T0201716

Introduction to Quantum Field Theory

Quantum field theory provides a theoretical framework for understanding fields and the particles associated with them, and is the basis of particle physics and condensed matter research. This graduate-level textbook provides a comprehensive introduction to quantum field theory, giving equal emphasis to operator and path-integral formalisms. It covers modern research such as helicity spinors, BCFW construction, and generalized unitarity cuts, as well as treating advanced topics including BRST quantization, loop equations, and finite-temperature field theory. Various quantum fields are described, including scalar and fermionic fields, abelian vector fields and quantum electrodynamics (QED), and finally non-abelian vector fields and quantum chromodynamics (QCD). Applications to scattering cross-sections in QED and QCD are also described. Each chapter ends with exercises and an important concepts section, allowing students to identify the key aspects of the chapter and test their understanding.

Horaţiu Năstase is a Researcher at the Institute for Theoretical Physics at the State University of São Paulo, Brazil. To date, his career has spanned four continents. As an undergraduate he studied at the University of Bucharest and Copenhagen University. He later completed his Ph.D. at the State University of New York, Stony Brook, before moving to the Institute for Advanced Study, Princeton University, New Jersey, where his collaboration with David Berenstein and Juan Maldacena defined the pp-wave correspondence. He has also held research and teaching positions at Brown University, Rhode Island and the Tokyo Institute of Technology. He has published three other books with Cambridge University Press: *Introduction to the AdS/CFT Correspondence* (2015), *String Theory Methods for Condensed Matter Physics* (2017), and *Classical Field Theory* (2019).

Introduction to Quantum Field Theory

HORAŢIU NĂSTASE

State University of São Paulo, Brazil

With material from unpublished
notes by Jan Ambjorn
and Jens Lyng Petersen

CAMBRIDGE
UNIVERSITY PRESS

CAMBRIDGE
UNIVERSITY PRESS

University Printing House, Cambridge CB2 8BS, United Kingdom

One Liberty Plaza, 20th Floor, New York, NY 10006, USA

477 Williamstown Road, Port Melbourne, VIC 3207, Australia

314–321, 3rd Floor, Plot 3, Splendor Forum, Jasola District Centre, New Delhi – 110025, India

79 Anson Road, #06–04/06, Singapore 079906

Cambridge University Press is part of the University of Cambridge.

It furthers the University's mission by disseminating knowledge in the pursuit of
education, learning, and research at the highest international levels of excellence.

www.cambridge.org
Information on this title: www.cambridge.org/9781108493994
DOI: 10.1017/9781108624992

First published 2020

Printed in the United Kingdom by TJ International Ltd. Padstow Cornwall

A catalogue record for this publication is available from the British Library.

Library of Congress Cataloging-in-Publication Data
Names: Nastase, Horatiu, 1972– author.
Title: Introduction to quantum field theory / Horatiu Nastase
(Universidade Estadual Paulista, Sao Paulo).
Description: Cambridge, United Kingdom ; New York, NY : Cambridge
University Press, 2020. | Includes bibliographical references and index.
Identifiers: LCCN 2019006491| ISBN 9781108493994 (alk. paper) |
ISBN 1108493998 (alk. paper)
Subjects: LCSH: Quantum field theory.
Classification: LCC QC174.45 .N353 2020 | DDC 530.14/3–dc23
LC record available at https://lccn.loc.gov/2019006491

ISBN 978-1-108-49399-4 Hardback

To the memory of my mother,
who inspired me to become a physicist

Contents

Preface

Quantum field theory is a subject that has been here for a while, and there are many books that teach it. However, I have found several reasons to write a book, based on lecture notes for a two-semester course I gave. One motivation is that, to my knowledge, there are no books that consistently treat together the operator and path-integral formalisms, on an equal footing, and using one or the other as is more convenient for the presentation. People usually have their favorite way of thinking about quantum field theory, either in the operator (usually for more phenomenological reasons) or the path-integral (usually for more theoretical reasons) formalism, and they almost always stick with it. But modern methods use both, and I think it important for students to be proficient in both also. There are many modern topics that don't make it into a quantum field theory book, but since the subject is constantly evolving, it is worth knowing the most important recent developments.

Another reason is that in most physics departments in the USA and Europe, quantum field theory is taught directly after classical mechanics and quantum mechanics, which is oftentimes a tough transition for a graduate student. At our institute, the Institute for Theoretical Physics at UNESP, we teach one semester of classical field theory, followed by two of quantum field theory, which makes the transition smoother, and is easier for the students to follow. For that reason, I have already published a classical field theory book (also with Cambridge University Press), as an extended version of the corresponding course I taught. The present quantum field theory book is conceived as a continuation of that classical field theory course, though I have tried to make it as self-consistent as possible. Notions of classical field theory are thus just reviewed, not treated in great detail.

Having decided to include path-integral treatment alongside the operator treatment (more heavily used in standard textbooks) early on, when deciding how to use path integrals in the exposition of most topics, I could think of no better way than the one I learned just before graduate school, in an exchange program at Niels Bohr Institute: a course following unpublished lecture notes by Jan Ambjorn and Jens Lyng Petersen, which were available from the NBI secretariat. So I used a lot of material from those notes, which are unpublished so far, in the building of my exposition, though of course following a different logic of exposition, in the end forming a significant part of the book, maybe a quarter or up to a third. When deciding to publish my notes, I asked Jan Ambjorn and Jens Lyng Petersen for permission to use the material in their notes as I did, with a clarification added in the subtitle. There are many quantum field theory books out there, among which I would highlight perhaps the ones by Peskin and Schroeder, Schwartz, Ramond, Weinberg, Zinn-Justin, Banks, Zee, and Srednicki, but I believe that none exactly match the requirements I had set out to fulfill, as stated above.

Acknowledgments

I want to thank everybody that shaped who I am as a physicist and helped me along the way, starting first and foremost with my mother Ligia, the first physicist I knew and the person who introduced me to the wonderful world of physics. Next, my high-school physics teacher Iosif Sever Georgescu made me realize that I could make a career out of physics, and made me see the path to it. Poul Olesen was my student exchange advisor at the Niels Bohr Institute (NBI), and he introduced me to string theory, which is still my field of study. Also at the NBI I had the wonderful quantum field theory course by Jan Ambjorn, based on the notes by him and Jens Lyng Petersen, which, I mentioned in the Preface, were used a lot in the presentation of the path-integral formalism in this book. Jens Lyng Petersen taught elementary particle physics, also forming my understanding of some of the issues in this book. My Ph.D. advisor, Peter van Nieuwenhuizen, taught me how to be a complete theoretical physicist, the beauty of calculations, and the value of rigor, and from him I learned many of the topics presented in this book, in particular various advanced topics in quantum field theory.

The book is based on a two-semester course I gave (twice) at the Institute for Theoretical Physics in São Paulo, so I would like to thank all the students in my class for their questions, input, and corrections to the notes that were then expanded into this book.

I would like to thank all my collaborators, and especially those with which I worked on various topics that ended up being presented in this book: Howard Schnitzer, Stephen Naculich, Juan Maldacena, David Berenstein, and Jeff Murugan. I want to thank my wife Antonia for her patience when I wrote this book, in the evenings at home, and her encouragement to continue. I want to thank my students and postdocs for accepting the reduced time I spent with them while working on the book. A big thanks to my editor at Cambridge University Press, Simon Capelin, for his encouragement and for helping me get this book, as well as my previous ones, published. To all the staff at Cambridge University Press, thanks for helping me with this book and my previous ones, and for making sure that it is as good as it can be.

Introduction

This book is meant as a two-semester course in quantum field theory, skipping some material that can be studied independently. The chapters with an asterisk I have not taught in my class, and they can be skipped in a first reading, or when teaching the material. The book and the corresponding course is supposed to follow a course in classical field theory; however, I have tried to make the book self-contained. This means that only a thorough knowledge of classical mechanics, quantum mechanics, and electromagnetism is really needed, though it is preferable to have first classical field theory. I will only review classical field theory, without going into great detail.

Quantum field theory represents the union of quantum mechanics and classical field theory, which itself is but a generalization of classical mechanics to an infinite number of degrees of freedom. As such, one needs to understand both quantum mechanics and classical mechanics (and perhaps its field theory generalization), so I will start by reviewing the necessary concepts. There are two main formalisms for treating quantum field theory, the operator formalism, and the more modern path-integral formalism, a generalization of the path integral for quantum mechanics, unfortunately not often taught in quantum mechanics. I will introduce them in parallel, and then use one or the other as is more convenient for the subject being treated.

In Part I, corresponding to the first semester of the course, I will introduce the general formalism and use it in "tree" processes, which are processes in the quantum mechanics form of classical field theory, but with no quantum field theory corrections. I will also give examples of physical processes treated with the quantum field theory formalism in the tree approximation, calculating scattering cross-sections for them. I will also describe briefly the modern formalism of helicity spinors for amplitudes of given helicity, and the BCFW iterative construction of amplitudes.

In Part II, corresponding to the second semester of the course, with several added chapters, I will describe "loops," which are true quantum field theory corrections, and the formalism of renormalization, which is a way to "absorb" the infinities of quantum field theory in the redefinition of parameters of the model, and the basis of the standard treatment. I will also treat nonabelian gauge theories, in particular QCD, IR divergences and anomalies, as well as many advanced topics that are usually less taught, like BRST quantization, the Makeenko–Migdal loop equation and order parameters, the Froissart unitarity bound, renormalization of spontaneously broken gauge theories, background field method, and finite-temperature quantum field theory. I also include more modern topics, like the generalized unitarity cut, polylogs and symbology, amplitudes in twistor space, and dual conformal invariance.

I

QUANTUM FIELDS, GENERAL FORMALISM, AND TREE PROCESSES

1 Review of Classical Field Theory: Lagrangians, Lorentz Group and its Representations, Noether Theorem

In this book, as I have mentioned, I will assume a knowledge of classical field theory and quantum mechanics, and I will only review a few notions from them, immediately useful, in the first two chapters. In this first chapter, I will start by describing what quantum field theory is, and after that I will review a few things about classical field theory. In the next chapter, a few relevant notions of quantum mechanics, not always taught, will be described.

Conventions   I will use theorist's conventions throughout, with $\hbar = c = 1$, which means that, for example, $[E] = [1/x] = 1$. I will also use the *mostly plus* metric, for instance in $3 + 1$ dimensions with signature $- + ++$.

1.1 What is and Why Do We Need Quantum Field Theory?

Quantum mechanics deals with the quantization of particles, and is a nonrelativistic theory: time is treated as special, and for the energy we use nonrelativistic formulas.

On the contrary, we want to apply *quantum field theory*, which is an application of quantum mechanics, to *fields* instead of particles, and this has the property of being *relativistic* as well.

Quantum field theory is often called (when derived from first principles) *second quantization*, the idea being that:

- The *first* quantization is when we have a single particle and we quantize its behavior (its motion) in terms of a wavefunction describing probabilities.
- The *second* quantization is when we quantize the wavefunction itself (instead of a function now we have an operator), the quantum object now being the number of particles the wavefunction describes, which is an arbitrary (variable) quantum number. Therefore, the field is now a description of an arbitrary number of particles (and *antiparticles*), and this number can *change* (i.e. it is not a constant).

People have tried to build a *relativistic quantum mechanics*, but it was quickly observed that if we do that, we cannot describe a single particle:

- First, the relativistic relation $E = mc^2$, together with the existence (experimentally confirmed) of *antiparticles* which annihilate with particles giving only energy (photons), means that if we have an energy $E > m_p c^2 + m_{\bar{p}} c^2$, we can create a particle–antiparticle pair, and therefore the number of particles cannot be a constant in a relativistic theory.

- Second, even if $E < m_p c^2 + m_{\bar{p}} c^2$, the particle–antiparticle pair can still be created for a short time. Indeed, Heisenberg's uncertainty principle in the (E, t) sector (as opposed to the usual (x, p) sector) means that $\Delta E \cdot \Delta t \sim \hbar$, meaning that for a short time $\Delta t \sim \hbar / \Delta E$ we can have an uncertainty in the energy ΔE, for instance such that $E + \Delta E > m_p c^2 + m_{\bar{p}} c^2$. This means that we can create a pair of *virtual particles*, that is particles which are forbidden by energy and momentum conservation to exist as asymptotic particles, but can exist as quantum fluctuations for a short time.

- Third, causality is violated by a single particle propagating via usual quantum mechanics formulas, even with the relativistic formula for the energy, $E = \sqrt{\vec{p}^2 + m^2}$.

The amplitude for propagation from $\vec{x}_0$ to $\vec{x}$ in a time t in quantum mechanics is

$$U(t) = \langle \vec{x} | e^{-iHt} | \vec{x}_0 \rangle, \tag{1.1}$$

and replacing E, the eigenvalue of H, by $\sqrt{\vec{p}^2 + m^2}$, we obtain

$$U(t) = \langle \vec{x} | e^{-it\sqrt{\vec{p}^2 + m^2}} | \vec{x}_0 \rangle = \frac{1}{(2\pi)^3} \int d^3 \vec{p}\, e^{-it\sqrt{\vec{p}^2 + m^2}} e^{i\vec{p} \cdot (\vec{x} - \vec{x}_0)}. \tag{1.2}$$

But

$$\int d^3 \vec{p}\, e^{i\vec{p} \cdot \vec{x}} = \int p^2 dp \int 2\pi \sin\theta d\theta e^{ipx\cos\theta}$$

$$= \int p^2 dp \left[\frac{2\pi}{ipx} (e^{ipx} - e^{-ipx}) \right] = \int p^2 dp \left[\frac{4\pi}{px} \sin(px) \right], \tag{1.3}$$

and therefore

$$U(t) = \frac{1}{2\pi^2 |\vec{x} - \vec{x}_0|} \int_0^\infty p\, dp \sin(p|\vec{x} - \vec{x}_0|) e^{-it\sqrt{p^2 + m^2}}. \tag{1.4}$$

For $x^2 \gg t^2$, we use a saddle-point approximation, which is the idea that the integral $I = \int dx e^{f(x)}$ can be approximated by the Gaussian around the saddle point (i.e. $I \simeq e^{f(x_0)} \int d\delta x e^{f''(x_0)\delta x^2} \simeq e^{f(x_0)} \sqrt{\pi/f''(x_0)}$, where x_0 is the saddle point) at whose position we have $f'(x_0) = 0$. Generally, if we are interested in leading behavior in some large parameter, the function $e^{f(x_0)}$ dominates $\sqrt{\pi/f''(x_0)}$ and we can just approximate $I \sim e^{f(x_0)}$.

In our case, we obtain

$$\frac{d}{dp} \left(ipx - it\sqrt{p^2 + m^2} \right) = 0 \Rightarrow x = \frac{tp}{\sqrt{p^2 + m^2}} \Rightarrow p = p_0 = \frac{imx}{\sqrt{x^2 - t^2}}. \tag{1.5}$$

Since we are at $x^2 \gg t^2$, we obtain

$$U(t) \propto e^{ip_0 x - it\sqrt{p_0^2 + m^2}} \sim e^{-\sqrt{x^2 - t^2}} \neq 0. \tag{1.6}$$

So we see that even much outside the lightcone, at $x^2 \gg t^2$, we have nonzero probability for propagation, meaning a breakdown of causality.

However, we will see that this problem is fixed in quantum field theory, which will be causal.

In quantum field theory, the fields describe many particles. One example of this fact that is easy to understand is the case of the electromagnetic field, $(\vec{E}, \vec{B}) \rightarrow F_{\mu\nu}$, which describes many photons. Indeed, we know from the correspondence principle of quantum mechanics that a classical state is equivalent to a state with many photons, and also that the number of photons is not a constant in any sense: we can define a (quantum) average number of photons that is related to the classical intensity of an electromagnetic beam, but the number of photons is not a classically measurable quantity.

We will describe processes involving many particles by *Feynman diagrams*, which will be an important part of this book. In quantum mechanics, a particle propagates forever, so its "Feynman diagram" is always a single line, as in Figure 1.1.

In quantum field theory, however, we will derive the mathematical form of Feynman diagrams, but the simple physical interpretation for which Feynman introduced them is that we can have processes where, for instance, a particle splits into two (or more) (see Figure 1.1(a)), two (or more) particles merge into one (see Figure 1.1(b)), or two (or more) particles of one type disappear and another type is created, like for instance in the annihilation of an e^+ (positron) with an e^- (electron) into a photon (γ) as in Figure 1.1(c), and so on.

Moreover, we can have (as we mentioned) virtual processes, like a photon γ creating an e^+e^- pair, which lives for a short time Δt and then annihilates into a γ, creating an e^+e^- virtual loop inside the propagating γ, as in Figure 1.2. Of course, $E, \vec{p}$ conservation means that $(E, \vec{p})$ is the same for the γ before and after the loop.

Next, we should review a few notions of classical field theory.

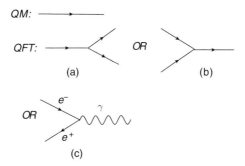

(a)   (b)

(c)

Figure 1.1 Quantum mechanics: particle goes on forever. Quantum field theory: particles can split (a), join (b), and particles of different types can appear and disappear, like in the quantum electrodynamics process (c).

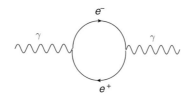

Figure 1.2 Virtual particles can appear for a short time in a loop. Here a photon creates a virtual electron–positron pair, which then annihilates back into the photon.

1.2 Classical Mechanics

Before doing that, however, we begin with an even quicker review of **classical mechanics**.

In classical mechanics, the description of a system is in terms of a Lagrangian $L(q_i, \dot{q}_i)$ for the variables $q_i(t)$, and the corresponding action

$$S = \int_{t_1}^{t_2} dt L(q_i(t), \dot{q}_i(t)). \tag{1.7}$$

By varying the action with fixed boundary values for the variables $q_i(t)$ (i.e. $\delta S = 0$), we obtain the Euler–Lagrange equations (or equations of motion)

$$\frac{\partial L}{\partial q_i} - \frac{d}{dt}\frac{\partial L}{\partial \dot{q}_i} = 0. \tag{1.8}$$

We can also do a Legendre transformation from the Lagrangian $L(q_i, \dot{q}_i)$ to the Hamiltonian $H(q_i, p_i)$ in the usual way, by

$$H(p, q) = \sum_i p_i \dot{q}_i - L(q_i, \dot{q}_i), \tag{1.9}$$

where

$$p_i \equiv \frac{\partial L}{\partial \dot{q}_i} \tag{1.10}$$

is the momentum canonically conjugate to the coordinate q_i.

Differentiating the Legendre transformation formula, we get the first-order Hamilton equations (instead of the second-order Lagrange equations)

$$\frac{\partial H}{\partial p_i} = \dot{q}_i,$$

$$\frac{\partial H}{\partial q_i} = -\frac{\partial L}{\partial q_i} = -\dot{p}_i. \tag{1.11}$$

1.3 Classical Field Theory

The generalization of classical mechanics to **field theory** is obtained by considering instead of a set $\{q_i(t)\}_i$, which is a collection of given particles, fields $\phi(\vec{x}, t)$, where $\vec{x}$ is a generalization of i, and not a coordinate of a particle.

We will be interested in *local* field theories, which means all objects are integrals over $\vec{x}$ of functions defined at a point, in particular the Lagrangian is written as

$$L(t) = \int d^3\vec{x} \mathcal{L}(\vec{x}, t). \tag{1.12}$$

Here $\mathcal{L}$ is called the *Lagrange density*, but by an abuse of notation, one usually refers to it also as the Lagrangian.

We are also interested in *relativistic field theories*, which means that $\mathcal{L}(\vec{x}, t)$ is a relativistically invariant function of fields and their derivatives:

$$\mathcal{L}(\vec{x}, t) = \mathcal{L}(\phi(\vec{x}, t), \partial_\mu \phi(\vec{x}, t)). \tag{1.13}$$

Considering also several fields ϕ_a, we have an action written as

$$S = \int L dt = \int d^4 x \mathcal{L}(\phi_a, \partial_\mu \phi_a), \tag{1.14}$$

where $d^4 x = dt d^3 \vec{x}$ is the relativistically invariant volume element for spacetime.

The Euler–Lagrange equations are obtained in the same way, as

$$\frac{\partial \mathcal{L}}{\partial \phi_a} - \partial_\mu \left[\frac{\partial \mathcal{L}}{\partial(\partial_\mu \phi_a)} \right] = 0. \tag{1.15}$$

Note that one could think of $L(q_i)$ as a discretization over $\vec{x}$ of $\int d^3 \vec{x} \mathcal{L}(\phi_a)$, but that is not particularly useful.

In the Lagrangian we have relativistic fields, that is fields that have a well-defined transformation property under Lorentz transformations

$$x'^\mu = \Lambda^\mu{}_\nu x^\nu, \tag{1.16}$$

namely

$$\phi'_i(x') = R_i^j \phi_i(x), \tag{1.17}$$

where i is some index for the fields, related to its Lorentz properties. We will come back to this later, but for now let us just observe that for a scalar field there is no i and $R \equiv 1$ (i.e. $\phi'(x') = \phi(x)$).

In this book I will use the convention for the spacetime metric with "mostly plus" on the diagonal, that is the Minkowski metric is

$$\eta_{\mu\nu} = diag(-1, +1, +1, +1). \tag{1.18}$$

Note that this is the convention that is the most natural in order to make heavy use of Euclidean field theory via Wick rotation, as we will do (by just redefining the time t by a factor of i), and so is very useful if we work with the functional formalism, where Euclidean field theory is essential.

On the contrary, for various reasons, people connected with phenomenology and making heavy use of the operator formalism often use the "mostly minus" metric ($\eta_{\mu\nu} = diag(+1, -1, -1, -1)$), for instance the standard textbook of Peskin and Schroeder [1] does so, so one has to be very careful when translating results from one convention to the other.

With this metric, the Lagrangian for a scalar field is generically

$$\begin{aligned}
\mathcal{L} &= -\frac{1}{2} \partial_\mu \phi \partial^\mu \phi - \frac{1}{2} m^2 \phi^2 - V(\phi) \\
&= \frac{1}{2} \dot{\phi}^2 - \frac{1}{2} |\vec{\nabla}\phi|^2 - \frac{1}{2} m^2 \phi^2 - V(\phi),
\end{aligned} \tag{1.19}$$

so is of the general type $\dot{q}^2/2 - \tilde{V}(q)$, as it should be (where the terms $1/2|\vec{\nabla}\phi|^2 + m^2\phi^2/2$ are also part of $\tilde{V}(q)$).

To go to the Hamiltonian formalism, we must first define the momentum canonically conjugate to the field $\phi(\vec{x})$ (remembering that $\vec{x}$ is a label like i):

$$p(\vec{x}) = \frac{\partial L}{\partial\dot{\phi}(\vec{x})} = \frac{\partial}{\partial\dot{\phi}(\vec{x})}\int d^3\vec{y}\mathcal{L}(\phi(\vec{y}), \partial_\mu\phi(\vec{y})) = \pi(\vec{x})d^3\vec{x}, \qquad (1.20)$$

where

$$\pi(\vec{x}) = \frac{\delta\mathcal{L}}{\delta\dot{\phi}(\vec{x})} \qquad (1.21)$$

is a conjugate momentum density, but by an abuse of notation again will just be called conjugate momentum.

Then the Hamiltonian is

$$H = \sum_{\vec{x}} p(\vec{x})\dot{\phi}(\vec{x}) - L$$

$$\rightarrow \int d^3\vec{x}[\pi(\vec{x})\dot{\phi}(\vec{x}) - \mathcal{L}] \equiv \int d^3\vec{x}\mathcal{H}, \qquad (1.22)$$

where $\mathcal{H}$ is a Hamiltonian density.

1.4 Noether Theorem

The statement of the Noether theorem is that for every symmetry of the Lagrangian L, there is a corresponding conserved charge.

The best known examples are the time translation $t \rightarrow t+a$, corresponding to conserved energy E, and the space translation $\vec{x} \rightarrow \vec{x} + \vec{a}$, corresponding to conserved momentum $\vec{p}$, together making the spacetime translation $x^\mu \rightarrow x^\mu + a^\mu$, corresponding to conserved 4-momentum P^μ. The *Noether currents* corresponding to these charges form the energy–momentum tensor $T_{\mu\nu}$.

Consider the symmetry $\phi(x) \rightarrow \phi'(x) = \phi(x) + \alpha\Delta\phi$ that transforms the Lagrangian density as

$$\mathcal{L} \rightarrow \mathcal{L} + \alpha\partial_\mu J^\mu, \qquad (1.23)$$

such that the action $S = \int d^4x\mathcal{L}$ is invariant, if the fields vanish on the boundary, usually considered at $t = \pm\infty$, since the boundary term

$$\int d^4x\partial_\mu J^\mu = \oint_{bd} dS_\mu J^\mu = \int d^3\vec{x}J^0|_{t=-\infty}^{t=+\infty} \qquad (1.24)$$

is then zero. In this case, there exists a conserved current j^μ, that is

$$\partial_\mu j^\mu(x) = 0, \qquad (1.25)$$

where

$$j^{\mu}(x) = \frac{\partial \mathcal{L}}{\partial(\partial_{\mu}\phi)}\Delta\phi - J^{\mu}. \tag{1.26}$$

For linear symmetries (symmetry transformations linear in ϕ), we can define

$$(\alpha\Delta\phi)^i \equiv \alpha^a (T^a)^i{}_j \phi^j \tag{1.27}$$

such that, if $J^{\mu} = 0$, we have the Noether current

$$j^{\mu,a} = \frac{\partial \mathcal{L}}{\partial(\partial_{\mu}\phi)}(T^a)^i{}_j \phi^j. \tag{1.28}$$

Applying this general formalism to translations, $x^{\mu} \rightarrow x^{\mu} + a^{\mu}$, we obtain, for an infinitesimal parameter a^{μ}:

$$\phi(x) \rightarrow \phi(x+a) = \phi(x) + a^{\mu}\partial_{\mu}\phi, \tag{1.29}$$

which are the first terms in the Taylor expansion around x. The corresponding conserved current is therefore

$$T^{\mu}{}_{\nu} \equiv \frac{\partial \mathcal{L}}{\partial(\partial_{\mu}\phi)}\partial_{\nu}\phi - \mathcal{L}\delta^{\mu}_{\nu}, \tag{1.30}$$

where we have added a term $J^{\mu}_{(\nu)} = \mathcal{L}\delta^{\mu}_{\nu}$ to get the conventional definition of the *energy–momentum tensor* or *stress–energy tensor*. The conserved charges are integrals of the energy–momentum tensor (i.e. P^{μ}). Note that the above translation can be considered as also giving the term $J^{\mu}_{(\nu)}$ from the general formalism, since we can check that for $\alpha^{\nu} = a^{\nu}$, the Lagrangian changes by $\partial_{\mu}J^{\mu}_{(\nu)}$.

1.5 Fields and Lorentz Representations

The Lorentz group is $SO(1,3)$, that is an orthogonal group that generalizes $SO(3)$, the group of rotations in the (Euclidean) three spatial dimensions.

Its basic objects in the fundamental representation, defined as the representation that acts on coordinates x^{μ} (or rather dx^{μ}), are called $\Lambda^{\mu}{}_{\nu}$, and thus

$$dx'^{\mu} = \Lambda^{\mu}{}_{\nu}dx^{\nu}. \tag{1.31}$$

If η is the matrix $\eta_{\mu\nu}$, the Minkowski metric $diag(-1, +1, +1, +1)$, the orthogonal group $SO(1,3)$ is the group of elements Λ that satisfy

$$\Lambda\eta\Lambda^T = \eta. \tag{1.32}$$

Note that the usual rotation group $SO(3)$ is an orthogonal group satisfying

$$\Lambda\Lambda^T = \mathbf{1} \Rightarrow \Lambda^{-1} = \Lambda^T, \tag{1.33}$$

but we should actually write this as

$$\Lambda\mathbf{1}\Lambda^T = \mathbf{1}, \tag{1.34}$$

which admits a generalization to $SO(p,q)$ groups as

$$\Lambda g \Lambda^T = g, \tag{1.35}$$

where $g = diag(-1,\ldots,-1,+1,\ldots,+1)$ with p minuses and q pluses. In the above, Λ satisfies the group property, namely if Λ_1, Λ_2 belong in the group, then

$$\Lambda_1 \cdot \Lambda_2 \equiv \Lambda \tag{1.36}$$

is also in the group.

General representations are a generalization of (1.31), namely instead of acting on x, the group acts on a vector space ϕ^a by

$$\phi'^a(\Lambda x) = R(\Lambda)^a{}_b \phi^b(x), \tag{1.37}$$

such that it respects the group property, that is

$$R(\Lambda_1)R(\Lambda_2) = R(\Lambda_1 \cdot \Lambda_2). \tag{1.38}$$

Group elements are represented for infinitesimally small parameters β^a as exponentials of the *Lie algebra generators* in the R representation $t_a^{(R)}$, that is

$$R(\beta) = e^{i\beta^a t_a^{(R)}}. \tag{1.39}$$

The statement that $t_a^{(R)}$ form a Lie algebra is the statement that we have a relation

$$[t_a^{(R)}, t_b^{(R)}] = i f_{ab}{}^c t_c^{(R)} \tag{1.40}$$

where $f_{ab}{}^c$ are called the *structure constants*. Note that the factor of i is conventional, with this definition we can have Hermitian generators, for which $\text{Tr}(t_a t_b) = \delta_{ab}$; if we redefine t_a by an i we can remove it from there, but then $\text{Tr}(t_a t_b)$ can be put only to $-\delta_{ab}$ (anti-Hermitian generators).

The representations of the Lorentz group are:

- Bosonic. Scalars ϕ for which $\phi'(x') = \phi(x)$; vectors like the electromagnetic field $A_\mu = (\phi, \vec{A})$ that transform as ∂_μ (covariant) or dx^μ (contravariant), and representations which have products of indices, like for instance the electromagnetic field strength $F_{\mu\nu}$ which transforms as

$$F'_{\mu\nu}(\Lambda x) = \Lambda_\mu{}^\rho \Lambda_\nu{}^\sigma F_{\rho\sigma}(x), \tag{1.41}$$

  where $\Lambda_\mu{}^\nu = \eta_{\mu\rho} \eta^{\nu\sigma} \Lambda^\rho{}_\sigma$. For fields with more indices, $B^{\nu_1\ldots\nu_j}_{\mu_1\ldots\mu_k}$, it transforms as the appropriate products of Λ.
- Fermionic. Spinors, which will be treated in more detail later on in the book. For now, let us just say that fundamental spinor representations ψ are acted upon by gamma matrices γ^μ.

The Lie algebra of the Lorentz group $SO(1,3)$ is

$$[J_{\mu\nu}, J_{\rho\sigma}] = -i\eta_{\mu\rho} J_{\nu\sigma} + i\eta_{\mu\sigma} J_{\nu\rho} - i\eta_{\nu\sigma} J_{\mu\rho} + i\eta_{\nu\rho} J_{\mu\sigma}. \tag{1.42}$$

Note that if we denote $a \equiv (\mu\nu), b \equiv (\rho\sigma)$, and $c \equiv (\lambda\pi)$, we then have

$$f_{ab}{}^{c} = -\eta_{\mu\rho}\delta^{\lambda}_{[\nu}\delta^{\pi}_{\sigma]} + \eta_{\mu\sigma}\delta^{\lambda}_{[\nu}\delta^{\pi}_{\rho]} - \eta_{\nu\sigma}\delta^{\lambda}_{[\mu}\delta^{\pi}_{\rho]} + \eta_{\nu\rho}\delta^{\lambda}_{[\mu}\delta^{\pi}_{\sigma]}, \qquad (1.43)$$

so (1.42) is indeed of the Lie algebra type.

The Lie algebra $SO(1,3)$ is (modulo some global subtleties) the same as the product of two $SU(2)$s (i.e. $SU(2) \times SU(2)$), which can be seen by first defining

$$J_{0i} \equiv K_i; \quad J_{ij} \equiv \epsilon_{ijk}J_k, \qquad (1.44)$$

where $i, j, k = 1, 2, 3$, and then redefining

$$M_i \equiv \frac{J_i + iK_i}{2}; \quad N_i \equiv \frac{J_i - iK_i}{2}, \qquad (1.45)$$

after which we obtain

$$[M_i, M_j] = i\epsilon_{ijk}M_k,$$
$$[N_i, N_j] = i\epsilon_{ijk}N_k,$$
$$[M_i, N_j] = 0, \qquad (1.46)$$

which we leave as an exercise to prove.

Important Concepts to Remember

- Quantum field theory is a relativistic quantum mechanics, which necessarily describes an arbitrary number of particles.
- Particle–antiparticle pairs can be created and disappear, both as real (energetically allowed) and virtual (energetically disallowed, only possible due to Heisenberg's uncertainty principle).
- If we use the usual quantum mechanics rules, even with $E = \sqrt{p^2 + m^2}$, we have causality breakdown: the amplitude for propagation is nonzero even much outside the lightcone.
- Feynman diagrams represent the interaction processes of creation and annihilation of particles.
- When generalizing classical mechanics to field theory, the label i is generalized to $\vec{x}$ in $\phi(\vec{x}, t)$, and we have a Lagrangian density $\mathcal{L}(\vec{x}, t)$, conjugate momentum density $\pi(\vec{x}, t)$, and Hamiltonian density $\mathcal{H}(\vec{x}, t)$.
- For relativistic and local theories, $\mathcal{L}$ is a relativistically invariant function defined at a point x^{μ}.
- The Noether theorem associates a conserved current ($\partial_{\mu}j^{\mu} = 0$) with a symmetry of the Lagrangian L, in particular the energy–momentum tensor T^{μ}_{ν} with translations $x^{\mu} \rightarrow x^{\mu} + a^{\mu}$.
- Lorentz representations act on the fields ϕ^a, and are the exponentials of Lie algebra generators.
- The Lie algebra of $SO(1,3)$ splits into two $SU(2)$s.

Further Reading

See, for instance, sections 2.1 and 2.2 in [1] and chapter 1 in [2].

Exercises

1. Prove that for the Lie algebra of the Lorentz group

$$[J_{\mu\nu}, J_{\rho\sigma}] = -\left(-i\eta_{\mu\rho}J_{\nu\sigma} + i\eta_{\mu\sigma}J_{\nu\rho} - i\eta_{\nu\sigma}J_{\mu\rho} + i\eta_{\nu\rho}J_{\mu\sigma}\right), \qquad (1.47)$$

if we define

$$J_{0i} \equiv K_i; \quad J_{ij} \equiv \epsilon_{ijk}J_k,$$
$$M_i \equiv \frac{J_i + iK_i}{2}; \quad N_i \equiv \frac{J_i - iK_i}{2}, \qquad (1.48)$$

we obtain that the M_i and N_i satisfy

$$[M_i, M_j] = i\epsilon_{ijk}M_k,$$
$$[N_i, N_j] = i\epsilon_{ijk}N_k,$$
$$[M_i, N_j] = 0. \qquad (1.49)$$

2. Consider the action in Minkowski space

$$S = \int d^4x \left(-\frac{1}{4}F_{\mu\nu}F^{\mu\nu} - \bar{\psi}(\slashed{D} + m)\psi - (D_\mu\phi)^*D^\mu\phi\right), \qquad (1.50)$$

where $D_\mu = \partial_\mu - ieA_\mu$, $\slashed{D} = D_\mu\gamma^\mu$, $F_{\mu\nu} = \partial_\mu A_\nu - \partial_\nu A_\mu$, $\bar{\psi} = \psi^\dagger i\gamma_0$, ψ is a spinor field and ϕ is a scalar field, and γ_μ are the gamma matrices, satisfying $\{\gamma_\mu, \gamma_\nu\} = 2\eta_{\mu\nu}$. Consider the electromagnetic $U(1)$ transformation

$$\psi'(x) = e^{ie\lambda(x)}\psi(x); \quad \phi'(x) = e^{ie\lambda(x)}\phi(x); \quad A'_\mu(x) = A_\mu(x) + \partial_\mu\lambda(x). \qquad (1.51)$$

Calculate the Noether current.

3. Find the invariances of the model for N real scalars Φ^I, with Lagrangian

$$\mathcal{L} = g_{IJ}(\Phi^I\Phi^I)\partial_\mu\Phi^I\partial^\mu\Phi^I \qquad (1.52)$$

in the case of a general metric g_{IJ}, and in the particular case of $g_{IJ} = \eta_{IJ}$ (at least in a local neighborhood in scalar space).

4. Calculate the equations of motion of the Dirac–Born–Infeld (DBI) scalar Lagrangian

$$\mathcal{L} = -\frac{1}{L^4}\sqrt{1 + L^4[g(\phi)(\partial_\mu\phi)^2 + m^2\phi^2]}. \qquad (1.53)$$

Quantum Mechanics: Harmonic Oscillator and Quantum Mechanics in Terms of Path Integrals

The career of a young theoretical physicist consists of treating the harmonic oscillator in ever-increasing levels of abstraction

Sidney Coleman

In this chapter, I will review some facts about the harmonic oscillator in quantum mechanics, and then show how to do quantum mechanics in terms of path integrals, something that should be taught in a standard quantum mechanics course, though it does not always happen.

As the quote above shows, understanding the harmonic oscillator really well is crucial: we understand everything if we understand this simple example really well, such that we can generalize it to more complicated systems. Similarly, most of the issues of quantum field theory in path-integral formalism can be described by using the simple example of the quantum-mechanical path integral.

2.1 The Harmonic Oscillator and its Canonical Quantization

The harmonic oscillator is the simplest possible nontrivial quantum system, with a quadratic potential, that is with the Lagrangian

$$L = \frac{\dot{q}^2}{2} - \omega^2 \frac{q^2}{2}, \tag{2.1}$$

giving the Hamiltonian

$$H = \frac{1}{2}(p^2 + \omega^2 q^2). \tag{2.2}$$

Using the definition

$$a = \frac{1}{\sqrt{2\omega}}(\omega q + ip),$$
$$a^\dagger = \frac{1}{\sqrt{2\omega}}(\omega q - ip), \tag{2.3}$$

inverted as

$$p = -i\sqrt{\frac{\omega}{2}}(a - a^\dagger),$$
$$q = \frac{1}{\sqrt{2\omega}}(a + a^\dagger), \tag{2.4}$$

we can write the Hamiltonian as

$$H = \frac{\omega}{2}(aa^\dagger + a^\dagger a), \tag{2.5}$$

where, even though we are now at the classical level, we have been careful to keep the order of $a, a^\dagger$ as is. Of course, classically we could then write

$$H = \omega a^\dagger a. \tag{2.6}$$

In **classical mechanics**, one can define the Poisson bracket of two functions $f(p, q)$ and $g(p, q)$ as

$$\{f, g\}_{P.B.} \equiv \sum_i \left(\frac{\partial f}{\partial q_i} \frac{\partial g}{\partial p_i} - \frac{\partial f}{\partial p_i} \frac{\partial g}{\partial q_i} \right). \tag{2.7}$$

With this definition, we can immediately check that

$$\{p_i, q_j\}_{P.B.} = -\delta_{ij}. \tag{2.8}$$

The Hamilton equations of motion then become

$$\dot{q}_i = \frac{\partial H}{\partial p_i} = \{q_i, H\},$$

$$\dot{p}_i = -\frac{\partial H}{\partial q_i} = \{p_i, H\}. \tag{2.9}$$

Then, **canonical quantization** is simply the procedure of replacing the c-number variables (q, p) with the operators $(\hat{q}, \hat{p})$, and replacing the Poisson brackets $\{, \}_{P.B.}$ with $1/(i\hbar)[,]$ (commutator).

In this way, in theoretical physicist's units, with $\hbar = 1$, we have

$$[\hat{p}, \hat{q}] = -i. \tag{2.10}$$

Substituting in p, q the definition of $a, a^\dagger$, we find also

$$[\hat{a}, \hat{a}^\dagger] = 1. \tag{2.11}$$

One thing that is not obvious from the above is the picture we are in. We know that we can describe quantum mechanics in the Schrödinger picture, with operators independent of time, or in the Heisenberg picture, where operators depend on time. There are other pictures, in particular the interaction picture that will be relevant for us later, but these are not important at this time.

In the Heisenberg picture, we can translate the classical Hamilton equations in terms of Poisson brackets into equations for the time evolution of the Heisenberg picture operators, obtaining

$$i\hbar \frac{d\hat{q}_i}{dt} = [\hat{q}_i, H],$$

$$i\hbar \frac{d\hat{p}_i}{dt} = [\hat{p}_i, H]. \tag{2.12}$$

For the quantum Hamiltonian of the harmonic oscillator, we write, from (2.5):

$$\hat{H}_{qu} = \frac{\hbar\omega}{2} \left(\hat{a}\hat{a}^\dagger + \hat{a}^\dagger\hat{a} \right) = \hbar\omega \left(\hat{a}^\dagger\hat{a} + \frac{1}{2} \right), \tag{2.13}$$

where we have reintroduced $\hbar$ just so we remember that $\hbar\omega$ is an energy. The operators $\hat{a}$ and $\hat{a}^\dagger$ are called destruction (annihilation) or lowering, and creation or raising operators, since the eigenstates of the harmonic oscillator Hamiltonian are defined by an occupation number n, such that

$$\hat{a}^\dagger|n\rangle \propto |n+1\rangle; \quad \hat{a}|n\rangle \propto |n-1\rangle, \tag{2.14}$$

meaning that

$$\hat{a}^\dagger\hat{a}|n\rangle \equiv \hat{N}|n\rangle = n|n\rangle. \tag{2.15}$$

This means that in the vacuum, for occupation number $n = 0$, we still have an energy

$$E_0 = \frac{\hbar\omega}{2}, \tag{2.16}$$

called the vacuum energy or zero-point energy. On the contrary, the remainder is called the *normal ordered Hamiltonian*

$$: \hat{H} := \hbar\omega\hat{a}^\dagger\hat{a}, \tag{2.17}$$

where we define the *normal order* such that $\hat{a}^\dagger$ is to the left of $\hat{a}$, that is

$$: \hat{a}^\dagger\hat{a} := \hat{a}^\dagger\hat{a}; \quad : \hat{a}\hat{a}^\dagger := \hat{a}^\dagger\hat{a}. \tag{2.18}$$

2.2 The Feynman Path Integral in Quantum Mechanics in Phase Space

We now turn to the discussion of the Feynman path integral, not always taught in quantum mechanics classes, though essential for the generalization to field theory.

Given a position q at time t, an important quantity is the amplitude for the probability of finding the particle at point q' and time t':

$$M(q',t';q,t) = {}_H\langle q',t'|q,t\rangle_H, \tag{2.19}$$

where $|q,t\rangle_H$ is the state, eigenstate of $\hat{q}(t)$ at time t, in the Heisenberg picture.

Let us remember a bit about pictures in quantum mechanics. There are more pictures, but for now we will be interested in the two basic ones, the Schrödinger picture and the Heisenberg picture. In the Heisenberg picture, operators depend on time, in particular we have $\hat{q}_H(t)$, and the state $|q,t\rangle_H$ is independent of time, and t is just a label. This means that the state is an eigenstate of $\hat{q}_H(T)$ at time $T = t$, that is

$$\hat{q}_H(T = t)|q,t\rangle_H = q|q,t\rangle_H, \tag{2.20}$$

and it is *not* an eigenstate for $T \neq t$. The operator in the Heisenberg picture $\hat{q}_H(t)$ is related to that in the Schrödinger picture $\hat{q}_S$ by

$$\hat{q}_H(t) = e^{i\hat{H}t} \hat{q}_S e^{-i\hat{H}t}, \qquad (2.21)$$

and the Schrödinger picture state is related to the Heisenberg picture state by

$$|q\rangle = e^{-i\hat{H}t} |q, t\rangle_H, \qquad (2.22)$$

and is an eigenstate of $\hat{q}_S$, that is

$$\hat{q}_S |q\rangle = q |q\rangle. \qquad (2.23)$$

In terms of the Schrödinger picture, we then have the probability amplitude

$$M(q', t'; q, t) = \langle q' | e^{-i\hat{H}(t'-t)} | q \rangle. \qquad (2.24)$$

From now on we will drop the index H and S for states, since it is obvious, if we write $|q, t\rangle$ we are in the Heisenberg picture, if we write $|q\rangle$ we are in the Schrödinger picture.

Let us now derive the path-integral representation.

Divide the time interval between t and t' into a large number $n + 1$ of equal intervals, and denote

$$\epsilon \equiv \frac{t' - t}{n + 1}; \qquad t_0 = t, t_1 = t + \epsilon, \dots, t_{n+1} = t'. \qquad (2.25)$$

But at any fixed t_i, the set $\{|q_i, t_i\rangle_H | q_i \in \mathbb{R}\}$ is a complete set, meaning that we have the completeness relation

$$\int dq_i |q_i, t_i\rangle \langle q_i, t_i| = \mathbf{1}. \qquad (2.26)$$

We then introduce n factors of $\mathbf{1}$, one for each t_i, $i = 1, \dots, n$, in the amplitude $M(q', t'; q, t)$ in (2.19), obtaining

$$M(q, t'; q, t) = \int dq_1 \dots dq_n \langle q', t' | q_n, t_n \rangle \langle q_n, t_n | q_{n-1}, t_{n-1} \rangle \dots | q_1, t_1 \rangle \langle q_1, t_1 | q, t \rangle, \quad (2.27)$$

where the discrete positions $q_i \equiv q(t_i)$ give us a *regularized path* between q and q'.

But note that this is not a classical path, since at any of the times t_i, the position q can be anything ($q_i \in \mathbb{R}$), independent of q_{i-1}, and independent of how small ϵ is, whereas classically we have a continuous path, meaning as ϵ gets smaller, $q_i - q_{i-1}$ can only be smaller and smaller. But integrating over all $q_i = q(t_i)$ means we integrate over these quantum paths, where q_i is arbitrary (independent on q_{i-1}), as in Figure 2.1. Therefore, we denote

$$\mathcal{D}q(t) \equiv \prod_{i=1}^{n} dq(t_i), \qquad (2.28)$$

and this is an "integral over all paths," or "path integral."

In contrast, considering that

$$|q\rangle = \int \frac{dp}{2\pi} |p\rangle \langle p|q\rangle,$$

$$|p\rangle = \int dq |q\rangle \langle q|p\rangle \qquad (2.29)$$

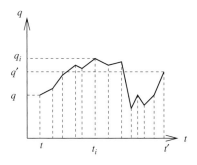

In the quantum-mechanical path integral, we integrate over discretized paths. The paths are not necessarily smooth, as classical paths are: we divide the path into a large number of discrete points, and then integrate over the positions of these points.

(note the factor of 2π, which is necessary, since $\langle q|p\rangle = e^{ipq}$ and $\int dq e^{iq(p-p')} = 2\pi\delta(p-p')$), we have

$$_H\langle q(t_i), t_i|q(t_{i-1}), t_{i-1}\rangle_H = \langle q(t_i)|e^{-i\epsilon\hat{H}}|q(t_{i-1})\rangle = \int \frac{dp(t_i)}{2\pi}\langle q(t_i)|p(t_i)\rangle\langle p(t_i)|e^{-i\epsilon\hat{H}}|q(t_{i-1})\rangle.$$
(2.30)

Now we need a technical requirement on the quantum Hamiltonian: it has to be ordered such that all the $\hat{p}$s are to the left of the $\hat{q}$s.

Then, to first order in ϵ, we can write

$$\langle p(t_i)|e^{-i\epsilon\hat{H}(\hat{p},\hat{q})}|q(t_{i-1})\rangle = e^{-i\epsilon H(p(t_i),q(t_{i-1}))}\langle p(t_i)|q(t_{i-1})\rangle = e^{-i\epsilon H(p(t_i),q(t_{i-1}))}e^{-ip(t_i)q(t_{i-1})},$$
(2.31)

since $\hat{p}$ will act on the left on $\langle p(t_i)|$ and $\hat{q}$ will act on the right on $|q(t_{i-1})\rangle$. Of course, to higher order in ϵ, we have $\hat{H}(\hat{p},\hat{q})\hat{H}(\hat{p},\hat{q})$, which will have terms like $\hat{p}\hat{q}\hat{p}\hat{q}$, which are more complicated. But since we have $\epsilon \to 0$, we only need the first order in ϵ.

Then we get

$$M(q',t';q,t) = \int \prod_{i=1}^{n}\frac{dp(t_i)}{2\pi}\prod_{j=1}^{n}dq(t_j)\langle q(t_{n+1}|p(t_{n+1})\rangle\langle p(t_{n+1})|e^{-i\epsilon\hat{H}}|q(t_n)\rangle \dots$$

$$\dots \langle q(t_1)|p(t_1)\rangle\langle p(t_1)|e^{-i\epsilon\hat{H}}|q(t_0)\rangle$$

$$= \mathcal{D}p(t)\mathcal{D}q(t)\exp\left\{i\left[p(t_{n+1})(q(t_{n+1}) - q(t_n)) + \dots + p(t_1)(q(t_1) - q(t_0))\right.\right.$$

$$\left.\left. -\epsilon(H(p(t_{n+1}),q(t_n)) + \dots + H(p(t_1),q(t_0)))\right]\right\}$$

$$= \mathcal{D}p(t)\mathcal{D}q(t)\exp\left\{i\int_{t_0}^{t_{n+1}}dt[p(t)\dot{q}(t) - H(p(t),q(t))]\right\},$$
(2.32)

where we have used the fact that $q(t_{i+1}) - q(t_i) \to dt\dot{q}(t_i)$. The above expression is called the *path integral in phase space*.

We note that this was derived rigorously (rigorously for a physicist, of course... a mathematician would disagree). But we would like a path integral in configuration space. For that, however, we need one more technical requirement: we need the Hamiltonian to be quadratic in momenta, that is

$$H(p, q) = \frac{p^2}{2} + V(q). \tag{2.33}$$

If this is not true, we have to start in phase space and see what we get in configuration space. But if we have only quadratic terms in momenta, we can use Gaussian integration to derive the path integral in configuration space. Therefore we will make a mathematical interlude to define some Gaussian integration formulas that will be useful here and later.

2.3 Gaussian Integration

The basic Gaussian integral is

$$I = \int_{-\infty}^{+\infty} e^{-\alpha x^2} dx = \sqrt{\frac{\pi}{\alpha}}. \tag{2.34}$$

Squaring this integral formula, we obtain also

$$I^2 = \int dx dy e^{-(x^2 + y^2)} = \int_0^{2\pi} d\phi \int_0^\infty r dr e^{-r^2} = \pi. \tag{2.35}$$

We can generalize this as

$$\int d^n x e^{-\frac{1}{2} x_i A_{ij} x_j} = (2\pi)^{n/2} (\det A)^{-1/2}, \tag{2.36}$$

which can be proven, for instance, by diagonalizing the matrix A, and since $\det A = \prod_i \alpha_i$, with α_i the eigenvalues of A, we get the above.

Finally, consider the object

$$S = \frac{1}{2} x^T A x + b^T x \tag{2.37}$$

(which later on in the book will be used as a discretized form for an action, hence the name S). Considering it as an action, the classical solution will be $\partial S / \partial x_i = 0$:

$$x_c = -A^{-1} b, \tag{2.38}$$

and then

$$S(x_c) = -\frac{1}{2} b^T A^{-1} b, \tag{2.39}$$

which means that we can write

$$S = \frac{1}{2} (x - x_c)^T A (x - x_c) - \frac{1}{2} b^T A^{-1} b, \tag{2.40}$$

and thus we find

$$\int d^n x e^{-S(x)} = (2\pi)^{n/2} (\det A)^{-1/2} e^{-S(x_c)} = (2\pi)^{n/2} (\det A)^{-1/2} e^{+\frac{1}{2} b^T A^{-1} b}. \tag{2.41}$$

2.4 Path Integral in Configuration Space

We are now ready to go to configuration space. The Gaussian integration we need to do is then the one over $\mathcal{D}p(t)$, which is

$$\int \mathcal{D}p(\tau)e^{i\int_t^{t'} d\tau[p(\tau)\dot{q}(\tau)-\frac{1}{2}p^2(\tau)]},\tag{2.42}$$

which when discretized becomes

$$\prod_i \frac{dp(t_i)}{2\pi}\exp\left[i\Delta\tau\left(p(t_i)\dot{q}(t_i)-\frac{1}{2}p^2(\tau_i)\right)\right],\tag{2.43}$$

and therefore in our mathematical notation for the Gaussian integral, we have $x_i = p(t_i)$, $A_{ij} = i\Delta\tau\delta_{ij}$, $b = -i\Delta\tau\dot{q}(t_i)$, giving

$$\int \mathcal{D}p(\tau)e^{i\int_t^{t'} d\tau[p(\tau)\dot{q}(\tau)-\frac{1}{2}p^2(\tau)]} = \mathcal{N}e^{i\int_t^{t'} d\tau \frac{\dot{q}(\tau)^2}{2}},\tag{2.44}$$

where $\mathcal{N}$ contains constant factors of $2, \pi, i, \Delta\tau$, which we will see are irrelevant.

Then the probability amplitude in configuration space is

$$M(q',t';q,t) = \mathcal{N}\int \mathcal{D}q\exp\left\{i\int_t^{t'} d\tau\left[\frac{\dot{q}^2(\tau)}{2}-V(q)\right]\right\}$$

$$= \mathcal{N}\int \mathcal{D}q\exp\left\{i\int_t^{t'} d\tau L(q(\tau),\dot{q}(\tau))\right\}$$

$$= \mathcal{N}\int \mathcal{D}q e^{iS[q]}.\tag{2.45}$$

This is the path integral in configuration space that we were seeking. But we have to remember that this formula is valid only if the Hamiltonian is quadratic in momenta, otherwise we need to redo the calculation starting from phase space.

2.5 Correlation Functions

We have found how to write the probability of transition between (q,t) and (q',t'), and that is good. But there are other observables that we can construct which are of interest, for instance the correlation functions.

The simplest one is the one-point function

$$\langle q',t'|\hat{q}(t_a)|q,t\rangle,\tag{2.46}$$

where we can make it such that t_a coincides with one of the t_is in the discretization of the time interval. The calculation now proceeds as before, but in the step (2.27) we introduce the **1**s such that we have (besides the usual products) also the expectation value

$$\langle q_{i+1},t_{i+1}|\hat{q}(t_a)|q_i,t_i\rangle = q(t_a)\langle q_{i+1},t_{i+1}|q_i,t_i\rangle,\tag{2.47}$$

since $t_a = t_i \Rightarrow q(t_a) = q_i$. Then the calculation proceeds as before, since the only new thing is the appearance of $q(t_a)$, leading to

$$\langle q', t' | \hat{q}(t_a) | q, t \rangle = \int \mathcal{D}q \, e^{iS[q]} q(t_a). \tag{2.48}$$

Consider next the two-point function

$$\langle q', t' | \hat{q}(t_b) \hat{q}(t_a) | q, t \rangle. \tag{2.49}$$

If we have $t_a < t_b$, we can proceed as before. Indeed, remember that in (2.27) the **1**s were introduced in time order, such that we can do the rest of the calculation and get the path integral. Therefore, if $t_a < t_b$, we can choose $t_a = t_i, t_b = t_j$, such that $j > i$, and then we have

$$\ldots \langle q_{j+1}, t_{j+1} | \hat{q}(t_b) | q_j, t_j \rangle \ldots \langle q_{i+1}, t_{i+1} | \hat{q}(t_a) | q_i, t_i \rangle \ldots$$
$$= \ldots q(t_b) \langle q_{j+1}, t_{j+1} | q_j, t_j \rangle \ldots q(t_a) \langle q_{i+1}, t_{i+1} | q_i, t_i \rangle \ldots, \tag{2.50}$$

besides the usual products, leading to

$$\int \mathcal{D}q \, e^{iS[q]} q(t_b) q(t_a). \tag{2.51}$$

Then, reversely, the path integral leads to the two-point function where the $q(t)$ are ordered according to time (time ordering), that is

$$\int \mathcal{D}q \, e^{iS[q]} q(t_a) q(t_b) = \langle q', t' | T\{\hat{q}(t_a)\hat{q}(t_b)\} | q, t \rangle, \tag{2.52}$$

where time ordering is defined as

$$T\{\hat{q}(t_a)\hat{q}(t_b)\} = \hat{q}(t_a)\hat{q}(t_b) \quad \text{if } t_a > t_b,$$
$$= \hat{q}(t_b)\hat{q}(t_a) \quad \text{if } t_b > t_a, \tag{2.53}$$

which has an obvious generalization to

$$T\{\hat{q}(t_{a_1}) \ldots \hat{q}(t_{a_N})\} = \hat{q}(t_{a_1}) \ldots \hat{q}(t_{a_N}) \quad \text{if } t_{a_1} > t_{a_2} > \ldots > t_{a_N}, \tag{2.54}$$

and otherwise they are ordered in the order of their times.

Then we similarly find that the *n-point function* or *correlation function* is

$$G_n(t_{a_1}, \ldots, t_{a_n}) \equiv \langle q', t' | T\{\hat{q}(t_{a_1}) \ldots q(t_{a_n})\} | q, t \rangle = \int \mathcal{D}q \, e^{iS[q]} q(t_{a_1}) \ldots q(t_{a_n}). \tag{2.55}$$

In mathematics, for a set $\{a_n\}_n$, we can define a *generating function*

$$f(x) \equiv \sum_n \frac{1}{n!} a_n x^n, \tag{2.56}$$

such that we can find a_n from its derivatives:

$$a_n = \frac{d^n}{dx^n} f(x) \Big|_{x=0}. \tag{2.57}$$

Similarly, we can now define a generating functional

$$Z[J] = \sum_{N \geq 0} \int dt_1 \ldots \int dt_N \frac{i^N}{N!} G_N(t_1, \ldots, t_N) J(t_1) \ldots J(t_N). \tag{2.58}$$

As we see, the difference is that now we have $G_N(t_1, \ldots, t_N)$ instead of a_N, so we needed to integrate over dt, and instead of x, introduce $J(t)$, and i was conventional. Using (2.55), the integrals factorize, and we obtain just a product of the same integral:

$$Z[J] = \int \mathcal{D}q e^{iS[q]} \sum_{N \geq 0} \frac{1}{N!} \left[\int dt i q(t) J(t) \right]^N, \tag{2.59}$$

so finally

$$Z[J] = \int \mathcal{D}q e^{iS[q,J]} = \int \mathcal{D}q e^{iS[q]+i \int dt J(t) q(t)}. \tag{2.60}$$

We then find that this object indeed generates the correlation functions by

$$\frac{\delta^N}{i\delta J(t_1) \ldots i J(t_N)} Z[J] \bigg|_{J=0} = \int \mathcal{D}q e^{iS[q]} q(t_1) \ldots q(t_N) = G_N(t_1, \ldots, t_N). \tag{2.61}$$

Important Concepts to Remember

- For the harmonic oscillator, the Hamiltonian in terms of $a, a^\dagger$ is $H = \omega/2(aa^\dagger + a^\dagger a)$.
- Canonical quantization replaces classical functions with quantum operators, and Poisson brackets with commutators, $\{,\}_{P.B.} \to 1/(i\hbar)[,]$.
- At the quantum level, the harmonic oscillator Hamiltonian is the sum of a normal ordered part and a zero-point energy part.
- The transition probability from (q, t) to (q', t') is a path integral in phase space, $F(q', t'; q, t) = \int \mathcal{D}q \mathcal{D}p e^{i \int (p\dot{q}-H)}$.
- If the Hamiltonian is quadratic in momenta, we can go to the path integral in configuration space and find $F(q', t'; q, t) = \int \mathcal{D}q e^{iS}$.
- The n-point functions or correlation functions, with insertions of $q(t_i)$ in the path integral, give the expectation values of time-ordered $\hat{q}(t)$s.
- The n-point functions can be found from the derivatives of the generating function $Z[J]$.

Further Reading

See section 1.3 in [3] and chapter 2 in [4].

Exercises

1. Consider the Lagrangian

$$L(q, \dot{q}) = \frac{\dot{q}^2}{2} - \frac{\lambda}{4!} q^4. \tag{2.62}$$

   Write down the *Hamiltonian* equations of motion and the path integral *in phase space* for this model.

2. Consider the generating functional

$$\ln Z[J] = \int dt \frac{J^2(t)}{2} f(t) + \lambda \int dt \frac{J^3(t)}{3!} + \tilde{\lambda} \int dt \frac{J^4(t)}{4!}. \tag{2.63}$$

   Calculate the three-point function and the four-point function.

3. Repeat the change from phase-space path integral to configuration-space path integral for a Hamiltonian of the type

$$H = \frac{p^2}{2} + \alpha p + \beta + V(q). \tag{2.64}$$

4. Calculate the generating functional $Z[J]$ for the case of a quadratic action, with $V[q] = \omega^2 q^2/2$, that is for the harmonic oscillator

$$L = \frac{\dot{q}^2}{2} - \omega^2 \frac{q^2}{2}. \tag{2.65}$$

3   Canonical Quantization of Scalar Fields

In this chapter, we will learn how to quantize classical scalar fields, similarly to the procedure in the quantum mechanics of a finite number of degrees of freedom.

As we saw, in quantum mechanics, for a particle with Hamiltonian $H(p, q)$, we replace the Poisson bracket

$$\{f, g\}_{P.B.} = \sum_i \left(\frac{\partial f}{\partial q_i} \frac{\partial g}{\partial p_i} - \frac{\partial f}{\partial p_i} \frac{\partial g}{\partial q_i} \right) \tag{3.1}$$

with $\{p_i, q_j\}_{P.B.} = -\delta_{ij}$, with the commutator, $\{,\}_{P.B.} \to \frac{1}{i\hbar}[,]$, and all functions of (p, q) become quantum operators, in particular $[\hat{p}_i, \hat{q}_j] = -i\hbar$, and for the harmonic oscillator we have

$$[\hat{a}, \hat{a}^\dagger] = 1. \tag{3.2}$$

3.1 Quantizing Scalar Fields: Kinematics

Then, in order to generalize to field theory, we must first define the Poisson brackets. As we already have a definition in the case of a set of particles, we will discretize *space*, in order to use that definition. Therefore we consider the coordinates, and their canonically conjugate momenta

$$\begin{aligned} q_i(t) &= \sqrt{\Delta V} \phi_i(t), \\ p_i(t) &= \sqrt{\Delta V} \pi_i(t), \end{aligned} \tag{3.3}$$

where $\phi_i(t) \equiv \phi(\vec{x}_i, t)$ and similarly for $\pi_i(t)$. We should also define how to go from the derivatives in the Poisson brackets to functional derivatives. The recipe is

$$\begin{aligned} \frac{1}{\Delta V} \frac{\partial f_i(t)}{\partial \phi_j(t)} &\to \frac{\delta f(\phi(\vec{x}_i, t), \pi(\vec{x}_i, t))}{\delta \phi(\vec{x}_j, t)}, \\ \Delta V &\to d^3x, \end{aligned} \tag{3.4}$$

where *functional derivatives* are defined such that, for instance

$$H(t) = \int d^3x \frac{\phi^2(\vec{x}, t)}{2} \Rightarrow \frac{\delta H(t)}{\delta \phi(\vec{x}, t)} = \phi(\vec{x}, t), \tag{3.5}$$

that is by dropping the integral sign and then taking normal derivatives.

Replacing these definitions in the Poisson brackets (3.1), we get

$$\{f,g\}_{P.B.} = \int d^3x \left[\frac{\delta f}{\delta \phi(\vec{x},t)} \frac{\delta g}{\delta \pi(\vec{x},t)} - \frac{\delta f}{\delta \pi(\vec{x},t)} \frac{\delta g}{\delta \phi(\vec{x},t)} \right], \tag{3.6}$$

and then we immediately find

$$\{\phi(\vec{x},t), \pi(\vec{x}',t)\}_{P.B.} = \delta^3(\vec{x} - \vec{x}'),$$
$$\{\phi(\vec{x},t), \phi(\vec{x}',t)\}_{P.B.} = \{\pi(\vec{x},t), \pi(\vec{x}',t)\} = 0, \tag{3.7}$$

where we note that these are *equal time commutation relations*, in the same way that we had in quantum mechanics, more precisely, $\{q_i(t), p_j(t)\}_{P.B.} = \delta_{ij}$.

We can now easily do *canonical quantization* of this scalar field. We just replace classical fields $\phi(\vec{x},t)$ with quantum Heisenberg operators $\phi_H(\vec{x},t)$ (we will drop the H, understanding that if there is time dependence we are in the Heisenberg picture and if we don't have time dependence we are in the Schrödinger picture), and $\{,\}_{P.B.} \to 1/(i\hbar)[,]$, obtaining the fundamental equal time commutation relations

$$[\phi(\vec{x},t), \pi(\vec{x}',t)] = i\hbar\delta^3(\vec{x} - \vec{x}'),$$
$$[\phi(\vec{x},t), \phi(\vec{x}',t)] = [\pi(\vec{x},t), \pi(\vec{x}',t)] = 0. \tag{3.8}$$

We further define the Fourier transforms

$$\phi(\vec{x},t) = \int \frac{d^3p}{(2\pi)^3} e^{i\vec{x}\cdot\vec{p}} \phi(\vec{p},t)$$

$$\Rightarrow \phi(\vec{p},t) = \int d^3x e^{-i\vec{p}\cdot\vec{x}} \phi(\vec{x},t),$$

$$\pi(\vec{x},t) = \int \frac{d^3p}{(2\pi)^3} e^{i\vec{x}\cdot\vec{p}} \pi(\vec{p},t)$$

$$\Rightarrow \pi(\vec{p},t) = \int d^3x e^{-i\vec{p}\cdot\vec{x}} \pi(\vec{x},t). \tag{3.9}$$

We also define, using the same formulas used for the harmonic oscillator, but now for each of the momentum modes of the fields:

$$a(\vec{k},t) = \sqrt{\frac{\omega_k}{2}} \phi(\vec{k},t) + \frac{i}{\sqrt{2\omega_k}} \pi(\vec{k},t),$$

$$a^\dagger(\vec{k},t) = \sqrt{\frac{\omega_k}{2}} \phi^\dagger(\vec{k},t) - \frac{i}{\sqrt{2\omega_k}} \pi^\dagger(\vec{k},t). \tag{3.10}$$

We will see later that we have $\omega_k = \sqrt{\vec{k}^2 + m^2}$, but for the moment we will only need that $\omega_k = \omega(|\vec{k}|)$. Then, replacing these definitions in ϕ and π, we obtain

$$\phi(\vec{x},t) = \int \frac{d^3p}{(2\pi)^3} \frac{1}{\sqrt{2\omega_p}} (a(\vec{p},t)e^{i\vec{p}\cdot\vec{x}} + a^\dagger(\vec{p},t)e^{-i\vec{p}\cdot\vec{x}})$$

$$= \int \frac{d^3p}{(2\pi)^3} \frac{1}{\sqrt{2\omega_p}} e^{i\vec{p}\cdot\vec{x}} (a(\vec{p},t) + a^\dagger(-\vec{p},t)),$$

$$\pi(\vec{x},t) = \int \frac{d^3p}{(2\pi)^3} \left(-i\sqrt{\frac{\omega_p}{2}}\right) (a(\vec{p},t)e^{i\vec{p}\cdot\vec{x}} - a^\dagger(\vec{p},t)e^{-i\vec{p}\cdot\vec{x}})$$

$$= \int \frac{d^3p}{(2\pi)^3} \left(-i\sqrt{\frac{\omega_p}{2}} \right) (a(\vec{p}, t) - a^\dagger(-\vec{p}, t)) e^{i\vec{p}\cdot\vec{x}}. \tag{3.11}$$

In terms of $a(\vec{p}, t)$ and $a^\dagger(\vec{p}, t)$, we obtain the commutators

$$[a(\vec{p}, t), a^\dagger(\vec{p}', t)] = (2\pi)^3 \delta^3(\vec{p} - \vec{p}'),$$
$$[a(\vec{p}, t), a(\vec{p}', t)] = [a^\dagger(\vec{p}, t), a^\dagger(\vec{p}', t)] = 0. \tag{3.12}$$

As a consistency check, we can check, for instance, that the $[\phi, \pi]$ commutator gives the correct result, given the above $a, a^\dagger$ commutators:

$$[\phi(\vec{x}, t), \pi(\vec{x}', t)] = \int \frac{d^3p}{(2\pi)^3} \frac{d^3p'}{(2\pi)^3} \left(-\frac{i}{2} \sqrt{\frac{\omega_{p'}}{\omega_p}} \right) ([a^\dagger(-\vec{p}, t), a(\vec{p}', t)]$$
$$- [a(\vec{p}, t), a^\dagger(-\vec{p}', t)]) e^{i(\vec{p}\cdot\vec{x} + \vec{p}'\cdot\vec{x}')}$$
$$= i\delta^3(\vec{x} - \vec{x}'). \tag{3.13}$$

We now note that the calculation above was independent of the form of ω_p ($= \sqrt{p^2 + m^2}$), we only used the fact that $\omega_p = \omega(|\vec{p}|)$, but otherwise we just used the definitions adapted from the harmonic oscillator. We have also not written any explicit time dependence, it was left implicit through $a(\vec{p}, t), a^\dagger(\vec{p}, t)$. Yet, we obtained the same formulas as for the harmonic oscillator.

3.2 Quantizing Scalar Fields: Dynamics and Time Evolution

We should now understand the dynamics, which will give us the formula for ω_p.

We therefore go back, and start systematically. We work with a *free scalar field* (i.e. one with $V = 0$) with Lagrangian

$$\mathcal{L} = -\frac{1}{2}\partial_\mu\phi\partial^\mu\phi - \frac{m^2}{2}\phi^2 \tag{3.14}$$

and action $S = \int d^4x \mathcal{L}$. Partially integrating $-1/2 \int \partial_\mu\phi\partial^\mu\phi = +1/2 \int \phi\partial_\mu\partial^\mu\phi$, we obtain the *Klein–Gordon (KG)* equation of motion

$$(\partial_\mu\partial^\mu - m^2)\phi = 0 \Rightarrow (-\partial_t^2 + \partial_{\vec{x}}^2 - m^2)\phi = 0. \tag{3.15}$$

Going to momentum ($\vec{p}$) space via a Fourier transform, we obtain

$$\left[\frac{\partial^2}{\partial t^2} + (\vec{p}^2 + m^2) \right] \phi(\vec{p}, t) = 0, \tag{3.16}$$

which is the equation of motion for a harmonic oscillator with $\omega = \omega_p = \sqrt{p^2 + m^2}$.

This means that the Hamiltonian is

$$H = \frac{1}{2}(p^2 + \omega_p^2\phi^2) \tag{3.17}$$

and we can use the transformation

$$\phi = \frac{1}{\sqrt{2\omega}}(a + a^\dagger); \quad p = -i\sqrt{\frac{\omega}{2}}(a - a^\dagger), \tag{3.18}$$

after which we obtain as usual $[a, a^\dagger] = 1$. Therefore, we can now justify *a posteriori* the transformations that we did on the field.

We should also calculate the Hamiltonian. As explained in Chapter 1, using the discretization of space, we write

$$H = \sum_{\vec{x}} p(\vec{x}, t)\dot{\phi}(\vec{x}, t) - L$$

$$= \int d^3x [\pi(\vec{x}, t)\dot{\phi}(\vec{x}, t) - \mathcal{L}] \equiv d^3x \mathcal{H}, \tag{3.19}$$

and from the Lagrangian (3.14) we obtain

$$\pi(\vec{x}, t) = \dot{\phi}(\vec{x}, t) \Rightarrow$$

$$\mathcal{H} = \frac{1}{2}\pi^2 + \frac{1}{2}(\vec{\nabla}\phi)^2 + \frac{1}{2}m^2\phi^2. \tag{3.20}$$

Substituting ϕ, π inside the Hamiltonian, we obtain

$$H = \int d^3x \int \frac{d^3p}{(2\pi)^3}\frac{d^3p'}{(2\pi)^3} e^{i(\vec{p}+\vec{p}')\vec{x}} \left\{ -\frac{\sqrt{\omega_p\omega_{p'}}}{4}(a(\vec{p}, t) - a^\dagger(-\vec{p}, t))(a(\vec{p}', t) - a^\dagger(-\vec{p}', t)) \right.$$

$$\left. + \frac{-\vec{p}\cdot\vec{p}' + m^2}{4\sqrt{\omega_p\omega_{p'}}}(a(\vec{p}, t) + a^\dagger(-\vec{p}, t))(a(\vec{p}', t) + a^\dagger(-\vec{p}', t)) \right\}$$

$$= \int \frac{d^3p}{(2\pi)^3}\frac{\omega_p}{2}(a^\dagger(\vec{p}, t)a(\vec{p}, t) + a(\vec{p}, t)a^\dagger(\vec{p}, t)), \tag{3.21}$$

where in the last line we have first integrated over $\vec{x}$, obtaining $\delta^3(\vec{p} + \vec{p}')$, and then integrated over $\vec{p}'$, obtaining $\vec{p}' = -\vec{p}$. We have finally reduced the Hamiltonian to an infinite (continuum, even) sum over harmonic oscillators.

We have dealt with the first issue stated earlier, about the dynamics of the theory. We now address the second, of the explicit time dependence.

We have Heisenberg operators, for which the time evolution is

$$i\frac{d}{dt}a(\vec{p}, t) = [a(\vec{p}, t), H]. \tag{3.22}$$

Calculating the commutator from the above Hamiltonian (using the fact that $[a, aa^\dagger] = [a, a^\dagger a] = a$), we obtain

$$i\frac{d}{dt}a(\vec{p}, t) = \omega_p a(\vec{p}, t). \tag{3.23}$$

More generally, the time evolution of the Heisenberg operators in field theories is given by

$$\mathcal{O}(x) = \mathcal{O}_H(\vec{x}, t) = e^{iHt}\mathcal{O}(\vec{x})e^{-iHt}, \tag{3.24}$$

which is equivalent to

$$i\frac{\partial}{\partial t}\mathcal{O}_H = [\mathcal{O}, H] \tag{3.25}$$

via

$$i\frac{d}{dt}(e^{iAt}Be^{-iAt}) = [B, A]. \tag{3.26}$$

The solution of (3.23) is

$$a(\vec{p}, t) = a_{\vec{p}}e^{-i\omega_p t},$$
$$a^\dagger(\vec{p}, t) = a_{\vec{p}}^\dagger e^{+i\omega_p t}. \tag{3.27}$$

Replacing in $\phi(\vec{x}, t)$ and $\pi(\vec{x}, t)$, we obtain

$$\phi(\vec{x}, t) = \int \frac{d^3p}{(2\pi)^3} \frac{1}{\sqrt{2E_p}}(a_{\vec{p}}e^{ipx} + a_{\vec{p}}^\dagger e^{-ipx})|_{p^0=E_p},$$
$$\pi(\vec{x}, t) = \frac{\partial}{\partial t}\phi(\vec{x}, t), \tag{3.28}$$

so we have formed the Lorentz invariants $e^{\pm ipx}$, though we haven't written an explicitly Lorentz-invariant formula. We will do this in Chapter 4. Here we have denoted $E_p = \sqrt{\vec{p}^2 + m^2}$, remembering that it is the relativistic energy of a particle of momentum $\vec{p}$ and mass m.

Finally, of course, if we want the Schrödinger picture operators, we have to remember the relation between the Heisenberg and Schrödinger pictures:

$$\phi_H(\vec{x}, t) = e^{iHt}\phi(\vec{x})e^{-iHt}. \tag{3.29}$$

3.3 Discretization

Continuous systems are hard to understand, so it would be better if we could find a rigorous way to discretize the system. Luckily, there is such a method, namely we consider a space of finite volume V (i.e. we "put the system in a box"). Obviously, this doesn't discretize space, but it does discretize *momenta*, since in a direction z of length L_z, allowed momenta will be only $k_n = 2\pi n/L_z$.

Then, the discretization is defined, as always, by

$$\int d^3k \rightarrow \frac{1}{V}\sum_{\vec{k}},$$
$$\delta^3(\vec{k} - \vec{k}') \rightarrow V\delta_{\vec{k}\vec{k}'}, \tag{3.30}$$

to which we add the redefinition

$$a_{\vec{k}} \rightarrow \sqrt{V(2\pi)^3}\alpha_{\vec{k}}, \tag{3.31}$$

which allows us to keep the usual orthonormality condition in the discrete limit, $[\alpha_{\vec{k}}, \alpha_{\vec{k}'}^\dagger] = \delta_{\vec{k}\vec{k}'}$.

Using these relations, and replacing the time dependence, which cancels out of the Hamiltonian, we get the Hamiltonian of the free scalar field in a box of volume V:

$$H = \sum_{\vec{k}} \frac{(\hbar)\omega_k}{2}(\alpha_{\vec{k}}^\dagger \alpha_{\vec{k}} + \alpha_{\vec{k}} \alpha_{\vec{k}}^\dagger) = \sum_{\vec{k}} h_{\vec{k}}, \tag{3.32}$$

where $h_{\vec{k}}$ is the Hamiltonian of a single harmonic oscillator:

$$h_{\vec{k}} = \omega_{\vec{k}} \left(N_{\vec{k}} + \frac{1}{2}\right), \tag{3.33}$$

$N_{\vec{k}} = \alpha_{\vec{k}}^\dagger \alpha_{\vec{k}}$ is the number operator for mode $\vec{k}$, with eigenstates $|n_{\vec{k}}>$:

$$N_{\vec{k}}|n_{\vec{k}}\rangle = n_{\vec{k}}|n_{\vec{k}}\rangle, \tag{3.34}$$

and the orthonormal eigenstates are

$$|n\rangle = \frac{1}{\sqrt{n!}}(\alpha^\dagger)^n|0\rangle; \quad \langle n|m\rangle = \delta_{mn}. \tag{3.35}$$

Here, $\alpha_{\vec{k}}^\dagger$ = raising/creation operator and $\alpha_{\vec{k}}$ = lowering/annihilation (destruction) operator, named since they create and annihilate a particle, respectively, that is

$$\alpha_{\vec{k}}|n_{\vec{k}}\rangle = \sqrt{n_{\vec{k}}}|n_{\vec{k}} - 1\rangle,$$
$$\alpha_{\vec{k}}^\dagger|n_{\vec{k}}\rangle = \sqrt{n_{\vec{k}} + 1}|n_{\vec{k}} + 1\rangle,$$
$$h_{\vec{k}}|n\rangle = \omega_{\vec{k}}\left(n + \frac{1}{2}\right)|n\rangle. \tag{3.36}$$

Therefore, as we know from quantum mechanics, there is a ground state $|0\rangle$, $\Leftrightarrow n_{\vec{k}} \in \mathbf{N}_+$, in which case $n_{\vec{k}}$ is called the occupation number, or number of particles in the state $\vec{k}$.

3.4 Fock Space and Normal Ordering for Bosons

3.4.1 Fock Space

The Hilbert space of states in terms of eigenstates of the number operator is called *Fock space*, or *Fock space representation*. The Fock space representation for the states of a single harmonic oscillator is $\mathcal{H}_{\vec{k}} = \{|n_{\vec{k}}>\}$.

Since the total Hamiltonian is the sum of the Hamiltonians for each mode, the total Hilbert space is the direct product of the Hilbert spaces of the Hamiltonians for each mode, $\mathcal{H} = \otimes_{\vec{k}}\mathcal{H}_{\vec{k}}$. Its states are then

$$|\{n_{\vec{k}}\}\rangle = \prod_{\vec{k}} |n_{\vec{k}}\rangle = \left(\prod_{\vec{k}} \frac{1}{\sqrt{n_{\vec{k}}!}}(\alpha_{\vec{k}}^\dagger)^{n_{\vec{k}}}\right)|0\rangle. \tag{3.37}$$

Note that we have defined a unique vacuum for all the Hamiltonians, $|0\rangle$, such that $a_{\vec{k}}|0\rangle = 0$, $\forall \vec{k}$, instead of denoting it as $\prod_{\vec{k}} |0\rangle_{\vec{k}}$.

3.4.2 Normal Ordering

For a single harmonic oscillator mode, the ground-state energy, or zero-point energy, is $\hbar\omega_{\vec{k}}/2$, which we may think could have some physical significance. But for a free scalar field, even one in a box ("discretized"), the total ground-state energy is $\sum_{\vec{k}} \hbar\omega_{\vec{k}}/2 = \infty$, and since an observable of infinite value doesn't make sense, we have to consider that it is *unobservable*, and put it to zero.

In this simple model, that's no problem, but consider the case where this free scalar field is coupled to gravity. In a gravitational theory, energy is equivalent to mass, and gravitates (i.e. it can be measured by its gravitational effects). So how can we drop a constant piece from the energy? Are we allowed to do that? In fact, this is part of one of the biggest problems of modern theoretical physics, the *cosmological constant problem*, and the answer to this question is far from obvious. At this level, however, we will not bother with this question anymore, and drop the infinite constant.

But it is also worth mentioning that while infinities are of course unobservable, the finite difference between two infinite quantities might be observable, and in fact one such case was already measured. If we consider the difference in the zero-point energies between fields in two different boxes, one of volume V_1 and another of volume V_2, that *is* measurable, and leads to the so-called Casimir effect, which we will discuss at the beginning of Part II.

We are then led to define the *normal ordered Hamiltonian*

$$: H := H - \frac{1}{2} \sum_{\vec{k}} \hbar\omega_{\vec{k}} = \sum_{\vec{k}} \hbar\omega_{\vec{k}} N_{\vec{k}} \tag{3.38}$$

by dropping the infinite constant. The *normal order* is to always have $a^{\dagger}$ before a, that is $: a^{\dagger}a := a^{\dagger}a$, $: aa^{\dagger} := a^{\dagger}a$. Since as commute among themselves, as do $a^{\dagger}s$, and operators from different modes, in case these appear, we don't need to bother with their order. For instance then, $: aa^{\dagger}a^{\dagger}aaaa^{\dagger} := a^{\dagger}a^{\dagger}a^{\dagger}aaa$.

We then consider that only normal ordered operators have physical expectation values, for instance

$$\langle 0| : \mathcal{O} : |0\rangle \tag{3.39}$$

is measurable.

One more observation to make is that in the expansion of ϕ we have

$$(a_{\vec{p}}e^{ipx} + a_{\vec{p}}^{\dagger}e^{-ipx})_{p^0=E_p} \tag{3.40}$$

and here $E_p = +\sqrt{\vec{p}^2 + m^2}$, but we note that the second term has *positive frequency* (energy), $a_{\vec{p}}^{\dagger}e^{-iE_pt}$, whereas the first has *negative frequency* (energy), ae^{+iE_pt}, that is we create $E > 0$ and destroy $E < 0$, which means that in this context we have only positive energy excitations. But we will see in Chapter 4 that in the case of the complex scalar field, we create both $E > 0$ and $E < 0$ and similarly destroy, leading to the concept of *antiparticles*. At this time, however, we don't have that.

3.4.3 Bose–Einstein Statistics

Since $[a_{\vec{k}}^{\dagger}, a_{\vec{k}'}^{\dagger}] = 0$, for a general state defined by a wavefunction $\psi(\vec{k}_1, \vec{k}_2)$:

$$
\begin{aligned}
|\psi\rangle &= \sum_{\vec{k}_1, \vec{k}_2} \psi(\vec{k}_1, \vec{k}_2) \alpha_{\vec{k}_1}^{\dagger} \alpha_{\vec{k}_2}^{\dagger} |0\rangle \\
&= \sum_{\vec{k}_1, \vec{k}_2} \psi(\vec{k}_2, \vec{k}_1) \alpha_{\vec{k}_1}^{\dagger} \alpha_{\vec{k}_2}^{\dagger} |0\rangle,
\end{aligned}
\tag{3.41}
$$

where in the second line we have commuted the two $\alpha^{\dagger}$s and then renamed $\vec{k}_1 \leftrightarrow \vec{k}_2$.

We then obtain the *Bose–Einstein statistics*

$$
\psi(\vec{k}_1, \vec{k}_2) = \psi(\vec{k}_2, \vec{k}_1),
\tag{3.42}
$$

that is for indistinguishable particles (permuting them, we obtain the same state).

As an aside, note that the Hamiltonian of the free (bosonic) oscillator is $1/2(aa^{\dagger} + a^{\dagger}a)$ (and of the free fermionic oscillator is $1/2(b^{\dagger}b - bb^{\dagger})$), so in order to have a well-defined system we must have $[a, a^{\dagger}] = 1$, $\{b, b^{\dagger}\} = 1$. In turn, $[a, a^{\dagger}] = 1$ leads to Bose–Einstein statistics, as above.

Important Concepts to Remember

- The commutation relations for scalar fields are defined at equal time. The Poisson brackets are defined in terms of integrals of functional derivatives.
- The canonical commutation relations for the free scalar field imply that we can use the same redefinitions as for the harmonic oscillator, for the momentum modes, to obtain the $[a, a^{\dagger}] = 1$ relations.
- The Klein–Gordon equation for the free scalar field implies the Hamiltonian of the free harmonic oscillator for each of the momentum modes.
- Putting the system in a box, we find a sum over discrete momenta of harmonic oscillators, each with a Fock space.
- The Fock space for the free scalar field is the direct product of the Fock space for each mode.
- We must use normal ordered operators, for physical observables, in particular for the Hamiltonian, in order to avoid unphysical infinities.
- The scalar field is quantized in terms of the Bose–Einstein statistics.

Further Reading

See section 2.3 in [1] and sections 2.1 and 2.3 in [2].

Exercises

1. Consider the classical Hamiltonian

$$H = \int d^3x \left\{ \frac{\pi^2(\vec{x}, t)}{2} + \frac{1}{2}(\vec{\nabla}\phi)^2 + \frac{\lambda}{3!}\phi^3(\vec{x}, t) + \frac{\tilde{\lambda}}{4!}\phi^4(\vec{x}, t) \right\}. \tag{3.43}$$

   Using the Poisson brackets, write the Hamiltonian equations of motion. Quantize canonically the *free* system and compute the equations of motion for the Heisenberg operators.
2. Write down the Hamiltonian above in terms of $a(\vec{p}, t)$ and $a^\dagger(\vec{p}, t)$ *at the quantum level*, and then write down the *normal ordered* Hamiltonian.
3. Consider the scalar Dirac–Born–Infeld (DBI) Lagrangian

$$\mathcal{L} = -\frac{1}{L^4}\sqrt{1 + L^4(\partial_\mu\phi)^2}. \tag{3.44}$$

   Calculate the Hamiltonian, and the Hamiltonian equations of motion, using the Poisson brackets.
4. For the model in Exercise 3, quantize canonically using the fundamental Poisson brackets, and write the Hamiltonian in terms of a and $a^\dagger$ defined as in (3.10).

Propagators for Free Scalar Fields

After having defined canonical quantization, in this chapter we will learn how to construct propagators, fundamental objects in quantum field theory, that solve the Klein–Gordon (KG) equation with delta function source, and describe propagation of the field. In order to define that, however, we will first construct a relativistically invariant version of canonical quantization, and then learn to quantize a *complex* scalar field.

4.1 Relativistic Invariant Canonical Quantization

We need to understand the relativistic invariance of the quantization of the free scalar field, which was described in Chapter 3 in nonrelativistic form. The first issue is the normalization of the states. We saw that in the discrete version of the quantized scalar field, we had in each mode state

$$|n_{\vec{k}}\rangle = \frac{1}{\sqrt{n_k!}}(\alpha_{\vec{k}}^{\dagger})^{n_{\vec{k}}}|0\rangle, \tag{4.1}$$

normalized as $\langle m|n \rangle = \delta_{mn}$. In discretizing, we had $a_{\vec{k}} \to \sqrt{V}\alpha_{\vec{k}}$ and $\delta^3(\vec{k} - \vec{k}') \to V\delta_{\vec{k}\vec{k}'}$.

However, we want to have a *relativistic normalization*

$$\langle \vec{p}|\vec{q}\rangle = 2E_{\vec{p}}(2\pi)^3\delta^3(\vec{p} - \vec{q}), \tag{4.2}$$

or, in general, for occupation numbers in all momentum modes

$$\langle \{\vec{k}_i\}|\{\vec{q}_j\}\rangle = \sum_{\pi(j)} \prod_i 2\omega_{\vec{k}_i}(2\pi)^3\delta^3(\vec{k}_i - \vec{q}_{\pi(j)}), \tag{4.3}$$

where $\{\pi(j)\}$ are permutations of the $\{j\}$ indices. We see that we are missing a factor of $2\omega_k V$ in each mode in order to get $2\omega_k\delta^3(\vec{k} - \vec{k}')$ instead of $\delta_{\vec{k}\vec{k}'}$. We therefore take the normalized states

$$\prod_{\vec{k}} \frac{1}{\sqrt{n_{\vec{k}}!}} \left(\sqrt{2\omega_{\vec{k}}}\sqrt{V(2\pi)^3}\alpha_{\vec{k}}\right)^{n_{\vec{k}}}|0\rangle \to \prod_{\vec{k}} \frac{1}{\sqrt{n_{\vec{k}}!}}[a_{\vec{k}}^{\dagger}\sqrt{2\omega_{\vec{k}}}]^{n_{\vec{k}}}|0\rangle \equiv |\{\vec{k}_i\}\rangle. \tag{4.4}$$

We now prove that we have a *relativistically invariant* formula.

First, we look at the normalization. It is obviously invariant under rotations, so we need only look at boosts:

$$p_3' = \gamma(p_3 + \beta E); \quad E' = \gamma(E + \beta p_3). \tag{4.5}$$

Since

$$\delta(f(x) - f(x_0)) = \frac{1}{|f'(x_0)|}\delta(x - x_0),\tag{4.6}$$

and a boost acts only on p_3, but not on p_1, p_2, we have

$$\delta^3(\vec{p} - \vec{q}) = \delta^3(\vec{p}' - \vec{q}')\frac{dp_3'}{dp_3} = \delta(\vec{p}' - \vec{q}')\gamma\left(1 + \beta\frac{dE_3}{dp_3}\right) = \delta^3(\vec{p}' - \vec{q}')\frac{\gamma}{E}(E + \beta p_3)$$

$$= \delta^3(\vec{p}' - \vec{q}')\frac{E'}{E}.\tag{4.7}$$

This means that $E\delta^3(\vec{p} - \vec{q})$ is relativistically invariant, as we wanted.

Also, the expansion of the scalar field

$$\phi(\vec{x}, t) = \int \frac{d^3p}{(2\pi)^3}\frac{1}{\sqrt{2E_p}}(a_{\vec{p}}e^{ipx} + a_{\vec{p}}^\dagger e^{-ipx})|_{p^0=E_p}\tag{4.8}$$

contains the relativistic invariants $e^{\pm ipx}$ and $\sqrt{2E_p}a_{\vec{p}}$ (since they create a relativistically invariant normalization, these operators are relativistically invariant), but we also have a relativistically invariant measure

$$\int \frac{d^3p}{(2\pi)^3}\frac{1}{2E_p} = \int \frac{d^4p}{(2\pi)^4}(2\pi)\delta(p^2 + m^2)|_{p^0>0}\tag{4.9}$$

(since $\delta(p^2 + m^2) = \delta(-(p^0)^2 + E_p^2)$, and then we use (4.6), allowing us to write

$$\phi(x) \equiv \phi(\vec{x}, t) = \int \frac{d^4p}{(2\pi)^4}(2\pi)\delta(p^2 + m^2)|_{p^0>0}\left(\sqrt{2E_p}a_{\vec{p}}e^{ipx} + \sqrt{2E_p}a_{\vec{p}}^\dagger e^{-ipx}\right)|_{p^0=E_p}.$$
$$\tag{4.10}$$

4.2 Canonical Quantization of the Complex Scalar Field

We now turn to the quantization of the complex scalar field, needed in order to understand the physics of propagation in quantum field theory.

The Lagrangian for the complex scalar field is

$$\mathcal{L} = -\partial_\mu\phi\partial^\mu\phi^* - m^2|\phi|^2 - U(|\phi|^2).\tag{4.11}$$

This Lagrangian has a $U(1)$ global symmetry $\phi \to \phi e^{i\alpha}$, or in other words ϕ is *charged under the U(1) symmetry*.

Note the absence of the factor $1/2$ in the kinetic term for ϕ with respect to the real scalar field. The reason is that we treat ϕ and ϕ^* as independent fields. Then the equation of motion of ϕ is $(\partial_\mu\partial^\mu - m^2)\phi^* = \partial U/\partial\phi$. We could write the Lagrangian as a sum of two real scalars, but then with a factor of $1/2$, $-\partial_\mu\phi_1\partial^\mu\phi_1/2 - \partial_\mu\phi_2\partial^\mu\phi_2/2$, since then we get $(\partial_\mu\partial^\mu - m^2)\phi_1 = \partial U/\partial\phi_1$.

Exactly paralleling the discussion of the real scalar field, we obtain an expansion in terms of a and $a^\dagger$ operators, just that now we have complex fields, with twice as many degrees of freedom, so we have $a_\pm$ and $a_\pm^\dagger$, with half the degrees of freedom in ϕ and half in $\phi^\dagger$:

$$\phi(\vec{x},t) = \int \frac{d^3p}{(2\pi)^3} \frac{1}{\sqrt{2\omega_p}} \left(a_+(\vec{p},t)e^{i\vec{p}\cdot\vec{x}} + a_-^\dagger(\vec{p},t)e^{-i\vec{p}\cdot\vec{x}} \right),$$

$$\phi^\dagger(\vec{x},t) = \int \frac{d^3p}{(2\pi)^3} \frac{1}{\sqrt{2\omega_p}} \left(a_+^\dagger(\vec{p},t)e^{-i\vec{p}\cdot\vec{x}} + a_-(\vec{p},t)e^{i\vec{p}\cdot\vec{x}} \right),$$

$$\pi(\vec{x},t) = \int \frac{d^3p}{(2\pi)^3} \left(-i\sqrt{\frac{\omega_p}{2}} \right) \left(a_-(\vec{p},t)e^{i\vec{p}\cdot\vec{x}} - a_+^\dagger(\vec{p},t)e^{-i\vec{p}\cdot\vec{x}} \right),$$

$$\pi^\dagger(\vec{x},t) = \int \frac{d^3p}{(2\pi)^3} \left(i\sqrt{\frac{\omega_p}{2}} \right) \left(a_-^\dagger(\vec{p},t)e^{-i\vec{p}\cdot\vec{x}} - a_+(\vec{p},t)e^{i\vec{p}\cdot\vec{x}} \right). \tag{4.12}$$

As before, this ansatz is based on the harmonic oscillator, whereas the form of ω_p comes out of the Klein–Gordon (KG) equation. Substituting this ansatz inside the canonical quantization commutator

$$[\phi(\vec{x},t), \pi(\vec{x}',t)] = i\hbar\delta^3(\vec{x} - \vec{x}') \tag{4.13}$$

and its complex conjugate, we find as before

$$[a_\pm(\vec{p},t), a_\pm^\dagger(\vec{p}',t)] = (2\pi)^3\delta^3(\vec{p} - \vec{p}'), \tag{4.14}$$

and the rest being zero (the proof is left as an exercise). Again, we note the equal time for the commutators. Also, the time dependence is the same as before.

We can calculate the $U(1)$ charge operator (left as an exercise), obtaining

$$Q = \int \frac{d^3k}{(2\pi)^3} \left[a_{+\vec{k}}^\dagger a_{+\vec{k}} - a_{-\vec{k}}^\dagger a_{-\vec{k}} \right]. \tag{4.15}$$

Thus, as expected from the notation used, a_+ has charge $+$ and a_- has charge $-$, and therefore we have

$$Q = \int \frac{d^3k}{(2\pi)^3} [N_{+\vec{k}} - N_{-\vec{k}}] \tag{4.16}$$

(the number of $+$ charges minus the number of $-$ charges).

We then see that ϕ creates $-$ charge and annihilates $+$ charge, and $\phi^\dagger$ creates $+$ charge and annihilates $-$ charge.

Since in this simple example there are no other charges, we see that $+$ and $-$ particles are *particle–antiparticle pairs* (i.e. pairs which are equal in everything, except they have opposite charges). As promised in Chapter 3, we have now introduced the concept of an antiparticle, and it is related to the existence of positive and negative frequency modes.

We also see now that for a real field, the particle is its own antiparticle, since $\phi = \phi^\dagger$ identifies "creating $+$ charge with creating $-$ charge" (really, there is no charge now).

4.3 Two-Point Functions and Propagators

In this section, we consider the object

$$\langle 0|\phi^\dagger(x)\phi(y)|0\rangle \tag{4.17}$$

corresponding to propagation from $y = (t_y, \vec{y})$ to $x = (t_x, \vec{x})$, the same way that in quantum mechanics $\langle q', t'|q, t\rangle$ corresponds to propagation from (q, t) to (q', t'). This object corresponds to a measurement of the field ϕ at y, then of $\phi^\dagger$ at x.

For simplicity, we will analyze the real scalar field, and we will use the complex scalar only for interpretation. Substituting the expansion of $\phi(x)$, since $a|0\rangle = \langle 0|a^\dagger = 0$, and for $\phi = \phi^\dagger$, we have

$$\langle 0|\phi^\dagger\phi|0\rangle \sim \langle 0|(a + a^\dagger)(a + a^\dagger)|0\rangle, \tag{4.18}$$

only

$$\langle 0|a_{\vec{p}}a_{\vec{q}}^\dagger|0\rangle e^{i(\vec{p}\cdot\vec{x} - \vec{q}\cdot\vec{y})} \tag{4.19}$$

survives in the sum, and, as $\langle 0|aa^\dagger|0\rangle = \langle 0|a^\dagger a|0\rangle + [a, a^\dagger]\langle 0|0\rangle$, we get $(2\pi)^3\delta(\vec{p} - \vec{q})$ from the expectation value. Then finally we obtain, for the scalar propagation from y to x:

$$D(x - y) \equiv \langle 0|\phi(x)\phi(y)|0\rangle = \int \frac{d^3p}{(2\pi)^3}\frac{1}{2E_p}e^{ip(x-y)}. \tag{4.20}$$

We now analyze what happens for varying $x - y$. By Lorentz transformations, we have only two cases to analyze.

(a) For timelike separation, we can put $t_x - t_y = t$ and $\vec{x} - \vec{y} = 0$. In this case, using $d^3p = d\Omega p^2 dp$ and $dE/dp = p/\sqrt{p^2 + m^2}$:

$$D(x - y) = 4\pi \int_0^\infty \frac{p^2 dp}{(2\pi)^3}\frac{1}{2\sqrt{p^2 + m^2}}e^{-i\sqrt{p^2 + m^2}t}$$

$$= \frac{1}{4\pi^2}\int_m^\infty dE\sqrt{E^2 - m^2}e^{-iEt} \stackrel{t\to\infty}{\propto} e^{-imt}, \tag{4.21}$$

which is oscillatory (i.e. it doesn't vanish). But that's fine, since in this case, we remain in the same point as time passes, so the probability should be large.

(b) For spacelike separation, $t_x = t_y$ and $\vec{x} - \vec{y} = \vec{r}$, we obtain

$$D(\vec{x} - \vec{y}) = \int \frac{d^3p}{(2\pi)^3}\frac{1}{2E_p}e^{i\vec{p}\cdot\vec{r}} = 2\pi \int_0^\infty \frac{p^2 dp}{2E_p(2\pi)^3}\int_{-1}^1 d(\cos\theta)e^{ipr\cos\theta}$$

$$= 2\pi \int_0^\infty \frac{p^2 dp}{2E_p(2\pi)^3}\left[\frac{e^{ipr} - e^{-ipr}}{ipr}\right] = \frac{-i}{(2\pi)^2 2r}\int_{-\infty}^{+\infty} p dp\frac{e^{ipr}}{\sqrt{p^2 + m^2}}, \tag{4.22}$$

where in the last line we have redefined in the second term $p \to -p$, and then added up $\int_0^\infty$ to $\int_{-\infty}^0$.

In the last form, we have e^{ipr} multiplying a function with poles at $p = \pm im$, so we know, by a theorem from complex analysis, that we can consider the integral in the complex p plane, and add for free the integral on an infinite semicircle in the upper half plane, since

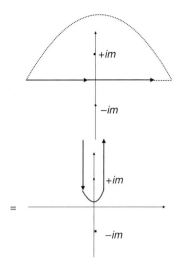

Figure 4.1 By closing the contour in the upper half plane, we pick up the residue of the pole in the upper half plane, at $+im$.

then $e^{ipr} \propto e^{-\mathrm{Im}(p)r} \to 0$. Thus, closing the contour, we can use the residue theorem and say that our integral equals the residue in the upper half plane (i.e. at $+im$), see Figure 4.1. Looking at the residue for $r \to \infty$, its leading behavior is

$$D(\vec{x} - \vec{y}) \propto e^{i(im)r} = e^{-mr}. \tag{4.23}$$

But for spacelike separation, at $r \to \infty$, we are much outside the lightcone (in no time, we move in space), and yet, propagation gives a small but nonzero amplitude, which is not fine.

But the relevant question is, will measurements be affected? We will see later why, but the only relevant issue is whether the commutator $[\phi(x), \phi(y)]$ is nonzero for spacelike separation. We thus compute

$$
\begin{aligned}
[\phi(x), \phi(y)] &= \int \frac{d^3p}{(2\pi)^3} \int \frac{d^3q}{(2\pi)^3} \frac{1}{\sqrt{2E_p 2E_q}} [(a_{\vec{p}} e^{ipx} + a_{\vec{p}}^{\dagger} e^{-ipx}), (a_{\vec{q}} e^{iqy} + a_{\vec{q}}^{\dagger} e^{-iqy})] \\
&= \int \frac{d^3p}{(2\pi)^3} \frac{1}{2E_p} (e^{ip(x-y)} - e^{ip(y-x)}) \\
&= D(x-y) - D(y-x). \tag{4.24}
\end{aligned}
$$

But if $(x - y)^2 > 0$ (spacelike), $(x - y) = (0, \vec{x} - \vec{y})$ and we can make a Lorentz transformation (a rotation, really) $(\vec{x} - \vec{y}) \to -(\vec{x} - \vec{y})$, leading to $(x - y) \to -(x - y)$. But since $D(x - y)$ is Lorentz invariant, it follows that for spacelike separation we have $D(x - y) = D(y - x)$, and therefore

$$[\phi(x), \phi(y)] = 0, \tag{4.25}$$

and we have causality. Note that this is due to the existence of negative frequency states (e^{ipx}) in the scalar field expansion. In contrast, we should also check that for timelike separation we have a nonzero result. Indeed, for $(x - y)^2 < 0$, we can set $(x - y) =$

$(t_x - t_y, 0)$ and so $-(x - y) = (-(t_x - t_y), 0)$ corresponds to time reversal, so is not a Lorentz transformation, therefore we have $D(-(x - y)) \neq D(x - y)$, and so $[\phi(x), \phi(y)] \neq 0$.

4.4 Propagators: Retarded and Feynman

4.4.1 Klein–Gordon Propagators

We are finally ready to describe the propagator.

Consider the c-number

$$[\phi(x), \phi(y)] = \langle 0 | [\phi(x), \phi(y)] | 0 \rangle = \int \frac{d^3 p}{(2\pi)^3} \frac{1}{2E_p} (e^{ip(x-y)} - e^{-ip(x-y)})$$

$$= \int \frac{d^3 p}{(2\pi)^3} \left[\frac{1}{2E_p} e^{ip(x-y)} \big|_{p^0 = E_p} + \frac{1}{-2E_p} e^{ip(x-y)} \big|_{p^0 = -E_p} \right]. \quad (4.26)$$

For $x^0 > y^0$, we can write it as

$$\int \frac{d^3 p}{(2\pi)^3} \int_C \frac{dp^0}{2\pi i} \frac{1}{p^2 + m^2} e^{ip(x-y)}, \quad (4.27)$$

where the contour C is on the real line, except it avoids slightly above the two poles at $p^0 = \pm E_p$, and then, in order to select both poles, we need to close the contour below, with an infinite semicircle in the lower half plane, as in Figure 4.2. Closing the contour below works, since for $x^0 - y^0 > 0$, we have $e^{-ip^0(x^0 - y^0)} = e^{Im(p^0)(x^0 - y^0)} \to 0$. Note that this way we get a contour closed clockwise, hence its result is minus the residue (plus the residue is for a contour closed anticlockwise), giving the extra minus sign for the contour integral to reproduce the right result.

In contrast, for this contour, if we have $x^0 < y^0$ instead, the same reason above says we need to close the contour above (with an infinite semicircle in the upper half plane). In this case, there are no poles inside the contour, therefore the integral is zero.

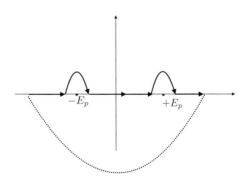

Figure 4.2   For the retarded propagator, the contour is such that closing it in the lower half plane picks up both poles at $\pm E_p$.

4.4.2 Retarded Propagator

We can then finally write for the *retarded propagator* (i.e. one that vanishes for $x^0 < y^0$)

$$D_R(x - y) \equiv \theta(x^0 - y^0)\langle 0|[\phi(x), \phi(y)]|0\rangle = \int \frac{d^3p}{(2\pi)^3} \frac{dp^0}{2\pi i} \frac{1}{p^2 + m^2} e^{ip(x-y)}$$

$$= \int \frac{d^4p}{(2\pi)^4} \frac{-i}{p^2 + m^2} e^{ip(x-y)}, \qquad (4.28)$$

where $\theta(x)$ is the Heaviside function.

The object above is a Green's function for the KG operator. This is easier to see in momentum space. Indeed, making a Fourier transform

$$D_R(x - y) = \int \frac{d^4p}{(2\pi)^4} e^{ip(x-y)} D_R(p), \qquad (4.29)$$

we obtain

$$D_R(p) = \frac{-i}{p^2 + m^2} \Rightarrow (p^2 + m^2)D_R(p) = -i, \qquad (4.30)$$

which means that in x space (the Fourier transform of 1 is $\delta(x)$, and the Fourier transform of p^2 is $-\partial^2$):

$$(\partial^2 - m^2)D_R(x - y) = i\delta^4(x - y) \leftrightarrow -i(\partial^2 - m^2)D_R(x - y) = \delta^4(x - y). \qquad (4.31)$$

4.4.3 Feynman Propagator

Consider now a different "$i\epsilon$ prescription" for the contour of integration C. Consider a contour that avoids slightly below the $-E_p$ pole and avoids slightly above the $+E_p$ pole, as in Figure 4.3. This is equivalent to the Feynman prescription for the propagator, changing D_R into D_F, defined as

$$D_F(x - y) = \int \frac{d^4p}{(2\pi)^4} \frac{-i}{p^2 + m^2 - i\epsilon} e^{ip(x-y)}. \qquad (4.32)$$

Since $p^2 + m^2 - i\epsilon = -(p^0)^2 + E_p^2 - i\epsilon = -(p^0 + E_p - i\epsilon/2)(p^0 - E_p + i\epsilon/2)$, we have poles at $p^0 = \pm(E_p - i\epsilon/2)$, so in this form we have a contour along the real line, but the poles are modified instead ($+E_p$ is moved below, and $-E_p$ is moved above the contour).

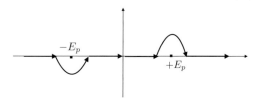

Figure 4.3   The contour for the Feynman propagator avoids $-E_p$ from below and $+E_p$ from above.

For this contour, as for D_R (for the same reasons), for $x^0 > y^0$ we close the contour below (with an infinite semicircle in the lower half plane). The result for the integration is then the residue inside the closed contour (i.e. the residue at $+E_p$). But we saw that the clockwise residue at $+E_p$ is $D(x-y)$ (and the clockwise residue at $-E_p$ is $-D(y-x)$).

For $x^0 < y^0$, we need to close the contour above (with an infinite semicircle in the upper half plane), and then we get the anticlockwise residue at $-E_p$, therefore $+D(y-x)$. The final result is then

$$D_F(x-y) = \theta(x^0 - y^0)\langle 0|\phi(x)\phi(y)|0\rangle + \theta(y^0 - x^0)\langle 0|\phi(y)\phi(x)|0\rangle \equiv \langle 0|T(\phi(x)\phi(y))|0\rangle.$$

(4.33)

This is then the Feynman propagator, which again is a Green's function for the KG operator, with the *time ordering operator* T in the two-point function, the same that we defined in the n-point functions in quantum mechanics (e.g. $\langle q', t'|T[\hat{q}(t_1)\hat{q}(t_2)]|q, t\rangle$).

As suggested by the quantum-mechanical case, the Feynman propagator will appear in the Feynman rules, and has a physical interpretation as propagation of the particle excitations of the quantum field.

Important Concepts to Remember

- The expansion of the free scalar field in quantum fields is relativistically invariant, as is the relativistic normalization.
- The complex scalar field is quantized in terms of $a_\pm$ and $a_\pm^\dagger$, which correspond to $U(1)$ charge ± 1. ϕ creates minus particles and destroys plus particles, and $\phi^\dagger$ creates plus particles and destroys minus particles.
- The plus and minus particles are particle–antiparticle pairs, since they only differ by their charge.
- The object $D(x-y) = \langle 0|\phi(x)\phi(y)|0\rangle$ is nonzero much outside the lightcone, however $[\phi(x), \phi(y)]$ is zero outside the lightcone, and since only this object leads to measurable quantities, quantum field theory is causal.
- $D_R(x-y)$ is a retarded propagator and corresponds to a contour of integration that avoids the poles $\pm E_p$ from slightly above, and is a Green's function for the KG operator.
- $D_F(x-y) = \langle 0|T[\phi(x)\phi(y)]|0\rangle$, the Feynman propagator, corresponds to the $-i\epsilon$ prescription (i.e. avoids $-E_p$ from below and $+E_p$ from above), is also a Green's function for the KG operator, and will appear in Feynman diagrams. It has the physical interpretation of propagation of particle excitations of the quantum field.

Further Reading

See section 2.4 in [1] and section 2.4 in [2].

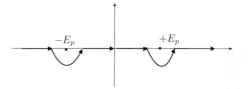

Figure 4.4   The contour for the advanced propagator avoids both poles at $\pm E_p$ from below.

Exercises

1. For the Lagrangian

$$\mathcal{L} = -\partial_\mu \phi \partial^\mu \phi^* - m^2 |\phi|^2 - U(|\phi|^2), \tag{4.34}$$

   calculate the Noether current for the $U(1)$ symmetry $\phi \to \phi e^{i\alpha}$ in terms of $\phi(\vec{x}, t)$ and show that it then reduces to the expression in the text:

$$Q = \int \frac{d^3 k}{(2\pi)^3} [a^\dagger_{+\vec{k}} a_{+\vec{k}} - a^\dagger_{-\vec{k}} a_{-\vec{k}}]. \tag{4.35}$$

2. Calculate the *advanced* propagator $D_A(x-y)$ by using the integration contour that avoids the $\pm E_p$ poles from below, instead of the contours for D_R, D_F, as in Figure 4.4.
3. Show that the canonical quantization equal time commutation relations for the complex scalar field give the same $a, a^\dagger$ relations

$$[a_\pm(\vec{p}, t), a^\dagger_\pm(\vec{p}', t)] = (2\pi)^3 \delta^3(\vec{p} - \vec{p}'), \tag{4.36}$$

   and the rest are zero.
4. Show that the Feynman propagator, and the advanced propagator defined in Exercise 2, are also Green's functions for the KG equation.

5 Interaction Picture and Wick Theorem for $\lambda\phi^4$ in Operator Formalism

Until now, we have considered the Schrödinger picture and the Heisenberg picture for the operators, but we must consider more general pictures, in particular the interaction (or Dirac) picture, in order to consider perturbation theory. Next, we lay the ground for perturbation theory in the operator formalism, by proving the Feynman theorem, relating correlation functions of Heisenberg operators in the true vacuum, in terms of an expansion in correlators of interaction picture operators in the free (unperturbed) vacuum. Finally, we prove the Wick theorem, for calculating the latter correlators.

5.1 Quantum Mechanics Pictures

Even though the Schrödinger and Heisenberg pictures are the most known, there is a continuum of quantum-mechanical descriptions. Indeed, one can transform states and operators (similar to the canonical transformations in classical mechanics) with unitary operators W ($W^\dagger = W^{-1}$), by

$$|\psi_W\rangle = W|\psi\rangle,$$
$$A_W = WAW^{-1}. \tag{5.1}$$

Then the Hamiltonian in the new formulation is

$$\mathcal{H}' = \mathcal{H}_W + i\hbar\frac{\partial W}{\partial t}W^\dagger, \tag{5.2}$$

where $\mathcal{H}_W$ stands for the original Hamiltonian, transformed as an operator by W. The statement is that $\mathcal{H}'$ generates the new time evolution, so in particular we obtain for the time evolution of the transformed operators

$$i\hbar\frac{\partial A_W}{\partial t} = i\hbar\left(\frac{\partial A}{\partial t}\right)_W + \left[i\hbar\frac{\partial W}{\partial t}W^\dagger, A_W\right], \tag{5.3}$$

and that the new evolution operator is

$$U'(t, t_0) = \hat{W}(t)U(t, t_0)W^\dagger(t_0) \neq \hat{U}_W \tag{5.4}$$

(note that on the left we have time t and on the right t_0, so we don't obtain U_W). Here the various pictures are assumed to coincide at time $t = 0$.

With this general formalism, we now review the usual pictures.

5.1.1 Schrödinger Picture (Usual)

In the Schrödinger picture, $\psi_S(t)$ depends on time, and the usual operators don't, $\partial \hat{A}_S / \partial t = 0$. We could have at most a c-number (non-operator) time dependence, like $A_s(t) \cdot \mathbf{1}$, maybe with $A_s(t)$ a c-number function.

5.1.2 Heisenberg Picture

In the Heisenberg picture, wavefunctions (states) are independent of time, $\partial \psi_H / \partial t = 0$, which is the same as saying that $\mathcal{H}' = 0$, the Hamiltonian generating the evolution of states vanishes. This means that we also have

$$\mathcal{H}_H = \mathcal{H}_S = -i\hbar \frac{\partial W}{\partial t} W^\dagger, \tag{5.5}$$

giving the time evolution of operators

$$i\hbar \frac{\partial}{\partial t} \hat{A}_H(t) = [\hat{A}_H(t), H]. \tag{5.6}$$

If, at t_0, the Schrödinger picture equals the Heisenberg picture, then

$$W(t) = U_S^{-1}(t, t_0) = U_S(t_0, t), \tag{5.7}$$

where $U_S(t, t_0)$ is the evolution operator (giving the time evolution of states) in the Schrödinger picture.

5.1.3 Dirac (Interaction Picture)

The need for the general theory of quantum-mechanical pictures was so that we could define the picture used in perturbation theory, the Dirac, or interaction, picture. We need it when we have a split

$$\hat{H} = \hat{H}_0 + \hat{H}_1, \tag{5.8}$$

where $\hat{H}_0$ is a free quadratic piece and $\hat{H}_1$ is an interacting piece. Then, we would like to write formulas like the ones already written, for the quantization of fields. Therefore we need the time evolution of operators to be only in terms of $\hat{H}_0$, leading to

$$i\hbar \frac{\partial}{\partial t} |\psi_I(t)\rangle = \hat{H}_{1I} |\psi_I(t)\rangle,$$

$$i\hbar \frac{\partial}{\partial t} \hat{A}_I(t) = [\hat{A}_I(t), \hat{H}_0], \tag{5.9}$$

where $H_{0,I} = H_{0,S}$. Note that here we have denoted the interaction piece by H_1 instead of H_i, as we will later on, in order not to confuse the subscript with I, meaning in the interaction picture. Once the distinction becomes clear, we will go back to the usual notation H_i.

5.2 Physical Scattering Set-up and Interaction Picture

Why is this picture, where operators (for instance the quantum fields) evolve with the free part ($\hat{H}_0$) only, preferred? Because of the usual physical set-up. The interaction region, where there is actual interaction between fields (or particles), like for instance in a scattering experiment, say at CERN, is finite both in space and time. This means that at $t = \pm\infty$, we can treat the states as free. And therefore, it would be useful if we could take advantage of the fact that asymptotically, the states are the free states considered in the previous chapters.

We must distinguish then between the true vacuum of the interacting theory, a state that we will call $|\Omega\rangle$, and the vacuum of the free theory, which we will call $|0\rangle$ as before, that satisfies $a_{\vec{p}}|0\rangle = 0, \forall \vec{p}$. For the full theory we should use $|\Omega\rangle$, but to use perturbation theory we will relate to $|0\rangle$.

The basic objects to study are correlators, like the two-point function $\langle\Omega|T[\phi(x)\phi(y)]|\Omega\rangle$, where $\phi(x)$ are Heisenberg operators. The interaction picture field is related to the field at the reference time t_0, where we define that the interaction picture field is the same as the Heisenberg and the Schrödinger picture fields, by

$$\phi_I(t, \vec{x}) = e^{iH_0(t-t_0)}\phi(t_0, \vec{x})e^{-iH_0(t-t_0)}$$

$$= \int \frac{d^3p}{(2\pi)^3}\frac{1}{\sqrt{2E_{\vec{p}}}}(a_{\vec{p}}e^{ipx} + a_{\vec{p}}^\dagger e^{-ipx})|_{x^0=t-t_0,p^0=E_{\vec{p}}}, \qquad (5.10)$$

since, as explained previously, $i\hbar\partial A/\partial t = [A, H_0]$ is equivalent to

$$A(t) = e^{iH_0(t-t_0)}A(t_0)e^{-iH_0(t-t_0)}, \qquad (5.11)$$

so the first line relates $\phi_I(t)$ with $\phi_I(t_0)$, and then by the definition of the reference point t_0 as the one point of equality of pictures, we can replace with the Schrödinger picture operator, with no time dependence and the usual definition of space dependence in terms of a and $a^\dagger$ (as in Chapter 3). Then the action of H_0 is replaced with the usual (free) p^0 in the exponent of e^{ipx} (again as in Chapter 3).

5.2.1 $\lambda\phi^4$ Theory

We will specialize to $\lambda\phi^4$ theory, with the free KG piece, plus an interaction

$$H = H_0 + H_1 = H_0 + \int d^3x \frac{\lambda}{4!}\phi^4 \qquad (5.12)$$

for simplicity.

5.3 Evolution Operator and the Feynman Theorem

We consider the interaction picture evolution operator (note that here t_0 is the reference time when the pictures are all equal; cf. (5.4))

$$U_I(t, t_0) = e^{iH_0(t-t_0)}e^{-iH(t-t_0)}, \tag{5.13}$$

where $e^{-iH(t-t_0)}$ is the Schrödinger picture evolution operator $U_S(t, t_0)$, that is

$$|\psi_S(t)\rangle = U_S(t, t_0)|\psi_S(t_0)\rangle,$$
$$|\psi_I(t)\rangle = U_I(t, t_0)|\psi_I(t_0)\rangle. \tag{5.14}$$

Since the Heisenberg operator has $\phi(t, \vec{x}) = e^{iH(t-t_0)}\phi(t_0, \vec{x})e^{-iH(t-t_0)}$, we have

$$\phi_I(t, \vec{x}) = e^{iH_0(t-t_0)}e^{-iH(t-t_0)}\phi(t, \vec{x})e^{iH(t-t_0)}e^{-iH_0(t-t_0)} \Rightarrow$$
$$\phi(t, \vec{x}) = U_I^\dagger(t, t_0)\phi_I(t, \vec{x})U_I(t, t_0). \tag{5.15}$$

The goal is to find an expression for $U_I(t, t_0)$, but to do that, we will first prove a differential equation

$$i\frac{\partial}{\partial t}U_I(t, t_0) = H_{i,I}U_I(t, t_0). \tag{5.16}$$

When we take the derivative on U_I, we choose to put the resulting H and H_0 in the same place, in the middle, as

$$i\frac{\partial}{\partial t}U_I(t, t_0) = e^{iH_0(t-t_0)}(H - H_0)e^{-iH(t-t_0)}. \tag{5.17}$$

But $H - H_0 = H_{i,S}$ (in the Schrödinger picture), and by the relation between Schrödinger and interaction picture operators:

$$H_{i,I} = e^{iH_0(t-t_0)}H_{i,S}e^{-iH_0(t-t_0)}, \tag{5.18}$$

and therefore

$$i\frac{\partial}{\partial t}U_I(t, t_0) = H_{i,I}e^{iH_0(t-t_0)}e^{-iH(t-t_0)} = H_{i,I}U_I(t, t_0). \tag{5.19}$$

q.e.d.

We then also note that

$$H_{i,I} = e^{iH_0(t-t_0)}H_{i,S}e^{-iH_0(t-t_0)} = \int d^3x \frac{\lambda}{4!}\phi_I^4. \tag{5.20}$$

We now try to solve this differential equation for something like $U \sim e^{-iH_I t}$ by analogy with the Schrödinger picture. However, we now need to be more careful. We can write for the first few terms in the solution

$$U_I(t, t_0) = 1 + (-i)\int_{t_0}^t dt_1 H_{i,I}(t_1) + (-i)^2\int_{t_0}^t dt_1 \int_{t_0}^{t_1} dt_2 H_{i,I}(t_1)H_{i,I}(t_2) + \dots \tag{5.21}$$

and naming these the zeroth term, the first term, and the second term, we can easily check that we have

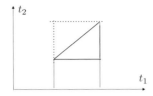

Figure 5.1   The integration in (t_1, t_2) is over a triangle, but can be written as half the integration over the rectangle.

$$i\partial_t(\text{first term}) = H_{i,I}(t) \times \text{zeroth term},$$

$$i\partial_t(\text{second term}) = H_{i,I}(t) \times \text{first term}, \qquad (5.22)$$

and this is like what we have when we write for c-numbers for instance $\partial_x e^{ax} = ae^{ax}$, and expanding $e^{ax} = \sum_n (ax)^n/n!$, we have a similar relation.

But the integration region is a problem. In the (t_1, t_2) plane, the integration domain is the lower right-angle triangle in between t_0 and t (i.e. half the rectangle bounded by t_0 and t). So we can replace the triangular domain by half the rectangular domain, see Figure 5.1. But we have to be careful, since $[H_I(t), H_I(t')] \neq 0$, and since in the integral expression above, we actually have time ordering $(t_1 > t_2)$. To obtain equality it suffices to put time ordering in front, that is

$$U_I(t, t_0) = 1 + (-i) \int_{t_0}^{t} dt_1 H_{i,I}(t_1) + \frac{(-i)^2}{2!} \int_{t_0}^{t} dt_1 \int_{t_0}^{t} dt_2 T\{H_{i,I}(t_1)H_{i,I}(t_2)\} + \dots \quad (5.23)$$

and the higher-order terms can be written in the same way, leading finally to

$$U_I(t, t_0) = T\left\{ \exp\left[-i \int_{t_0}^{t} dt' H_{i,I}(t') \right] \right\}. \qquad (5.24)$$

Now that we have an expression for U_I, we still want an expression for the vacuum of the full theory, $|\Omega\rangle$. We want to express it in terms of the free vacuum, $|0\rangle$. Consider $\langle\Omega|H|\Omega\rangle = E_0$ (the energy of the full vacuum), and we also have states $|n\rangle$ of higher energy, $E_n > E_0$. In contrast, for the free vacuum, $H_0|0\rangle = 0$. The completeness relation for the full theory is then

$$\mathbf{1} = |\Omega\rangle\langle\Omega| + \sum_{n \neq 0} |n\rangle\langle n|, \qquad (5.25)$$

and introducing it in between e^{-iHT} and $|0\rangle$ in their product, we obtain

$$e^{-iHT}|0\rangle = e^{-iE_0 T}|\Omega\rangle\langle\Omega|0\rangle + \sum_{n \neq 0} e^{-iE_n T}|n\rangle\langle n|0\rangle. \qquad (5.26)$$

Then consider $T \to \infty(1-i\epsilon)$, where we take a slightly complex time in order to dampen away the higher $|n\rangle$ modes, since then $e^{-iE_n T} \to e^{-iE_n \infty} \times e^{-E_n \infty \epsilon}$ and $e^{-E_n \infty \epsilon} \ll e^{-E_0 \infty \epsilon}$, so we can drop the terms with $n \neq 0$, with $e^{-iE_n T}$. Thus we obtain

$$|\Omega\rangle = \lim_{T \to \infty(1-i\epsilon)} \frac{e^{-iH(T+t_0)}|0\rangle}{e^{-iE_0(T+t_0)}\langle\Omega|0\rangle}$$

$$= \lim_{T \to \infty(1-i\epsilon)} \frac{e^{-iH(T+t_0)}e^{iH_0(T+t_0)}|0\rangle}{e^{-iE_0(T+t_0)}\langle\Omega|0\rangle}$$

$$= \lim_{T \to \infty(1-i\epsilon)} \frac{U_I(t_0, -T)|0\rangle}{e^{-iE_0(T+t_0)}\langle\Omega|0\rangle}, \qquad (5.27)$$

where in the first line we have isolated $T \to T + t_0$, in the second line we have introduced a term for free, due to $H_0|0\rangle = 0$, and in the third line we have formed the evolution operator. Similarly, we obtain for $\langle\Omega|$:

$$\langle\Omega| = \lim_{T\to\infty(1-i\epsilon)} \frac{\langle 0|U_I(T, t_0)}{e^{-iE_0(T-t_0)}\langle 0|\Omega\rangle}. \tag{5.28}$$

We now have all the ingredients to calculate the two-point function. Consider in the case $x^0 > y^0 > t_0$ the quantity

$$\langle\Omega|\phi(x)\phi(y)|\Omega\rangle = \lim_{T\to\infty(1-i\epsilon)} \frac{\langle 0|U_I(T, t_0)}{e^{-iE_0(T-t_0)}\langle 0|\Omega\rangle} U_I^\dagger(x^0, t_0)\phi_I(x)U_I(x^0, t_0)$$

$$\times U_I^\dagger(y^0, t_0)\phi(y)U_I(y^0, t_0)\frac{U_I(t_0, -T)|0\rangle}{e^{-iE_0(T+t_0)}\langle\Omega|0\rangle}$$

$$= \lim_{T\to\infty(1-i\epsilon)} \frac{\langle 0|U_I(T, x^0)\phi_I(x)U_I(x^0, y^0)\phi_I(y)U_I(y^0, -T)|0\rangle}{e^{-iE_0(2T)}|\langle 0|\Omega\rangle|^2}, \tag{5.29}$$

where we have used $U^\dagger(t, t') = U^{-1}(t, t') = U(t', t)$ and $U(t_1, t_2)U(t_2, t_3) = U(t_1, t_3)$.

In contrast, multiplying the expression we have for $|\Omega\rangle$ and $\langle\Omega|$, we get

$$1 = \langle\Omega|\Omega\rangle = \frac{\langle 0|U_I(T, t_0)U_I(t_0, -T)|0\rangle}{|\langle 0|\Omega\rangle|^2 e^{-iE_0(2T)}}. \tag{5.30}$$

We then divide (5.29) by it, obtaining

$$\langle\Omega|\phi(x)\phi(y)|\Omega\rangle = \lim_{T\to\infty(1-i\epsilon)} \frac{\langle 0|U_I(T, x^0)\phi_I(x)U_I(x^0, y^0)\phi_I(y)U_I(y^0, -T)|0\rangle}{\langle 0|U_I(T, -T)|0\rangle}. \tag{5.31}$$

We now observe that, since we used $x^0 > y^0 > t_0$ in our calculation, both the left-hand side and the right-hand side are time ordered, so we can generalize this relation with T symbols on both the left-hand side and the right-hand side. But then, since $T(AB) = T(BA)$ (the time ordering operators order temporally, independent of their initial order inside it), we can permute how we want the operators inside $T()$, in particular by extracting the field operators to the left and then multiplying the resulting evolution operators, $U_I(T, x^0)U_I(x^0, y^0)U_I(y^0, -T) = U_I(T, -T)$. We then use the exponential expression for the evolution operator, to finally get our desired result:

$$\langle\Omega|T\{\phi(x)\phi(y)\}|\Omega\rangle = \lim_{T\to\infty(1-i\epsilon)} \frac{\langle 0|T\left\{\phi_I(x)\phi_I(y)\exp\left[-i\int_{-T}^{T} dtH_I(t)\right]\right\}|0\rangle}{\langle 0|T\left\{\exp\left[-i\int_{-T}^{T} dtH_I(t)\right]\right\}|0\rangle}. \tag{5.32}$$

This is called **Feynman's theorem** and we can generalize it to a product of any number of ϕs, and in fact any insertions of Heisenberg operators $\mathcal{O}_H$, so we can write

$$\langle\Omega|T\{\mathcal{O}_H(x_1)\ldots\mathcal{O}_H(x_n)\}|\Omega\rangle = \lim_{T\to\infty(1-i\epsilon)} \frac{\langle 0|T\left\{\mathcal{O}_I(x_1)\ldots\mathcal{O}_I(x_n)\exp\left[-i\int_{-T}^{T} dtH_I(t)\right]\right\}|0\rangle}{\langle 0|T\left\{\exp\left[-i\int_{-T}^{T} dtH_I(t)\right]\right\}|0\rangle}. \tag{5.33}$$

5.4 Wick's Theorem

From Feynman's theorem, we see that we need to calculate $\langle 0|T\{\phi_I(x_1)\ldots\phi_I(x_n)\}|0\rangle$ (from explicit insertions, or the expansion of the exponential of $H_I(t)$s). To do that, we split

$$\phi_I(x) = \phi_I^+(x) + \phi_I^-(x), \tag{5.34}$$

where $+$ and $-$ refer to positive and negative frequency terms:

$$\phi_I^+ = \int \frac{d^3p}{(2\pi)^3} \frac{1}{\sqrt{2E_p}} a_{\vec{p}} e^{ipx},$$

$$\phi_I^- = \int \frac{d^3p}{(2\pi)^3} \frac{1}{\sqrt{2E_p}} a_{\vec{p}}^\dagger e^{-ipx}. \tag{5.35}$$

Note that then, $\phi_I^+|0\rangle = 0 = \langle 0|\phi_I^-$. We defined *normal order*, $:\,()\,:$, which we will denote here (since in the Wick's theorem case it is usually denoted like this) by $N()$, which means ϕ_I^- to the left of ϕ_I^+.

Consider then, in the $x^0 > y^0$ case:

$$T(\phi_I(x)\phi_I(y)) = \phi_I^+(x)\phi_I^+(y) + \phi_I^-(x)\phi_I^-(y) + \phi_I^-(x)\phi_I^+(y) + \phi_I^-(y)\phi_I^+(x) + [\phi_I^+(x), \phi_I^-(y)]$$

$$= N(\phi_I(x)\phi_I(y)) + [\phi_I^+(y), \phi_I^-(x)]. \tag{5.36}$$

In the $y^0 > x^0$ case, we need to change $x \leftrightarrow y$.

We now define the *contraction*

$$\overline{\phi_I(x)\phi_I(y)} = [\phi_I^+(x), \phi_I^-(y)], \quad \text{for } x^0 > y^0,$$

$$= [\phi_I^+(y), \phi_I^-(x)], \quad \text{for } y^0 > x^0. \tag{5.37}$$

Then we can write

$$T[\phi_I(x)\phi_I(y)] = N[\phi_I(x)\phi_I(y) + \overline{\phi_I(x)\phi_I(y)}]. \tag{5.38}$$

But since $[\phi^+, \phi^-]$ is a c-number, we have

$$[\phi^+, \phi^-] = \langle 0|[\phi^+, \phi^-]|0\rangle = \langle 0|\phi^+\phi^-|0\rangle = D(x - y), \tag{5.39}$$

where we have also used $\langle 0|\phi^- = \phi^+|0\rangle = 0$. Then we get

$$\overline{\phi(x)\phi(y)} = D_F(x - y). \tag{5.40}$$

Generalizing (5.38), we get *Wick's theorem*

$$T\{\phi_I(x_1)\ldots\phi_I(x_n)\} = N\{\phi_I(x_1)\ldots\phi_I(x_n) + \text{all possible contractions}\}, \tag{5.41}$$

where all possible contractions means not only full contractions, but also partial contractions. Note that for instance, if we contract 1 with 3, we obtain

$$N(\overline{\phi_1\phi_2\phi_3}\phi_4) = D_F(x_1 - x_3)N(\phi_2\phi_4). \tag{5.42}$$

Why is the Wick theorem useful? Since for any nontrivial operator, $\langle 0|N(anything)|0\rangle = 0$, as $\langle 0|\phi^- = \phi^+|0\rangle = 0$, so only the *fully contracted* terms (c-numbers) remain in $\langle 0|N()|0\rangle$, giving a simple result. Consider, for instance, the simplest nontrivial term

$$T(\phi_1\phi_2\phi_3\phi_4) = N(\phi_1\phi_2\phi_3\phi_4 + \overbrace{\phi_1\phi_2}\phi_3\phi_4 + \overbrace{\phi_1\phi_2\phi_3}\phi_4 + \overbrace{\phi_1\phi_2\phi_3\phi_4}$$

$$+ \phi_1\overbrace{\phi_2\phi_3}\phi_4 + \phi_1\overbrace{\phi_2\phi_3\phi_4} + \phi_1\phi_2\overbrace{\phi_3\phi_4}$$

$$+ \overbrace{\phi_1\phi_2}\,\overbrace{\phi_3\phi_4} + \phi_1\phi_2\phi_3\phi_4 + \phi_1\phi_2\phi_3\phi_4). \tag{5.43}$$

This then gives, under the vacuum expectation value:

$$\langle 0|T(\phi_1\phi_2\phi_3\phi_4)|0\rangle = D_F(x_1 - x_2)D_F(x_3 - x_4) + D_F(x_1 - x_4)D_F(x_2 - x_3)$$
$$+ D_F(x_1 - x_3)D_F(x_2 - x_4). \tag{5.44}$$

For the proof of Wick's theorem, we use induction. We have proved the initial step, for $n = 2$, so it remains to prove the step for n, assuming the step for $n - 1$. This is left as an exercise.

Important Concepts to Remember

- Unitary operators transform states and operators in quantum mechanics, changing the picture. The new Hamiltonian is different from the transformed Hamiltonian.
- In quantum field theory, we use the Dirac or interaction picture, where operators evolve in time with H_0, and states evolve with $H_{i,I}$.
- Perturbation theory is based on the existence of asymptotically free states (at $t = \pm\infty$ and/or spatial infinity), where we can use canonical quantization of the free theory, and use the vacuum $|0\rangle$ to perturb around, while the full vacuum is $|\Omega\rangle$.
- In the interacting theory, the interaction picture field $\phi_I(t, \vec{x})$ has the same expansion as the Heisenberg field in the free KG case studied before.
- The time-evolution operator of the interaction picture is written as $U_I(t, t_0) = T \exp[-i \int dt\, H_{i,I}(t)]$.
- Feynman's theorem expresses correlators (n-point functions) of the full theory as the ratio of vacuum expectation values *in the free vacuum of T* of interaction picture operators, with an insertion of $U_I(T, -T)$, divided by the same thing without the insertion of operators.
- The contraction of two fields gives the Feynman propagator.
- Wick's theorem relates the time order of operators with the normal order of a sum of terms: the operators, plus all their possible contractions.
- In $\langle 0|T(\ldots)|0\rangle$, which appears in Feynman's theorem as the thing to calculate, only the full contractions give a nonzero result.

Further Reading

See sections 4.2 and 4.3 in [1] and section 2.5 in [2].

Exercises

1. For an interaction Hamiltonian

$$H_I = \int d^3x \frac{\lambda}{4!} \phi_I^4, \tag{5.45}$$

   write down the *explicit* form of the Feynman theorem for the operators $\mathcal{O}_I(x)\mathcal{O}_I(y)\mathcal{O}_I(z)$, where $\mathcal{O}_I(x) = \phi^2(x)$, in order λ^2 for the numerator, and then the Wick theorem for the numerator (not all the terms, just a few *representative* ones).

2. Complete the induction step of Wick's theorem and then write all the *nonzero* terms of the Wick theorem for

$$\langle 0|T\{\phi_1\phi_2\phi_3\phi_4\phi_5\phi_6\}|0\rangle. \tag{5.46}$$

3. Consider two different ways of splitting the Hamiltonian into a free and an interaction piece, $H = H_{0,1} + H_{i,i}$ and $H = H_{0,2} + H_{i,2}$. Relate the operators and wavefunctions in the two corresponding Dirac pictures.

4. Write explicitly the expansion of the evolution operator $U_I(t, t_0)$ up to fourth order in the Hamiltonian.

6 Feynman Rules for $\lambda\phi^4$ from the Operator Formalism

In this chapter, we define and derive the Feynman rules, for calculating perturbatively the correlation functions, from the operator formalism.

We saw that the Feynman theorem relates the correlators to a ratio of vacuum expectation values (VEVs) in the free theory, of time orderings of the interaction picture fields, that is things like

$$\langle 0|T\{\phi(x_1)\ldots\phi(x_n)|0\rangle, \tag{6.1}$$

which can then be evaluated using Wick's theorem, as the sum of all possible contractions (products of Feynman propagators).

This includes insertions of

$$\left[-i\int dt H_{i,I}(t)\right]^n \tag{6.2}$$

inside the time ordering.

6.1 Diagrammatic Representation of Free Four-Point Function

We can make a diagrammatic representation for

$$\langle 0|T(\phi_1\phi_2\phi_3\phi_4)|0\rangle = D_F(x_1-x_2)D_F(x_3-x_4) + D_F(x_1-x_4)D_F(x_2-x_3)$$
$$+ D_F(x_1-x_3)D_F(x_2-x_4) \tag{6.3}$$

as the sum of three terms: one where the pairs of points $(12),(34)$ are connected by lines, the others for $(13),(24)$ and $(14),(23)$, as in Figure 6.1. We call these the *Feynman diagrams*, and we will see later how to make this more general.

These quantities that we are calculating are not yet physical, we will later relate it to some physical scattering amplitudes, but we can still use a physical interpretation of particles propagating, being created, and annihilated.

We will later define more precisely the so-called S-matrices that relate to physical scattering amplitudes, and define their relation to the correlators we compute here, but for the moment we will just note that we have something like $S_{fi} = \langle f|S|i\rangle$, where $S = U_I(+\infty, -\infty)$. So in some sense this corresponds to propagating the state $|i\rangle$ from $-\infty$ to $+\infty$, and then computing the probability of ending up in the state $|f\rangle$. For now, we will just study the abstract story, and leave the physical interpretation for later.

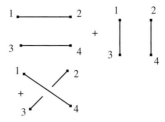

Feynman diagrams for free processes for the four-point function. We can join the external points in three different ways.

6.2 Interacting Four-Point Function: First-Order Result and its Diagrammatic Representation

Let us consider a term with a $H_{i,I}$ in it, for instance the first-order term (order λ term) in $H_{i,I}$ for the two-point functions:

$$\langle 0|T\left\{\phi(x)\phi(y)\left[-i\int dtH_{i,I}\right]\right\}|0\rangle, \tag{6.4}$$

where $\int dtH_{i,I}(t) = \int d^4z\lambda/4!\,\phi_I^4$. We then write this term as

$$\langle 0|T\left\{\phi(x)\phi(y)\left(-i\frac{\lambda}{4!}\right)\int d^4z\phi(z)\phi(z)\phi(z)\phi(z)\right\}|0\rangle. \tag{6.5}$$

In this, there are two independent Wick contractions we can consider:

- Contracting x with y and zs among themselves: $\overline{\phi(x)\phi(y)}$ and $\overline{\phi(z)\phi(z)}\,\overline{\phi(z)\phi(z)}$.
- Contracting x with z and y with z: $\overline{\phi(x)\phi(z)}$, $\overline{\phi(y)\phi(z)}$, $\overline{\phi(z)\phi(z)}$.

The first contraction can be done in three ways, because we can choose in three ways which is the contraction partner for the first $\phi(z)$ (and the last contraction is then fixed).

The second contraction can be done in $4 \times 3 = 12$ ways, since we have four choices for the $\phi(z)$ to contract with $\phi(x)$, then three remaining choices for the $\phi(z)$ to contract with $\phi(y)$. In total we have $3 + 12 = 15$ terms, corresponding to having the six ϕs, meaning five ways to make the first contraction between them, three ways to make the second, and one way to make the last.

The result is then

$$\langle 0|T\left\{\phi(x)\phi(y)\left[-i\int dtH_{i,I}\right]\right\}|0\rangle = 3\left(-i\frac{\lambda}{4!}\right)D_F(x-y)\int d^4zD_F(z-z)D_F(z-z)$$

$$+12\left(-i\frac{\lambda}{4!}\right)\int d^4zD_F(x-z)D_F(y-z)D_F(z-z). \tag{6.6}$$

The corresponding Feynman diagrams for these terms are: a line between x and y, and a figure eight with the middle point being z, and for the second term, a line between

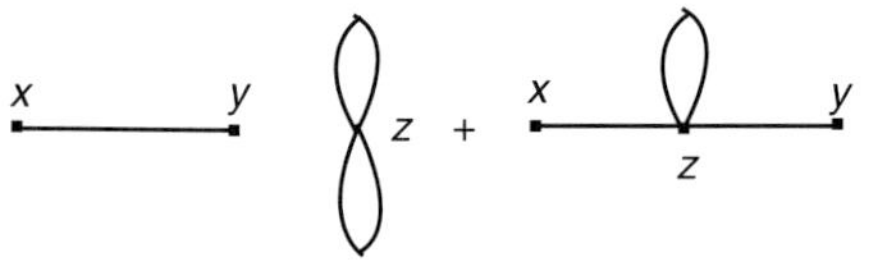

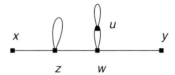

Figure 6.2   The two Feynman diagrams at order λ in the expansion (first order in the interaction vertex).

Figure 6.3   A Feynman diagram at order λ^3, forming three loops.

x and y with a z in the middle, and an extra loop starting and ending again at z, as in Figure 6.2.

6.3 Other Contractions and Diagrams

Consider now a more involved contraction: for instance, in the $\mathcal{O}(\lambda^3)$ term

$$\langle 0|T\left\{\phi(x)\phi(y)\frac{1}{3!}\left(-i\frac{\lambda}{4!}\right)^3\int d^4z\phi(z)\phi(z)\phi(z)\phi(z)\int d^4w\phi(w)\phi(w)\phi(w)\phi(w)\int d^4u\right.$$

$$\left.\times\,\phi(u)\phi(u)\phi(u)\phi(u)\right\}|0\rangle, \tag{6.7}$$

contract $\phi(x)$ with a $\phi(z)$, $\phi(y)$ with a $\phi(w)$, a $\phi(z)$ with a $\phi(w)$, two $\phi(w)$s with two $\phi(u)$s, and the remaining $\phi(z)$s between them, the remaining $\phi(u)$s between them. This gives the result

$$\frac{1}{3!}\left(\frac{-i\lambda}{4!}\right)^3\int d^4z d^4w d^4u D_F(x-z)D_F(z-z)D_F(z-w)D_F(w-u)D_F(w-u)D_F(u-u)D_F(w-y). \tag{6.8}$$

The corresponding Feynman diagram then has the points x and y with a line in between them, on which we have the points z and w, with an extra loop starting and ending at z, two lines forming a loop between w and an extra point u, and an extra loop starting and ending at u, as in Figure 6.3.

Let us count the number of identical contractions for this diagram: there are 3! ways of choosing which to call z, w, u (permutations of three objects). Then in $\int d^4z\phi\phi\phi\phi$, we can choose the contractions with $\phi(x)$ and with $\phi(w)$ in 4×3 ways, in $\int d^4w\phi\phi\phi\phi$ we can choose the contractions with z, y and u, u in $4\times3\times2\times1$ ways, and then choose the contractions in $\int d^4u\phi\phi\phi\phi$ with $\phi(w)\phi(w)$ in 4×3 ways. But then, we overcounted in choosing the two $w-u$ contractions, by counting a 2 in both w and u, so we must divide by 2, for a total of

$$3!\times(4\times3)\times(4\times3\times2)\times(4\times3)\times\frac{1}{2}=\frac{3!\times(4!)^3}{8}. \tag{6.9}$$

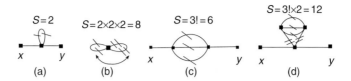

Figure 6.4   Examples of symmetry factors. (a) $S = 2$, (b) $S = 8$, (c) $S = 6$, and (d) $S = 12$.

We note that this almost cancels the $1/(3! \times (4!)^3)$ in front of the Feynman diagram result. We therefore define the *symmetry factor*, in general

$$\frac{\text{denominator}}{\text{no. of contractions}} = \frac{p!\,(4!)^p}{\#\ \text{contractions}}, \qquad (6.10)$$

which in fact equals the number of symmetries of the Feynman diagram. Here the symmetry factor is

$$2 \times 2 \times 2 = 8, \qquad (6.11)$$

which comes about because we can interchange the two ends of $D_F(z - z)$ in the Feynman diagram, obtaining the same diagram, then we can interchange the two ends of $D_F(u - u)$ in the same way, and we can also interchange the two $D_F(w - u)$s, obtaining the same diagram.

Let us see a few more examples. Consider the one-loop diagram with a line between x and y, with a point z in the middle, with an extra loop starting and ending at the same z, see Figure 6.4(a). We can interchange the two ends of $D_F(z - z)$, obtaining the same diagram, thus the symmetry factor is $S = 2$. Consider then the figure-eight diagram (vacuum bubble) with a central point z, see Figure 6.4(b). It has a symmetry factor of $S = 2 \times 2 \times 2 = 8$, since we can interchange the ends of one of the $D_F(z - z)$, also of the other $D_F(z - z)$, and we can also interchange the two $D_F(z - z)$s between them. Consider then the "setting sun" diagram, see Figure 6.4(c), with $D_F(x - z)$, three $D_F(z - w)$s, and then $D_F(w - y)$. We can permute in 3! ways the three $D_F(z - w)$s, obtaining the same diagram, thus $S = 3! = 6$. Finally, consider the diagram with $D_F(x - z)$, $D_F(z - y)$, $D_F(z - u)$, $D_F(z - w)$, and three $D_F(u - w)$s, see Figure 6.4(d). It has a symmetry factor of $S = 3! \times 2 = 12$, since there are 3! ways of interchanging the three $D_F(u - w)$s, and we can rotate the subfigure touching at point z, thus interchanging $D_F(z - u)$ with $D_F(z - w)$ and u with w.

6.4 *x*-Space Feynman Rules for $\lambda\phi^4$

We are now ready to state the *x*-space Feynman rules for the numerator of the Feynman theorem:

$$\langle 0 | T \left\{ \phi_I(x_1) \ldots \phi_I(x_n) \exp\left[-i \int H_{i,I}(t) \right] \right\} | 0 \rangle, \qquad (6.12)$$

see Figure 6.5.

- For the *propagator* we draw a line between x and y, corresponding to $D_F(x - y)$.
- For the *vertex* we draw a point z with four lines coming out of it, and it corresponds to $(-i\lambda) \int d^4z$.

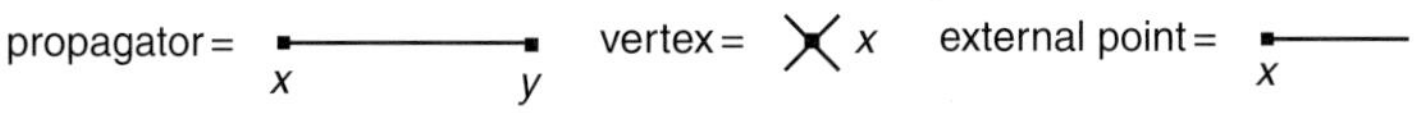

propagator = •——————•   vertex = ✕ x   external point = •——————
             x   y   x

Figure 6.5   Pictorial representations for Feynman rules in x space.

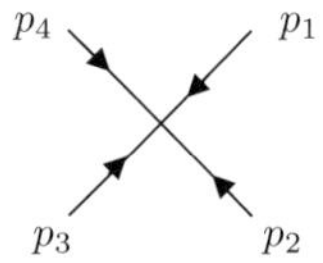

Figure 6.6   Examples of momenta at a vertex, for definition of convention.

- For the *external point* (line) at x, we draw a line ending at x, corresponding to a factor of 1 (nothing new).
- After drawing the Feynman diagram according to the above, we divide the result by the symmetry factor.
- Then, summing over all possible diagrams of all orders in λ, we get the full result for the numerator above.

We next turn to p **space**.

The Feynman propagator is written as

$$D_F(x - y) = \int \frac{d^4p}{(2\pi)^4} \frac{-i}{p^2 + m^2 - i\epsilon} e^{ip(x-y)}, \tag{6.13}$$

so the p-space propagator is

$$D_F(p) = \frac{-i}{p^2 + m^2 - i\epsilon}. \tag{6.14}$$

We must choose a direction for the propagator, and we choose it arbitrarily to be from x to y (note that $D_F(x - y) = D_F(y - x)$, so the order doesn't matter). Then, for instance, the line going into a point y corresponds to a factor of e^{-ipy} (see the expression above). Consider that at the four-point vertex z, for instance p_4 goes out, and p_1, p_2, p_3 go in, as in Figure 6.6. Then we have for the factors depending on z:

$$\int d^4z\, e^{-ip_1 z} e^{-ip_2 z} e^{-ip_3 z} e^{+ip_4 z} = (2\pi)^4 \delta^{(4)}(p_1 + p_2 + p_3 - p_4), \tag{6.15}$$

that is *momentum conservation*.

6.5 p-Space Feynman Rules and Vacuum Bubbles

Thus, the Feynman rules in p space are (see Figure 6.7):

- For the *propagator*, we write a line with an arrow in the direction of the momentum, corresponding to $D_F(p)$.
- For the *vertex*, we write four lines going into a point, corresponding to a factor of $-i\lambda$.

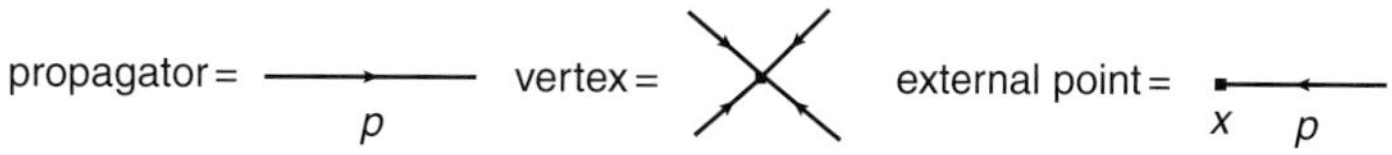

Figure 6.7 Pictorial representation of Feynman rules in p space.

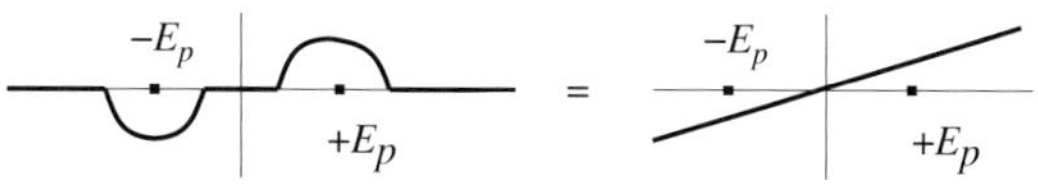

Figure 6.8 The Feynman contour can be understood in two different ways. On the right-hand side, the contour is a straight line.

- For the *external point* (line), we write an arrow going into the point x, with momentum p, corresponding to a factor of e^{-ipx}.
- We then impose momentum conservation at each vertex.
- We integrate over the internal momenta, $\int d^4p/(2\pi)^4$.
- We divide by the symmetry factor.

We should note something: why do we have the *Feynman* propagator (i.e. the Feynman contour)? The Feynman contour avoids the $+E_p$ pole from above and the $-E_p$ pole from below, and is equivalent to the $-i\epsilon$ prescription, which takes $p^0 \propto 1 + i\epsilon$, see Figure 6.8.

It is needed since without it, due to $T \to \infty(1 - i\epsilon)$, in $\int_{-T}^{T} dz^0 d^3z\, e^{i(p_1+p_2+p_3-p_4)\cdot z}$, we would have a factor of $e^{\pm(p_1^0+p_2^0+p_3^0-p_4^0)\epsilon}$, which blows up on one of the sides (positive or negative p^0). But with it, we have in the exponent $(1+i\epsilon)(1-i\epsilon) = 1+\mathcal{O}(\epsilon^2)$, so we have no factor to blow up in order ϵ.

6.5.1 Canceling of the Vacuum Bubbles in Numerator vs. Denominator in Feynman Theorem

Vacuum bubbles give infinities, for instance the p-space double figure-eight vacuum diagram, with momenta p_1 and p_2 between points z and w, and also a loop of momentum p_3 going out and then back inside z and another loop of momentum p_4 going out and then back inside w, as in Figure 6.9. Then momentum conservation at z and w, together with the integration over p_1, gives

$$\int \frac{d^4p_1}{(2\pi)^4}(2\pi)^4\delta^4(p_1+p_2) \times (2\pi)^4\delta^4(p_1+p_2) = (2\pi)^4\delta^4(0) = \infty, \qquad (6.16)$$

which can also be understood as $\int d^4w(\text{const.}) = 2T \cdot V$ (time $T \to \infty$ and volume $V \to \infty$).

But these unphysical infinities cancel out between the numerator and the denominator of the Feynman theorem: they factorize and exponentiate in the same way in the two. Let us see that.

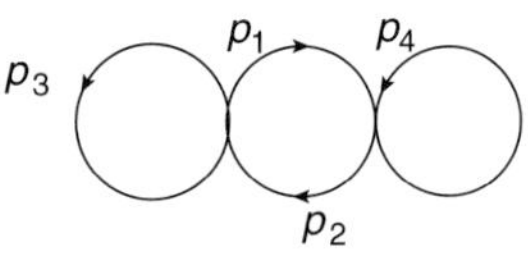

Figure 6.9   Vacuum bubble diagram, giving infinities.

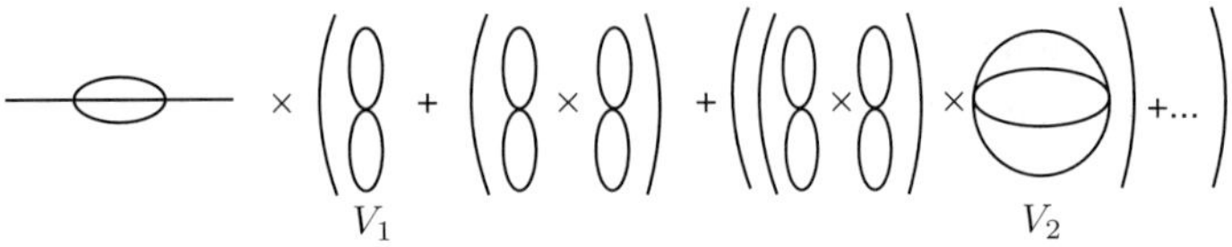

Figure 6.10   The infinite vacuum bubbles factorize in the calculation of n-point functions (here, two-point function).

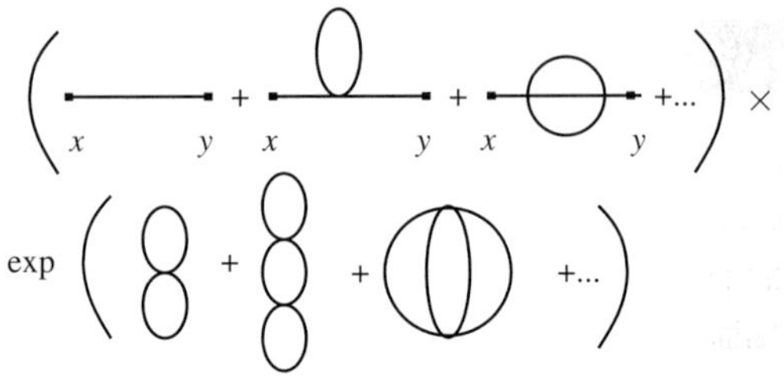

Figure 6.11   The final result for the factorization of the vacuum bubble diagrams in the two-point function, with an exponential that cancels against the same one in the denominator.

In the numerator, for the two-point function with $\phi(x)$ and $\phi(y)$, we have the connected pieces, like the setting sun diagram, factorized, multiplied by a sum of all the possible vacuum bubbles (i.e. with no external points x and y): terms like the figure-eight bubble, that we will call V_1, a product of two figure eights, $(V_1)^2$, the same times a setting sun bubble (four propagators between two internal points) called V_2, and so on. See Figure 6.10.

Then, in general, we have (the $1/n_i!$ is for permutations of identical diagrams)

$$\lim_{T\to\infty(1-i\epsilon)} \langle 0|T\left\{\phi_I(x)\phi_I(y)\exp\left[-i\int_{-T}^{T} dt H_I(t)\right]\right\}|0\rangle$$

$$= \text{connected piece} \times \sum_{\text{all }\{n_i\}\text{ sets}} \prod_i \frac{1}{n_i!}(V_i)^{n_i}$$

$$= \text{connected pieces} \times \left(\sum_{n_1} \frac{1}{n_1!}(V_1)^{n_1}\right)\left(\sum_{n_2} \frac{1}{n_2!}(V_2)^{n_2}\right) \times \dots$$

$$= \text{connected pieces} \times \prod_i \left(\sum_{n_i} \frac{1}{n_i!}(V_i)^{n_i}\right)$$

$$= \text{connected pieces} \times e^{\sum_i V_i}, \qquad (6.17)$$

as in Figure 6.11.

But the same thing happens for the denominator, obtaining

$$\lim_{T \to \infty(1-i\epsilon)} \langle 0|T\left\{\exp\left[-i\int_{-T}^{T}dtH_I(t)\right]\right\}|0\rangle = e^{\sum_i V_i}, \qquad (6.18)$$

and cancels with the numerator, obtaining that

$$\langle\Omega|T\{\phi(x)\phi(y)\}|\Omega\rangle \qquad (6.19)$$

is the sum of only the *connected diagrams*.

Thus, in the same way, in general

$$\langle\Omega|T\{\phi(x_1)\dots\phi(x_n)\}|\Omega\rangle = \text{sum of all connected diagrams.} \qquad (6.20)$$

This is the last thing to add to the Feynman rules: for the correlators, take only connected diagrams.

Important Concepts to Remember

- Feynman diagrams are pictorial representations of the contractions of the perturbative terms in the Feynman theorem. They have the interpretation of particle propagation, creation, and annihilation, but at this point, it is a bit abstract, since we have not yet related to the physical S-matrices for particle scattering.
- The symmetry factor is given by the number of symmetric diagrams (by a symmetry, we get the same diagram).
- The Feynman rules in x space have $D_F(x-y)$ for the propagator, $-i\lambda \int d^4z$ for the vertex, and 1 for the external point, dividing by the symmetry factor at the end.
- The Feynman rules in p space have $D_F(p)$ for the propagator, $-i\lambda$ for the vertex, e^{-ipx} for the external line going in x, momentum conservation at each vertex, integration over internal momenta, and dividing by the symmetry factor.
- Vacuum bubbles factorize, exponentiate, and cancel between the numerator and the denominator of the Feynman theorem, leading to the fact that the n-point functions have only connected diagrams.

Further Reading

See section 4.4 in [1].

Exercises

1. Apply the x-space Feynman rules to write down the expression for the Feynman diagram in Figure 6.12.

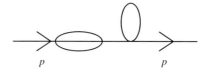

Figure 6.12 x-Space Feynman diagram.

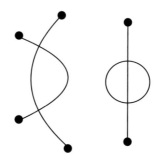

Figure 6.13 p-Space Feynman diagram.

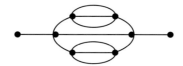

Figure 6.14   Feynman diagram for six-point function in ϕ^4 theory.

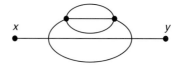

Figure 6.15   Feynman diagram in ϕ^4 theory.

2. Similarly for the p-space diagram in Figure 6.13.
3. Is the diagram in Figure 6.14 one that contributes to the six-point correlator?
4. Write the integral expression for the diagram in Figure 6.15 in both x space and p space using the Feynman rules.

The Driven (Forced) Harmonic Oscillator

In this chapter we start the analysis of the path-integral formalism for quantization. Before considering the field theory case, here we treat the simplest case, the (forced) harmonic oscillator, which, as before, has all the required features that will be generalized to field theory, once we understand it well enough.

7.1 Set-up

We have seen that for a quantum-mechanical system, we can write the transition amplitudes as path integrals, via

$$M(q', t'; q, t) \equiv {}_H\langle q', t'|q, t\rangle_H = \langle q'|e^{-i\hat{H}(t'-t)}|q\rangle$$
$$= \int \mathcal{D}p(t)\mathcal{D}q(t) \exp\left\{i\int_{t_0}^{t_{n+1}} dt[p(t)\dot{q}(t) - H(q(t), q(t))]\right\} \quad (7.1)$$

and, if the Hamiltonian is quadratic in momenta, $H(p, q) = p^2/2 + V(q)$, then

$$M(q', t'; q, t) = \mathcal{N} \int \mathcal{D}q e^{iS[q]}, \quad (7.2)$$

where $\mathcal{N}$ is a constant.

The important objects to calculate in general (in quantum mechanics $M(q't; q't)$ would be sufficient, but not in quantum field theory) are the correlators or n-point functions

$$G_N(\bar{t}_1, \ldots, \bar{t}_N) = \langle q', t'|T\{\hat{q}(\bar{t}_1)\ldots\hat{q}(\bar{t}_N)\}|q, t\rangle$$
$$= \int \mathcal{D}q(t)e^{iS[q]}q(\bar{t}_1)\ldots q(\bar{t}_N). \quad (7.3)$$

We can compute all the correlators from their generating functional

$$Z[J] = \int \mathcal{D}q e^{iS[q;J]} \equiv \int \mathcal{D}q e^{iS[q]+i\int dtJ(t)q(t)} \quad (7.4)$$

by

$$G_N(t_1, \ldots, t_N) = \frac{\delta}{i\delta J(t_1)} \cdots \frac{\delta}{i\delta J(t_N)} Z[J]\bigg|_{J=0}. \quad (7.5)$$

In the above, the object $J(t)$ was just a mathematical artifice, useful only to obtain the correlators through derivatives of $Z[J]$. But actually, considering that we have the action

$S[q; J] = S[q] + \int dt J(t) q(t)$, we see that a nonzero $J(t)$ acts as a source term for the classical $q(t)$, that is an external driving force for the harmonic oscillator, since its equation of motion is now

$$0 = \frac{\delta S[q; J]}{\delta q(t)} = \frac{\delta S[q]}{\delta q(t)} + J(t). \tag{7.6}$$

So in the presence of nonzero $J(t)$, we have a driven harmonic oscillator.

We can consider then as a field theory primer the (free) harmonic oscillator driven by an external force, with action

$$S[q; J] = \int dt \left\langle \left[\frac{1}{2}(\dot{q}^2 - \omega^2 q^2) + J(t) q(t) \right] \right\rangle, \tag{7.7}$$

which is quadratic in q, therefore the path integral is Gaussian, of the type we already performed in Chapter 2 (when we went from the phase-space path integral to the configuration-space path integral). This means that we will be able to compute the path integral exactly.

But there is an important issue of boundary conditions for $q(t)$. We will first make a naive treatment, then come back to do it better.

7.2 Sloppy Treatment

If we can partially integrate $\dot{q}^2/2$ in the action without boundary terms (not quite correct, see later), then we have

$$S[q; J] = \int dt \left\{ -\frac{1}{2}q(t) \left[\frac{d^2}{dt^2} + \omega^2 \right] q(t) + J(t) q(t) \right\}, \tag{7.8}$$

so then the path integral is of the form

$$Z[J] = \mathcal{N} \int \mathcal{D}q e^{-\frac{1}{2}q\Delta^{-1}q + iJ \cdot q}, \tag{7.9}$$

where we have defined $iS = -1/2 q \Delta^{-1} q + \ldots$, so that

$$\Delta^{-1} q(t) \equiv i \left[\frac{d^2}{dt^2} + \omega^2 \right] q(t),$$

$$J \cdot q \equiv \int dt J(t) q(t). \tag{7.10}$$

Here, Δ is the *propagator*. Then, remembering the general Gaussian integration formula

$$S = \frac{1}{2}x^T A x + b^T x \Rightarrow$$

$$\int d^n x e^{-S(x)} = (2\pi)^{n/2} (\det A)^{-1/2} e^{\frac{1}{2}b^T A^{-1} b}, \tag{7.11}$$

where now $b = -iJ(t)$ and $A = \Delta^{-1}$, we obtain

$$Z[J] = \mathcal{N}' e^{-\frac{1}{2}J \cdot \Delta \cdot J}, \tag{7.12}$$

where $\mathcal{N}'$ contains, besides $\mathcal{N}$ and factors of 2π, also $(\det \Delta)^{1/2}$, which is certainly nontrivial, however it is J-independent, so it is put as part of the overall constant. Also,

$$J \cdot \Delta \cdot J \equiv \int dt \int dt' J(t) \Delta(t, t') J(t'). \tag{7.13}$$

Here we find

$$\Delta(t, t') = i \int \frac{dp}{2\pi} \frac{e^{-ip(t-t')}}{p^2 - \omega^2}, \tag{7.14}$$

since

$$\Delta^{-1} \Delta(t, t') = i\left[\frac{d^2}{dt^2} + \omega^2\right] i \int \frac{dp}{2\pi} \frac{e^{-ip(t-t')}}{p^2 - \omega^2} = -\int \frac{dp}{2\pi} \frac{-p^2 + \omega^2}{p^2 - \omega^2} e^{-ip(t-t')}$$

$$= \int \frac{dp}{2\pi} e^{-ip(t-t')} = \delta(t - t'). \tag{7.15}$$

But we note that there is a singularity at $p^2 = \omega^2$. In fact, we have seen before that we avoided these singularities using a certain integration contour in complex p^0 space, giving the various propagators. Let us think better what this means. The question is, in the presence of this singularity, is the operator Δ^{-1} invertible, so that we can write down the above formula for Δ? An operator depends also on the space of functions on which it is defined (i.e. in the case of quantum mechanics, on the Hilbert space of the theory).

If we ask whether Δ^{-1} is invertible on the space of *all* the functions, then the answer is obviously no. Indeed, there are zero modes (i.e. eigenfunctions with eigenvalue zero for Δ^{-1}), namely ones satisfying

$$\left[\frac{d^2}{dt^2} + \omega^2\right] q_0(t) = 0, \tag{7.16}$$

and since $\Delta^{-1} q_0 = 0$, obviously on q_0 we cannot invert Δ^{-1}. Moreover, these zero modes are not even pathological, so that we can say we neglect them, but rather these are the classical solutions for the free oscillator!

So in order to find an invertible operator Δ^{-1}, we must exclude these zero modes from the Hilbert space *on which* Δ^{-1} *acts* (they are classical solutions, so obviously they exist in the theory), by imposing some boundary conditions that exclude them.

We will argue that the correct result for the propagator (inverse of Δ^{-1}) is

$$\Delta_F(t, t') = i \int \frac{dp}{2\pi} \frac{e^{-ip(t-t')}}{p^2 - \omega^2 + i\epsilon}, \tag{7.17}$$

which is the Feynman propagator (in $0 + 1$ dimensions, i.e. with only time, but no space). Indeed, check that, with $p \cdot p = -(p^0)^2$ and p^0 called simply p in quantum mechanics, we get this formula from the previously defined Feynman propagator.

Note that if Δ_F^{-1} has $\{q_i(t)\}$ eigenfunctions with $\lambda_i \neq 0$ eigenvalues, such that the eigenfunctions are orthonormal, that is

$$q_i^* \cdot q_j \equiv \int dt q_i(t)^* q_j(t) = \delta_{ij}, \tag{7.18}$$

then we can write

$$\Delta_F^{-1}(t,t') = \sum_i \lambda_i q_i(t) q_i(t)^*, \tag{7.19}$$

inverted to

$$\Delta_F(t,t') = \sum_i \frac{1}{\lambda_i} q_i(t) q_i(t')^*. \tag{7.20}$$

More generally, for an operator A with eigenstates $|q\rangle$ and eigenvalues a_q, that is $A|q\rangle = a_q|q\rangle$, with $\langle q|q'\rangle = \delta_{qq'}$, then

$$A = \sum_q a_q |q\rangle\langle q|. \tag{7.21}$$

Our operator is roughly of this form, but not quite. We have something like $q_i(t) \sim e^{-ipt}$, $\sum_i \sim \int dp/(2\pi)$, and $\lambda_i \sim (p^2 - \omega^2 + i\epsilon)$, but we must be more precise. Note that if at $t \to +\infty$, *all* $q_i(t) \to$ same function, then $\Delta_F(t \to +\infty)$ gives the same, and analogously for $t \to -\infty$.

We can in fact do the integral in Δ_F as in previous chapters, since we have poles at $p = \pm(\omega - i\epsilon)$, corresponding to the Feynman contour for the p integration. For $t - t' > 0$, we can close the contour below, since we have exponential decay in the lower half plane, and pick up the pole at $+\omega$, whereas for $t - t' < 0$ we close the contour above and pick up the pole at $-\omega$, all in all giving

$$\Delta_F(t,t') = \frac{1}{2\omega} e^{-i\omega|t-t'|}. \tag{7.22}$$

We see that for $t \to \infty$, $\Delta_F \sim e^{-i\omega t}$, whereas for $t \to -\infty$, $\Delta_F \sim e^{+i\omega t}$, therefore the boundary conditions for the functions $q(t)$ on which we define the inverse propagator are

$$\begin{aligned} q(t) &\sim e^{-i\omega t}, \quad t \to \infty, \\ q(t) &\sim e^{+i\omega t}, \quad t \to -\infty. \end{aligned} \tag{7.23}$$

Note that $[d^2/dt^2 + \omega^2]e^{\pm i\omega t} = 0$, though of course $[d^2/dt^2 + \omega^2]q(t) \neq 0$ (since it is nonzero at finite time).

This means that we must define the space of functions as functions satisfying (7.23), and we do the path integral on this space.

So, a *better definition of the path integral* is in terms of

$$q(t) = q_{cl}(t;J) + \delta q(t), \tag{7.24}$$

where the classical solution $q_{cl}(t;J)$ satisfies the correct boundary conditions, and the quantum fluctuation $\delta q(t)$ satisfies zero boundary conditions, so that it doesn't modify those of $q_{cl}(t;J)$. But we know that we have

$$S(x;J) = \frac{1}{2}Ax^2 + J \cdot x \Rightarrow$$

$$S(x;J) = S(x_0;J) + \frac{1}{2}A(x - x_0)^2 = S(x_0;J) + S(x - x_0;0), \tag{7.25}$$

where $x_0 = x_0(J)$ is the extremum. Therefore, in our case we have

$$S[q;J] = S[q_{cl};J] + S[\delta q;0],\qquad(7.26)$$

and the path integral is

$$Z[q;J] = \int \mathcal{D}q e^{iS[q;J]} = e^{iS[q_{cl};J]} \int \mathcal{D}\delta q e^{iS[\delta q;0]},\qquad(7.27)$$

meaning the path integral over δq is now J-independent (i.e. part of $\mathcal{N}$).

The classical equation of motion

$$\Delta^{-1}q_{cl}(t;J) = iJ(t)\qquad(7.28)$$

is solved by

$$q_{cl}(t;J) = q_{cl}(t;0) + i(\Delta \cdot J)(t),\qquad(7.29)$$

where $q_{cl}(t;0)$ is a zero mode, satisfying

$$\Delta^{-1}q_{cl}(t;0) = 0.\qquad(7.30)$$

Note the appearance of the zero mode $q_{cl}(t;0)$ that doesn't satisfy the correct boundary conditions. But the quantum solutions $q(t)$ on which we invert Δ_F^{-1} have correct boundary conditions instead, and do not include these zero modes. This is so since $q_{cl}(t;J)$ has the boundary conditions of the $i(\Delta \cdot J) = i\int dt' \Delta(t,t')J(t')$ term (i.e. of $\Delta(t,t')$, (the only nontrivial t dependence), as opposed to $q_{cl}(t;0)$, which has trivial boundary conditions.

Then we have

$$\frac{\delta S[q_{cl};J]}{\delta J(t)} = \int dt' \left[\frac{\delta_{\text{partial}}S}{\delta q(t')}\bigg|_{q=q_{cl}} \frac{\delta q_{cl}(t;J)}{\delta J(t)} + \frac{\delta_{\text{partial}}S}{\delta J(t)} \right],\qquad(7.31)$$

and the first term is zero by the classical equations of motion, and the second is equal to $q_{cl}(t;J)$, giving

$$\frac{\delta S[q_{cl};J]}{\delta J(t)} = q_{cl}(t;J) = q_{cl}(t;0) + i(\Delta \cdot J)(t) \Rightarrow$$

$$S[q_{cl}(J);J] = S[q_{cl}(0);0] + q_{cl}(0) \cdot J + \frac{i}{2}J \cdot \Delta \cdot J,\qquad(7.32)$$

where in the second line we have integrated over J the first line.

Then we obtain for the path integral

$$Z[J] = \mathcal{N}'' e^{-\frac{1}{2}J \cdot \Delta \cdot J + iq_{cl}(0) \cdot J},\qquad(7.33)$$

where the second term is new, and it depends on the fact that we have nontrivial boundary conditions. Indeed, this was a zero mode that we needed to introduce in q_{cl}, as it has the correct boundary conditions, as opposed to the quantum fluctuations which have zero boundary conditions.

All of the above however was more suggestive than rigorous, as we were trying to define the boundary conditions for our functions, but we didn't really need the precise boundary conditions.

7.3 Correct Treatment: Harmonic Phase Space

The correct treatment gives the path integral in terms of a modified phase-space path integral, called harmonic phase space. We now derive it.

The classical Hamiltonian of the free harmonic oscillator is

$$H(p,q) = \frac{p^2}{2} + \frac{\omega^2 q^2}{2}. \tag{7.34}$$

Making the usual definitions of the phase-space variables in terms of $a, a^\dagger$:

$$q(t) = \frac{1}{\sqrt{2\omega}}[a(t) + a^\dagger(t)],$$

$$p(t) = -i\sqrt{\frac{\omega}{2}}[a(t) - a^\dagger(t)], \tag{7.35}$$

inverted as

$$a(t) = \frac{1}{\sqrt{2\omega}}[\omega q - ip] = a(0)e^{-i\omega t},$$

$$a^\dagger(t) = \frac{1}{\sqrt{2\omega}}[\omega q + ip] = a^\dagger(0)e^{+i\omega t}, \tag{7.36}$$

the Hamiltonian is

$$H(a, a^\dagger) = \omega a^\dagger a. \tag{7.37}$$

In quantum mechanics, we quantize by $[\hat{a}, \hat{a}^\dagger] = 1$, and we can write the Fock space representation, in terms of states

$$|n\rangle = \frac{1}{\sqrt{n!}}(\hat{a}^\dagger)^n|0\rangle, \tag{7.38}$$

which are eigenstates of the Hamiltonian.

But we can also define *coherent states*

$$|\alpha\rangle = e^{\alpha\hat{a}^\dagger}|0\rangle = \sum_{n\geq 0}\frac{\alpha^n}{n!}(\hat{a}^\dagger)^n|0\rangle, \tag{7.39}$$

which are a linear combination of all the Fock space states.

These states are eigenstates of $\hat{a}$, since

$$\hat{a}|\alpha\rangle = [\hat{a}, e^{\alpha\hat{a}^\dagger}]|0\rangle = \alpha e^{\alpha\hat{a}^\dagger}|0\rangle = \alpha|\alpha\rangle, \tag{7.40}$$

where we have used

$$[\hat{a}, e^{\alpha\hat{a}^\dagger}] = \sum_{n\geq 0}\frac{\alpha^n}{n!}([\hat{a}, \hat{a}^\dagger](\hat{a}^\dagger)^{n-1} + \hat{a}^\dagger[\hat{a}, \hat{a}^\dagger](\hat{a}^\dagger)^{n-2} + \ldots (\hat{a}^\dagger)^{n-1}[\hat{a}, \hat{a}^\dagger])$$

$$= \alpha \sum_{n\geq 0}\frac{\alpha^{n-1}}{(n-1)!}(\hat{a}^\dagger)^{n-1} = \alpha e^{\alpha a^\dagger}. \tag{7.41}$$

Thus, these coherent states are eigenstates of a, with eigenvalue α.

Similarly, defining the bra state

$$\langle \alpha^* | \equiv \langle 0 | e^{\alpha^* \hat{a}}, \qquad (7.42)$$

we have

$$\langle \alpha^* | \hat{a}^\dagger = \langle \alpha^* | \alpha^*. \qquad (7.43)$$

For the inner product of coherent states, we obtain

$$\langle \alpha^* | \alpha \rangle = \langle 0 | e^{\alpha^* \hat{a}} | \alpha \rangle = e^{\alpha^* \alpha} \langle 0 | \alpha \rangle = e^{\alpha^* \alpha}, \qquad (7.44)$$

since $\langle 0 | \alpha \rangle = \langle 0 | e^{\alpha \hat{a}^\dagger} | 0 \rangle = \langle 0 | 0 \rangle = 1$, as $\langle 0 | a^\dagger = 0$. We also have the completeness relation

$$\mathbf{1} = \int \frac{d\alpha d\alpha^*}{2\pi i} e^{-\alpha \alpha^*} | \alpha \rangle \langle \alpha^* |, \qquad (7.45)$$

which is left as an exercise (note that the constant depends on the definition of the complex integration measure $dz d\bar{z}$).

We will compute the transition amplitude between the Heisenberg states $| \alpha, t \rangle_H$ and $_H \langle \alpha^*, t' |$:

$$M(\alpha^*, t'; \alpha, t) = {}_H \langle \alpha^*, t' | \alpha, t \rangle_H = \langle \alpha^* | e^{-i\hat{H}(t'-t)} | \alpha \rangle, \qquad (7.46)$$

where $| \alpha \rangle$ and $\langle \alpha^* |$ are Schrödinger states.

The Hamiltonian in the presence of a driving force is

$$H(a^\dagger, a; t) = \omega a^\dagger a - j(t) a^\dagger - \bar{j}(t) a. \qquad (7.47)$$

The sources (driving forces) $\gamma, \bar{\gamma}$ in the Hamiltonian formalism are related to the sources J defined before in the Lagrangian formalism by

$$j(t) = \frac{J(t)}{\sqrt{2\omega}}; \quad \bar{j}(t) = \frac{\bar{J}(t)}{\sqrt{2\omega}}. \qquad (7.48)$$

Indeed, we can check that for real J (i.e. $J = \bar{J}$), the coupling is $j(t) a^\dagger + \bar{j}(t) a = J(t) q(t)$. We need one more formula before we compute $M(\alpha^*, t'; \alpha, t)$, namely

$$\langle \alpha^* | \hat{H}(\hat{a}^\dagger, \hat{a}; t) | \beta \rangle = H(\alpha^*, \beta; t) \langle \alpha^* | \beta \rangle = H(\alpha^*, \beta; t) e^{\alpha^* \beta}. \qquad (7.49)$$

Now, like in the case of the phase-space path integral, we divide the path into $N+1$ small pieces, with $\epsilon \equiv (t' - t)/(N + 1)$, and times $t_0 = t, t_1, \ldots, t_n, t_{n+1} = t'$. We then insert the identity as the completeness relation (7.45) at each point t_i, $i = 1, \ldots, n$, dividing $e^{-i\hat{H}(t'-t)} = e^{-i\epsilon \hat{H}} \times \ldots \times e^{-i\epsilon \hat{H}}$. We obtain

$$M(\alpha^*, t'; \alpha, t) = \int \prod_i \left[\frac{d\alpha(t_i) d\alpha^*(t_i)}{2\pi i} e^{-\alpha^*(t_i)\alpha(t_i)} \right] \langle a^*(t') | e^{-i\epsilon \hat{H}} | \alpha(t_n) \rangle$$

$$\times \langle \alpha^*(t_n) | e^{-i\epsilon \hat{H}} | \alpha(t_{n-1}) \rangle \langle \alpha^*(t_{n-1}) | \ldots \langle \alpha^*(t_1) | e^{-i\epsilon \hat{H}} | \alpha(t) \rangle. \quad (7.50)$$

Since

$$\langle \alpha^*(t_{i+1}) | e^{-i\epsilon \hat{H}} | \alpha(t_i) \rangle = e^{-i\epsilon H(\alpha^*(t_{i+1}), \alpha(t_i))} e^{\alpha^*(t_{i+1}) \alpha(t_i)}, \qquad (7.51)$$

when we collect all the terms we obtain

$$\int \prod_i \left[\frac{d\alpha(t_i) d\alpha^*(t_i)}{2\pi i} \right] \exp\left[\alpha^*(t')\alpha(t_n) - \alpha^*(t_n)\alpha(t_n) + \alpha^*(t_n)\alpha(t_{n-1}) - \alpha^*(t_{n-1})\alpha(t_{n-1}) \right.$$

$$\left. + \ldots + \alpha^*(t_1)\alpha(t) - i \int_t^{t'} d\tau H(\alpha^*(\tau), \alpha(\tau); \tau) \right]. \tag{7.52}$$

We see that we have pairs with alternating signs, with $\alpha(t_{i+1})\alpha(t_i) - \alpha^*(t_i)\alpha(t_i)$ being the discrete version of $d\tau \dot{\alpha}^*(\tau)\alpha(\tau)$, and the last term remaining uncoupled, giving finally

$$\int \prod_i \left[\frac{d\alpha(t_i) d\alpha^*(t_i)}{2\pi i} \right] \exp\left\{ \int_t^{t'} d\tau [\dot{\alpha}^*(\tau)\alpha(\tau) - iH(\alpha^*(\tau), \alpha(\tau); \tau)] + \alpha^*(t)\alpha(t) \right\},$$
$$\tag{7.53}$$

so that we obtain

$$M(\alpha^*, t'; \alpha, t) = \int \mathcal{D}\alpha \mathcal{D}\alpha^* \exp\left\{ i \int_t^{t'} d\tau \left[\frac{\dot{\alpha}^*(\tau)}{i}\alpha(\tau) - H \right] + \alpha^*(t)\alpha(t) \right\}. \tag{7.54}$$

Important Concepts to Remember

- The source term J in the generating functional $Z[J]$ acts as a driving force, or source, for the classical action.
- At the naive level, the Feynman propagator appears as the inversion of the kinetic operator, but we need to avoid the singularities.
- To avoid the singularities, we need to define boundary conditions for the space of functions. We find that the correct boundary conditions are $e^{-i\omega t}$ at $+\infty$ and $e^{+i\omega t}$ at $-\infty$, which means that the classical driven field has these boundary conditions, but the quantum fluctuations have zero boundary conditions. In this way, the zero modes do not appear in the space on which we invert Δ^{-1}.
- With the boundary conditions, we also get a term linear in J in the exponential, $Z[J] = \mathcal{N}'' e^{-1/2 J \cdot \Delta J + i q_{cl}(0) J}$.
- The correct treatment is in harmonic phase space, in terms of coherent states $|\alpha\rangle$, which are eigenstates of $\hat{a}$.
- The transition amplitude between Heisenberg states of $|\alpha\rangle$ is written as a path integral over $\alpha(t)$ and $\alpha^*(t)$.

Further Reading

See section 1.4 in [3] and section 2.3 in [4].

Exercises

1. Using the $Z[J]$ for the free driven harmonic oscillator calculated in the text, calculate the four-point function $G_4(t_1, t_2, t_3, t_4)$ for the free driven harmonic oscillator.

2. Prove that

$$\mathbf{1} = \int \frac{d\alpha \, d\alpha^*}{2\pi i} e^{-\alpha\alpha^*} |\alpha\rangle \langle \alpha^*|, \tag{7.55}$$

and calculate

$$e^{i[\hat{a}^\dagger \hat{a} + \lambda(\hat{a} + \hat{a}^\dagger)^3]} |\alpha\rangle. \tag{7.56}$$

3. Consider the driven harmonic oscillator action (7.7), and calculate the two-point function $G_2(t_1, t_2)$, but without putting $J = 0$ at the end (as a function of $J(t)$).

4. Calculate

$$e^{\lambda(\hat{a} + \hat{a}^\dagger)^3} - : e^{\lambda(\hat{a} + \hat{a}^\dagger)^3} : \tag{7.57}$$

up to (including) $\mathcal{O}(\lambda^3)$.

8 Euclidean Formulation and Finite-Temperature Field Theory

In this chapter we will define the Euclidean formulation of quantum mechanics, to be extended in the next chapter to quantum field theory, and the finite-temperature version associated with it, making the connection to statistical mechanics.

8.1 Phase-Space and Configuration-Space Path Integrals and Boundary Conditions

But first, we finish the discussion of the path integral in harmonic phase space, and relate it to the boundary conditions for the fields in the path integral. We have seen that the transition amplitude between Heisenberg states $|\alpha, t\rangle$ and $\langle \alpha^*, t'|$ is given as a path integral:

$$M(\alpha^*, t'; \alpha, t) = \int \mathcal{D}\alpha \mathcal{D}\alpha^* \exp \left\{ i \int_t^{t'} d\tau \left[\frac{\dot{\alpha}^*(\tau)}{i} \alpha(\tau) - H \right] + \alpha^*(t)\alpha(t) \right\}. \quad (8.1)$$

But we need to understand the boundary conditions for the path integral, related to the boundary conditions for the transition amplitude.

Classically, if we have two variables, like $\alpha(t)$ and $\alpha^*(t)$, obeying linear differential equations, or one variable obeying a quadratic differential equation, we could choose to impose their values (or the values of the function and its derivative in the second case) at a given initial time t, or the same at a final time t'. But in quantum mechanics, it is not so simple. We know that the eigenvalues of two operators that don't commute cannot be given (measured with infinite precision) at the same time (in the same state).

The precise statement is that if we have operators $\hat{A}$ and $\hat{B}$ such that $[\hat{A}, \hat{B}] \neq 0$, then defining ΔA by $(\Delta A)^2 = \langle \psi | (\hat{A} - A\mathbf{1})^2 | \psi \rangle$, where $|\psi\rangle$ is some given state, we have (by a theorem in quantum mechanics) the generalized Heisenberg's uncertainty principle

$$(\Delta A)(\Delta B) \geq \frac{1}{2} |\langle \psi | i[\hat{A}, \hat{B}] | \psi \rangle|. \quad (8.2)$$

For instance, since $[\hat{q}, \hat{p}] = i\hbar$, we get $(\Delta q)(\Delta p) \geq \hbar/2$ in a given state, for instance at a given time t for a single particle in its evolution.

This means that now, since $[\hat{a}, \hat{a}^\dagger] = 1$, we cannot specify their eigenvalues α and α^* at the same time (with infinite precision). So the only possibility for boundary conditions is something that from the point of view of classical mechanics looks strange, but is fine in quantum mechanics, namely to define

- at time t, $\alpha(t) = \alpha$ and $\alpha^*(t)$ is unspecified (free), and
- at time t', $\alpha^*(t') = \alpha^*$ and $\alpha(t)$ is unspecified (free).

The equations of motion (i.e. the equations obtained from the stationarity of the path integral, by varying the exponential with respect to $\alpha(\tau)$ and $\alpha^*(\tau)$) are

$$\dot{\alpha} + i\frac{\partial H}{\partial \alpha^*} = 0 \rightarrow \dot{\alpha} + i\omega\alpha - i\,j(t) = 0,$$

$$\dot{\alpha}^* - i\frac{\partial H}{\partial \alpha} = 0 \rightarrow \dot{\alpha}^* - i\omega\alpha^* + i\,\bar{j}(t) = 0. \tag{8.3}$$

The classical solutions of this system of equations are

$$\alpha_{cl}(\tau) = \alpha e^{i\omega(t-\tau)} + i\int_t^\tau e^{i\omega(s-\tau)}j(s)ds,$$

$$\alpha^*_{cl}(\tau) = \alpha^* e^{i\omega(\tau-t')} + i\int_\tau^{t'} e^{i\omega(\tau-s)}\bar{j}(s)ds. \tag{8.4}$$

On these solutions (i.e. using the above equations of motion), we can compute that the object in the exponent of the path integral is

$$= \alpha^*(t)\alpha(t) + i\int_t^{t'} d\tau j(\tau)\alpha^*(\tau)$$

$$= \alpha^*\alpha e^{-i\omega(t'-t)} + i\int_t^{t'} ds[\alpha e^{i\omega(t-s)}\bar{j}(s) + \alpha^* e^{i\omega(s-t')}j(s)]$$

$$- \frac{1}{2}\int_t^{t'} ds \int_t^{t'} ds' j(s)\bar{j}(s')e^{-i\omega|s'-s|}. \tag{8.5}$$

In the above we have skipped some steps, which are left as an exercise.

Then, like we did in Chapter 7 for the configuration-space path integral, we can perform the Gaussian integral by shifting $\alpha(t) = \alpha_{cl}(t) + \tilde{\alpha}(t)$ and $\alpha^*(t) = \alpha^*_{cl}(t) + \tilde{\alpha}^*(t)$, where $\alpha_{cl}(t), \alpha^*_{cl}(t)$ are the above classical solutions, and the tilde quantities are quantum fluctuations. Then, by shifting the path integral, we get that (due to the quadratic nature of the exponential) the object in the exponent is written as

$$E[\alpha(t), \alpha^*(t); \gamma, \bar{\gamma}] = E[\alpha_{cl}, \alpha^*_{cl}(t); \gamma, \bar{\gamma}] + E[\tilde{\alpha}(t), \tilde{\alpha}^*(t); 0], \tag{8.6}$$

resulting in the path integral being a constant times the exponential of the classical exponent:

$$\mathcal{N}e^{E[\alpha_{cl}(t), \alpha^*_{cl}(t); \gamma, \bar{\gamma}]}. \tag{8.7}$$

We see that this object has the same structure as (7.33), since we can absorb the constant in (8.5) in $\mathcal{N}$, and then we are left with a term linear in J and a term quadratic in J in the exponent, like in the case of the configuration-space path integral.

But we still need to relate the above harmonic phase-space path integral with the configuration-space path integral of Chapter 7. First, we would need to have vacuum states at the two ends. Since $|\alpha\rangle = e^{\alpha a^\dagger}|0\rangle$, choosing $\alpha = 0$ gives $|\alpha\rangle = |0\rangle$, and choosing $\alpha^* = 0$ gives $\langle\alpha^*| = \langle 0|$. Next, we need to take $t \rightarrow -\infty$ and $t' \rightarrow \infty$, since this is the same as was required for the configuration-space path integral. In this case, in the exponent in (8.5), the constant and linear terms disappear, and we are left with the object

$$Z[J] \equiv \langle 0, +\infty | 0, -\infty \rangle_J = \exp \left\{ -\frac{1}{2} J \cdot \Delta_F \cdot J \right\} \langle 0|0 \rangle_0, \qquad (8.8)$$

where $|0, -\infty \rangle$ means $|\alpha = 0, t = -\infty \rangle$, and we have written the overall constant as $\langle 0|0 \rangle_0$, since it is indeed what we obtain for the path integral if we put $J = 0$. Note that here we have

$$\Delta_F = \frac{1}{2\omega} e^{-i\omega |s-s'|} \qquad (8.9)$$

in the quadratic piece in the critical exponent, as we can see by substituting $\gamma = J/\sqrt{2\omega}$. This, as we saw, is the Feynman propagator.

Now the boundary condition is $\alpha(t) = 0$ at $t = -\infty$, and $\alpha^*(t)$ free, in other words, pure creation part (α^* is an eigenvalue of $a^\dagger$) at $t = -\infty$, and $\alpha^*(t) = 0$ at $t = +\infty$, and $\alpha(t)$ free, in other words, pure annihilation part (α is an eigenvalue of a). Since

$$q(t) = \frac{1}{\sqrt{2\omega}} [a e^{-i\omega t} + a^\dagger e^{+i\omega t}], \qquad (8.10)$$

we see that this is consistent with the previously defined (unrigorous) boundary condition, $q(t = -\infty) \sim e^{+i\omega t}$ and $q(t = +\infty) \sim e^{-i\omega t}$. But now we have finally defined rigorously the boundary condition and the resulting path integral.

8.2 Wick Rotation to Euclidean Time and Connection with Statistical Mechanics Partition Function

But while the path integral is well defined now, its calculation in relevant cases of interest is not. A useful approximation for any path integral is what is known as a "saddle point approximation," which is the Gaussian integral around a classical solution, an extremum of the action

$$S = S_{cl} + \frac{1}{2} \delta q_i S_{ij} \delta q_j + \mathcal{O}((\delta q)^3), \qquad (8.11)$$

where $S_{cl} = S[q_{cl} : \delta S/\delta q(q_{cl}) = 0]$.

If we have a free action, this is exact and not an approximation. We have performed the Gaussian integration as if it were correct, but in reality it is only correct for real integrals, which decay exponentially, $\int_{-\infty}^{+\infty} dx e^{-\alpha x^2}$, whereas for imaginary integrals, when the integral is over a phase (purely oscillatory), $\int dx e^{-i\alpha x^2}$ is much less well defined (since if we take $\int_{-\Lambda}^{+\Lambda}$ or $\int_{-\Lambda-C}^{+\Lambda+C}$ where $\Lambda \to \infty$ and C finite, the two integrals differ by a finite contribution that is highly oscillatory in the value of C, $\sim 2 \int_0^C dx e^{-2i\alpha \Lambda x}$). And when it is a path integral instead of a single integral, it becomes even more obvious this is not so well defined.

So if we could have somehow e^{-S} instead of e^{iS}, that would solve our problems. Luckily, this is what happens when we go to *Euclidean space*.

Consider a time-independent Hamiltonian $\hat{H}$, with a complete set of eigenstates $\{|n\rangle\}$ (so that $\mathbf{1} = \sum_n |n\rangle\langle n|$) and eigenvalues $E_n > 0$. Then the transition amplitude can be written as

$$_H\langle q', t'|q, t\rangle_H = \langle q'|e^{-i\hat{H}(t'-t)}|q\rangle = \sum_n \sum_m \langle q'|n\rangle\langle n|e^{-i\hat{H}(t'-t)}|m\rangle\langle m|q\rangle$$

$$= \sum_n \langle q'|n\rangle\langle n|q\rangle e^{-iE_n(t'-t)} = \sum_n \psi_n(q')\psi_n^*(q)e^{-iE_n(t'-t)}, \quad (8.12)$$

where we have used $\langle n|e^{-i\hat{H}(t'-t)}|m\rangle = \delta_{nm}e^{-iE_n(t'-t)}$ and $\langle q|n\rangle = \psi_n(q)$. This expression is analytic in $\Delta t = t' - t$.

Now consider the analytical continuation to Euclidean time, called *Wick rotation*, $\Delta t \to -i\beta$. Then we obtain

$$\langle q', \beta|q, 0\rangle = \sum_n \psi_n(q')\psi_n^*(q)e^{-\beta E_n}. \quad (8.13)$$

We observe that if we specialize to the case $q' = q$ and integrate over this value of q, we obtain the statistical mechanics partition function of a system at a temperature T, with $kT = 1/\beta$, thus obtaining a relation to statistical mechanics!

Indeed, we then obtain

$$\int dq\langle q, \beta|q, 0\rangle = \int dq \sum_n |\psi_n(q)|^2 e^{-\beta E_n} = \text{Tr}\{e^{-\beta\hat{H}}\} = Z[\beta]. \quad (8.14)$$

This corresponds in the path integral to taking closed paths of Euclidean time length $\beta = 1/(kT)$, since $q' \equiv q(t_E = \beta) = q(t_E = 0) \equiv q$.

Let us see how we write the path integral. The Minkowski space Lagrangian is

$$L(q, \dot{q}) = \frac{1}{2}\left(\frac{dq}{dt}\right)^2 - V(q), \quad (8.15)$$

meaning the exponent in the path integral becomes

$$iS[q] = i\int_0^{t_E=\beta}(-idt_E)\left[\frac{1}{2}\left(\frac{dq}{d(-it_E)}\right)^2 - V(q)\right] \equiv -S_E[q], \quad (8.16)$$

where by definition, the *Euclidean action* is

$$S_E[q] = \int_0^\beta dt_E\left[\frac{1}{2}\left(\frac{dq}{dt_E}\right)^2 + V(q)\right] = \int dt_E L_E(q, \dot{q}). \quad (8.17)$$

We finally obtain the *Feynman–Kac formula*

$$Z(\beta) = \text{Tr}\{e^{-\beta\hat{H}}\} = \int \mathcal{D}q e^{-S_E[q]}|_{q(t_E+\beta)=q(t_E)}, \quad (8.18)$$

where the path integral is then taken over *all closed paths of Euclidean time length β*.

8.3 Quantum-Mechanical Statistical Partition Function and Correlation Functions

As we know, the partition function in statistical mechanics contains all the relevant information about the system, so using this formalism we can extract any statistical mechanics quantity of interest.

To do so, we can introduce currents $J(t)$ as usual and calculate the correlation functions. That is, we define

$$Z[\beta;J] = \int \mathcal{D}q e^{-S_E(\beta) + \int_0^\beta J_E(\tau) q_E(\tau)} d\tau. \tag{8.19}$$

Note that here, the current term is the usual one we encountered before, since

$$i \int dt J(t) q(t) = i \int d(-it_E) J(-it_E) q(-it_E) \equiv \int dt_E J_E(t_E) q_E(t_E). \tag{8.20}$$

Let's see a simple, but very useful, quantity, the *propagator in imaginary (Euclidean) time*. We immediately find in the usual way

$$\frac{1}{Z(\beta)} \frac{\delta^2 Z[\beta;J]}{\delta J(\tau_1) \delta J(\tau_2)} = \frac{1}{Z(\beta)} \int \mathcal{D}q(\tau) q(\tau_1) q(\tau_2) e^{-S_E(\beta)}. \tag{8.21}$$

Note that here in $J(\tau_1), J(\tau_2), q(\tau_1), q(\tau_2)$ we have Euclidean time (τ), but since the formula for $Z[\beta;J]$ was obtained by analytical continuation, this is equal to

$$\langle \Omega | T\{\hat{q}(-i\tau_1)\hat{q}(-i\tau_2)\} | \Omega \rangle_\beta = \frac{1}{Z(\beta)} \text{Tr}[e^{-\beta \hat{H}} T\{\hat{q}(-i\tau_1)\hat{q}(-i\tau_2)\}]. \tag{8.22}$$

Also note that here $Z(\beta)$ is a normalization constant, since it is J-independent (now we have a new parameter β, but except for this dependence, it is constant). In the above, the Heisenberg operators are also Wick rotated, that is

$$\hat{q}(t) = e^{i\hat{H}t} \hat{q} e^{-i\hat{H}t} \Rightarrow$$
$$\hat{q}(-i\tau) = e^{\hat{H}\tau} \hat{q} e^{-\hat{H}\tau}. \tag{8.23}$$

We could ask: why is it not the Euclidean correlator of the vacuum expectation value (VEV) of the time-ordered Euclidean operators? The answer is that in Euclidean space, space and time are the same, so we can't define the "time ordering operator." In a certain sense, what we obtain is then just the VEV of the product of qs in Euclidean space, for the above $\langle \Omega | \hat{q}(\tau_1) \hat{q}(\tau_2) | \Omega \rangle$, just that it is better defined as the continuation from Minkowski space of the VEV of the time-ordered product.

We can now extend the definition of the free propagator $\Delta(\tau)$ to the interval $[-\beta, \beta]$ by the periodicity of $q(\tau)$, $q(\tau + \beta) = q(\tau)$. Next we can define

$$\Delta(\tau) = \langle \Omega | T\{\hat{q}(-i\tau)\hat{q}(0)\} | \Omega \rangle_\beta, \tag{8.24}$$

such that

$$\Delta(\tau - \beta) = \Delta(\tau). \tag{8.25}$$

We could of course compute it from the path integral in principle, but consider reversely the following problem. The propagator equation

$$\left[-\frac{d^2}{dt^2} + \omega^2\right] K(\tau, \tau') = \delta(\tau, \tau'), \tag{8.26}$$

with the above periodicity, where $K(\tau, \tau') = \Delta_{free}(\tau - \tau')$, has a *unique* solution: if $\tau \in [0, \beta]$, the solution is

$$\Delta_{free}(\tau) = \frac{1}{2\omega}[(1 + n(\omega))e^{-\omega\tau} + n(\omega)e^{\omega\tau}], \tag{8.27}$$

where

$$n(\omega) = \frac{1}{e^{\beta|\omega|} - 1} \tag{8.28}$$

is the Bose–Einstein distribution. The proof of the above statement is left as an exercise.

We can also check that for $\beta \to \infty$:

$$\Delta_{free}(\tau > 0) \to \Delta_F(\tau > 0). \tag{8.29}$$

So we obtain the Feynman propagator in the limit of zero temperature ($\beta = 1/(kT) \to \infty$).

Moreover, in this zero-temperature limit, in which we have an infinite period in Euclidean time, we have

$$\langle q', \beta | q, 0 \rangle = \sum_n \psi_n(q')\psi_n^*(q)e^{-\beta E_n} \to \psi_0(q')\psi_0(q)e^{-\beta E_0}, \tag{8.30}$$

that is we only get the *vacuum contribution*.

In conclusion, we can define the sources $J(t)$ to be nonzero on a finite time interval, and take infinitely long periodic Euclidean time. Then we obtain a definition of the *vacuum functional*, where the initial and final states are the vacuum $|\Omega\rangle$, like we defined before in Minkowski space, and this functional is given by the path integral in Euclidean space.

Because of the statistical mechanics connection, we call $Z[J] = Z[\beta \to \infty; J]$ the *partition function*, and we will continue to call it that from now on. Later we will also call $-\ln Z[J] = W[J]$ the free energy, since the same happens in statistical mechanics.

Note that all we have said here is for quantum mechanics, and the corresponding statistical mechanics with the Feynman–Kac formula. But we can generalize this formalism to quantum field theory, and we will do so in Chapter 9. We can also generalize statistical mechanics to finite-temperature field theory, but we will do so only at the end of the book.

8.4 Example: Driven Harmonic Oscillator

Let us now return to our basic example for everything, the harmonic oscillator. The Euclidean partition function is

$$Z_E[J] = \int \mathcal{D}q \exp\left\{-\frac{1}{2}\int dt\left[\left(\frac{dq}{dt}\right)^2 + \omega^2 q^2\right] + \int dt J(t)q(t)\right\}. \tag{8.31}$$

We saw that in Minkowski space the partial integration of the kinetic term can introduce problematic boundary terms related to the boundary conditions. But now, in Euclidean space, we have only closed paths, so there are no boundary terms!

Therefore, we can write as before

$$Z_E[J] = \int \mathcal{D}q \exp\left\{-\frac{1}{2}\int dt q(t)\left[-\frac{d^2}{dt^2}+\omega^2\right]q(t) + \int dt J(t)q(t)\right\}$$
$$= \mathcal{N} \exp\left\{\frac{1}{2}\int ds \int ds' J(s)\Delta_E(s,s')J(s')\right\}, \qquad (8.32)$$

where the *Euclidean propagator* is defined by $-S_E = -1/2q\Delta^{-1}q + \ldots$, giving

$$\Delta_E(s,s') = \left(-\frac{d^2}{ds^2}+\omega^2\right)^{-1}(s,s') = \int \frac{dE_E}{2\pi}\frac{e^{-iE_E(s-s')}}{E_E^2+\omega^2}. \qquad (8.33)$$

Note first that the Gaussian integration above is now well defined, as we explained, since we don't have oscillatory terms anymore, but we have the usual integral of a real decaying exponential, $\int dx e^{-\alpha x^2}$. Second, the Euclidean propagator is well defined, since in the integral, we don't have any singularities anymore, as $E_E^2+\omega^2 > 0$, so we don't need to choose a particular integration contour avoiding them, as in the case of the Minkowski space propagator.

Let us summarize what we did: we Wick rotated the Minkowski space theory to Euclidean space in order to better define it. In Euclidean space we have no ambiguities anymore: the path integral is well defined, Gaussian integration and the propagators are also well defined. Then, in order to relate to physical quantities, we need to Wick rotate the calculations in Euclidean space back to Minkowski space.

Let's see that for the propagator. We have $t = -is$, where s is Euclidean time. But then we want to have the object Et be Wick rotation invariant, $Et = E_E s$, so $E_E = -iE$.

The Euclidean space propagator is well defined, but of course when we do the above Wick rotation to Minkowski space, we find the usual poles, since the Euclidean propagator has imaginary poles $E_E = \pm i\omega$, corresponding to poles at $E = \pm\omega$. Therefore we can't do the full Wick rotation, corresponding to a rotation of the integration contour with $\pi/2$ in the complex energy plane, since then we cross (or rather touch) the poles. To avoid that, we must rotate the contour only by $\pi/2 - \epsilon$:

$$E_E \rightarrow e^{-i(\frac{\pi}{2}-\epsilon)}E = -i(E+i\epsilon'). \qquad (8.34)$$

Doing this Wick rotation, we obtain

$$\Delta_E(s = it) = -i\int_{-\infty}^{+\infty}\frac{dE}{2\pi}\frac{e^{-iEt}}{-E^2+\omega^2-i\epsilon} = \Delta_F(t), \qquad (8.35)$$

that is the Feynman propagator. We see that the $\pi/2 - \epsilon$ Wick rotation corresponds to the Feynman prescription for the integration contour to avoid the poles. It comes naturally from the Euclidean space construction (any other integration contour could not be obtained from any smooth deformation of the Euclidean contour), one more reason to consider the Feynman propagator as the relevant object for path integrals.

In conclusion, let us mention that the puristic view of quantum field theory is that the Minkowski space theory is not well defined, and to define it well, we must work in Euclidean space and then analytically continue back to Minkowski space.

Also, we saw that we can have a statistical mechanics interpretation of the path integral, which makes the formalism for quantum field theory and statistical mechanics the same. A number of books take this seriously, and treat the two subjects together, but we will continue with just quantum field theory.

Important Concepts to Remember

- Relating the harmonic phase-space path integral to the configuration-space path integral, we obtain the boundary conditions that we have a pure creation part at $t = -\infty$ and a pure annihilation part at $t = +\infty$, compatible with the configuration-space boundary condition $q(t = -\infty) \sim e^{i\omega t}$, $q(t = +\infty) \sim e^{-i\omega t}$. This defines rigorously the path integral.
- The Feynman–Kac formula relates the partition function in statistical mechanics, $Z(\beta) = \text{Tr}\{e^{-\beta \hat{H}}\}$, to the Euclidean-space path integral over all closed paths in Euclidean time, with length β.
- The correlators in imaginary time (like the imaginary-time propagator) are the analytical continuation of Minkowski-space correlators.
- The free Euclidean propagator at periodicity β involves the Bose–Einstein distribution.
- The vacuum functional, or partition function $Z[J]$, is given by the Euclidean path integral of infinite periodicity.
- For the harmonic oscillator in Euclidean space, we can rigorously perform Gaussian integration to obtain the exact solution for $Z[J]$, in terms of a well-defined Euclidean propagator.
- The analytical continuation (Wick rotation) of the Euclidean propagator, by smoothly deforming the integration contour to avoid the poles in the complex plane, uniquely selects the Feynman propagator.

Further Reading

See section 1.5 in [3] and section 3.7 in [4].

Exercises

1. Complete the omitted steps in the proof for going from

$$F(\alpha^*, t'; \alpha, t) = \int \mathcal{D}\alpha \mathcal{D}\alpha^* \exp\left\{ i \int_t^{t'} d\tau \left[\frac{\dot{\alpha}^*(\tau)}{i}\alpha(\tau) - H \right] + \alpha^*(t)\alpha(t) \right\} \quad (8.36)$$

to

$$Z[J] \equiv \langle 0, +\infty | 0, -\infty \rangle_J = \exp\left\{ -\frac{1}{2} J \cdot \Delta_F \cdot J \right\} \langle 0 | 0 \rangle_0. \tag{8.37}$$

2. Prove that

$$\left[-\frac{d^2}{d\tau^2} + \omega^2 \right] K(\tau, \tau') = \delta(\tau - \tau'), \tag{8.38}$$

where $K(\tau, \tau') = \Delta_{free}(\tau - \tau')$ and $\Delta(\tau - \beta) = \Delta(\tau)$ has a *unique* solution: if $\tau \in [0, \beta]$, the solution is

$$\Delta_{free}(\tau) = \frac{1}{2\omega}[(1 + n(\omega))e^{-\omega\tau} + n(\omega)e^{\omega\tau}], \tag{8.39}$$

where

$$n(\omega) = \frac{1}{e^{\beta|\omega|} - 1}. \tag{8.40}$$

3. Wick rotate the formula

$$I(E_E) = \int \frac{dE_{1,E}}{2\pi} \int \frac{dE_{2,E}}{2\pi} \frac{1}{(E_{1,E}^2 + \omega_1^2)(E_{2,E}^2 + \omega_2^2)((E_{1,E} + E_{2,E} - E_E)^2 + \omega_3^2)}. \tag{8.41}$$

4. For the driven harmonic oscillator in Euclidean space, (8.31), calculate the four-point function $G_{4,E}(t_1, t_2, t_3, t_4)$.

9

The Feynman Path Integral for a Scalar Field

In this chapter we generalize what we have learned from quantum mechanics to the case of quantum field theory. We start by defining a Euclidean formulation, after which we define perturbation theory through the Dyson formula, and then the Wick theorem in path-integral formulation. Coupled with the solution of the free theory, this fully defines perturbation theory.

9.1 Euclidean Formulation

We have seen that in order to better define the theory, we must do a Wick rotation to Euclidean space, $t = -it_E$. If we choose closed (periodic) paths, $q(t_E + \beta) = q(\beta)$ in Euclidean time, we obtain the statistical mechanics partition function, $Z(\beta) = \text{Tr}\{e^{-\beta \hat{H}}\}$, given as a path integral $\int \mathcal{D}q e^{-S_E[q]}|_{q(t_E+\beta)=q(t_E)}$ (the Feynman–Kac formula). The Euclidean action is positive definite if the Minkowski space Hamiltonian was. To obtain the vacuum functional (transition between vacuum states), we take $\beta \to \infty$. In Euclidean space we obtain well-defined path integrals; for instance, the driven harmonic oscillator can now be easily solved. Partial integration has no problems, since periodic paths have no boundary, and we obtain $Z_E[J] = \mathcal{N}e^{\frac{1}{2}J \cdot \Delta_E \cdot J}$, where the Euclidean propagator is well defined (has no poles). Wick rotation, however, will give the same problem: a full Wick rotation with $\pi/2$ rotation of the integration contour will touch the poles, so we must rotate only with $\pi/2 - \epsilon$, obtaining the Feynman propagator. No other propagator can arise from a *smooth* deformation of the integration contour.

To generalize to field theory, as usual we think of replacing $q_i(t) \to \phi_{\vec{x}}(t) \equiv \phi(\vec{x}, t) \equiv \phi(x)$. Of course, there are issues with the correct regularization involved in this (some type of discretization of space, we have discussed that a bit already), but we will ignore this, since we only want to see how to translate quantum mechanics results into quantum field theory results.

In *Minkowski space*, the action is

$$S[\phi] = \int d^4x \left[-\frac{1}{2}\partial_\mu\phi\partial^\mu\phi - \frac{1}{2}m^2\phi^2 - V(\phi) \right], \tag{9.1}$$

and the Minkowski space n-point functions, or Green's functions, are

$$G_n(x_1, \ldots, x_n) = \langle 0|T\{\hat{\phi}(x_1)\ldots\hat{\phi}(x_n)\}|0\rangle = \int \mathcal{D}\phi e^{iS[\phi]}\phi(x_1)\ldots\phi(x_n). \tag{9.2}$$

To define the theory better, we go to *Euclidean space*, where we take only periodic paths with infinite period, obtaining the *vacuum functional*. The Euclidean action is

$$S_E[\phi] = \int d^4x \left[\frac{1}{2} \partial_\mu \phi \partial_\mu \phi + \frac{1}{2} m^2 \phi^2 + V(\phi) \right], \tag{9.3}$$

where, since we are in Euclidean space, $a_\mu b_\mu = a_\mu b^\mu = a_\mu b_\nu \delta^{\mu\nu}$, and time is defined as $t_M \equiv x^0 = -x_0 = -it_E$, $t_E = x_4 = x^4$, and so $x^4 = ix^0$.

The Euclidean space Green's functions are

$$G_n^{(E)}(x_1, \ldots, x_n) = \int \mathcal{D}\phi\, e^{-S_E[\phi]} \phi(x_1) \ldots \phi(x_n). \tag{9.4}$$

We can write their generating functional, the *partition function*, so called because of the connection with statistical mechanics which stays the same: at finite periodicity β in Euclidean time

$$Z[\beta, J] = \text{Tr}\{e^{-\beta \hat{H}_J}\} = \int \mathcal{D}\phi\, e^{-S_E[\phi] + J \cdot \phi}|_{\phi(\vec{x}, t_E + \beta) = \phi(\vec{x}, t_E)}, \tag{9.5}$$

so in the vacuum

$$Z[J] = \int \mathcal{D}\phi\, e^{-S_E[\phi] + J \cdot \phi} \equiv {}_J\langle 0 | 0 \rangle_J, \tag{9.6}$$

where in d dimensions

$$J \cdot \phi \equiv \int d^d x J(x) \phi(x). \tag{9.7}$$

So the Green's functions are obtained from derivatives as usual:

$$G_n(x_1, \ldots, x_n) = \frac{\delta}{\delta J(x_1)} \cdots \frac{\delta}{\delta J(x_n)} \int \mathcal{D}\phi\, e^{-S_E + J \cdot \phi} \Bigg|_{J=0}. \tag{9.8}$$

Note, however, the absence of the factors of i in the denominators, since we now have $J \cdot \phi$ instead of $iJ \cdot \phi$. The partition function sums the Green's functions as usual:

$$Z[J] = \sum_{n \geq 0} \frac{1}{n!} \int \prod_{i=1}^n d^d x_i G(x_1, \ldots, x_n) J(x_1) \ldots J(x_n). \tag{9.9}$$

9.2 Perturbation Theory

We now move to analyzing the perturbation theory, that is when $S[\phi] = S_0[\phi] + S_I[\phi]$, where S_0 is a free (quadratic) part and S_I is an interaction piece. As before, we will try to perturb in the interaction piece. We can generalize the harmonic oscillator case and calculate transition amplitudes between states. But even though in physical situations we need to calculate transitions between physical states, which are usually wavefunctions over particle states of given momentum, it is much easier from a theoretical point of view to calculate the Green's functions (transitions between vacuum states).

We will see, however, that the two are related. Namely, we can consider the Green's functions in momentum space, and the S-matrix elements, of the general type $S_{fi} = \langle f | S | i \rangle$, where $S = U_I(+\infty, -\infty)$. Then the S-matrix elements will be related in a more precise manner later with the *residues at all the mass-shell poles* ($p_i^2 = -m_i^2$ for external lines) of the momentum space Green's functions. The proof of this formula (which will be just written down later in the book), called the *Lehmann–Symanzik–Zimmermann (LSZ) reduction formula*, will be given in Part II, since it requires the use of renormalization.

The momentum space Green's functions are

$$\tilde{G}_n(p_1, \ldots, p_n) = \int d^d x_1 \ldots d^d x_n e^{i(p_1 x_1 + \ldots + p_n x_n)} G_n(x_1, \ldots, x_n). \tag{9.10}$$

But due to translational invariance (any good theory must be invariant under translations, $x_i \to x_i + a$):

$$G_n(x_1, \ldots, x_n) = G_n(x_1 - X_0, x_2 - X_0, \ldots, x_n - X_0). \tag{9.11}$$

Choosing $X_0 = x_1$ and changing integration variables $x_i \to x_i + X_0$, for $i = 2, \ldots, n$, we get

$$\tilde{G}_n(p_1, \ldots, p_n) = \left[\int d^d x_1 e^{i x_1 (p_1 + \ldots + p_n)} \right] \int d^d x_2 \ldots d^d x_n e^{i(x_2 p_2 + \ldots + x_n p_n)} G_n(0, x_2, \ldots, x_n)$$

$$= (2\pi)^d \delta^d(p_1 + \ldots + p_n) G_n(p_1, \ldots, p_n), \tag{9.12}$$

and we usually calculate $G_n(p_1, \ldots, p_n)$, where these external momenta already satisfy momentum conservation.

As we said, it will be much easier theoretically to work with Green's functions rather than S-matrices, and the two are related.

9.3 Dyson Formula for Perturbation Theory

This is a formula that was initially derived in the operator formalism, but in the path-integral formalism it is kind of trivial. It is useful in order to define the perturbation theory.

Working in the Euclidean case, we define $|0\rangle$ as the vacuum of the *free theory*, and we consider VEVs in the free theory of some operators, which as usual are written as path integrals:

$$\langle 0 | \mathcal{O}[\{\hat{\phi}\}] | 0 \rangle = \int \mathcal{D}\phi e^{-S_0[\phi]} \mathcal{O}[\{\phi\}]. \tag{9.13}$$

Now, considering the particular case of $\mathcal{O} = e^{-S_I[\phi]}$, we obtain

$$\int \mathcal{D}\phi e^{-S_0[\phi] - S_I[\phi]} = \langle 0 | e^{-S_I[\hat{\phi}]} | 0 \rangle. \tag{9.14}$$

If the operator $\mathcal{O}$ contains also a product of fields, we obtain the Green's functions, that is

$$G_n(x_1, \ldots, x_n) = \langle 0 | \hat{\phi}(x_1) \ldots \hat{\phi}(x_n) e^{-S_I[\hat{\phi}]} | 0 \rangle = \int \mathcal{D}\phi e^{-S_0[\phi]} \phi(x_1) \ldots \phi(x_n) e^{-S_I[\phi]}. \tag{9.15}$$

We note here again that in Euclidean space, there is no distinction between time and space, so we cannot define a "Euclidean time ordering," so in a certain sense we have the usual product. Just when we analytically continue (Wick rotate) from Minkowski space, we can define the time ordering implicitly (via the Minkowski time ordering).

Finally, for the generating functional of the Green's functions, the partition function, we can similarly write, by summing the Green's functions:

$$Z[J] = \langle 0|e^{-S_I[\hat{\phi}]}e^{\int d^dx J(x)\hat{\phi}(x)}|0\rangle = \int \mathcal{D}\phi e^{-S_0[\phi]+J\cdot\phi}e^{-S_I[\phi]}, \qquad (9.16)$$

which is called *Dyson's formula*, and as we can see, it is rather trivial in this path-integral formulation, but for S-matrices in the operator formalism in Minkowski space, where it was originally derived, it is less so.

9.4 Solution of the Free Field Theory

We now solve for the free field theory, $S_I[\phi] = 0$. We have

$$Z_0[J] = \int \mathcal{D}\phi e^{-S_0[\phi]+J\cdot\phi} = \int \mathcal{D}\phi \left\{ -\frac{1}{2}\int d^dx[\partial_\mu\phi\partial_\mu\phi + m^2\phi^2] + \int d^dx J(x)\phi(x)\right\}. \qquad (9.17)$$

In Euclidean space we have no boundary terms, so we can partially integrate the kinetic term to obtain

$$Z_0[J] = \int \mathcal{D}\phi \exp\left\{ -\frac{1}{2}\int d^dx\phi[-\partial_\mu\partial_\mu + m^2]\phi + \int d^dx J(x)\phi(x)\right\}, \qquad (9.18)$$

where $[-\partial_\mu\partial_\mu + m^2] \equiv \Delta^{-1}$, and we can then shift the integration variable by first writing (using the formal notation with scalar products)

$$Z_0[J] = \int \mathcal{D}\phi \exp\left\{ -\frac{1}{2}[\phi - J\cdot\Delta]\Delta^{-1}[\phi - \Delta\cdot J] + \frac{1}{2}J\cdot\Delta\cdot J\right\}$$
$$= e^{\frac{1}{2}J\cdot\Delta\cdot J}\int \mathcal{D}\phi' e^{-\frac{1}{2}\phi'\cdot\Delta^{-1}\phi'}, \qquad (9.19)$$

and now the path integral that remains is just a normalization constant, giving finally

$$Z_0[J] = e^{\frac{1}{2}J\cdot\Delta\cdot J}\langle 0|0\rangle_0 \equiv e^{\frac{1}{2}\int d^dx\int d^dy J(x)\Delta(x,y)J(y)}\langle 0|0\rangle_0, \qquad (9.20)$$

and we can put $Z_0[0] = \langle 0|0\rangle_0$ to 1. Note that here the propagator comes from $\Delta^{-1} = -\partial_\mu\partial_\mu + m^2$ and is given by

$$\Delta(x,y) = \int \frac{d^dp}{(2\pi)^d}\frac{e^{ip(x-y)}}{p^2 + m^2}, \qquad (9.21)$$

and now again it is nonsingular, since in the denominator $p^2+m^2 > 0$. But again, analytical continuation back to Minkowski space (Wick rotating back) can lead to the usual poles at $|p| = \pm im$, so we need to rotate the integration contour in the complex p^0 plane with $\pi/2 - \epsilon$ instead of $\pi/2$. Then we obtain the Feynman propagator, since $p_E^2 + m^2 \rightarrow$

$p_M^2 + m^2 - i\epsilon$. Note that here, $p_E^2 = p_\mu^E p_\nu^E \delta^{\mu\nu}$, but $p_M^2 = p_\mu^M p_\nu^M \eta^{\mu\nu} = -(p^0)^2 + \vec{p}^2$. And again, after the rotation (smooth deformation of the contour):

$$\Delta(s = it, \vec{x}) = \Delta_F(t, \vec{x}), \tag{9.22}$$

uniquely chosen.

9.5 Wick's Theorem

In the path-integral formalism, Wick's theorem is both simple to state, and easy to prove, though it will take us some examples in Chapter 10 in order to figure out why it is the same as in the operator formalism.

We start with an example, to be able to see the generalization. Consider, for example, the function $f[\{\phi\}] = \phi(x_1)\phi^3(x_2)\phi^2(x_3)$. Its VEV is a path integral, as we saw, which we can write as

$$\langle 0|f[\{\phi\}]|0\rangle = \int \mathcal{D}\phi e^{-S_0[\phi]} \phi(x_1)\phi^3(x_2)\phi^2(x_3)$$

$$= \left(\frac{\delta}{\delta J(x_1)}\right)\left(\frac{\delta}{\delta J(x_2)}\right)^3 \left(\frac{\delta}{\delta J(x_3)}\right)^2 \int \mathcal{D}\phi e^{-S_0+J\cdot\phi}\bigg|_{J=0}. \tag{9.23}$$

From this example, it is easy to see that, in general, for an arbitrary function f, and in the presence of an arbitrary source J:

$$_J\langle 0|f[\{\phi\}]|0\rangle_J = f\left[\left\{\frac{\delta}{\delta J}\right\}\right] Z_0[J]\bigg|_J. \tag{9.24}$$

In particular, we can use Dyson's formula, that is apply it for

$$f[\{\phi\}] = e^{-\int d^d x V(\phi(x))}, \tag{9.25}$$

to obtain

$$Z[J] = e^{-\int d^d x V\left(\frac{\delta}{\delta J(x)}\right)} Z_0[J] = e^{-\int d^d x V\left(\frac{\delta}{\delta J(x)}\right)} e^{\frac{1}{2}J\cdot\Delta\cdot J}$$

$$= e^{-\int d^d x V\left(\frac{\delta}{\delta J(x)}\right)} e^{\frac{1}{2}\int d^d x \int d^d y J(x)\Delta(x,y)J(y)}, \tag{9.26}$$

which is the form of Wick's theorem in the path-integral formalism.

This formula seems very powerful. Indeed, we seem to have solved completely the interacting theory, as we have a closed formula for the partition function, from which we can derive everything.

Of course, we can't get anything for free: if it was tricky in the operator formalism, it means it is tricky here also. The point is that this formal expression is in general not well defined: there are divergences (infinite integrals that will appear) and apparent singularities (like what happens, for instance, when several derivatives at the same point act at the same time). The above formula then has to be understood as a *perturbative* expansion: we expand the exponential, and then calculate the derivatives term by term. We will look at

Important Concepts to Remember

- The generalization from quantum mechanics to quantum field theory is obvious: the vacuum functional is the path integral in Euclidean space over periodic paths of infinite period, with a positive definite Euclidean action.
- Green's functions are obtained from a partition function with sources.
- In perturbation theory, we will study Green's functions because they are simpler, but one can relate them to the S-matrix elements (real scattering amplitudes) via the LSZ formula, whose precise form will be given later.
- For momentum-space Green's functions, one usually factorizes the overall momentum conservation, and talks about the Green's functions with momenta that already satisfy momentum conservation.
- Dyson's formula relates $\langle 0|e^{-S_I[\hat{\phi}]}e^{J\cdot\hat{\phi}}|0\rangle$ to the partition function (path integral over the full action, with sources).
- In free field theory, $Z_0[J] = e^{\frac{1}{2}J\cdot\Delta\cdot J}$.
- Wick's theorem gives the partition function of the interacting theory as the formal expression $Z[J] = e^{-\int V[\delta/\delta J]}Z_0[J]$.

some examples in Chapter 10, then formulate Feynman rules, which will be the same ones as we have found in the operator formalism.

Further Reading

See sections 2.1 and 2.2 in [3] and sections 3.1 and 3.2 in [4].

Exercises

1. Using the formulas given in the text, compute the four-point function in momentum space, in free field theory.
2. If $S_I = \int \lambda\phi^4/4!$, write down a path-integral expression for the four-point function $G_4(x_1, \ldots, x_4)$ up to (including) order λ^2.
3. For a $\lambda\phi^3/3!$ model, expand the Wick theorem formula in λ, to write an explicit form of the partition function up to λ^3.
4. Wick rotate (write everything explicitly!) back to Minkowski space the solution of the free field theory partition function in Euclidean space.

10

Wick's Theorem for Path Integrals and Feynman Rules Part I

In this chapter we continue the definition of perturbation theory in the path-integral formalism. We continue defining the Wick theorem, doing examples, and defining a second form for it, and then we define Feynman rules that will construct perturbation theory from the Wick theorem.

In Chapter 9 we saw the Euclidean formulation of scalar field theories. We have written Dyson's formula for the partition function:

$$Z[J] = \langle 0|e^{-S_I[\hat{\phi}]}e^{\int d^d x J(x)\hat{\phi}(x)}|0\rangle = \int \mathcal{D}\phi e^{-S_0[\phi]+J\cdot\phi}e^{-S_I[\phi]}, \tag{10.1}$$

and calculated the partition function of the free theory:

$$Z_0[J] = e^{\frac{1}{2}J\cdot\Delta\cdot J}\langle 0|0\rangle = e^{\frac{1}{2}J\cdot\Delta\cdot J}, \tag{10.2}$$

where the Euclidean propagator is

$$\Delta(x,y) = \int \frac{d^d p}{(2\pi)^d} \frac{e^{ip(x-y)}}{p^2+m^2}. \tag{10.3}$$

Finally we wrote Wick's theorem for path integrals, which is

$$Z[J] = e^{-\int d^d x V\left(\frac{\delta}{\delta J(x)}\right)}Z_0[J] = e^{-\int d^d x V\left(\frac{\delta}{\delta J(x)}\right)}e^{J\cdot\Delta\cdot J}. \tag{10.4}$$

10.1 Examples

It is not completely obvious why this is the same Wick's theorem as the operator formalism, so we will investigate this by looking at explicit examples.

We consider first the theory at zeroth order in the coupling constant (i.e. the free theory), $Z_0[J]$.

We will denote the Green's functions by $G_n^{(p)}(x_1,\ldots,x_n)$ for the n-point function at order p in the coupling.

We start with the one-point function. At nonzero J:

$$G_1^{(0)}(x_1)_J = \frac{\delta}{\delta J(x_1)}e^{\frac{1}{2}J\cdot\Delta\cdot J} = \Delta\cdot J(x_1)e^{\frac{1}{2}J\cdot\Delta\cdot J}, \tag{10.5}$$

where $\Delta \cdot J(x_1) = \int d^d x \Delta(x_1, x) J(x)$. Putting $J = 0$, the resulting one-point function is zero:

$$G_1^{(0)}(x_1) = 0. \tag{10.6}$$

We can easily see that this generalizes for all the odd n-point functions, since we have even numbers of Js in $Z_0[J]$, so by taking an odd number of derivatives and then putting $J = 0$ we get zero:

$$G_{2k+1}^{(0)}(x_1, \ldots, x_{2k+1}) = 0. \tag{10.7}$$

The next case is the two-point function

$$G_2^{(0)}(x_1, x_2)_J = \frac{\delta}{\delta J(x_1)} \frac{\delta}{\delta J(x_2)} e^{\frac{1}{2} J \cdot \Delta \cdot J} = \frac{\delta}{\delta J(x_1)} \left[\Delta \cdot J(x_2) e^{\frac{1}{2} J \cdot \Delta \cdot J} \right]$$

$$= [\Delta(x_1, x_2) + (\Delta \cdot J(x_2))(\Delta \cdot J(x_1))] e^{\frac{1}{2} J \cdot \Delta \cdot J}, \tag{10.8}$$

and by putting $J = 0$ we find

$$G_2^{(0)}(x_1, x_2) = \Delta(x_1, x_2). \tag{10.9}$$

The corresponding Feynman diagram is obtained by drawing a line connecting the external points x_1 and x_2.

For the next nontrivial case, the four-point function is

$$G_4^{(0)}(x_1, x_2, x_3, x_4) = \prod_{i=1}^{4} \frac{\delta}{\delta J(x_1)} e^{\frac{1}{2} J \cdot \Delta \cdot J} \bigg|_{J=0}$$

$$= \Delta(x_1, x_2) \Delta(x_3, x_4) + \Delta(x_1, x_3) \Delta(x_2, x_4) + \Delta(x_1, x_4) \Delta(x_2, x_3), \tag{10.10}$$

which can be represented as the sum of Feynman diagrams with lines connecting (12) and (34); (13) and (24); (14) and (23), as in Figure 10.1. The (simple) details are left as an exercise.

We then get the general rule for computing $G_n^{(0)}(x_1, \ldots, x_n)$. Write all the Feynman diagrams by connecting pairwise all external points in all possible ways.

We now move to the first nontrivial example of interaction. Consider the theory with $V(\phi) = \lambda \phi^3 / 3!$. Of course, such a theory is not so good: the Hamiltonian is unbounded from below (the energy becomes arbitrarily negative for large enough negative ϕ), so the system is unstable. But we just want to use this as a simple example of how to calculate Feynman diagrams.

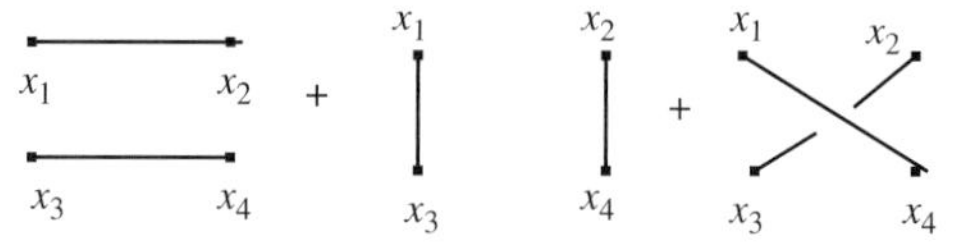

Figure 10.1 Free contractions for the four-point function.

Consider the theory at first order in λ, so replace

$$e^{-\int d^d x \frac{\lambda}{3!}\left(\frac{\delta}{\delta J(x)}\right)^3} \rightarrow -\int d^d x \frac{\lambda}{3!}\left(\frac{\delta}{\delta J(x)}\right)^3. \tag{10.11}$$

Then we have

$$\begin{aligned}
Z^{(1)}[J] &= -\int d^d x \frac{\lambda}{3!}\left(\frac{\delta}{\delta J(x)}\right)^3 e^{\frac{1}{2}J\cdot\Delta\cdot J} = -\frac{\lambda}{3!}\int d^d x \left(\frac{\delta}{\delta J(x)}\right)^2 \left[\Delta \cdot J(x) e^{\frac{1}{2}J\cdot\Delta\cdot J}\right] \\
&= -\frac{\lambda}{3!}\int d^d x \frac{\delta}{\delta J(x)}[\Delta(x,x) + (\Delta\cdot J(x))^2]e^{\frac{1}{2}J\cdot\Delta\cdot J} \\
&= -\frac{\lambda}{3!}\int d^d x[3\Delta(x,x)(\Delta\cdot J)(x) + (\Delta\cdot J(x))^3]e^{\frac{1}{2}J\cdot\Delta\cdot J}. \tag{10.12}
\end{aligned}$$

From this expression, which contains only odd powers of J, we see that the even n-point functions at order 1 are zero:

$$G_{2k}^{(1)}(x_1,\ldots,x_{2k}) = 0. \tag{10.13}$$

The one-point function is

$$G_1^{(1)}(x_1) = \left.\frac{\delta}{\delta J(x_1)}Z^{(1)}[J]\right|_{J=0} = -\frac{\lambda}{2}\int d^d x \Delta(x,x)\Delta(x,x_1), \tag{10.14}$$

which diverges due to $\Delta(x,x)$, and has as Feynman diagram a propagator from the external point x_1 to the point x to be integrated over, followed by a loop starting and ending at x, as in Figure 10.2.

Next, we calculate the three-point function:

$$\begin{aligned}
G_3^{(1)}(x_1,x_2,x_3) &= \left.\frac{\delta}{\delta J(x_1)}\frac{\delta}{\delta J(x_2)}\frac{\delta}{\delta J(x_3)}Z^{(1)}[J]\right|_{J=0} \\
&= -\lambda\int d^d x \left\{\Delta(x,x_1)\Delta(x,x_2)\Delta(x,x_3) + \frac{1}{2}\Delta(x,x)[\Delta(x,x_1)\Delta(x_2,x_3)\right. \\
&\quad \left. + \Delta(x,x_2)\Delta(x_1,x_3) + \Delta(x,x_3)\Delta(x_1,x_2)]\right\}, \tag{10.15}
\end{aligned}$$

which can be represented as four Feynman diagrams: one with a 3-vertex connected with the three external points, one with x_2 connected with x_3 and a $G_1^{(1)}(x_1)$ contribution (i.e. a line with a loop at the end connected to x_1), and the two permutations for it, as in Figure 10.3.

We now exemplify the diagrams at second order with the zero-point function (vacuum bubble) at second order $\mathcal{O}(\lambda^2)$ (i.e. $Z^{(2)}[J=0]$). We have

$$Z^{(2)}[J] = \frac{1}{2!}\int d^d x \frac{\lambda}{3!}\left(\frac{\delta}{\delta J(x)}\right)^3 \int d^d y \frac{\lambda}{3!}\left(\frac{\delta}{\delta J(y)}\right)^3 e^{\frac{1}{2}J\cdot\Delta\cdot J}. \tag{10.16}$$

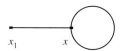

x_1 x

Figure 10.2   One-point function tadpole diagram.

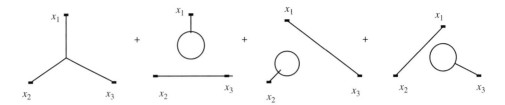

Diagrams for the three-point function at order λ.

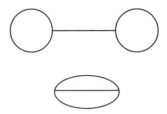

Diagrams for the vacuum bubble (zero-point function) at order λ^2.

But when we put $J = 0$, only terms with six Js contribute, so we have

$$Z^{(2)}[J = 0] = \frac{1}{2!} \int d^d x \frac{\lambda}{3!} \left(\frac{\delta}{\delta J(x)} \right)^3 \int d^d y \frac{\lambda}{3!} \left(\frac{\delta}{\delta J(y)} \right)^3 \frac{1}{3!} \left(\frac{1}{2} \right)^3 \int d^d z_1 d^d z_2 d^d z_3$$

$$d^d z_1' d^d z_2' d^d z_3' J(z_1) \Delta(z_1, z_1') J(z_1') J(z_2) \Delta(z_2, z_2')$$

$$J(z_2') J(z_3) \Delta(z_3, z_3') J(z_3') \big|_{J=0} .$$

(10.17)

When we do the derivatives, we obtain a result that is the sum of two Feynman diagrams:

$$Z^{(2)}[J = 0] = \frac{\lambda^2}{2^3} \int d^d x d^d y \Delta(x, x) \Delta(x, y) \Delta(y, y) + \frac{\lambda^2}{2 \times 3!} \int d^d x d^d y \Delta^3(x, y),$$

(10.18)

the two Feynman diagrams being one with two loops (circles) connected by a line, and one with three propagators connecting the same two points (integrated over), x and y, as in Figure 10.4. The details (which derivatives give which diagram) are left as an exercise.

10.2 Wick's Theorem: Second Form

We now formulate a second form for the Wick theorem:

$$Z[J] = e^{\frac{1}{2} \frac{\delta}{\delta\phi} \cdot \Delta \cdot \frac{\delta}{\delta\phi}} \left\{ e^{-\int d^d x V(\phi) + J \cdot \phi} \right\} \Big|_{\phi=0}$$

$$= \exp \left[\frac{1}{2} \int d^d x d^d y \Delta(x - y) \frac{\delta}{\delta\phi(x)} \frac{\delta}{\delta\phi(x)} \right] \left\{ e^{-\int d^d x V(\phi) + J \cdot \phi} \right\} \Big|_{\phi=0} .$$

(10.19)

To prove it, we use the previous form of the Wick theorem, and the following:

Lemma 1 (Coleman)   Consider two functions of multi-variables

$$f(x) = f(x_1, \ldots, x_n); \quad g(y) = g(y_1, \ldots, y_n).$$

(10.20)

Then we have

$$f\left(\frac{\partial}{\partial x}\right)g(x) = g\left(\frac{\partial}{\partial y}\right)\{f(y)e^{x\cdot y}\}_{y=0}. \tag{10.21}$$

To prove it, because of the Fourier decomposition theorem, we only have to prove the lemma for

$$f(x) = e^{a\cdot x}; \quad g(y) = e^{b\cdot y}. \tag{10.22}$$

Let's do that, first on the left-hand side of the lemma:

$$f\left(\frac{\partial}{\partial x}\right)g(x) = e^{a\cdot\frac{\partial}{\partial x}}e^{b\cdot x} = \sum_m \frac{1}{m!}\left(a\cdot\frac{\partial}{\partial x}\right)^m e^{b\cdot x}$$

$$= \sum_m \frac{1}{m!}(a\cdot b)^m e^{b\cdot x} = e^{a\cdot b}e^{b\cdot x} = e^{b\cdot(a+x)}, \tag{10.23}$$

and then on the right-hand side of the lemma:

$$g\left(\frac{\partial}{\partial y}\right)\{f(y)e^{x\cdot y}\}_{y=0} = e^{b\cdot\frac{\partial}{\partial y}}\{e^{a\cdot y}e^{x\cdot y}\}_{y=0}. \tag{10.24}$$

Using the result for the left-hand side of the lemma, we have

$$g\left(\frac{\partial}{\partial y}\right)\{f(y)e^{x\cdot y}\}_{y=0} = e^{b\cdot(a+x)}e^{(a+x)\cdot y}|_{y=0} = e^{b\cdot(a+x)}, \tag{10.25}$$

which is the same as the left-hand side of the lemma.   *q.e.d.*

We can now apply this lemma to the previous form of Wick's theorem, with $x \to J(x)$ and $y \to \phi(x)$, so we generalize between a discrete and a finite number of variables in the above to a continuum of variables. Then we obtain

$$Z[J] = e^{-\int d^d x V\left[\frac{\delta}{\delta J}\right]}e^{\frac{1}{2}J\cdot\Delta\cdot J} = e^{\frac{1}{2}\frac{\delta}{\delta\phi}\cdot\Delta\cdot\frac{\delta}{\delta\phi}}\{e^{-\int d^d x V(\phi)+J\cdot\phi}\}_{\phi=0}. \tag{10.26}$$

q.e.d.

10.3 Feynman Rules in x Space

Consider the general polynomial potential

$$V(\phi) = \lambda\phi^p. \tag{10.27}$$

We write down the Feynman rules for the n-point functions, to order λ^N. But note that we will now consider Feynman diagrams that give the same result as different. This is an algorithmic method, so when we can't compute the symmetry factors for diagrams, we can use this construction. We will return to the usual Feynman rules in Chapter 11.

The rules are: write down the n external points, $x_1,\ldots,x_n$, each with a line (leg) sticking out, and then N vertices (i.e. internal, to be integrated over), points $y_1,\ldots,y_N$, each with p legs sticking out. Then connect all the lines sticking out in all possible ways by propagators, thus constructing all the Feynman diagrams. Then:

- For each p-legged vertex, write a factor of $-\lambda$, see Figure 10.5. Note the missing factor of i in this Euclidean-space case compared to Minkowski space.
- For each propagator from z to w, write a $\Delta(z-w)$.
- The resulting expression $I_D(x_1,\ldots,x_n;y_1,\ldots,y_N)$ is integrated over the vertices to obtain the result for the Feynman diagram:

$$F_{diag}^{(N)}(x_1,\ldots,x_n) = \int d^d y_1 \ldots d^d y_N I_{diag}(x_1,\ldots,x_n;y_1,\ldots,y_N). \qquad (10.28)$$

- Then the n-point function is given by

$$G_n(x_1,\ldots,x_n) = \sum_{N\geq 0} G_n^{(N)}(x_1,\ldots,x_n) = \sum_{N\geq 0} \frac{1}{N!} \sum_{diag} F_{diag}^{(N)}(x_1,\ldots,x_n). \qquad (10.29)$$

As an **example**, we take the two-point function to order λ^1 for a potential with $p = 4$. Then we draw the two points x_1, x_2 with a line sticking out and a vertex y with four legs sticking out, as in Figure 10.6(a).

We can connect x_1 with x_2, and we have three ways of connecting the four legs of the vertex to each other, as in Figure 10.6(b)–(d), leading to

$$F_{D1} = F_{D2} = F_{D3} = -\lambda \int d^d y \Delta(x_1 - x_2) \Delta(y - y) \Delta(y - y). \qquad (10.30)$$

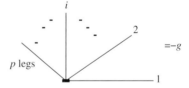

Figure 10.5 Feynman diagram for the p vertex.

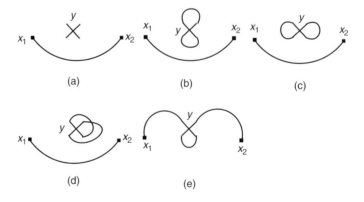

Figure 10.6 We can draw all possible Feynman diagrams at order λ in the two-point function by drawing the vertex and the external lines (a). Then we can connect them in all possible ways, obtaining diagrams (b)–(e).

We can also connect each of the x_1 and x_2 with one of the legs of the vertex, and the remaining two legs of the vertex to each other, as in Figure 10.6(e). This gives 12 diagrams, since we have six ways of choosing two lines out of the four, and two ways of choosing which of the two to connect with x_1. Thus we have

$$F_{D4} = \ldots = F_{D15} = -\lambda \int d^d y \Delta(x_1 - y)\Delta(x_2 - y)\Delta(y - y). \qquad (10.31)$$

Important Concepts to Remember

- In free field theory (zeroth order in perturbation), the odd n-point functions are zero, $G_{2k+1}^{(0)}(x_1, \ldots, x_{2k+1}) = 0$.
- For $G_{2k}^{(0)}(x_1, \ldots, x_{2k})$, we write all the Feynman diagrams by connecting pairwise all external legs in all possible ways.
- We can write down a second form of the Wick theorem, using Coleman's lemma, as $Z[J] = e^{\frac{1}{2} \frac{\delta}{\delta \phi} \cdot \Delta \cdot \frac{\delta}{\delta \phi}} \left\{ e^{-\int V(\phi) + J \cdot \phi} \right\} \Big|_{\phi=0}$.
- The Feynman rules in x space, for $G_n^{(N)}$ in $\lambda \phi^p$ theory, the long form, are: write n external points, with a leg sticking out, and N vertices, with p legs sticking out, then connect all the legs in all possible ways. For a vertex, write $-\lambda$ and for a propagator, write $\Delta(x, y)$. Then integrate over the vertices, sum over diagrams (now we will have many diagrams giving the same result), and sum over N with $1/N!$.
- This prescription is equivalent to the usual one of writing only inequivalent diagrams, and writing a symmetry factor S.

Further Reading

See sections 2.3 and 4.1 in [3].

Exercises

1. *Prove* that $G_4^{(0)}(x_1, \ldots, x_4)$ is the sum of the three diagrams with connections of (12), (34); (13), (24); (14), (23) in Figure 10.1 and then generalize to *write down* $G_6^{(0)}(x_1, \ldots, x_6)$.
2. Explain which derivatives give which Feynman diagram in the calculation of the zero-point function to $\mathcal{O}(\lambda^2)$ in $\lambda \phi^3$ theory (two circles connected by a line; and two points connected by three propagators as in Figure 10.4), and then write down the Feynman diagrams for $G_4^{(2)}(x_1, \ldots, x_4)$ in $\lambda \phi^4$ theory.

3. Consider $\lambda\phi^4$ theory. Write an expression for the "setting sun" diagram (like the third diagram in Figure 10.4, just with two external lines coming out from the two vertices) for the two-point function, using the Feynman rules. Then explain which term in the expression of the two-point function, using the Wick theorem, this diagram comes from.

4. Write all the Feynman diagrams up to (including) order λ^2 in $\lambda\phi^4$ theory contributing to the two-point function.

11

Feynman Rules in x Space and p Space

In this chapter we continue with the definition of perturbation theory via the Feynman rules. We first prove the x-space Feynman rules derived in Chapter 10, and understand the role of the symmetry factor (statistical weight factor). Then we define the rules in momentum space, and finally consider (formally) the rules for the most general bosonic field theory.

11.1 Proof of the Feynman Rules

To prove the Feynman rules, we start with the definition of the perturbative expansion defined previously: we consider the second form of the Wick theorem in the path-integral formalism (using the form of the free partition function) for the partition function, and then differentiate to get the Green's functions.

We thus compute the Green's functions as

$$
\begin{aligned}
G(x_1, \ldots, x_n) &= \left. \frac{\delta^n Z[J]}{\delta J(x_1) \ldots \delta J(x_n)} \right|_{J=0} \\
&= \left. e^{\frac{1}{2} \frac{\delta}{\delta \phi} \cdot \Delta \cdot \frac{\delta}{\delta \phi}} \left\{ \phi(x_1) \ldots \phi(x_n) e^{-\int d^d x [V(\phi(x)) - J(x)\phi(x)]} \right\} \right|_{\phi=0} \right|_{J=0}.
\end{aligned} \quad (11.1)
$$

Note that these Green's functions are the Green's functions corresponding to the VEV $\langle 0 | T\{\phi(x_1) \ldots \phi(x_n) e^{i \int H_{int}}\} | 0 \rangle$ (in the *free* vacuum of the theory) in Minkowski space, and as a result they still contain vacuum bubbles. The vacuum bubbles will be seen later to factorize exactly as in the operator formalism, and then we will calculate the Green's functions corresponding to $\langle \Omega | T\{\phi(x_1) \ldots \phi(x_n)\} | \Omega \rangle$ from $W = -\ln Z[J]$.

We next note that in the above we can drop the $J \cdot \phi$ term, since by taking all the derivatives we will have only terms linear in J from this contribution, and putting $J = 0$ these terms vanish. The result is then

$$
G(x_1, \ldots, x_n) = \left. e^{\frac{1}{2} \frac{\delta}{\delta \phi} \cdot \Delta \cdot \frac{\delta}{\delta \phi}} \left\{ \phi(x_1) \ldots \phi(x_n) e^{-\int d^d x V(\phi)} \right\} \right|_{\phi=0}. \quad (11.2)
$$

This means that in order N (i.e. $\mathcal{O}(\lambda^N)$), we find

$$
\begin{aligned}
G^{(N)}(x_1, \ldots, x_n) = e^{\frac{1}{2} \frac{\delta}{\delta \phi} \cdot \Delta \cdot \frac{\delta}{\delta \phi}} \Bigg\{ \phi(x_1) \ldots \phi(x_n) \frac{(-\lambda)^N}{N!} \\
\left. \int d^d y_1 \ldots d^d y_N \phi^p(y_1) \ldots \phi^p(y_N) \right\} \Bigg|_{\phi=0}.
\end{aligned} \quad (11.3)
$$

Consider now that there are $Q = n + pN$ ϕs, on which we act with the derivatives on the exponential. If we have more derivatives than fields, we obviously get zero, but we also get zero if we have fewer derivatives, because at the end we must put $\phi = 0$, so if we have ϕs left over, we also get zero. Therefore, we only get a nonzero result if $Q = 2q$ (even), and then the result is

$$G^{(N)}(x_1, \ldots, x_n) = \frac{1}{q! \, 2^q} \int d^d z_1 d^d w_1 \ldots d^d z_q d^d w_q$$

$$\times \frac{\delta}{\delta\phi(z_1)} \Delta(z_1 - w_1) \frac{\delta}{\delta\phi(w_1)} \cdots \frac{\delta}{\delta\phi(z_q)} \Delta(z_q - w_q) \frac{\delta}{\delta\phi(w_q)}$$

$$\times \left\{ \phi(x_1) \ldots \phi(x_n) \frac{(-\lambda)^N}{N!} \int d^d y_1 \ldots d^d y_N \phi^p(y_1) \ldots \phi^p(y_N) \right\}. \quad (11.4)$$

Therefore, when acting with a factor of $\int d^d z d^d w \frac{\delta}{\delta\phi(z)} \Delta(z-w) \frac{\delta}{\delta\phi(w)}$ on some $\phi(x)\phi(z)$, we obtain a Wick contraction $\overline{\phi(x)\phi(y)}$ (i.e. we replace it with $\Delta(x-y)$). But with a factor of 2 in front, since it comes from $\int d^d z d^d w [\delta(z-x)\delta(w-y) + \delta(z-y)\delta(w-x)]\Delta(z-w)$. Since all the q factors above give the same factor of 2, in total we get a 2^q canceling the $1/2^q$. Also, we have q factors as above with which we can act on the $Q = 2q$ ϕs, and so by permuting them we get the same contribution. Concluding, the resulting factor of $q!$ will cancel the $1/q!$ in front.

11.2 Statistical Weight Factor (Symmetry Factor)

As explained before, from most diagrams, we get a lot of contributions that are identical, almost giving a $p!$ factor for each vertex, and an overall $N!$, so it is customary to redefine $\lambda = \lambda_p/p!$ so as to (almost) cancel this factor. Then, we have in front of the result for a Feynman diagram a factor of $1/S$, where by definition the *statistical weight factor, or symmetry factor S*, is

$$S = \frac{N! \, (p!)^N}{\# \text{ equivalent diagrams}}. \quad (11.5)$$

Then we can construct only the topologically inequivalent diagrams, and associate a statistical weight factor, or symmetry factor S, corresponding to all possible symmetries of the diagram that leave it invariant. If there is no symmetry, then $S = 1$.

For tree diagrams $S_{tree} = 1$ always. For instance, we can write a Feynman diagram in ϕ^4 theory, like in Figure 11.1, and explicitly check that $S = 1$ for it.

There are $N!$ ways to label the vertices as $y_1, \ldots, y_N$. Many topologically equivalent diagrams obtained by connecting the $p = 4$ legs on the vertices with other legs from vertices, or from the external lines, can be thought of as just the above relabeling the vertices. Then there are $p! = 4!$ ways of attaching given propagators at the $p = 4$ legs of the vertex. In total, we cancel the $1/(N! \, (p!)^N)$ factor in front, giving $S_{tree} = 1$.

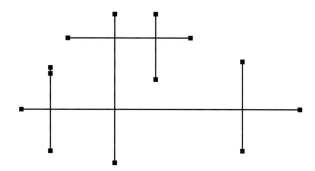

Figure 11.1 Example of a tree diagram. It has $S = 1$.

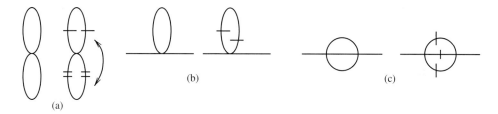

(a)   (b)   (c)

Figure 11.2 Computing the symmetry factors for three loop diagrams (a), (b), and (c).

We can find more symmetry factors. The figure-eight vacuum bubble diagram of Figure 11.2(a) has a symmetry factor of 8, which matches with the number of diagrams. One vertex with four legs has three ways of connecting them 2 by 2, giving $3 \times 1/(1!\,(4!\,)^1) = 1/8$.

The propagator from x_1 to x_2 with a loop at an intermediate point x, as in Figure 11.2(b), has a symmetry factor of 2. From the vertex, there are four ways to connect one leg to x_1, then three ways to connect to x_2, and one way to connect the remaining two legs between themselves, for a total of $4 \times 3/(1!\,(4!\,)^1) = 1/2$.

The "setting sun" diagram, propagating from x_1 to x_2, with three propagators between two intermediate points x and y, as in Figure 11.2(c), has a symmetry factor of 3!, from permuting the three propagators between x and y. In contrast, we draw two vertices with four legs, and two external lines. For the first contraction there are eight ways to connect x to a vertex, then four ways to connect y to the other vertex, and then $3! = 3 \times 2 \times 1$ to connect the three internal propagators, for a total of $8 \times 4 \times 3!\,/(2!\,(4!\,)^2) = 1/3!$, as expected.

11.3 Feynman Rules in p Space

As already explained, we can define the Fourier transform of the x-space Green's function, but by translational invariance we can always factor out a delta function, with the remaining

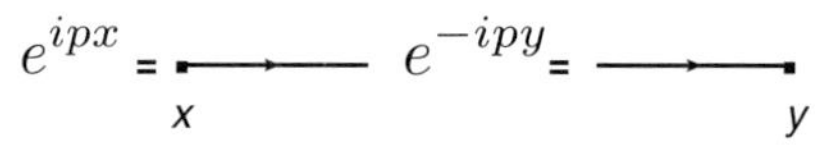

$$e^{ipx} = \underset{x}{\underset{\longrightarrow}{\bullet \quad}} \qquad e^{-ipy} = \underset{y}{\underset{\longrightarrow}{\quad \bullet}}$$

Figure 11.3   Momentum convention for external lines.

Green's function, already satisfying momentum conservation, being the one we usually calculate:

$$\tilde{G}(p_1,\ldots,p_n) = \int \prod_i d^d x_i e^{i\sum x_j p_j} G(x_1,\ldots,x_n) = (2\pi)^d \delta^d(p_1 + \ldots + p_n) G(p_1,\ldots,p_n),$$

(11.6)

where in $G(p_1,\ldots,p_n)$ the momenta already satisfy momentum conservation.

Since the Euclidean propagator in x space is

$$\Delta(x-y) = \int \frac{d^d p}{(2\pi)^d} \frac{e^{ip(x-y)}}{p^2 + m^2},$$

(11.7)

in momentum space we have

$$\Delta(p) = \frac{1}{p^2 + m^2}.$$

(11.8)

By convention, a momentum p going out of a point x means a factor of e^{ipx} and going in means a factor of e^{-ipx}, as in Figure 11.3.

We write a version of the Feynman diagrams where we already include the momentum conservation. But for that, we need to know the number of *loops* (i.e. independent loops, circles or closed paths) in our diagrams, or the number of integrations left over after we use the momentum conservation delta functions.

We denote by V the number of vertices, I the number of internal lines, E the number of external lines, and L the number of loops. There are I momentum variables (one for each internal line), but there are V constraints on them (momentum conservation delta functions). However, one of these is the overall constraint which we put outside G to form $\tilde{G}$. So, for the calculation of G, we have $L = I - V + 1$ loops (independent integrations). But we must express this in terms of things that are independent of the diagram, and depend only on having $G^{(N)}(p_1,\ldots,p_n)$, so we must find another formula. If we cut all propagators (or equivalently, consider the diagram before we make the connections of the p legs on vertices with other legs on them and on external lines), the legs on the vertices come as follows. For each external line, one of the two pieces of propagator stays on the external point, but one comes from a vertex. For internal lines, however, both come from the vertex, so all in all, $E + 2I = pV$. Replacing this formula in the above for L, we finally find

$$L = V\left(\frac{p}{2} - 1\right) - \frac{E}{2} + 1.$$

(11.9)

So, after using all the momentum conservation delta functions, we are left with L independent integrations, and the Feynman rules become:

- Draw all the topologically inequivalent Feynman diagrams with n external lines and N vertices.
- Label external momenta $p_1, \ldots, p_n$ with arrows for them, introduce $l_1, \ldots, l_L$ independent loop momenta, integrated over with the measure $\int d^d l_1/(2\pi)^d \ldots d^d l_L/(2\pi)^d$, and label the propagator lines q_j, using the l_i and momentum conservation.
- For an external line (between two points), put a propagator $1/(p_i^2 + m^2)$.
- For an internal line (between two points), put a propagator $1/(q_j^2 + m^2)$.
- For a vertex, we have a factor of $-\lambda_p$.
- Calculate the statistical weight factor (symmetry factor) S and multiply the diagram by $1/S$.
- Sum over Feynman diagrams.

To understand better the p-space rules, consider the x-space "setting sun" diagram above, and Fourier transform it to p space. This gives $\tilde{G}(p_1, p_2)$, with two external points, but without x labels, which therefore has a $\Delta(p_1)$ and a $\Delta(p_2)$ factor. The external line rule for a momentum p going out of an external point labeled x, giving a factor of e^{ipx}, is just a wavefunction for a more physical case: the only way to introduce an x dependence into a p space object is via an explicit wavefunction for the external states of momenta p_i. If there is no label x, we have no wavefunction. But also, if there is no external point (what is known as an *amputated diagram*, to be discussed later in the book), we have no corresponding external propagator. The final thing to observe is of course that in general, we are not interested in $\tilde{G}(p_1, p_2)$ but in $G(p_1, p_2) = G(p_1, -p_1) \equiv G(p_1)$, in which momentum conservation is already taken into account.

For instance, consider the free propagator, a line between the two external points. We could in principle consider momentum p_1 going in at one point, and p_2 going in at the other. But then

$$
\begin{aligned}
\tilde{G}(p_1, p_2) &= \int d^d x_1 d^d x_2 e^{i(p_1 x_1 + p_2 x_2)} G(x_1, x_2) \\
&= \int d^d x_1 d^d x_2 e^{i(p_1 x_1 + p_2 x_2)} \int \frac{d^d p}{(2\pi)^d} \frac{e^{ip(x_1 - x_2)}}{p^2 + m^2} \\
&= \int d^d x_1 e^{ip_1 x_1} \frac{e^{ip_2 x_1}}{p_2^2 + m^2} = (2\pi)^d \delta^d(p_1 + p_2) \frac{1}{p_2^2 + m^2} \\
&= (2\pi)^d \delta^d(p_1 + p^2) G(p_2).
\end{aligned}
\tag{11.10}
$$

11.4 Most General Bosonic Field Theory

We now deal with a general bosonic field theory. Consider fields $\phi_p(x)$, where p is some general label, which could signify some Lorentz index, like on a vector field $A_\mu(x)$, or some label for several scalar fields, or a combination of both.

Consider the kinetic term

$$S_0 = \frac{1}{2} \int d^d x \sum_{p,q} \phi_p(x) \Delta_{p,q}^{-1} \phi_q(x), \tag{11.11}$$

where $\Delta_{p,q}^{-1}$ is some differential operator. We invert this object, obtaining the propagator $\Delta_{p,q}$. If this is not possible, it means that there are redundancies, and we must find the independent fields. For instance, for a gauge field $A_\mu(x)$, there is a gauge invariance, $\delta A_\mu = \partial_\mu \lambda(x)$, which means that we must introduce a gauge fixing (and also introduce "ghosts," to be defined later in the book) in order to define independent fields. Then, we can invert the propagator.

We next derive a generalized Wick theorem, for a general interaction term

$$S_{p_1...p_n} = \int d^d x V_{p_1...p_n} \phi_{p_1}(x) ... \phi_{p_n}(x), \tag{11.12}$$

where $V_{p_1...p_n}$ can contain couplings and derivatives, meaning that it is more general than the previous construction of the ϕ^n theory.

Then, the vertex is

$$- \int d^d x_1 ... d^d x_n e^{i(k_1 x_1 + ... + k_n x_n)} \frac{\delta}{\delta \phi_{p_1}(x_1)} \cdots \frac{\delta}{\delta \phi_{p_n}(x_n)} S_{p_1...p_n}. \tag{11.13}$$

(Note that we don't really need to define $V_{p_1...p_n}$, all we need is the form of the interaction term in the action, $S_{p_1...p_n}$.) Let's check that this gives the correct result for the $\lambda_4/4! \, \phi^4$ theory. Indeed, we then have the interaction term

$$S_I = \frac{\lambda_4}{4!} \int d^d x \phi^4(x). \tag{11.14}$$

We easily find that, applying the above formula, we get

$$- (2\pi)^d \delta^d(k_1 + ... + k_4) \lambda_4, \tag{11.15}$$

which is the correct vertex (it has the $-\lambda$ factor, as well as the momentum conservation at the vertex that we need to take into account).

Important Concepts to Remember

- We can write the Feynman rules in p space directly in terms of independent integrations (= loop momenta). The number of loops, or independent integrations, is $L = V(p/2 - 1) - E/2 + 1$.
- The Feynman rules then consist of labeling the internal lines using the loop momenta and momentum conservation, with $\Delta(p_i)$ for external lines (ending at an unlabeled point), $\Delta(q_j)$ for internal lines, and $-\lambda_p$ for vertices. Then we must divide by the statistical weight factor (symmetry factor for the diagram).
- For the most general bosonic theory, we need to invert the kinetic term to find the propagator (if there are redundancies, like for gauge fields, we must first find the correct kinetic term for the independent fields). For the interaction term $S_{r_1...r_p}$, we find the vertex as the Fourier transform of $-\delta^n/\delta \phi_{r_1} ... \phi_{r_p} S_{r_1...r_p}$.

Further Reading

See sections 2.4.2, 2.4.3, and 2.7 in [3], section 4.1 in [4], and section 3.4 in [2].

Exercises

1. Write down the statistical weight factor (symmetry factor) for the Feynman diagram in Figure 11.4 in x space, and then write down an integral expression for it, applying the Feynman rules.

2. Consider the interaction term

$$S_I = \int d^d x \sum_{i_1, i_2 = 1}^{N} (\partial^2 \phi^{i_1})(\partial_\mu \phi^{i_2}) \phi_{i_1}(\partial^\mu \phi_{i_2}), \qquad (11.16)$$

   for N scalars ϕ_i, $i = 1, \ldots, N$, each with the usual massless kinetic term. Write down the Feynman rules for this model.

3. Consider the massive Dirac–Born–Infeld (DBI) scalar Lagrangian

$$\mathcal{L} = -\frac{1}{L^4}\sqrt{1 + L^4[(\partial_\mu \phi)^2 + m^2 \phi^2]}. \qquad (11.17)$$

   Expand it in L up to order L^8 (keeping two nontrivial orders beyond the kinetic term). Considering the extra terms as interaction terms, write the resulting Feynman rules.

4. Consider the Feynman diagram in Figure 11.5 for ϕ^5 theory in p space. Calculate the statistical weight factor for the diagram, and write an integral expression for it using the Feynman rules.

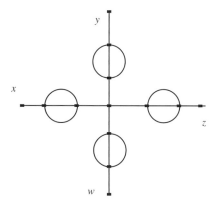

Figure 11.4 x-Space Feynman diagram.

Figure 11.5 p-Space Feynman diagram.

12

Quantization of the Dirac Field and Fermionic Path Integral

Until now we have considered bosonic theories. In particular, we considered scalar fields, but at the end of Chapter 11 we considered the most general bosonic field theory, in the case without redundancies. So any theory other than a gauge theory can be analyzed this way, while gauge theories will be studied in detail later on.

In this chapter we consider fermionic field theories instead, specifically the spin-1/2 fields. The spin-3/2 field, called a gravitino, is only relevant for supergravity theories, which are too specialized, so will not be considered in this book, and spins higher than that have no simple interactions, so are usually not considered.

We start with a review of the classical Dirac field, which is taught in a classical field theory course. After describing the Dirac equation and Weyl spinors, we consider solutions of the Dirac equation. Then, we quantize the Dirac field, and define the fermionic path integral.

12.1 The Dirac Equation

We start by reviewing the classical Dirac field and the equation governing it, which was defined in classical field theory. Spinors were also defined, by analyzing the possible representations of the Poincaré group, as half-integer representations. But there is a more relevant definition of spin-1/2 spinors as follows.

Spinors are fields in a Hilbert space acted on by gamma matrices. The gamma matrices are objects satisfying the *Clifford algebra*

$$\{\gamma^\mu, \gamma^\nu\} = 2g^{\mu\nu}\mathbf{1}. \tag{12.1}$$

Therefore, the spinors are representations of the Clifford algebra (Hilbert spaces on which we represent the abstract algebra). Since in this case the objects

$$S_{\mu\nu} = \frac{i}{4}[\gamma_\mu, \gamma_\nu] \tag{12.2}$$

satisfy the Lorentz algebra

$$[J^{\mu\nu}, J^{\rho\sigma}] = i[g^{\nu\rho}J^{\mu\sigma} - g^{\mu\rho}J^{\nu\sigma} - g^{\nu\sigma}J^{\mu\rho} + g^{\mu\sigma}J^{\nu\rho}], \tag{12.3}$$

which is left as an exercise to prove, the spinors are also a representation of the Lorentz algebra, called the spinor representation.

For instance, in three Euclidean dimensions, the gamma matrices are the same as the Pauli matrices, $\gamma^i = \sigma^i$, since indeed then $\{\gamma^i, \gamma^j\} = 2\delta^{ij}$. Remember that the Pauli matrices

$$\sigma^1 = \begin{pmatrix} 0 & 1 \\ 1 & 0 \end{pmatrix}; \quad \sigma^2 = \begin{pmatrix} 0 & -i \\ i & 0 \end{pmatrix}; \quad \sigma^3 = \begin{pmatrix} 1 & 0 \\ 0 & -1 \end{pmatrix} \tag{12.4}$$

satisfy the relation

$$\sigma^i \sigma^j = \delta^{ij} + i\epsilon^{ijk}\sigma^k, \tag{12.5}$$

from which it follows that they do indeed satisfy the Clifford algebra in three Euclidean dimensions.

We can make linear transformations on the spinor space (Hilbert space for the Clifford algebra) via $\psi \to S\psi$ and $\gamma^\mu \to S\gamma^\mu S^{-1}$, and thus choose various representations for the gamma matrices. A particularly useful such representation is called the Weyl (or chiral) representation, which is defined by

$$\gamma^0 = -i\begin{pmatrix} \mathbf{0} & \mathbf{1} \\ \mathbf{1} & \mathbf{0} \end{pmatrix}; \quad \gamma^i = -i\begin{pmatrix} \mathbf{0} & \sigma^i \\ -\sigma^i & \mathbf{0} \end{pmatrix}; \quad i = 1, 2, 3, \tag{12.6}$$

where $\mathbf{1}$ and $\mathbf{0}$ are 2×2 matrices. For this representation, we can immediately check that $(\gamma^0)^\dagger = -\gamma^0$ and $(\gamma^i)^\dagger = \gamma^i$.

We can also define the 4-vector 2×2 matrices

$$\sigma^\mu = (\mathbf{1}, \sigma^i); \quad \bar{\sigma}^\mu = (\mathbf{1}, -\sigma^i), \tag{12.7}$$

such that in the Weyl representation we can write

$$\gamma^\mu = -i\begin{pmatrix} \mathbf{0} & \sigma^\mu \\ \bar{\sigma}^\mu & \mathbf{0} \end{pmatrix}. \tag{12.8}$$

Moreover, we define the matrix

$$\gamma_5 = -i\gamma^0\gamma^1\gamma^2\gamma^3, \tag{12.9}$$

which in the Weyl representation becomes just

$$\gamma_5 = \begin{pmatrix} \mathbf{1} & \mathbf{0} \\ \mathbf{0} & -\mathbf{1} \end{pmatrix}. \tag{12.10}$$

This is not a coincidence. In reality the Weyl representation was chosen such that γ_5, the product of gamma matrices, has this form. Also, the notation γ_5 is not coincidental: if we go to five dimensions, the first four gamma matrices are the same, and the fifth is γ_5.

Also, as a side remark, the pattern of obtaining the gamma matrices as tensor products of σ^i and $\mathbf{1}$ continues to be valid in all dimensions. The gamma matrices in two Euclidean dimensions can be chosen to be σ^1, σ^2 (note that then $\sigma^3 = -i\sigma^1\sigma^2$) and in higher dimensions we can always build the gamma matrices in terms of them.

We now write down the Dirac equation for a Dirac spinor:

$$(\gamma^\mu \partial_\mu + m)\psi = 0. \tag{12.11}$$

We can define the *Dirac conjugate*

$$\bar{\psi} = \psi^\dagger \beta; \quad \beta = i\gamma^0, \tag{12.12}$$

which is defined such that $\bar{\psi}\psi$ is a Lorentz invariant.

Note that there are several conventions possible for β (for instance, others use $\beta = i\gamma_0 = -i\gamma^0$, or $\beta = \gamma^0$ or γ_0). With these conventions, the Dirac action is written as

$$S_\psi = -\int d^4x\, \bar{\psi}(\gamma^\mu \partial_\mu + m)\psi. \tag{12.13}$$

We will use the notation $\slashed{\partial} \equiv \gamma^\mu \partial_\mu$, which is common. Since the Dirac field ψ is complex, by varying the action with respect to $\bar{\psi}$ (considered as independent of ψ), we get the Dirac equation. Note that shortly we will find for Majorana (real) spinors, for which $\bar{\psi}$ is related to ψ, that there is a factor of $1/2$ in front of the action, since by varying with respect to ψ we get two terms (one where we vary ψ proper, and one where we vary ψ from $\bar{\psi}$, giving the same result).

12.2 Weyl Spinors

In four Minkowski dimensions, the Dirac representation is reducible as a representation of the Lorentz algebra (or of the Clifford algebra). The irreducible representation (irrep) is found by imposing a constraint on it, and the spinors are called Weyl (or chiral) spinors. In the Weyl representation for the gamma matrices, the Dirac spinors split simply as

$$\psi_D = \begin{pmatrix} \psi_L \\ \psi_R \end{pmatrix}, \tag{12.14}$$

which is why the Weyl representation for gamma matrices was chosen as such. In general, we have

$$\psi_L = \frac{1 + \gamma_5}{2}\psi_D \Rightarrow \frac{1 - \gamma_5}{2}\psi_L = 0,$$

$$\psi_R = \frac{1 - \gamma_5}{2}\psi_D \Rightarrow \frac{1 + \gamma_5}{2}\psi_R = 0, \tag{12.15}$$

and note that we chose the Weyl representation for gamma matrices such that we have the projectors onto the Weyl spinor representations

$$\frac{1 + \gamma_5}{2} = \begin{pmatrix} 1 & 0 \\ 0 & 0 \end{pmatrix}; \quad \frac{1 - \gamma_5}{2} = \begin{pmatrix} 0 & 0 \\ 0 & 1 \end{pmatrix}. \tag{12.16}$$

Another possible choice for irreducible representation, completely equivalent (in four Minkowski dimensions) to the Weyl representation, is the *Majorana representation*. Majorana spinors satisfy the reality condition

$$\bar{\psi} = \psi^C \equiv \psi^T C, \tag{12.17}$$

where ψ^C is called the Majorana conjugate (thus the reality condition is "Dirac conjugate equals Majorana conjugate"), and C is a matrix called the *charge conjugation matrix*. Note that since ψ^T is just another ordering of ψ, whereas $\bar{\psi}$ contains $\psi^\dagger = (\psi^*)^T$, this is indeed a reality condition, of the type $\psi^* = (\ldots)\psi$.

The charge conjugation matrix in four Minkowski dimensions satisfies

$$C^T = -C; \quad C\gamma^\mu C^{-1} = -(\gamma^\mu)^T. \tag{12.18}$$

In other dimensions and/or signatures, the definition is more complicated, and it can involve other signs on the right-hand side of the two equations above.

In the Weyl representation, we can choose

$$C = \begin{pmatrix} -\epsilon^{\alpha\beta} & \mathbf{0} \\ \mathbf{0} & \epsilon^{\alpha\beta} \end{pmatrix}, \tag{12.19}$$

where we have

$$\epsilon^{\alpha\beta} = \begin{pmatrix} 0 & 1 \\ -1 & 0 \end{pmatrix} = i\sigma^2, \tag{12.20}$$

and so

$$C = \begin{pmatrix} -i\sigma^2 & \mathbf{0} \\ \mathbf{0} & i\sigma^2 \end{pmatrix} = -i\gamma^0\gamma^2. \tag{12.21}$$

Then we also have

$$C^{-1} = \begin{pmatrix} i\sigma^2 & \mathbf{0} \\ \mathbf{0} & -i\sigma^2 \end{pmatrix} = -C. \tag{12.22}$$

We can now check explicitly that this C is indeed a representation for the C-matrix (i.e. that it satisfies (12.18)).

As we mentioned, the action for Majorana fields, with $\bar{\psi}$ related to ψ, is

$$S_\psi = -\frac{1}{2} \int d^4x \, \bar{\psi}(\slashed{\partial} + m)\psi. \tag{12.23}$$

The Dirac equation implies the KG equation; more precisely, the Dirac equation is a sort of square root of the KG equation, which is roughly how Dirac thought about deriving it. Let's see this. First, we note that

$$\slashed{\partial}\slashed{\partial} = \gamma^\mu\gamma^\nu \partial_\mu \partial_\nu = \frac{1}{2}\{\gamma^\mu, \gamma^\nu\}\partial_\mu \partial_\nu = g^{\mu\nu}\partial_\nu \partial_\nu = \Box. \tag{12.24}$$

Then, it follows that

$$(\slashed{\partial} - m)(\slashed{\partial} + m)\psi = (\partial^2 - m^2)\psi, \tag{12.25}$$

so indeed, the KG operator is the product of the Dirac operator with $+m$ and the Dirac operator with $-m$.

12.3 Solutions of the Free Dirac Equation

Since a solution of the Dirac equation is also a solution of the KG equation, solutions of the Dirac equation have to be solutions of the KG equation, that is $e^{\pm ipx}$, with $p^2 + m^2 = 0$, times some matrices (column vectors) depending on p.

One set of solutions can then be written as

$$\psi(x) = u(p)e^{ipx}, \tag{12.26}$$

where $p^2 + m^2 = 0$ and $p^0 > 0$. Since this is basically a Fourier transform, $u(p)$ satisfies the Fourier-transformed equation

$$(i\not{p} + m)u(p) = 0, \tag{12.27}$$

where $\not{p} \equiv \gamma^\mu p_\mu$ (note that $p_\mu = g_{\mu\nu}p^\nu$, and for the same p^μ, the p_μ changes by a sign between our "mostly plus" metric convention and the "mostly minus" one). The two solutions for the above equation for $u(p)$ can be written compactly and formally as

$$u^s(p) = \begin{pmatrix} \sqrt{-p \cdot \sigma}\,\xi^s \\ \sqrt{-p \cdot \bar\sigma}\,\xi^s \end{pmatrix}, \tag{12.28}$$

where $s = 1, 2$ and $\xi^1 = \begin{pmatrix} 1 \\ 0 \end{pmatrix}, \xi^2 = \begin{pmatrix} 0 \\ 1 \end{pmatrix}$. Here $\sqrt{-p \cdot \sigma} = \sqrt{-p_\mu \sigma^\mu}$ is understood in matrix sense (if $A^2 = B$, then $A = \sqrt{B}$).

In case the matrix is diagonal, we can take the square root of the diagonal elements, but in general we can't. For instance, in the rest frame, where $\vec{p} = 0$ and $p^0 = m$:

$$u^s(p) \propto \begin{pmatrix} 0 \\ 1 \\ 0 \\ 1 \end{pmatrix} \quad \text{or} \quad \begin{pmatrix} 1 \\ 0 \\ 1 \\ 0 \end{pmatrix}. \tag{12.29}$$

The other two solutions are of type

$$\psi(x) = v(p)e^{-ipx}, \tag{12.30}$$

where $p^2 + m^2 = 0$, $p^0 > 0$, and $v(p)$ satisfies

$$(-i\not{p} + m)v(p) = 0. \tag{12.31}$$

The two $v^s(p)$ can be similarly written as

$$v^s(p) = \begin{pmatrix} \sqrt{-p \cdot \sigma}\,\xi^s \\ -\sqrt{-p \cdot \bar\sigma}\,\xi^s \end{pmatrix}, \tag{12.32}$$

and in the rest frame we have

$$v^s(p) \propto \begin{pmatrix} 0 \\ 1 \\ 0 \\ -1 \end{pmatrix} \quad \text{or} \quad \begin{pmatrix} 1 \\ 0 \\ -1 \\ 0 \end{pmatrix}. \tag{12.33}$$

The normalization conditions for the $u(p)$ and $v(p)$ are written in Lorentz-invariant form as

$$\bar{u}^r(p)u^s(p) = 2m\delta^{rs},$$
$$\bar{v}^r(p)v^s(p) = -2m\delta^{rs}, \tag{12.34}$$

or using $u^\dagger$ and $v^\dagger$ as

$$u^{r\dagger}(p)u^s(p) = 2E_p\delta^{rs},$$

$$v^{r\dagger}(p)v^s(p) = 2E_p\delta^{rs}. \tag{12.35}$$

The u and v solutions are orthogonal, that is

$$\bar{u}^r(p)v^s(p) = \bar{v}^r(p)u^s(p) = 0. \tag{12.36}$$

Note that now

$$u^{r\dagger}(p)v^s(p) \neq 0; \quad v^{r\dagger}(p)u^s(p) \neq 0, \tag{12.37}$$

however, we also have

$$u^{r\dagger}(\vec{p})v^s(-\vec{p}) = v^{r\dagger}(-\vec{p})u^s(\vec{p}) = 0. \tag{12.38}$$

12.4 Quantization of the Dirac Field

The spin-statistics theorem says that fields of spin $S = (2k + 1)/2$ ($k \in \mathbb{N}$) obey Fermi statistics, which means that in their quantization we must use anticommutators, $\{,\}$, instead of commutators, $[,]$. In contrast, for $S = k$, we have the Bose–Einstein statistics, with commutators $[,]$ for their quantization.

The Lagrangian for the Dirac field is

$$\mathcal{L} = -\bar{\psi}\gamma^\mu\partial_\mu\psi - m\bar{\psi}\psi + \ldots = -\psi^\dagger i\gamma^0\gamma^\mu\partial_\mu\psi + \ldots = +i\psi^\dagger\partial_0\psi + \ldots \tag{12.39}$$

(since $(\gamma^0)^2 = -1$). Then the canonical conjugate to ψ is

$$p_\psi = \frac{\partial\mathcal{L}}{\partial\dot{\psi}} = i\psi^\dagger. \tag{12.40}$$

The Hamiltonian is

$$H = p\dot{q} - L = i\int d^3x\,\psi^\dagger(+\gamma^0\gamma^i\partial_i + m\gamma^0)\psi. \tag{12.41}$$

When we quantize, we write anticommutation relations at equal time, $\{,\}_{P.B.} \to 1/i\hbar\{,\}$, namely (after canceling the i)

$$\{\psi_\alpha(\vec{x},t), \psi_\beta^\dagger(\vec{y},t)\} = \delta^3(\vec{x} - \vec{y})\delta_{\alpha\beta},$$

$$\{\psi_\alpha(\vec{x},t), \psi_\beta(\vec{y},t)\} = \{\psi_\alpha^\dagger(\vec{x},t), \psi_\beta^\dagger(\vec{y},t)\} = 0 \tag{12.42}$$

are the *equal-time anticommutation relations*.

The quantization proceeds exactly as for the complex scalar field, just that instead of the properly normalized solution of the KG equation, $e^{\pm ipx}$, we have the above solutions of the Dirac equation, and again, their coefficients are harmonic oscillator operators. Therefore, we have

$$\psi(x) = \int \frac{d^3p}{(2\pi)^3} \frac{1}{\sqrt{2E_p}} \sum_s (a_{\vec{p}}^s u^s(p)e^{ipx} + b_{\vec{p}}^{s\dagger} v^s(p)e^{-ipx}),$$

$$\bar{\psi}(x) = \int \frac{d^3p}{(2\pi)^3} \frac{1}{\sqrt{2E_p}} \sum_s (b_{\vec{p}}^s \bar{v}^s(p)e^{ipx} + a_{\vec{p}}^{s\dagger} \bar{u}^s(p)e^{-ipx}), \qquad (12.43)$$

where the as, $a^\dagger$s and bs, $b^\dagger$s are the annihilation/creation operators obeying the anticommutation relations of the *fermionic* harmonic oscillator, that is

$$\{a_{\vec{p}}^r, a_{\vec{q}}^{\dagger s}\} = \{b_{\vec{p}}^r, b_{\vec{q}}^{\dagger s}\} = (2\pi)^3 \delta^3(\vec{p} - \vec{q})\delta_{rs}, \qquad (12.44)$$

and the rest of the anticommutators are zero.

There is a Fock vacuum $|0\rangle$, satisfying

$$a_{\vec{p}}^s |0\rangle = b_{\vec{p}}^s |0\rangle = 0, \qquad (12.45)$$

and then Fock states are created by acting with $a_{\vec{p}}^{s\dagger}$ and $b_{\vec{p}}^{s\dagger}$ on it. But, due to the anticommutation relations, which mean in particular that $(a_{\vec{p}}^{s\dagger})^2 = (b_{\vec{p}}^{s\dagger})^2 = 0$, we can't have two excitations in the same Fock state:

$$(a_{\vec{p}}^{s\dagger})^2 |\psi\rangle = (b_{\vec{p}}^{s\dagger})^2 |\psi\rangle = 0, \qquad (12.46)$$

which is a manifestation of the Pauli exclusion principle.

The Hamiltonian is then written as

$$H = \int \frac{d^3p}{(2\pi)^3} \sum_s E_p(a_{\vec{p}}^{\dagger s} a_{\vec{p}}^s - b_{\vec{p}}^s b_{\vec{p}}^{\dagger s})$$

$$= \int \frac{d^3p}{(2\pi)^3} \sum_s E_p(a_{\vec{p}}^{\dagger s} a_{\vec{p}}^s + b_{\vec{p}}^{s\dagger} b_{\vec{p}}^s - \{b_{\vec{p}}^{\dagger s}, b_{\vec{p}}^s\}), \qquad (12.47)$$

where the anticommutator in the integral is equal to 1, giving a negative infinite constant. As in the bosonic case, the infinite constant is removed by *normal ordering*.

The difference is that now, since $a_{\vec{p}}^r a_{\vec{q}}^s = -a_{\vec{q}}^s a_{\vec{p}}^r$, $a_{\vec{p}}^{r\dagger} a_{\vec{q}}^{s\dagger} = -a_{\vec{q}}^{s\dagger} a_{\vec{p}}^{r\dagger}$, and similar for the bs, we have

$$: a_{\vec{p}}^r a_{\vec{q}}^s := a_{\vec{p}}^r a_{\vec{q}}^s = -a_{\vec{q}}^s a_{\vec{p}}^r, \quad \text{etc.} \qquad (12.48)$$

and so also

$$: a_{\vec{p}}^r a_{\vec{q}}^{s\dagger} := -a_{\vec{q}}^{s\dagger} a_{\vec{p}}^r. \qquad (12.49)$$

So the net effect of normal ordering is to remove the anticommutator terms, moving all the creation operators to the right with the anticommutation rules. Note that since the fermions have a negative infinite constant, whereas the bosons have a positive infinite constant, the infinite constant can cancel in a theory with an equal number of bosonic and fermionic modes, which is called a *supersymmetric theory*. These supersymmetric theories therefore possess better behavior with respect to unphysical infinities.

Also, note that we can interpret $a_{\vec{p}}^\dagger$ as creating positive-energy particles and $b_{\vec{p}}$ as the creation operator for a negative-energy particle, as the second term in the first line of (12.47) shows (compare with the first line). Dirac introduced the concept of a Dirac sea, which says that there is a full "sea" of occupied states of negative energy (remember that for fermions, a state can only be occupied by a single particle). Then $b_p^\dagger$ destroys a particle of negative energy, which is equivalent to creating a "hole" in the Dirac sea (unoccupied state in a sea of occupied states), which acts as a state of positive energy (consider the second line in (12.47)).

12.5 The Fermionic Path Integral

How do we write a path integral? The path integral is used to describe quantum mechanics, but it involves classical objects (functions), just going over all possible paths instead of just the classical path. These classical objects are obtained in the classical limit of quantum operators (i.e. when $\hbar \to 0$). For bosons, for instance for the harmonic oscillators, $[a, a^\dagger] = \hbar \to 0$, so in the path integral over the harmonic phase space we used the classical objects α and α^*, corresponding to the quantum objects a and $a^\dagger$.

But now, we have anticommutation relations, in particular for the fermionic harmonic oscillators, $\{\hat{a}, \hat{a}^\dagger\} = \hbar \to 0$. So in the classical limit for these operators we don't obtain the usual functions, but we obtain objects that anticommute, forming what is known as a *Grassmann algebra*. Considering the $a, a^\dagger$ of a fermionic harmonic oscillator, we have

$$\{a, a^\dagger\} = \{a, a\} = \{a^\dagger, a^\dagger\} = 0. \tag{12.50}$$

The objects $a, a^\dagger$ are known as the "odd" part of the algebra, since they anticommute, whereas products of odd objects are called "even," since they commute. For instance, we see that $[aa^\dagger, aa^\dagger] = 0$. Thus in general we have *bose* × *bose* = *bose*, *fermi* × *fermi* = *bose*, and *bose* × *fermi* = *fermi*, where *bose* stands for even and *fermi* stands for odd.

However, what is the meaning of a "classical" Grassmann field if even in the classical $\hbar \to 0$ limit we can't put more than one particle in the same state: $\{a^\dagger, a^\dagger\} = 0 \Rightarrow a^\dagger a^\dagger |0\rangle = 0$, where $a^\dagger$ is an object that appears in the expansion of the "classical" Grassmann field $\psi(x)$. The meaning is not clear, but this is one of those instances where, in quantum mechanics, there are questions which are meaningless. The point is that this is a formalism that works for the path integral (i.e. we obtain the right results, which agree with experiments and also with the operator formalism of quantum field theory), and the path integral is quantum mechanical in any case, even if it is written as an integral over "classical" Grassmann fields.

12.5.1 Definitions

We now define better the calculus with Grassmann fields. The general Grassmann algebra of N objects x_i, $i = 1, \ldots, N$, is $\{x_i, x_j\} = 0$, together with the element **1**, which commutes with the rest, $[x_i, \mathbf{1}] = 0$, and with complex number coefficients (i.e. it is an algebra over $\mathbb{C}$).

Since $(x_i)^2 = 0$, the Fourier expansion of a general function stops after a finite number of terms, specifically after N terms:

$$f(\{x_i\}) = f(x_0) + \sum_i f_i^{(1)} x_i + \sum_{i<j} f_{ij}^{(2)} x_i x_j + \sum_{i<j<k} f_{ijk}^{(3)} x_i x_j x_k + \ldots + f_{12\ldots N}^{(N)} x_1 \ldots x_N. \quad (12.51)$$

Only for an infinite number of generators x_i is the Taylor expansion infinite in extent. Note that here, since we cannot add bosons (even objects) with fermions (odd objects), the functions must be either even, in which case there is always an even number of xs in the expansion, or odd, in which case there is an odd number of xs in the expansion.

However, we will often think of only a subset of the xs in making the expansion. For instance, we can expand in terms of a single x (even though there are more objects in the Grassmann algebra), and write for a general even function of x the expansion

$$f(x) = a + bx, \quad (12.52)$$

where, since x is odd, a is even and b is odd (fermionic). This is possible, since there is at least one other Grassmann object, for instance called y, so we could have $b = cy$ (i.e. $f(x) = a + cyx$).

In fact, we will often consider (at least) an even number of xs, half of which are used for the expansion, and half for the coefficients, allowing us to write the general expansion (12.51) with coefficients $f^{(2k)}$ being even (bosonic) but $f^{(2k+1)}$ being odd (fermionic). (This is what one does in the case of supersymmetry, where one expands a "superfield" Φ in terms of auxiliary θs like the xs above, with odd coefficients which are still Grassmann functions corresponding to spinor fields.)

We define differentiation by first writing

$$\left(\frac{\partial}{\partial x_i} x_j\right) = \delta_{ij}, \quad (12.53)$$

but since differential operators must also be Grassmann objects, we have

$$\frac{\partial}{\partial x_i}(x_j \ldots) = \delta_{ij}(\ldots) - x_j \frac{\partial}{\partial x_i}(\ldots). \quad (12.54)$$

Note that for bosons we would have a plus sign in the second term, but now $\partial/\partial x_i$ anticommutes past x_j.

As an example, we have

$$\frac{\partial}{\partial x_i}(x_1 x_2 \ldots x_n) = \delta_{i1} x_2 x_3 \ldots x_n - \delta_{i2} x_1 x_3 \ldots x_n + \delta_{i3} x_1 x_2 x_4$$

$$\ldots x_n + \ldots + (-)^{n-1} \delta_{in} x_1 x_2 \ldots x_{n-1}. \quad (12.55)$$

Then, for instance, differentiation of a product of even functions $f(x)$ and $g(x)$ is done in the usual way:

$$\frac{\partial}{\partial x_i}(f(x)g(x)) = \left(\frac{\partial}{\partial x_i} f(x)\right) g(x) + f(x)\left(\frac{\partial}{\partial x_i} g(x)\right), \quad (12.56)$$

but if $f(x)$ is an odd function, we have the modified rule

$$\frac{\partial}{\partial x_i}(f(x)g(x)) = \left(\frac{\partial}{\partial x_i} f(x)\right) g(x) - f(x)\left(\frac{\partial}{\partial x_i} g(x)\right). \quad (12.57)$$

As another example, consider the function $e^{\sum_i x_i}$, which is defined in the usual way, just that we now substitute $(x_i)^2 = 0$ in the Taylor expansion of the exponential. Then we have

$$\frac{\partial}{\partial x_k} e^{\sum_i x_i y_i} = y_k e^{\sum_i x_i y_i},$$

$$\frac{\partial}{\partial y_k} e^{\sum_i x_i y_i} = -x_k e^{\sum_i x_i y_i}. \tag{12.58}$$

Next, we define integration, which is not defined as a Riemann sum in the usual way for functions over complex numbers. Indeed, due to the Grassmann properties, we cannot do this, but rather must define it as a linear operator, and in particular we can only define the indefinite integral, not a definite one.

Since the two basic elements of the algebra with an element x are **1** and x, we must define what happens for these two elements. We define

$$\int dx\, \mathbf{1} = 0; \quad \int dx\, x = 1. \tag{12.59}$$

For several x_is, we define the anticommutation relations for $i \neq j$:

$$\{dx_i, dx_j\} = 0; \quad \{x_i, dx_j\} = 0. \tag{12.60}$$

For instance, we have

$$\int dx_1 dx_2\, x_1 x_2 = -\int dx_1\, x_1 \int dx_2\, x_2 = -1. \tag{12.61}$$

Then the integral operation is translational invariant, since

$$\int dx\, f(x+a) = \int dx[f^0 + f^1(x+a)] = \int dx\, f^1 x = \int dx\, f(x). \tag{12.62}$$

In view of the above relations, we see that *integration is the same as differentiation* (satisfies the same rules). For instance:

$$\int dx_2\, x_1 x_2 x_3 x_4 = -\int dx_2\, x_2 x_1 x_3 x_4 = -x_1 x_3 x_4, \tag{12.63}$$

which is the same as

$$\frac{\partial}{\partial x_2} x_1 x_2 x_3 x_4 = -\frac{\partial}{\partial x_2} x_2 x_1 x_3 x_4 = -x_1 x_3 x_4. \tag{12.64}$$

We now define the delta function on the Grassmann space. We have

$$\delta(x) = x. \tag{12.65}$$

To prove this, consider a general even function of x, $f(x) = f^0 + f^1 x$, where f^0 is even and f^1 is odd. Then

$$\int dx\delta(x-y)f(x) = \int dx(x-y)(f^0 + f^1 x) = \int dx(xf^0 - yf^1 x)$$

$$= f^0 + y \int dx f^1 x = f^0 - yf^1 = f^0 + f^1 y = f(y). \tag{12.66}$$

Finally, let us consider the change of variables from the Grassmann variable x to the variable $y = ax$, where $a \in \mathbb{C}$. But then we must have

$$1 = \int dx\, x = \int dy\, y = a \int dy\, x, \qquad (12.67)$$

so it follows that

$$dy = \frac{1}{a} dx, \qquad (12.68)$$

which is not like the usual integration, but is rather like differentiation.

Important Concepts to Remember

- Spinor fields are representations of the Clifford algebra $\{\gamma^\mu, \gamma^\nu\} = 2g^{\mu\nu}$, and spinorial representations of the Lorentz algebra.
- In the Weyl (chiral) representation, γ_5 is diagonal and has ± 1 on the diagonal.
- The Dirac action is $-\int \bar{\psi}(\partial\!\!\!/ + m)\psi$, with $\bar{\psi} = \psi^\dagger i\gamma^0$.
- The irreducible spinor representations are, in four-dimensional Minkowski space, either Weyl spinors, or Majorana spinors. Weyl spinors satisfy $(1 \pm \gamma_5)/2\psi = 0$ and Majorana spinors satisfy $\bar{\psi} = \psi^T C$, with C the charge conjugation matrix.
- The Dirac operator is a kind of square root of the KG operator, since $(\partial\!\!\!/ - m)(\partial\!\!\!/ + m) = \Box - m^2$.
- The solutions of the Dirac equation are $u(p)e^{ipx}$ and $v(p)e^{-ipx}$, with $u(p)$ and $v(p)$ orthonormal, and in canonical quantization we expand in these solutions, with fermionic harmonic oscillator coefficients.
- The momentum conjugate to ψ is $i\psi^\dagger$, giving the nontrivial anticommutator $\{\psi_\alpha(\vec{x}, t), \psi_\beta^\dagger(\vec{y}, t)\} = \delta(\vec{x} - \vec{y})\delta_{\alpha\beta}$.
- The infinite constant (zero-point energy) in the Hamiltonian is negative.
- We can interpret as: $a^\dagger$ creates positive-energy states, b creates negative-energy states, and $b^\dagger$ destroys negative-energy states, thus effectively creating "holes" in the Dirac sea of positive energy.
- For the fermionic path integral, we use Grassmann algebra-valued objects.
- The Taylor expansion of a function of Grassmann algebra objects ends at order N, the number of x_is.
- Grassmann differentiation is also anticommuting with others and with xs.
- Grassmann integration is the same as Grassmann differentiation and $\delta(x) = x$ for Grassmann variables.

Further Reading

See sections 3.1.1, 3.1.2, and 3.3 in [3], section 7.1 in [2], sections 3.2 and 3.3 in [1], and sections 5.2 and 5.3 in [4].

Exercises

1. Consider the Rarita–Schwinger action for a vector–spinor field $\psi_{\mu\alpha}$ in Minkowski space

$$S = -\frac{1}{2} \int d^4x \bar{\psi}_\mu \gamma^{\mu\nu\rho} \partial_\nu \psi_\rho, \qquad (12.69)$$

where $(\psi_\mu)_\alpha$ is a *Majorana* spinor and $\gamma^{\mu\nu\rho} = \gamma^{[\mu}\gamma^\nu\gamma^{\rho]}$. Calculate the variation δS under a variation $\delta\psi$, and then the equations of motion, in x and p space.

2. Consider the Lagrangian for a *Dirac* field in Minkowski space

$$S = -\int d^4x [\bar{\psi}(\slashed{\partial} + m)\psi + \alpha(\bar{\psi}\psi)^2], \qquad (12.70)$$

and use the free-field quantization in the text. Compute the quantum Hamiltonian in terms of $a, a^\dagger, b, b^\dagger$ and then the normal ordered Hamiltonian.

3. Show that $S_{\mu\nu} = \frac{i}{2}\gamma_{\mu\nu} = \frac{i}{4}[\gamma_\mu, \gamma_\nu]$ satisfies the Lorentz algebra

$$[J^{\mu\nu}, J^{\rho\sigma}] = i[g^{\nu\rho}J^{\mu\sigma} - g^{\mu\rho}J^{\nu\sigma} - g^{\nu\sigma}J^{\mu\rho} + g^{\mu\sigma}J^{\nu\rho}]. \qquad (12.71)$$

4. Show that $C = -i\gamma^0\gamma^2$ is a C-matrix (i.e. it satisfies (12.18)).

5. Find the normalization constants in (12.29) and (12.33), such that we have the normalization conditions (12.34) and (12.35).

13 Wick Theorem, Gaussian Integration, and Feynman Rules for Fermions

In Chapter 12 we saw that the path integral for fermions is done using integrals of objects in a Grassmann algebra. We defined the calculus on the Grassmann algebra, defining an anti-commuting differential operator and an integral operator (defined as a linear operation, not a Riemann sum), which turned out to have the same properties as the differential operator.

We then saw that changing variables in the Grassmann algebra, between x and $y = ax$, where a is a c-number, gives us $dy = dx/a$, like for the differential operator. The obvious generalization to an n-dimensional Grassmann algebra is $y = A \cdot x$ implies $d^n y = d^n x / \det A$.

In this chapter, we continue to define perturbation theory in the path-integral formalism for fermions. We first define Gaussian integration for fermions, then define the path integral for the fermionic harmonic oscillator, in a similar manner to the bosonic case. We finally derive the Wick theorem for fermions, which allows us to construct perturbation theory with fermions, and define the resulting Feynman rules for the Yukawa interaction.

13.1 Gaussian Integration for Fermions

13.1.1 Gaussian Integration – The Real Case

We now define Gaussian integration over a Grassmann algebra over real numbers.

Theorem 13.1 Consider a real $n \times n$ antisymmetric matrix $\{A_{ij}\}$, such that the (symmetric) matrix A^2 has negative, nonzero eigenvalues. Then $n = 2m$, and for $x_1, \ldots, x_n$ Grassmann variables:

$$\int d^n x e^{x^T A x} = 2^m \sqrt{\det A}. \tag{13.1}$$

Proof Consider first the case $m = 1$ (i.e. $n = 2$). Then

$$A = \begin{pmatrix} 0 & \lambda \\ -\lambda & 0 \end{pmatrix} \Rightarrow A^2 = \begin{pmatrix} -\lambda^2 & 0 \\ 0 & -\lambda^2 \end{pmatrix}, \tag{13.2}$$

and $\det A = \lambda^2 > 0$, whereas $x^T A x = \lambda(x_1 x_2 - x_2 x_1) = 2\lambda x_1 x_2$. Since $(x_i)^2 = 0$, we then have

$$e^{x^T A x} = 1 + 2\lambda x_1 x_2 \Rightarrow \int d^2 x e^{x^T A x} = \int dx_2 dx_1 [1 + 2\lambda x_1 x_2] = 2\lambda. \tag{13.3}$$

Note that here we had to choose a specific order for the integrals in d^2x, namely $dx_2 dx_1$, in order to get a plus sign.

Next, we consider a version of the *Jordan lemma* for the Grassmann algebra, namely that there is a transformation matrix B (real, orthogonal) acting on the Grassmann algebra space such that $B^T A B$ is block-diagonal in Jordan blocks, that is

$$
B^T A B = \begin{pmatrix} 0 & \lambda_1 & & & & & \\ -\lambda_1 & 0 & & & & & \\ & & 0 & \lambda_2 & & & \\ & & -\lambda_2 & 0 & & & \\ & & & & \ddots & & \\ & & & & & 0 & \lambda_m \\ & & & & & -\lambda_m & 0 \end{pmatrix}, \tag{13.4}
$$

which then gives $\det A = \det(B^T A B) = \lambda_1^2 \ldots \lambda_m^2$.

Proof of lemma We need to show that there is a basis of vectors $\{\vec{e}_1, \vec{e}_{-1}, \vec{e}_2, \vec{e}_{-2}, \ldots, \vec{e}_m, \vec{e}_{-m}\}$ such that

$$
A\vec{e}_k = \lambda_k \vec{e}_k; \quad A\vec{e}_{-k} = -\lambda_k \vec{e}_k. \tag{13.5}
$$

Then we have $A^2 \vec{e}_{\pm k} = -\lambda_k^2 \vec{e}_{\pm k}$. We see that it is sufficient to take a basis of eigenvectors of A^2, since we already know A^2 has nonzero negative eigenvalues. We need to check that there is a twofold degeneracy for the basis, which is true, since if $\vec{e}$ has eigenvalue $-\lambda^2$, so does $\vec{e}' = A\vec{e}/\lambda$, as we can easily check: $A^2 \vec{e}' = -\lambda A\vec{e} = -\lambda^2 \vec{e}'$. The basis of vectors is linearly independent, since if there were a number f such that $\vec{e}' = f\vec{e}$, on the one hand, we could write what $\vec{e}'$ is and obtain $A\vec{e} = \lambda f \vec{e}$, applying this twice to obtain $A^2 \vec{e} = \lambda^2 f^2 \vec{e}$. On the other hand, we know that $A^2 \vec{e} = -\lambda^2 \vec{e}$, but since f is real, $f^2 > 0$, giving a contradiction. Therefore $\vec{e}$ and $\vec{e}'$ are linearly independent, and we have found the Jordan basis for the matrix A. Finally, $\vec{e} \cdot \vec{e}' = \frac{1}{\lambda} \vec{e} A\vec{e} = 0$, since A is antisymmetric, hence the eigenvectors are also orthogonal. *q.e.d.* lemma

Then we can write

$$
\int d^n x e^{x^T A x} = \int d^n x' e^{(B^T x)^T B^T A B (B^T x)} = \prod_{i=1}^{m} [m = 1 \ \text{case}]_i = 2^m \sqrt{\det A}, \tag{13.6}
$$

where $x' = B^T x$. *q.e.d.* Theorem 13.1

Theorem 13.2 Complex Gaussian integration We write independent complex variables x_i, y_i, without putting $y_i = x_i^*$, and an $n \times n$ antisymmetric matrix A, and then we have

$$
\int d^n x d^n y e^{y^T A x} = \det A. \tag{13.7}
$$

We consider this since in the path integral we have something exactly of this form, for instance $\int \mathcal{D}\psi \mathcal{D}\bar{\psi} e^{-\bar{\psi}(\slashed{\partial}+m)\psi}$.

Proof In $e^{y^T Ax}$, the terms that are nonzero under the integration are only terms of order n (since $(x_i)^2 = 0$, so for $> n$ terms necessarily one of the xs is repeated, and for $< n$ terms we get some $\int dx_i 1 = 0$). Specifically, these terms will be

$$\frac{1}{n!} \sum_P \sum_Q (y_{P(1)} A_{P(1)Q(1)} x_{Q(1)}) \ldots (y_{P(n)} A_{P(n)Q(n)} x_{Q(n)})$$

$$= \frac{1}{n!} \sum_P \sum_Q (y_1 A_{1,QP^{-1}(q)} x_{QP^{-1}(1)}) \ldots (y_n A_{n,QP^{-1}(n)} x_{QP^{-1}(n)})$$

$$= 1 \times \sum_{Q'} (y_1 A_{1,Q'(q)} x_{Q'(1)}) \ldots (y_n A_{n,Q'(n)} x_{Q'(n)})$$

$$= \epsilon(y_1 \ldots y_n)(x_1 \ldots x_n) \sum_Q \epsilon_Q A_{1Q(1)} \ldots A_{nQ(n)}$$

$$= \epsilon(y_1 \ldots y_n)(x_1 \ldots x_n) \det A, \tag{13.8}$$

where in the first line we wrote the sums over permutations P, Q of $1, 2, \ldots, n$, in the second line we rewrote the sum over permutations, such that the $\sum_P$ is done trivially as a factor of $n!$, then we commuted the xs and the ys, obtaining $y_1 x_1 y_2 x_2 \ldots y_n x_n = \epsilon y_1 \ldots y_n x_1 \ldots x_n$, with ϵ is a sign that depends on n, but not on Q. We can fix the sign of the integral $d^n x d^n y (y_1 \ldots y_n)(x_1 \ldots x_n)$ to be ϵ by properly defining the order of dxs and dys, in which case we obtain just $\det A$ for the Gaussian integral.   *q.e.d.* Theorem 13.2

13.1.2 Real vs. Complex Integration

We can also check that complex Gaussian integration is consistent with the real integration case. If the complex objects x_i and y_i are written as $x_i = a_i + ib_i$ and $y_i = a_i - ib_i$, then $y^T Ax = a^T Aa + b^T Ab$. The Jacobian of the transformation $d^n x d^n y = J d^n a d^n b$ is $J = 2^{-m}$ (for the usual complex numbers we can check that it would be 2^m, and the Grassmann integral works as the inverse). Then the result of the integral done as the double real integral over $d^n a d^n b$ is $J 2^m (\sqrt{\det A})^2 = \det A$, as we want.

13.2 The Fermionic Harmonic Oscillator and Generalization to Field Theory

For the fermionic harmonic oscillator, the quantum hamiltonian is

$$\hat{H}_F = \omega \left(\hat{b}^\dagger \hat{b} - \frac{1}{2} \right). \tag{13.9}$$

We consider then, as in the bosonic case, coherent states $|\beta\rangle$ defined by (β is a Grassmann-valued number)

$$|\beta\rangle = e^{\hat{b}^\dagger \beta}|0\rangle = (1 + \hat{b}^\dagger \beta)|0\rangle = (1 - \beta \hat{b}^\dagger)|0\rangle \Rightarrow \hat{b}|\beta\rangle = \beta|0\rangle = \beta(1 - \beta \hat{b}^\dagger)|0\rangle = \beta|\beta\rangle$$
$$(13.10)$$

that satisfy the completeness relation

$$1 = \int d\beta^* d\beta |\beta\rangle\langle\beta^*|e^{-\beta^*\beta}. \qquad (13.11)$$

Then consider the Hamiltonian with currents

$$H(b^\dagger, b; t) = \omega b^\dagger b - b^\dagger \eta(t) - \bar{\eta}(t)b, \qquad (13.12)$$

and define, as in the bosonic case, the transition amplitude

$$M(\beta^*, t'; \beta, t) = \langle \beta^*, t'|\beta, t\rangle, \qquad (13.13)$$

for which we carry out the same steps as in the bosonic case and find the path integral

$$M(\beta^*, t'; \beta, t) = \int \mathcal{D}\beta^* \mathcal{D}\beta \exp\left\{ i \int_t^{t'} d\tau [-i\dot{\beta}^*(\tau)\beta(\tau) - H] + \beta^*(t)\beta(t) \right\}. \quad (13.14)$$

The equations of motion of the Hamiltonian with sources are

$$\dot{\beta}^* - i\omega\beta^* + i\bar{\eta} = 0,$$
$$\dot{\beta} + i\omega\beta - i\eta = 0, \qquad (13.15)$$

with solutions

$$\beta(\tau) = \beta e^{i\omega(t-\tau)} + i \int_t^\tau e^{i\omega(s-\tau)}\eta(s)ds,$$

$$\beta^*(\tau) = \beta^* e^{i\omega(\tau-t')} + i \int_\tau^{t'} e^{i\omega(\tau-s)}\bar{\eta}(s)ds. \qquad (13.16)$$

For the case $\beta = \beta^* = 0$, and $t \to -\infty$, $t' \to +\infty$, we obtain the vacuum functional (vacuum expectation value).

In the usual manner, we obtain

$$Z[\eta, \bar{\eta}] = \langle 0|0\rangle \exp\left\{ -\int_{-\infty}^{+\infty} d\tau \int_\tau^{+\infty} ds\, e^{i\omega(\tau-s)}\bar{\eta}(s)\eta(\tau) \right\}. \qquad (13.17)$$

The details are left as an exercise.

Renaming $\beta \to \psi$ and $\beta^* \to \bar{\psi}$ for easy generalization to the field theory case, and writing $\int \dot{\beta}^*\beta \to \int \partial_t\bar{\psi}\psi = -\int \bar{\psi}\partial_t\psi$, the path integral becomes

$$Z[\eta, \bar{\eta}] = \int \mathcal{D}\bar{\psi}\mathcal{D}\psi \exp\left\{ i \int dt[\bar{\psi}(i\partial_t - \omega)\psi + \bar{\psi}\eta + \bar{\eta}\psi] \right\}$$

$$= Z[0, 0] \exp\left\{ -\int ds d\tau\, \bar{\eta}(s)D_F(s, \tau)\eta(\tau) \right\}. \qquad (13.18)$$

Here we have defined the Feynman propagator such that (as usual) $iS^M = -\bar{\psi}\Delta^{-1}\psi$, that is

$$D_F(s, \tau) = (-i(i\partial_t - \omega))^{-1}(s, \tau) = i \int \frac{dE}{2\pi} \frac{e^{-iE(s-\tau)}}{E - \omega + i\epsilon} = \theta(s - \tau)e^{-i\omega(s-\tau)}, \quad (13.19)$$

where we have first written the Fourier transform, then done the complex integral with the residue theorem: there is now a single pole at $E = \omega - i\epsilon$. We get a nonzero result only if we close the contour below (in the lower half plane), which we can only do if $s - \tau > 0$, so that $-iE(s - \tau) = -|\text{Im}E|(s - \tau) + \text{imaginary}$, otherwise the result is zero. Substituting this $D_F(s, \tau)$ we indeed obtain (13.17).

In the Euclidean case, we write $t_M = -it_E$ as usual, and substituting in $Z[\eta, \bar\eta]$ we obtain

$$Z_E[\eta, \bar\eta] = \int \mathcal{D}\bar\psi \mathcal{D}\psi \exp\left[-\int dt_E \bar\psi \left(\frac{\partial}{\partial t_E} + \omega\right)\psi + \int dt_E(\bar\psi \eta + \bar\eta \psi)\right]$$

$$= Z[0, 0] \exp\left\{\int d\tau ds \bar\eta(s)D(s, \tau)\eta(\tau)\right\}, \tag{13.20}$$

where we define $-S_E = -\bar\psi D^{-1}\psi + \ldots$, that is

$$D(s, \tau) = (\partial_{\tau_E} + \omega)^{-1} = i\int \frac{dE}{2\pi} \frac{e^{-iE(s-\tau)}}{E + i\omega}, \tag{13.21}$$

and as usual this Euclidean propagator is well defined.

The Wick rotation to Minkowski space also proceeds as usual, namely $\tau_E \to i\tau_M$, and in order to have the same $E\tau$ product, $E_E = (-i + \epsilon)E_M$, where we only rotate by $\pi/2 - \epsilon$ to avoid crossing the Minkowski-space pole. Then we obtain the Feynman propagator as usual:

$$D(is_M, i\tau_M) = i\int \frac{dE_M}{2\pi} \frac{e^{-iE_M(s_M - \tau_M)}}{E_M - \omega + i\epsilon'} = D_F(s_M, \tau_M). \tag{13.22}$$

We are now ready to generalize to field theory. The Minkowski-space Lagrangian is

$$\mathcal{L}_F^{(M)} = -\bar\psi(\slashed{\partial} + m)\psi. \tag{13.23}$$

The Euclidean action is defined by the Wick rotation $t_E = it_M$, with $iS^M = -S^E$ as usual, with $\int dt_M = -i\int dt_E$, thus giving

$$\mathcal{L}_F^{(E)} = +\bar\psi(\slashed{\partial} + m)\psi. \tag{13.24}$$

Note, however, that since the gamma matrices must satisfy the Clifford algebra $\{\gamma^\mu, \gamma^\nu\} = 2g^{\mu\nu}$, in the Minkowski case we had $(\gamma^0)^2 = -1$, but in the Euclidean case we have $(\gamma^4)^2 = +1$, which is satisfied by choosing the Wick rotation $\gamma^4 = i\gamma^0$ (same rotation as for t), and as a result we also write

$$\bar\psi = \psi^\dagger i\gamma^0 = \psi^\dagger \gamma^4. \tag{13.25}$$

Also, in Euclidean space with $\gamma^4 = i\gamma^0$ we can check that $\gamma^\mu = \gamma_\mu = (\gamma^\mu)^\dagger$.

Then the free fermionic partition function in field theory is

$$Z_F^{(0)}[\bar\eta, \eta] = \int \mathcal{D}\bar\psi \mathcal{D}\psi \exp\left\{-\int d^4x \bar\psi(\slashed{\partial} + m)\psi + \int d^4x(\bar\eta\psi + \bar\psi\eta)\right\}$$

$$= Z[0, 0]e^{\bar\eta(\slashed{\partial} + m)^{-1}\eta}. \tag{13.26}$$

The Euclidean propagator is defined by $-S_E = -\bar\psi \Delta^{-1}\psi$, giving

$$S_F(x, y) = (\slashed{\partial} + m)^{-1} = i\int \frac{d^4p}{(2\pi)^4} \frac{e^{ip(x-y)}}{-\slashed{p} + im}. \tag{13.27}$$

Note that then

$$(\slashed{\partial} + m)_x S_F = \int \frac{d^4p}{(2\pi)^4}(i\slashed{p} + m)e^{ip(x-y)}\frac{i}{-\slashed{p} + im} = \delta^4(x - y), \tag{13.28}$$

and note also that

$$\frac{1}{-\slashed{p} + im} = \frac{-\slashed{p} - im}{p^2 + m^2}. \tag{13.29}$$

The relation to Minkowski-space operators is also as in the bosonic case, namely consider the general $N + M$ point function

$$\langle 0|T\{\hat{\psi}_{\alpha_1}(x_1)\ldots\hat{\psi}_{\alpha_N}(x_N)\hat{\bar{\psi}}_{\beta_1}(y_1)\ldots\hat{\bar{\psi}}_{\beta_M}(y_M)\}|0\rangle$$

$$\int \mathcal{D}\bar{\psi}\mathcal{D}\psi e^{iS_F^{(M)}}\{\psi_{\alpha_1}(x_1)\ldots\psi_{\alpha_N}(x_N)\bar{\psi}_{\beta_1}(y_1)\ldots\bar{\psi}_{\beta_M}(y_M)\}. \tag{13.30}$$

In particular, the propagator in the free theory is

$$S_F^{(M)}(x - y) = \langle 0|T\{\psi(x)\bar{\psi}(y)\}|0\rangle = i\int \frac{d^4p}{(2\pi)^4}\frac{e^{ip(x-y)}}{-i\slashed{p} - m + i\epsilon}$$

$$= \int \frac{d^4p}{(2\pi)^4}\frac{-\slashed{p} - im}{p^2 + m^2 - i\epsilon}e^{ip(x-y)}. \tag{13.31}$$

As usual, by Wick rotation ($t_E = it_M, E_E = (-i + \epsilon)E_M$), we get

$$S_F^{(E)}(it_M - it'_M; \vec{x} - \vec{y}) = iS_F^{(M)}(x - y). \tag{13.32}$$

13.3 Wick Theorem for Fermions

The **Wick theorem for Fermi fields**, for instance for some interaction with a scalar ϕ with source J and interaction term $S_I[\bar{\psi}, \psi, \phi]$, is

$$Z[\bar{\eta}, \eta, J] = e^{-S_I\left(-\frac{\delta}{\delta\eta}, \frac{\delta}{\delta\bar{\eta}}, \frac{\delta}{\delta J}\right)}Z_F^{(0)}[\bar{\eta}, \eta]Z_\phi^{(0)}[J]. \tag{13.33}$$

Note that the only difference from the bosonic case is the minus sign for $\delta/\delta\eta$, due to the fact that we have the source term $\int d^4x[\bar{\eta}\psi + \bar{\psi}\eta + \phi J]$, so we need to commute the $\delta/\delta\eta$ past $\bar{\psi}$ to act on η.

Coleman's lemma for fermions is also the same, except for the same minus sign, that is

$$F\left(-\frac{\delta}{\delta\eta}, \frac{\delta}{\delta\bar{\eta}}\right)Z[\bar{\eta}, \eta] = Z\left[-\frac{\delta}{\delta\psi}, \frac{\delta}{\delta\bar{\psi}}\right]\left(F(\bar{\psi}, \psi)e^{\bar{\psi}\eta + \bar{\eta}\psi}\right)|_{\bar{\psi}=\psi=0}. \tag{13.34}$$

Then, applying it to the first form of the Wick theorem, we get the second form of the Wick theorem

$$Z[\bar{\eta}, \eta, J] = e^{-S_I\left(-\frac{\delta}{\delta\eta}, \frac{\delta}{\delta\bar{\eta}}, \frac{\delta}{\delta J}\right)} e^{\bar{\eta} S_F \eta} Z_\phi^{(0)}[J]$$

$$= e^{-\frac{\delta}{\delta\psi} S_F \frac{\delta}{\delta\bar{\psi}}} e^{-S_I\left(\bar{\psi}, \psi, \frac{\delta}{\delta J}\right) + \int(\bar{\psi}\eta + \bar{\eta}\psi)} Z_\phi^{(0)}[J]\Big|_{\psi=\bar{\psi}=\phi=0}. \tag{13.35}$$

13.4 Feynman Rules for Yukawa Interaction

We are now ready to write down the Feynman rules for the Yukawa interaction between two fermions and a scalar:

$$\mathcal{L}_Y = g\bar{\psi}\psi\phi. \tag{13.36}$$

The second form of the Wick theorem is

$$Z[\bar{\eta}, \eta, J] = e^{\frac{1}{2}\frac{\delta}{\delta\phi}\Delta\frac{\delta}{\delta\phi}} e^{-\frac{\delta}{\delta\psi} S_F \frac{\delta}{\delta\bar{\psi}}} e^{-g\int d^4x \bar{\psi}\psi\phi + \int d^4x[\bar{\psi}\eta + \bar{\eta}\psi + J\phi]}\Big|_{\psi=\bar{\psi}=\phi=0}. \tag{13.37}$$

For the free two-point function, we must take two derivatives, and consider only the order $1 = g^0$ term, that is

$$\langle 0|\psi(x)\bar{\psi}(y)|0\rangle = \frac{\delta}{\delta\bar{\eta}(x)}\left(-\frac{\delta}{\delta\eta(y)}\right) Z[\bar{\eta}, \eta, J]|_{\eta=\bar{\eta}=J=0}$$

$$= -\frac{\delta}{\delta\psi} S_F \frac{\delta}{\delta\bar{\psi}} \{\psi(x)\bar{\psi}(y)\}|_{\psi=\bar{\psi}=\phi=0} = S_F(x - y). \tag{13.38}$$

Therefore the first Feynman rule is that for a propagator we have $S_F(x - y)$, represented by a solid line with an arrow between y ($\bar{\psi}(y)$) and x ($\psi(x)$), see Figure 13.1. Note that now, because fermions anticommute, the order matters, and if we interchange two fermions, for instance these ψ and $\bar{\psi}$ (interchange the order of the arrow), then we have a minus sign.

A scalar is now represented by a dotted line without an arrow on it.

Also, the vertex is standard. Since all the fields in the interaction term are different (no field is repeated), we have just a factor of $-g$ for the vertex (in the pure scalar case, for a $\phi^n/n!$ interaction we have a factor of $n!$ in front, which comes from permuting various ϕs, but now we don't have this).

The interchange of two fermions means also the only real difference from the bosonic rules. If we have a fermion loop with N lines sticking out of it (in this case, ϕ lines), this means that we must act with $-\frac{\delta}{\delta\psi} S_F \frac{\delta}{\delta\bar{\psi}}$ on factors of $\psi\bar{\psi}$, obtaining S_F as above. But the

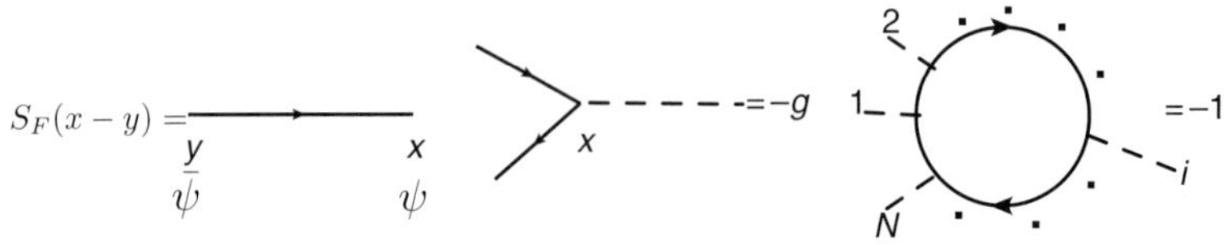

$$S_F(x - y) = \underset{\substack{y \\ \bar{\psi}}}{\overset{}{\longrightarrow}} \underset{\substack{x \\ \psi}}{}$$

Figure 13.1 x-Space Feynman rules for Yukawa interaction.

factors of $\psi\bar{\psi}$ come from expanding the $e^{-g\int \bar{\psi}\psi\phi}$, and since the $-g$ is part of a vertex, the important part is

$$\left(-\frac{\delta}{\delta\psi}S_F\frac{\delta}{\delta\bar{\psi}}\right)\ldots\left(-\frac{\delta}{\delta\psi}S_F\frac{\delta}{\delta\bar{\psi}}\right)[(\bar{\psi}\psi)\ldots(\bar{\psi}\psi)]. \qquad (13.39)$$

But in order to form the $\psi\bar{\psi}$ combinations giving us the S_F propagators, which by our assumption are cyclically linked (form a loop) as $S_F(x_N-x_1)S_F(x_1-x_2)\ldots S_F(x_{N-1}-x_N)$, we must commute the last $\psi(x_N)$ past the other $2(N-1)+1$ fermions, to put it in front, giving a minus sign.

Therefore, *a fermion loop gives a minus sign.*

The p-**space Feynman rules** are then:

- The scalar propagator is, as usual

$$\frac{1}{p^2+M^2}. \qquad (13.40)$$

- The fermion propagator is

$$\frac{1}{i\not{p}+m}=\frac{i}{-\not{p}+im}=\frac{-i\not{p}+m}{p^2+m^2}. \qquad (13.41)$$

- The vertex between two fermions and a scalar is $-g$.
- There is a minus sign for a fermion loop, and for interchanging a ψ with a $\bar{\psi}$.

For a **general interaction between fermions and a number of scalars**, with interaction term

$$S_I(\bar{\psi},\psi,\phi_1,\ldots,\phi_n), \qquad (13.42)$$

the vertex is found exactly as in the pure bosonic case, with the exception of the usual minus sign in the functional derivative for fermions, namely

$$\int d^4x d^4y d^4z_1\ldots d^4z_n e^{-i(px+p'y+q_1z_1+\ldots+q_nz_n)}$$

$$\times\frac{\delta}{\delta\phi_1(z_1)}\cdots\frac{\delta}{\delta\phi_n(z_n)}\left(-\frac{\delta}{\delta\bar{\psi}(y)}\frac{\delta}{\delta\psi(x)}\right)\{-S_I[\bar{\psi},\psi,\phi_1,\ldots,\phi_n]\}. \qquad (13.43)$$

The previous rules were in Euclidean space.

The **Minkowski-space Feynman rules in p space** are (see Figure 13.2):

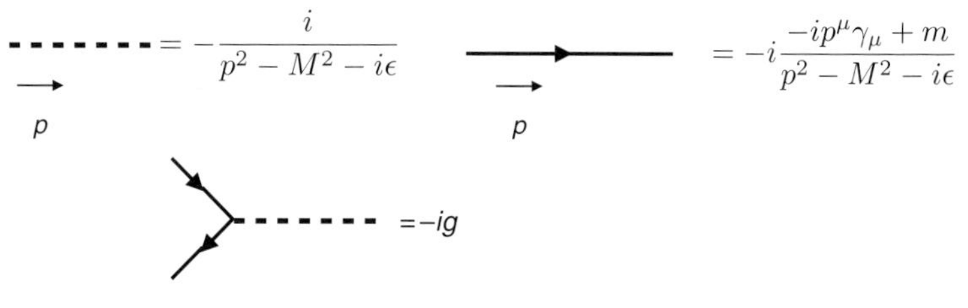

$$- - - - - - - = -\frac{i}{p^2-M^2-i\epsilon} \qquad\qquad \longrightarrow = -i\frac{-ip^\mu\gamma_\mu+m}{p^2-M^2-i\epsilon}$$

$$p \qquad\qquad\qquad\qquad\qquad\qquad\qquad\qquad\qquad p$$

$$= -ig$$

Figure 13.2 p-space Feynman rules in Minkowski space.

- The scalar propagator is

$$\frac{-i}{p^2 + M^2 - i\epsilon}.$$ (13.44)

- The fermion propagator is

$$-i\frac{-i\not{p} + m}{p^2 + m^2 - i\epsilon}.$$ (13.45)

- The vertex is $-ig$.
- There is a minus sign for a fermion loop or for changing a ψ with a $\bar{\psi}$.

This concludes our analysis of perturbation theory for fermions in the path-integral formulation.

Important Concepts to Remember

- Real Gaussian integration over n Grassmann variables gives $2^{n/2}\sqrt{\det A}$, and complex Gaussian integration, $\int d^n x d^n y e^{y^T A x}$, gives $\det A$.
- For the fermionic harmonic oscillator, the vacuum functional (vacuum expectation value) gives the same expression as in the bosonic case, just in terms of the fermion propagator $D_F(s, \tau)$.
- Again the Euclidean propagator is Wick rotated to the Minkowski-space propagator.
- The generalization to field theory contains no new facts.
- We can now have $N + M$ point functions of ψs and $\bar{\psi}$s, again given by the usual path integrals.
- In the Wick theorem for fermions, the only new thing is that we replace $\bar{\psi}$ by $-\delta/\delta\eta$, due to the anticommuting nature of the sources and fields.
- In the Feynman rules for Yukawa theory, the only new thing is the fact that for a fermion loop we have a minus sign, as we have for interchanging a ψ with a $\bar{\psi}$, or equivalently, the direction of the arrow. The vertex factor for $-g \int \bar{\psi}\psi\phi$ is $-g$.

Further Reading

See sections 3.1.3, 3.1.4, 3.3.2, and 3.3.3 in [3], section 7.2 in [2], section 5.3 in [4], and section 4.7 in [1].

Exercises

1. (a) Consider θ^α, $\alpha = 1, 2$, Grassmann variables and the even function

$$\Phi(x, \theta) = \phi(x) + \sqrt{2}\theta^\alpha \psi_\alpha(x) + \theta^2 F(x),$$ (13.46)

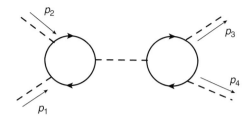

Figure 13.3   Feynman diagram in Yukawa theory.

where $\theta^2 \equiv \epsilon_{\alpha\beta}\theta^\alpha\theta^\beta$. Calculate

$$\int d^2\theta(a_1\Phi + a_2\Phi^2 + a_3\Phi^3), \tag{13.47}$$

where

$$d^2\theta = -\frac{1}{4}\int d\theta^\alpha d\theta^\beta \epsilon_{\alpha\beta}, \tag{13.48}$$

and $x, a_1, a_2, a_3 \in \mathbb{R}$.

    (b) Fill in the details missed in the text for the calculation of

$$Z[\bar{\eta}, \eta] = Z[0,0]\exp\left\{-i\int d\tau ds\bar{\eta}(s)D_F(s,\tau)\eta(\tau)\right\}. \tag{13.49}$$

2. Write down an expression for the Feynman diagram in Yukawa theory in Figure 13.3.
    Note: There are no external points on the p_i lines.

3. Consider the 4-fermi interaction term

$$S_I = \int d^4x \left[\bar{\psi}(x)\gamma_\mu\psi(x)\right]\left[\bar{\psi}(x)\gamma^\mu\psi(x)\right]. \tag{13.50}$$

Calculate the expression for the vertex in the Feynman rules, and use it to write down the formula for the one-loop diagram with two such vertices connected by two fermion lines.

4. Expand the second form of the Wick theorem for the Yukawa interaction up to (including) third order in g for $Z[\bar{\eta}, \eta, J]$.

14 Spin Sums, Dirac Field Bilinears, and C, P, T Symmetries for Fermions

In the previous two chapters we have defined the quantization, path integrals, and Feynman rules for fermions. This chapter is dedicated to some technical details for fermions that we will use later on: spin sums, bilinears in the Dirac field, and the action of symmetries of C, P, T type in theories with fermions.

14.1 Spin Sums

Often when preparing Feynman diagrams we will need to do sums over polarizations of fermions. For instance, if we do an experiment where we can't measure the spin of external particles, we should sum over this spin. It is therefore important to compute the relevant spin sums.

The first one is

$$
\sum_{s=1,2} u^s(p)\bar{u}^s(p) = \sum_s \begin{pmatrix} \sqrt{-p\cdot\sigma}\xi^s \\ \sqrt{-p\cdot\bar{\sigma}}\xi^s \end{pmatrix} \begin{pmatrix} \xi^{s\dagger} \sqrt{-p\cdot\bar{\sigma}} & \xi^{s\dagger} \sqrt{-p\cdot\sigma} \end{pmatrix}
$$

$$
= \begin{pmatrix} \sqrt{-p\cdot\sigma}\sqrt{-p\cdot\bar{\sigma}} & \sqrt{-p\cdot\sigma}\sqrt{-p\cdot\sigma} \\ \sqrt{-p\cdot\bar{\sigma}}\sqrt{-p\cdot\bar{\sigma}} & \sqrt{-p\cdot\bar{\sigma}}\sqrt{-p\cdot\sigma} \end{pmatrix}
$$

$$
= \begin{pmatrix} m & -p\cdot\sigma \\ -p\cdot\bar{\sigma} & m \end{pmatrix} = -i\slashed{p} + m, \tag{14.1}
$$

where we have used $\bar{u} = u^\dagger i\gamma^0$, $\sum_{s=1,2} \xi^s\xi^{s\dagger} = \mathbf{1}$, and $\{\sigma_i, \sigma_j\} = 2\delta_{ij}$, implying that $\{\sigma^{(\mu}, \bar{\sigma}^{\nu)}\} = -2\eta^{\mu\nu}$ and so

$$
(p\cdot\sigma)(p\cdot\bar{\sigma}) = p_\mu p_\nu \frac{1}{2}\{\sigma^\mu, \bar{\sigma}^\nu\} = -p^2 = m^2. \tag{14.2}
$$

The second sum is over vs, giving similarly

$$
\sum_s v^s(p)\bar{v}^s(p) = -i\slashed{p} - m. \tag{14.3}
$$

14.2 Dirac Field Bilinears

A single fermion is classically unobservable (it is a Grassmann object, which cannot be measured experimentally). Classically observable objects are fermion bilinears, which are commuting variables, and can have a VEV that one can measure experimentally. Even in quantum theory, the fermion bilinears are objects that appear very often, so it is important to understand their properties, in particular their properties under Lorentz transformations.

We have defined $\bar{\psi}$ by the condition that $\bar{\psi}\psi$ is a Lorentz scalar, so this bilinear is given. The spinorial representations of spin $1/2$ are: the fundamental, or $1/2$, to which ψ belongs, and the conjugate representation, or $\overline{1/2}$, to which $\bar{\psi}$ belongs, such that $\bar{\psi}\psi$ is a scalar.

The Lorentz transformation of ψ is

$$x'^{\mu} = \Lambda^{\mu}{}_{\nu}x^{\nu} \Rightarrow \psi'^{\alpha} = S^{\alpha}{}_{\beta}(\Lambda)\psi^{\beta}. \tag{14.4}$$

The next bilinear in order of complexity is obtained by inserting a γ_{μ} matrix (i.e. $\bar{\psi}\gamma^{\mu}\psi$). γ^{μ} is a vector under Lorentz transformations, meaning that

$$S^{\alpha}{}_{\gamma}(\gamma^{\mu})^{\gamma}{}_{\delta}(S^{-1})^{\delta}{}_{\beta} = \Lambda^{\mu}{}_{\nu}(\gamma^{\nu})^{\alpha}{}_{\beta}, \tag{14.5}$$

and therefore $\bar{\psi}\gamma^{\mu}\psi$ is a vector, that is

$$\bar{\psi}\gamma^{\mu}\psi \rightarrow \Lambda^{\mu}{}_{\nu}\bar{\psi}\gamma^{\nu}\psi. \tag{14.6}$$

What other bilinears can we form? The most general would be $\bar{\psi}\Gamma\psi$, where Γ is a general 4×4 matrix, which can be decomposed into a basis of 16, 4×4 matrices. Such a basis is given by

$$\mathcal{O}_i = \left\{ \mathbf{1}, \gamma^{\mu}, \gamma_5, \gamma^{\mu}\gamma_5, \gamma^{\mu\nu} = \frac{1}{2}\gamma^{[\mu}\gamma^{\nu]} \right\}, \tag{14.7}$$

which is a total number of $1 + 4 + 1 + 4 + 4 \times 3/2 = 16$ matrices. Then this basis is normalized as

$$\mathrm{Tr}\{\mathcal{O}^i\mathcal{O}_j\} = 4\delta^i_j, \tag{14.8}$$

since for instance $\mathrm{Tr}\{\mathbf{1} \times \mathbf{1}\} = 4$ and

$$\mathrm{Tr}\{\gamma^{\mu}\gamma_{\nu}\} = \frac{1}{2}\mathrm{Tr}\{\gamma^{\mu}, \gamma_{\nu}\} = 4\delta^{\mu}_{\nu}. \tag{14.9}$$

Moreover, this basis is complete, satisfying the completeness relation

$$\delta^{\beta}_{\alpha}\delta^{\delta}_{\gamma} = \frac{1}{4}(\mathcal{O}_i)^{\delta}_{\alpha}(\mathcal{O}_i)^{\beta}_{\gamma}. \tag{14.10}$$

This is indeed a completeness relation, since on multiplying by M^{γ}_{β} we get

$$M^{\delta}{}_{\alpha} = \frac{1}{4}\mathrm{Tr}(M\mathcal{O}_i)(\mathcal{O}_i)^{\delta}{}_{\alpha}, \tag{14.11}$$

which is indeed the expansion in a complete basis. The coefficient is correct, since on multiplying again by $\mathcal{O}_j{}^{\alpha}_{\delta}$ we get an identity.

Now, multiplying (14.10) by arbitrary spinors χ_β, $\bar\psi^\gamma$, and $N\phi_\delta$, where N is a matrix, we obtain the relation (after multiplying it by yet another matrix M)

$$M\chi(\bar\psi N\phi) = -\frac{1}{4}\sum_j M\mathcal{O}_j N\phi(\bar\psi \mathcal{O}_j \chi), \qquad (14.12)$$

which is called the *Fierz identity* or *Fierz recoupling*. Note the minus sign, which appears because we interchanged the order of the fermions. The Fierz identity is useful for simplifying fermionic terms, allowing us to "recouple" the fermions and hopefully obtain something simpler.

We might think that we could introduce other Lorentz structures made up from gamma matrices, but for instance

$$\gamma_{[\mu\nu\rho\sigma]} = \gamma_{[\mu}\gamma_\nu\gamma_\rho\gamma_{\sigma]} \propto \epsilon_{\mu\nu\rho\sigma}\gamma_5, \qquad (14.13)$$

since $\gamma_5 = -i\gamma^0\gamma^1\gamma^2\gamma^3$, and the gamma matrix products are antisymmetric by construction, since gamma matrices anticommute.

Then we also have

$$\gamma_{[\mu\nu\rho]} \propto \epsilon_{\mu\nu\rho\sigma}\gamma^\sigma\gamma_5 \qquad (14.14)$$

and there are no higher antisymmetric bilinears, since the antisymmetric product of five γ_μ is zero (there are only four different values for indices).

As terminology, $\mathbf{1}$ is a scalar, γ_5 is a pseudoscalar, γ^μ is a vector, $\gamma^\mu\gamma_5$ is a pseudovector, and $\gamma^{\mu\nu}$ is an antisymmetric tensor. Here, for instance, pseudoscalar means that it transforms like a scalar under the usual Lorentz transformations, except under parity it gets an extra minus. Similarly, pseudo-something means it transforms usually under Lorentz transformations, except under parity it gets an extra minus.

If $\psi(x)$ satisfies the Dirac equation, then the *vector current* $j^\mu = \bar\psi(x)\gamma^\mu\psi(x)$ is conserved:

$$\partial_\mu j^\mu = (\partial_\mu\bar\psi)\gamma^\mu\psi + \bar\psi\gamma^\mu\partial_\mu\psi = (m\bar\psi)\psi + \bar\psi(-m\psi) = 0 \qquad (14.15)$$

and the *axial vector current* $j^{\mu 5} = \bar\psi(x)\gamma^\mu\gamma_5\psi(x)$ is conserved:

$$\partial_\mu j^{\mu 5} = (\partial_\mu\bar\psi)\gamma^\mu\gamma_5\psi + \bar\psi\gamma^\mu\gamma_5\partial_\mu\psi = m\bar\psi\gamma_5\psi - \bar\psi\gamma_5\slashed\partial\psi = 2m\bar\psi\gamma_5\psi, \qquad (14.16)$$

$\partial_\mu j^{\mu 5} = 0$ if $m = 0$ only.

14.3 C, P, T Symmetries for Fermions

The continuous Lorentz group $L_+^\uparrow$, the proper orthochronous group, continuously connected with the identity, is a symmetry of relativistic field theories.

But we also have the parity and time-reversal transformations

$$P : (t, \vec{x}) \to (t, -\vec{x}),$$
$$T : (t, \vec{x}) \to (-t, \vec{x}), \qquad (14.17)$$

which can also be symmetries, but not necessarily (there is no physical principle which demands it).

Besides $L_+^\uparrow$, there are three other disconnected groups, $L_-^\uparrow$, $L_+^\downarrow$, and $L_+^\downarrow$. L_+ is called a proper group, and L_- improper, and the two are related by P, whereas $L^\uparrow$ is called orthocronous and $L^\downarrow$ nonorthochronous, and the two are related by T.

Besides these transformations, we also have C, the charge conjugation transformation, which transforms particles into antiparticles.

For a long time, it was thought that C, P, T must be symmetries of physics, for instance since:

- Gravitational, electromagnetic, and strong interactions respect $C, P,\ T$.
- Weak interactions however break C and P separately, but preserve CP. The Nobel Prize was awarded for the discovery of this breaking of parity, thought to have been fundamental.
- By now there is good experimental evidence for CP breaking also, which would point to interactions beyond the Standard Model.
- However, there is a theorem that, under very general conditions (like locality and unitarity), CPT must be preserved always, so CP breaking implies T breaking.

In the quantum theory, the action of the C, P, T symmetries is encoded in operators C, P, T that act on other operators. The c-numbers are generally not modified (though see T below), even if they contain $t, \vec{x}$.

14.3.1 Parity

Parity is defined as being the same as reflection in a mirror. If we look in a mirror at a momentum vector with a spin, or helicity, represented by a rotation around the momentum axis, we see the momentum inverted, but not the spin, hence parity flips the momentum $\vec{p}$, but not the spin s, see Figure 14.1.

This means that the action of P on the annihilation/creation operators in the expansion of the quantum field ψ is

$$P a_{\vec{p}}^s P^{-1} = \eta_a a_{-\vec{p}}^s,$$
$$P b_{\vec{p}}^s P^{-1} = \eta_b b_{-\vec{p}}^s, \tag{14.18}$$

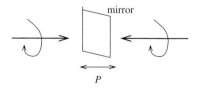

Figure 14.1   Parity is reflection in a mirror. The momentum gets inverted, but not the spin.

where η_a and η_b are possible phases. Since two parity operations revert the system to its original, we have $P^2 = \mathbf{1}$, so $P^{-1} = P$, and then also $\eta_a^2 = \eta_b^2 = \pm 1$, since observables are fermion bilinears, so a $\pm$ sign doesn't make any difference.

Defining the momenta $\tilde{p} = (p^0, -\vec{p})$, we have

$$
u(p) = \begin{pmatrix} \sqrt{-p \cdot \sigma}\,\xi \\ \sqrt{-p \cdot \bar{\sigma}}\,\xi \end{pmatrix} = \begin{pmatrix} \sqrt{-\tilde{p} \cdot \bar{\sigma}}\,\xi \\ \sqrt{-\tilde{p} \cdot \sigma}\,\xi \end{pmatrix} = i\gamma^0 u(\tilde{p}),
$$

$$
v(p) = \begin{pmatrix} \sqrt{-p \cdot \sigma}\,\xi \\ -\sqrt{-p \cdot \bar{\sigma}}\,\xi \end{pmatrix} = \begin{pmatrix} \sqrt{-\tilde{p} \cdot \bar{\sigma}}\,\xi \\ -\sqrt{-\tilde{p} \cdot \sigma}\,\xi \end{pmatrix} = -i\gamma^0 v(\tilde{p}), \tag{14.19}
$$

where

$$
\gamma^0 = -i \begin{pmatrix} \mathbf{0} & \mathbf{1} \\ \mathbf{1} & \mathbf{0} \end{pmatrix}. \tag{14.20}
$$

Then

$$
P\psi(x)P = \int \frac{d^3p}{(2\pi)^3} \frac{1}{\sqrt{2E_{\tilde{p}}}} \sum_s (\eta_a a^s_{-\vec{p}} u^s(p)e^{ipx} + \eta_b^* b^{s\dagger}_{-\vec{p}} v^s(p)e^{-ipx})
$$

$$
= \int \frac{d^3\tilde{p}}{(2\pi)^3} \frac{1}{\sqrt{2E_{\tilde{p}}}} \sum_s (\eta_a a^s_{\tilde{p}} i\gamma^0 u^s(\tilde{p})e^{i\tilde{p}(t,-\vec{x})} - \eta_b^* b^{s\dagger}_{\tilde{p}} i\gamma^0 v^s(\tilde{p})e^{-i\tilde{p}(t,-\vec{x})}), \tag{14.21}
$$

where in the second line we have changed the integration variable from p to $\tilde{p}$, and then used the relation between $u(p)$ and $u(\tilde{p})$ and similarly for v. Finally, we see that we need to have $\eta_b^* = -\eta_a$ to relate to another ψ field. In that case, we obtain

$$
P\psi(x)P = \eta_a i\gamma^0 \psi(t, -\vec{x}). \tag{14.22}
$$

Then we also have

$$
(P\psi(x)P)^\dagger = -i\eta_a^* \psi^\dagger(t, -\vec{x})\gamma^{0\dagger} = i\eta_a^* \psi^\dagger(t, -\vec{x})\gamma^0, \tag{14.23}
$$

and so

$$
P\bar{\psi}(x)P = (P\psi(t,\vec{x})P)^\dagger i\gamma^0 = i\eta_a^* \bar{\psi}(t, -\vec{x})\gamma^0. \tag{14.24}
$$

We can now also compute the behavior of fermion bilinears under parity. We obtain

$$
P\bar{\psi}\psi P = |\eta_a|^2 \bar{\psi}(t, -\vec{x})\gamma^0(-\gamma^0)\psi(t, -\vec{x}) = \bar{\psi}\psi(t, -\vec{x}), \tag{14.25}
$$

that is it transforms as a scalar. Here we have used the fact that η_a is a phase, so $|\eta_a|^2 = 1$ and $-(\gamma^0)^2 = 1$.

Next, we have

$$
P\bar{\psi}\gamma_5\psi P = \bar{\psi}(t, -\vec{x})\gamma^0\gamma_5(-\gamma^0)\psi(t, -\vec{x}) = -\bar{\psi}\gamma_5\psi(t, -\vec{x}), \tag{14.26}
$$

which is a pseudoscalar.

Also

$$
P\bar{\psi}\gamma^\mu\psi P = \bar{\psi}\gamma^0\gamma^\mu(-\gamma^0)\psi(t, -\vec{x}) = (-1)^\mu \bar{\psi}\gamma^\mu\psi(t, -\vec{x}), \tag{14.27}
$$

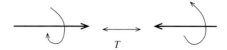

Figure 14.2   Time reversal inverts both momentum and spin.

where $(-1)^\mu = +1$ for $\mu = 0$ and $= -1$ for $\mu \neq 0$, so it transforms like a vector. Further:

$$P\bar{\psi}\gamma^\mu\gamma_5\psi P = \bar{\psi}\gamma^0\gamma^\mu\gamma_5(-\gamma^0)\psi(t, -\vec{x}) = -(-1)^\mu\bar{\psi}\gamma^\mu\gamma_5\psi(t, -\vec{x}), \tag{14.28}$$

so it transforms as a pseudovector.

14.3.2 Time Reversal

Consider now the time-reversal transformation. Under this, the momentum changes direction (time flows opposite), and also the spin changes, since it rotates back, see Figure 14.2. However, time reversal cannot be made into a linear operator like P, but it can be made into an *antilinear operator*

$$T(\text{c-number}) = (\text{c-number})^* T. \tag{14.29}$$

Then we have

$$Ta_p^s T = a_{-p}^{-s},$$

$$Tb_p^s T = b_{-p}^{-s}, \tag{14.30}$$

and since the T operator is antilinear, for the full Dirac equation modes we have

$$Ta_p^s u^s(p)e^{ipx}T = a_{-p}^{-s}[u^s(p)]^* e^{-ipx},$$

$$Tb_p^{s\dagger} v^s(p)e^{-ipx}T = b_{-p}^{-s\dagger}[v^s(p)]^* e^{ipx}. \tag{14.31}$$

We now consider a general spin basis. Instead of $\begin{pmatrix} 1 \\ 0 \end{pmatrix}$ and $\begin{pmatrix} 0 \\ 1 \end{pmatrix}$, corresponding to spin oriented along the z direction, we take a spin direction of arbitrary spherical angles θ, ϕ, giving

$$\xi(\uparrow) = R(\theta, \phi)\begin{pmatrix} 1 \\ 0 \end{pmatrix} = \begin{pmatrix} \cos\theta/2 \\ e^{i\phi}\sin\theta/2 \end{pmatrix},$$

$$\xi(\downarrow) = R(\theta, \phi)\begin{pmatrix} 0 \\ 1 \end{pmatrix} = \begin{pmatrix} -e^{-i\phi}\sin\theta/2 \\ \cos\theta/2 \end{pmatrix}, \tag{14.32}$$

and we have $\xi^s = (\xi(\uparrow), \xi(\downarrow))$, then $(\vec{\sigma}\sigma^2 = \sigma^2(-\vec{\sigma}^*) \Rightarrow$ if $\vec{n} \cdot \vec{\sigma}\xi = +\xi$, so $(\vec{n} \cdot \vec{\sigma})(-i\sigma^2\xi^*) = -(-i\sigma^2\xi^*))$

$$\xi^{-s} = (\xi(\downarrow), -\xi(\uparrow)) = -i\sigma^2(\xi^s)^*. \tag{14.33}$$

Analogously, we write for the annihilation operators $a_{\vec{p}}^{-s} = (a_{\vec{p}}^2, -a_{\vec{p}}^1)$ and $b_{\vec{p}}^{-s} = (b_{\vec{p}}^2, -b_{\vec{p}}^1)$.

For the column vectors we have

$$u^{-s}(\tilde{p}) = \begin{pmatrix} \sqrt{-\tilde{p} \cdot \sigma}(-i\sigma^2 \xi^{s*}) \\ \sqrt{-\tilde{p} \cdot \bar{\sigma}}(-i\sigma^2 \xi^{s*}) \end{pmatrix} = \begin{pmatrix} -i\sigma^2 \sqrt{-p \cdot \sigma^*} \xi^{s*} \\ -i\sigma^2 \sqrt{-p \cdot \bar{\sigma}^*} \xi^{s*} \end{pmatrix}$$

$$= -i \begin{pmatrix} \sigma^2 & 0 \\ 0 & \sigma^2 \end{pmatrix} [u^s(p)]^* = \gamma^1 \gamma^3 [u^s(p)]^*. \tag{14.34}$$

Here we have used the relation $\sqrt{\tilde{p} \cdot \sigma}\sigma^2 = \sigma^2 \sqrt{p \cdot \sigma^*}$, which we can prove by expanding the square root for small $\vec{p}$, and then $(\vec{p} \cdot \vec{\sigma})\sigma_2 = -\sigma_2(\vec{p} \cdot \vec{\sigma}^*)$, true since $\sigma_2^* = -\sigma_2, \sigma_{1,3}^* = \sigma_{1,3}$.

We can now compute the effect on the quantum field operator, and obtain

$$T\psi(t, \vec{x})T = -\gamma^1 \gamma^3 \psi(-t, \vec{x}). \tag{14.35}$$

For the fermion bilinears, we then obtain

$$T\bar{\psi}\psi T = +\bar{\psi}\psi(-t, \vec{x}),$$
$$T\bar{\psi}\gamma_5\psi T = -\bar{\psi}\gamma_5\psi(-t, \vec{x}),$$
$$T\bar{\psi}\gamma^\mu\psi T = (-1)^\mu \bar{\psi}\gamma^\mu\psi(-t, \vec{x}), \tag{14.36}$$

which we leave as an exercise to check.

14.3.3 Charge Conjugation

Charge conjugation is the operation that takes particles to antiparticles. On the classical spinor space, it acts not on the classical space $(t, \vec{x})$, but by the C-matrix (charge conjugation matrix) defined before. We need to define how it acts on the quantum field operators.

As we said before, due to the fact that $\psi \sim a_{\vec{p}}^s u^s(p)e^{ipx} + b_{\vec{p}}^{s\dagger} v^s(p)e^{-ipx}$, and in the energy we have $\int \omega(a^\dagger a - bb^\dagger)$, we can understand $a^\dagger$ as a creation operator for fermions of positive energy (e.g. electrons e^-) and $b^\dagger$ as an annihilation operator for fermions of negative energy in the fully occupied "Dirac sea," equivalent to a creation operator for a "hole" of effective positive energy, or *antiparticle* (e.g. positron e^+).

Therefore, the C operator maps as into bs:

$$Ca_{\vec{p}}^s C^{-1} = b_{\vec{p}}^s,$$
$$Cb_{\vec{p}}^s C^{-1} = a_{\vec{p}}^s. \tag{14.37}$$

Note that we have ignored possible phases here. Next, we use the fact that

$$(v^s(p))^* = \begin{pmatrix} 0 & -i\sigma^2 \\ i\sigma^2 & 0 \end{pmatrix} \begin{pmatrix} \sqrt{-p \cdot \sigma} \xi^s \\ \sqrt{-p \cdot \bar{\sigma}} \xi^s \end{pmatrix} = \gamma^2 \begin{pmatrix} \sqrt{-p \cdot \sigma} \xi^s \\ \sqrt{-p \cdot \bar{\sigma}} \xi^s \end{pmatrix}, \tag{14.38}$$

where again we have used $\sqrt{-p \cdot \sigma}\sigma^2 = \sigma^2 \sqrt{-p \cdot \bar{\sigma}^*}$. Then we have for the column vectors

$$u^s(p) = \gamma^2 (v^s(p))^*,$$
$$v^s(p) = \gamma^2 (u^s(p))^*. \tag{14.39}$$

Finally, using the linearity property of the C operator and the above properties, we can calculate the effect on the quantum field operator:

$$C\psi(x)C^{-1} = \gamma^2 \psi^*(x) = \gamma^2 (\psi^\dagger)^T. \tag{14.40}$$

Therefore, the charge conjugation operator interchanges ψ and $\psi^\dagger$ up to some matrix. For the fermion bilinears we find

$$C\bar{\psi}\psi C^{-1} = \bar{\psi}\psi,$$
$$C\bar{\psi}\gamma_5\psi C^{-1} = \bar{\psi}\gamma_5\psi,$$
$$C\bar{\psi}\gamma^\mu\psi C^{-1} = -\bar{\psi}\gamma^\mu\psi,$$
$$C\bar{\psi}\gamma^\mu\gamma_5\psi C^{-1} = \bar{\psi}\gamma^\mu\gamma_5\psi. \tag{14.41}$$

The full transformations under P, T, C and CPT, for various bilinears and for the derivatives ∂_μ, are

$$\bar{\psi}\psi : +1, +1, +1, +1,$$
$$\bar{\psi}\gamma_5\psi : -1, -1, +1, +1,$$
$$\bar{\psi}\gamma^\mu\psi : (-1)^\mu, (-1)^\mu, -1, -1,$$
$$\bar{\psi}\gamma^\mu\gamma_5\psi : -(-1)^\mu, (-1)^\mu, +1, -1,$$
$$\bar{\psi}\gamma^{\mu\nu}\psi : (-1)^\mu(-1)^\nu, -(-1)^\mu(-1)^\nu, -1, +1,$$
$$\partial_\mu : (-1)^\mu, -(-1)^\mu, +1, -1, \tag{14.42}$$

where of course under P and T we also transform the position for the fields.

Important Concepts to Remember

- The spin sums are $\sum_s u^s(p)\bar{u}^s(p) = -i\not{p} + m$ and $\sum_s v^s(p)\bar{v}^s(p) = -i\not{p} - m$.
- The complete basis of 4×4 matrices is $\mathcal{O}_i = \{\mathbf{1}, \gamma^\mu, \gamma_5, \gamma^\mu\gamma_5, \gamma^{\mu\nu}\}$, and fermion bilinears are $\bar{\psi}\mathcal{O}_i\psi$.
- The presence of γ_5 means a pseudoscalar/pseudovector, and so on (i.e. an extra minus under parity).
- Fierz recoupling follows from the completeness of the $\mathcal{O}_i$ basis.
- Vector currents are conserved, and axial vector currents are conserved only if $m = 0$.
- T, C, P act on quantum operators.
- Time reversal T is antilinear and charge conjugation C changes particles to antiparticles.

Further Reading

See sections 3.4 and 3.6 in [1].

Exercises

1. Using the Fierz identity, prove that we have for Majorana spinors λ^a

$$(\bar{\lambda}^a \gamma_\mu \lambda^c)(\bar{\epsilon} \gamma^\mu \lambda^b) f_{abc} = 0, \tag{14.43}$$

where f_{abc} is totally antisymmetric, and also using

$$\gamma_\mu \gamma_\rho \gamma^\mu = -2\gamma_\rho,$$
$$\gamma_\mu \gamma_{\rho\sigma} \gamma^\mu = 0, \tag{14.44}$$

and the fact that Majorana spinors satisfy

$$\bar{\epsilon} \chi = \bar{\chi} \epsilon,$$
$$\bar{\epsilon} \gamma_\mu \chi = -\bar{\chi} \gamma_\mu \epsilon,$$
$$\bar{\epsilon} \gamma_5 \chi = \bar{\chi} \gamma_5 \epsilon,$$
$$\bar{\epsilon} \gamma_\mu \gamma_5 \chi = \bar{\chi} \gamma_\mu \gamma_5 \epsilon. \tag{14.45}$$

2. Prove that the transformations of $\bar{\psi} \gamma^\mu \gamma_5 \psi$ under T, C and CPT are with $(-1)^\mu$, $+1$, and -1, respectively.

3. Calculate the quantities in terms of the basis $\mathcal{O}_i$:

$$\gamma_\mu \gamma^{\mu\nu}, \quad \gamma_\mu \gamma^{\nu\rho\sigma} \gamma^\mu, \quad \gamma_\mu \gamma^{\nu\rho} \gamma_5 \gamma^\mu. \tag{14.46}$$

4. Prove that under time reversal, we have

$$T\bar{\psi}\psi T = +\bar{\psi}\psi(-t, \vec{x}),$$
$$T\bar{\psi}\gamma_5 \psi T = -\bar{\psi}\gamma_5 \psi(-t, \vec{x}),$$
$$T\bar{\psi}\gamma^\mu \psi T = (-1)^\mu \bar{\psi}\gamma^\mu \psi(-t, \vec{x}). \tag{14.47}$$

15 Dirac Quantization of Constrained Systems

In this chapter we will describe the quantization of constrained systems developed by Dirac. I will mostly follow the presentation of P. A. M. Dirac, in the book *Lectures on Quantum Mechanics* [5], which is a series of four lectures given at Yeshiva University, published in 1964, and is possibly one of the most influential books in physics, per number of pages. The first two lectures describe the procedure now known as Dirac quantization, which is very important for modern theoretical physics, and even now, there is little that needs to be changed in the presentation of Dirac. The question we want to answer is, how do we deal with constraints in quantum mechanics? Naively, it looks like they could become some operator constraints, but the truth is more complicated. We will see that constraints can be split into what are known as first-class, and second-class, constraints. The latter can be used to define some modified Poisson brackets between fields (in effect, "solving the constraints" to find an independent space), called Dirac brackets, that we can then quantize.

15.1 Set-up and Hamiltonian Formalism

So what is the specific problem that we want to solve in quantum field theory, for which we need Dirac quantization? We want to quantize gauge fields (we will do so in Chapter 16), which have an action

$$S = -\frac{1}{4} \int d^4x F_{\mu\nu}^2, \tag{15.1}$$

where the field strength is written in terms of the gauge field A_μ as

$$F_{\mu\nu} = \partial_\mu A_\nu - \partial_\nu A_\mu. \tag{15.2}$$

This means that we have a gauge invariance

$$\delta A_\mu = \partial_\mu \Lambda, \tag{15.3}$$

which leaves $F_{\mu\nu}$, and therefore the action, invariant. This means that there is a redundancy in the system, there are fewer degrees of freedom in the system than variables used to describe it (we can use Λ to put one component, for instance A_0, to zero). To quantize only the physical degrees of freedom, we must fix a gauge, for instance the Lorenz, or covariant, gauge $\partial^\mu A_\mu = 0$. That imposes a constraint on the system, so we must learn how to quantize in the presence of constraints, which is the subject that Dirac tackled.

We can deal with this in the Hamiltonian formalism, leading to the Dirac quantization of constrained systems. Even though we will not use much of the Dirac formalism later, it is important theoretically, to understand the concepts involved better, hence we started with it.

In the Lagrangian formulation of classical mechanics, constraints are introduced in the Lagrangian by multiplying them by Lagrange multipliers. We will see that in the Hamiltonian formulation, we also add the constraints to the Hamiltonian with some coefficients, with some notable differences.

In the Hamiltonian formalism, we first define conjugate momenta as

$$p_n = \frac{\partial L}{\partial \dot{q}_n}, \tag{15.4}$$

and then the Hamiltonian is

$$H = \sum_n (p_n \dot{q}_n) - L. \tag{15.5}$$

The Hamilton equations of motion are

$$\dot{q}_n = \frac{\partial H}{\partial p_n}; \quad \dot{p}_n = -\frac{\partial H}{\partial q_n}. \tag{15.6}$$

We define the Poisson brackets of two functions of p and q, $f(p,q)$ and $g(p,q)$, as

$$\{f, g\}_{P.B.} = \sum_n \left[\frac{\partial f}{\partial q_n} \frac{\partial g}{\partial p_n} - \frac{\partial f}{\partial p_n} \frac{\partial g}{\partial q_n} \right]. \tag{15.7}$$

Then the Hamilton equations of motion are

$$\dot{q}_n = \{q_n, H\}_{P.B.}; \quad \dot{p}_n = \{p_n, H\}_{P.B.}, \tag{15.8}$$

and more generally, we can write for the time evolution (equation of motion) of some function on phase space, $g(q,p)$,

$$\dot{g} = \{g, H\}_{P.B.}. \tag{15.9}$$

15.2 System with Constraints in Hamiltonian Formalism: Primary/Secondary and First/Second-Class Constraints

But now assume that there are M constraints on the phase space:

$$\phi_m(q, p) = 0, \quad m = 1, \ldots, M. \tag{15.10}$$

We will call these the *primary constraints* of the Hamiltonian formalism, and see why shortly.

We will define $\approx$, called *weak equality*, as equality only after using the constraints $\phi_m = 0$. Note that we must use the constraints only *at the end of a calculation*, for instance when using Poisson brackets, since the Poisson brackets are defined by assuming a set of independent variables q_n, p_n (so that partial derivatives mean when we keep all the other variables fixed; this would not be possible if there is a relation between all the qs and ps). So only when all the dynamical calculations are done, can we use $\phi_m = 0$.

The weak equality means, by definition

$$\phi_m \approx 0, \tag{15.11}$$

since otherwise ϕ_m is some function of qs and ps, for which we can calculate Poisson brackets, and is not *identically* zero.

Note that as far as the time evolution is concerned, there is no difference between H and $H + u_m \phi_m$ (they are indistinguishable), since $\phi_m \approx 0$. Then the equations of motion are

$$\dot{q}_n = \frac{\partial H}{\partial p_n} + u_m \frac{\partial \phi_m}{\partial p_n} \approx \{q_n, H + u_m \phi_m\}_{P.B.},$$

$$\dot{p}_n = -\frac{\partial H}{\partial q_n} - u_m \frac{\partial \phi_m}{\partial q_n} \approx \{p_n, H + u_m \phi_m\}_{P.B.}, \tag{15.12}$$

and in general we write

$$\dot{g} \approx \{g, H_T\}_{P.B.}, \tag{15.13}$$

where the *total Hamiltonian* H_T is

$$H_T = H + u_m \phi_m. \tag{15.14}$$

Note that in the above we have used the fact that the Poisson bracket of the coefficient u_m (which is in principle a function of (q, p)) does not contribute, since it is multiplied by ϕ_m, so $\{g, u_m\}\phi_m \approx 0$.

Let us now apply the above time evolution to the particular case of the constraints ϕ_m themselves. Physically, we know that if ϕ_m are good constraints, then the time evolution should keep us within the constraint hypersurface, so we must have $\dot{\phi}_m \approx 0$ (which means that the time variation of ϕ_m must be proportional to some ϕs itself). This gives the equations

$$\{\phi_m, H\}_{P.B.} + u_n \{\phi_m, \phi_n\}_{P.B.} \approx 0. \tag{15.15}$$

This in turn will give other, potentially independent, constraints, called *secondary constraints*. We can then apply the time variation again, and gain new constraints, and so on, until we get nothing new, finally obtaining the full set of secondary constraints, that we will call ϕ_k, $k = M + 1, \ldots, M + K$. Together, all the constraints, primary and secondary, form the set

$$\phi_j \approx 0, \quad j = 1, \ldots, M + K. \tag{15.16}$$

While there is not much difference between primary and secondary constraints, as we will see, they both act in the same way, there is one distinction that is useful. We will call a function $R(p, q)$ on phase space *first class* if

$$\{R, \phi_j\}_{P.B.} \approx 0 \quad (\text{i.e.} = r_{jj'} \phi_{j'}), \tag{15.17}$$

for all $j = 1, \ldots, M+K$, and *second class* if $\{R, \phi_j\}_{P.B.}$ is not ≈ 0 for some ϕ_j. In particular, we have for the time evolution of all the constraints, by definition (all the set of constraints satisfies that their time evolution is also a constraint)

$$\{\phi_j, H_T\}_{P.B.} \approx \{\phi_j, H\}_{P.B.} + u_m\{\phi_j, \phi_m\}_{P.B.} \approx 0 \tag{15.18}$$

(here again we dropped the term $\{u_m, \phi_j\}\phi_m \approx 0$).

These equations first tell us that H_T is first class, and second give us a set of $M + K$ equations for the M coefficients u_m, which are in principle functions of q and p (i.e. $u_m(q, p)$). Even if the system looks overconstrained, these equations need to have at least a solution from physics consistency, since if not, it would mean that there is no possible consistent time evolution of constraints, which is clearly wrong. But in general, the solution is not unique. If U_m is a particular solution of (15.18), then the general solution is

$$u_m = U_m + v_a V_{am}, \tag{15.19}$$

where v_a are arbitrary numbers, and V_{am} satisfy the equation

$$V_{am}\{\phi_j, \phi_m\}_{P.B.} \approx 0. \tag{15.20}$$

Then we can split the total Hamiltonian, which is first class as we saw, as

$$\begin{aligned} H_T &= H + U_m\phi_m + v_a V_{am}\phi_m \\ &= H' + v_a\phi_a, \end{aligned} \tag{15.21}$$

where

$$H' = H + U_m\phi_m \tag{15.22}$$

is first class, by the definition of U_m as a particular solution of (15.18), and

$$\phi_a = V_{am}\phi_m \tag{15.23}$$

are first-class primary constraints, because of the definition of V_{am} as satisfying (15.20).

Theorem 15.1   If R and S are first class (i.e. $\{R, \phi_j\}_{P.B.} = r_{jj'}\phi_{j'}$, $\{S, \phi_j\}_{P.B.} = s_{jj'}\phi_{j'}$), then $\{R, S\}_{P.B.}$ is first class.

Proof   We first use the Jacobi identity for antisymmetric brackets (like the commutator and the Poisson brackets), which is an identity (type $0 = 0$) proved by writing explicitly the brackets

$$\{\{R, S\}, P\} + \{\{P, R\}, S\} + \{\{S, P\}, R\} = 0, \tag{15.24}$$

to get

$$\begin{aligned} \{\{R, S\}_{P.B.}, \phi_j\}_{P.B.} &= \{\{R, \phi_j\}_{P.B.}, S\}_{P.B.} - \{\{S, \phi_j\}_{P.B.}, R\}_{P.B.} \\ &= \{r_{jj'}\phi_{j'}, S\}_{P.B.} - \{s_{jj'}\phi_{j'}, R\}_{P.B.} \\ &= -r_{jj'}s_{j'j''}\phi_{j''} + \{r_{jj'}, S\}_{P.B.}\phi_{j'} \\ &\quad -s_{jj'}(-r_{j'j''}\phi_{j''}) - \{s_{jj'}, R\}_{P.B.}\phi_{j'} \approx 0. \end{aligned} \tag{15.25}$$

q.e.d.

Finally, we can add the first-class secondary constraints $\phi_{a'}$ as well to the Hamiltonian, since there is no difference between the first-class primary and first-class secondary constraints, obtaining the *extended Hamiltonian*

$$H_E = H_T + v'_{a'}\phi_{a'}. \tag{15.26}$$

However, the second-class constraints ϕ_w are different. We cannot add them to the Hamiltonian, since $\{\phi_w, \phi_j\}_{P.B.}$ is not ≈ 0, so the time evolution of ϕ_j, $\dot{\phi}_j = \{\phi_j, H\}_{P.B.}$, would be modified.

Therefore, we have the physical interpretation that first-class constraints generate motion tangent to the constraint hypersurface (adding them to the Hamiltonian, the new term is ≈ 0, i.e. in the constraint hypersurface $\phi_m = 0$). In contrast, second-class constraints generate motion away from the constraint hypersurface, since adding them to the Hamiltonian adds terms which are not ≈ 0 to the time evolution of ϕ_j.

15.3 Quantization and Dirac Brackets

Now we finally come to the issue of quantization. Normally, we substitute the Poisson bracket $\{,\}_{P.B.}$ with $1/i\hbar[,]$ when quantizing. But now there is a subtlety, namely the constraints.

The simplest possibility is to impose the constraints on states (wavefunctions), that is to put

$$\hat{\phi}_j|\psi\rangle = 0. \tag{15.27}$$

But that leads to a potential problem. For consistency, applying this twice we get $[\hat{\phi}_j, \hat{\phi}_{j'}]|\psi\rangle = 0$. But since the ϕ_j are supposed to be *all* the constraints, the commutator must be a linear combination of the constraints themselves, so

$$[\hat{\phi}_j, \hat{\phi}_{j'}] = c_{jj'j''}\hat{\phi}_{j''}. \tag{15.28}$$

Note the fact that in quantum mechanics the cs could in principle not commute with the $\hat{\phi}_j$, so we must put the cs to the left of the $\hat{\phi}_j$s, to have the commutator on the left-hand side equal zero.

Then the time evolution of the constraints should also be a constraint, so

$$[\hat{\phi}_j, H] = b_{jj'}\hat{\phi}_{j'}. \tag{15.29}$$

If there are no second-class constraints, then $\{\phi_j, \phi_{j'}\}_{P.B.} \approx 0$ (the condition for all the constraints to be first class), which in the quantum case turns into the algebra of constraints (15.28), and we have no problem.

But if there are second-class constraints, we have a problem, since then the fact that $\{\phi_w, \phi_{j'}\}_{P.B.}$ is not ≈ 0 contradicts the consistency condition (15.28).

Let's see what happens in an example. Consider a system with several degrees of freedom, and the constraints $q_1 \approx 0$ and $p_1 \approx 0$. But then their Poisson brackets give

$\{q_1, p_1\}_{P.B.} = 1 \neq 0$, which means that they are not first class. This means that we can't quantize the constraints by imposing them on states as

$$\hat{q}_1 |\psi\rangle = 0 = \hat{p}_1 |\psi\rangle, \qquad (15.30)$$

since then on the one hand applying the constraints twice we get $[\hat{q}_1, \hat{p}_1] |\psi\rangle = 0$, but on the other hand the commutator gives

$$[\hat{q}_1, \hat{p}_1] \psi = i\hbar |\psi\rangle, \qquad (15.31)$$

which is a contradiction.

Therefore one of the two assumptions we used is invalid: either $\hat{\phi}_j |\psi\rangle = 0$ is not a good way to impose the constraint, or the quantization map $\{,\}_{P.B.} \rightarrow 1/i\hbar[,]$ is invalid. It turns out that the latter is the case. We need to modify the Poisson bracket in order to map to $1/i\hbar[,]$. In the example above, the solution is obvious. Since the constraints are $q_1 = p_1 = 0$, we can just drop them from the phase space, and write the modified Poisson bracket

$$\{f, g\}_{P.B.'} = \sum_{n=2}^{N} \left(\frac{\partial f}{\partial q_n} \frac{\partial g}{\partial p_n} - \frac{\partial f}{\partial p_n} \frac{\partial g}{\partial q_n} \right). \qquad (15.32)$$

But what do we do in general?

In general, we must introduce the *Dirac brackets* as follows. We consider the *independent* second-class constraints (i.e. taking out the linear combinations which are first class, leaving only the constraints which have all linear combinations second class), called χ_s, and define $c_{ss'}$ by

$$c_{ss'} \{\chi_{s'}, \chi_{s''}\}_{P.B.} = \delta_{ss''}, \qquad (15.33)$$

that is we have the inverse matrix of Poisson brackets, $c_{ss'} = \{\chi_s, \chi_{s'}\}_{P.B.}^{-1}$. Then we define the Dirac brackets as

$$[f, g]_{D.B.} = \{f, g\}_{P.B.} - \{f, \chi_s\}_{P.B.} c_{ss'} \{\chi_{s'}, g\}_{P.B.}. \qquad (15.34)$$

If we use the Dirac brackets instead of the Poisson brackets, the time evolution is not modified, since

$$[g, H_T]_{D.B.} = \{g, H_T\}_{P.B.} - \{g, \chi_s\} c_{ss'} \{\chi_{s'}, H_T\}_{P.B.} \approx \{g, H_T\}_{P.B.} \approx \dot{g}, \qquad (15.35)$$

where we have used the fact that H_T is first class, so $\{\chi_{s'}, H_T\}_{P.B.} \approx 0$.

On the contrary, the Dirac bracket of the second-class constraints with any function on the phase space is zero, since

$$[f, \chi_{s''}]_{D.B.} = \{f, \chi_{s''}\}_{P.B.} - \{f, \chi_s\}_{P.B.} c_{ss'} \{\chi_{s'}, \chi_{s''}\}_{P.B.} = 0 \qquad (15.36)$$

(equal to 0 strongly!), where we have used the definition of $c_{ss'}$ as the inverse matrix of constraints. This means that in classical mechanics we can put $\chi_s = 0$ *strongly* if we use Dirac brackets. In quantum mechanics, we again impose it on states as

$$\hat{\chi}_s |\psi\rangle = 0, \qquad (15.37)$$

but now we don't have any more contradictions, since the Dirac bracket, turning into a commutator in quantum mechanics, is zero for the second-class constraints with anything. Therefore, $[\hat{f}, \hat{\chi}_s]|\psi\rangle = 0$ gives no contradictions now.

Finally, note that in quantum mechanics the difference between the primary and secondary constraints is irrelevant, whereas the difference between first and second-class constraints is important, since second-class constraints modify the Dirac bracket.

15.4 Example: Electromagnetic Field

We now return to the example which started the discussion, the electromagnetic field with action

$$S = -\int d^4x \frac{F_{\mu\nu}^2}{4} = \int \left[-\frac{1}{4} F_{ij} F^{ij} - \frac{1}{2} F_{0i} F^{0i} \right], \tag{15.38}$$

where $F_{\mu\nu} = \partial_\mu A_\nu - \partial_\nu A_\mu$. We also denote derivatives as $\partial_\mu B = B_{,\mu}$. Then the momenta conjugate to the fields A_μ are

$$P^\mu = \frac{\delta \mathcal{L}}{\delta A_{\mu,0}} = F^{\mu 0}. \tag{15.39}$$

Since $F^{00} = 0$ by antisymmetry, it means that we have the primary constraint

$$P^0 \approx 0. \tag{15.40}$$

The basic Poisson brackets of the fields A_μ with their conjugate momenta are as usual

$$\{A_\mu(\vec{x}, t), P^\nu(\vec{x}', t)\}_{P.B.} = \delta_\mu^\nu \delta^3(\vec{x} - \vec{x}'). \tag{15.41}$$

The Hamiltonian is then

$$\begin{aligned} H &= \int d^3x P^\mu A_{\mu,0} - L = \int d^3x \left[F^{i0} A_{i,0} + \frac{1}{4} F^{ij} F_{ij} + \frac{1}{2} F^{i0} F_{i0} \right] \\ &= \int d^3x \left[F^{i0} A_{0,i} + \frac{1}{4} F_{ij} F^{ij} - \frac{1}{2} F^{i0} F_{i0} \right] \\ &= \int d^3x \left[\frac{1}{4} F_{ij} F^{ij} + \frac{1}{2} P^i P^i - A_0 P^i_{,i} \right], \end{aligned} \tag{15.42}$$

where in the third equality we used $F_{i0} = A_{0,i} - A_{i,0}$ and in the last equality we used $P^i = F^{0i}$ and partial integration.

Next we compute the secondary constraints. First we compute the time evolution of the primary constraint, giving a secondary constraint

$$\{P^0, H\}_{P.B.} = P^i_{,i} \approx 0. \tag{15.43}$$

The time evolution of the secondary constraint is in turn trivial:

$$\{P^i_{,i}, H\}_{P.B.} = 0, \tag{15.44}$$

so there are no other secondary constraints.

All the constraints are first class, since we have

$$\{P^0, P^0\}_{P.B.} = 0 = \{P^0, P^i{}_{,i}\}_{P.B.} = \{P^i{}_{,i}, P^j{}_{,j}\}_{P.B.} = 0. \tag{15.45}$$

This means that there is no need for Dirac brackets. We have $H = H'$, and we add the first-class primary constraint P^0 with an arbitrary coefficient, giving the total Hamiltonian

$$H_T = H' + \int vP^0 = \int \left(\frac{1}{4}F^{ij}F_{ij} + \frac{1}{2}P^iP^i\right) - \int A_0 P^i{}_{,i} + \int v(x)P^0(x). \tag{15.46}$$

The extended Hamiltonian is found by adding also the secondary first-class constraint with an arbitrary coefficient function:

$$H_E = H_T + \int d^3x\, u(x)P^i{}_{,i}(x). \tag{15.47}$$

But A_0 and P^0 contain no relevant information. We can put $P^0 = 0$ in H_T since that only makes the time evolution of A_0 trivial, $\dot{A}_0 = 0$, and then we can get rid of A_0 by redefining $u'(x) = u(x) - A_0(x)$, obtaining the Hamiltonian

$$H_E = \int d^3x \left(\frac{1}{4}F^{ij}F_{ij} + \frac{1}{2}P^iP^i\right) + \int d^3x\, u'(x)P^i{}_{,i}(x). \tag{15.48}$$

Finally, we should note that the Dirac formalism is very important in modern theoretical physics. It is the beginning of more powerful formalisms: Becchi–Rouet–Stora–Tyutin (BRST) quantization, Batalin–Vilkovisky (BV) formalism, field–antifield formalism, used for gauge theories and string theory, among others. However, we will not explain them now.

Important Concepts to Remember

- Dirac quantization of a constrained system starts in the Hamiltonian formalism with the primary constraints $\phi_m(p, q) = 0$.
- We write weak equality $F(q, p) \approx 0$ if, at the end of the calculation, we use $\phi_m = 0$.
- The primary constraints are added to the Hamiltonian, forming the total Hamiltonian $H_T = H + u_m\phi_m$.
- The secondary constraints are obtained from the time evolution of the primary constraints, $\{\phi_m, H_T\}_{P.B.}$, and iterated until nothing new is found. In total, $\phi_j, j = 1, \ldots, M + K$, are all the constraints.
- First-class quantities R satisfy $\{R, \phi_j\} \approx 0$, and second-class quantities don't.
- H_T is written as $H_T = H' + v_a\phi_a$, where H' is first class, ϕ_a are first-class primary constraints, and v_a are numbers.
- The extended Hamiltonian is obtained by adding the first-class secondary constraints $H_E = H_T + v'_{a'}\phi_{a'}$.
- When quantizing, we must impose the constraints on states, $\hat{\phi}_j|\psi\rangle = 0$, but to avoid inconsistencies in the presence of second-class constraints χ_s, we must introduce Dirac brackets (to be replaced by the quantum commutators), by subtracting $\{f, \chi_s\}_{P.B.}\{\chi_s, \chi_{s'}\}^{-1}_{P.B.}\{\chi_{s'}, g\}_{P.B.}$.
- The electromagnetic field has primary constraint $P^0 \approx 0$ and secondary constraint $P^i{}_{,i} = F^{0i}{}_{,i} \approx 0$, but in the extended Hamiltonian we can drop A_0 and P^0.

Further Reading

See Dirac's book [5].

Exercises

1. Consider the Dirac action for a Dirac spinor in Minkowski space:

$$S_{cl} = \int d^4x(-\psi^\dagger i\gamma^0\gamma^\mu\partial_\mu\psi), \qquad (15.49)$$

with eight independent variables, ψ^A and $\psi^*_A = (\psi^A)^*$, where $A = 1,\dots,4$ is a spinor index. Calculate p_A and p^{*A} and the eight resulting primary constraints. Note that classical fermions are anticommuting, so we define p by taking the derivatives from the left, for example $\frac{\partial}{\partial\psi}(\psi\chi) = \chi$, so that $\{p^A, q_B\} = -\delta^A_B$. In general, we must define

$$\{f,g\}_{P.B.} = -(\partial f/\partial p^\alpha)\left(\frac{\partial}{\partial q_\alpha}g\right) + (-)^{fg}(\partial g/\partial p^\alpha)\left(\frac{\partial}{\partial q_\alpha}f\right), \qquad (15.50)$$

where $\partial f/\partial p^\alpha$ is the right derivative, for example $\partial/\partial\psi(\chi\psi) = \chi$, and $(-)^f\, g = -1$ for f and g being both fermionic and $+1$ otherwise (if f and/or g is bosonic). This bracket is antisymmetric if f and g are bose–bose or bose–fermi, and symmetric if f and g are both fermionic. Then compute H_T by adding to the classical Hamiltonian H the eight primary constraints with coefficients u_m. Check that there are no secondary constraints, and then from

$$\{\phi_m, H_T\}_{P.B.} \approx 0, \qquad (15.51)$$

solve for u^A, v_A.

2. (Continuation) Show that all constraints are second class, thus finding that

$$H_T = H' = H_E. \qquad (15.52)$$

Write Dirac brackets. Using the Dirac brackets we can now put $p^* = 0$ and replace $i\psi^*$ by $-p$. Show that finally, we have

$$H_T = -\int d^3x(p\gamma^0\gamma^k\partial_k\psi), \qquad (15.53)$$

and

$$[\psi^A, p_B]_{D.B.} = \{\psi^A, p_B\}_{P.B.} = -\delta^B_A\delta^3(\vec{x} - \vec{y}). \qquad (15.54)$$

Observation: Note that the analysis for Majorana (real) spinors is somewhat different, and there one finds

$$[\psi^A, p_B]_{D.B.} = -\frac{1}{2}\delta^A_B\delta^4(\vec{x} - \vec{y}). \qquad (15.55)$$

3. Consider the Lagrangian for the variable $q(t)$:

$$L = q\dot{q} - \alpha q^2. \tag{15.56}$$

Find the primary and secondary constraints, and check if they are first or second class. Then write Dirac brackets for the quantization of the system.

4. Write the extended Hamiltonian H_E and the total Hamiltonian H_T for the model in Exercise 3.

16

Quantization of Gauge Fields, their Path Integral, and the Photon Propagator

As we saw in Chapter 15, for gauge fields there are redundancies in the description due to the gauge invariance $\delta A_\mu = \partial_\mu \lambda$, which means that there are components which are not degrees of freedom. For instance, since $\lambda(x)$ is an arbitrary function, we could put $A_0 = 0$ by a gauge transformation. So we must impose a gauge in order to quantize.

- The simplest choice perhaps would be to impose a **physical gauge** condition, like the Coulomb gauge $\vec{\nabla} \cdot \vec{A} = 0$, where only the physical modes propagate (no redundant modes are present), and impose quantization of only these modes.
- But we will see that it is better to impose a **covariant gauge** condition, like the Lorenz gauge $\partial^\mu A_\mu = 0$, and quantize all the modes, imposing some constraints. This will be called covariant quantization.

16.1 Physical Gauge

We start with a discussion of the quantization in Coulomb, or radiation, gauge. The equation of motion for the action

$$S = -\frac{1}{4} \int d^4 x F_{\mu\nu}^2 = -\frac{1}{2} \int [(\partial_\mu A_\nu) \partial^\mu A^\nu - (\partial_\mu A_\nu) \partial^\nu A^\mu] \tag{16.1}$$

is

$$\Box A^\nu - \partial^\nu (\partial^\mu A_\mu) = 0 \tag{16.2}$$

and is difficult to solve. However, we see that in the *Lorenz, or covariant, gauge* $\partial^\mu A_\mu = 0$, the equation simplifies to just the Klein–Gordon (KG) equation

$$\Box A_\mu = 0, \tag{16.3}$$

which has solution of the type $A_\mu \sim \epsilon_\mu(k) e^{\pm ikx}$, with $k^2 = 0$ and $k^\mu \epsilon_\mu(k) = 0$ (from the gauge condition $\partial^\mu A_\mu = 0$ in momentum space).

But note that the Lorenz gauge does not fix completely the gauge, there is a residual gauge symmetry, namely $\delta A_\mu = \partial_\mu \lambda$, with λ satisfying (since $\partial^\mu \delta A_\mu = 0$)

$$\partial^\mu \delta A_\mu = \Box \lambda = 0. \tag{16.4}$$

We can use this λ to also fix the gauge $A_0 = 0$, by transforming with $\partial_0 \lambda = -A_0$, since the KG equation for A_μ, (16.2), gives

$$\Box A_0 = 0 \Rightarrow \partial_0 \Box \lambda = 0, \qquad (16.5)$$

so it is fine to use a λ restricted to $\Box \lambda = 0$ only. If $A_0 = 0$, the Lorenz gauge reduces to $\vec{\nabla} \cdot \vec{A} = 0$.

Note that this $A_0 = 0$ gauge is consistent only with $J^\mu = 0$ (no current, since otherwise $A_0 = \int J_0$), but we will consider such a case (no classical background). Therefore we consider the full gauge condition as

$$A_0 = 0; \qquad \vec{\nabla} \cdot \vec{A} = 0, \qquad (16.6)$$

known as *radiation, or Coulomb, gauge*, and familiar from electrodynamics (see e.g. Jackson's 1998 book *Classical Electrodynamics*). In this gauge there are only two propagating modes, the physical degrees of freedom (corresponding to the two polarizations of the electromagnetic field, transverse to the direction of propagation, or circular and anticircular), since we removed A_0 and $\partial^\mu A_\mu$ from the theory.

The classical solution is

$$\vec{A}(x) = \int \frac{d^3 k}{(2\pi)^3 \sqrt{2E_k}} \sum_{\lambda=1,2} \vec{\epsilon}^{(\lambda)}(k) \left[e^{ikx} a^{(\lambda)}(k) + a^{(\lambda)\dagger}(k) e^{-ikx} \right], \qquad (16.7)$$

supplemented with $k^2 = 0$ (from the KG equation) and $\vec{k} \cdot \vec{\epsilon}^{(\lambda)}(k) = 0$ (from the Lorenz gauge condition in momentum space). The normalization for the solutions was chosen, as in previous cases, so that under quantization, a and $a^\dagger$ act as creation and annihilation operators for the harmonic oscillator. We can also choose the *polarization vectors* $\vec{\epsilon}^{(\lambda)}(k)$ such that they are orthogonal:

$$\vec{\epsilon}^{(\lambda)}(\vec{k}) \cdot \vec{\epsilon}^{(\lambda')}(k) = \delta^{\lambda\lambda'}. \qquad (16.8)$$

16.2 Quantization in Physical Gauge

We now turn to quantization of the system. As we saw in Chapter 15, in Dirac formalism, we can drop A_0 and P^0 from the system entirely, which is now needed since we want to use the gauge $A_0 = 0$, and impose the remaining gauge condition $\vec{\nabla} \cdot \vec{A} = 0$ as an operatorial condition.

Consider the conjugate momenta

$$P^i(= \Pi^i) = F^{0i} = E^i. \qquad (16.9)$$

We would be tempted to impose the canonical equal-time quantization conditions

$$[A^i(\vec{x}, t), E^j(\vec{x}', t)] = i \int \frac{d^3 k}{(2\pi)^3} \delta^{ij} e^{i\vec{k}(\vec{x}-\vec{x}')} = i\delta^{ij}\delta^3(\vec{x} - \vec{x}'), \qquad (16.10)$$

but note that if we do so, applying $\nabla_{\vec{x},i}$ on the above relation does not give zero as it should (since we should have $\vec{\nabla} \cdot \vec{A} = 0$). So we generalize $\delta^{ij} \rightarrow \Delta^{ij}$ and from the condition $k_i \Delta^{ij} = 0$, we get

$$[A^i(\vec{x},t), E^j(\vec{x}',t)] = i \int \frac{d^3k}{(2\pi)^3} \Delta^{ij} e^{i\vec{k}(\vec{x}-\vec{x}')} = i \int \frac{d^3k}{(2\pi)^3} \left(\delta^{ij} - \frac{k^i k^j}{k^2} \right) e^{i\vec{k}(\vec{x}-\vec{x}')}$$

$$= i \left(\delta^{ij} - \frac{\partial_i \partial_j}{\vec{\nabla}^2} \right) \delta^3(\vec{x} - \vec{x}'), \qquad (16.11)$$

and the rest of the commutators are zero:

$$[A^i(\vec{x},t), A^j(\vec{x}',t)] = [E^i(\vec{x},t), E^j(\vec{x}',t)] = 0. \qquad (16.12)$$

Replacing the mode decomposition of $\vec{A}$ in the above commutators, we obtain the usual harmonic oscillator creation/annihilation operator algebra

$$[a^{(\lambda)}(k), a^{(\lambda')\dagger}(k')] = (2\pi)^3 \delta_{\lambda\lambda'} \delta^{(3)}(\vec{k} - \vec{k}'),$$

$$[a^{(\lambda)}(k), a^{(\lambda')}(k')] = 0 = [a^{(\lambda)\dagger}(k), a^{(\lambda')\dagger}(k')]. \qquad (16.13)$$

We can then compute the energy and obtain the same sum over harmonic oscillator Hamiltonians as in previous cases:

$$E = H = \frac{1}{2} \int d^3x (\vec{E}^2 + \vec{B}^2)$$

$$= \sum_\lambda \int \frac{d^3k}{(2\pi)^3} \frac{k^0}{2} \left[a^{(\lambda)\dagger}(k) a^{(\lambda)}(k) + a^{(\lambda)}(k) a^{(\lambda)\dagger}(k) \right]. \qquad (16.14)$$

The proof is similar to previous cases, so is left as an exercise. Again, as in previous cases, we define normal ordering and find for the normal ordered Hamiltonian

$$: H := \sum_\lambda \int \frac{d^3k}{(2\pi)^3} k^0 a^{(\lambda)\dagger}(k) a^{(\lambda)}(k). \qquad (16.15)$$

The quantization described here is good, but cumbersome, since while the classical theory was Lorentz invariant, the quantization procedure isn't manifestly so, since we did it using the Lorentz symmetry-breaking radiation gauge condition (16.6). This means that in principle, quantum corrections (in quantum field theory) could break Lorentz invariance, so we need to check Lorentz invariance explicitly at each step.

Therefore a better choice is to use a formalism that doesn't break the manifest Lorentz invariance, like the next case to be considered.

16.3 Lorenz Gauge (Covariant) Quantization

If we only impose the KG equation, but no other condition, we still have all four polarization modes, so we have the classical solution

$$A_\mu(x) = \int \frac{d^3k}{(2\pi)^3 \sqrt{2E_k}} \sum_{\lambda=0}^{3} \epsilon_\mu^{(\lambda)}(k) \left[e^{ikx} a^{(\lambda)}(k) + a^{(\lambda)\dagger}(k) e^{-ikx} \right], \qquad (16.16)$$

where again $k^2 = 0$ from the solution of the KG equation, but since the KG equation is obtained only in the Lorenz gauge, we still need to impose the Lorenz gauge in an operatorial fashion, as we will see later.

Let us fix the coordinate system such that the momentum solving the KG equation $k^2 = 0$ is in the third direction, $k^\mu = (k, 0, 0, k)$, and define the four polarizations in the directions of the four axes of coordinates, that is $\epsilon_\mu^\lambda = \delta_\mu^\lambda$, or

$$\epsilon^{(\lambda)} = \begin{pmatrix} 1 \\ 0 \\ 0 \\ 0 \end{pmatrix}; \begin{pmatrix} 0 \\ 1 \\ 0 \\ 0 \end{pmatrix}; \begin{pmatrix} 0 \\ 0 \\ 1 \\ 0 \end{pmatrix}; \begin{pmatrix} 0 \\ 0 \\ 0 \\ 1 \end{pmatrix} \text{ for } \lambda = 0, 1, 2, 3. \qquad (16.17)$$

Then for $\lambda = 1, 2$ we obtain *transverse polarizations*, that is physical polarizations, transverse to the direction of the momentum:

$$k^\mu \epsilon_\mu^{(1,2)} = 0, \qquad (16.18)$$

and for the unphysical modes $\lambda = 0$ (*timelike polarization*) and $\lambda = 3$ (*longitudinal polarization*, i.e. parallel to the momentum):

$$k^\mu \epsilon_\mu^{(0)} = E_k = k^\mu \epsilon_\mu^{(3)}(k). \qquad (16.19)$$

These modes are unphysical since they don't satisfy the gauge condition $\partial^\mu A_\mu = 0$, which would also be needed in order to obtain the KG equation for them. This means that these modes must somehow cancel out of the physical calculations, by imposing some operatorial gauge condition on physical states.

The natural guess would be to impose the Lorenz gauge condition on states, $\partial^\mu A_\mu |\psi\rangle = 0$, which is what Fermi tried to do first. But in view of the mode expansion (16.16), this is not a good idea, since it contains also creation operators, so $\partial^\mu A_\mu$ would not be zero even on the vacuum $|0\rangle$, since $a^\dagger|0\rangle \neq 0$ (there are some details, but we can check that it is impossible). Instead, the good condition was found by Gupta and Bleuler, namely that *only the positive-frequency part* of the Lorenz condition is imposed on states:

$$\partial^\mu A_\mu^{(+)}(x)|\psi\rangle = 0 \qquad (16.20)$$

(note the somewhat confusing notation: the plus means positive frequency, not dagger, in fact only the annihilation part must be zero on states), known as the *Gupta–Bleuler* quantization condition.

Therefore the equal-time commutation relations in the covariant gauge are

$$[A_\mu(\vec{x}, t), \Pi_\nu(\vec{x}', t)] = ig_{\mu\nu}\delta^{(3)}(\vec{x} - \vec{x}'),$$
$$[A_\mu(\vec{x}, t), A_\nu(\vec{x}', t) = [\Pi_\mu(\vec{x}, t), \Pi_\nu(\vec{x}', t)] = 0. \qquad (16.21)$$

Note that since from the Lagrangian for electromagnetism we found $\Pi_0 = 0$, this contradicts the commutation relation above (and within this context we can't impose $\Pi_0 = 0$ on states, and even if we do, we would still obtain a contradiction with the commutation relation). One solution is to change the Lagrangian, and we will see later that we can in fact do exactly that.

Substituting the expansion (16.16) at the quantum level, we obtain *almost* the usual commutation relations for harmonic oscillator creation and annihilation operators, namely

$$[a^{(\lambda)}(k), a^{(\lambda')\dagger}(k')] = g^{\lambda\lambda'}(2\pi)^3\delta^{(3)}(\vec{k} - \vec{k}'),$$
$$[a^{(\lambda)}(k), a^{(\lambda')}(k')] = [a^{(\lambda)\dagger}(k), a^{(\lambda')\dagger}(k')] = 0. \qquad (16.22)$$

For the $\lambda = 1, 2, 3$ modes that is fine, but for the $\lambda = 0$ mode not, since

$$[a^{(0)}(k), a^{(0)\dagger}(k')] = -(2\pi)^3\delta^{(3)}(\vec{k} - \vec{k}'). \qquad (16.23)$$

This means that these modes have negative norm, since

$$||a^\dagger|0\rangle||^2 \equiv \langle 0|aa^\dagger|0\rangle = -\langle 0|a^\dagger a|0\rangle\langle 0|0\rangle = -1, \qquad (16.24)$$

so they are clearly unphysical.

The Gupta–Bleuler condition means that, substituting the form of A_μ and $k^\mu\epsilon_\mu^{(1,2)} = 0$:

$$(k^\mu\epsilon_\mu^{(0)}(k)a^{(0)}(k) + k^\mu\epsilon_\mu^{(3)}a^{(3)}(k))|\psi\rangle = 0 \Rightarrow$$
$$[a^{(0)}(k) + a^{(3)}(k)]|\psi\rangle = 0. \qquad (16.25)$$

Let us see how this looks on a state which is a linear combination of one (0) and one (3) state:

$$[a^{(0)}(k) + a^{(3)}(k)](a^{(0)\dagger}(k) + a^{(3)\dagger}(k))|0\rangle$$
$$= ([a^{(0)}(k), a^{(0)\dagger}(k)] + [a^{(3)}(k), a^{(3)\dagger}(k)])|0\rangle = 0, \qquad (16.26)$$

so it is indeed a physical state. In general, we see that we need to have the same number of (0) and (3) modes in the state (as we can check).

This means that in general the contribution of $a^{(0)}$s (timelike modes) and $a^{(3)}$s (longitudinal modes) cancels each other; for instance, in loops, these modes will cancel each other, as we will see in Part II of the book.

Note that by imposing only the positive-frequency part of the Lorenz gauge condition on $|\psi\rangle$s, we impose the full gauge condition on expectation values:

$$\langle\psi|\partial^\mu A_\mu|\psi\rangle = \langle\psi|(\partial^\mu A_\mu^{(+)} + \partial^\mu A_\mu^{(-)})|\psi\rangle = 0, \qquad (16.27)$$

since $(\partial^\mu A_\mu^{(+)}|\psi\rangle)^\dagger = \langle\psi|\partial^\mu A_\mu^{(-)}$.

We have defined the above two methods of quantization, since they were historically first. But now, there are also other methods of quantization:

- In other gauges, like lightcone gauge quantization.
- More powerful (modern) covariant quantization: Becchi–Rouet–Stora–Tyutin (BRST) quantization. To describe this, however, we would need to understand BRST symmetry, and this will be considered in Part II of the book.
- Path integral procedures:
  - We can write a Hamiltonian path integral based on Dirac's formalism from Chapter 15. This path is taken for instance in Ramond's book [4], but we will not continue it here.
  - We can use a procedure more easily generalized to the nonabelian case, namely Fadeev–Popov quantization. This is what we will describe here.

First, we come back to the observation that we need to modify the Lagrangian in order to avoid getting $\Pi_0 = 0$. It would be nice to find just the KG equation (what we have in the covariant gauge) from this Lagrangian, for instance. We try

$$\mathcal{L} = -\frac{1}{4}F_{\mu\nu}^2 - \frac{\lambda}{2}(\partial_\mu A^\mu)^2, \tag{16.28}$$

and indeed, its equations of motion are

$$\Box A^\mu - (1-\lambda)\partial^\mu(\partial_\lambda A^\lambda) = 0, \tag{16.29}$$

and so for $\lambda = 1$ we obtain the KG equation $\Box A_\mu = 0$. $\lambda = 1$ is called the "Feynman gauge," although this is a misnomer, since it is not really a gauge, just a choice. The Lagrangian above was just something that gives the right equation of motion from the Lorenz gauge, but we will see that it is what we will obtain in the Fadeev–Popov quantization procedure.

16.4 Fadeev–Popov Path-Integral Quantization

To properly define the path-integral quantization, we must work in Euclidean space, as we saw. The Wick rotation is $x^0 = t \to -ix_4$. But now, since we have the vector A_μ that should behave like x_μ, we must also Wick rotate A_0 like $x_0 = -t$. Then we obtain

$$E_i^{(E)} = F_{4i} = \frac{\partial}{\partial x_4}A_i - \frac{\partial}{\partial x_i}A_4 = -iE_i^{(M)}, \tag{16.30}$$

since $F_{0i} \to iF_{4i}$, like $x_0 \to ix_4$.

Since the action is Wick rotated as $iS^{(M)} \to -S^{(E)}$, and we have $i\int dt(-F_{ij}^2) = -\int dx_4 F_{ij}^2$, we obtain for the Lagrangian in Euclidean space

$$\mathcal{L}_{em}^{(E)}(A) = +\frac{1}{4}F_{\mu\nu}^{(E)}F_{\mu\nu}^{(E)} = \frac{1}{2}\left[(E_i^{(E)})^2 + (B_i^{(E)})^2\right], \tag{16.31}$$

and for the action

$$S_{em}^{(E)} = \int d^4x \mathcal{L}_{em}^{(E)}(A). \tag{16.32}$$

Since we are in Euclidean space, we can do partial integrations without boundary terms (as we saw, the Euclidean theory is defined for periodic time, on a circle with an infinite radius, hence there are no boundary terms), so we can write

$$S_{em}[A] = \frac{1}{2} \int d^d x A_\mu(x)(-\partial^2 \delta_{\mu\nu} + \partial_\mu \partial_\nu) A_\nu(x). \tag{16.33}$$

This means that we would be tempted to think that we can write as usual

$$
\begin{aligned}
Z[J] &= \int \mathcal{D}A_\mu(x) \exp\left\{ -\frac{1}{2} \int d^d x [A_\mu(-\partial^2 \delta_{\mu\nu} + \partial_\mu \partial_\nu) A_\nu] \right\} \\
&\text{`` = ''} Z[0] \exp\left\{ \frac{1}{2} \int d^d x d^d y J_\mu(x) G_{\mu\nu}(x,y) J_\nu(y) \right\},
\end{aligned}
\tag{16.34}
$$

where $G_{\mu\nu}$ is the inverse of the operator $(-\partial^2 \delta_{\mu\nu} + \partial_\mu \partial_\nu)$. However, that is not possible, since the operator has zero modes, $A_\mu = \partial_\mu \lambda$:

$$(-\partial^2 \delta_{\mu\nu} + \partial_\mu \partial_\nu)\partial_\nu \lambda = 0, \tag{16.35}$$

and an operator with zero modes (zero eigenvalues) cannot be inverted.

This fact is a consequence of gauge invariance, since these zero modes are exactly the pure gauge modes, $\delta A_\mu = \partial_\mu \chi$. So in order to be able to invert the operator, we need to get rid of these zero modes, or gauge transformations, from the path integral. Since physical quantities are related to ratios of path integrals, it will suffice if we can *factorize* the integration over the gauge invariance:

$$Vol(G_{inv}) = \prod_{x \in \mathbb{R}^d} \int d\lambda(x), \tag{16.36}$$

or the volume of the local $U(1)$ symmetry group:

$$G_{inv} = \prod_{x \in \mathbb{R}^d} U_x(1). \tag{16.37}$$

To do this, we will consider the set of more general covariant gauge conditions

$$\partial_\mu A_\mu = c(x), \tag{16.38}$$

instead of the Euclidean Lorenz gauge $\partial_\mu A_\mu = 0$.

Consider an arbitrary gauge field A and the *orbit of A, $Or(A)$*, obtained from all possible gauge transformations of this A, see Figure 16.1. Then consider also the space of all possible gauge conditions, $\mathcal{M}$. We should have that there is only one point at the intersection of $Or(A)$ and $\mathcal{M}$; that is there is a unique gauge transformation of A, with $\lambda^{(A)}$, that takes us into the gauge condition. We will suppose this is the case in the following (in fact, there is some issue with this assumption in the nonabelian case at least: it is not true for *large* nonabelian gauge transformations; there exist what are called *Gribov copies*, which arise for large differences between As, but we will ignore these here, since anyway we are dealing with the abelian case). Consider then the definition of λ^A:

$$\partial^2 \lambda^{(A)}(x) = -\partial_\mu A_\mu(x) + c(x). \tag{16.39}$$

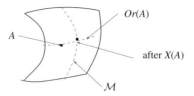

The gauge-fixed configuration is at the intersection of the orbit of A, the gauge transformations of a gauge field configuration A, and the space $\mathcal{M}$ of all possible gauge conditions.

There is a unique solution in Euclidean space, considering that $c(x)$ falls off sufficiently fast at infinity, and similarly for A and $\chi^{(A)}$.

Define the gauge field A transformed by some gauge transformation λ as

$$^{\lambda}A_{\mu}(x) \equiv A_{\mu}(x) + \partial_{\mu}\lambda(x). \tag{16.40}$$

Then we have the following:

Lemma (identity)

$$\int \prod_{x \in \mathbb{R}^d} d\lambda(x) \prod_{y \in \mathbb{R}^d} \delta(-\partial_{\mu}(^{\lambda}A_{\mu}(y)) + c(y)) = \frac{1}{\det(-\partial^2)}. \tag{16.41}$$

Proof   We have

$$-\partial_{\mu}(^{\lambda}A_{\mu}) + c = -\partial^2\lambda - \partial_{\mu}A_{\mu} + c = -\partial^2\lambda + \partial^2\chi^{(A)}, \tag{16.42}$$

so we shift the integration from $\int d\lambda$ to $\lambda \to \lambda - \lambda^{(A)}$, absorbing the second term above. Then we obtain

$$\int \prod_{x \in \mathbb{R}^d} d\lambda(x) \prod_{y \in \mathbb{R}^d} \delta(-\partial_{\mu}(^{\lambda}A_{\mu}(y)) + c(y)) = \int \prod_{x} d\lambda(x) \prod_{y} \delta(-\partial^2\lambda(y)). \tag{16.43}$$

But note that this is a continuum version of the discrete relation

$$\int \prod_{i=1}^{n} d\lambda_i \prod_{j=1}^{n} \delta(\Delta_{ij}\lambda_j) = \frac{1}{\det \Delta}, \tag{16.44}$$

for the operator

$$-\partial^2(x, y) \equiv -\partial_x^2 \delta^{(d)}(x - y). \tag{16.45}$$

So we have proved the lemma.   *q.e.d.*

We can then write the lemma with the determinant on the other side as

$$\text{``1''} = \int \prod_{x} d\lambda(x) \det(-\partial^2) \prod_{y} \delta(-\partial_{\mu}(^{\lambda}A_{\mu}(y)) + c(y)), \tag{16.46}$$

and we can do a Gaussian integration over the gauge conditions $c(x)$ of this result as

$$
\text{``}1(\alpha)\text{''} = \int \mathcal{D}c(x) e^{-\frac{1}{2\alpha}\int d^d x c^2(x)} \text{``}1\text{''}
$$

$$
= \int \prod_x d\lambda(x) \det(-\partial^2) e^{-\frac{1}{2\alpha}\int d^d x (\partial_\mu(^\lambda A_\mu(x)))^2}, \tag{16.47}
$$

where in the second equality we used the lemma and used the delta function in it. Note that "$1(\alpha)$" means that it is a function of the arbitrary number α only, so it is not important, since this will cancel out in the ratio of path integrals which defines observables.

Then in the path integral for an observable $\mathcal{O}(A_\mu)$, we obtain

$$
\int \mathcal{D}A_\mu e^{-S[A_\mu]} \mathcal{O}[A_\mu] \text{``}1(\alpha)\text{''} = \det(-\partial^2) \int \prod_x d\lambda(x) \int \mathcal{D}A_\mu e^{-S[A_\mu]-\frac{1}{2\alpha}\int d^d x (\partial^\lambda A_\mu)^2} \mathcal{O}(A_\mu). \tag{16.48}
$$

Now we change variables in the path integral from A_μ to $A_\mu - \partial_\mu \lambda$, so as to turn $\partial_\mu(^\lambda A_\mu) \rightarrow \partial_\mu A_\mu$. Then the dependence on λ disappears from inside the integral, and the integration over the gauge modes finally factorizes:

$$
\int \mathcal{D}A_\mu e^{-S[A_\mu]} \mathcal{O}[A_\mu] \text{``}1(\alpha)\text{''} = \left[\det(-\partial^2) \int \prod_x d\lambda(x) \right]
$$

$$
\times \int \mathcal{D}A_\mu e^{-S[A_\mu]-\frac{1}{2\alpha}\int d^d x (\partial_\mu A_\mu)^2} \mathcal{O}(A_\mu). \tag{16.49}
$$

This means that in correlator variables, defined as ratios of the path integral with insertion to the one without, to cancel out the bubble diagrams, the integration over the gauge modes cancels:

$$
\langle \mathcal{O}(A_\mu \ldots) \rangle = \frac{\int \mathcal{D}(A_\mu \ldots) \mathcal{O}(A_\mu \ldots) e^{-S(A_\mu)} \text{``}1(\alpha)\text{''}}{\int \mathcal{D}(A_\mu \ldots) e^{-S(A_\mu)} \text{``}1(\alpha)\text{''}}
$$

$$
= \frac{\int \mathcal{D}(A_\mu \ldots) \mathcal{O}(A_\mu \ldots) e^{-S_{\text{eff}}(A_\mu, \ldots)}}{\int \mathcal{D}(A_\mu \ldots) e^{-S_{\text{eff}}(A_\mu \ldots)}}. \tag{16.50}
$$

Here the effective Lagrangian is

$$
\mathcal{L}_{\text{eff}} = \frac{1}{4} F_{\mu\nu}^2 + \frac{1}{2\alpha} (\partial_\mu A_\mu)^2, \tag{16.51}
$$

where the extra term is called the *gauge-fixing term*, as it is not gauge invariant, and the remaining path integral is without zero modes.

16.5 Photon Propagator

We can again partially integrate and obtain, for the effective action

$$
S_{\text{eff}}(A) = \frac{1}{2} \int d^d x A_\mu \left(-\partial^2 \delta_{\mu\nu} + \left(1 - \frac{1}{\alpha}\right) \partial_\mu \partial_\nu \right) A_\nu, \tag{16.52}
$$

where the operator in brackets, $(G^{(0)})^{-1}_{\mu\nu}$, is now invertible. We write in momentum space

$$S_{\text{eff}}(A) = \frac{1}{2}\int \frac{d^d k}{(2\pi)^d} A_\mu(-k)(G^{(0)})^{-1}_{\mu\nu}(k)A_\nu(k). \tag{16.53}$$

Since

$$(G^{(0)})^{-1}_{\mu\nu}(k) = k^2\delta_{\mu\nu} - \left(1 - \frac{1}{\alpha}\right)k_\mu k_\nu, \tag{16.54}$$

we have

$$\left(k^2\delta_{\mu\nu} - \left(1 - \frac{1}{\alpha}\right)k_\mu k_\nu\right)G^{(0)}_{\nu\lambda} = \delta_{\mu\lambda}, \tag{16.55}$$

meaning that finally the photon propagator is

$$G^{(0)}_{\mu\nu}(k) = \frac{1}{k^2}\left(\delta_{\mu\nu} - (1-\alpha)\frac{k_\mu k_\nu}{k^2}\right). \tag{16.56}$$

Then in the case of $\alpha = 1$ ($\lambda = 1/\alpha = 1$), the "Feynman gauge" (a misnomer, as we said), we have the KG propagator

$$G^{(0)}_{\mu\nu}(k;\alpha = 1) = \frac{1}{k^2}\delta_{\mu\nu}. \tag{16.57}$$

Important Concepts to Remember

- The physical gauge is $A_0 = 0$ and $\partial^\mu A_\mu = 0$, or $\vec{\nabla}\cdot\vec{A} = 0$, and here the equation of motion is KG.
- Quantization in the physical gauge is done for only the two physical transverse polarizations, for which we expand in the usual harmonic oscillator creation and annihilation operators.
- Covariant gauge quantization is done keeping all four polarizations of the gauge field, but imposing $\partial^\mu A^{(+)}_\mu|\psi\rangle = 0$ on physical states.
- Timelike modes have negative norm, but they cancel in calculations against longitudinal modes. In particular, the physical condition says that the longitudinal and timelike modes should match inside physical states.
- Fadeev–Popov path-integral quantization factorizes the gauge modes and leaves an effective action which contains a gauge-fixing term.
- The photon propagator in Feynman gauge is the KG propagator.

Further Reading

See sections 6.1, 6.2, and 6.3.1 in [3], section 7.3 in [2], and section 7.1 in [4].

Exercises

1. Prove that for quantization in physical gauge, the quantum energy is

$$E = \frac{1}{2} \int d^3x (\vec{E}^2 + \vec{B}^2) = \sum_{\lambda=1,2} \int \frac{d^3k}{(2\pi)^3} \frac{k^0}{2} \left[a^{(\lambda)\dagger}(k) a^{(\lambda)}(k) + a^{(\lambda)}(k) a^{\dagger(\lambda)}(k) \right].$$

(16.58)

2. Consider the state in covariant quantization

$$|\psi\rangle = \left(a^{(1)\dagger}(k_1) - a^{(2)\dagger}(k_3) \right) \left(a^{(0)\dagger}(k_2) + a^{(3)\dagger}(k_4) \right) |0\rangle. \qquad (16.59)$$

   Is it physical? Why? Write a physical state with two physical particles.
3. Using the effective action S_{eff} (with gauge-fixing term), for $k^\mu = (k, 0, 0, k)$, write down the equation of motion in momentum space, *separately* for the longitudinal, timelike, and transverse modes, at arbitrary "gauge" α.
4. Consider a complex scalar minimally coupled to the photon, and write the one-loop diagram with two external scalars and a gauge loop. Write a formula for the difference between the result in "Feynman gauge" $\alpha = 1$ and the "gauge" $\alpha = 0$ for the one-loop diagram. How do you explain the fact that this difference is nonzero?

17

Generating Functional for Connected Green's Functions and the Effective Action (1PI Diagrams)

In this chapter we will consider several types of n-point functions and their generating functionals in the path-integral formalism. We saw that the partition function $Z[J]$ is the generating functional of the full Green's functions. Here we will consider the free energy $W[J]$, the generating functional of connected Green's functions, the effective action $\Gamma[\Phi]$, the generating functional of the one-particle irreducible (1PI) Green's functions, and the classical action $S[\Phi]$, the generating functional of tree diagrams.

17.1 Generating Functional of Connected Green's Functions

As we saw, $Z[J]$ is the generating functional of the full (connected and disconnected) Green's functions.

But now we will prove that the generating functional of connected Green's functions is $-W[J]$, where

$$Z[J] = e^{-W[J]}. \tag{17.1}$$

We define the n-point Green's functions in the presence of a nonzero source J as

$$G_n(x_1, \ldots, x_n)_J = \frac{\delta^n}{\delta J(x_1) \ldots \delta J(x_n)} Z[J] \tag{17.2}$$

(before we had defined the Green's functions at nonzero J). We will denote them by a box with a J inside it, and n lines ending on points $x_1, \ldots, x_n$ coming out of it, as in Figure 17.1. A box without any lines from it is the zero-point function (i.e. $Z[J]$, see Figure 17.1).

For the one-point function in ϕ^4 theory, writing the perturbative expansion as we already did (before we put $J = 0$), we have

$$G(x)_J = \int d^d y \Delta(x - y) J(y) - g \int d^d z d^d y_1 d^d y_2 d^d y_3$$
$$\Delta(x - z) \Delta(y_1 - z) \Delta(y_2 - z) \Delta(y_3 - z) J(y_1) J(y_2) J(y_3) + \ldots, \tag{17.3}$$

and the Feynman diagrams that correspond to it are: line from x to a cross representing the source, line from x to a vertex from which three other lines end on crosses. There is also: line from x to a cross with a loop on it (setting sun), and so on, see Figure 17.1(a) for the diagrammatic form of the above equation.

A cross is $\int J(x)$, or $\sum_i J_i$ in the discretized version. A propagator is a line, and so a line between a point and a cross is $\Delta_{ij} J_j = \int d^d y \Delta(x - y) J(y)$, see Figure 17.1.

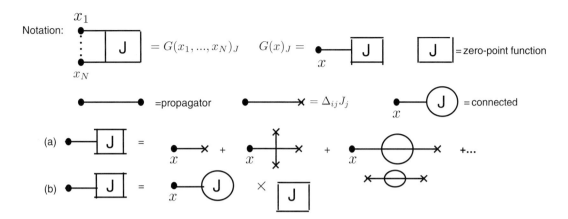

Figure 17.1 Notation for Green's functions, followed by the diagrammatic expansion for the one-point function with source J (a), which can be seen diagrammatically to factorize into the connected part times the vacuum bubbles (b).

Figure 17.2 For the two-point function with a source, the factorization doesn't work anymore, since we can have disconnected pieces without vacuum bubbles, as exemplified here.

As we said, a box with J inside and a line ending on an external point is the one-point function $G(x)_J$. We also draw the connected piece by replacing the box with a circle (Figure 17.1).

Then, the full one-point function is the connected one-point function times the zero-point function (vacuum bubbles, or $Z[J]$), in exactly the same way as we showed in the operator formalism, see Figure 17.1(b). Note that this only works for the one-point functions, since for instance for the two-point function we also have contributions like x_1 connected with a cross, and x_2 connected with another cross, as in Figure 17.2.

The above diagrammatic equation is written as

$$\frac{\delta Z[J]}{\delta J(x)} = -\frac{\delta W[J]}{\delta J(x)} Z[J], \qquad (17.4)$$

where here by definition $-W[J]$ is the generating functional of connected diagrams. The solution of this equation is

$$Z[J] = \mathcal{N}e^{-W[J]}, \qquad (17.5)$$

as we said it should be. Here, $W[J]$ is called *free energy* since, exactly like in thermodynamics, the partition function is the exponential of minus the free energy.

17.2 Effective Action and 1PI Green's Functions

Another functional of interest is the *effective action*, which we will define soon, and which we will find is the generating functional of 1PI Green's functions. One-particle irreducible means that the diagrams cannot be separated into two disconnected pieces by cutting a single propagator. An example of a one-particle reducible diagram would be two lines between two points, then one line to another, and then another two to the last point, see Figure 17.3. To make it 1PI we would need to add another line for the points in the middle.

The effective action we will find is a Legendre transform of the free energy, exactly like we take a Legendre transform of the free energy in thermodynamics to obtain (Gibbs) potentials like $G = F - Q\Phi$.

We need to define first the object conjugate to the source J (just like the electric potential Φ is conjugate to the charge Q for G above). This is the *classical field* $\phi_{cl}[J]$ in the presence of an external current J. As the name suggests, it is the properly normalized VEV of the quantum field, which replaces the field of classical field theory at the quantum level:

$$\phi_{cl}[J] \equiv \frac{\langle 0|\hat{\phi}(x)|0\rangle_J}{\langle 0|0\rangle_J} = \frac{1}{Z[J]}\int \mathcal{D}\phi e^{-S[\phi]+J\cdot\phi}\phi(x) = \frac{1}{Z[J]}\frac{\delta Z[J]}{\delta J(x)} = \frac{\delta(-W[J])}{\delta J(x)} = G_1^c(x;J),$$

(17.6)

that is the connected one-point function in the presence of the external source.

17.2.1 Example: Free Scalar Field Theory in the Discretized Version

Consider the action

$$S_0 - J\cdot\phi = \frac{1}{2}\int d^d x[(\partial_\mu \phi)^2 + m^2\phi^2] - \int d^d x J(x)\phi(x)$$

$$= \frac{1}{2}\int d^d x \int d^d y \phi(x)\Delta^{-1}(x,y)\phi(y) - \int d^d x J(x)\phi(x)$$

$$\equiv \frac{1}{2}\phi_i\Delta_{ij}^{-1}\phi_j - J_k\phi_k.$$

(17.7)

The classical equation of motion for this, and its solution, are

$$\Delta_{ij}^{-1}\phi_j - J_i = 0 \Rightarrow \phi_i = \Delta_{ij}J_j,$$

(17.8)

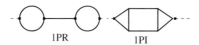

1PR   1PI

Figure 17.3 One-particle reducible graph (left): the middle propagator can be cut, separating the graph into two pieces vs. one-particle irreducible graph (right): one cannot separate the graph into two pieces by cutting a single propagator.

which is the Coulomb law. In the continuum, we have

$$\phi(x) = \int d^d x \Delta(x, y) J(y). \tag{17.9}$$

In contrast, we saw that the free partition function is $Z_0 = e^{\frac{J_i \Delta_{ij} J_j}{2}}$, so that the free energy is

$$- W_0[J] = \frac{J_i \Delta_{ij} J_j}{2}, \tag{17.10}$$

implying that in the free theory the classical field is

$$\phi_i^{cl(0)} = \frac{\delta}{\delta J_i}(-W_0[J]) = \Delta_{ij} J_j, \tag{17.11}$$

that is the same as the solution of the classical equation of motion.

This is so since free fields are *classical*; no interactions means there are no quantum fluctuations.

The classical field is represented by a propagator from the point i to a circle with a J inside it.

17.2.2 1PI Green's Functions

Some of the diagrams for the classical field in ϕ^4 theory are as follows: line from i to a cross; line from i to a vertex, then from the vertex three lines to crosses; the same just adding a setting sun on propagators; the vertex diagram with a circle surrounding the vertex and crossing all four propagators; and so on, as in Figure 17.4.

If we think about how to write this in general, and also what to write for a general theory instead of ϕ^4, denoting 1PI diagrams by a shaded blob, the relevant diagrams can be written in a self-consistent sort of way as: line from i to a cross; plus propagator from i to a 1PI blob with one external point; plus propagator from i to a 1PI blob with two external points,

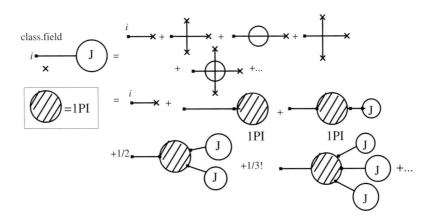

Figure 17.4 Diagrammatic expansion of the classical field. It can be reorganized into a self-consistent equation, in terms of the classical field and the 1PI n-point functions.

and the external point connected to the same classical field, plus 1/2 times the same with three-point 1PI with two external classical fields; and so on, as in Figure 17.4.

We then obtain the following self-consistency equation for the classical field:

$$
\phi_i^{cl} = \Delta_{ij} \left[J_j - \left(\Gamma_j + \Pi_{jk}\phi_k^{cl} + \frac{1}{2}\Gamma_{jkl}\phi_k^{cl}\phi_l^{cl} \right. \right.
$$
$$
\left. \left. + \ldots + \frac{1}{(n-1)!}\Gamma_{ji_1\ldots i_{n-1}}\phi_{i_1}^{cl}\ldots\phi_{i_{n-1}}^{cl} + \ldots \right) \right]. \qquad (17.12)
$$

Note that only Π_{jk} was not called Γ_{jk}, the rest of the 1PI n-point functions were called $\Gamma_{i_1\ldots i_n}$, we will see why shortly.

We define the generating functional of 1PI Green's functions

$$
\hat{\Gamma}(\phi^{cl}) = \Gamma_i\phi_i^{cl} + \frac{1}{2}\Pi_{ij}\phi_i^{cl}\phi_j^{cl} + \frac{1}{3!}\Gamma_{ijk}\phi_i^{cl}\phi_j^{cl}\phi_k^{cl} + \ldots \qquad (17.13)
$$

such that the 1PI n-point functions are given by its multiple derivatives

$$
\Gamma_{i_1\ldots i_n} = \frac{\delta^n}{\delta\phi_{i_1}^{cl}\ldots\delta\phi_{i_n}^{cl}}\hat{\Gamma}(\phi^{cl})|_{\phi^{cl}=0} \qquad (17.14)
$$

(only for two points do we have Π_{ij}). Then the self-consistency equation (17.12) is written as

$$
-\frac{\delta W}{\delta J_i} \equiv \phi_i^{cl} = \Delta_{ij}\left(J_j - \frac{\delta\hat{\Gamma}}{\delta\phi_j^{cl}}\right). \qquad (17.15)
$$

In turn, this is rewritten as

$$
\frac{\delta\hat{\Gamma}}{\delta\phi_i^{cl}} + \Delta_{ij}^{-1}\phi_j^{cl} = J_i \Rightarrow \frac{\delta}{\delta\phi_i^{cl}}\left[\hat{\Gamma}(\phi^{cl}) + \frac{1}{2}\phi_k^{cl}\Delta_{kj}^{-1}\phi_j^{cl}\right] = J_i. \qquad (17.16)
$$

This means that we can define the *effective action*

$$
\Gamma(\phi^{cl}) = \hat{\Gamma}(\phi^{cl}) + \frac{1}{2}\phi_k^{cl}\Delta_{kj}^{-1}\phi_j^{cl}, \qquad (17.17)
$$

such that we have

$$
\frac{\delta\Gamma(\phi^{cl})}{\delta\phi_i^{cl}} = J_i, \qquad (17.18)
$$

similar to the classical equation of motion

$$
\frac{\delta S}{\delta\phi_i} = J_i. \qquad (17.19)
$$

In other words, ϕ^{cl} plays the role of the field ϕ in the classical field theory, and instead of the classical action S, we have the effective action Γ, which contains all the quantum corrections. For the effective action we define

$$
\Gamma(\phi^{cl}) = \sum_{N\geq 1}\frac{1}{N!}\Gamma_{i_1\ldots i_N}\phi_{i_1}^{cl}\ldots\phi_{i_N}^{cl}, \qquad (17.20)
$$

$$\Gamma_{ij} = \text{[diagram]} + \left(\text{[diagram]} \right)^{-1}$$

The effective action is not technically diagrammatic, because of the inverse free propagator.

where the only difference from the generating functional of the 1PI diagrams is in the two-point function, where

$$\Gamma_{ij} = \Pi_{ij} + \Delta_{ij}^{-1}, \tag{17.21}$$

which does not have a diagrammatic interpretation (we have Δ^{-1}, not Δ), see Figure 17.5. Otherwise, the effective action contains only 1PI diagrams.

Theorem 17.1   The effective action is the Legendre transform of the free energy, that is

$$e^{-\Gamma(\phi^{cl})+J_i\phi_i^{cl}} = \int \mathcal{D}\phi \, e^{-S[\phi]+J_i\phi_i} = e^{-W[J]}, \tag{17.22}$$

or

$$\Gamma[\phi^{cl}] = W[J] + J_k\phi_k^{cl}. \tag{17.23}$$

Proof   We take the derivative with respect to ϕ_i^{cl} of the required relation above, and obtain

$$\frac{\delta\Gamma}{\delta\phi_i^{cl}} = J_i = \frac{\delta W}{\delta J_k}\frac{\delta J_k}{\delta\phi_i^{cl}} + \frac{\delta J_k}{\delta\phi_i^{cl}}\phi_k^{cl} + J_i = -\phi_k^{cl}\frac{\delta J_k}{\delta\phi_i^{cl}} + \frac{\delta J_k}{\delta\phi_i^{cl}}\phi_k^{cl} + J_i = J_i, \tag{17.24}$$

that is we obtain an identity, as wanted.   *q.e.d.*

The effective action contains all the information about the full quantum field theory, so in principle, if we were able to find the exact effective action, it would be equivalent to solving the quantum field theory.

17.3 The Connected Two-Point Function

The connected one-point function is

$$G_i^C[J] = -\frac{\delta W[J]}{\delta J_i} = \phi_i^{cl}[J], \tag{17.25}$$

and we obtain the connected two-point function by taking a derivative

$$G_{ij}^C[J] = -\frac{\delta}{\delta J_i}\frac{\delta}{\delta J_j}W[J] = \frac{\delta\phi_i^{cl}}{\delta J_j}. \tag{17.26}$$

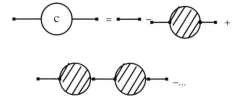

Figure 17.6   Diagrammatic equation relating the connected two-point function and the 1PI two-point function.

Then we can substitute (17.12) and write

$$G_{ij}^C[J] = \Delta_{ij} - \Delta_{ik}\Pi_{kl}\frac{\delta\phi_l^{cl}}{\delta J_j} + \ldots, \tag{17.27}$$

where the terms that were dropped vanish when $J = 0$, with the assumption that $\phi_i^{cl}[J = 0] = 0$, which is a reasonable requirement in quantum field theory (classically, at zero source we should have no field, and so quantum mechanically under the same condition we should have no field VEV, or rather, this should be removed by renormalization, which we will learn about in Part II of the book).

Then at $J = 0$, the two-point function obeys the equation

$$G_{ij}^C = \Delta_{ij} - \Delta_{ik}\Pi_{kl}G_{lj}^C, \tag{17.28}$$

which gives

$$(\delta_{ik} + \Delta_{il}\Pi_{lk})G_{kj}^c = \Delta_{ij} \Rightarrow$$
$$(\Delta_{mk}^{-1} + \Pi_{mk})G_{kj}^C = \delta_{mj}, \tag{17.29}$$

meaning that the connected two-point function is the inverse of the two-point term in the effective action:

$$\Gamma_{ik}G_{kj}^C = \delta_{ij}, \tag{17.30}$$

or explicitly:

$$G^C = (1 + \Delta\Pi)^{-1}\Delta = (1 - \Delta\Pi + \Delta\Pi\Delta\Pi - \Delta\Pi\Delta\Pi\Delta\Pi + \ldots)\Delta \Rightarrow$$
$$G_{ij}^C = \Delta_{ij} - \Delta_{il}\Pi_{lk}\Delta_{kj} + \Delta_{il}\Pi_{lk}\Delta_{km}\Pi_{mn}\Delta_{nj} - \ldots \tag{17.31}$$

So, the connected two-point function is a propagator; minus a propagator, two-point 1PI, propagator again; plus a propagator, two-point 1PI, propagator, two-point 1PI, propagator; minus..., as in Figure 17.6.

So apart from signs, it is obvious from diagrammatics.

17.4 Classical Action as Generating Functional of Tree Diagrams

In the classical $\hbar \to 0$ limit, the effective action turns into the classical action (all the quantum corrections vanish). In the same limit, all the 1PI diagrams disappear and we are left with only the vertices in the action (e.g. the four-point vertex in the case of the ϕ^4 theory). Therefore, from all the diagrams we are left with only the tree diagrams.

But note that having only tree diagrams is still *quantum mechanics*, since we still calculate amplitudes for quantum transition, probabilities, and can have interference, and so on. The point, however, is that we don't have *quantum field theory* ("*second* quantization"), but we still have quantum mechanics. For the transition amplitudes we still have quantum mechanics rules, and we know for instance that some of the early successes of quantum electrodynamics come from tree diagram calculations. The point is, of course, that when we talk about external particles, and transitions for them, that is a quantum mechanics process: the classical field is a collection of *many* particles, but if we have a single particle we use quantum mechanics.

As an example, let's consider the classical ϕ^4 theory:

$$S[\phi] = \frac{1}{2}\phi_i \Delta_{ij}^{-1}\phi_j + \frac{\lambda}{4!}\sum_i \phi_i^4, \tag{17.32}$$

with equation of motion

$$\Delta_{ij}^{-1}\phi_j + \frac{\lambda}{3!}\phi_i^3 = J_i, \tag{17.33}$$

having the solution

$$\phi_i = \Delta_{ij}J_j - \frac{\lambda}{3!}\Delta_{ij}\phi_j^3. \tag{17.34}$$

This is a self-consistent solution (like for ϕ_{cl} at the quantum level), written as the classical field (line connected with a circle with J in the middle, at classical level) = line ending in cross $-\lambda/3! \times$ line ending in vertex connected with three classical fields.

It can be solved perturbatively, by replacing the classical field first with the line ending in a cross, substituting on the right-hand side of the equation, then the right-hand side of the resulting equation is reintroduced instead of the classical field, and so on. In this way, we obtain that the classical field (at classical level) equals the line ending in a cross; $-\lambda/3! \times$ line ending in a vertex, with three lines ending in a cross; $+3(-\lambda/3!)^2 \times$ a tree with one external point and five crosses; and so on, as in Figure 17.7 (in which we have considered the more general ϕ^n interaction case). In this way we obtain all possible trees with one external point and many external crosses. This is also what we would get from ϕ^{cl} by keeping only the tree diagrams.

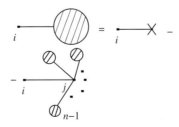

Figure 17.7 Equation satisfied by ϕ_i in ϕ^n theory, represented diagrammatically.

- The generating functional of connected diagrams is $-W[J] = \ln Z[J]$, known as the free energy by analogy with thermodynamics.
- The classical field, defined as the normalized VEV of the quantum field, is the connected one-point function in the presence of a source J.
- The generating functional of 1PI diagrams is $\hat{\Gamma}(\phi^{cl})$, and the effective action Γ is the same, just in the two-point object we need to add Δ^{-1}, that is, we have $\Gamma_{ij} = \Pi_{ij} + \Delta_{ij}^{-1}$.
- The effective action $\Gamma[\phi^{cl}]$ is the Legendre transform of the free energy $W[J]$.
- The connected two-point function at $J = 0$, G_{ij}^C, is the inverse of the two-point effective action Γ_{ij}.
- The classical action is the generating functional of tree diagrams. Tree diagrams can still be used for quantum mechanics, but we have no quantum field theory.

Further Reading

See sections 2.5 and 4.1 in [3] and sections 3.1 and 3.3 in [4].

Exercises

1. Consider the hypothetical effective action in p space (here $\Delta^{-1} = p^2$, as usual for massless scalars)

$$\Gamma(\phi^{cl}) = \int \frac{d^4p}{(2\pi)^4} \phi^{cl}(p)(p^2 + Gp^4)\phi^{cl}(-p)$$

$$+ \exp\left\{\lambda \int \left(\prod_{i=1}^{4} \frac{d^4p_i}{(2\pi)^4}\right) \delta^4(p_1 + p_2 + p_3 + p_4)\phi^{cl}\right.$$

$$\left. \times (p_1)\phi^{cl}(p_2)\phi^{cl}(p_3)\phi^{cl}(p_4)\right\}. \tag{17.35}$$

Calculate the 1PI four-point Green's function $\Gamma_{p_1 p_2 p_3 p_4}$.

2. For the model in Exercise 1, calculate the 1PI two-point function $\Pi_{p,-p}$ and the connected two-point function.

3. Consider the theory

$$S[\phi] = \frac{1}{2}\phi_i \Delta_{ij}^{-1} \phi_j + \frac{\lambda}{3!}\sum_i \phi_i^3, \tag{17.36}$$

in the presence of sources J_i. Write down the perturbative diagrams for the classical equation of motion up to (including) order λ^3.

4. Consider the hypothetical discretized partition function

$$Z[J] = \exp\left[\frac{1}{2}J_i\Delta_{ij}J_j + \lambda_{ijk}J_iJ_jJ_k\right]. \tag{17.37}$$

Calculate the connected three-point function. Restrict to the case $\lambda_{ijk} = \lambda\delta_{ij}\delta_{jk}$.

18   Dyson–Schwinger Equations and Ward Identities

In this chapter, we consider identities satisfied by the Green's functions, obtained in the path-integral formalism. We start with the Dyson–Schwinger equations, obtained as a quantum-mechanical version of the equations of motion. Iterating these equations allows us to obtain the perturbative expansion of the Green's functions. Next we review the Noether theorem, with the purpose of obtaining quantum versions for it, more precisely for the conservation of the global current. These will be the Ward identities.

18.1 Dyson–Schwinger Equations

At the classical level, we have the classical equations of motion

$$\frac{\delta S[\phi]}{\delta \phi_i} - J_i = 0. \tag{18.1}$$

But we know from Ehrenfest's theorem for quantum mechanics that in the quantum theory these classical equations should hold on the quantum average, or VEV (note that we are talking about the quantum average of the equation for the field, not the classical equation for the quantum average of the field!). In the path-integral formalism, these turn out to be the Dyson–Schwinger equations in a very easy (almost trivial) way. Historically, these equations were found in the operator formalism, and it was quite nontrivial to find them.

Consider the following trivial observation:

$$\int_{-\infty}^{+\infty} dx \frac{d}{dx} f(x) = 0 \tag{18.2}$$

if $f(\pm\infty) = 0$, and generalize it to the path-integral case. In the path integral in Euclidean space, we have something similar, just that the boundary condition is even simpler: instead of fields going to zero at $t = \pm\infty$, we use periodic boundary conditions (Euclidean time is periodic, namely is a circle of radius $R \to \infty$), hence we have

$$0 = \int \mathcal{D}\phi \frac{\delta}{\delta \phi_i} e^{-S[\phi] + J \cdot \phi}. \tag{18.3}$$

Writing explicitly, we have

$$
0 = \int \mathcal{D}\phi \left[-\frac{\delta S[\phi]}{\delta \phi_i} + J_i \right] e^{-S[\phi] + J \cdot \phi}
$$

$$
= \left[-\frac{\delta S}{\delta \phi_i} \bigg|_{\phi = \frac{\delta}{\delta J_i}} + J_i \right] Z[J], \tag{18.4}
$$

where in the second equality we used the usual fact that we have, for example, $\int \mathcal{D}\phi \, \phi_i e^{-S[\phi] + J \cdot \phi} = \int \mathcal{D}\phi (\delta/\delta J_i) e^{-S[\phi] + J \cdot \phi} = \delta/\delta J_i Z[J]$.

This equation is the Dyson–Schwinger equation. Now it appears trivial, but it was originally derived from an analysis of Feynman diagrams, where it looks nontrivial. Now, we will derive the relation for Feynman diagrams from this.

To see it, we first specialize for a theory with quadratic kinetic term, which is a pretty general requirement. Therefore consider

$$
S[\phi] = \sum_{ij} \frac{1}{2} \phi_i \Delta_{ij}^{-1} \phi_j + S_I[\phi]. \tag{18.5}
$$

Then we obtain (by differentiating and multiplying by Δ_{li})

$$
\sum_i \Delta_{li} \frac{\delta S[\phi]}{\delta \phi_i} = \phi_l + \sum_i \Delta_{li} \frac{\delta S_I[\phi]}{\delta \phi_i}, \tag{18.6}
$$

and substituting it in the Dyson–Schwinger equation (18.4), we get

$$
\left[-\frac{\delta}{\delta J_l} - \sum_i \Delta_{li} \frac{\delta S_I}{\delta \phi_i} \bigg|_{\phi = \frac{\delta}{\delta J}} + \sum_i \Delta_{li} J_i \right] Z[J] = 0, \tag{18.7}
$$

so that we get the Dyson–Schwinger equation for $Z[J]$:

$$
\frac{\delta}{\delta J_l} Z[J] = \sum_i \Delta_{li} J_i Z[J] - \sum_i \Delta_{li} \frac{\delta S_I[\phi]}{\delta \phi_i} \bigg|_{\phi = \frac{\delta}{\delta J}} Z[J]. \tag{18.8}
$$

From this, we can derive a Dyson–Schwinger equation for the full Green's functions (with a diagrammatic interpretation) if we choose a particular interaction term.

18.1.1 Specific Interaction

We consider a ϕ^3 plus ϕ^4 interaction

$$
S_I[\phi] = \frac{g_3}{3!} \sum_i \phi_i^3 + \frac{g_4}{4!} \sum_i \phi_i^4. \tag{18.9}
$$

Then the Dyson–Schwinger equation (18.8) becomes

$$
\frac{\delta}{\delta J_l} Z[J] = \sum_i \Delta_{li} J_i Z[J] - \sum_i \Delta_{li} \left[\frac{g_3}{2!} \frac{\delta^2}{\delta J_i \delta J_i} + \frac{g_4}{3!} \frac{\delta^3}{\delta J_i \delta J_i \delta J_i} \right] Z[J]. \tag{18.10}
$$

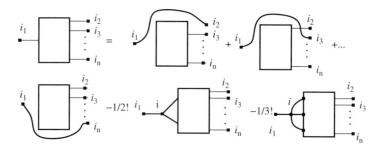

Figure 18.1 Dyson–Schwinger equation for the n-point function for the ϕ^3 plus ϕ^4 theory.

We put $l = i_1$ and then take $(\delta/\delta J_{i_2}) \ldots (\delta/\delta J_{i_n})$ and finally put $J = 0$ to obtain a relation between the (full) Green's functions

$$
\begin{aligned}
G^{(n)}_{i_1 \ldots i_n} = \; & \Delta_{i_1 i_2} G^{(n-2)}_{i_3 \ldots i_n} + \Delta_{i_1 i_3} G^{(n-2)}_{i_2 i_4 \ldots i_n} + \cdots \\
& + \Delta_{i_1 i_n} G^{(n-2)}_{i_2 i_3 \ldots i_{n-1}} - \sum_i \Delta_{i_1 i} \left[\frac{g_3}{2!} G^{(n+1)}_{i i i_2 \ldots i_n} + \frac{g_4}{3!} G^{(n+2)}_{i i i i_2 \ldots i_n} \right].
\end{aligned} \tag{18.11}
$$

We can write a diagrammatic form of this equation. The full Green's function is represented by a box with n external points, $i_1, \ldots, i_n$. We need to choose one special point, i_1 above, and we can write it on the left of the box, and the other on the right of the box. There are $n - 1$ terms where we connect i_1 with one of the other points by a propagator, disconnected from a box with the remaining $n - 2$ external points, minus $1/2! \times$ a term where the propagator from i_1 reaches first an i, where it splits into two before reaching the box, minus $1/3! \times$ a term where the propagator from i_1 reaches first an i where it splits into three before reaching the box, see Figure 18.1 for more details.

18.2 Iterating the Dyson–Schwinger Equation

The whole perturbative expansion can be obtained by *iterating* the Dyson–Schwinger equation repeatedly.

18.2.1 Example

We will see this in the example of the two-point function in the above theory. We first write the Dyson–Schwinger equation for it as above:

$$
G^{(2)}_{i_1 i_2} = \Delta_{i_1 i_2} G^{(0)} - \frac{g_3}{2!} \sum_i \Delta_{i_1 i} G^{(3)}_{i i i_2} - \frac{g_4}{3!} \sum_i \Delta_{i_1 i} G^{(4)}_{i i i i_2}, \tag{18.12}
$$

and then substitute in it the Dyson–Schwinger equations for $G^{(3)}_{i i i_2}$ and $G^{(4)}_{i i i i_2}$, with the first index on each considered as the special one (note that we cannot write a Dyson–Schwinger equation for $G^{(0)}$):

iterate:

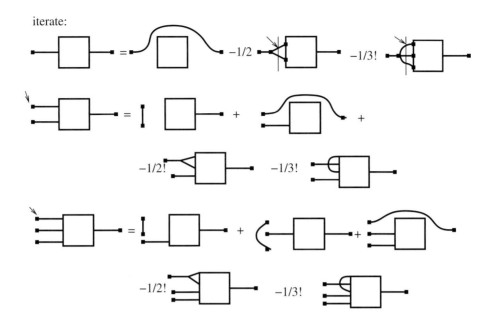

Ingredients for iterating the Dyson–Schwinger equation (step 1). Consider the first iteration for the two-point function, and choose a specific leg on the right-hand side. The Dyson–Schwinger equations for the three-point and four-point functions with a special leg as chosen above are written below.

$$
G^{(3)}_{iii_2} = \Delta_{ii} G^{(1)}_{i_2} + \Delta_{ii_2} G^{(1)}_{i} - \frac{g_3}{2!} \sum_j \Delta_{ij} G^{(4)}_{jjii_2} - \frac{g_4}{3!} \sum_j \Delta_{ij} G^{(5)}_{jjjii_2},
$$

$$
G^{(4)}_{iiii_2} = \Delta_{ii} G^{(2)}_{ii_2} + \Delta_{ii} G^{(2)}_{ii_2} + \Delta_{ii_2} G^{(2)}_{ii} - \frac{g_3}{2!} \sum_j \Delta_{ij} G^{(5)}_{jjiii_2} - \frac{g_4}{3!} \sum_j \Delta_{ij} G^{(6)}_{jjjiii_2}. \tag{18.13}
$$

These are represented in Figure 18.2.
We then obtain

$$
G^{(2)}_{ij} = \Delta_{ij} G^{(0)} - \frac{g_3}{2!} \sum_i \Delta_{i_1 i} \left[\Delta_{ii} G^{(1)}_{i_2} + \Delta_{ii_2} G^{(1)}_{i} - \frac{g_3}{2!} \sum_j \Delta_{ij} G^{(4)}_{jjii_2} - \frac{g_4}{3!} \sum_j \Delta_{ij} G^{(5)}_{jjjii_2} \right]
$$

$$
- \frac{g_4}{3!} \sum_i \Delta_{i_1 i} \left[\Delta_{ii} G^{(2)}_{ii_2} + \Delta_{ii} G^{(2)}_{ii_2} + \Delta_{ii_2} G^{(2)}_{ii} \right.
$$

$$
\left. - \frac{g_3}{2!} \sum_j \Delta_{ij} G^{(5)}_{jjiii_2} - \frac{g_4}{3!} \sum_j \Delta_{ij} G^{(6)}_{jjjiii_2} \right]. \tag{18.14}
$$

This is represented in Figure 18.3. By iterating it further, we can obtain the whole perturbative expansion.

Let us consider the expansion to order $(g_3)^0 (g_4)^1$ of the above second iteration. Since, for $G^{(0)}$, the nontrivial terms need to have at least two vertices, it only contributes to order

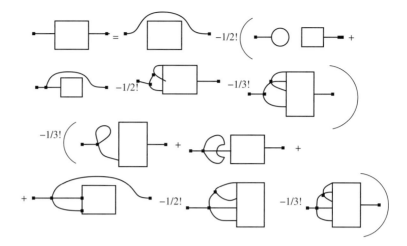

Figure 18.3   The second iteration of the Dyson–Schwinger equation, using the ingredients from step 1.

Figure 18.4   The result of iterating the Dyson–Schwinger equation for the two-point function, at order $(g_3)^0(g_4)^1$.

0+1 (i.e. as $= 1 + g_4 \Delta_{kk} \Delta_{kk}$). Also, the first square bracket does not contribute, since all the terms have at least one g_3 in them. In the second bracket, the last two terms are of orders $(g_3)^1(g_4)^1$ and $(g_4)^2$, respectively, so also don't contribute. Thus, besides the trivial propagator term, we only have the first three terms in the second bracket contributing. But in it, the iterated two-point functions must be trivial to contribute to order $(g_3)^0(g_4)^1$ to $G_{ij}^{(2)}$ (i.e. must be replaced by 1).

We then obtain that to order $(g_3)^0(g_4)^1$, the two-point function $G_{ij}^{(2)}$ is the propagator Δ_{ij} plus three times $1/3! \times$ the propagator self-correction, that is

$$G_{ij}^{(2)} = \Delta_{ij}(1 + g_4 \Delta_{kk} \Delta_{kk}) + \frac{3g_4}{3!} \sum_k \Delta_{ik} \Delta_{kk} \Delta_{kj} + \dots \tag{18.15}$$

The connected part of this is written diagrammatically in Figure 18.4.

Note that in this way we have obtained the correct symmetry factor for the one-loop diagram, $S = 2$, since $1/S = 3/3!$. This means that the iterated Dyson–Schwinger equation is an algorithmic way to compute the symmetry factors in cases where they are very complicated and we are not sure of the algebra.

Note also that we wrote Dyson–Schwinger equations for the full Green's functions, but we can also write such equations for the connected and 1PI Green's functions, and these are also iterative and self-consistent. We will, however, not do it here.

We next turn to a study of symmetries and Ward identities.

18.3 Noether's Theorem

First, we will review Noether's theorem. Consider a global symmetry

$$\delta\phi^i = \epsilon^a(iT^a)_{ij}\phi^j, \tag{18.16}$$

where ϵ^a are constant symmetry parameters. We have

$$0 = \delta S = \int d^4x \left[\frac{\partial\mathcal{L}}{\partial\phi^i}\delta\phi^i + \frac{\partial\mathcal{L}}{\partial(\partial_\mu\phi^i(x))}\partial_\mu\delta\phi^i \right]$$

$$= \int d^4x \left[\left(\frac{\partial\mathcal{L}}{\partial\phi^i} - \partial_\mu\left(\frac{\partial\mathcal{L}}{\partial(\partial_\mu\phi^i(x))}\right) \right)\delta\phi^i + \partial_\mu\left(\frac{\partial\mathcal{L}}{\partial(\partial_\mu\phi^i(x))}\delta\phi^i\right) \right]. \tag{18.17}$$

If the equations of motion are satisfied, the first bracket is zero, and then substituting for $\delta\phi^i$ the transformation under the global symmetry, we obtain that classically (if the equations of motion are satisfied)

$$(\delta\mathcal{L})^{\text{class}}_{\text{symm}} = i\sum_a \epsilon^a\partial_\mu\left[\sum_{ij}\frac{\partial\mathcal{L}}{\partial(\partial_\mu\phi^i(x))}(T^a)_{ij}\phi^j(x)\right]. \tag{18.18}$$

We then obtain that the current is

$$j^a_\mu = \sum_{ij}\frac{\partial\mathcal{L}}{\partial(\partial^\mu\phi^i(x))}(T^a)_{ij}\phi^j(x), \tag{18.19}$$

and is classically *conserved on-shell*, that is

$$\partial^\mu j^a_\mu = 0, \tag{18.20}$$

which is the statement of the Noether theorem.

Since $Q^a = \int d^3x j^a_0$, current conservation means that $dQ/dt = 0$, that is the charge is conserved in time.

Under a symmetry transformation, but off-shell (*if the equations of motion are not satisfied*), we still have $\delta S = 0$, which in turn means that

$$0 = \delta\mathcal{L} = i\epsilon^a(T^a)_{ij}\left[\left(\frac{\partial\mathcal{L}}{\partial\phi^i} - \partial_\mu\left(\frac{\partial\mathcal{L}}{\partial(\partial_\mu\phi^i(x))}\right)\right)\phi^j + \partial_\mu\left(\frac{\partial\mathcal{L}}{\partial(\partial_\mu\phi^i(x))}\phi^j\right)\right]. \tag{18.21}$$

We now promote the global transformation (18.16) to a local one, with parameter $\epsilon^a(x)$. Obviously then, it is not a symmetry anymore, so under it, δS is not zero anymore, but rather

$$\delta S = \int d^4x i\epsilon^a(x)(T^a)_{ij}\left[\left(\frac{\partial\mathcal{L}}{\partial\phi^i} - \partial_\mu\left(\frac{\partial\mathcal{L}}{\partial(\partial_\mu\phi^i(x))}\right)\right)\phi^j + \partial_\mu\left(\frac{\partial\mathcal{L}}{\partial(\partial_\mu\phi^i(x))}\phi^j\right)\right]$$

$$+ i\int d^4x \sum_{a,i,j}(\partial_\mu\epsilon^a)(T^a)_{ij}\left[\frac{\partial\mathcal{L}}{\partial(\partial_\mu\phi^i(x))}\phi^j(x)\right]$$

$$= i\sum_a\int d^4x(\partial^\mu\epsilon^a(x))j^a_\mu(x) = -i\sum_a\int d^4x\epsilon^a(x)(\partial^\mu j^a_\mu(x)), \tag{18.22}$$

where in the second equality we have used (18.21) and in the last step we have assumed that $\epsilon^a(x)$ "vanish at infinity," so that we can partially integrate without boundary terms.

Therefore we have the variation under the local version of the global symmetry

$$\delta S = i \sum_a \int d^4x (\partial^\mu \epsilon^a(x)) j_\mu^a(x) = -i \sum_a \int d^4x \epsilon^a(x)(\partial^\mu j_\mu^a(x)). \tag{18.23}$$

This formula is valid off-shell as we said, so we can use it inside the path integral. This gives a way to identify the current as the coefficient of $\partial_\mu \epsilon^a$ in the variation of the action under the local version of the global symmetry.

18.4 Ward Identities

Since classically (on the classical equations of motion) we have $\partial^\mu j_\mu^a = 0$, by Ehrenfest's theorem we expect to have $\partial^\mu j_\mu^a = 0$ as a VEV (as a quantum average), and that will be the Ward identity.

However, it can happen that there are *quantum anomalies*, meaning that the classical symmetry is not respected by quantum corrections, and we then get that the classical equation $\partial^\mu j_\mu^a = 0$ is not satisfied as an average at the quantum level.

We consider the local version of the symmetry transformation

$$\delta \phi^i(x) = i \sum_{a,j} \epsilon^a(x)(T^a)_{ij}\phi^j(x), \tag{18.24}$$

where $\phi' = \phi + \delta\phi$. Changing integration variables from ϕ to ϕ' (renaming the variable, really) does nothing, so

$$\int \mathcal{D}\phi' e^{-S[\phi']} = \int \mathcal{D}\phi e^{-S[\phi]}. \tag{18.25}$$

However, now comes a crucial assumption: *if the Jacobian from $\mathcal{D}\phi'$ to $\mathcal{D}\phi'$ is 1*, then $\mathcal{D}\phi' = \mathcal{D}\phi$, so

$$0 = \int \mathcal{D}\phi \left[e^{-S[\phi']} - e^{-S[\phi]}\right] = -\int \mathcal{D}\phi \delta S[\phi] e^{-S[\phi]}. \tag{18.26}$$

This assumption, however, is not true in general. In the path-integral formalism, quantum anomalies appear exactly as anomalous Jacobians for the change of variables in the measure, and lead to breakdown of the symmetries at the quantum level. However, these will be described in the second part of the book, so we will not discuss them further here.

Now we can use the previously derived form for δS under a local version of a global symmetry, (18.23), obtaining

$$0 = \int d^4x i\epsilon^a(x) \int \mathcal{D}\phi e^{-S[\phi]} \partial^\mu j_\mu^a(x). \tag{18.27}$$

But since the parameters $\epsilon^a(x)$ are arbitrary, we can also derive that

$$\int \mathcal{D}\phi e^{-S[\phi]} \partial^\mu j_\mu^a(x) = 0, \tag{18.28}$$

which is indeed the quantum-averaged version of the conservation law, namely the Ward identity.

Note that here, j_μ^a is a function of ϕ also (the object integrated over), so it cannot be taken out of the path integral (this is an equation of the type $\int dx f(x) = 0$, from which nothing further can be deduced).

We can derive more general Ward identities by considering a general operator $A = A(\{\phi^i\})$, such that

$$\delta A(\{\phi^i\}) = \int d^4x \frac{\delta A(\{\phi^i\})}{\delta \epsilon^a(x)} \epsilon^a(x). \tag{18.29}$$

Then analogously we get

$$0 = \int \mathcal{D}\phi \, \delta \left[e^{-S[\phi]} A(\{\phi^i\}) \right] = i \int d^4x \epsilon^a(x) \int \mathcal{D}\phi \, e^{-S[\phi]} \left[\partial^\mu j_\mu^a(x) A - i \frac{\delta A(\{\phi^i\})}{\delta \epsilon^a(x)} \right], \tag{18.30}$$

so that we have the *more general Ward identities*

$$\int \mathcal{D}\phi \, e^{-S[\phi]} (\partial^\mu j_\mu^a(x)) A = i \int \mathcal{D}\phi \, e^{-S[\phi]} \frac{\delta A(\{\phi^i\})}{\delta \epsilon^a(x)}. \tag{18.31}$$

There are also other forms of Ward identities we can write. For instance, choosing $A = e^{J \cdot \phi}$, or in other words adding a current J, we obtain

$$0 = \int \mathcal{D}\phi \frac{\delta}{\delta \epsilon^a(x)} e^{-S+J\cdot\phi} = i(T^a)_{ij} \int \mathcal{D}\phi \left[-\frac{\delta S}{\delta \phi^i(x)} \phi^j(x) + J_i \phi^j(x) \right] e^{-S+J\cdot\phi}. \tag{18.32}$$

Now using the usual trick of replacing ϕ^i inside the path integral with $\delta/\delta J$, we obtain the Ward identities

$$i(T^a)_{ij} \left[-\frac{\delta S}{\delta \phi^i(x)} \left\{ \phi = \frac{\delta}{\delta J} \right\} + J_i(x) \right] \frac{\delta}{\delta J_j(x)} Z[J] = 0. \tag{18.33}$$

Taking further derivatives with respect to J and then putting $J = 0$, we obtain Ward identities between the Green's functions, in the same way as happened with the Dyson–Schwinger equation. Also in this case we can write Ward identities for connected or 1PI Green's functions as well.

Ward identities play a very important role in the discussion of symmetries, since they constrain the form of the Green's functions.

For example, we will see later that in quantum electrodynamics, the transversality condition $k^\mu A_\mu = 0$, coming from the Lorenz gauge condition, that fixes (part of) the (local) gauge invariance, gives a similar constraint on Green's functions. In particular, for the 1PI two-point function $\Pi_{\mu\nu}$ we can write

$$k^\mu \Pi_{\mu\nu} = 0, \tag{18.34}$$

which implies that the Lorentz structure of $\Pi_{\mu\nu}$ is fixed, namely

$$\Pi_{\mu\nu}(k) = (k^2 \delta_{\mu\nu} - k_\mu k_\nu) \Pi(k^2). \tag{18.35}$$

This constraint is then a local analogue of the Ward identities for global symmetries.

Important Concepts to Remember

- The Dyson–Schwinger equation is the quantum version of the classical equation of motion $\delta S/\delta\phi^i = J_i$, namely its quantum average (under the path integral).
- It can be written as an operator acting on $Z[J]$, and from it we can deduce a relation between Green's functions, for ϕ^k interactions relating $G^{(n)}$ with $G^{(n-2)}$ and $G^{(n+k-2)}$.
- By iterating the Dyson–Schwinger equation we can get the full perturbative expansion, with the correct symmetry factors. It therefore gives an algorithmic way to calculate the symmetry factors.
- The Noether conservation gives on-shell conservation of the current j^a_μ associated with a global symmetry.
- By making $\epsilon \to \epsilon(x)$, the variation of the action is $\delta S = i\int(\partial^\mu\epsilon^a)j^a_\mu$.
- Ward identities are quantum-averaged versions of the classical (on-shell) current conservation equations $\partial^\mu j^a_\mu = 0$, perhaps with an operator A inserted, and its variation on the right-hand side. However, there can be quantum anomalies which spoil it, manifesting themselves as anomalous Jacobians for the transformation of the path-integral measure.
- Ward identities can also be written as an operator acting on $Z[J]$, and from it deriving relations between Green's functions.

Further Reading

See sections 4.2 and 4.3 in [3], section 3.1 in [6], and section 9.6 in [1].

Exercises

1. Consider the scalar theory with interaction term

$$S_I(\phi) = \frac{g_5}{5!}\sum_i \phi_i^5. \tag{18.36}$$

Write down the Dyson–Schwinger equation for the n-point function in this theory (equation and diagrammatic). Iterate it once more for the two-point function.

2. Consider the Euclidean action for N real scalars ϕ_i:

$$S = \int d^dx \left[\frac{1}{2}\sum_{i=1}^N(\partial_\mu\phi_j)^2 + \frac{\lambda}{4!}\left(\sum_i\phi_i^2\right)^2\right], \tag{18.37}$$

invariant under rotations of the N scalars. Write down the explicit forms of the two types of Ward identities:

- for $A = \sum_i(\phi_i)^2$;
- for the partition function $Z[J]$.

3. Consider a system of N massless abelian vectors, with action

$$S = -\frac{1}{4} \int d^4x \sum_{I=1}^{N} (F^I_{\mu\nu})^2.$$ (18.38)

   Write the Noether current associated with the continuous global symmetry, and the corresponding Ward identity.

4. Consider the sine-Gordon action in four dimensions:

$$S = \int d^4x \left[-\frac{1}{2}(\partial_\mu\phi)^2 - V_0[1 - \cos(a\phi)] \right].$$ (18.39)

   Write down the Dyson–Schwinger equation for this model.

19

Cross-Sections and the S-Matrix

In this chapter, we turn to experimentally measurable quantities, cross-sections and decay rates, and their relations to S-matrices and amplitudes.

We have already said that the S-matrix is defined roughly as $S_{fi} = \langle f|S|i \rangle$, with $S = U_I(+\infty, -\infty)$ (note that here the evolution operator is in the interaction picture; however, in this chapter we will use the Heisenberg picture, in the next one we will use the interaction picture), and that the LSZ formula relates S-matrices to residues at all the $p_i^2 = -m_i^2$ external line poles of the momentum space Green's functions $\tilde{G}_n(p_1, \dots, p_n)$. Now we define these facts better and relate them to experimental observables, in particular the cross-section and decay rates.

19.1 Cross-Sections and Decay Rates

In almost all experiments, we scatter two objects, usually projectiles off a target (in the "laboratory reference frame"), or (like at the LHC collider at CERN) in a center of mass frame, collide two particles off each other.

If we have a box of target particles A, with density ρ_A and length l_A along the direction of impact, and a box of projectiles B, with density ρ_B and length l_B along the direction of impact, and with common cross-sectional area (area of impact) A, as in Figure 19.1, the number of scattering events (when the projectiles are scattered by the target) is easily seen to be proportional to all these, ρ_A, l_A, ρ_B, l_B, and A. We can then define the *cross-section* σ as

$$\sigma = \frac{\text{No. scattering events}}{\rho_A l_A \rho_B l_B A}. \tag{19.1}$$

This is the definition used in some books, like for instance by Peskin and Schroeder [1], but we have more intuitive ways to define it. First, let us notice that the dimension of σ is $[\sigma] = 1/[(L/L^3)(L/L^3)L^2] = L^2$ (i.e. that of area). The physical interpretation of the cross-section is the effective (cross-sectional) area of interaction per target particle, that is within such an area around one target, incident projectiles are scattered.

The cross-section should be familiar from classical mechanics, where one can do the same. For instance, in the famous Rutherford experiment, proving that the atoms are not continuous media (the "jellium model"), but rather a nucleus surrounded by empty space and then orbiting electrons, one throws charged particles (e.g. electrons) against a target, and observes the scattered objects. The simple model which agrees with the

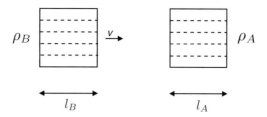

Figure 19.1 Scattering usually involves shooting a moving projectile at a fixed target.

experiment ("Rutherford scattering") is: one scatters the classical pointlike particle off a central potential $V(r) = Ze^2/r$, and calculates the classical cross-section.

The cross-section definition for scattering *on one single target* in that case is, more intuitively:

$$\sigma = \frac{\Delta N_{scatt}/\Delta t}{\phi_0} = \frac{\Delta N_{scatt}/\Delta t}{\Delta N_{in}/(\Delta tA)} = \frac{\Delta N_{scatt}}{n_B}, \tag{19.2}$$

that is the ratio of scattered particles per time, over the incident flux, or particles per time and unit area. Here, n_B is the number of incident particles per area. Also, the flux is

$$\phi_0 = \frac{\Delta N_{in}}{\Delta tA} = \frac{\rho_B(v\Delta t)A}{\Delta tA} = \rho_B v, \tag{19.3}$$

since the incident volume in time Δt is $(v\Delta t)A$.

Now consider the case of N targets, like above, where $N = \rho_A l_A A$. Then to find the cross-section (defined *per target*) we must divide by N, getting

$$\sigma = \frac{\Delta N_{scatt}/\Delta t}{\phi_0 N} = \frac{\Delta N_{scatt}}{(v\Delta t)\rho_B(\rho_A l_A A)} = \frac{\Delta N_{scatt}}{\rho_B l_B \rho_A l_A A}, \tag{19.4}$$

as above. Note that, in general, if both the target and projectiles are moving, v refers to the relative velocity, $v = v_{rel} = |\vec{v}_1 - \vec{v}_2|$.

The above *total* cross-section is a useful quantity, but in experiments we generally measure the momenta (or directions, at least) of scattered particles as well, so a more useful quantity is the *differential cross-section*, given momenta $\vec{p}_1, \ldots, \vec{p}_2$ (almost) well defined in the final state:

$$\frac{d\sigma}{d^3p_1 \ldots d^3p_n}. \tag{19.5}$$

An important case, for instance for the classical Rutherford scattering above, is of $n = 2$ (i.e. two final states). This means six momentum components, constrained by the 4-momentum conservation delta functions, leaving two independent variables, which can be taken to be two angles θ, ϕ. In the case of Rutherford scattering, these were the scattering angle θ relative to the incoming momentum, and the angle ϕ of rotation around the incoming momentum. Together, these two angles form a solid angle Ω. Therefore, in the case of $2 \rightarrow 2$ scattering, after using 4-momentum conservation, one defines the differential cross-section

$$\frac{d\sigma}{d\Omega}. \tag{19.6}$$

19.1.1 Decay Rate

There is one more process that is useful, besides the $2 \to n$ scattering.

Consider the decay of an unstable particle at rest into a final state made up of several particles. Then we define the decay rate

$$\Gamma = \frac{\text{\# decays/time}}{\text{\# particles}} = \frac{dN}{Ndt}. \tag{19.7}$$

The lifetime τ of a particle with decay rates Γ_i into various channels is

$$\sum_i \Gamma_i = \frac{1}{\tau}. \tag{19.8}$$

Consider an unstable atomic state in nonrelativistic quantum mechanics. In those cases, we know that the decay of the unstable state is found as a Breit–Wigner resonance in the scattering amplitude, that is

$$f(E) \propto \frac{1}{E - E_0 + i\Gamma/2}, \tag{19.9}$$

which means that the cross-section for decay, or probability, is proportional to $||^2$ of the above:

$$\sigma \propto \frac{1}{(E - E_0)^2 + \Gamma^2/4}. \tag{19.10}$$

That is the resonance appears as a bell curve with width Γ, centered around E_0.

In relativistic quantum mechanics, or quantum field theory, the same happens. Initial particles combine to form an unstable (or metastable) particle, which then decays into others. In the amplitude, this is reflected in a relativistically invariant generalization of the Breit–Wigner formula, namely an amplitude proportional to

$$\frac{1}{p^2 + m^2 - im\Gamma} \simeq \frac{-1}{2E_p(p^0 - E_p + \frac{im}{E_p}\frac{\Gamma}{2})}, \tag{19.11}$$

so of the same form near the mass shell and for Γ small.

19.2 In and Out States, the S-Matrix, and Wavefunctions

In quantum mechanics, we first need to define the states that scatter.

- We can consider states whose wavepackets are well isolated at $t = -\infty$, so we can then consider them noninteracting. But these (Heisenberg) states will overlap, and therefore interact at finite t. We call these Heisenberg states the in states:

$$|\{\vec{p}_i\}\rangle_{in}. \tag{19.12}$$

- We can also consider states whose wavepackets are well isolated at $t = +\infty$, so we can consider them noninteracting there, but they will overlap at finite t. We call these Heisenberg states the out states:

$$|\{\vec{p}_i\}\rangle_{out}. \tag{19.13}$$

Then the in states, after the interaction, at $t = +\infty$, look very complicated, and reversely, the out states, tracked back to $-\infty$, look very complicated. But *all* the in states, as well as *all* the out states, form complete sets:

$$\sum |\{\vec{p}_i\}; in\rangle \langle \{\vec{p}_i\}; in| = \sum |\{\vec{p}_i\}; out\rangle \langle \{\vec{p}_i\}; out| = 1. \tag{19.14}$$

This means that we can expand one in the other, so the amplitude for an (isolated) out state, given an (isolated) in state is

$$\begin{aligned}
_{out}\langle \vec{p}_1, \vec{p}_2, \ldots | \vec{k}_A, \vec{k}_B \rangle_{in} &= \lim_{T \to \infty} \langle \vec{p}_1, \vec{p}_2, \ldots (T) | \vec{k}_A, \vec{k}_B(-T) \rangle \\
&= \lim_{T \to \infty} \langle \vec{p}_1, \vec{p}_2, \ldots | e^{-iH(2T)} | \vec{k}_A, \vec{k}_B \rangle,
\end{aligned} \tag{19.15}$$

where in the last form the states are defined at the same time, which we can define as the time when Heisenberg=Schrödinger, thus think of these states as Schrödinger states.

Then the *S-matrix* is

$$\langle \vec{p}_1, \vec{p}_2, \ldots | S | \vec{k}_A, \vec{k}_B \rangle = {}_{out}\langle \vec{p}_1, \vec{p}_2, \ldots | \vec{k}_A, \vec{k}_B \rangle_{in}, \tag{19.16}$$

which means that

$$S = e^{-iH(2T)} \tag{19.17}$$

is the *Heisenberg picture evolution operator*.

19.2.1 Wavefunctions

The one-particle states are always isolated, so in that case

$$|\vec{p}_{in}\rangle = |\vec{p}_{out}\rangle = |\vec{p}\rangle \left(= \sqrt{2E_p} a_p^\dagger |0\rangle \right), \tag{19.18}$$

where in brackets we wrote the free theory result, and we can construct the one-particle states with wavefunctions

$$|\phi\rangle = \int \frac{d^3k}{(2\pi)^3} \frac{1}{\sqrt{2E_k}} \phi(\vec{k}) |\vec{k}\rangle. \tag{19.19}$$

With this normalization, the wavefunctions give the probabilities, since

$$\int \frac{d^3k}{(2\pi)^3} |\phi(\vec{k})|^2 = 1, \tag{19.20}$$

and so $\langle \phi | \phi \rangle = 1$. A wavefunction can be something like $e^{i\vec{k}\cdot\vec{x}}$, giving an $\vec{x}$ dependence.

In the case of two-particle states, we can write the in states as

$$|\phi_A \phi_B\rangle_{in} = \int \frac{d^3k_A}{(2\pi)^3} \int \frac{d^3k_B}{(2\pi)^3} \frac{\phi_A(\vec{k}_A)\phi_B(\vec{k}_B)}{\sqrt{2E_A}\sqrt{2E_B}} e^{-i\vec{b}\cdot\vec{k}_B} |\vec{k}_A \vec{k}_B\rangle_{in}. \tag{19.21}$$

Note that we could have absorbed the factor $e^{-i\vec{k}_B \cdot \vec{b}}$ in the B wavefunction, but we have written it explicitly since if the wavefunctions are centered around a momentum, like in a classical case, we have a separation between particles. In the Rutherford experiment, the imaginary line of the nondeflected incoming projectile passes at a $\vec{b}$ minimum distance close to the target. Note then that $\vec{b}$ is perpendicular to the collision direction (i.e. is transversal).

The out state of several momenta is defined as usual:

$$_{out}\langle \phi_1, \phi_2 \dots | = \prod_f \left(\frac{d^3 p_f}{(2\pi)^3} \frac{\phi_f(p_f)}{\sqrt{2E_f}} \right) {}_{out}\langle p_1 p_2 \dots |. \tag{19.22}$$

The S-matrix of states with wavefunctions is

$$S_{\beta\alpha} = \langle \beta_{out} | \alpha_{in} \rangle. \tag{19.23}$$

Because this matrix gives probabilities, and the S operator is, as we saw above, an evolution operator, it corresponds to a unitary operator, so

$$SS^\dagger = S^\dagger S = \mathbf{1}. \tag{19.24}$$

But the operator S contains the case where particles go on without interacting (i.e. the identity $\mathbf{1}$). To take that out, we define

$$S = 1 + iT. \tag{19.25}$$

Note that the i is conventional. Moreover, amplitudes always contain a momentum conservation delta function. Therefore, in order to define nonsingular and finite objects, we define the *invariant (or reduced) matrix element* $\mathcal{M}$ by

$$\langle \vec{p}_1, \vec{p}_2, \dots | iT | \vec{k}_A, \vec{k}_B \rangle = (2\pi)^4 \delta^4 \left(k_A + k_B - \sum p_f \right) i\mathcal{M}(k_A, k_B \to p_f). \tag{19.26}$$

19.3 The Reduction Formula (Lehmann, Symanzik, Zimmermann)

The LSZ formula relates S-matrices to Green's functions, as mentioned. We will not prove it here, since for that we need renormalization, which will be carried out in Part II of the book. We will just state it.

We also introduce an operator $\mathcal{A}$ to make the formula a bit more general, but this is not really needed. Define the momentum-space Green's functions

$$\tilde{G}_{n+m}^{(A)}(p_i^\mu, k_j^\mu) = \int \prod_{i=1}^n \int d^4 x_i e^{-ip_i \cdot x_i} \prod_{j=1}^m \int d^4 y_j e^{ik_j \cdot y_j}$$

$$\times \langle \Omega | T\{\phi(x_1) \dots \phi(x_n) \mathcal{A}(0) \phi(y_1) \dots \phi(y_m)\} | \Omega \rangle. \tag{19.27}$$

Then we have

$$_{out}\langle\{p_i\}n|\mathcal{A}(0)|\{k_j\}_m\rangle_{in}$$

$$= \lim_{p_i^2\to -m_i^2, k_j^2\to -m_j^2} \frac{1}{(-i\sqrt{Z})^{m+n}} \prod_{i=1}^{n}(p_i^2 + m^2 - i\epsilon)\prod_{j=1}^{m}(k_j^2 + m^2 - i\epsilon)\tilde{G}_{n+m}^{(A)}(p_i^\mu, k_j^\mu).$$

$$(19.28)$$

For $\mathcal{A} = 1$, we obtain a formula for the S-matrix as the multiple residue at all the external line poles of the momentum-space Green's functions, dividing by the extra factors of Z.

The full two-point function behaves near the pole as

$$G_2(p) = \int d^4x e^{-ip\cdot x}\langle\Omega|T\{\phi(x)\phi(0)\}|\Omega\rangle \sim \frac{-iZ}{p^2 + m^2 - i\epsilon}. \qquad (19.29)$$

In other words, to find the S-matrix, we put the external lines on shell, and divide by the full propagators corresponding to all the external lines (but note that Z belongs to two external lines, hence the $\sqrt{Z}$). This implies a diagrammatic procedure called *amputation*, which will be explained in Chapter 20.

Note that the factor Z has a kind of physical interpretation, since we can define it as

$$Z \equiv |\langle\Omega|\phi(0)|\vec{p}\rangle|^2. \qquad (19.30)$$

In other words, it is the probability of creating a state from the vacuum. Note that the factor $Z = 1 + \mathcal{O}(g^2)$, but the "correction" is an infinite loop correction (this is the oddity of renormalization, to be understood in Part II of the book). However, at tree level, $Z = 1$.

19.4 Cross-Sections from Amplitudes $\mathcal{M}$

The probability of going from $|\phi_A\phi_B\rangle$ to a state within the infinitesimal interval $d^3p_1 \ldots d^3p_n$ is

$$\mathcal{P}(AB \to 12\ldots n) = \left(\prod_f \frac{d^3p_f}{(2\pi)^3 2E_f}\right)|_{out}\langle p_1 p_2 \ldots |\phi_A\phi_B\rangle_{in}|^2, \qquad (19.31)$$

since it must be proportional to the infinitesimal interval, and with the $||^2$ of the amplitude, and the rest is the correct normalization.

For one target (i.e. $N_A = 1$) and for n_B particles coming in per unit of transverse area, the number of scattered particles is

$$\Delta N = \int d^2b n_B \mathcal{P}(\vec{b}). \qquad (19.32)$$

If n_B is constant, we can take it out of the integral. Then the cross-section is (as we saw in (19.2))

$$d\sigma = \frac{\Delta N}{n_B} = \int d^2b \mathcal{P}(\vec{b}). \qquad (19.33)$$

Replacing the form of $\mathcal{P}$ and on the states in the amplitudes there, we get

$$
d\sigma = \left(\prod_f \frac{d^3 p_f}{(2\pi)^3} \frac{1}{2E_f} \right) \int d^2 b \prod_{i=A,B} \int \frac{d^3 k_i}{(2\pi)^3} \frac{\phi_i(\vec{k}_i)}{\sqrt{2E_i}} \int \frac{d^3 \bar{k}_i}{(2\pi)^3} \frac{\phi_i^*(\vec{\bar{k}}_i)}{\sqrt{2\bar{E}_i}}
$$

$$
\times e^{i\vec{b}\cdot(\vec{\bar{k}}_B - \vec{k}_B)} \left({}_{out}\langle\{p_f\}|\{k_i\}\rangle_{in} \right) \left({}_{out}\langle\{p_f\}|\{\bar{k}_i\}\rangle \right)^* . \tag{19.34}
$$

To compute it, we use

$$
\int d^2\vec{b}\, e^{i\vec{b}\cdot(\vec{\bar{k}}_B - \vec{k}_B)} = (2\pi)^2 \delta^{(2)}(k_B^\perp - \bar{k}_B^\perp),
$$

$$
{}_{out}\langle\{p_f\}|\{k_i\}\rangle_{in} = i\mathcal{M}(2\pi)^4 \delta^{(4)} \left(\sum k_i - \sum p_f \right),
$$

$$
{}_{out}\langle\{p_f\}|\{\bar{k}_i\}\rangle_{in}^* = -i\mathcal{M}^*(2\pi)^4 \delta^{(4)} \left(\sum \bar{k}_i - \sum p_f \right). \tag{19.35}
$$

We also use

$$
\int d^3\bar{k}_A \int d^3\bar{k}_B \delta^{(4)} \left(\sum \bar{k}_i - \sum p_f \right) \delta^{(2)}(k_B^\perp - \bar{k}_B^\perp)
$$

$$
= (k_i^\perp = \bar{k}_i^\perp) \times \int d\bar{k}_A^z d\bar{k}_B^z \delta(\bar{k}_A^z + \bar{k}_B^z - \sum p_f^z) \delta \left(\bar{E}_A + \bar{E}_B - \sum E_f \right)
$$

$$
= \int d\bar{k}_A^z \, \delta \left(\sqrt{\bar{k}_A^2 + m_A^2} + \sqrt{\bar{k}_B^2 + m_B^2} - \sum E_f \right) \Bigg|_{\bar{k}_B^z = \sum p_f^z - \bar{k}_A^z}
$$

$$
= \frac{1}{\left| \frac{\bar{k}_A^z}{\bar{E}_A} - \frac{\bar{k}_B^z}{\bar{E}_B} \right|} = \frac{1}{|v_A - v_B|}. \tag{19.36}
$$

In the last line we have used the fact that (since $E = \sqrt{k^2 + m^2}$ and there we have $\bar{k}_B^z = \sum p_f - \bar{k}_A^z$)

$$
\frac{d\bar{E}_A}{d\bar{k}_A} = \frac{\bar{k}_A}{\bar{E}_A}; \quad \frac{d\bar{E}_B}{d\bar{k}_B} = \frac{\bar{k}_B}{\bar{E}_B} = -\frac{d\bar{E}_B}{d\bar{k}_A}, \tag{19.37}
$$

and the fact that $\int dx \delta(f(x) - f(x_0)) = 1/|f'(x_0)|$.

Putting everything together, we find

$$
d\sigma = \left(\prod_f \frac{d^3 p_f}{(2\pi)^3} \frac{1}{2E_f} \right) \int \frac{d^3 k_A}{(2\pi)^3} \int \frac{d^3 k_B}{(2\pi)^3}
$$

$$
\frac{|\mathcal{M}(k_A, k_B \to \{p_f\})|^2}{2E_A 2E_B |v_A - v_B|} |\phi_A(k_A)|^2 |\phi_B(k_B)|^2 (2\pi)^4 \delta^{(4)} \left(k_A + k_B - \sum p_f \right). \tag{19.38}
$$

For states with wavefunctions centered sharply on a given momentum $\delta(\vec{k} - \vec{p})$, we obtain

$$
d\sigma = \frac{1}{2E_A 2E_B |v_A - v_B|} \left(\prod_f \frac{d^3 p_f}{(2\pi)^3} \frac{1}{2E_f} \right) |\mathcal{M}(k_A, k_B \to \{p_f\})|^2 (2\pi)^4 \delta^{(4)}
$$

$$
\times \left(k_A + k_B - \sum p_f \right). \tag{19.39}
$$

In this formula, $|\mathcal{M}|^2$ is Lorentz invariant.

$$\int d\Pi_n = \left(\prod_f \frac{d^3 p_f}{(2\pi)^3} \frac{1}{2E_f}\right)(2\pi)^4 \delta^{(4)}\left(k_A + k_B - \sum p_f\right) \qquad (19.40)$$

is relativistic invariant as well, and is called *relativistically invariant n-body phase space*, however

$$\frac{1}{E_A E_B |v_A - v_B|} = \frac{1}{|E_B p_A^z - E_A p_B^z|} = \frac{1}{|\epsilon_{\mu x y \nu} p_A^\mu p_B^\nu|} \qquad (19.41)$$

is therefore not relativistically invariant, meaning the cross-section is not relativistically invariant either.

However, if $\vec{k}_A \| \vec{k}_B$ (for instance in the center of mass frame, or the laboratory frame), this can be written as

$$\frac{1}{\sqrt{(p_1 \cdot p_2)^2 - m_1^2 m_2^2}}. \qquad (19.42)$$

So if we adopt this form in all reference frames, that is replace in the formula for the differential cross-section

$$\frac{1}{E_A E_B |v_A - v_B|} \rightarrow \frac{1}{\sqrt{(p_1 \cdot p_2)^2 - m_1^2 m_2^2}}, \qquad (19.43)$$

then we obtain the *relativistically invariant cross-section*. This is a theoretical concept, which is useful since we can write this cross-section in a simple way in terms of Mandelstam variables s, t (to be defined later). But it is not measurable, so for comparison with experiment, we should remember to go back to the usual form (or just use the center of mass frame or laboratory frame).

To understand the relativistically invariant n-body phase space better, consider the important case of $n = 2$ (i.e. $2 \rightarrow 2$ scattering) and consider working in the center of mass frame, so $\vec{p}_{total} = 0$. Then from the momenta $\vec{p}_1, \vec{p}_2$, the delta function over 3-momenta imposes $\vec{p}_2 = -\vec{p}_1$, and we are left with

$$\int d\Pi_2 = \int \frac{dp_1}{(2\pi)^3} \frac{p_1^2}{2E_1} \frac{d\Omega}{2E_2} 2\pi \delta(E_{CM} - E_1 - E_2)$$

$$= \int d\Omega \frac{p_1^2}{16\pi^2 E_1 E_2} \frac{1}{\frac{p_1}{E_1} + \frac{p_1}{E_2}}\Bigg|_{p_1 = -p_2} = \int d\Omega \frac{|p_1|}{16\pi^2 E_{CM}}. \qquad (19.44)$$

Here, $E_{CM} = E_1 + E_2$ and $E_1 = \sqrt{p_1^2 + m_1^2}, E_2 = \sqrt{p_1^2 + m_2^2}$.

Therefore, finally we obtain that in the center of mass frame

$$\left(\frac{d\sigma}{d\Omega}\right)_{CM} = \frac{1}{2E_A 2E_B |v_A - v_B|} \frac{|\vec{p}_1|}{16\pi^2 E_{CM}} |\mathcal{M}(p_A, p_B \rightarrow p_1, p_2)|^2. \qquad (19.45)$$

In the case of identical masses for all the particles (A, B, 1, 2), we have

$$\frac{E_{CM}}{2} = E_A = E_B; \quad |p_A| = |p_B| = |p_1| = |p_2|, \qquad (19.46)$$

and substituting above (together with $E_A E_B |v_A - v_B| = |E_A p_B - E_B p_A|$), we get

$$\left(\frac{d\sigma}{d\Omega}\right)_{CM} = \frac{|\mathcal{M}|^2}{64\pi^2 E_{CM}^2}. \tag{19.47}$$

19.4.1 Particle Decay

We can now calculate particle decay very easily. Formally, all we have to do is keep only the k_As and drop the k_Bs in the calculation above. Dropping the integrals $\int d^3 k_B/(2\pi)^3$ and $\int d^3 \bar{k}_B/(2\pi)^3$ implies that in the above we don't have $\int d\bar{k}_B^z$ anymore in (19.36), only

$$\int d\bar{k}_A^z \delta\left(\bar{k}_A^z - \sum p_f^z\right) \delta\left(\bar{E}_A - \sum E_f\right) = \delta\left(\bar{E}_A - \sum E_f\right), \tag{19.48}$$

so in effect, we need to remove the factor $1/|v_A - v_B|$ from the calculation. We can then immediately write the result for the decay rate in the center of mass system:

$$d\Gamma|_{CM} = \frac{1}{2m_A}\left(\prod_f \frac{d^3 p_f}{(2\pi)^3}\frac{1}{2E_f}\right)|\mathcal{M}(m_A \to \{p_f\})|^2(2\pi)^4\delta^4\left(p_A - \sum p_f\right). \tag{19.49}$$

We only have a problem of interpretation: what does this represent, since the state is unstable, so how do we define the asymptotic (noninteracting) state A? Nevertheless, the result is correct.

Important Concepts to Remember

- The cross-section is the number of scattering events per time, divided by the incoming flux and the number of targets, and it measures the effective area of interaction around a single target.
- The decay rate is the number of decays per unit time, divided by the number of particles decaying. The sum of the decay rates gives the inverse of the lifetime.
- In a cross-section, a resonance appears via a relativistic generalization of the Breit–Wigner formula, with amplitudes proportional to $1/(p^2 + m^2 - im\Gamma)$ for the resonance.
- In states are well-separated states at $t = -\infty$, out states are well-separated states at $+\infty$, and their overlap gives the S-matrix.
- The S operator is $1 + iT$, and extracting from the matrix elements of T the overall momentum conservation delta functions, we obtain the finite reduced matrix element $\mathcal{M}$.
- The LSZ formula says that the S-matrix is the residue at all the external poles of the momentum space Green's function, divided by the Z factors, or the Green's function divided by the full external propagators near the mass shell.
- The differential cross-section equals $|\mathcal{M}|^2$ times the relativistically invariant n-body phase space times $1/(E_A E_B |v_A - v_B|)$. If we replace the last factor with the relativistically invariant formula $1/\sqrt{(p_1 \cdot p_2)^2 - m_1^2 m_2^2}$, we obtain the *relativistically invariant cross-section*, a useful theoretical concept.

Further Reading

See section 4.5 in [1], and sections 2.5, 4.2, and 4.3 in [2].

Exercises

1. Consider the scattering of a massive field off a massless field ($m_1 = m_3 = m; m_2 = m_4 = 0$) in a theory where $i\mathcal{M} \simeq i\lambda$ =constant (e.g. Fermi's 4-fermion theory for weak interactions). Calculate the *total, relativistically invariant* cross-section $\sigma_{tot,rel}$ as a function of the center of mass energy.

2. Consider the theory

$$\mathcal{L} = -\frac{1}{2}(\partial_\mu \phi)^2 - \frac{1}{2}(\partial_\mu \Phi)^2 - \frac{1}{2}M^2\Phi^2 - \frac{1}{2}m^2\phi^2 - \mu\Phi\phi\phi. \tag{19.50}$$

Then, if $M > 2m$, the first-order amplitude for Φ to decay in two ϕs is $|\mathcal{M}| = \mu$. Calculate the lifetime τ for Φ.

3. Consider the model for a real scalar field Φ and a complex scalar ϕ, with Lagrangian

$$\mathcal{L} = -\frac{1}{2}|\partial_\mu\phi|^2 - \frac{1}{2}(\partial_\mu\Phi)^2 - \frac{1}{2}M^2\Phi^2 - \frac{1}{2}m^2\phi^2 - \lambda\Phi^2|\phi|^2. \tag{19.51}$$

Calculate the first-order differential cross-section in the center of mass reference frame.

4. Integrate the first-order differential cross-section in the center of mass reference frame in Exercise 3, to find the total cross-section. Compare with the *relativistically invariant* cross-section obtained in a similar manner.

The S-Matrix and Feynman Diagrams

In Chapter 19 we saw how to connect observable cross-sections to S-matrix elements. In this chapter we develop the perturbation theory for these S-matrices, by finding the Feynman and Wick theorems for them, and then defining Feynman diagrams.

20.1 Perturbation Theory for S-Matrices: Feynman and Wick

We now want to go from the Heisenberg picture used in Chapter 19 to the interaction picture, the same way as we did for Green's functions. In Chapter 19 we wrote for the S-matrix

$$\langle \vec{p}_1 \vec{p}_2 \ldots | S | \vec{k}_A \vec{k}_B \rangle = \lim_{T \to \infty} \langle \vec{p}_1 \vec{p}_2 \ldots | e^{-iH(2T)} | \vec{k}_A \vec{k}_B \rangle, \tag{20.1}$$

where the states are in the *full* theory (not free states), that is they are eigenfunctions of H, and are defined at the same time.

But, as for the Green's functions, we want to replace them by *free* states (i.e. eigenfunctions of H_0). Before, we wrote

$$|\Omega\rangle = \lim_{T \to \infty(1-i\epsilon)} (e^{-iE_0 T} \langle \Omega|0\rangle)^{-1} e^{-iHT} |0\rangle = \lim_{T \to \infty(1-i\epsilon)} (e^{-iE_0 T} \langle \Omega|0\rangle)^{-1} U_I(0, -T) |0\rangle \tag{20.2}$$

(by introducing a complete set of full states $|n\rangle\langle n|$, with $|\Omega\rangle$ having $H|\Omega\rangle = E_0|\Omega\rangle$ and the $T \to \infty(1 - i\epsilon)$ guaranteeing only the lowest-energy mode, i.e. the vacuum, remains). Now we want to write in a similar way

$$|\vec{k}_A \vec{k}_B\rangle \propto \lim_{T \to \infty(1-i\epsilon)} e^{-iHT} |\vec{k}_A \vec{k}_B\rangle_0, \tag{20.3}$$

where we wrote a proportionality sign, since proving the relation and getting the constant is now complicated: the external state is not a ground state anymore. Using this, we rewrite the right-hand side of (20.1) as

$$\lim_{T \to \infty(1-i\epsilon)} {}_0\langle \vec{p}_1 \ldots \vec{p}_n | e^{-iH(2T)} | \vec{p}_A \vec{p}_B \rangle_0$$

$$\propto \lim_{T \to \infty(1-i\epsilon)} {}_0\langle \vec{p}_1 \ldots \vec{p}_n | T \left\{ \exp\left[-i \int_{-T}^{T} dt H_I(t) \right] \right\} | \vec{p}_A \vec{p}_B \rangle_0, \tag{20.4}$$

therefore

$$S_{fi}|_{int} = \langle f | U_I(+\infty, -\infty) | i \rangle. \tag{20.5}$$

The proportionality factors cancel out in the Green's functions case by using the Feynman and Wick theorems, and only the connected pieces remain. Here a similar thing happens, although the proof is harder. We will explain later the result in terms of the LSZ formalism. The correct formula is

$$\langle \vec{p}_1 \dots \vec{p}_n | iT | \vec{p}_A \vec{p}_B \rangle = (\sqrt{Z})^{n+2} \lim_{T \to \infty(1-i\epsilon)} \left({}_0\langle \vec{p}_1 \dots \vec{p}_n | T \right.$$
$$\left. \times \left\{ \exp\left[-i \int_{-T}^{T} dt H_I(t) \right] \right\} | \vec{p}_A \vec{p}_B \rangle_0 \right)_{\text{connected,amputated}} .$$
(20.6)

We have iT instead of S, since the formula gives only the matrix elements where the initial and final states are different. To understand the meaning of "connected, amputated" we will look at some examples. We first note the factors of Z equal to 1 at tree level, and we will understand the need to put them there later, from the LSZ formula.

20.2 Example: ϕ^4 Theory in Perturbation Theory and First-Order Differential Cross-Section

We first consider the free term on the right-hand side of (20.6). It is given by

$$_0\langle \vec{p}_1 \vec{p}_2 | \vec{p}_A \vec{p}_B \rangle_0 \equiv \sqrt{2E_A 2E_B 2E_1 2E_2} \langle 0 | a_1 a_2 a_A^\dagger a_B^\dagger | 0 \rangle$$
$$= 2E_A 2E_B (2\pi)^6 \left(\delta^{(3)}(\vec{p}_A - \vec{p}_1)\delta^{(3)}(\vec{p}_B - \vec{p}_2) + \delta^{(3)}(\vec{p}_A - \vec{p}_2)\delta^{(3)}(\vec{p}_B - \vec{p}_1) \right).$$
(20.7)

This has the diagrammatic interpretation of propagator lines connecting $(1A)(2B)$ and $(1B)(2A)$, see Figure 20.1, and corresponds to the $\mathbf{1}$ in $S = \mathbf{1} + iT$, therefore should be excluded from the right-hand side of (20.6). It is excluded, as the two lines are not connected among each other.

We next consider the first-order term in λ:

$$_0\langle \vec{p}_1 \vec{p}_2 | T \left\{ -i \frac{\lambda}{4!} \int d^4x \phi_I^4(x) \right\} | \vec{p}_A \vec{p}_B \rangle_0$$
$$= {}_0\langle \vec{p}_1 \vec{p}_2 | N \left\{ -i \frac{\lambda}{4!} \int d^4x \phi_I^4(x) + \text{contractions} \right\} | \vec{p}_A \vec{p}_B \rangle_0,$$
(20.8)

where in the equality we used Wick's theorem.

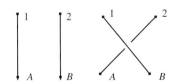

Figure 20.1 Free term for the scattering matrix: it is excluded (only interacting pieces considered).

But now we don't have $|0\rangle$ anymore, but rather $|\vec{p}_A\vec{p}_B\rangle_0$, so the "non-contracted" pieces give a nonzero result. Better said, there is another type of contraction.

Consider the annihilation part of ϕ_I, ϕ_I^+, acting on an external state:

$$\phi_I^+(x)|\vec{p}\rangle_0 = \left(\int \frac{d^3k}{(2\pi)^3}\frac{1}{\sqrt{2E_k}}a_{\vec{k}}e^{ikx}\right)\left(\sqrt{2E_p}a_{\vec{p}}^\dagger|0\rangle\right)$$

$$= \int \frac{d^3k}{(2\pi)^3}e^{ikx}\left((2\pi)^3\delta^{(3)}(\vec{k}-\vec{p})\right)|0\rangle = e^{ipx}|0\rangle. \qquad (20.9)$$

Inside $\langle..|N(\phi^k)|..\rangle = \langle..|(\phi^-)^n(\phi^+)^{k-n}|..\rangle$, we have ϕ^+s acting on a state on the right and ϕ^-s acting on the left, so we define the contraction with an external state:

$$\overline{\phi_I(x)|\vec{p}\rangle} = e^{ipx}|0\rangle, \qquad (20.10)$$

and the conjugate

$$\overline{\langle\vec{p}|\phi_I(x)} = \langle0|e^{-ipx}. \qquad (20.11)$$

We now analyze the first-order term in the S-matrix (20.8). It contains a term $\overline{\phi\phi}\phi\phi$, a term $\phi\phi\phi\phi$, and a term $\overline{\phi\phi}\;\overline{\phi\phi}$.

Consider the last term, with two $\phi\phi$ contractions. It is

$$-i\frac{\lambda}{4!}\int d^4x\,\overline{\phi\phi}\;\overline{\phi\phi}\;{}_0\langle\vec{p}_1\vec{p}_2|\vec{p}_A\vec{p}_B\rangle_0. \qquad (20.12)$$

This gives the figure-eight vacuum bubble times the free term (zeroth order) in Figure 20.1, so it is a disconnected piece where the initial state and the final state are the same. Therefore, again this is a piece that contributes to $\mathbf{1}$, not to iT, and is disconnected, consistent with (20.6).

Next consider the second term, with a single contraction. After applying the normal ordering N operator, we have $\sim a^\dagger a^\dagger + 2a^\dagger a + aa$, but only the term with the same number of $a^\dagger$ and a contributes in the expectation value (e.g. $\langle p_1 p_2|a^\dagger a^\dagger|p_A p_B\rangle \sim \langle0|aa(a^\dagger)^4|0\rangle \sim [a,a^\dagger]^2\langle0|a^\dagger a^\dagger|0\rangle = 0$), that is

$$-i\frac{\lambda}{4!}\int d^4x\,2\overline{\phi\phi}_0\langle\vec{p}_1\vec{p}_2|\phi^-\phi^+|\vec{p}_A\vec{p}_B\rangle_0. \qquad (20.13)$$

Doing the contractions of the ϕs with the external states we obtain the same zeroth-order term with a loop on one of the legs (thus a total of four terms), see Figure 20.2. Contractions with external states are as in Figure 20.3. Since the $\int d^4x$ interval gives a delta function, on the leg with a loop we also have the same momentum on both sides, so

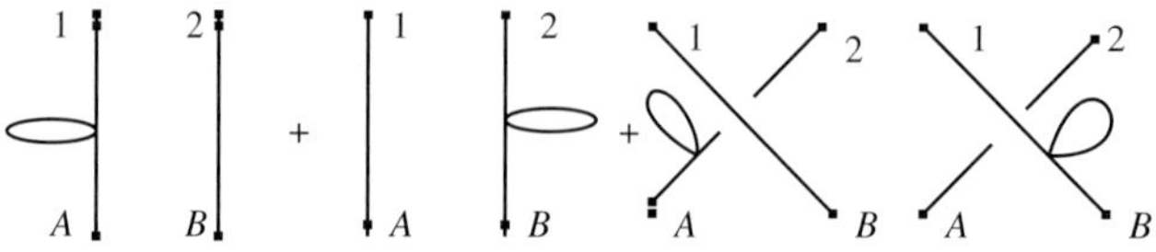

Figure 20.2   Other disconnected terms, this time with nontrivial contractions (loops), excluded from the scattering matrix.

$$\overline{\phi_I(x)|p\rangle} = \underset{x \quad p}{\longrightarrow} \qquad \overline{\langle p|\phi_I(x)\rangle} = \longleftarrow$$

Figure 20.3 Feynman diagrammatic representation for contractions with external states.

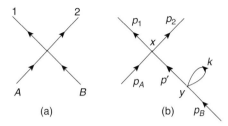

Figure 20.4 (a) The first-order contribution to the S-matrix. (b) Even among the connected diagrams there are some that do not contribute to the S-matrix, like this one. It is one which will be *amputated*.

this is again a contribution to the identity **1**, and not to iT, and is disconnected, consistent with (20.6).

Finally, the only nonzero term is the one where there are no contractions between the ϕs, and only contractions with the external states, of the type

$$_0\langle \vec{p}_1\vec{p}_2|\phi^-\phi^-\phi^+\phi^+(x)|\vec{p}_A\vec{p}_b\rangle_0, \tag{20.14}$$

where all the ϕs are at the same point, so this gives a term $e^{i(p_A+p_B-p_1-p_2)x}$. But this comes from $N(\phi^4(x))$, with $\phi = \phi^+ + \phi^-$, so there are $4! = 24$ such terms as the above. From the first ϕ, we can pick either ϕ^- or ϕ^+ to contract with any one of the four external states $(\vec{p}_1, \vec{p}_2, \vec{p}_A, \vec{p}_B)$, the next ϕ can be contracted with any of the three remaining (uncontracted) external states, and so on, giving $4 \times 3 \times 2 \times 1$ terms. The total result is then

$$4! \times \left(-i\frac{\lambda}{4!}\right) \int d^4x\, e^{i(p_A+p_B-p_1-p_2)x} = -i\lambda(2\pi)^4\delta^{(4)}(p_A + p_B - p_1 - p_2), \tag{20.15}$$

which is just a vertex connecting the four external lines, as in Figure 20.4(a).

Since iT is $(2\pi)^4\delta(\ldots) \times i\mathcal{M}$, we have $\mathcal{M} = -\lambda$. Replacing in the formula for the differential cross-section in the center of mass frame from Chapter 19, we get

$$\left(\frac{d\sigma}{d\Omega}\right)_{CM} = \frac{\lambda^2}{64\pi^2 E_{CM}^2}. \tag{20.16}$$

This is independent of the angles θ, ϕ, therefore it integrates trivially, by multiplying with $\Omega = 4\pi$, giving finally

$$\sigma_{tot} = \frac{\lambda^2}{32\pi E_{CM}^2}. \tag{20.17}$$

Here we added a factor of $1/2$ since we have identical particles in the final state, so it is impossible to differentiate experimentally between 1 and 2.

20.3 Second-Order Perturbation Theory and Amputation

So we need to consider only *fully connected* diagrams, where all the external lines are connected as well. But even among these diagrams there are some that need to be excluded. To see this, consider the second-order diagram obtained as follows: a vertex diagram with a loop on one of the external states (p_B), with k in the loop and p' after it, as in Figure 20.4(b). It gives

$$\frac{1}{2} \int \frac{d^4 p'}{(2\pi)^4} \frac{-i}{p'^2 + m^2} \int \frac{d^4 k}{(2\pi)^4} \frac{-i}{k^2 + m^2} (-i\lambda)(2\pi)^4 \delta^{(4)}(p_A + p' - p_1 - p_2)$$

$$\times (-i\lambda)(2\pi)^4 \delta^{(4)}(p_B - p'), \tag{20.18}$$

where the delta functions come from the integration over vertices. But doing the p' integration we get

$$\left. \frac{1}{p'^2 + m^2} \right|_{p'=p_B} = \frac{1}{p_B^2 + m^2} = \frac{1}{0}, \tag{20.19}$$

so the result is badly defined. This means that this diagram must be excluded from the physical calculation, so we must find a procedure that excludes this kind of diagram.

We now define *amputation* as follows. From the tip of an external leg, find the furthest point in the connected diagram where we can disconnect the external line from the rest by cutting just one propagator, and remove that whole piece (including the external line) from the diagrams. See examples of diagrams in Figure 20.5.

Therefore we can say that

$$i\mathcal{M}(2\pi)^4 \delta^{(4)} \left(p_A + p_B - \sum p_f \right) = \left(\sum \text{connected, amputated Feynman diagram} \right)$$
$$\times (\sqrt{Z})^{n+2}. \tag{20.20}$$

These pieces of diagrams that we are excluding are contributions to the two-point functions for the external lines (quantum corrections to the propagators). We will see the interpretation of this shortly.

Figure 20.5 Examples for the diagrammatic representation of amputation, to obtain the Feynman diagram contributions to the S-matrix.

20.4 Feynman Rules for S-Matrices

First we consider the **Feynman diagrams in x space** (see Figure 20.6(a)):

- propagator $D_F(x - y)$
- vertex $-i\lambda \int d^4x$
- external line e^{ipx}
- divide by the symmetry factor

and the **Feynman diagrams in p space** (see Figure 20.6(b)):

- propagator $D_F(p) = -i/(p^2 + m^2 - i\epsilon)$
- vertex $-i\lambda$
- external line $= 1$
- momentum conservation at vertices
- $\int d^4p/(2\pi)^4$ for loops
- divide by the symmetry factor

Given the LSZ formula (which we only wrote, we didn't derive), we can understand the above rule (20.20). Indeed, we can rewrite the LSZ formula as

$$S_{fi} - \delta_{fi} = \langle f|iT|i\rangle = (2\pi)^4 \delta^{(4)}\left(p_A + p_B - \sum p_f\right) i\mathcal{M}$$

$$= \lim_{p_i^2 + m^2 \to 0, k_j^2 + m^2 \to 0} (\sqrt{Z})^{m+n} \prod_{i=1}^{n}\left(\frac{p_i^2 + m^2 - i\epsilon}{-iZ}\right) \prod_{j=1}^{m}\left(\frac{k_j^2 + m^2 - i\epsilon}{-iZ}\right)$$

$$\times \tilde{G}_{n+m}(p_i, k_j). \tag{20.21}$$

But in $\tilde{G}_{n+m}$ we have the connected Feynman diagrams, and the factors in the brackets are, near their mass shell, the inverse of the full propagators for the external lines:

$$G_2(p) \sim \frac{-iZ}{p^2 + m^2 - i\epsilon}. \tag{20.22}$$

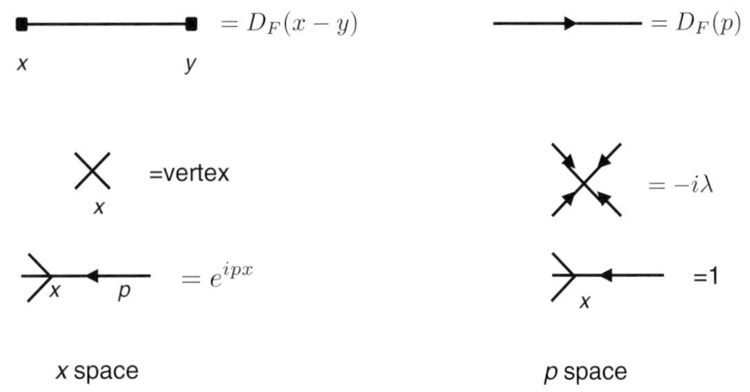

Figure 20.6   Diagrammatic representation of the relevant Feynman rules for S-matrices in (a) x space and (b) p space.

So amputation is just the removal (dividing by) these full on-shell two-point functions for the external lines. Thus we obtain exactly the rule (20.20).

Important Concepts to Remember

- In the interaction picture, $S_{fi} \propto \langle f | U_I(+\infty, -\infty) | i \rangle$.
- The matrix elements of iT (nontrivial S-matrices) are the sum over connected, amputated diagrams.
- Amputation means removing the parts that contribute to the (on-shell) two-point functions of the external lines.
- The S-matrix diagrammatic rules follow from the LSZ formula.

Further Reading

See section 4.6 in [1], and sections 4.1 and parts of 4.4 in [2].

Exercises

1. Consider the second-order term in the S-matrix

$$_0\langle \vec{p}_1 \vec{p}_2 | T \left\{ \left(\frac{-i\lambda}{4!}\right)^2 \int d^4x \phi_I^4(x) \int d^4y \phi_I^4(y) \right\} | \vec{p}_A \vec{p}_B \rangle_0. \tag{20.23}$$

Using the Wick contractions and theorem, show only the Feynman diagrams which contribute to the S-matrix.

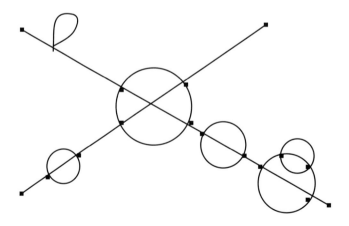

Figure 20.7   Example of Feynman diagram.

2. Consider the Feynman diagram in Figure 20.7 in $\lambda\phi^4/4!$ theory. Write down the Feynman diagram corresponding to this in $_0\langle\vec{p}_1\vec{p}_2|S|\vec{p}_3\vec{p}_4\rangle_0$ and write the integral expression for it using the Feynman rules.

3. Consider a $\lambda\phi^4/4!$ theory. Draw all the Feynman diagrams contributing to the S-matrix up to (including) order $\mathcal{O}(\lambda^3)$.

4. Use the S-matrix Feynman rules to write down the formula for the S-matrix "setting sun" diagram.

In this chapter we describe the optical theorem, defining unitarity of the quantum mechanics in the Feynman diagrammatic perturbation theory, that is, the fact that the sum of probabilities equals one. This has the implication of a relation between the imaginary part of some amplitudes, and the scattering cross-section for the same initial external states. But further, one can consider it in Feynman perturbation theory as a relation between diagrams, one "cut" one, and one "half" of the cut diagram squared. Cutting refers to an operation that can be applied to Feynman diagrams, and will be defined more precisely using the rules found by Cutkovsky.

21.1 The Optical Theorem: Formulation

The optical theorem is a straightforward consequence of the *unitarity of the S-matrix*. In a good quantum theory, the operator defining the evolution of the system, in our case $S_{fi} = \langle f|U(+\infty, -\infty)|i\rangle$, has to be unitary, in order to conserve probability (there are no "sources or sinks of probability"). Therefore, with $S = 1 + iT$, we have

$$S^\dagger S = 1 = SS^\dagger \Rightarrow -i(T - T^\dagger) = T^\dagger T, \tag{21.1}$$

which is the essence of the optical theorem.

But the point is that in our definition of quantum field theory, and more importantly in our perturbative definition of S-matrices, there are many unknowns and approximations, so it is by no means obvious that the S-matrix is unitary. We have to *prove it perturbatively (order by order in perturbation theory)*, and from the above, that is equivalent to proving the optical theorem.

To obtain the usual formulation of the optical theorem, we evaluate (21.1) between two two-particle states, $|\vec{p}_1\vec{p}_2\rangle$ and $|\vec{k}_1\vec{k}_2\rangle$. Moreover, in between $T^\dagger T$, we introduce the identity expressed by considering the completeness relation for external states:

$$1 = \sum_n \prod_{i=1}^{n} \frac{d^3 q_i}{(2\pi)^3} \frac{1}{2E_i} |\{\vec{q}_i\}\rangle \langle\{\vec{q}_i\}|, \tag{21.2}$$

obtaining

$$\langle\vec{p}_1\vec{p}_2|T^\dagger T|\vec{k}_1\vec{k}_2\rangle = \sum_n \prod_{i=1}^{n} \frac{d^3 q_i}{(2\pi)^3} \frac{1}{2E_i} \langle\vec{p}_1\vec{p}_2|T^\dagger|\{\vec{q}_i\}\rangle \langle\{\vec{q}_i\}|T|\vec{k}_1\vec{k}_2\rangle. \tag{21.3}$$

But

$$\langle\{\vec{p}_f\}|T|\vec{k}_1\vec{k}_2\rangle = \mathcal{M}(k_1,k_2 \to \{p_f\})(2\pi)^4\delta^{(4)}\left(k_1+k_2-\sum p_f\right), \quad (21.4)$$

so by replacing in the above and taking out a common factor $(2\pi)^4\delta^{(4)}(k_1+k_2-p_1-p_2)$, we obtain

$$-i[\mathcal{M}(k_1,k_2 \to p_1,p_2) - \mathcal{M}^*(p_1,p_2 \to k_1,k_2)]$$

$$= \sum_n \left(\prod_{i=1}^n \frac{d^3q_i}{(2\pi)^3}\frac{1}{2E_i}\right)\mathcal{M}^*(p_1p_2 \to \{q_i\})\mathcal{M}(k_1k_2 \to \{q_i\})(2\pi)^4\delta^{(4)}\left(k_1+k_2-\sum_i q_i\right),$$

$$(21.5)$$

or more schematically (here $d\Pi_f$ is the n-body relativistically invariant phase space defined in Chapter 19)

$$-i[\mathcal{M}(a \to b) - \mathcal{M}^*(b \to a)] = \sum_f \int d\Pi_f \mathcal{M}^*(b \to f)\mathcal{M}(a \to f), \quad (21.6)$$

which is the general statement of the optical theorem on states.

An important particular case is when the initial and final states are the same ($a = b$), or *forward scattering*, $p_i = k_i$, in which case the right-hand side of (21.5) contains the total cross-section

$$\text{Im}\mathcal{M}(k_1k_2 \to k_1k_2) = 2E_{CM}p_{CM}\sigma_{\text{tot}}(k_1k_2 \to \text{anything}) \quad (21.7)$$

or *the imaginary part of the forward scattering amplitude is proportional to the total cross-section for the two particles to go to anything*, see Figure 21.1.

But, as we supposed that the S-matrix is unitary, and *a priori* it cannot be, we need to prove this in perturbative quantum field theory. It was proved for Feynman diagrams to all orders, first in $\lambda\phi^4$ theory by Cutkovsky, then in quantum electrodynamics by Feynman, and in gauge theories later. In particular, the proof for spontaneously broken gauge theories was done by 't Hooft and Veltman, which together with their proof of renormalizability of the same meant that spontaneously broken gauge theories are well defined, and for which they got the Nobel Prize. That proof was done using a more general formalism, of the *largest time equation*, which will be treated in Chapter 22. Here we just present the one-loop case in $\lambda\phi^4$.

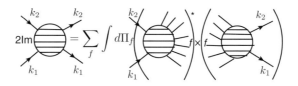

Figure 21.1 The optical theorem, diagrammatic representation.

21.2 Unitarity: Optical Theorem at One Loop in $\lambda\phi^4$ Theory

The unitarity proof is done by proving directly the optical theorem for Feynman diagrams.

Note that above the reduced-amplitude $\mathcal{M}$ is defined formally, but we can define it purely by Feynman diagrams, and that is the definition we will use below, since we want to prove unitarity for Feynman diagrams. Then, $\mathcal{M}$ is defined at some center of mass energy, with $s = E_{CM}^2 \in \mathbb{R}_+$. But we analytically continue to an analytic function $\mathcal{M}(s)$ defined on the complex s plane.

Consider $\sqrt{s_0}$ the threshold energy for production of the lightest multi-particle state. Then, if $s < s_0$ and real, it means that any possible intermediate state created from the initial state cannot go on-shell, and can only be virtual. In turn, this means that any propagators in the expression of $\mathcal{M}$ in terms of Feynman diagrams are always nonzero, leading to an $\mathcal{M}(s)$ real for $s < s_0$ real, that is

$$\mathcal{M}(s) = [\mathcal{M}(s^*)]^*. \tag{21.8}$$

We then analytically continue to $s > s_0$ real, and since both sides of the above relation are analytical functions, we apply it for $s \pm i\epsilon$ with $s > s_0$ real, giving (since now the propagators can be on-shell, so we need to avoid the poles by moving s a bit away from the real axis)

$$\mathrm{Re}\mathcal{M}(s + i\epsilon) = \mathrm{Re}\mathcal{M}(s - i\epsilon); \quad \mathrm{Im}\mathcal{M}(s + i\epsilon) = -\mathrm{Im}\mathcal{M}(s - i\epsilon). \tag{21.9}$$

This means that for $s > s_0$ and real there is a branch cut in s (by moving accross it, we go between two Riemann sheets), with discontinuity

$$\mathrm{Disc}\mathcal{M}(s) = 2i\mathrm{Im}\mathcal{M}(s + i\epsilon). \tag{21.10}$$

It is much easier to compute the discontinuity only rather than the full amplitude for complex s, taking the discontinuity.

The discontinuity of Feynman diagrams with loops, across cuts on loops (equal from the above to the imaginary parts of the same diagrams) will be related to $\int d\Pi |\bar{\mathcal{M}}|^2$, that is the optical theorem formula, see Figure 21.2. We will obtain the so-called "cutting rules," ironically found by Cutkovsky, so also called the Cutkovsky rules.

We consider the $2 \to 2$ one-loop diagram in $\lambda\phi^4$ where k_1, k_2 go to two intermediate lines, then to two final lines, called k_3, k_4. Define $k = k_1 + k_2$, and in the two intermediate lines define the momenta as $k/2 + q$ and $k/2 - q$, where q is the (integrated over) loop

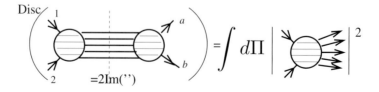

Figure 21.2 On the left-hand side of the optical theorem, represent the imaginary part of diagrams as a discontinuity ("cut").

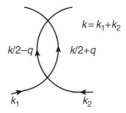

Figure 21.3 The optical theorem is valid diagram by diagram. Consider this one-loop diagram, to derive the Cutkovsky rules.

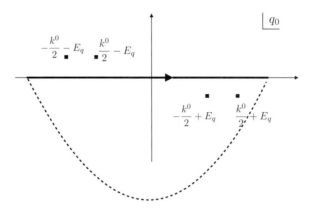

Figure 21.4 Closing the contour downwards, we pick up the poles in the lower half plane.

momentum, see Figure 21.3. Since this diagram has symmetry factor $S = 2$, we have (we call it $\delta\mathcal{M}$ since it could be part of a larger diagram)

$$i\delta\mathcal{M} = \frac{\lambda^2}{2} \int \frac{d^4q}{(2\pi)^4} \frac{1}{(k/2 - q)^2 + m^2 - i\epsilon} \frac{1}{(k/2 + q)^2 + m^2 - i\epsilon}. \qquad (21.11)$$

This amplitude has a discontinuity for $k^0 > 2m$ (in the physical region, since with center of mass energy k^0 we can create an on-shell two-particle intermediate state of energy $2m$ only if $k^0 > 2m$), but we compute it for $k^0 < 2m$ and then analytically continue.

In the center of mass system, $k = (k^0, \vec{0})$, where $s = (k^0)^2$, the poles of (21.11) are at $|k^0/2 \pm q| = \vec{q}^2 + m^2 \equiv E_q^2$, or at

$$q^0 = \frac{k^0}{2} \pm (E_q - i\epsilon) \text{ and}$$

$$= -\frac{k^0}{2} \pm (E_q - i\epsilon). \qquad (21.12)$$

We now close the contour downwards, with a semicircle in the lower half plane, because of the prescription $T = \infty(1 - i\epsilon)$, therefore picking the poles in the lower half planes, see Figure 21.4. We will see later that the pole at $+k^0/2 + E_q - i\epsilon$ does not contribute to the discontinuity, so only the pole at $-k^0/2 + E_q - i\epsilon$ does. But picking up the residue at the pole is equivalent to replacing

$$\frac{1}{(k/2+q)^2 + m^2 - i\epsilon} \simeq \frac{1}{q^0 + k^0/2 - E_q + i\epsilon} \frac{-1}{2E_q}$$

$$\to 2\pi i \delta((k/2+q)^2 + m^2) = 2\pi i \frac{\delta(q^0 + k/2 - E_q)}{2E_q}. \tag{21.13}$$

This gives

$$i\delta\mathcal{M} = -2\pi i \frac{\lambda^2}{2} \int \frac{d^3q}{(2\pi)^4} \frac{1}{2E_q} \frac{1}{(k^0 - E_q)^2 - E_q^2}$$

$$= -2\pi i \frac{\lambda^2}{2} \frac{4\pi}{(2\pi)^4} \int_m^\infty dE_q E_q |q| \frac{1}{2E_q} \frac{1}{k^0(k_0 - 2E_q)}. \tag{21.14}$$

In the first line we used the above delta function, and $(k/2-q)^2 + m^2 = -[(k^0 - E_q)^2 - E_q^2]$, and in the second line $\int d^3q = \int d\Omega q^2 dq$, with $\int d\Omega = 4\pi$ and $\int q^2 dq = \int q E_q dE_q$, since $E_q^2 = q^2 + m^2$ (which also gives $E_q \geq m$). Note that if $k^0 < 2m$, the denominator in the last line has no pole on the integral path, as $k^0 < 2m < 2E_q$, so δM is real (has no discontinuity). (Also note that the pole in q^0 neglected, at $+k^0/1 + E_q - i\epsilon$, indeed does not contribute, since its residue has now no pole for the E_q integral, as we can easily check.) But if $k^0 > 2m$, we have a pole on the path of integration, so there is a discontinuity between $k^2 + i\epsilon$ and $k^2 - i\epsilon$. To find it, we can write

$$\frac{1}{k^0 - 2E_q \pm i\epsilon} = P\frac{1}{k^0 - 2E_q} \mp i\pi \delta(k^0 - 2E_q), \tag{21.15}$$

where P stands for the principal part.

Therefore for the discontinuity, equivalently, we can also replace the second propagator with the delta function

$$\frac{1}{(k/2-q)^2 + m^2 - i\epsilon} = \frac{1}{-(k^0 - E_q)^2 + E_q^2 - i\epsilon} = \frac{1}{k^0(k^0 - 2E_q)}$$

$$\to 2\pi i\delta((k/2-q)^2 + m^2) = 2\pi i\delta(k^0(k^0 - 2E_q)) = 2\pi i \frac{\delta(k^0 - 2E_q)}{k^0}. \tag{21.16}$$

We now go back, to see what we did, and rewrite the original loop integration as the integration over momenta on the propagators, with a vertex delta function ($k/2 - q = p_1, k/2 + q = p_2$):

$$\int \frac{d^4q}{(2\pi)^4} = \int \frac{d^4p_1}{(2\pi)^4} \int \frac{d^4p_2}{(2\pi)^4} (2\pi)^4 \delta^{(4)}(p_1 + p_2 - k). \tag{21.17}$$

In this formulation, the discontinuity is obtained by replacing the propagators with delta functions for them, that is

$$\frac{1}{p_i^2 + m_i^2 - i\epsilon} \to 2\pi i\delta(p_i^2 + m^2). \tag{21.18}$$

So by doing the integrals over the zero component of the momenta $\int d^4p_1 \int d^4p_2$, we put the momenta on-shell, and we finally get

$$\mathrm{Disc}\mathcal{M}(k) = 2i\mathrm{Im}\mathcal{M}(k) = \frac{i}{2}\int \frac{d^3p_1}{(2\pi)^3}\frac{1}{2E_1}\int \frac{d^3p_2}{(2\pi)^3}\frac{1}{2E_2}|\tilde{\mathcal{M}}(k)|^2(2\pi)^4\delta^{(4)}(p_1 + p_2 - k),$$

$$(21.19)$$

where the $1/2$ factor is interpreted as the symmetry factor for two identical bosons, $|\tilde{\mathcal{M}}(k)|^2 = \lambda^2$ corresponds in general to $\tilde{\mathcal{M}}(k_1, k_2 \to p_1, p_2)\tilde{\mathcal{M}}^*(k_3, k_4 \to p_1, p_2)$.

On the left-hand side the discontinuity is in the one-loop amplitude $\mathcal{M}(k_1, k_2 \to k_3, k_4)$ (we wrote above for forward scattering $k_1 = k_3, k_2 = k_4$), where we cut the intermediate propagators, putting them on-shell (i.e. we replace the propagator with the delta function for the on-shell condition, as in the left-hand side of Figure 21.2). On the right-hand side we have the right-hand side of Figure 21.2.

21.3 General Case and the Cutkovsky Cutting Rules

Therefore we have verified the optical theorem at one loop, and by the discussion at the beginning of the chapter, we have thus verified unitarity at one loop in $\lambda\phi^4$. We can also write some one-loop diagrams in quantum electrodynamics, cut them, and write the corresponding optical theorem at one loop, see Figure 21.5. One can prove unitarity for them equivalently to the above.

Finally, Cutkovsky gave the **general cutting rules**, true in general (at all loops):

1. Cut the diagram in all possible ways such that cut propagators can be put *simultaneously* on-shell.
2. For each cut, replace

$$\frac{1}{p_i^2 + m_i^2 - i\epsilon} \to 2\pi i\delta(p^2 + m^2) \qquad (21.20)$$

   in the cut propagators, and then do the integral.
3. Sum over all possible cuts.

Then, he proved unitarity (equivalent to the optical theorem) in perturbation theory.

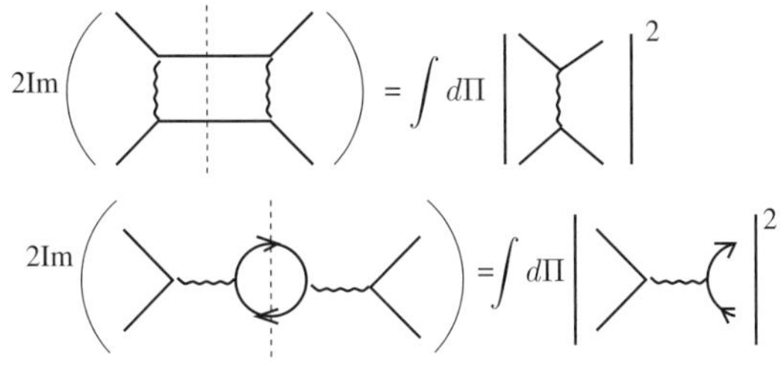

Figure 21.5 Examples of the optical theorem at one loop in QED.

There exists a more general method, the method of the *largest time equation* (which is a generalization of the above method), which was used by 't Hooft and Veltman to prove the perturbative unitarity of spontaneously broken gauge theories, which we will consider in Chapter 22.

Important Concepts to Remember

- The optical theorem is the statement $-i(T - T^\dagger) = T^\dagger T$ on states, which follows from unitarity of the S-matrix.
- The diagrammatic interpretation for it is that we cut over all possible intermediate states the amplitude for $2 \to 2$ scattering, and this equals the cut amplitude parts $\mathcal{M}(a \to f)\mathcal{M}^*(b \to f)$, integrated over the final phase space.
- It implies that the imaginary part of the forward scattering amplitude is proportional to the total cross-section $\sigma_{\text{tot}}(k_1 k_2 \to \text{anything})$.
- In $\lambda\phi^4$ at one loop, the discontinuity for $\mathcal{M}(k)$ is found by replacing the intermediate cut propagators with the delta function of the mass shell condition, leading to the one-loop optical theorem (defined diagrammatically), which is equivalent to one-loop unitarity.
- The general cutting rules say that we should cut a diagram in all possible ways such that the cut propagators can be put simultaneously on-shell, then replace the cut propagators with their mass shell delta function, and sum over all possible cuts.
- Using this method, and its generalization, the largest time equation, perturbative unitarity was proven for $\lambda\phi^4$ theory, gauge theories, and spontaneously broken gauge theories.

Further Reading

See section 7.3 in [1].

Exercises

1. (**"Black disk eikonal"**) Consider $2 \to 2$ scattering of massless particles $k_1, k_2 \to k_3, k_4$, with all momenta defined as incoming, as in Figure 21.6, and $s = -(k_1 + k_2)^2$, $t = -\vec{q}^2 = -(k_1 + k_3)^2$, and the S-matrix

$$S = e^{i\delta(b,s)}, \tag{21.21}$$

where $\delta(b, s)$ satisfies

$$\text{Re}[\delta(b, s)] = 0,$$

$$\text{Im}[\delta(b, s)] = 0 \ \text{ for } \ b > b_{max}(s); \quad \text{Im}[\delta(b, s)] = \infty \ \text{ for } \ b < b_{max}(s), \tag{21.22}$$

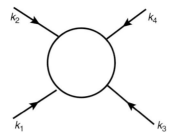

Figure 21.6 $2 \rightarrow 2$ Scattering with incoming momenta.

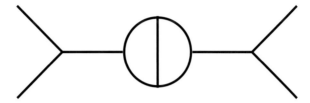

Figure 21.7 Two-loop diagram in ϕ^3 theory.

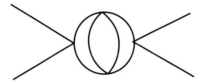

Figure 21.8 Three-loop diagram in ϕ^4 theory.

and

$$\frac{1}{s}\mathcal{M}(s,t) = \int d^2b\, e^{i\vec{q}\cdot\vec{b}} T(b,s), \tag{21.23}$$

where $\vec{b}$ is the impact parameter. Calculate $\sigma_{tot}(k_1, k_2 \rightarrow \text{anything})$.

2. Consider the two-loop Feynman diagram in $\lambda\phi^3$ theory (two incoming external lines merge into an intermediate one, which splits into a circle with a vertical line in it, then mirror symmetric: intermediate line splitting into the final two external states) in Figure 21.7.

    Write down the unitarity relation for it using the cutting rules. Which cuts depend on whether all lines have the same mass, and which ones don't?

3. Consider the three-loop diagram in a massive ϕ^4 theory (with a single massive scalar) for the four-point function, with two external lines on each side, one scalar loop, and two propagators connecting points on the upper and lower propagators, as in Figure 21.8.

Write all the cut diagrams, and then the unitarity relation using the cutting rules.

4. Write down all the Feynman diagrams for the four-point function up to (including) two loops in massive $\lambda\phi^4$ theory (for a single scalar), and then write the full unitarity relation for this sum of diagrams, giving a scattering cross-section, up to a corresponding order in λ.

22*   Unitarity and the Largest Time Equation

In this chapter, we give an alternative derivation of perturbative unitarity, one based on the "largest time equation," introduced by Veltman, and later used by 't Hooft and Veltman for their unitarity proof for spontaneously broken gauge theories. More precisely, instead of the approach based on analytically continued complex momentum amplitudes, we will now decompose the Feynman propagator into Δ^+ and Δ^-, and use real momenta. There is a third, and more modern approach to perturbative unitarity, and that is the one based on the Becchi–Rouet–Stora–Tyutin (BRST) symmetry. We will not deal with it here, since BRST will be introduced in Part II of the book. There we show that one can restrict the sum over all states in the unitarity relation to a sum over only the physical states, due to properties of the BRST operator, thus proving unitarity, given that the S-matrix is unitary in the space of all the states.

22.1 The Largest Time Equation for Scalars: Propagators

Like in Chapter 21, we will start from an observation about Feynman diagrams. This time, consider the *integrand* I of a Feynman diagram. Then, we have the statement that $I - I^*$ (the imaginary part of the integrand) can be written as a sum of cut diagrams, just like in Chapter 21. But the intermediate state will be larger than the physical states. When we do the integrations, however, we obtain energy conservations at each vertex, which in turn will restrict to the physical states only.

We start with the basic unitarity equation (21.1) (condition for the S-matrix to be unitary), sandwiched between general initial $|i\rangle$ and final $|f\rangle$ states, with an insertion of the identity on the right-hand side, as a sum over a complete set of physical states $|n\rangle$:

$$\langle f|(iT)|i\rangle + \langle f|(iT)^\dagger|i\rangle = -\sum_n{}' \langle f|(iT)^\dagger|n\rangle\langle n|(iT)|i\rangle, \qquad (22.1)$$

where a prime refers to a sum only on physical states (in the Hilbert space) of the theory (if, for instance in a gauge theory, we have unphysical states like longitudinal, negative norm states).

Note that, as we saw in Chapter 21, this implies the optical theorem (21.6), which is roughly speaking $\text{Im}\,\mathcal{M} = \int |\mathcal{M}|^2$, since the amplitude $\mathcal{M}$ is the matrix element of the T-matrix, less a momentum conservation delta function.

197

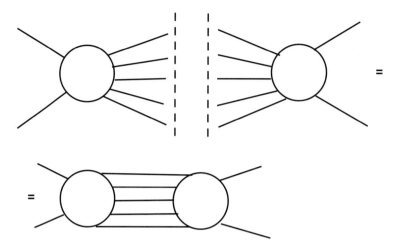

Figure 22.1 Optical theorem for a Feynman diagram.

It can be represented graphically by considering a Feynman diagram with a cut through the legs on the right (for the final state), plus the complex conjugate graph, with a cut through the legs on the left (now the same "final" state), equaling the two graphs put together, with a cut through their joining, representing all the states summed over, as in Figure 22.1.

To prove unitarity, we consider the types of propagators we have in the Feynman diagrams. We have seen in Chapter 4 that the Feynman propagator is written as

$$D_F(x - y) = \theta(x^0 - y^0)\langle 0|\phi(x)\phi(y)|0\rangle + \theta(y^0 - x^0)\langle 0|\phi(y)\phi(x)|0\rangle = \langle 0|T(\phi(x)\phi(y))|0\rangle,$$
$$(22.2)$$

where T is the time-ordering operator, and the propagator $D(x - y)$ is

$$D(x - y) = \langle 0|\phi(x)\phi(y)|0\rangle = \int \frac{d^3k}{(2\pi)^3 2E_k} e^{ik(x-y)}. \qquad (22.3)$$

For the purposes of writing the largest time equation, we change the notation a bit, and rename $D(x - y)$ as $\Delta^+(x - y)$, and then note that

$$(\Delta^+(x - y))^* = \Delta^+(y - x) \equiv \Delta^-(x - y), \qquad (22.4)$$

such that

$$\Delta^\pm(x - y) = \int \frac{d^3k}{(2\pi)^3 2E_k} e^{\pm ik(x-y)}. \qquad (22.5)$$

Then we can write the Feynman propagator as

$$D_F(x - y) = \theta(x^0 - y^0)\Delta^+(x - y) + \theta(y^0 - x^0)\Delta^-(x - y). \qquad (22.6)$$

Moreover, we can now define the anti-time-ordering operator $T^\dagger$, such that

$$T^\dagger(\phi(x)\phi(y)) = \theta(y^0 - x^0)\phi(x)\phi(y) + \theta(x^0 - y^0)\phi(y)\phi(x). \qquad (22.7)$$

Then similarly, we obtain the anti-Feynman propagator as

$$D_F^*(x-y) \equiv \langle 0|T^\dagger(\phi(x)\phi(y))|0\rangle = \theta(y^0 - x^0)\Delta^+(x-y) + \theta(x^0 - y^0)\Delta^-(x-y). \quad (22.8)$$

22.2 Cut Diagrams

We can now define "cut diagrams" as diagrams where, on the left of the cut, we use the Feynman propagator D_F, on the right we use the anti-Feynman propagator D_F^*, and for the cut lines we use either Δ^+ or Δ^-. For such diagrams, we will derive an identity that will be called the largest time equation. The use of the Feynman propagator will be for the part $\langle f|(iT)^\dagger$, that of the anti-Feynman propagator for $(iT)|i\rangle$, and the cut lines for the sum over states in the middle.

Consider a Feynman diagram with N points $x_1, x_2, \ldots, x_N$, with integrand $I(x_1, \ldots, x_N)$ and composed of Feynman propagators $D_F(x_i - x_j)$ connecting nodes i and j (remember that $D_F(x) = D_F(-x)$), and a factor i at each vertex (there is also a $-\lambda$ in addition, but we are not interested in that for the proof below). At the end, we will need to integrate over the positions of the vertices.

From the unintegrated Feynman graph $I(x_1, \ldots, x_N)$ above, we can construct another $2^N - 1$ "cut" graphs as follows:

- Draw circles around some of the vertices, which will be related to complex conjugation.
- Replace the i by $-i$ in the circled graphs (see above).
- The propagators between two vertices are (see Figure 22.2):

  - $D_F(x_i - x_j)$ for lines connecting uncircled vertices (usual Feynman propagator).
  - $D_F^*(x_i - x_j)$ for lines connecting circled vertices (anti-Feynman propagator for the complex conjugate situation).
  - $\Delta^+(x_i - x_j)$ for circled x_i to uncircled x_j (equivalent to replacing $D_F(x-y) = \theta(x^0 - y^0)\Delta^+(x-y) + \theta(y^0 - x^0)\Delta^-(x-y)$ with $\theta(x^0 - y^0)\Delta^+(x-y) + \theta(y^0 - x^0)(\Delta^-(x-y))^*$).
  - $\Delta^-(x_i - x_j)$ for uncircled x_i to circled x_j (equivalent to replacing $D_F(x-y) = \theta(x^0 - y^0)\Delta^+(x-y) + \theta(y^0 - x^0)\Delta^-(x-y)$ with $\theta(x^0 - y^0)(\Delta^+(x-y))^* + \theta(y^0 - x^0)\Delta^-(x-y)$).

There are a total of 2^N diagrams, since for each vertex we can either circle it or not.

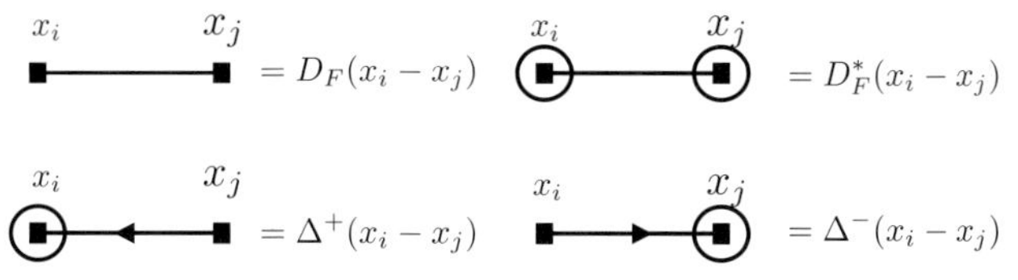

Figure 22.2 Propagators.

Note that complex conjugation of the cut Feynman diagram would replace the circled vertices with uncircled ones, and vice versa (use the fact that $(\Delta^+(x))^* = \Delta^-(-x)$). In particular, the complex conjugate of the uncut Feynman diagram (no circles) is the Feynman diagram with only circles.

22.3 The Largest Time Equation for Scalars: Derivation

We now derive the largest time equation, which is an equation for $I(x_1,\ldots,x_N) + I(x_1,\ldots,x_N)^*$. For the given graph, and in a certain reference frame, consider that x_l has the largest time among the x_is (i.e. $x_l^0 > x_i^0, \forall i$). Consider, among the cut graphs, two graphs that differ only by the fact that x_l is uncircled in one, and circled in another, and consider that x_l is connected by a propagator to x_i. Then the sum of the two graphs vanishes, since:

- If x_i is not circled, we have the sum of a Feynman propagator $D_F(x_l-x_i)$ (not circled x_i to not circled x_l) and a Δ^+ propagator (circled x_l to uncircled x_i) $\Delta^+(x_l-x_i)$, but the circled vertex in the second has an extra minus sign. In contrast, for $x_l^0 > x_i^0$, $D_F(x_l - x_i) = \Delta^+(x_l - x_i)$, so the sum of the two diagrams is zero.
- If x_i is circled, we have the sum of a $\Delta^-(x_l-x_i)$ propagator (uncircled x_l to circled x_i) and a $D_F^*(x_l-x_i)$ propagator (circled x_l to circled x_i), but again there is an extra minus sign in between them, due to the vertex factor. In contrast, for $x_l^0 > x_i^0$, $D_F^*(x_l-x_i) = \Delta^-(x_l-x_i)$, so the sum of the two diagrams is zero.

This argument used essentially that x_l is the largest time, so it is valid for all x_is. Note that the argument above didn't rely on there being a single propagator from x_l, there could be several connecting them to several points x_i. Then, as above, the propagators from x_l to any x_i are the same, so the only difference between the two graphs is the minus sign at the x_l vertex, so again they cancel out.

In conclusion, all the graphs will cancel pairwise, since there will be 2^{N-1} pairs of graphs, where in the pair one vertex is either circled, or uncircled, for a total result of zero. But we rewrite this zero by keeping on one side the graph with no circles (i.e. $I(x_1,\ldots,x_N)$) and the graph with only circles (i.e. $I^*(x_1,\ldots,x_N)$) and moving the rest (graphs with between one and $N-1$ circles) on the right-hand side, called $\tilde{I}(x_1,\ldots,x_N)$, to obtain

$$I(x_1,\ldots,x_N) + I^*(x_1,\ldots,x_N) = -\tilde{I}(x_1,\ldots,x_N). \tag{22.9}$$

This is called the *largest time equation*. We now show that from it we can obtain unitarity, specifically that when we integrate over the vertices, with wave factors for the external lines, the sum of terms on the right-hand side can be equated to a sum over physical states on the right-hand side of the unitarity equation (22.1). Since the integrated left-hand side of the largest time equation obviously gives the left-hand side of the unitarity equation, this is enough to prove it.

To understand the general proof, we start with the simplest nontrivial example, of a Born graph in a $V = \lambda\phi^3/3!$ theory, so with $\mathcal{L}_{\rm int} = -V = -\lambda\phi^3/3!$. Two external lines come into a vertex x, propagate to a vertex y, and then two other external lines. The largest time equation for this graph is, dividing by a common factor of $(-\lambda)^2$:

$$(i)D_F(x-y)(i) + (-i)D_F^*(x-y)(-i) = -(i)\Delta^-(x-y)(-i) - (-i)\Delta^+(x-y)(i), \quad (22.10)$$

where the two terms on the right-hand side correspond to the graphs where y is circled, and x is circled, respectively.

To obtain Feynman diagrams in p space (for the S-matrix), we multiply the external lines by the factors e^{ipx} for momenta p going into the vertex x, and integrate over the vertices x. But then on the right-hand side of the largest time equation above, only one of the two terms contributes.

For concreteness, consider that energy and momentum flows in from the left (vertex x) and flows out from the right (vertex y). Then the first graph, with circled x and uncircled y, thus with $\Delta^+(x-y)$, doesn't contribute. Indeed:

$$\Delta^+(x-y) = \int \frac{d^3k}{(2\pi)^3 2E_k} e^{ik(x-y)} = \int \frac{d^3}{(2\pi)^3 2E_k} e^{i\vec{k}(\vec{x}-\vec{y})} e^{-iE_k(x^0-y^0)}, \quad (22.11)$$

and the external factors are $e^{ip(x-y)} = e^{-ip^0(x^0-y^0)} e^{i\vec{p}(\vec{x}-\vec{y})}$, where p is the total momentum entering the x vertex. The integration over x^0 then gives a factor of $\delta(p^0 + E_k) = 0$. On the contrary, in the $\Delta^-(x-y)$ term, coming from the second graph, with uncircled x and circled y, we similarly obtain $\delta(p^0 - E_k) \neq 0$. Moreover, it is clear that the integration over $\vec{x}$ will also give momentum conservation, in effect putting the intermediate line on-shell (since the external lines were on-shell).

This means that only the graph where external energy flows from the uncircled vertex to the circled vertex contributes, and the other one doesn't.

We can now generalize this argument to any more complicated graph, with loops. We draw arrows on the lines with one circled and one uncircled end, from the uncircled to the circled. This amounts to an arrow from x_j to x_i for $\Delta^+(x_i - x_j)$ and an arrow from x_i to x_j for $\Delta^-(x_i - x_j)$. Lines with both uncircled (or both circled) ends, corresponding to $D_F(x_i - x_j)$ and $D_F(x_j - x_i)$, are given no arrows, since they contain both terms.

22.4 General Case

In order to construct "cut" diagrams, we draw a line (a cut) through all the lines with arrows, but not through the lines without them. Examples are given in Figure 22.3.

From the simple example above, we see that if the arrow points in the opposite direction to the external energy flow, the contribution to the graph will be zero. Indeed, in that case, we will have factors at the endpoints of the line that are in contradiction with the ones consistent with the arrow (from Δ^+ or Δ^-), and thus the integration over the vertices gives zero. More precisely, if all arrows flow into a vertex, and external energy also flows into it, then we get $\delta(\sum_a p_a^0 + \sum_i E_{k_i}) = 0$, so the graph vanishes.

This means that we are instructed to keep only the graphs where on one side of the cut we have only uncircled vertices, and on the other only circled ones, and the external energy

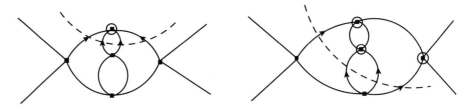

Figure 22.3   Propagators.

flows from the uncircled side to the circled side. These graphs have a side that is of type
$\mathcal{M}$, and the other of type $\mathcal{M}^*$ (complex conjugate), and the lines between them (the "cut"
lines) are lines that will be put on-shell by the vertex integrations, thus becoming physical
states. The rest of the graphs will vanish, as we saw.

Therefore, indeed, we have a contribution to

$$-\sum_{n}\langle f|(iT)^{\dagger}|n\rangle\langle n|(iT)|i\rangle, \tag{22.12}$$

where the $|n\rangle$ states are multi-particle states, which correspond to as many particles as lines
are cut in the diagram (for a cut that goes through two lines, we have a two-particle state,
for a cut that goes through three lines, a three-particle state, etc.).

We have thus shown that the largest time equation leads to a proof of unitarity for scalars,
and it is a proof that goes *diagram by diagram*.

For a full proof of perturbative unitarity, we would need to regularize and renormalize,
which are concepts that will be dealt with in Part II of the book. But we can already say that
in dimensional regularization (which will be defined rigorously at the beginning of Part II),
the proof will hold, since in an arbitrary dimension, we still have $D_F = \theta\Delta^+ + \theta\Delta^-$,
as in 3+1 dimensions, so we continue to have the largest time equation, and everything
else we said here. We can, in fact, extend to any reasonable regularization method. Then
renormalization will introduce more vertices, and one has to reanalyze the proof, but it will
continue to hold.

One can show that the proof can be extended easily to spin-1/2 fields (fermions). For
gauge fields, and for spontaneously broken gauge theories, the proof is more involved,
since we have the added complication of the existence of unphysical states, but one can
also extend it here.

Important Concepts to Remember

- The Feynman propagator D_F splits into a piece with Δ^+ and a piece with Δ^-, where $\Delta^+(x-y) = \Delta^-(y-x)$.
- Cut diagrams are diagrams with D_F to the left of the cut, D_F^* to the right, and Δ^+ or Δ^- on the cut.
- We draw circles on some vertices, by complex conjugating them, and D_F connects uncircled vertices, D_F^* circled vertices, Δ^+ goes from uncircled to circled vertices, and Δ^- from circled to uncircled.

- The largest time equation relates the sum of the fully circled and the fully uncircled integrands for the graphs to minus the sum of the rest, $I + I^* = -\sum \tilde{I}$. By integrating it, we obtain the unitarity equation, and it is a proof that works diagram by diagram.
- General cut diagrams are obtained by cutting all the lines with arrows, which are lines on which we have Δ^+ and Δ^-. But then only the diagrams in which the arrows point in the direction of the flow of external energy contribute.

Further Reading

See the original works of 't Hooft and Veltman on perturbative unitarity, in particular Veltman's paper [7].

Exercises

1. Consider the diagram in Figure 22.4. Write all the possible circled diagrams for it, and the corresponding largest time equation.
2. For the case in Exercise 1, draw the arrows, and show the only remaining diagrams, and the corresponding unitarity equation.
3. For the case in Exercises 1 and 2, show that the Cutkovsky rules give the same result as the unitarity equation, and show explicitly what makes it so (so that it can be generalized).
4. How many circled diagrams are there up to (including) order λ^2 in $\lambda\phi^3$ theory?

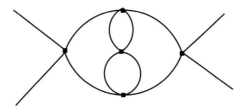

Figure 22.4   Diagram for the largest time equation in ϕ^4 theory.

23

QED: Definition and Feynman Rules; Ward–Takahashi Identities

We now turn to applications in the real world, in particular to quantum electrodynamics (QED). This is given by a gauge field (electromagnetic field) A_μ and a fermion (the electron field) ψ coupled minimally to it. For generality, we couple also minimally to a complex scalar field ϕ. In this chapter, we will define the theory and its path integral, derive the Feynman rules, and describe the Ward–Takahashi identities, which are local forms of the Ward identities (in the case of QED), giving examples for them.

23.1 QED: Definition

The Lagrangian for QED, with a complex scalar, in *Minkowski space* is

$$\mathcal{L}_{\text{QED,M}}(A, \psi, \phi) = -\frac{1}{4}F_{\mu\nu}^2 - \bar\psi(\slashed{D} + m)\psi - (D_\mu\phi)^* D^\mu\phi - V(\phi^*\phi)$$
$$= \mathcal{L}(A) + \mathcal{L}(\psi, A) + \mathcal{L}(\phi, A). \tag{23.1}$$

Here, $F_{\mu\nu} = \partial_\mu A_\nu - \partial_\nu A_\mu$, $\slashed{D} = D_\mu\gamma^\mu$ and $D_\mu = \partial_\mu - ieA_\mu$, and $\bar\psi = \psi^\dagger i\gamma^0$, $\{\gamma_\mu, \gamma_\nu\} = 2g_{\mu\nu}$.

The Lagrangian possesses a local $U(1)$ gauge invariance, under which the fields transform as

$$\psi(x) \to \psi'(x) = \psi(x)e^{ie\chi(x)},$$
$$\phi(x) \to \phi'(x) = \phi(x)e^{ie\chi(x)},$$
$$A_\mu(x) \to A'_\mu(x) = A_\mu(x) + \partial_\mu\chi(x). \tag{23.2}$$

We note that the Lagrangian at $A_\mu = 0$ has only global $U(1)$ invariance, and by the condition to make the invariance local (to "gauge it"), we need to introduce the gauge field $A_\mu(x)$ with the transformation law above.

In *Euclidean space*, since as usual $iS_M \to -S$, we get for the Euclidean fermion Lagrangian

$$\mathcal{L}_E[\psi] = \bar\psi(\gamma^\mu\partial_\mu + m)\psi, \tag{23.3}$$

where now $\{\gamma_\mu, \gamma_\nu\} = 2\delta_{\mu\nu}$. We also saw that for the gauge field we have

$$\mathcal{L}_E[A] = +\frac{1}{4}F_{\mu\nu}^{(E)}F^{\mu\nu(E)}. \tag{23.4}$$

Therefore the QED Lagrangian in Euclidean space is

$$\mathcal{L}_E(A, \psi, \phi) = +\frac{1}{4}F_{\mu\nu}^2 + \bar{\psi}(\slashed{D} + m)\psi + (D_\mu\phi)^* D^\mu \phi + V(\phi^*\phi). \qquad (23.5)$$

23.2 QED Path Integral

As we saw, defining the path integral by quantizing, we use a procedure like the Fadeev–Popov one, leading to a gauge-fixing term, with

$$\mathcal{L}_{\text{eff}} = \mathcal{L}(A) + \mathcal{L}_{\text{gauge fix}} = \frac{1}{4}F_{\mu\nu}^2 + \frac{1}{2\alpha}(\partial^\mu A_\mu)^2. \qquad (23.6)$$

The action with the gauge-fixing term can then be written as

$$S_{\text{eff}}[A] = \frac{1}{2}\int d^d x A_\mu \left(-\partial^2 \delta_{\mu\nu} + \left(1 - \frac{1}{\alpha}\right)\partial_\mu \partial_\nu\right)A_\nu = \frac{1}{2}\int d^d x A_\mu G_{\mu\nu}^{(0)-1} A_\nu. \qquad (23.7)$$

Then in momentum space the photon propagator is

$$G_{\mu\nu}^{(0)} = \frac{1}{k^2}\left(\delta_{\mu\nu} - (1 - \alpha)\frac{k_\mu k_\nu}{k^2}\right). \qquad (23.8)$$

From now on, we talk about pure QED, without the complex scalar ϕ.

The Euclidean space partition function is

$$Z[J_\mu, \xi, \bar{\xi}] = \int \mathcal{D}A \mathcal{D}\psi \mathcal{D}\bar{\psi} \, e^{-S_{\text{eff}}[A, \psi, \bar{\psi}, J, \xi, \bar{\xi}]}, \qquad (23.9)$$

where in the presence of sources

$$S_{\text{eff}} = \int d^d x [\mathcal{L}_{\text{eff}}(A, \psi, \bar{\psi}) - J_\mu A_\mu - \bar{\xi}\psi - \bar{\psi}\xi]. \qquad (23.10)$$

The VEV of an operator $\mathcal{O}(A, \bar{\psi}, \psi)$ is

$$\langle \mathcal{O}(A, \bar{\psi}, \psi)\rangle = \frac{\int \mathcal{D}A \mathcal{D}\bar{\psi} \mathcal{D}\psi \, \mathcal{O} e^{-S_{\text{eff}}}}{\int \mathcal{D}A \mathcal{D}\bar{\psi} \mathcal{D}\psi \, e^{-S_{\text{eff}}}}. \qquad (23.11)$$

The free energy F is defined as usual by

$$Z[J, \xi, \bar{\xi}] = e^{-F[J, \xi, \bar{\xi}]}. \qquad (23.12)$$

The effective action is, also as usual, the Legendre transform of the free energy

$$\Gamma[A^{cl}, \bar{\psi}^{cl}, \psi^{cl}] = F[J, \xi, \bar{\xi}] + \int d^d x [J_\mu A_\mu^{cl} + \bar{\xi}\psi^{cl} + \bar{\psi}^{cl}\xi]. \qquad (23.13)$$

By taking derivatives with respect to $J_\mu, \xi, \bar{\xi}, A_\mu^{cl}, \psi^{cl}, \bar{\psi}^{cl}$ of the Legendre transform relation, we obtain

$$\frac{\delta F}{\delta J_\mu} = -A_\mu^{cl}; \quad \frac{\delta F}{\delta \bar{\xi}} = -\psi^{cl}; \quad \frac{\delta F}{\delta \xi} = \bar{\psi}^{cl}$$

$$\frac{\delta \Gamma}{\delta A_\mu^{cl}} = J_\mu; \quad \frac{\delta \Gamma}{\delta \psi^{cl}} = -\bar{\xi}; \quad \frac{\delta \Gamma}{\delta \bar{\psi}^{cl}} = \xi. \qquad (23.14)$$

23.3 QED Feynman Rules

23.3.1 Feynman Rules for Green's Functions in Euclidean Momentum Space

It is easier to calculate Green's functions in Euclidean space, hence we first write the Feynman rules in Euclidean space (see Figure 23.1(a)):

- The ψ propagator from α to β (arrow from α to β), with momentum p, is

$$\left(\frac{1}{i\not{p}+m}\right)_{\alpha\beta} = \frac{(-i\not{p}+m)_{\alpha\beta}}{p^2+m^2}. \tag{23.15}$$

- The photon propagator, from μ to ν, is

$$\frac{1}{k^2}\left(\delta_{\mu\nu} - (1-\alpha)\frac{k_\mu k_\nu}{k^2}\right). \tag{23.16}$$

- Vertex $+ie(\gamma^\mu)_{\alpha\beta}$ for a photon with μ to come out of a line from α to β. This is because the interaction term in the action is $S_{\text{int}} = -ie\int\bar{\psi}\gamma^\mu A_\mu\psi$, and we know that for an interaction $+g\int\prod_i\phi_i$, the vertex is $-g$.
- A minus sign for a fermion loop.

23.3.2 Feynman Rules for S-Matrices in Minkowski Space

S-matrices can only be defined in Minkowski space, since we need to have external states, satisfying $p^2 + m^2 = 0$, but that equation has no real solution in Euclidean space, only in Minkowski space. For the graphical representation of these Feynman rules, see Figure 23.1(b), (c). They are:

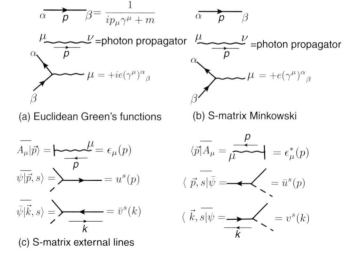

(a) Euclidean Green's functions   (b) S-matrix Minkowski

(c) S-matrix external lines

Figure 23.1   Relevant Feynman rules for Green's functions and S-matrices.

- The fermion propagator can be found from the Euclidean space one by the replacement $i\not{p} + m \to i\not{p} + m$, $p^2 + m^2 \to p^2 + m^2 - i\epsilon$, and an extra $-i$ in front (from the x-space continuation of t in the action), giving

$$-i\frac{-i\not{p} + m}{p^2 + m^2 - i\epsilon} = \frac{-(\not{p} + im)}{p^2 + m^2 - i\epsilon}. \tag{23.17}$$

- The photon propagator. The same continuation rules give

$$-\frac{i}{k^2 - i\epsilon}\left(g_{\mu\nu} - (1 - \alpha)\frac{k_\mu k_\nu}{k^2 - i\epsilon}\right). \tag{23.18}$$

- The vertex is $+e(\gamma^\mu)_{\alpha\beta}$ (there is an i difference between the Euclidean and Minkowski vertices).
- A minus sign for fermion loops.
- For the external fermion lines: for the $\overline{\psi\,|\vec{p}, s} >$ contraction, $= u^s(p)$, for the $\overline{\overline{\psi}\,|\vec{p}, s} >$ contraction, $= \bar{v}^s(p)$, and for the bars, similar: $< \overline{\vec{p}, s|\psi} = \bar{u}^s(p)$ and $< \overline{\vec{p}, s|\overline{\psi}} = v^s(p)$.
- For the external photon lines, $\overline{A_\mu\,|\vec{p}} >= \epsilon_\mu(p)$ and $< \overline{\vec{p}|A_\mu} = \epsilon^*(p)$.
- We can also see that if we exchange the order of the external fermion line contractions, we get a minus sign, so we deduce that if we exchange identical external fermion lines in otherwise identical diagrams, we get a minus sign.

23.4 Ward–Takahashi Identities

These are a type of Ward identities for QED, that is a local version of the global Ward identities we studied before. Consider an infinitesimal local gauge-invariance transformation

$$\delta\psi(x) = ie\epsilon(x)\psi(x),$$
$$\delta A_\mu(x) = \partial_\mu\epsilon(x). \tag{23.19}$$

If we consider this as just a change (renaming) of integration variables in the path integral, the path integral should not change. In contrast, *assuming that the regularization of the path integral respects gauge invariance*, we also expect that there is no nontrivial Jacobian, and $\delta(\mathcal{D}A\mathcal{D}\psi\mathcal{D}\phi) = 0$.

Note that this assumption is in principle nontrivial. There are classical symmetries that are broken by the quantum corrections. In the path integral formalism, the way this happens is that the regularization of the path integral measure (i.e. defining better $\mathcal{D}\phi$s as $\prod_i d\phi(x_i)$s) does not respect the symmetries (the only other ingredient of the path integral is the classical action, which is invariant). These are called *quantum anomalies*. But as we will see in Part II of the book, anomalies in gauge symmetries are actually bad, and make the theory sick, so should be absent from a well-defined regularization. Therefore we will assume that the regularization of the measure respects gauge invariance.

Since, as we said, the change is just a redefinition of the path integral, we also have

$$\delta Z[J, \xi, \bar{\xi}] = 0. \tag{23.20}$$

208
23 QED: Definition and Feynman Rules

Using the form of the partition function, and the fact that both the classical action and the measure are invariant, we only need to evaluate the change due to the gauge-fixing and source terms. This gives

$$0 = \delta Z = \int \mathcal{D}A\mathcal{D}\psi\mathcal{D}\bar{\psi} e^{-S_{\text{eff}}} \int d^d x \left[J^\mu \partial_\mu \epsilon + ie\epsilon(\bar{\xi}\psi - \bar{\psi}\xi) - \frac{1}{\alpha}(\partial_\mu A_\mu)\partial^2\epsilon \right],$$
(23.21)

where the last term comes from varying the gauge-fixing term, and the others from varying the source terms.

Taking the derivative with respect to ϵ, and then expressing A, ψ, $\bar{\psi}$ as derivatives of S_{eff} (e.g. $A_\mu e^{-S_{\text{eff}}} = \delta/\delta J_\mu e^{-S_{\text{eff}}}$), taking out of the path integral the remaining objects that are not integrated over, we are left with an operator that acts on Z, namely

$$0 = \frac{\delta Z}{\delta\epsilon(x)} = -\frac{1}{\alpha}\partial^2\partial_\mu \frac{\delta Z}{\delta J_\mu} - (\partial_\mu J_\mu)Z + ie\left(\bar{\xi}\frac{\delta Z}{\delta\bar{\xi}} + \frac{\delta Z}{\delta\xi}\xi\right).$$
(23.22)

Dividing by Z and using the fact that $\ln Z = -F$, we get

$$0 = \frac{1}{\alpha}\partial^2\partial_\mu \frac{\delta F}{\delta J_\mu} - \partial_\mu J_\mu - ie\left(\bar{\xi}\frac{\delta F}{\delta\bar{\xi}} + \frac{\delta F}{\delta\xi}\xi\right).$$
(23.23)

Then replacing the equations (23.14) in the above, we finally get

$$-\frac{1}{\alpha}\partial^2\partial_\mu A_\mu^{cl} - \partial_\mu \frac{\delta\Gamma}{\delta A_\mu^{cl}} - ie\left(\frac{\delta\Gamma}{\delta\psi^{cl}}\psi^{cl} + \bar{\psi}^{cl}\frac{\delta\Gamma}{\delta\bar{\psi}^{cl}}\right) = 0.$$
(23.24)

These are called *generalized Ward–Takahashi identities*, or (for the 1PI functions, of which we will shortly derive some examples) the *Lee–Zinn-Justin (LZJ) identities*.

23.4.1 Example 1: Photon Propagator

By taking derivatives of the LZJ identities with respect to $A_\mu^{cl}(y)$ and putting $A^{cl}, \psi^{cl}, \bar{\psi}^{cl}$ to zero, that is

$$\frac{\delta}{\delta A_\nu^{cl}(y)}(L - Z - J)\bigg|_{A^{cl}=\psi^{cl}=\bar{\psi}^{cl}=0},$$
(23.25)

we get

$$-\frac{1}{\alpha}\partial_x^2\partial_{x^\nu}\delta(x-y)-\partial_{x^\mu}\frac{\delta^2\Gamma}{\delta A_\mu^{cl}(x)\delta A_\nu^{cl}(y)}\bigg|_{A^{cl}=\psi^{cl}=\bar{\psi}^{cl}=0} = -\frac{1}{\alpha}\partial_x^2\partial_{x^\nu}\delta(x-y)-\partial_{x^\mu}G_{\mu\nu}^{-1}(x,y) = 0,$$
(23.26)

where $G_{\mu\nu}^{-1}$ is the inverse of the full connected propagator.

Indeed, as we saw before, this equals the two-point effective action (i.e. the second derivative of the effective action). Taking the Fourier transform, we obtain

$$k^\mu\left[\frac{k^2}{\alpha}\delta_{\mu\nu} - G_{\mu\nu}^{-1}(k)\right] = 0.$$
(23.27)

On the contrary, we have

$$G_{\mu\nu}^{-1}(k) = \Gamma_{\mu\nu}(k) = G_{\mu\nu}^{(0)-1} + \Pi_{\mu\nu},$$
(23.28)

and since

$$G_{\mu\nu}^{(0)-1}(k) = k^2 \delta_{\mu\nu} - k_\mu k_\nu + \frac{1}{\alpha} k_\mu k_\nu, \tag{23.29}$$

we have

$$k^\mu G_{\mu\nu}^{(0)-1}(k) = \frac{k^2 k_\nu}{\alpha}. \tag{23.30}$$

Replacing it in (23.27), we get

$$k^\mu \Pi_{\mu\nu}(k) = 0. \tag{23.31}$$

That is, the 1PI corrections to the photon propagator are transverse (perpendicular to the momentum k^μ). This means that we can write them in terms of a single (scalar) function:

$$\Pi_{\mu\nu}(k) = (k^2 \delta_{\mu\nu} - k_\mu k_\nu)\Pi(k). \tag{23.32}$$

23.4.2 Example 2: n-Photon Vertex Function for $n \geq 3$

We now take $n - 1$ derivatives with respect to A_μ^{cl}, that is we consider

$$\left. \frac{\delta^{n-1}}{\delta A_{\mu_2}^{cl}(x_2) \ldots \delta A_{\mu_n}^{cl}(x_n)}(LZJ) \right|_{A^{cl}=\psi^{cl}=\bar\psi^{cl}=0} \tag{23.33}$$

and find

$$\left. \frac{\partial}{\partial x_{\mu_1}} \frac{\delta^{(n)}\Gamma}{\delta A_{\mu_1}^{cl}(x_1) \ldots \delta A_{\mu_n}^{cl}(x_n)} \right|_{A^{cl}=\psi^{cl}=\bar\psi^{cl}=0} = 0. \tag{23.34}$$

Going to momentum space, we then find

$$k^{\mu_1} \Gamma_{\mu_1 \ldots \mu_n}^{(n)}(k^{(1)}, \ldots, k^{(n)}) = 0. \tag{23.35}$$

That is, the n-point 1PI photon vertex functions (from the effective action) are also transverse, like the 1PI two-point function above.

23.4.3 Example 3: Original Ward–Takahashi Identity

We take the derivatives with respect to ψ^{cl} and $\bar\psi^{cl}$, that is we consider

$$\left. \frac{\delta^2}{\delta\psi_\beta^{cl}(z)\delta\bar\psi_\alpha^{cl}(y)}(L - Z - J) \right|_{A^{cl}=\psi^{cl}=\bar\psi^{cl}=0}, \tag{23.36}$$

to obtain

$$0 = \left[-\frac{\partial}{\partial x^\mu} \frac{\delta^3\Gamma}{\delta\psi_\beta^{cl}(z)\delta\bar\psi^{cl}(y)\delta A_\mu(x)} - ie\delta^{(d)}(x-y)\frac{\delta^2\Gamma}{\delta\psi_\beta^{cl}(z)\delta\bar\psi_\alpha^{cl}(x)} \right.$$

$$\left. + ie\delta^{(d)}(x-z)\frac{\delta^2\Gamma}{\delta\psi_\beta^{cl}(x)\delta\bar\psi_\alpha^{cl}(y)} \right]_{A^{cl}=\psi^{cl}=\bar\psi^{cl}=0}. \tag{23.37}$$

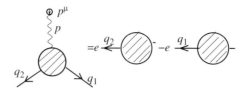

Figure 23.2   Diagrammatic representation of the original Ward–Takahashi identity.

Replacing the derivatives with the corresponding n-point effective action terms, we get

$$-\frac{\partial}{\partial x^\mu}\Gamma_{\mu;\alpha\beta}(x;y,z) = -ie\delta^{(d)}(x-z)(S_F^{-1})_{\alpha\beta}(y-x)+ie\delta^{(d)}(x-y)(S_F^{-1})_{\alpha\beta}(x-z), \quad (23.38)$$

where S_F^{-1} is the inverse of the full connected fermion propagator (i.e. the 1PI two-point effective action).

In momentum space, this relation becomes

$$p^\mu\Gamma_{\mu\alpha\beta}(p;q_2,q_1) = e(S_F^{-1}(q_2)_{\alpha\beta} - S_F^{-1}(q_1)_{\alpha\beta}), \quad (23.39)$$

which is the original Ward–Takahashi identity: the photon–fermion–antifermion vertex contracted with the momentum gives the difference of the inverse fermion propagator (or 1PI two-point function) for q_2 minus that for q_1, where $q_{1,2}$ are the fermion momenta, as in Figure 23.2.

Important Concepts to Remember

- QED is a gauge field (electromagnetic field) coupled to a fermion (generalized to include a coupling to a complex scalar).
- The generalized Ward–Takahashi identities or Lee–Zin-Justin identities, are identities between n-point effective actions, or 1PI n-point functions, resulting from local versions of Ward identities.
- The 1PI corrections to the photon propagator are transverse, $p^\mu\Pi_{\mu\nu}(k) = 0$.
- The 1PI n-photon vertex functions for $n \geq 3$ are also transverse, $k^{\mu_1}\Gamma_{\mu_1...\mu_n} = 0$.
- The original Ward–Takahashi identity relates the photon–fermion–antifermion vertex, contracted with the photon momentum, to the difference of the inverse photon propagator (two-point effective action) for the two fermions.

Further Reading

See section 4.8 in [1], section 6.3 in [3], and section 8.1 in [2].

Exercises

1. Write down the expression for the QED Feynman diagram in Figure 23.3 for the S-matrix $e^-(p_1)e^+(p_2) \rightarrow e^-(p_3)e^+(p_4)$.
2. Using the Lee–Zinn-Justin identities, derive a Ward identity for the vertex $\Gamma_{\mu_1\mu_2\alpha\beta}$.
3. Write down the one-loop diagram for photon-on-photon scattering $\gamma + \gamma \rightarrow \gamma + \gamma$ in QED, and use the S-matrix Feynman rules to write down the resulting integral expression.
4. Write down all the Ward-type identities obtained by taking derivatives with respect to gauge fields A_μ of the original Ward–Takahashi identity.

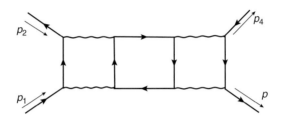

Figure 23.3   Feynman diagram for the S-matrix in QED.

24

Nonrelativistic Processes: Yukawa Potential, Coulomb Potential, and Rutherford Scattering

In this chapter we will rederive some formulas for classical, nonrelativistic processes using the quantum field theory formalism we have developed, to check that our formalism works, and understand how it is applied. In particular, we will calculate the Yukawa and Coulomb potentials from our formalism, and derive the differential cross-section for Rutherford scattering.

24.1 Yukawa Potential

Even though we are more interested in the QED example of the Coulomb potential, we will start with the simplest example, the Yukawa potential, as a warm-up.

In Yukawa theory, the interaction of two fermions, $\psi(p) + \psi(k) \rightarrow \psi(p') + \psi(k')$, happens via the exchange of a scalar particle ϕ, of nonzero mass m_ϕ. We want to see that this indeed gives the well-known Yukawa potential.

There are two diagrams that in principle contribute to $i\mathcal{M}$. One where the scalar with momentum $q = p - p'$ is exchanged between the two fermions (Figure 24.1(a)) and one where the final fermions are interchanged (the fermion with momentum p' comes out of the one with k, and the fermion with momentum k' comes out of the one with p) (Figure 24.1(b)). But we will work with *distinguishable fermions* (i.e. we will assume that there is some property which we can measure that will distinguish between them, for instance, we could consider one to be an e^-, and another to be a μ^-). In this case, the second diagram will not contribute. Since this is what happens classically, it is a good assumption in order to get a classical potential.

In the nonrelativistic limit, we have

$$p \simeq (m, \vec{p}); \quad p' \simeq (m, \vec{p}'); \quad k = (m, \vec{k}); \quad k' = (m, \vec{k}'),$$

$$(p - p')^2 \simeq (\vec{p} - \vec{p}')^2,$$

$$u^s(p) \simeq \sqrt{m} \begin{pmatrix} \xi^s \\ \xi^s \end{pmatrix}, \tag{24.1}$$

where $\xi^s = \begin{pmatrix} 1 \\ 0 \end{pmatrix}$ and $\begin{pmatrix} 0 \\ 1 \end{pmatrix}$ for $s = 1, 2$.

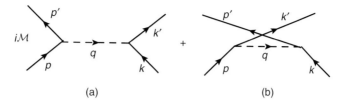

Figure 24.1 The two Feynman diagrams for the potential in the case of Yukawa interaction. For distinguishable particles, only diagram (a) contributes.

The last property means that

$$\bar{u}^{s'}(p')u^s(p) = u^{\dagger s'}(i\gamma^0)u^s = m\left(\xi^{\dagger s'} \quad \xi^{\dagger s'}\right)\begin{pmatrix}0 & 1\\ 1 & 0\end{pmatrix}\begin{pmatrix}\xi^s\\ \xi^s\end{pmatrix} = 2m\xi^{\dagger s'}\xi^s = 2m\delta^{ss'}. \quad (24.2)$$

Then, writing both terms in $i\mathcal{M}$ for completeness, we have

$$i\mathcal{M} = (-ig)\bar{u}(p')u(p)\frac{-i}{(p-p')^2 + m_\phi^2}(-ig)\bar{u}(k')u(k)$$
$$-(-ig)\bar{u}(p')u(k)\frac{-i}{(p-k')^2 + m_\phi^2}(-ig)\bar{u}(k')u(p), \quad (24.3)$$

but as we said, we will drop the second term since we consider distinguishable fermions. Some comments are in order here. We have considered a Yukawa coupling g, which means a vertex factor $-ig$ in Minkowski space. The fermions are contracted along each of the fermion lines separately. The overall sign for the fermions is conventional, only the relative sign between diagrams is important, and that can be found from the simple rule that when we interchange two fermions between otherwise identical diagrams (like interchanging $u(p)$ with $u(k)$ in the two terms above), we add a minus sign.

But we can choose a convention. We can define

$$|\vec{p},\vec{k}\rangle \sim a_{\vec{p}}^\dagger a_{\vec{k}}^\dagger|0\rangle; \quad \langle\vec{p}',\vec{k}'| = (|\vec{p}',\vec{k}'\rangle)^\dagger \sim \langle 0|a_{\vec{k}'}a_{\vec{p}'}, \quad (24.4)$$

so that in the matrix element

$$\langle\vec{p}',\vec{k}'|(\bar{\psi}\psi)_x(\bar{\psi}\psi)_y|\vec{p},\vec{k}\rangle \sim \langle 0|a_{\vec{k}'}a_{\vec{p}'}(\bar{\psi}\psi)_x(\bar{\psi}\psi)_y a_{\vec{p}}^\dagger a_{\vec{k}}^\dagger|0\rangle, \quad (24.5)$$

the contraction of $\vec{k}'$ with $\bar{\psi}_x$, $\vec{p}'$ with $\bar{\psi}_y$, $\vec{p}$ with ψ_y, and $\vec{k}$ with ψ_x, corresponding to the first diagram (the one we keep), gives a plus sign (we can count how many jumps we need for the fermions to be able to contract them, and it is an even number).

Then we can easily see that the second diagram, where we now contract $\vec{k}'$ with $\bar{\psi}_y$ and $\vec{p}'$ with $\bar{\psi}_x$ instead, has a minus sign. As we said, the simple Feynman rule above accounts for this relative sign.

We then finally have for the contribution of the first diagram (for distinguishable fermions)

$$i\mathcal{M} \simeq +\frac{ig^2}{|\vec{p}-\vec{p}'|^2 + m_\phi^2}(2m\delta^{ss'})(2m\delta^{rr'}). \quad (24.6)$$

But we can then compare with the Born approximation to the scattering amplitude in nonrelativistic quantum mechanics, written in terms of the potential in momentum space:

$$\langle \vec{p}' | iT | \vec{p} \rangle = -iV(\vec{q}) 2\pi \delta(E_{p'} - E_p), \tag{24.7}$$

where $\vec{q} = \vec{p}' - \vec{p}$.

When comparing, we should remember a few facts: the $2m$ factors are from the relativistic normalization we used for the fermions, whereas the Born approximation above uses nonrelativistic normalization, so we should drop them in the comparison. Then $iT = 2\pi \delta(E_f - E_i) i\mathcal{M}$, so we drop the $2\pi \delta(E_{p'} - E_p)$ in the comparison; and also the formula is at $s = s', r = r'$. We finally get

$$V(\vec{q}) = \frac{-g^2}{|\vec{q}|^2 + m_\phi^2}. \tag{24.8}$$

We should also comment on the logic: it is impossible to directly measure the potential $V(\vec{q})$ in the scattering, since this is a classical concept, so we can only measure things that depend on it. But for expedience's sake, we used the comparison with nonrelativistic quantum mechanics, where we have a quantum mechanics calculation, yet in a classical potential, in order to directly extract the result for $V(\vec{q})$.

The x-space potential is (here $|\vec{x}| \equiv r$)

$$\begin{aligned}
V(\vec{x}) &= \int \frac{d^3q}{(2\pi)^3} \frac{-g^2}{|\vec{q}|^2 + m_\phi^2} e^{i\vec{q} \cdot \vec{x}} \\
&= -\frac{g^2}{(2\pi)^3} 2\pi \int_0^\infty q^2 dq \frac{e^{iqr} - e^{-iqr}}{iqr} \frac{1}{q^2 + m_\phi^2} \\
&= -\frac{g^2}{4\pi^2 ir} \int_{-\infty}^{+\infty} dq \frac{q e^{iqr}}{q^2 + m_\phi^2}.
\end{aligned} \tag{24.9}$$

Here, in the second equality we have used $\int_0^\pi \sin\theta d\theta e^{iqr\cos\theta} = \int_0^1 d(\cos\theta) e^{(iqr)\cos\theta} = (e^{iqr} - e^{-iqr})/(iqr)$ and in the last equality we have rewritten the second term as an integral from $\int_{-\infty}^0$. The integral has complex poles at $q = \pm im_\phi$, so we can close the contour over the real axis with an infinite semicircle in the upper half plane, since then e^{iqr} decays exponentially at infinity. We then pick up the residue of the pole $q = im_\phi$, giving finally

$$V(r) = \frac{-g^2}{4\pi r} e^{-m_\phi r}, \tag{24.10}$$

the well-known attractive Yukawa potential.

24.2 Coulomb Potential

We can now move to the case of nonrelativistic scattering in QED, as in Figure 24.2. The difference is now that we have a vector exchanged between the two fermions, with a

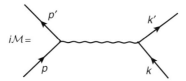

Figure 24.2 Feynman diagram for the Coulomb interaction between distinguishable fermions.

(massless) vector propagator, in the Feynman gauge just $g_{\mu\nu}$ times the KG propagator, and a vertex factor $+e\gamma^\mu$ in Minkowski space.

We then get

$$i\mathcal{M} = (+e)\bar{u}(p')\gamma^\mu u(p)\frac{-ig_{\mu\nu}}{(p-p')^2}(+e)\bar{u}(k')\gamma^\nu u(k). \qquad (24.11)$$

But in the nonrelativistic limit we can check that the only nonzero components are the $\mu = 0$ ones (as expected), so we only need to calculate

$$\bar{u}(p')\gamma^0 u(p) = u^\dagger(p')i(\gamma^0)^2 u(p) \simeq -im\left(\xi^{\dagger s'}\quad \xi^{\dagger s'}\right)\begin{pmatrix}\xi^s\\\xi^s\end{pmatrix} = -2im\delta^{ss'} \qquad (24.12)$$

(since $(\gamma^0)^2 = -1$, from $\{\gamma^\mu,\gamma^\nu\} = 2g^{\mu\nu}$), finally obtaining

$$i\mathcal{M} \simeq \frac{ie^2 g_{00}}{|\vec{p}-\vec{p}'|^2}(2m\delta^{ss'})(2m\delta^{rr'}) = -\frac{ie^2}{|\vec{p}'-\vec{p}|^2}(2m\delta^{ss'})(2m\delta^{rr'}). \qquad (24.13)$$

Thus there is a sign change with respect to the Yukawa potential, meaning that the Coulomb potential is repulsive. Also, the mass is now zero. To go to the configuration space, it is easier to take the zero-mass limit of the Yukawa potential, getting

$$V(r) = \frac{+e^2}{4\pi r} = \frac{\alpha}{r}. \qquad (24.14)$$

Note that $\alpha = e^2/(4\pi) \simeq 1/137$ is the electromagnetic coupling, which is very weak.

24.3 Particle–Antiparticle Scattering

24.3.1 Yukawa Potential

We now consider scattering a fermion off an antifermion in Yukawa theory. Since now the second fermion line is antifermionic, with momentum k in and momentum k' out, as in Figure 24.3, we have a few changes from the calculation before. First, we replace the us with vs for the second line, that is

$$\bar{u}^{s'}(k')u^s(k) \to \bar{v}^s(k)v^{s'}(k') \simeq m\left(\xi^{\dagger s}\quad -\xi^{\dagger s}\right)\begin{pmatrix}0 & 1\\1 & 0\end{pmatrix}\begin{pmatrix}\xi^{s'}\\-\xi^{s'}\end{pmatrix} = -2m\delta^{ss'}, \qquad (24.15)$$

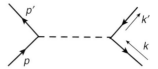

Figure 24.3   Feynman diagram for the Yukawa potential interaction between particle and antiparticle.

which gives a minus sign with respect to the fermion–fermion calculation. But also changing the fermion into an antifermion amounts to exchanging the fermion lines with momenta k and k', as we can check. Another way to say this is that if we still do the contraction like in the fermion–fermion case (i.e. such that the order is k' first, then k), we obtain $v(k')\bar{v}(k)$, and in order to obtain the Lorentz invariant $\bar{v}(k)v(k')$, we need to change the order of the fermions.

All in all, we obtain two minus signs (i.e. a plus). Thus the potential between fermion and antifermion is the same as that between two fermions: scalar particle exchange gives a universally attractive potential.

24.3.2 Coulomb Potential

Considering now fermion–antifermion scattering in QED, we again have the same minus sign for exchanging the fermions, and exchanging the us with vs in the second fermion line gives

$$\bar{v}(k)\gamma^0 v(k') = v^\dagger(k)i(\gamma^0)^2 v(k) = -im\left(\xi^{\dagger s} - \xi^{\dagger s}\right)\begin{pmatrix} \xi^s \\ -\xi^s \end{pmatrix} = -2im\delta^{ss'}, \qquad (24.16)$$

instead of

$$\bar{u}(k)\gamma^0 u(k') = -2im\delta^{ss'}, \qquad (24.17)$$

so no change here. Thus, overall, we have a minus sign with respect to fermion–fermion scattering (i.e. the fermion–antifermion potential is now *attractive*). Indeed, e^+e^- attract, whereas e^-e^- repel.

In conclusion, for the exchange of a vector particle (electromagnetic field in this case), like charges repel, and opposite charges attract.

We note that the repulsive nature of the fermion–fermion potential was due to the presence of g_{00} in (24.13).

We can now guess also what happens for an exchange of a tensor (i.e. a spin-2 particle, the graviton), without doing the full calculation. Since a tensor has two indices, we will have propagation between a point with indices $\mu\nu$ and one with indices $\rho\sigma$, which in some appropriate gauge (the equivalent of the Feynman gauge) should be proportional to $g_{\mu\rho}g_{\nu\sigma} + g_{\mu\sigma}g_{\nu\rho}$, and in the nonrelativistic limit only the 00 components should contribute, giving $\propto (g_{00})^2 = +1$.

Therefore we can guess that gravity is attractive, and moreover universally attractive, like we know from experiments to be the case.

24.4 Rutherford Scattering

We now calculate an example of a nonrelativistic scattering cross-section, namely the case of Rutherford scattering, of a charged particle ("electron") off a fixed electromagnetic field ("field of a nucleus").

We should remember that in the classic experiment of Rutherford which determined that the atoms are made up of a positively charged nucleus, surrounded by electrons (as opposed to a "raisin pudding," a ball of constant density of positive charge filled with electrons), where charged particles were scattered on a fixed target made of a metal foil, Rutherford made a classical calculation for the scattering cross-section, which was confirmed by experiment, thus proving the assumption.

We will consider the scattering of an electron off the constant electromagnetic field of a fixed nucleus. We will thus treat the electromagnetic field classically, and just the fermion quantum mechanically.

The interaction Hamiltonian is

$$H_I = \int d^3x\, e\bar{\psi} i\gamma^\mu \psi A_\mu. \tag{24.18}$$

Then the S-matrix contribution, coming from $\sim e^{-i\int H_I dt}$, the first-order contribution is

$$\langle p'|iT|p\rangle = \langle p'|T\left\{-i\int d^4x\, \bar{\psi} ie\gamma^\mu \psi A_\mu\right\}|p\rangle$$

$$= -ie\bar{u}(p')i\gamma^\mu u(p)\int d^4x\, e^{ipx}e^{-ip'x}A_\mu(x)$$

$$= +e\bar{u}(p')\gamma^\mu u(p)A_\mu(p-p'). \tag{24.19}$$

But if we consider that $A_\mu(x)$ is time-independent, meaning that

$$A_\mu(p-p') = A_\mu(\vec{p}-\vec{p}')2\pi\delta(E_f - E_i), \tag{24.20}$$

and since the $\mathcal{M}$ matrix is defined by

$$\langle p'|iT|p\rangle = i\mathcal{M}2\pi\delta(E_f - E_i), \tag{24.21}$$

we see that in the Feynman rules for $i\mathcal{M}$ we can add a rule for the interaction with a classical field, namely add $+e\gamma^\mu A_\mu(\vec{q})$, with $\vec{q} = \vec{p} - \vec{p}'$, with a diagram: off the fermion line starts a wavy photon line that ends on a circled cross, as in Figure 24.4.

We now calculate the cross-section for scattering of the electron off the classical field centered at a point situated at an impact parameter $\vec{b}$ from the line of the incoming electron

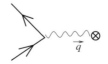

Figure 24.4   Feynman diagram representation for the interaction of a fermion with a classical potential.

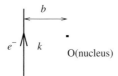

Figure 24.5 The impact parameter for scattering of an electron off a nucleus (fixed classical object).

(or rather, the impact parameter is the distance of the incoming electron from the nucleus), as in Figure 24.5. This is in principle a $2 \to 2$ scattering, even though we treat the nucleus and its electromagnetic field classically. As such, we must consider a situation similar to the one in $2 \to 2$ scattering, by writing incoming wavefunctions with impact parameter $\vec{b}$ as

$$\phi_i = \int \frac{d^3 k_i}{(2\pi)^3} \frac{\phi(k_i)}{\sqrt{2E_i}} e^{i\vec{k}_i \cdot \vec{b}}. \tag{24.22}$$

As usual, the probability for $i \to f$ is

$$dP(i \to f) = \frac{d^3 p_f}{(2\pi)^3 2E_f} |_{out}\langle p_f | \phi_A \rangle_{in}|^2, \tag{24.23}$$

and the cross-section element is

$$d\sigma = \int d^2 \vec{b} \, dP(\vec{b})$$

$$= \int d^2 b \frac{d^3 p_f}{(2\pi)^3 2E_f} \int \frac{d^3 k_i \phi(k_i)}{(2\pi)^3 \sqrt{2E_i}} \int \frac{d^3 \bar{k}_i \phi(\bar{k}_i)^*}{(2\pi)^3 \sqrt{2E_i}} e^{i\vec{b}(\vec{k}_i - \vec{\bar{k}}_i)} {}_{out}\langle p_f | k_i \rangle_{in} ({}_{out}\langle p_f | k_i \rangle_{in})^*. \tag{24.24}$$

We then use the relations

$$_{out}\langle p_f | k_i \rangle_{in} = i\mathcal{M}(i \to f) 2\pi \delta(E_f - E_i),$$

$$(_{out}\langle p_f | k_i \rangle_{in})^* = -i\mathcal{M}^*(i \to f) 2\pi \delta(E_f - \bar{E}_i),$$

$$\int d^2 b \, e^{i\vec{b}(\vec{k}_i - \vec{\bar{k}}_i)} = (2\pi)^2 \delta^2(k_i^\perp - \bar{k}_i^\perp),$$

$$\delta(E_f - E_i)\delta(E_f - \bar{E}_i) = \delta(E_f - E_i) \frac{\delta(\bar{k}_i^z - k_i^z)}{\left|\frac{\bar{k}_i^z}{\bar{E}_i}\right|},$$

$$\int \frac{d^3 p_f}{(2\pi)^3 2E_f} 2\pi \delta(E_f - E_i) = d\Omega \int \frac{dp_f p_f^2}{(2\pi)^3 2E_f} 2\pi \delta(E_f - E_i) = \frac{d\Omega}{8\pi^2} \int \frac{dp_f p_f^2}{E_f} \frac{E_f}{p_f} \delta(p_f - p_i)$$

$$= \frac{d\Omega}{8\pi^2} \int dp_f p_f \delta(p_f - p_i), \tag{24.25}$$

where we have used $\delta(\bar{E}_i - E_i) = \delta(\bar{k}_i^z - k_i^z)/|\bar{k}_i^z/E_i|$ and $\delta(E_f - E_i) = \delta(p_f - p_i)/|p_f/E_f|$. Using the wavefunctions peaked on the value p_i (i.e. $|\phi(k_i)|^2 = (2\pi)^3 \delta^3(k_i - p_i)$), and since $\bar{k}_i^z = v_i E_i$, putting everything together, we finally have

$$\frac{d\sigma}{d\Omega} = \frac{1}{16\pi^2} \frac{1}{v_i E_i} \int dp_f p_f \delta(p_f - p_i) |\mathcal{M}(p_i \to p_f)|^2, \tag{24.26}$$

where $E_i \simeq m$.

We now apply this to the case of the electric field of a nucleus of charge $+Ze$, that is we take

$$A_0 = \frac{+Ze}{4\pi r},$$ (24.27)

with Fourier transform

$$A_0(q) = \frac{Ze}{|\vec{q}|^2},$$ (24.28)

inside the matrix element

$$i\mathcal{M} = e\bar{u}(p_f)\gamma^\mu u(p_i)A_\mu(\vec{p}_f - \vec{p}_i).$$ (24.29)

In the nonrelativistic limit, as we saw (and summing over s', final-state helicities):

$$|\bar{u}(p_f)\gamma^\mu u(p_i)|^2 \simeq \sum_{s'} |2m(\xi^{\dagger s'}\xi^s)|^2 \delta^{\mu 0} = 4m^2 \delta^{\mu 0},$$ (24.30)

giving

$$|\mathcal{M}|^2 = e^2 4m^2 |A_0(\vec{p}_f - \vec{p}_i)|^2 = e^2 4m^2 \frac{Z^2 e^2}{|\vec{p}_f - \vec{p}_i|^4}.$$ (24.31)

Since $E_f = E_i$ (energy conservation), we have $|\vec{p}_f| = |\vec{p}_i|$, and if they have an angle θ between them, $|\vec{p}_f - \vec{p}_i| = 2p_i \sin\theta/2 \simeq 2mv_i \sin\theta/2$, see Figure 24.6.

Putting everything together, we find

$$\frac{d\sigma}{d\Omega} \simeq \frac{1}{8\pi^2} \frac{mv_i}{v_i 2m} \frac{e^2 4m^2 Z^2 e^2}{16m^4 v_i^4 \sin^4 \theta/2}$$

$$= \frac{Z^2 \alpha^2}{4m^2 v_i^4 \sin^4 \theta/2},$$ (24.32)

which is the Rutherford formula, originally found by Rutherford with a classical calculation.

Important Concepts to Remember

- For distinguishable fermions, fermion–fermion scattering has only one diagram, the exchange of the force particle, which can be scalar for Yukawa interaction, vector for Coulomb interaction, tensor for gravitational interaction, and so on.
- In the nonrelativistic limit, the scalar exchange diagram gives rise to the Yukawa potential. Since the potential is not directly measurable, we calculate it by matching with the nonrelativistic quantum-mechanical calculation of the Born approximation for scattering in a given potential. The Yukawa potential is attractive.

Figure 24.6   The scattering angle.

- In the Coulomb case, due to the $g_{00} = -1$ in the vector propagator, we have a repulsive process for like charges (fermion–fermion scattering).
- For fermion–antifermion scattering, we find the opposite sign in the potential for Coulomb, and the same sign in the potential for Yukawa.
- To derive the classical Rutherford formula for nonrelativistic scattering of an electron (or charged particle in general) off a fixed nucleus, we treat the electromagnetic field generated by the nucleus as a time-independent classical field, interacting with the quantum field of the incoming electron.

Further Reading

See sections 4.7, 4.8 and exercises in [1].

Exercises

1. Consider a Yukawa interaction with two scalars, ϕ_1, ϕ_2, with masses m_1, m_2. Using the quantum-mechanical argument in the text, calculate the potential $V(\vec{x})$. What happens for $m_1 \ll m_2$? How is the above result for $V(\vec{x})$ consistent with the linear principle of quantum mechanics (wavefunctions add, not probabilities)?

2. Integrate $d\sigma/d\Omega$ for Rutherford scattering to find σ_{tot} (comment on the result). The $d\sigma/d\Omega$ formula is the same as for classical scattering in a Coulomb potential (that's how Rutherford computed it). Yet we used a quantum field theory calculation. Argue how the various steps of the quantum field theory calculation would turn into steps of a classical calculation.

3. Consider a massive vector B_μ, with mass m (mass term $-m^2/2B_\mu B^\mu$) interacting with fermions like the gauge field A_μ of electromagnetism. Write the Feynman diagram corresponding to the interacting potential between two fermions (the "Coulomb" potential) and calculate the amplitude. Derive the corresponding interacting potential. *Hint: note that the equation of motion for B_μ implies that $\partial^\mu B_\mu = 0$, even though we have no gauge invariance to fix.*

4. Repeat Exercise 3 for the fermion–antifermion and antifermion–antifermion potential.

25 $e^+e^- \rightarrow l\bar{l}$ Unpolarized Cross-Section

In Chapter 24 we rederived some formulas for classical, nonrelativistic processes using the formalism of quantum field theory, in order to gain some confidence in the formalism. In this chapter we turn to the first new calculation, for the $e^+e^- \rightarrow l\bar{l}$ scattering at first order (tree level). This will be a relativistic calculation, and it is also quantum field theoretical, that is it has no classical or (nonrelativistic) quantum mechanics counterpart, since in this process an electron and a positron annihilate, to create a pair of lepton–antilepton, and this is a purely quantum field theoretic process.

25.1 $e^+e^- \rightarrow l\bar{l}$ Unpolarized Cross-Section: Set-up

The leptons are e^-, μ^-, τ^- and their antiparticles, together with the corresponding neutrinos (ν_e, ν_μ, ν_τ and their antiparticles), but the neutrinos have no charge, so within quantum electrodynamics (interaction only via photons, coupling to charged particles) there is no process creating neutrinos. The case when $l = e^-$, of *Bhabha scattering*, is special, and there we have more diagrams, but for $l = \mu^-$ or τ^- we have only one diagram: $e^-(p)e^+(p')$ annihilate into a photon, which then creates $l(k)\bar{l}(k')$. For definiteness we will assume that we have a μ^-. Note that we need sufficient energy to create the lepton pair, so only for $E_{CM} > 2m_l$ is the process possible. Therefore, for $E_{CM} = 2E_{e,CM} < 2m_\mu$, we can create only e^+e^-, for $2m_\mu < E_{CM} < 2m_\tau$ we can create e^+e^- or $\mu^+\mu^-$, and for $E_{CM} > 2m_\tau$ we can create all of $e^+e^-, \mu^+\mu^-, \tau^+\tau^-$.

Using the Feynman rules for Figure 25.1, we can write the amplitude (in Feynman gauge)

$$
\begin{aligned}
i\mathcal{M} &= \bar{v}^{s'}(p')(+e\gamma^\mu)u^s(p)\left(\frac{-ig_{\mu\nu}}{q^2}\right)\bar{u}^r(k)(+e\gamma^\nu)v^{r'}(k') \\
&= -i\frac{e^2}{q^2}\left(\bar{v}^{s'}(p')\gamma^\mu u^s(p)\right)\left(\bar{u}^s(k)\gamma_\mu v^{s'}(k')\right).
\end{aligned}
\tag{25.1}
$$

Note that the vector propagator is $-ig_{\mu\nu}/(q^2 - i\epsilon)$, but in this tree-level process q is not integrated, but rather $q = p + p'$, so $q^2 \neq 0$, therefore we can drop the $i\epsilon$.

For the large brackets, since they are numbers, their complex conjugates (equal to the adjoint for a number) give

$$
(\bar{v}i\gamma^\mu u)^* = u^\dagger(i\gamma^\mu)^\dagger(i\gamma^0)^\dagger v = u^\dagger i\gamma^0 i\gamma^\mu v = \bar{u}i\gamma^\mu v,
\tag{25.2}
$$

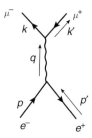

Figure 25.1   Feynman diagram for $e^+e^- \to \mu^+\mu^-$ scattering at tree level.

where we have used $(i\gamma^0)^\dagger = i\gamma^0$ and $(\gamma^\mu)^\dagger i\gamma^0 = -i\gamma^0(\gamma^\mu)$. Therefore, without the i, we have $(\bar{v}\gamma^\mu u)^* = -\bar{u}\gamma^\mu v$.

Then we can calculate $|\mathcal{M}|^2$ as

$$|\mathcal{M}|^2 = (-)^2 \frac{e^4}{q^4} \left[\left(\bar{v}^{s'}(p')\gamma^\mu u^s(p)\right)\left(\bar{u}^s(p)\gamma^\nu v^{s'}(p')\right)\right]\left[\left(\bar{u}^r(k)\gamma_\mu v^{r'}(k')\right)\left(\bar{v}^{r'}(k')\gamma_\nu u^r(k)\right)\right].$$
(25.3)

But in many experiments, we don't keep track of spins (helicities), usually because it is difficult to do so. Therefore, we must *average over the initial spins* (since we don't know the values of the initial spins), and *sum over the final spins* (since we measure the final particles no matter what their spin is). We call the resulting probabilities the *unpolarized scattering cross-section*.

This is obtained from the formula

$$\left(\frac{1}{2}\sum_s\right)\left(\frac{1}{2}\sum_{s'}\right)\sum_r\sum_{r'}|\mathcal{M}(s,s' \to r,r')|^2 = \frac{1}{4}\sum_{spins}|\mathcal{M}|^2.$$
(25.4)

But in order to do the sums over spins, we use the results derived in Chapter 14:

$$\sum_s u^{s\alpha}(p)\bar{u}^s_\beta(p) = -i\slashed{p}^\alpha{}_\beta + m\delta^\alpha_\beta; \qquad \sum_s v^{s\alpha}(p)\bar{v}^s_\beta(p) = -i\slashed{p}^\alpha{}_\beta - m\delta^\alpha_\beta.$$
(25.5)

This means that we have

$$\sum_{ss'}\bar{v}^{s'}_\alpha(p')(\gamma^\mu)^\alpha{}_\beta u^{s\beta}(p)\bar{u}^s_\delta(p)(\gamma^\nu)^\delta{}_\epsilon v^{s'\epsilon}(p') = (-i\slashed{p}' - m_e)^\epsilon{}_\alpha(\gamma^\mu)^\alpha{}_\beta(-i\slashed{p} + m_e)^\beta{}_\delta(\gamma^\nu)^\delta{}_\epsilon$$

$$= \text{Tr}[(-i\slashed{p}' - m_e)\gamma^\mu(-i\slashed{p} + m_e)\gamma^\nu].$$
(25.6)

Then similarly we obtain

$$\sum_{rr'}\left[\left(\bar{u}^r(k)\gamma_\mu v^{r'}(k')\right)\left(\bar{v}^{r'}(k')\gamma_\nu u^r(k)\right)\right] = \text{Tr}[(-i\slashed{k} + m_\mu)\gamma_\mu(-i\slashed{k}' - m_\mu)\gamma_\nu],$$
(25.7)

so that finally

$$\frac{1}{4}\sum_{spins}|\mathcal{M}|^2 = \frac{e^4}{4q^2}\text{Tr}[(-i\slashed{p}' - m_e)\gamma^\mu(-\slashed{p} + m_e)\gamma^\nu]\,\text{Tr}[(-i\slashed{k} + m_\mu)\gamma_\mu(-i\slashed{k}' - m_\mu)\gamma_\nu].$$
(25.8)

To calculate this further, we need to use gamma matrix identities, therefore we stop here and now write all possible gamma matrix identities which we will need here and later on.

25.2 Gamma Matrix Identities

First, we remember our conventions for gamma matrices, with the representation

$$\gamma^0 = -i\begin{pmatrix} \mathbf{0} & \mathbf{1} \\ \mathbf{1} & \mathbf{0} \end{pmatrix}; \quad \gamma^i = -i\begin{pmatrix} \mathbf{0} & \sigma^i \\ -\sigma^i & \mathbf{0} \end{pmatrix}; \quad \gamma^5 = \begin{pmatrix} \mathbf{1} & \mathbf{0} \\ \mathbf{0} & -\mathbf{1} \end{pmatrix}. \tag{25.9}$$

We then see that at least in this representation, we have

$$\text{Tr}[\gamma^\mu] = \text{Tr}[\gamma^5] = 0. \tag{25.10}$$

But this result is more general, since, using the fact that in general $(\gamma^5)^2 = 1$, and $\{\gamma^\mu, \gamma^5\} = 0$, we have

$$\text{Tr}[\gamma^\mu] = \text{Tr}[(\gamma^5)^2\gamma^\mu] = -\text{Tr}[\gamma^5\gamma^\mu\gamma^5] = -\text{Tr}[(\gamma^5)^2\gamma^\mu] = -\text{Tr}[\gamma^\mu], \tag{25.11}$$

and therefore $\text{Tr}[\gamma^\mu] = 0$. Here we have used the anticommutation of γ^5 with γ^μ, and then the cyclicity of the trace, to put back the γ^5 from the end of the trace to the beginning.

Using the same cyclicity of the trace, we have

$$\text{Tr}[\gamma^\mu\gamma^\nu] = \text{Tr}\left[\frac{1}{2}\{\gamma^\mu, \gamma^\nu\}\right] = g^{\mu\nu}\,\text{Tr}[\mathbf{1}] = 4g^{\mu\nu}. \tag{25.12}$$

We can also calculate that the trace of an odd number of gammas gives zero, by the same argument as for the single gamma above:

$$\text{Tr}[\gamma^{\mu_1}\ldots\gamma^{\mu_{2n+1}}] = \text{Tr}[(\gamma^5)^2\gamma^{\mu_1}\ldots\gamma^{\mu_{2n+1}}] = (-1)^{2n+1}\,\text{Tr}[\gamma^5\gamma^{\mu_1}\ldots\gamma^{\mu_{2n+1}}\gamma^5]$$
$$= -\text{Tr}[(\gamma^5)^2\gamma^{\mu_1}\ldots\gamma^{\mu_{2n+1}}] = -\text{Tr}[\gamma^{\mu_1}\ldots\gamma^{\mu_{2n+1}}], \tag{25.13}$$

and therefore

$$\text{Tr}[\gamma^{\mu_1}\ldots\gamma^{\mu_{2n+1}}] = 0. \tag{25.14}$$

In general, the method for calculating the traces of products of gamma matrices is to expand in the complete basis of 4×4 matrices:

$$\mathcal{O}_\mathcal{I} = \{\mathbf{1}, \gamma^5, \gamma^\mu, \gamma^\mu\gamma^5, \gamma^{\mu\nu}\}, \tag{25.15}$$

since for $\mathcal{O}_I \neq \mathbf{1}$, we can check that $\text{Tr}[\mathcal{O}_I] = 0$. Here and always, a gamma matrix with more than one index means that the indices are totally antisymmetrized, so $\gamma^{\mu\nu} = 1/2[\gamma^\mu, \gamma^\nu]$, $\gamma^{\mu\nu\rho} = 1/6[\gamma^\mu\gamma^\nu\gamma^\rho - 5 \text{ terms}]$, and so on. Indeed, we have

$$\text{Tr}[\gamma^{\mu\nu}] = \frac{1}{2}\text{Tr}[[\gamma^\mu, \gamma^\nu]] = 4g^{(\mu\nu)} = 0 \tag{25.16}$$

(the antisymmetric part of $g^{\mu\nu}$ is zero), and also by gamma matrix anticommutation

$$\text{Tr}[\gamma^\mu\gamma^5] = -\text{Tr}[\gamma^5\gamma^\mu], \tag{25.17}$$

therefore

$$\mathrm{Tr}[\gamma^\mu \gamma^5] = 0. \tag{25.18}$$

We also have the relations

$$\gamma^{\mu\nu}\gamma^5 = -\frac{i}{2}\epsilon^{\mu\nu\rho\sigma}\gamma_{\rho\sigma},$$
$$\gamma^{\mu\nu\rho} = -i\epsilon^{\mu\nu\rho\sigma}\gamma_\sigma\gamma_5,$$
$$\gamma^{\mu\nu\rho}\gamma_5 = -i\epsilon^{\mu\nu\rho\sigma}\gamma_\sigma,$$
$$\gamma^{\mu\nu\rho\sigma} = i\epsilon^{\mu\nu\rho\sigma}\gamma_5. \tag{25.19}$$

To prove this, it suffices to show them for some particular indices. Consider the first relation for $\mu = 0, \nu = 1$. Note that for gamma matrices with several indices, the indices must be different (the totally antisymmetric part is nonzero only if the indices are different). But for different indices, the product of gamma matrices is already antisymmetrized, due to the anticommutation property of the gamma matrices, $\gamma^\mu\gamma^\nu = -\gamma^\nu\gamma^\mu$, if $\mu \neq \nu$. Therefore we have, for example, $\gamma^{01} = \gamma^0\gamma^1$, and so on. On the left-hand side of the first relation, we have

$$\gamma^0\gamma^1\gamma^5 = \gamma^0\gamma^1(-i)\gamma^0\gamma^1\gamma^2\gamma^3 = -i\gamma^2\gamma^3 \tag{25.20}$$

(where we have used $\gamma^0\gamma^1 = -\gamma^1\gamma^0$ and $(\gamma^0)^2 = -1, (\gamma^1)^2 = 1$, which follow from $\{\gamma^\mu, \gamma^\nu\} = 2g^{\mu\nu}$) and on the right-hand side, we have $-i/2 \times 2\epsilon^{0123}\gamma_{23} = -i\gamma_2\gamma^3 = -i\gamma^2\gamma^3$, therefore the same result (the factor of 2 is because we can sum over (23) or (32)).

The second relation is proved by choosing, for example, $\mu = 0, \nu = 1, \rho = 2$, in which case on the left-hand side we have $\gamma^0\gamma^1\gamma^2$ and on the right-hand side we have

$$-i\epsilon^{0123}\gamma_3(-i)\gamma^0\gamma^1\gamma^2\gamma^3 = -\gamma^3\gamma^0\gamma^1\gamma^2\gamma^3 = +\gamma^0\gamma^1\gamma^2(\gamma^3)^2 = +\gamma^0\gamma^1\gamma^2. \tag{25.21}$$

The third relation follows from multiplication with γ^5 and using $(\gamma^5)^2 = 1$. The fourth relation can be proved by choosing the only nonzero case, $\mu = 0, \nu = 1, \rho = 2, \sigma = 3$ (of course, we can also have permutations of these), in which case the left-hand side is $\gamma^0\gamma^1\gamma^2\gamma^3$ and the right-hand side is

$$i\epsilon^{0123}(-i)\gamma^0\gamma^1\gamma^2\gamma^3 = \gamma^0\gamma^1\gamma^2\gamma^3. \tag{25.22}$$

To calculate the trace of a product of four gamma matrices, we decompose the product in the basis $\mathcal{O}_I$, which has Lorentz indices, times a corresponding constant Lorentz structure that gives a total of four Lorentz indices. But in four dimensions the only possibilities are $g^{\mu\nu}, \epsilon^{\mu\nu\rho\sigma}$, and products of them. This means that there are no constant Lorentz tensors with odd number of indices, and correspondingly in the product of four gamma matrices we cannot have γ^μ and $\gamma^\mu\gamma_5$ in the $\mathcal{O}_I$ decomposition, since it would need to be multiplied by a 3-index constant Lorentz structure. We can only have $\mathbf{1}, \gamma_5$, or $\gamma^{\mu\nu}$.

For $\mathbf{1}$ and γ_5 we can multiply with two possible Lorentz structures, $\epsilon^{\mu\nu\rho\sigma}, g^{\mu\nu}g^{\rho\sigma}$, and permutations. But we already know that $\gamma^{\mu\nu\rho\sigma}$ (which should certainly appear in the product of gammas) can be written as $i\epsilon^{\mu\nu\rho\sigma}\gamma_5$, and therefore the $g^{\mu\nu}g^{\rho\sigma}$s should multiply the $\mathbf{1}$ term. Then finally we can write the general formula

$$\gamma^\mu\gamma^\nu\gamma^\rho\gamma^\sigma = c_1\epsilon^{\mu\nu\rho\sigma}\gamma_5 + c_2 g^{\mu\nu}\gamma^{\rho\sigma} + c_3 g^{\mu\rho}\gamma^{\nu\sigma} + c_4 g^{\mu\sigma}\gamma^{\nu\rho} + c_5 g^{\nu\rho}\gamma^{\mu\sigma}$$
$$+ c_6 g^{\nu\sigma}\gamma^{\mu\rho} + c_7 g^{\rho\sigma}\gamma^{\mu\nu} + c_8 g^{\mu\nu}g^{\rho\sigma} + c_9 g^{\mu\rho}g^{\nu\sigma} + c_{10}g^{\mu\sigma}g^{\nu\rho}. \quad (25.23)$$

Since $\text{Tr}[\gamma_5] = \text{Tr}[\gamma_{\mu\nu}] = 0$, we have

$$\text{Tr}[\gamma^\mu\gamma^\nu\gamma^\rho\gamma^\sigma] = (c_8 g^{\mu\nu}g^{\rho\sigma} + c_9 g^{\mu\rho}g^{\nu\sigma} + c_{10}g^{\mu\sigma}g^{\nu\rho})\,\text{Tr}[\mathbf{1}]$$
$$= 4(c_8 g^{\mu\nu}g^{\rho\sigma} + c_9 g^{\mu\rho}g^{\nu\sigma} + c_{10}g^{\mu\sigma}g^{\nu\rho}). \quad (25.24)$$

To determine c_8, c_9, c_{10}, we consider indices for which only one structure gives a nonzero result. We can choose $\mu = \nu = 1, \rho = \sigma = 2$. Then c_1 does not contribute, and also $c_2 - c_7$, since they have symmetric $\times$ antisymmetric indices, unlike here, whereas c_9, c_{10} do not contribute due to the fact that $\rho = \sigma \neq \mu = \nu$. Then the left-hand side is $\gamma^1\gamma^1\gamma^2\gamma^2 = +1$, and the right-hand side is $+c_8$, meaning $c_8 = 1$. If we choose $\mu = \rho = 1, \nu = \sigma = 2$ instead, in the same way, we isolate c_9. Then the left-hand side is $\gamma^1\gamma^2\gamma^1\gamma^2 = -1$, and the right-hand side is $+c_9$, meaning $c_9 = -1$. If we choose $\mu = \sigma = 1, \nu = \rho = 2$, we isolate c_{10}. Then the left-hand side is $\gamma^1\gamma^2\gamma^2\gamma^1 = +1$, and the right-hand side is $+c_{10}$, meaning $c_{10} = +1$. Therefore

$$\text{Tr}[\gamma^\mu\gamma^\nu\gamma^\rho\gamma^\sigma] = 4(g^{\mu\nu}g^{\rho\sigma} - g^{\mu\rho}g^{\nu\sigma} + g^{\mu\sigma}g^{\nu\rho}). \quad (25.25)$$

This relation can be derived in another way as well, by commuting one of the gamma matrices past all the other ones, and then using cyclicity of the trace at the end:

$$\text{Tr}[\gamma^\mu\gamma^\nu\gamma^\rho\gamma^\sigma] = \text{Tr}[(2g^{\mu\nu} - \gamma^\nu\gamma^\mu)\gamma^\rho\gamma^\sigma] = \ldots$$
$$= \text{Tr}[2g^{\mu\nu}\gamma^\rho\gamma^\sigma - 2g^{\mu\rho}\gamma^\nu\gamma^\sigma + 2g^{\mu\sigma}\gamma^\nu\gamma^\rho] - \text{Tr}[\gamma^\mu\gamma^\nu\gamma^\rho\gamma^\sigma], \quad (25.26)$$

which implies the same relation as above. We have calculated it via decomposition in $\mathcal{O}_I$ since the method generalizes easily to other cases.

In particular, we can now use it directly to calculate $\text{Tr}[\gamma^\mu\gamma^\nu\gamma^\rho\gamma^\sigma\gamma_5]$, by multiplying (25.23) with γ_5. Then again c_2–c_7 do not contribute, since $\gamma_{\mu\nu}\gamma_5 \propto \gamma_{\rho\sigma}$ as we proved above, so it has zero trace. But now c_8–c_{10} are multiplied by γ_5, so also have zero trace, and the only nonzero result in the trace comes from c_1, since it multiplies $(\gamma^5)^2 = 1$. Then

$$\text{Tr}[\gamma^\mu\gamma^\nu\gamma^\rho\gamma^\sigma\gamma_5] = 4c_1\epsilon^{\mu\nu\rho\sigma}. \quad (25.27)$$

To calculate c_1, we isolate indices that only contribute to c_1, namely $\mu = 0, \nu = 1, \rho = 2, \sigma = 3$. Then the left-hand side of (25.23) is $\gamma^0\gamma^1\gamma^2\gamma^3$, and the right-hand side is $c_1\epsilon^{0123}(-i)\gamma^0\gamma^1\gamma^2\gamma^3 = -ic_1\gamma^0\gamma^1\gamma^2\gamma^3$, and therefore $c_1 = i$, giving finally

$$\text{Tr}[\gamma^\mu\gamma^\nu\gamma^\rho\gamma^\sigma\gamma_5] = 4i\epsilon^{\mu\nu\rho\sigma}. \quad (25.28)$$

We will also use contractions of ϵ tensors, so we write the results for them here:

$$\epsilon^{\alpha\beta\gamma\delta}\epsilon_{\alpha\beta\gamma\delta} = 4!\,\epsilon^{0123}\epsilon_{0123} = -24,$$
$$\epsilon^{\alpha\beta\gamma\mu}\epsilon_{\alpha\beta\gamma\nu} = 3!\,\delta^\mu_\nu\epsilon^{0123}\epsilon_{0123} = -6\delta^\mu_\nu,$$
$$\epsilon^{\alpha\beta\mu\nu}\epsilon_{\alpha\beta\rho\sigma} = 2!\,(\delta^\mu_\rho\delta^\nu_\sigma - \delta^\mu_\sigma\delta^\nu_\rho)\epsilon^{0123}\epsilon_{0123} = -2(\delta^\mu_\rho\delta^\nu_\sigma - \delta^\mu_\sigma\delta^\nu_\rho). \quad (25.29)$$

In the first relation we have 4! terms, for the permutations of 0123, in the second we have $\mu \neq \alpha\beta\gamma$ and also $\nu \neq \alpha\beta\gamma$, which means that $\mu = \nu$, and for given $\mu = \nu$, $\alpha\beta\gamma$ run over the remaining three values, with 3! permutations. For the third relation, we have $\mu\nu \neq \alpha\beta$, and also $\rho\sigma \neq \alpha\beta$, meaning $\mu\nu = \rho\sigma$, or $\delta^\mu_\rho \delta^\nu_\sigma - \delta^\mu_\sigma \delta^\nu_\rho$, and for given $\mu\nu$, $\alpha\beta$ run over two values, giving 2! terms.

We also have

$$\mathrm{Tr}[\gamma^\mu \gamma^\nu \ldots] = \mathrm{Tr}[C\gamma^\mu C^{-1} C\gamma^\nu C^{-1} \ldots C^{-1}] = (-1)^n \, \mathrm{Tr}[(\gamma^\mu)^T (\gamma^\nu)^T \ldots]$$
$$= (-1)^n [\mathrm{Tr}[(\ldots \gamma^\nu \gamma^\mu)^T] = (-1)^n \, \mathrm{Tr}[\ldots \gamma^\nu \gamma^\mu], \tag{25.30}$$

but since in any case for odd n the trace is zero, it means that the trace of many gammas is equal to the trace of the gammas in the opposite order:

$$\mathrm{Tr}[\gamma^\mu \gamma^\nu \ldots] = \mathrm{Tr}[\ldots \gamma^\nu \gamma^\mu]. \tag{25.31}$$

Finally, we write the sums over $\gamma^\mu (\ldots) \gamma_\mu$ with some other gammas in the middle. First, since $\{\gamma^\mu, \gamma^\nu\} = 2g^{\mu\nu}$, it means that

$$\gamma^\mu \gamma_\mu = \delta^\mu_\mu = 4, \tag{25.32}$$

and further, using this, we get

$$\gamma^\mu \gamma^\nu \gamma_\mu = -\gamma^\nu \gamma^\mu \gamma_\mu + 2g^{\mu\nu} \gamma_\mu = -2\gamma^\nu. \tag{25.33}$$

In turn, using this, we can calculate the trace with two gammas in the middle:

$$\gamma^\mu \gamma^\nu \gamma^\rho \gamma_\mu = -\gamma^\nu \gamma^\mu \gamma^\rho \gamma_\mu + 2g^{\mu\nu} \gamma^\rho \gamma_\mu = 2\gamma^\nu \gamma^\rho + 2\gamma^\rho \gamma^\nu = 4g^{\nu\rho}. \tag{25.34}$$

And finally, using this, we can calculate the trace with three gammas in the middle:

$$\gamma^\mu \gamma^\nu \gamma^\rho \gamma^\sigma \gamma_\mu = -\gamma^\nu \gamma^\mu \gamma^\rho \gamma^\sigma \gamma_\mu + 2g^{\mu\nu} \gamma^\rho \gamma^\sigma \gamma_\mu$$
$$= -4\gamma^\nu g^{\rho\sigma} + 2(-\gamma^\sigma \gamma^\rho + 2g^{\rho\sigma}) = -2\gamma^\sigma \gamma^\rho \gamma^\nu. \tag{25.35}$$

25.3 Cross-Section for Unpolarized Scattering

We can now go back to the calculation of the cross-section. Since the trace of three gamma matrices gives zero, in the two traces in (25.8), only the traces of two and four gamma matrices contribute.

We then have for the first trace

$$\mathrm{Tr}[(-i\not{p}' - m_e)\gamma^\mu (-i\not{p} + m_e)\gamma^\nu] = -4p'_\rho p_\sigma (g^{\rho\mu} g^{\nu\sigma} - g^{\rho\sigma} g^{\mu\nu} + g^{\nu\rho} g^{\mu\sigma}) - m_e^2 4g^{\mu\nu}$$
$$= 4[-p'^\mu p^\nu - p'^\nu p^\mu - g^{\mu\nu}(-p \cdot p' + m_e^2)], \tag{25.36}$$

and similarly for the second trace

$$\mathrm{Tr}[(-i\not{k} + m_\mu)\gamma_\mu (-i\not{k}' - m_\mu)\gamma_\nu] = -4k_\rho k'_\sigma (g^{\mu\rho} g^{\nu\sigma} - g^{\rho\sigma} g^{\mu\nu} + g^{\rho\nu} g^{\mu\sigma}) - 4m_\mu^2 g_{\mu\nu}$$
$$= 4[-k_\mu k'_\nu - k_\nu k'_\mu - g_{\mu\nu}(-k \cdot k' + m_\mu^2)]. \tag{25.37}$$

But since $m_e/m_\mu \simeq 1/200$, we neglect m_e and only keep m_μ.

Then we obtain for the unpolarized $|\mathcal{M}|^2$:

$$\frac{1}{4}\sum_{spins}|\mathcal{M}|^2 = \frac{4e^4}{q^4}[p'^{\mu}p^{\nu} + p'^{\nu}p^{\mu} - p \cdot p'g^{\mu\nu}][k_{\mu}k'_{\nu} + k_{\nu}k'_{\mu} + g_{\mu\nu}(-k \cdot k' + m_{\mu}^2)]$$

$$= \frac{4e^4}{q^4}[2(p \cdot k)(p' \cdot k') + 2(p \cdot k')(p' \cdot k) + 2p \cdot p'(-k \cdot k' + m_{\mu}^2)$$

$$+2k \cdot k'(-p \cdot p') + 4(-p \cdot p')(-k \cdot k' + m_{\mu}^2)]$$

$$= \frac{8e^4}{q^4}[(p \cdot k)(p' \cdot k') + (p \cdot k')(p' \cdot k) - m_{\mu}^2 p \cdot p']. \tag{25.38}$$

25.4 Center of Mass Frame Cross-Section

Since muons are created, we need to have $2E_e > 2m_{\mu} \sim 400m_e$, therefore the electrons are ultrarelativistic, and can be considered massless. Therefore, for the electron and positron, we have $p = (E, E\hat{z})$ and $p' = (E, -E\hat{z})$. Since the $\mu^+\mu^-$ have the same mass, $E_e = E_{\mu}$, but they are not necessarily ultrarelativistic, so $k = (E, \vec{k})$ and $k' = (E, -\vec{k})$, and $|\vec{k}| = \sqrt{E^2 - m_{\mu}^2}$, see Figure 25.2. We also define the angle θ between the electrons and the muons (i.e. $\vec{k} \cdot \hat{z} = |\vec{k}| \cos\theta$).

Then also

$$q^2 = (p + p')^2 = -4E^2 = 2p \cdot p' \Rightarrow p \cdot p' = -2E^2. \tag{25.39}$$

We also have

$$p \cdot k = p' \cdot k' = -E^2 + E|\vec{k}|\cos\theta,$$
$$p \cdot k' = p' \cdot k = -E^2 - E|\vec{k}|\cos\theta. \tag{25.40}$$

Substituting these formulas in $|\mathcal{M}|^2$, we get

$$\frac{1}{4}\sum_{spins}|\mathcal{M}|^2 = \frac{8e^4}{16E^4}[E^2(E - |k|\cos\theta)^2 + E^2(E + |k|\cos\theta)^2 + 2E^2 m_{\mu}^2]$$

$$= \frac{e^4}{E^2}[E^2 + k^2 \cos^2\theta + m_{\mu}^2]$$

$$= e^4\left[1 + \frac{m_{\mu}^2}{E^2} + \left(1 - \frac{m_{\mu}^2}{E^2}\right)\cos^2\theta\right]. \tag{25.41}$$

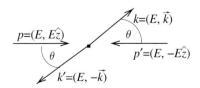

Figure 25.2   The kinematics of center of mass scattering for $e^+e^- \rightarrow \mu^+\mu^-$.

The differential cross-section is

$$\left(\frac{d\sigma}{d\Omega}\right)_{CM} = \frac{1}{2E_A 2E_B |v_A - v_B|} \frac{|p_1|}{(2\pi)^2 4E_{CM}} \frac{1}{4} \sum_{spins} |\mathcal{M}|^2. \tag{25.42}$$

But $2E_A = 2E_B = 2E = E_{CM}$, $\vec{v}_A = \vec{p}_A/E_A = \hat{z}$, and $\vec{v}_B = \vec{p}_B/E_B = -\hat{z}$, so $|v_A - v_B| = 2$. Then

$$\left(\frac{d\sigma}{d\Omega}\right)_{CM} = \frac{|k|}{2E_{CM}^2 16\pi^2} \frac{1}{2E} \frac{1}{4} \sum_{spins} |\mathcal{M}|^2$$

$$= \frac{\alpha^2}{4E_{CM}^2} \sqrt{1 - \frac{m_\mu^2}{E^2}} \left[1 + \frac{m_\mu^2}{E^2} + \left(1 - \frac{m_\mu^2}{E^2}\right)\cos^2\theta\right]. \tag{25.43}$$

In the ultrarelativistic limit $E \gg m_\mu$, it becomes

$$\left(\frac{d\sigma}{d\Omega}\right)_{CM,\text{ultrarel.}} = \frac{\alpha^2}{4E_{CM}^2}(1 + \cos^2\theta). \tag{25.44}$$

The total cross-section is found by integrating over the solid angle, $d\Omega = 2\pi \sin\theta d\theta = -2\pi d(\cos\theta)$, and using $\int_{-1}^{1} d(\cos\theta) = 2$ and $\int_{-1}^{1} d(\cos\theta)\cos^2\theta = 2/3$, we get

$$\sigma_{tot} = 2\pi \int_{-1}^{1} d(\cos\theta) \frac{d\sigma}{d\Omega}(\cos\theta)$$

$$= 2\pi \frac{\alpha^2}{4E_{CM}^2} \sqrt{1 - \frac{m_\mu^2}{E^2}} \left[\frac{8}{3} + \frac{4}{3}\frac{m_\mu^2}{E^2}\right]$$

$$= \frac{4\pi}{3} \frac{\alpha^2}{E_{CM}^2} \sqrt{1 - \frac{m_\mu^2}{E^2}} \left[1 + \frac{m_\mu^2}{2E^2}\right]. \tag{25.45}$$

In the ultrarelativistic limit, $E \gg m_\mu$, it becomes

$$\sigma_{tot} \to \frac{4\pi}{3} \frac{\alpha^2}{E_{CM}^2}. \tag{25.46}$$

Important Concepts to Remember

- The $e^+e^- \to l\bar{l}$ is quantum field theoretic, and for $l = \mu, \tau$, we have only one Feynman diagram.
- The complex conjugation changes $(\bar{u}\gamma v)$ into $\bar{v}\gamma u$, meaning that the sum over spins in $|\mathcal{M}|^2$ generates sums over spins $\sum_s v_s(p)\bar{v}^s(p)$, allowing us to convert all the us and vs into traces of gamma matrices.
- In calculations with gamma matrices, we use the Clifford algebra, anticommutations, Lorentz invariant structures, the complete set $\mathcal{O}_I$, and so on.
- The total unpolarized $e^+e^- \to l\bar{l}$ cross-section is finite, and given in the ultrarelativistic limit and in the center of mass frame by $4\pi/3\alpha^2/E_{CM}^2$.

Further Reading

See section 5.1 of [1] and section 8.1 of [2].

Exercises

1. Consider Bhabha scattering, $e^+e^- \to e^+e^-$. Write down the Feynman diagrams, and find the expression for $|\mathcal{M}|^2$ in terms of $u^s(p)$s and $v^s(p)$s.
2. Use the gamma matrix identities to calculate the unpolarized scattering cross-section for Bhabha scattering in Exercise 1.
3. Consider the gamma matrices in three Euclidean dimensions, γ^i, satisfying the Clifford algebra $\{\gamma^i, \gamma^j\} = 2\delta^{ij}$. Calculate $\mathrm{Tr}[\gamma^i\gamma^j]$, $\mathrm{Tr}[\gamma^i\gamma^j\gamma^k]$, and $\mathrm{Tr}[\gamma^i\gamma^j\gamma^k\gamma^l]$. You can use a specific representation for γ^i if you want (though it's not needed).
4. Using the method in the text, decompose

$$\gamma^\mu\gamma^\nu\gamma^\rho\gamma_5 \tag{25.47}$$

in the basis $\mathcal{O}_I$.

$e^+e^- \rightarrow l\bar{l}$ Polarized Cross-Section; Crossing Symmetry

In this chapter we show how to calculate the same process as in Chapter 25, but for given spin (i.e. the polarized cross-section). Then we will explore an important symmetry of quantum field theory processes called crossing symmetry and how it appears in the Mandelstam variables (which we will introduce).

26.1 $e^+e^- \rightarrow l\bar{l}$ Polarized Cross-Section

For simplicity we will work in the ultrarelativistic case $m_e, m_\mu \rightarrow 0$.

We have seen that in the Weyl representation for gamma matrices, we have

$$\gamma_5 = \begin{pmatrix} 1 & 0 \\ 0 & -1 \end{pmatrix}. \tag{26.1}$$

This means that $P_L = (1 + \gamma_5)/2$ projects a fermion ψ into its upper two components, called ψ_L, and $P_R = (1 - \gamma_5)/2$ projects the fermion into its lower two components, called ψ_R (i.e. $P_L \psi = \psi_L$, $P_R \psi = \psi_R$). P_L, P_R are (chiral) *projectors* (i.e. $P_L^2 = P_L, P_R^2 = P_R$) and $P_L P_R = P_R P_L = 0$ (i.e. $P_R \psi_L = 0$ and $P_L \psi_R = 0$).

Here ψ_L and ψ_R have given *helicities* (i.e. eigenvalues of the projection of the spin onto the direction of the momentum: $h = \vec{S} \cdot \vec{p}/|\vec{p}| = \pm 1/2$). Therefore, measuring a given spin, or more precisely a given helicity, means considering only ψ_L or ψ_R fermions.

Note that if $\psi_L = P_L \psi = (1 + \gamma_5)/2\psi$, then (since $\gamma_5^\dagger = \gamma_5$ and γ_5 anticommutes with the other gammas)

$$\overline{\psi_L} = \overline{\frac{1 + \gamma_5}{2}\psi} = \left(\frac{1 + \gamma_5}{2}\psi\right)^\dagger i\gamma^0 = \psi^\dagger \frac{1 + \gamma_5}{2} i\gamma^0 = \psi^\dagger i\gamma^0 \frac{1 - \gamma_5}{2}$$

$$= \bar{\psi} \frac{1 - \gamma_5}{2} = (\bar{\psi})_R, \tag{26.2}$$

so for instance $\overline{u_L} = (\bar{u})_R$.

Consider a Lorentz-invariant object like the one we calculated before, $\bar{v}(p')\gamma^\mu u(p)$. We want to calculate $\bar{v}(p')^{s'}\gamma^\mu u^s(p)$ with s, s' corresponding to u_R and $(\bar{v})_L = \overline{v_R}$. Since $P_R u_R = u_R$, we can introduce for free a P_R in front of u_R. Then, also on the bar side we have a nonzero element, since

$$\bar{v}(p')\gamma^\mu \frac{1 - \gamma_5}{2} u(p) = \bar{v}(p')\frac{1 + \gamma_5}{2}\gamma^\mu u(p) = \overline{\frac{1 - \gamma_5}{2}v(p')}\gamma^\mu u(p) = \bar{v}(p')\gamma^\mu u(p). \tag{26.3}$$

But since $P_R u_L = 0$, and also $(\bar{\psi})_R P_L = 0$, we can for free add the terms with the other spins (u_L and/or $(\bar{v})_R$), since they give zero when we have P_R inside, and therefore calculate the sum over spins for the e^+e^- factor in $|\mathcal{M}^2|$ (note that now it is a sum over spins, not an average, because *we do know the initial spin* – we don't average over it, we just choose to add contributions of spins which give zero in the sum)

$$\sum_{spins} |\bar{v}(p')\gamma^\mu \left(\frac{1-\gamma_5}{2}\right) u(p)|^2 = \sum_{spins} \bar{v}(p')\gamma^\mu \left(\frac{1-\gamma_5}{2}\right) u(p)\bar{u}(p)\gamma^\nu \left(\frac{1-\gamma_5}{2}\right) v(p').$$
(26.4)

Using the sums over spins, $\sum_s u^s(p)\bar{u}^s(p) = -i\slashed{p} + m \simeq -i\slashed{p}$ and $\sum_s v^s(p)\bar{v}^s(p) = -i\slashed{p} - m \simeq -i\slashed{p}$, we obtain

$$-\,\mathrm{Tr}\left[\slashed{p}'\gamma^\mu \left(\frac{1-\gamma_5}{2}\right)\slashed{p}\gamma^\nu \left(\frac{1-\gamma_5}{2}\right)\right] = -\,\mathrm{Tr}\left[\slashed{p}'\gamma^\mu \slashed{p}\gamma^\nu \frac{1-\gamma_5}{2}\right], \qquad (26.5)$$

where we have commuted the first P_R past two gammas, thus giving P_R again, and then used $P_R^2 = P_R$. Now we have two terms, one is the same as in the unpolarized cross-section, with four gammas inside the trace, and a new one that has an extra γ_5 (the trace with four gammas and a γ_5 was calculated in Chapter 25 also). Substituting the result of the traces, we get

$$-\frac{4}{2}[p'^\mu p^\nu + p'^\nu p^\mu - g^{\mu\nu} p \cdot p'] + \frac{1}{2}4ip'_\rho p_\sigma \epsilon^{\rho\mu\sigma\nu} \qquad (26.6)$$

$$= -2[p'^\mu p^\nu + p'^\nu p^\mu - g^{\mu\nu} p \cdot p' - i\epsilon^{\rho\mu\sigma\nu} p'_\rho p_\sigma].$$

Similarly, we can calculate the sum over spins in the $\mu^+\mu^-$ factor in $|\mathcal{M}|^2$, which gives

$$\sum_{spins} |\bar{u}(k)\gamma_\mu \left(\frac{1-\gamma_5}{2}\right) v(k')|^2 = -\,\mathrm{Tr}\left[\slashed{k}\gamma_\mu \left(\frac{1-\gamma_5}{2}\right)\slashed{k}'\gamma_\nu \left(\frac{1-\gamma_5}{2}\right)\right]$$

$$= -2\left[k_\mu k'_\nu + k_\nu k'_\mu - g_{\mu\nu} k \cdot k' - ik^\alpha k'^\beta \epsilon_{\alpha\mu\beta\nu}\right]. \qquad (26.7)$$

Then we get for $|\mathcal{M}|^2$ (where besides the two factors calculated above we have the factor e^4/q^4)

$$|\mathcal{M}(e_R^- e_L^+ \to \mu_R^- \mu_L^+)|^2 = \frac{4e^4}{q^4}[2(p\cdot k)(p'\cdot k') + 2(p\cdot k')(p'\cdot k) - \epsilon^{\rho\mu\sigma\nu}\epsilon_{\alpha\mu\beta\nu}p'_\rho p_\sigma k^\alpha k'^\beta]$$

$$= \frac{4e^4}{q^4}[2(p\cdot k)(p'\cdot k') + 2(p\cdot k')(p'\cdot k)$$

$$+2((p\cdot k')(p\cdot k') - (p\cdot k)(p'\cdot k'))]$$

$$= \frac{16e^4}{q^4}(p\cdot k')(p'\cdot k), \qquad (26.8)$$

where we have used $\epsilon^{\rho\mu\sigma\nu}\epsilon_{\alpha\mu\beta\nu} = -2(\delta^\rho_\alpha\delta^\sigma_\beta - \delta^\rho_\beta\delta^\sigma_\alpha)$. Using also (from Chapter 25) $q^2 = -4E^2$ and $p\cdot k' = p'\cdot k = -E(E + k \cdot \cos\theta)$, and in the ultrarelativistic limit $E = k$, we obtain

$$|\mathcal{M}(e_R^- e_L^+ \to \mu_R^- \mu_L^+)|^2 = e^4(1 + \cos\theta)^2. \qquad (26.9)$$

Then, since in the center of mass frame, in the ultrarelativistic limit we have (as we saw before)

$$\frac{d\sigma}{d\Omega}\bigg|_{CM,m_i \rightarrow 0} = \frac{|\mathcal{M}|^2}{64\pi^2 E_{CM}^2}, \qquad (26.10)$$

we finally obtain

$$\frac{d\sigma}{d\Omega}(e_R^- e_L^+ \rightarrow \mu_R^- \mu_L^+)\bigg|_{CM,m_i \rightarrow 0} = \frac{\alpha^2}{4E_{CM}^2}(1 + \cos\theta)^2. \qquad (26.11)$$

We can similarly calculate the cross-sections for other polarizations. For the process $e_R^- e_L^+ \rightarrow \mu_L^- \mu_R^+$, for instance, only one of the two factors has $\gamma_5 \rightarrow -\gamma_5$, meaning that there is a minus sign in front of the $\epsilon^{\rho\mu\sigma\nu}$ in the first term, therefore changing the relative sign in the next to last line in (26.8). The details are left as an exercise, but it follows that we have

$$\frac{d\sigma}{d\Omega}(e_R^- e_L^+ \rightarrow \mu_L^- \mu_R^+)\bigg|_{CM,m_i \rightarrow 0} = \frac{\alpha^2}{4E_{CM}^2}(1 - \cos\theta)^2. \qquad (26.12)$$

It is then clear that the other two nonzero processes give

$$\frac{d\sigma}{d\Omega}(e_L^- e_R^+ \rightarrow \mu_R^- \mu_L^+)\bigg|_{CM,m_i \rightarrow 0} = \frac{\alpha^2}{4E_{CM}^2}(1 - \cos\theta)^2,$$

$$\frac{d\sigma}{d\Omega}(e_L^- e_R^+ \rightarrow \mu_L^- \mu_R^+)\bigg|_{CM,m_i \rightarrow 0} = \frac{\alpha^2}{4E_{CM}^2}(1 + \cos\theta)^2. \qquad (26.13)$$

All the other helicity combinations give zero by the fact that $P_L P_R = P_R P_L = 0$.

There is another way of calculating the polarized cross-section that we will just sketch, but not follow through. Namely, we can use explicit forms for the spinors:

$$u(p) = \begin{pmatrix} \sqrt{-p \cdot \sigma}\xi \\ \sqrt{-p \cdot \bar{\sigma}}\xi \end{pmatrix} = \begin{pmatrix} \left[\sqrt{E + p^3}\left(\frac{1-\sigma^3}{2}\right) + \sqrt{E - p^3}\left(\frac{1+\sigma^3}{2}\right)\right]\xi \\ \left[\sqrt{E + p^3}\left(\frac{1+\sigma^3}{2}\right) + \sqrt{E - p^3}\left(\frac{1-\sigma^3}{2}\right)\right]\xi \end{pmatrix}$$

$$\overset{E \rightarrow \infty}{\rightarrow} \sqrt{2E}\begin{pmatrix} \frac{1}{2}(1 + \hat{p} \cdot \vec{\sigma})\xi \\ \frac{1}{2}(1 - \hat{p} \cdot \vec{\sigma})\xi \end{pmatrix}, \qquad (26.14)$$

and similarly

$$v(p) = \begin{pmatrix} \sqrt{-p \cdot \sigma}\xi \\ -\sqrt{-p \cdot \bar{\sigma}}\xi \end{pmatrix} \overset{E \rightarrow \infty}{\rightarrow} \sqrt{2E}\begin{pmatrix} \frac{1}{2}(1 + \hat{p} \cdot \vec{\sigma})\xi \\ -\frac{1}{2}(1 - \hat{p} \cdot \vec{\sigma})\xi \end{pmatrix}. \qquad (26.15)$$

One can then use the explicit form of the gamma matrices in the Weyl representation together with the above formulas to compute the $|\mathcal{M}|^2$s. We will not do it here, but the details can be found for instance in the book by Peskin and Schroeder [1].

26.2 Crossing Symmetry

Feynman diagrams contain more information than the specific processes we analyze, since by simply rotating the diagrams (considering time running in a different direction), we obtain a new process, related to the old one in a simple way, by a transformation called *crossing*. We will find that we can transform the momenta as well in order to obtain the same value for the amplitude $\mathcal{M}$, obtaining *crossing symmetry*. Of course, if we apply the transformation of momenta on the functional form of a given amplitude, in general we will not obtain a symmetry. We will discuss this more later.

In the case of the $e^+e^- \to \mu^+\mu^-$ process, the "crossed diagram" ("crossed process"), obtained by a 90° rotation, or by time running horizontally instead of vertically in the diagram, is $e^-(p_1)\mu^-(p_2) \to e^-(p_1')\mu^-(p_2')$, as in Figure 26.1, giving

$$i\mathcal{M} = \bar{u}(p_1')(+e\gamma^\mu)u(p_1)\frac{-ig_{\mu\nu}}{q^2}\bar{u}(p_2')(+e\gamma^\nu)u(p_2) = -i\frac{e^2}{q^2}\bar{u}(p_1')\gamma^\mu u(p_1)\bar{u}(p_2')\gamma_\mu u(p_2).$$
(26.16)

Then the unpolarized $|\mathcal{M}|^2$ gives

$$\frac{1}{4}\sum_{spins}|\mathcal{M}|^2 = \frac{e^4}{4q^4}\,\mathrm{Tr}[(-i\not{p}_1+m_e)\gamma^\nu(-i\not{p}_1'+m_e)\gamma^\mu]\,\mathrm{Tr}[(-i\not{p}_2'+m_\mu)\gamma_\mu(-i\not{p}_2+m_\mu)\gamma_\nu].$$
(26.17)

Here $q = p_1 - p_1'$.

This is the same result as for $e^+e^- \to \mu^+\mu^-$, except for renaming the momenta, and some signs. More precisely, when we change antiparticles to particles, we change $v^s(p)$s to $u^s(p)$s, and since $\sum_s u^s(p)\bar{u}^s(p) = -i\not{p}+m$, but $\sum_s v^s(p)\bar{v}^s(p) = -i\not{p}-m$, and we do this for two terms, changing $-i\not{p}+m \to -i\not{p}-m = -(i\not{p}+m)$. Thus we finally see that the rules for replacing the original diagram with the crossed diagram are

$$p \to p_1; \quad p' \to -p_1'; \quad k \to p_2'; \quad k' \to -p_2. \tag{26.18}$$

Thus we obtain

$$\frac{1}{4}\sum_{spins}|\mathcal{M}(e^-\mu^- \to e^-\mu^-)|^2 = \frac{8e^4}{q^4}[(p_1{\cdot}p_2')(p_1'{\cdot}p_2)+(p_1{\cdot}p_2)(p_1'{\cdot}p_2')+m_\mu^2 p_1{\cdot}p_1']. \tag{26.19}$$

In this process, in the center of mass frame, we have for the initial electron $p_1 = (k, k\hat{z})$, and for the initial muon we have opposite initial momentum (but higher energy),

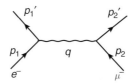

Figure 26.1 Feynman diagram for $e^-\mu^- \to e^-\mu^-$ scattering at tree level.

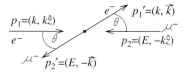

Figure 26.2 Kinematics of $e^-\mu^- \to e^-\mu^-$ scattering in the center of mass frame.

$p_2 = (E, -k\hat{z})$, with $E^2 = k^2 + m_\mu^2$. For the final electron we have $p_1' = (k, \vec{k})$ and for the final muon $p_2' = (E, -\vec{k})$, see Figure 26.2. The angle between the initial and final directions being θ, we have $\vec{k} \cdot \hat{z} = k \cos\theta$. We also have the center of mass energy $E_{CM} = E + k$ and for the various invariants

$$p_1 \cdot p_2 = p_1' \cdot p_2' = -k(E + k),$$
$$p_1' \cdot p_2 = p_1 \cdot p_2' = -k(E + k\cos\theta),$$
$$p_1 \cdot p_1' = -k^2(1 - \cos\theta),$$
$$q^2 = -2p_1 \cdot p_1' = 2k^2(1 - \cos\theta). \tag{26.20}$$

Then the sum over spins in the center of mass frame is

$$\frac{1}{4} \sum_{spins} |\mathcal{M}|^2 = \frac{2e^4}{k^2(1 - \cos\theta)^2}[(E + k)^2 + (E + k\cos\theta)^2 - m_\mu^2(1 - \cos\theta)], \tag{26.21}$$

and the differential cross-section is

$$\frac{d\sigma}{d\Omega}_{CM} = \frac{|\mathcal{M}|^2}{64\pi^2 E_{CM}^2}$$

$$= \frac{\alpha^2}{2k^2(1 - \cos\theta)^2(E + k)^2}[(E + k)^2 + (E + k\cos\theta)^2 - m_\mu^2(1 - \cos\theta)]. \tag{26.22}$$

In the ultrarelativistic limit, $E \gg m_\mu$, we get

$$\frac{d\sigma}{d\Omega}_{CM} \to \frac{\alpha^2}{2E_{CM}^2(1 - \cos\theta)^2}[4 + (1 + \cos\theta)^2] \overset{\theta \to 0}{\propto} \frac{1}{\theta^4}. \tag{26.23}$$

The fact that $d\sigma/d\Omega \to 1/\theta^4$ as $\theta \to 0$ means that there is a strong divergence at $\theta = 0$ (since $\int d\theta|_0 \sim \int d\theta/\theta^4 \to \infty$ in the massless limit $m \to 0$; remember that the exchanged photon is massless, as is the electron in the ultrarelativistic limit). This is indicative of something that will be dealt with in Part II of the book, infrared (IR) divergences for scattering of massless particles.

To complete this portion, we can now define the general crossed diagram by generalizing the case $e^+e^- \to \mu^+\mu^-$ into $e^-\mu^- \to e^-\mu^-$, and the corresponding transformation that leaves the amplitude invariant as

$$\mathcal{M}(\phi(p) + \ldots \to \ldots) = \mathcal{M}(\ldots \to \ldots + \bar{\phi}(k)). \tag{26.24}$$

Here ϕ is a particle, and $\bar{\phi}$ is the corresponding antiparticle (i.e. we change particle to antiparticle, exchanging in and out, and the momenta as $k = -p$, see Figure 26.3). This equality defines crossing symmetry. Note that since $k = -p$, we can't have $k^0 > 0$ and

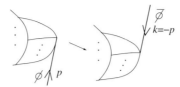

Figure 26.3 Crossing symmetry (includes a charge conjugation).

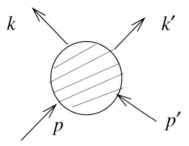

Figure 26.4 Mandelstam variables.

$p^0 > 0$ simultaneously, which means that when a particle is real, the crossed particle is virtual, and vice versa. Thus really, crossing symmetry also corresponds to an analytical continuation in energy.

26.3 Mandelstam Variables

Crossing symmetry is easier to describe in Mandelstam variables, so we will define them better now.

For a $2 \rightarrow 2$ scattering process, with incoming momenta p, p' and outgoing momenta k, k', as in Figure 26.4, we define the Mandelstam variables s, t, u by

$$
\begin{aligned}
s &= -(p + p')^2 = -(k + k')^2 = E_{CM}^2, \\
t &= -(k - p)^2 = -(k' - p')^2, \\
u &= -(k' - p)^2 = -(k - p')^2.
\end{aligned}
\tag{26.25}
$$

Note that $s = E_{CM}^2$. Also, in the more interesting case that the particles with momenta p and k are the same one (elastic scattering), $E_k = E_p$, meaning that $t = -(\vec{k} - \vec{p})^2$ is minus the invariant *momentum transfer squared* ($\vec{k} - \vec{p} = -(\vec{k}' - \vec{p}')$ is the momentum transferred between the scattering particles).

Now consider again the process $e^+ e^- \rightarrow \mu^+ \mu^-$. Then

$$
\begin{aligned}
s &= -(p + p')^2 = -q^2, \\
t &= -(k - p)^2 = 2p \cdot k = 2p' \cdot k', \\
u &= -(p - k')^2 = 2p \cdot k' = 2p' \cdot k,
\end{aligned}
\tag{26.26}
$$

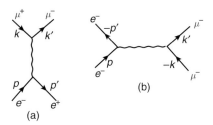

Figure 26.5   90° Rotation followed by mirror image of (a), giving crossed diagram (b).

and the sum over spins is

$$\frac{1}{4} \sum_{spins} |\mathcal{M}|^2 = \frac{8e^4}{s^2} \left[\left(\frac{t}{2}\right)^2 + \left(\frac{u}{2}\right)^2 \right]. \tag{26.27}$$

Consider the 90° rotation of $e^-(p)e^+(p') \rightarrow \mu^+(k)\mu^-(k')$ (followed by a mirror image exchanging left with right that doesn't change physics), as in Figure 26.5. By the above crossing rules, it gives for $e^-\mu^- \rightarrow e^-\mu^-$, $e^-(p)\mu^-(-k) \rightarrow e^-(-p)\mu^-(k')$. The Mandelstam variables for the process are

$$s = -(p-k)^2 = \bar{t},$$
$$t = -(p-(-p'))^2 = \bar{s},$$
$$u = -(p-k')^2 = \bar{u}, \tag{26.28}$$

where the quantities with a bar are defined using the original (uncrossed) diagram. Therefore, if we have a functional form $\mathcal{M}(s,t)$ for the original amplitude, the form of the amplitude for the crossed process is obtained by exchanging s with t and keeping u unchanged. This in general will not give the same result. In particular, in our case we obtain

$$\frac{1}{4} \sum_{spins} |\mathcal{M}|^2 = \frac{8e^4}{t^2} \left[\left(\frac{s}{2}\right)^2 + \left(\frac{u}{2}\right)^2 \right], \tag{26.29}$$

which is not the same as for the original process. If nevertheless we obtain the same formula, we say the amplitude is *crossing symmetric*.

The Mandelstam variables are not independent. In fact, a simple counting argument tells us that there are only two independent variables. There are four momenta, and for each three independent components (since $E^2 = \vec{p}^2 + m^2$), for a total of 12 components. But there are four momentum conservation conditions, leaving eight components. But there are Lorentz transformations $\Lambda^\mu{}_\nu$ that we can make on the reference frame, corresponding to an antisymmetric 4×4 matrix ($\mu, \nu = 1, \ldots, 4$), or equivalently three rotations (around the three axes of coordinates) and three Lorentz boosts (speed of the frame in three directions), for a total of six frame-dependent components, leaving only two Lorentz invariants. In fact, we can write a simple relation between s, t, u:

$$
\begin{aligned}
s + t + u &= -(p + p')^2 - (k - p)^2 - (k' - p)^2 \\
&= -3p^2 - p'^2 - k^2 - k'^2 - 2p \cdot p' + 2p \cdot k + 2p \cdot k' \\
&= 2p \cdot (k + k' - p') - 3p^2 - p'^2 - k^2 - k'^2 = -p^2 - p'^2 - k^2 - k'^2 \\
&= \sum_{i=1}^{4} m_i^2.
\end{aligned}
\tag{26.30}
$$

In the case of $2 \to 2$ scattering and a single virtual particle being exchanged between the two, we talk of "channels." Consider a particle–antiparticle pair of momenta p, p' annihilating and then creating a particle ϕ, which creates a different particle–antiparticle pair. Since then the momentum on ϕ is $p + p'$, the diagram has a propagator for ϕ with a $1/[(p + p')^2 + m_\phi^2] = -1/[s - m_\phi^2]$ factor, meaning

$$
\mathcal{M} \propto \frac{1}{s - m_\phi^2},
\tag{26.31}
$$

that is it has a pole in s. We thus refer to this as the "s-channel," see Figure 26.6 for details.

Consider now the crossed diagram with the particle with momentum p going into the particle with momentum k, and the ϕ particle being exchanged with the other particle. In this case, the propagator for ϕ has a factor of $1/[(p - k)^2 + m_\phi^2] = -1/[t - m_\phi^2]$, so

$$
\mathcal{M} \propto \frac{1}{t - m_\phi^2},
\tag{26.32}
$$

that is it has a pole in t. We refer to this as the "t-channel."

Finally, consider a different crossing, where the particle of momentum p goes into the particle with momentum k' instead, and ϕ is exchanged with the other particle. Then the ϕ is propagator gives a factor of $1/[(p - k')^2 + m_\phi^2] = -1/[u - m_\phi^2]$, and therefore

$$
\mathcal{M} \propto \frac{1}{u - m_\phi^2},
\tag{26.33}
$$

that is it has a pole in u. We call this the "u-channel."

We now define a relativistically invariant differential cross-section for elastic scattering (the same particles are in the initial and final states). We have defined $d\sigma/d\Omega$ before, even a relativistically invariant formula that becomes the usual $d\sigma/d\Omega$ in the center of mass frame.

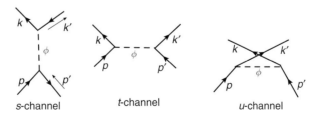

Figure 26.6 In $2 \to 2$ scattering, we can have the s-channel, the t-channel, and the u-channel, depending on whether the physical pole of the S-matrix is in s, t, or u, corresponding to effective Feynman diagrams as shown here.

But now we can define a formula only in terms of the independent relativistic invariants s and t. We can write $d\sigma/dt$. Consider that $t = -(p - p')^2$ and since $p = (E, \vec{p}_{CM})$ and $k = (E, \vec{k}_{CM})$ (if p and k correspond to the same particle, it has the same initial and final energy):

$$t = -(\vec{p}_{CM}^2 - \vec{k}_{CM})^2 = -(p_{CM}^2 + k_{CM}^2 - 2k_{CM}p_{CM}\cos\theta). \qquad (26.34)$$

This means that

$$\frac{dt}{d\cos\theta} = 2|k_{CM}||p_{CM}|. \qquad (26.35)$$

Consider then that the solid angle element defined by a fixed θ would be $d\phi\, d(\cos\theta)$, and integrating over the angle ϕ at fixed θ between the initial and final momenta, we have $2\pi\, d(\cos\theta)$, therefore

$$\frac{dt}{d\Omega} = \frac{dt}{2\pi\, d(\cos\theta)} = \frac{1}{\pi}|k_{CM}||p_{CM}|. \qquad (26.36)$$

We already defined the relativistically invariant formula

$$\left(\frac{d\sigma}{d\Omega}\right)_{rel.inv.} = \frac{1}{\sqrt{(p_1 \cdot p_2)^2 - m_1^2 m_2^2}} \frac{\vec{k}_{CM}}{64\pi^2 E_{CM}}|\mathcal{M}|^2, \qquad (26.37)$$

where $p_1 \to p, p_2 \to p'$, and since

$$s = -(p_1 + p_2)^2 = -p_1^2 - p_2^2 - 2p_1 \cdot p_2 = m_1^2 + m_2^2 - 2p_1 \cdot p_2$$

$$\Rightarrow (p_1 \cdot p_2)^2 - m_1^2 m_2^2 = \frac{1}{4}(s - m_1^2 - m_2^2) - m_1^2 m_2^2. \qquad (26.38)$$

In contrast, in the center of mass frame:

$$(p_1 \cdot p_2)^2 - m_1^2 m_2^2 = (E_1 E_2 + p_{CM})^2 - m_1^2 - m_2^2 = 2p_{CM}^4 + p_{CM}^2(m_1^2 + m_2^2) + 2E_1 E_2 p_{CM}^2$$

$$= p_{CM}^2(2p_{CM} + 2E_1 E_2 + m_1^2 + m_2^2) = p_{CM}^2 s, \qquad (26.39)$$

where we have used $s = -(p_1 + p_2)^2 = m_1^2 + m_2^2 - 2p_1 \cdot p_2 = m_1^2 + m_2^2 + 2E_1 E_2 + 2\vec{p}_{CM}^2$.

Then we have

$$\frac{d\sigma}{dt} = \frac{\pi}{|k_{CM}||p_{CM}|}\left(\frac{d\sigma}{d\Omega}\right)_{rel.inv.} = \frac{1}{64\pi |p_{CM}|\sqrt{s}\sqrt{(p_1 \cdot p_2)^2 - m_1^2 m_2^2}}|\mathcal{M}|^2$$

$$= \frac{1}{64\pi [(p_1 \cdot p_2)^2 - m_1^2 m_2^2]}|\mathcal{M}|^2. \qquad (26.40)$$

We can write for the total cross-section

$$\sigma_{tot} = \int dt \frac{d\sigma}{dt}(s, t) = \sigma_{tot}(s). \qquad (26.41)$$

Note that t is the momentum transfer, and as such it can in principle be anything, though of course we expect that for very large t (in which case at least one of the outgoing 3-momenta, and therefore also its energy, will be very large), the amplitude for this process would be very small or zero.

In the case of equal masses, $m_1 = m_2 = m$, (26.38) gives $s(s - 4m^2)/4$ and therefore finally

$$\frac{d\sigma}{dt} = \frac{1}{16\pi s(s - 4m^2)} |\mathcal{M}(s,t)|^2. \tag{26.42}$$

Important Concepts to Remember

- In order to calculate the polarized cross-section (i.e. for given helicities) we can introduce projectors onto helicities P_L, P_R for free, since $P_L \psi_L = \psi_L, P_R \psi_R = \psi_R$, and then sum over spins since the other spins have zero contributions, because $P_L \psi_R = P_R \psi_L = 0$.
- The crossed diagrams are diagrams where we rotate the direction of time, and reinterpret the diagram by changing particles into antiparticles.
- By changing the particle momenta into minus antiparticle momenta when crossing, we obtain the same value for the amplitude, hence crossing symmetry means $\mathcal{M}(\phi(p) + \ldots \to \ldots) = \mathcal{M}(\ldots \to \ldots \bar{\phi}(k))$, with $k = -p$.
- If we keep the functional form of the amplitude fixed, crossing symmetry corresponds in Mandelstam variables to exchanging s with t and keeping u fixed. If we obtain the same amplitude after the exchange, we say that the amplitude is crossing symmetric.
- In the $2 \to 2$ scattering, there are only two independent Lorentz invariants, s and t. We have $s + t + u = \sum_{i=1}^{4} m_i^2$.
- In the diagrams with only an intermediate particle ϕ, we talk of the s-channel, t-channel, and u-channel for the diagrams, which contain poles in s, t, and u, respectively, due to the ϕ propagator having s, t, u as $-p^2$.
- In $2 \to 2$ scattering, we can define the relativistic invariant differential cross-section $d\sigma/dt(s,t)$, which integrates to $\sigma_{tot}(s)$.

Further Reading

See sections 5.2 and 5.4 in [1], and section 8.1 in [2].

Exercises

1. Check that

$$\frac{d\sigma}{d\Omega}(e_L^- e_R^+ \to \mu_R^- \mu_L^+) = \frac{\alpha^2}{4E_{CM}^2}(1 - \cos\theta)^2, \tag{26.43}$$

as in the text (check all the details).

2. Consider the amplitude

$$\mathcal{M}(s,t) = \frac{\Gamma(-\alpha(s))\Gamma(-\alpha(t))}{\Gamma(-\alpha(s) - \alpha(t))} \qquad (26.44)$$

(Veneziano), where $\alpha(x) = a + bx$ $(a, b > 0)$. Show that it is crossing symmetric. Write it as an (infinite) sum of s-channel contributions (processes), and equivalently as a sum of t-channel contributions (processes).

3. Consider the crossing applied to the process $e^+e^- \rightarrow e^+e^-$ (Bhabha scattering). Show how the diagram(s), and their corresponding expressions, change under the operation.

4. Consider the unpolarized cross-section from the previous chapter. Write the relativistically invariant formula for it, first $d\sigma/dt$, and then the relativistically invariant total cross-section.

27

(Unpolarized) Compton Scattering

In this chapter we will study the most (calculationally) complicated process so far, the last example of applications of the quantum field theory formalism to quantum electrodynamics, namely Compton scattering, $e^- \gamma \to e^- \gamma$. Also, in this chapter, we will have a new ingredient compared to previous ones, the sum over photon polarizations. It is also the first example we calculate where we have a sum of diagrams. The scattering we will calculate is the unpolarized one, where we can't measure the spin of the particles.

27.1 Compton Scattering: Set-up

We have an electron of momentum p scattering with a photon of momentum k, resulting in a final electron of momentum p' and a final photon of momentum k'. There are two possible Feynman diagrams, with the photon being absorbed by the electron line before the final photon is emitted from the electron line (or we can write it as the intermediate electron line being vertical in the diagram), or first the final photon is emitted, and then the incoming photon is absorbed (or we can write it as the intermediate electron line being horizontal in the diagram), see Figure 27.1. In the first diagram the intermediate electron line has momentum $p + k$, in the second one it has momentum $p - k'$.

Using the Feynman rules, we can write the expression for the amplitude as (note that the two fermion lines are identical, so there is no relative minus sign between the diagrams)

$$
\begin{aligned}
i\mathcal{M} &= \bar{u}(p')(+e\gamma^\mu)\epsilon_\mu^*(k')(-)\frac{\not{p}+\not{k}+im}{(p+k)^2+m^2}(+e\gamma^\nu)\epsilon_\nu(k)u(p) \\
&\quad + \bar{u}(p')(+e\gamma^\nu)\epsilon_\nu(k)(-)\frac{\not{p}-\not{k}'+im}{(p-k')^2+m^2}(+e\gamma^\mu)\epsilon_\mu^*(k')u(p) \\
&= -e^2\epsilon_\mu^*(k')\epsilon_\nu(k)\bar{u}(p')\left[\frac{\gamma^\mu(\not{p}+\not{k}+im)\gamma^\nu}{(p+k)^2+m^2}+\frac{\gamma^\nu(\not{p}-\not{k}'+im)\gamma^\mu}{(p-k')^2+m^2}\right]u(p). \quad (27.1)
\end{aligned}
$$

But note that $(p+k)^2 + m^2 = p^2 + k^2 + 2p \cdot k + m^2 = 2p \cdot k$ (since $k^2 = 0$ for the external photon, and $p^2 + m^2 = 0$ for the external electron), and similarly $(p-k')^2 + m^2 = p^2 + k'^2 - 2p \cdot k' + m^2 = -2p \cdot k'$.

We now also use the fact that $u(p)$ and $v(p)$ satisfy the Dirac equation, that is $(\not{p} - im)u(p) = 0$ and $(\not{p} + im)v(p) = 0$. Then we have

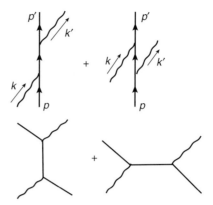

Figure 27.1 Feynman diagrams for Compton scattering, shown in two ways: above as scatterings off an electron line, and below as s-channel and t-channel diagrams.

$$(\not{p} + im)\gamma^\nu u(p) = (-\gamma^\nu \not{p} + 2g^{\mu\nu}p_\mu + im\gamma^\nu)u(p)$$
$$= 2p^\nu u(p) - \gamma^\nu(\not{p} - im)u(p) = 2p^\nu u(p), \qquad (27.2)$$

where we have used $\gamma^\mu \gamma^\nu = -\gamma^\nu \gamma^\mu + 2g^{\mu\nu}$ and the Dirac equation above. Substituting this formula in the square bracket in $\mathcal{M}$, we find

$$i\mathcal{M} = -e^2 \epsilon_\mu^*(k')\epsilon_\nu(k)\bar{u}(p')\left[\frac{\gamma^\mu \not{k}\gamma^\nu + 2\gamma^\mu p^\nu}{2p \cdot k} + \frac{-\gamma^\nu \not{k}'\gamma^\mu + 2\gamma^\nu p^\mu}{-2p \cdot k'}\right]u(p). \qquad (27.3)$$

As we saw in Chapter 26:

$$(\bar{u}\gamma^\mu \tilde{u})^* = -\bar{\tilde{u}}\gamma^\mu u, \qquad (27.4)$$

due to the fact that for a number, complex conjugation is the same as adjoint, and then we can write the transposed matrices in the opposite order, and use $(\gamma^\mu)^\dagger i\gamma^0 = -i\gamma^0 \gamma^\mu$ (where $\bar{u} = u^\dagger i\gamma^0$). Here $u, \tilde{u}$ can be either u or v. Then we can generalize this to

$$(\bar{u}\gamma^{\mu_1} \dots \gamma^{\mu_n}\tilde{u})^* = (-)^n \bar{\tilde{u}}\gamma^{\mu_n} \dots \gamma^{\mu_1} u. \qquad (27.5)$$

We then find

$$-i\mathcal{M}^* = +e^2 \epsilon_\rho(k')\epsilon_\sigma^*(k)\bar{u}(p)\left[\frac{\gamma^\sigma \not{k}\gamma^\rho + 2\gamma^\rho p^\sigma}{2p \cdot k} + \frac{-\gamma^\mu \not{k}'\gamma^\sigma + 2\gamma^\sigma p^\rho}{-2p \cdot k'}\right]u(p'). \qquad (27.6)$$

27.2 Photon Polarization Sums

We note that if we sum over polarizations, in $|\mathcal{M}|^2$ we have $\sum_{pol.} \epsilon_\mu^*(k)\epsilon_\nu(k)$. We then have the theorem that in $|\mathcal{M}|^2$, such sums over polarizations can be done with the replacement

$$\sum_{pol.} \epsilon_\mu^*(k)\epsilon_\nu(k) \to g_{\mu\nu}. \qquad (27.7)$$

More precisely, if

$$iM = \epsilon_\mu^*(k)\mathcal{M}^\mu(k), \tag{27.8}$$

then

$$\sum_{pol.} MM^* = \sum_{pol.} \epsilon_\mu^*(k)\epsilon_\nu(k)\mathcal{M}^\mu(k)\mathcal{M}_\nu^*(k) = g_{\mu\nu}\mathcal{M}^\mu(k)\mathcal{M}^{*\nu}(k). \tag{27.9}$$

Indeed, Ward identities imply that $k_\mu \mathcal{M}^\mu(k) = 0$. We have seen from the Ward–Takahashi identities that we have the same for the 1PI functions, $k^\mu \Pi_{\mu\nu}(k) = 0$, $k^{\mu_1}\Gamma_{\mu_1\ldots\mu_n} = 0$. The S-matrix corresponds to a connected, amputated n-point function, contracted with external lines for the states (e.g. for the photons with $\epsilon_\mu(k)$). As it is reasonable to assume that the connected and amputated n-point functions have the same property, we should have $k_\mu \mathcal{M}^\mu = 0$. Another approximate way to see this is that $\mathcal{M}^\mu$ is the Fourier transform of an electromagnetic current between some initial and final state:

$$\mathcal{M}^\mu(k) = \int d^4x e^{ikx} \langle f|j^\mu(x)|i\rangle, \tag{27.10}$$

where $j^\mu = \bar{\psi}\gamma^\mu\psi$. The reason is that external photons are created by the interaction term $e\int d^4x j^\mu A_\mu$. We can check this fact for our amplitude, where each half of a diagram (initial and final halves) have this property.

Then, consider an on-shell photon momentum ($k^2 = 0$), for instance $k^\mu = (k, 0, 0, k)$ (moving in the third direction). The two physical polarizations are then transverse to the momentum, and of unit norm, that is $\epsilon_{(1)}^\mu = (0, 1, 0, 0)$ and $\epsilon_{(2)}^\mu = (0, 0, 1, 0)$. Then $k_\mu \mathcal{M}^\mu(k) = 0$ is written as

$$-k\mathcal{M}^0(k) + k\mathcal{M}^3(k) = 0 \Rightarrow \mathcal{M}^3(k) = \mathcal{M}^0(k). \tag{27.11}$$

Thus, finally the sum over photon polarizations gives

$$\sum_{pol.} \epsilon_\mu^*(k)\epsilon_\nu(k)\mathcal{M}^\mu(k)\mathcal{M}^{*\nu}(k) = |\mathcal{M}^1|^2 + |\mathcal{M}^2|^2 = -|\mathcal{M}^0|^2 + |\mathcal{M}^1|^2 + |\mathcal{M}^2|^2 + |\mathcal{M}^3|^2$$

$$= g_{\mu\nu}\mathcal{M}^\mu(k)\mathcal{M}^{*\nu}(k). \tag{27.12}$$

$$q.e.d.$$

27.3 Cross-Section for Compton Scattering

We now go back to the Compton scattering calculation, and find for the unpolarized $|\mathcal{M}|^2$ (doing also the sum over electron spins in the usual manner):

$$\frac{1}{4} \sum_{spins} |\mathcal{M}|^2 = -\frac{e^4}{4} g_{\mu\rho} g_{\nu\sigma} \, \mathrm{Tr} \left\{ (-i\slashed{p}' + m) \left[\frac{\gamma^\mu \slashed{k} \gamma^\nu + 2\gamma^\mu p^\nu}{2p \cdot k} + \frac{-\gamma^\nu \slashed{k}' \gamma^\mu + 2\gamma^\nu p^\mu}{-2p \cdot k'} \right] \right.$$

$$\left. (-i\slashed{p} + m) \left[\frac{\gamma^\sigma \slashed{k} \gamma^\rho + 2\gamma^\rho p^\sigma}{2p \cdot k} + \frac{-\gamma^\rho \slashed{k}' \gamma^\sigma + 2\gamma^\sigma p^\rho}{-2p \cdot k'} \right] \right\}$$

$$\equiv -\frac{e^4}{4} \left[\frac{I}{(2p \cdot k)^2} + \frac{II}{(2p \cdot k)(2p \cdot k')} + \frac{III}{(2p \cdot k')(2p \cdot k)} + \frac{IV}{(2p \cdot k')^2} \right].$$
(27.13)

But we note that $II = III$ and that $I(k = -k') = IV$, so we only need to calculate I and II. We have

$$I = \mathrm{Tr} \left[(-i\slashed{p}' + m)(\gamma^\mu \slashed{k} \gamma^\nu + 2\gamma^\mu p^\nu)(-i\slashed{p} + m)(\gamma_\nu \slashed{k} \gamma_\mu + 2\gamma_\mu p_\nu) \right], \qquad (27.14)$$

and we can split it into eight nonzero traces. Indeed, there are $2^4 = 16$ terms, but as we already saw, only the trace of an even number of gammas is nonzero, so half of the terms are zero. We calculate each of the eight traces separately.

First, the longest trace is

$$I_1 = (-i)^2 \, \mathrm{Tr}[\slashed{p}' \gamma^\mu \slashed{k} \gamma^\nu \slashed{p} \gamma_\nu \slashed{k} \gamma_\mu] = \mathrm{Tr}[\gamma^\mu \slashed{p}' \gamma_\mu \slashed{k} (-2\slashed{p}) \slashed{k}] = -4 \, \mathrm{Tr}[\slashed{p}' \slashed{k} \slashed{p} \slashed{k}]$$

$$= -4 \, \mathrm{Tr}[\slashed{p}' \slashed{k} (-\slashed{k} \slashed{p} + 2p \cdot k)] = -8p \cdot k \, \mathrm{Tr}[\slashed{p}' \slashed{k}] = -32(p \cdot k)(p' \cdot k). \qquad (27.15)$$

Here we have used the fact that $\gamma^\mu \gamma_\nu \gamma_\mu = -2\gamma_\nu$, thus $\gamma^\nu \slashed{p} \gamma^\mu = -2\slashed{p}$, that the trace is cyclically symmetric, that $\gamma^\mu \gamma^\nu = -\gamma^\nu \gamma^\mu + 2g^{\mu\nu}$, thus $\slashed{p}\slashed{k} = -\slashed{k}\slashed{p} + 2p \cdot k$, and $\slashed{k}\slashed{k} = k^2 = 0$ (the photon is massless), and finally $\mathrm{Tr}[\gamma^\mu \gamma^\nu] = 4g^{\mu\nu}$, thus $\mathrm{Tr}[\slashed{p}' \slashed{k}] = 4p' \cdot k$.

Similarly:

$$I_2 = (-i)^2 \, \mathrm{Tr}[\slashed{p}' \gamma^\mu \slashed{k} 2p^\nu \gamma_\nu \slashed{p} \gamma_\mu] = -2 \, \mathrm{Tr}[\gamma_\mu \slashed{p}' \gamma^\mu \slashed{k} (-m^2)] = -4m^2 \, \mathrm{Tr}[\slashed{p}' \slashed{k}] = -16m^2 p' \cdot k,$$

$$I_3 = (-i)^2 \, \mathrm{Tr}[\slashed{p}' \gamma^\mu 2p^\nu \gamma_\nu \slashed{p} \slashed{k} \gamma_\mu] = -2 \, \mathrm{Tr}[\gamma_\mu \slashed{p}' \gamma^\mu (-m^2) \slashed{k}] = -4m^2 \, \mathrm{Tr}[\slashed{p}' \slashed{k}] = -16m^2 p' \cdot k,$$

$$I_4 = (-i)^2 (2p_\nu 2p^\nu) \, \mathrm{Tr}[\slashed{p}' \gamma^\mu \slashed{p} \gamma_\mu] = +4m^2 \, \mathrm{Tr}[\slashed{p}(-2\slashed{p})] = -32m^2 p \cdot p'. \qquad (27.16)$$

Here we have used also the fact that $p^2 = -m^2$ for the external electron, besides the others, already used. These are all the terms with the $(-i\slashed{p}')$ and $(-i\slashed{p})$ factors. There are also terms with $(-i\slashed{p}')m$ and with $m(-i\slashed{p})$, but these are zero, since they have an odd number of gammas in the trace. Therefore we only have the four terms with m^2 left, which give

$$I_5 = m^2 \, \mathrm{Tr}[\gamma^\mu \slashed{k} \gamma^\nu \gamma_\nu \slashed{k} \gamma_\mu] = 4m^2 \, \mathrm{Tr}[\gamma^\mu \gamma_\mu \slashed{k} \slashed{k}] = 0,$$

$$I_6 = 2m^2 \, \mathrm{Tr}[\gamma^\mu \slashed{k} p^\nu \gamma_\nu \gamma_\mu] = 8m^2 \, \mathrm{Tr}[\slashed{k}\slashed{p}] = 32m^2 k \cdot p,$$

$$I_7 = 2m^2 \, \mathrm{Tr}[\gamma^\mu \slashed{p} \slashed{k} \gamma_\mu] = 8m^2 \, \mathrm{Tr}[\slashed{p}\slashed{k}] = 32m^2 p \cdot k,$$

$$I_8 = 4m^2 p^\nu p_\nu \, \mathrm{Tr}[\gamma^\mu \gamma_\mu] = -64m^4. \qquad (27.17)$$

Then for the I term we have

$$I = \sum_{i=1}^{8} I_i = -32(p \cdot k)(p' \cdot k) - 32m^2 p' \cdot k - 32m^2 p \cdot p' + 64m^2 p \cdot k - 64m^4$$

$$= -32[(p \cdot k)(p' \cdot k) + m^2 p' \cdot k + m^2 p \cdot p' - 2m^2 p \cdot k + 2m^4]. \qquad (27.18)$$

Similarly, we find (the details are left as an exercise, similarly to the *I* term above)

$$II = III = -16[-2k \cdot k'p \cdot p' + 2k \cdot pp' \cdot p + m^2 p' \cdot k - 2k' \cdot pp \cdot p' - m^2 p' \cdot k' + m^2 p \cdot p'$$
$$-m^2 k \cdot k' + 2m^2 p \cdot k - 2m^2 p \cdot k' - m^4]. \tag{27.19}$$

We now translate into Mandelstam variables. We have

$$\begin{aligned}
s &= -(p+k)^2 = -p^2 - k^2 - 2p \cdot k = m^2 - 2p \cdot k \\
&= -(p'+k')^2 = -p'^2 - k'^2 - 2p' \cdot k' = m^2 - 2p' \cdot k', \\
t &= -(p-p')^2 = -p^2 - p'^2 + 2p \cdot p' = 2m^2 + 2p \cdot p' \\
&= -(k-k')^2 = -k^2 - k'^2 + 2k \cdot k' = 2k \cdot k', \\
u &= -(k'-p)^2 = -k'^2 - p^2 + 2k' \cdot p = m^2 + 2k' \cdot p \\
&= -(k-p')^2 = -k^2 - p'^2 + 2k \cdot p' = m^2 + 2k \cdot p',
\end{aligned} \tag{27.20}$$

which allow us to write all the inner products of external momenta in terms of s, t, u. We have also to remember that

$$s + t + u = \sum_{i=1}^{4} m_i^2 = 2m^2. \tag{27.21}$$

Further, to calculate *IV* we need to replace $k \leftrightarrow -k'$, which we easily see implies exchanging $s \leftrightarrow u$.

We then find

$$\begin{aligned}
I &= -32 \left[\frac{m^2 - s}{2} \frac{u - m^2}{2} + m^2 \frac{u - m^2}{2} + m^2 \frac{t - 2m^2}{2} - 2m^2 \frac{m^2 - s}{2} + 2m^4 \right] \\
&= -32 \left[-\frac{su}{4} + m^2 \left(\frac{3}{4} u + \frac{t}{2} + \frac{s}{4} \right) - \frac{3}{4} m^4 \right] \\
&= -32 \left[-\frac{su}{4} + \frac{m^2}{4}(3s + u) + \frac{m^4}{4} \right] \\
&= -16 \left[-\frac{1}{2}(s - m^2)(u - m^2) + m^2(s - m^2) + 2m^4 \right], \tag{27.22}
\end{aligned}$$

where in the next to last line we used $(s + t + u)/2 = m^2$.

Then we immediately find the expression for *IV*, just by switching $u \leftrightarrow s$:

$$IV = -16 \left[-\frac{1}{2}(s - m^2)(u - m^2) + m^2(u - m^2) + 2m^4 \right]. \tag{27.23}$$

Similarly, after similar algebra left as an exercise, we find

$$II = III = 16 \left[\frac{m^2}{2}(s - m^2) + \frac{m^2}{2}(u - m^2) + 2m^4 \right]. \tag{27.24}$$

Finally we have

$$\frac{1}{4}\sum_{spins}|\mathcal{M}|^2 = \frac{e^4}{4}\left\{\frac{16}{(s-m^2)^2}\left[-\frac{1}{2}(s-m^2)(u-m^2)+m^2(s-m^2)+2m^4\right]\right.$$

$$+\frac{16}{(u-m^2)^2}\left[-\frac{1}{2}(s-m^2)(u-m^2)+m^2(u-m^2)+2m^4\right]$$

$$\left.+\frac{2\cdot16}{(s-m^2)(u-m^2)}\left[\frac{m^2}{2}(s-m^2)+\frac{m^2}{2}(u-m^2)+2m^4\right]\right\}$$

$$= 4e^4\left\{-\frac{s-m^2}{2(u-m^2)}-\frac{u-m^2}{2(s-m^2)}\right.$$

$$\left.+\frac{2m^2}{u-m^2}+\frac{2m^2}{s-m^2}+2m^4\left(\frac{1}{u-m^2}+\frac{1}{s-m^2}\right)^2\right\}.$$

$$(27.25)$$

The relativistically invariant differential cross-section is

$$\frac{d\sigma}{dt} = \frac{1}{64\pi[(p_1\cdot p_2)^2-m_1^2 m_2^2]}\frac{1}{4}\sum_{spins}|\mathcal{M}|^2. \qquad (27.26)$$

But in our case, with $p_1 = p, p_2 = k$, we have

$$(p_1\cdot p_2)^2 - m_1^2 m_2^2 = \left(\frac{s-m^2}{2}\right)^2 - 0 = \frac{(s-m^2)^2}{4}, \qquad (27.27)$$

giving

$$\frac{d\sigma}{dt} = \frac{e^4}{4\pi(s-m^2)^2}\frac{1}{4}\sum_{spins}|\mathcal{M}|^2$$

$$= \frac{4\pi\alpha^2}{(s-m^2)^2}\left\{-\frac{s-m^2}{2(u-m^2)}-\frac{u-m^2}{2(s-m^2)}\right.$$

$$\left.+\frac{2m^2}{u-m^2}+\frac{2m^2}{s-m^2}+2m^4\left(\frac{1}{u-m^2}+\frac{1}{s-m^2}\right)^2\right\}. \qquad (27.28)$$

As we see, this formula is not that simple.

We can also analyze this formula in the laboratory frame, obtaining the *spin-averaged Klein–Nishina formula*

$$\frac{d\sigma}{d\cos\theta} = \frac{\pi\alpha^2}{m^2}\left(\frac{\omega'}{\omega}\right)^2\left[\frac{\omega'}{\omega}+\frac{\omega}{\omega'}-\sin^2\theta\right], \qquad (27.29)$$

where ω and ω' are the angular frequencies of the incoming and outgoing photons in the laboratory frame (where the electron target is fixed), related to each other by

$$\frac{\omega'}{\omega} = \frac{1}{1+\frac{\omega}{m}(1-\cos\theta)}. \qquad (27.30)$$

We will not prove the Klein–Nishina formula here, but rather will leave it as an exercise.

Important Concepts to Remember

- For sums over photon polarizations, if $i\mathcal{M} = \epsilon_\mu^*(k)\mathcal{M}^\mu(k)$, we replace $\sum_{pol.} \epsilon_\mu^*(k)\epsilon_\nu(k)$ by $g_{\mu\nu}$ in $|\mathcal{M}|^2$.
- When we have fermion propagators in Feynman diagrams, we can use the fact that the $u(p)$s and $v(p)$s satisfy the Dirac equation, $(\not{p} - im)u(p) = 0$ and $(\not{p} + im)v(p) = 0$.

Further Reading

See section 5.5 in [1] and section 8.2 in [2].

Exercises

1. Show that

$$\mathrm{Tr}\left[(-i\not{p}' + m)(\gamma^\mu \not{k}\gamma^\nu + 2\gamma^\mu p^\nu)(-i\not{p} + m)(\gamma^\mu \not{k}'\gamma^\nu - 2\gamma^\nu p^\mu)\right]$$
$$= 16\left[\frac{m^2}{2}(s - m^2) + \frac{m^2}{2}(u - m^2) + 2m^4\right]. \tag{27.31}$$

2. Using the formulas in the text, calculate the total relativistically invariant cross-section, $\sigma_{tot}(s)$.
3. Is the formula (27.28) crossing symmetric? Why do you think this is so?
4. Prove the Klein–Nishina formula (27.29), starting from (27.28).

28

The Helicity Spinor Formalism

Until now, we have considered quantum field theory amplitudes constructed via Feynman diagrams, and this is indeed the traditional way. A problem with this formalism, however, is that the number of Feynman diagrams increases exponentially not only with the number of loops L, but also with the number of external lines n. However, in the last 20 years or so, another formalism gained a lot of importance, since it was realized that it allows one to deal with much fewer terms, and in the case of tree diagrams only one: the spinor helicity formalism.

The goal is to write the amplitudes not for arbitrary helicities of external particles, but only for given helicities, most notably for what is called maximally helicity violating (MHV) amplitudes, and then the sum of the Feynman diagram (i.e. the full amplitude) has a very simple expression in terms of spinor helicities, which will be the Parke–Taylor formula. Moreover, the Britto–Cachazo–Feng–Witten (BCFW) construction allows one to construct larger tree amplitudes out of smaller ones.

While there has been some progress in the case of amplitudes with massive external states, the best understood case is still the one for massless external states, so that is what we will focus on.

28.1 Helicity Spinor Formalism for Spin 1/2

We can write momenta p^μ as matrices with spinor indices, by multiplication with the gamma matrices $(\gamma^\mu)^\alpha{}_\beta$. We make the split of the Dirac spinor indices into $\alpha = (a, \dot{a})$, where a and $\dot{a}$ refers to a Weyl spinor subspace, that is a spans the spinors defined by $\psi = (1 + \gamma_5)\psi/2$ and $\dot{a}$ the spinors defined by $\psi = (1 - \gamma_5)\psi/2$. This is the Weyl representation for the gamma matrices, where $\gamma_5 = \begin{pmatrix} 1 & 0 \\ 0 & -1 \end{pmatrix}$. Then we have

$$\not{p} = p_\mu \gamma^\mu = \begin{pmatrix} \mathbf{0} & p_\mu(\sigma^\mu)_{a\dot{b}} \\ p_\mu(\bar{\sigma}^\mu)^{\dot{a}b} & \mathbf{0} \end{pmatrix} \equiv \begin{pmatrix} \mathbf{0} & p_{a\dot{b}} \\ p^{\dot{a}b} & \mathbf{0} \end{pmatrix}, \tag{28.1}$$

and so

$$p_{a\dot{b}} = \begin{pmatrix} -p^0 + p^3 & p^1 - ip^2 \\ p^1 + ip^2 & -p^0 - p^3 \end{pmatrix} \Rightarrow \det p_{a\dot{b}} = -p_\mu p^\mu = m^2. \tag{28.2}$$

In particular, for massless spinors, $\det p_{a\dot{b}} = 0$, which means that it can be factorized as

$$p_{a\dot{b}} = \lambda_a \tilde{\lambda}_{\dot{b}}. \tag{28.3}$$

Indeed, the two columns of the $p_{a\dot{b}}$ matrix are then proportional to each other, and so are the two rows, implying zero determinant.

More precisely, we are interested in on-shell spinors, that is spinors which satisfy the massless Dirac equation $\not{p}\psi(p) = 0$, so we can use them to characterize massless external states.

Moreover, as we said, we are interested in eigenvectors of *helicity*, the projection of spin on the direction of the momentum:

$$H = \frac{\vec{p} \cdot \vec{S}}{|\vec{p}|}. \tag{28.4}$$

As an operator, the spin S^i is the diagonal $SU(2)$ in the decomposition of the Lorentz group into $SU(2) \times SU(2)$, as we learned in the classical field theory course (i.e. the operator we used to call J^i in Chapter 1). We then have

$$S^i = J^i = \frac{1}{2}\epsilon^{ijk}J_{jk} = \frac{1}{2}\epsilon^{ijk}\frac{i}{4}[\gamma_j, \gamma_k]. \tag{28.5}$$

In Chapters 12–14 we considered mostly massive fermions, since that is the physical case of electrons or quarks, and in that case there was no correlation between the chiral basis (eigenvectors of $P_{L/R} = (1 \pm \gamma_5)/2$, called Weyl spinors) and the helicity basis, but in this case there is.

Being interested in massless spinors, the equation of motion for both positive frequency modes $u_s(p)$ and negative frequency modes $v_s(p)$ is the same:

$$\not{p}u_s(p) = 0, \tag{28.6}$$

so we will only consider $u_s(p)$ (otherwise, we have $u_\pm(p) = v_\mp(p)$, denoting $s = +, -$). As we already know, there are two independent solutions, that were labeled $s = 1, 2$. We consider the Weyl representation, so that the chiral (Weyl) spinors correspond to the upper two components of the Dirac spinor, and the lower two components, respectively. We call the first type, corresponding to index a, chiral (eigenvalue $+1$ for γ_5) and the second, corresponding to index $\dot{a}$, antichiral (eigenvalue -1 for γ_5). Then the equation of motion is

$$\begin{pmatrix} \mathbf{0} & p \cdot \sigma \\ p \cdot \bar{\sigma} & \mathbf{0} \end{pmatrix} u_s(p) = 0. \tag{28.7}$$

Then, if the momentum is for instance moving in the third direction, $p^\mu = E(1, 0, 0, 1)$, the equation of motion becomes

$$-2E \begin{pmatrix} \mathbf{0} & \begin{pmatrix} 0 & 0 \\ 0 & 1 \end{pmatrix} \\ \begin{pmatrix} 1 & 0 \\ 0 & 0 \end{pmatrix} & \mathbf{0} \end{pmatrix} u_s(p) = 0, \tag{28.8}$$

so the two independent solutions would be (we change the notation from $s = 1, 2$ to $s = +, -$)

$$u_+(p) = \begin{pmatrix} 0 \\ 1 \\ 0 \\ 0 \end{pmatrix}, \quad u_-(p) = \begin{pmatrix} 0 \\ 0 \\ 1 \\ 0 \end{pmatrix}, \tag{28.9}$$

and we see that they have opposite chirality (eigenvalue of γ_5). In fact, we can check that we can arrange for any p for the two independent solutions to be chiral, of opposite chirality. Moreover, we can easily check in this case that the helicity operator is

$$H = \frac{\vec{p} \cdot \vec{S}}{E} = S_3 = -\frac{1}{2} \begin{pmatrix} \sigma_3 & \mathbf{0} \\ \mathbf{0} & \sigma_3 \end{pmatrix}, \tag{28.10}$$

and thus we have the relation

$$H u_\pm(p) = h u_\pm(p) = \pm \frac{1}{2} u_\pm(p). \tag{28.11}$$

The relation is in fact true for general p, and it means that $u_+(p)$ has helicity $+1/2$, and $u_-(p)$ has helicity $+1/2$. But we saw that $u_+(p)$ was chiral (or left-handed), with index a, and $u_-(p)$ was antichiral (or right-handed), with index $\dot{a}$. We therefore denote

$$u_+(p) = |\lambda^+\rangle = \lambda^a, \quad u_-(p) = |\tilde{\lambda}^-] = \tilde{\lambda}^{\dot{a}}, \tag{28.12}$$

or in column vector notation

$$u_+(p) = \begin{pmatrix} |\lambda^a\rangle \\ 0 \end{pmatrix}, \quad u_-(p) = \begin{pmatrix} 0 \\ |\tilde{\lambda}^{\dot{a}}] \end{pmatrix}, \tag{28.13}$$

where we have distinguished between the two by the use of either angle, or square brackets. At the moment this is a bit irrelevant, but it will soon become important.

We can repeat the analysis for the conjugate Dirac equation:

$$\bar{u}(p)\not{p} = 0, \tag{28.14}$$

and (again by using the Weyl representation and momentum in the third direction) we can easily prove that the solutions are

$$\bar{u}_+(p) = (0, [\tilde{\lambda}_{\dot{a}}|) \equiv \lambda_{\dot{a}}, \quad \bar{u}_-(p) = (\langle \lambda_a|, 0) \equiv \lambda_a. \tag{28.15}$$

Here indices were raised and lowered using the Levi–Civitta antisymmetric symbol, defined by

$$\epsilon^{12} = \epsilon^{\dot{1}\dot{2}} = +1 \tag{28.16}$$

and

$$\epsilon_{ab} = -\epsilon^{ab}, \quad \epsilon_{\dot{a}\dot{b}} = -\epsilon^{\dot{a}\dot{b}}, \tag{28.17}$$

as

$$\lambda_a = \epsilon_{ab}\lambda_b, \quad \lambda^a = \epsilon^{ab}\lambda_b, \quad \tilde{\lambda}_{\dot{a}} = \epsilon_{\dot{a}\dot{b}}\tilde{\lambda}^{\dot{b}}, \quad \tilde{\lambda}^{\dot{a}} = \epsilon^{\dot{a}\dot{b}}\tilde{\lambda}_{\dot{b}}. \tag{28.18}$$

Momenta are real numbers, but we will shortly see that it is often useful to consider complex values for the momenta, so in that case, the $\bar{u}_\pm(p)$ are not related to the $u_\pm(p)$, and λ and $\tilde{\lambda}$ are also independent. For real momenta, however, we can take the Dirac conjugate of $\not{p}u_\pm(p) = 0$ and, since $\bar{\psi} \equiv \psi^\dagger i\gamma^0 = \psi^\dagger \begin{pmatrix} \mathbf{0} & \mathbf{1} \\ \mathbf{1} & \mathbf{0} \end{pmatrix}$, which takes the complex conjugate and exchanges left and right chiral spinors, we obtain $\bar{u}_\pm(p)\not{p} = 0$ provided we make the identification

$$\lambda^a = (\tilde{\lambda}^{\dot{a}})^*. \tag{28.19}$$

In the case of several momenta, as we have in amplitudes, we will use the notation

$$u^+(k_i) = |i\rangle = \lambda_i^a , \quad u^-(k_i) = |i] = \tilde{\lambda}_i^{\dot{a}}. \tag{28.20}$$

Since λ and $\tilde{\lambda}$ define minimal spinor Lorentz representations, by contractions with the Levi–Civitta symbol we can get Lorentz invariants, the spinor products

$$\langle i\,j\rangle \equiv \langle i|j\rangle \equiv \epsilon^{ab}(\lambda_i)_a(\lambda_j)_b = \bar{u}^-(k_i)u^+(k_j),$$
$$[i\,j] \equiv [i|j] \equiv \epsilon^{\dot{a}\dot{b}}(\tilde{\lambda}_i)_{\dot{a}}(\tilde{\lambda}_j)_{\dot{b}} = \bar{u}^+(k_i)u^-(k_j), \tag{28.21}$$

or angle spinor bracket and square spinor bracket, respectively.

In the case of real momenta, we then have

$$\langle i\,j\rangle = [i\,j]^*. \tag{28.22}$$

We reconstruct the momenta from the helicity spinors, since

$$2p^\mu = p_\nu \, \text{Tr}[\sigma^\nu \sigma^\mu] = p_{a\dot{b}}(\sigma^\mu)^{\dot{b}a} = [\tilde{\lambda}|\gamma_\mu|\lambda\rangle = \tilde{\lambda}_{\dot{b}}(\sigma^\mu)^{\dot{b}a}\lambda_a. \tag{28.23}$$

But moreover, we can reconstruct generalized Mandelstam variables from the helicity spinors. We have seen in Chapter 26 that we can define Mandelstam variables s, t, u for $2 \to 2$ scattering such that, for massless external particles, $s + t + u = 0$. But more generally, we can define generalized Mandelstam variables for any n-point amplitude. In the convention that all the external momenta k_i^μ are incoming, the generalized variables are

$$s_{ij} = (k_i + k_j)^2 = 2k_i \cdot k_j, \tag{28.24}$$

where in the last equality we have used $k_i^2 = k_j^2 = 0$ for massless external particles.

In 1+1 dimensions, the σ^μ matrices are a complete set over the 2×2 matrices, which means that any matrix $M_{a\dot{a}}$ can be decomposed into them, $c_\mu(\sigma^\mu)_{a\dot{a}}$. Taking the trace on both sides, we find

$$M_{a\dot{a}} = \frac{1}{2}M_{b\dot{b}}(\sigma_\mu)^{\dot{b}b}(\sigma^\mu)_{a\dot{a}}. \tag{28.25}$$

This means that we have the completeness relation

$$(\sigma_\mu)^{\dot{b}b}(\sigma^\mu)_{a\dot{a}} = 2\delta_a^b \delta_{\dot{a}}^{\dot{b}}. \tag{28.26}$$

Then we can express the two momenta in s_{ij} in terms of the spinors and use the completeness relation above to find

$$s_{ij} = 2k_i \cdot k_j = \frac{1}{2}(\tilde{\lambda}_i)_{\dot{a}}(\sigma^\mu)^{\dot{a}a}(\lambda_i)^a \ (\tilde{\lambda}_j)_{\dot{b}}(\sigma^\mu)^{\dot{b}b}(\lambda_j)^b = (\tilde{\lambda}_i)_{\dot{a}}(\tilde{\lambda}_j)^{\dot{a}}(\lambda_j)_a(\lambda^i)^a = \langle i\,j\rangle[j\,i].$$

$$(28.27)$$

For real momenta, when $\langle i\,j\rangle = [i\,j]^*$, this then defines the angle and square brackets up to a phase, since then

$$\langle i\,j\rangle = \sqrt{-s_{ij}}\,e^{i\theta_{ij}}, \quad [i\,j] = -\sqrt{-s_{ij}}\,e^{-i\theta_{ij}}.$$

$$(28.28)$$

Note that among the properties of the brackets is antisymmetry, derived from the antisymmetry of the Levi–Civitta symbol in their definition:

$$\langle i\,j\rangle = -\langle j\,i\rangle, [i\,j] = -[j\,i].$$

$$(28.29)$$

We can get other useful identities for the helicity spinors. One is a completeness relation derived from the spin sum for helicity spinors. We have seen in Chapter 14 that we have the spin sum $\sum_s u^s(p)\bar{u}^s(p) = -i\not{p} + m$. In the massless spinor case, sandwiching it between $P_L = (1 + \gamma_5)/2$ (that projects onto helicity + states) and $P_R = (1 - \gamma_5)/2$ (that projects onto helicity − states), we find

$$\sum_s P_L u^s(p)\bar{u}^s(p)P_R = u^+(p)\bar{u}^-(p) = -iP_L\not{p}P_R = -ip_{ab},$$

$$(28.30)$$

which is just a statement of the decomposition $p_{ab} = \lambda_a\tilde{\lambda}_b$. In matrix form it can be written as $-iP_L\not{p}P_R = |\lambda\rangle[\tilde{\lambda}|$. But in an amplitude with momenta k_i, satisfying momentum conservation $\sum_i k_i^\mu = 0$, we can sum over i on both sides and obtain

$$\sum_{i=1}^n |i\rangle[i| = 0.$$

$$(28.31)$$

There is one more useful helicity spinor identity the Schouten identity. In general, the Schouten identity is the statement that if we completely antisymmetrize a product of $n + 1$ objects, when only n of them can be independent, we obtain zero. As such, one starts from a trivially obvious statement, but usually ends up with a useful identity. In our case, consider that the helicity spinors λ_i, as well as $\tilde{\lambda}_i$, are two-component, so there can only be two independent ones. Consider then the antisymmetrized product of three of them, where two of them are dotted into an angle product. The result should be zero:

$$\lambda_{[i}(\lambda_j \cdot \lambda_{k]}) = 0.$$

$$(28.32)$$

Written in terms of states and their products, and with an explicit sum, we obtain

$$|i\rangle\langle jk\rangle + |k\rangle\langle ij\rangle + |j\rangle\langle ki\rangle = 0.$$

$$(28.33)$$

We can also present it as an identity only between angle brackets, by taking the product with an $|l\rangle$:

$$\langle li\rangle\langle jk\rangle + \langle lk\rangle\langle ij\rangle + \langle lj\rangle\langle ki\rangle = 0.$$

$$(28.34)$$

An amplitude calculated with helicity spinor formalism will be an amplitude of given external helicity, so the external states will be defined by momenta and helicity + or − (i.e. a helicity spinor state). Before giving examples of amplitudes, we move on to spin 1.

28.2 Helicity Spinor Formalism for Spin 1

The full power of the helicity spinor formalism is realized in amplitudes of massless spin-1 and spin-2 particles (i.e. gluons and gravitons), hence we now describe the extension of the formalism to this case. In these cases, there are also two possible helicities, $h = \pm 1$ for massless vectors and $h = \pm 2$ for massless gravitons.

We can write the polarization vectors for massless gauge fields of momentum k_i and of definite helicities by using a reference momentum q of spinors λ_q^a and $\tilde{\lambda}_q^{\dot{a}}$ ($q^{a\dot{a}} = \lambda_q^a \tilde{\lambda}_q^{\dot{a}}$) as (remember that $|i]$ has negative helicity, $|i^-]$ and $|i\rangle$ has positive helicity, $|i^+\rangle$)

$$\epsilon_\mu^+(k_i; q) = \frac{\langle q|\gamma_\mu|i]}{\sqrt{2}\langle qi\rangle} , \quad \epsilon_\mu^-(k_i; q) = -\frac{[q|\gamma_\mu|i\rangle}{\sqrt{2}[qi]}. \tag{28.35}$$

In terms of spinors components, we have

$$\epsilon_{a\dot{a}}^+(k_i; q) = \frac{\sqrt{2}\lambda_q^a \tilde{\lambda}_i^{\dot{a}}}{\langle qi\rangle} , \quad \epsilon_{a\dot{a}}^-(k_i; q) = -\frac{\sqrt{2}\tilde{\lambda}_q^{\dot{a}}\lambda_i^a}{[qi]}, \tag{28.36}$$

where we remember that $\langle qi\rangle = \epsilon^{ab}(\lambda_q)_a(\lambda_i)_b$ and $[qi] = \epsilon^{\dot{a}\dot{b}}(\lambda_q)_{\dot{a}}(\lambda_i)_{\dot{b}}$.

Because the helicity spinors by definition satisfy the massless Dirac equation, that is

$$\slashed{k}|i^+\rangle = \slashed{k}|i^-] = \slashed{q}|q^+\rangle = \slashed{q}|q^-] = 0, \tag{28.37}$$

we find that the polarization vectors are transverse to the momentum, as to the reference momentum:

$$\epsilon^\pm(k_i; q) \cdot k_i = \epsilon^\pm(k_i; q) \cdot q = 0. \tag{28.38}$$

Moreover, we see that $\epsilon_\mu^+(k_i; q) \propto \tilde{\lambda}_i^{\dot{a}}/\lambda_i^b$ and $\epsilon_\mu^-(k_i; q) \propto \lambda_i^a/\tilde{\lambda}_i^{\dot{b}}$, and λ_i^a has helicity $-1/2$ and $\tilde{\lambda}_i^a$ has helicity $+1/2$, meaning that their ratios have helicities $+1$ and -1, respectively, as stated.

28.3 Amplitudes with External Spinors

As a simple example of the formalism for amplitudes with external spinors, we consider a case where there is a single tree-level Feynman diagram, of the $e\bar{e} \to l\bar{l}$ scattering from Chapter 24, just in the massless quark limit. Note that we put in bars instead of the sign of the charge, in order not to confuse with helicities. Indeed, we now want to study the amplitudes of given helicities, for which the result was (25.1), and in the notation of Chapter 27, we would study $e_R^-(-k_1)e_L^+(-k_2) \to l_L(k_3)\bar{l}_R(k_4)$. The convention we use is for all momenta k_i to be incoming, so $k_1 + k_2 + k_3 + k_4 = 0$. The amplitude is then

$$A_4 = A_4^{\text{tree}}(1_e^+, 2_e^-, 3_l^+ 4_l^-), \tag{28.39}$$

and by the formula (25.1) we find

$$A_4 = \frac{ie^2}{(k_1 + k_2)^2} \overline{v_-}(k_2)\gamma_\mu u_-(k_1)\overline{u_+}(k_3)\gamma^\mu v_+(k_4)$$

$$= \frac{ie^2}{s_{12}} \lambda_2^a (\sigma_\mu)_{a\dot{a}} \tilde{\lambda}_1^{\dot{a}}(\tilde{\lambda}_3)_{\dot{b}} (\sigma^\mu)^{\dot{b}b}(\lambda_4)_b, \tag{28.40}$$

so that finally

$$A_4 = A_4^{\text{tree}}(1_{\bar{e}}^+, 2_e^-, 3_l^+ 4_{\bar{l}}^-) = ie^2 \frac{\langle 24\rangle[13]}{s_{12}} = ie^2 \frac{\langle 24\rangle[13]}{\langle 12\rangle[21]}. \tag{28.41}$$

This amplitude exemplifies the general case: the result for the helicity amplitude is given just in terms of (generalized Mandelstam variables and) angle and square brackets for helicity spinors.

But as we said, the power of the formalism is that the *sum* of the Feynman diagrams can be expressed in a simple manner, one that can be calculated using other methods (for instance, the Parke–Taylor formula and the BCFW construction). This will be studied in Chapter 29.

Important Concepts to Remember

- For massless spinors, their momentum p_μ can be factorized as $p_{a\dot{b}} = \lambda_a \tilde{\lambda}_{\dot{b}}$.
- For the spinors $|i\rangle = \lambda_i^a$ and $|i]$, we have the Lorentz invariants $\langle i\,j\rangle = \langle i|j\rangle = \epsilon^{ab}\lambda_{ia}\lambda_{jb}$ and $[i\,j] = [i|j] = \epsilon^{\dot{a}\dot{b}}\tilde{\lambda}_{i\dot{a}}\tilde{\lambda}_{j\dot{b}}$, which are antisymmetric and satisfy $s_{ij} = \langle i\,j\rangle[j\,i]$.
- For spin-1 gauge fields, we can define the polarization vectors in terms of helicity spinors and a reference momentum q, by $\epsilon_\mu^+(k_i; q) = \frac{\langle q|\gamma_\mu|i]}{\sqrt{2}\langle qi\rangle}$ and $\epsilon_\mu^-(k_i; q) = -\frac{[q|\gamma_\mu|i\rangle}{\sqrt{2}[qi]}$.
- The four-point gauge field amplitude with two positive and two negative helicities is $A_4 = A_4^{\text{tree}}(1_{\bar{e}}^+, 2_e^-, 3_l^+ 4_{\bar{l}}^-) = ie^2 \frac{\langle 24\rangle[13]}{\langle 12\rangle[21]}$.

Further Reading

See the review [8] and the book [9].

Exercises

1. Show that if $u_\pm$ are the solutions to the Dirac equation, then $Hu_\pm(p) = \pm u_\pm(p)$ for general p, so the $\pm$ index refers to helicity.
2. How does the calculation of the electron–positron scattering amplitude $A_4^{\text{tree}}(1_{\bar{e}}^+, 2_e^-, 3_l^+ 4_{\bar{l}}^-)$ get modified if the final states are replaced by $e, \bar{e}$ instead of $l, \bar{l}$? Write the amplitude in terms of helicity spinors.

3. Using the properties of the helicity spinors, show that the amplitude

$$A_5 = -i\frac{\langle 2\,5\rangle}{s_{12}}\frac{\langle 1^+|(k_3+k_4)|q^+\rangle[4\,3]}{s_{34}\langle q\,4\rangle} + i\frac{[1\,3]}{s_{12}}\frac{\langle 2^-|(k_4+k_5)|4^-\rangle\langle q\,5\rangle}{s_{45}\langle q\,4\rangle} \qquad (28.42)$$

can be rewritten, by choosing a q that makes the second term vanish, as

$$A_5 = i\frac{\langle 2\,5\rangle^2}{\langle 1\,2\rangle\langle 3\,4\rangle\langle 4\,5\rangle}. \qquad (28.43)$$

4. Using the properties of the helicity spinors, show that the amplitude for the tree process $e_R^-(-k_1)e_L^+(-k_2) \to l_L(k_3)\bar{l}_R(k_4)$ in (28.41) can be rewritten as

$$A_4 = ie^2\frac{\langle 2\,4\rangle^2}{\langle 1\,2\rangle\langle 3\,4\rangle}. \qquad (28.44)$$

29

Gluon Amplitudes, the Parke–Taylor Formula, and the BCFW Construction

In this chapter we continue the study of amplitudes with helicity in the case of most interest: gluon amplitudes. We will show an important formula using helicity spinors, the Parke–Taylor formula, and a recursion relation, the Britto–Cachazo–Feng–Witten (BCFW) construction, that allows one, among other things, to prove simply the Parke–Taylor formula.

29.1 Amplitudes with External Gluons and Color-Ordered Amplitudes

The full power of the helicity spinor formalism is really reached in the case of nonabelian gauge theories or gravity theories, specifically for amplitudes with external gluons or gravitons. Even though we have not yet introduced the nonabelian gauge theory (nor gravity), we will rely on the little that is known from the classical field theory course in order to set up the problem in a way that can be understood.

In nonabelian gauge theories, the gauge fields, called gluons, are self-interacting, which means that there is a three-point vertex between three gluons. The gluons are in the adjoint representation, which means that they have a nonabelian label a, A_μ^a, and the three-point vertex is proportional to a factor f^{abc} for the color indices, the structure constant of the (local) symmetry group. There will also be a term of the type $\eta_{\mu\nu}p_1^\rho + cyclic$ for the coupling of $A_\mu^a(p_1), A_\nu^b(p_2), A_\rho^c(p_3)$, since it comes from the coupling of $(\partial_\mu A_\nu^a)f_{abc}A_\mu^b A_\nu^c$ type. One usually considers generators T_a for the group in the fundamental representation, $(T_a)_i^{\bar j}$, which are thus $N \times N$ matrices. One canonical normalization is $\text{Tr}_{\text{adj}}[T^a T^b] = \delta^{ab}$ in the adjoint representation, in which case the normalization in the fundamental representation is with a factor of 1/2, $\text{Tr}_{\text{fd}}[T^a T^b] = \frac{1}{2}\delta^{ab}$. But in the context of spinor helicity formalism, it is more common to define it with factor 1 in the fundamental representation:

$$\text{Tr}_{\text{fd}}[T^a T^b] = \delta^{ab}, \tag{29.1}$$

which has however the effect of rescaling the T^as by $\sqrt 2$ with respect to the usual definition, so the Lie algebra with this normalization is taken to be

$$[T^a, T^b] = i\sqrt 2 f^{abc} T_c, \tag{29.2}$$

where f^{abc} is as usual the three-point gluon coupling (the four-point one is related to it). But then in amplitudes, we first define $\tilde{f}^{abc} = i\sqrt{2}f^{abc}$, and then find from the Lie algebra relation and the normalization that

$$\tilde{f}^{abc} \equiv i\sqrt{2}f^{abc} = \text{Tr}[T^a T^b T^c] - \text{Tr}[T^a T^c T^b]. \tag{29.3}$$

This means that in the Feynman diagrams, we can exchange the couplings for traces of the T_a factors.

Moreover, if we have external quarks (i.e. fields charged under the fundamental representation of the gauge group) of the type q^i, the coupling of a quark, an antiquark, and a gluon is given by the matrix $(T_a)_i{}^j$, which means that besides traces of T_as, we will also find some products of T^as with external indices (untraced over). Then we can also use, in the case of an $SU(N)$ gauge group, the most common application, the completeness relation

$$(T_a)_i{}^j (T^a)_{i'}{}^{j'} = \delta_i^{j'} \delta_{i'}^j - \frac{1}{N}\delta_i^j \delta_{i'}^{j'}. \tag{29.4}$$

In any case, the result is that for amplitudes with only external gluons, we can factorize completely the dependence on gluon indices via traces of T_as, and be left with reduced amplitudes that are independent of them, and moreover with a single ordering of the color indices, the *color-ordered amplitude*. This *color decomposition* at tree level takes the form (for an n-gluon amplitude)

$$\mathcal{A}_n^{\text{tree}}(12\ldots n) = g^{n-2} \sum_{\sigma \in S_n/Z_n} \text{Tr}[T^{a_{\sigma(1)}} \ldots T^{a_{\sigma(n)}}] A_n^{\text{tree}}(\sigma(1), \ldots, \sigma(n)). \tag{29.5}$$

Here, $\mathcal{A}_n^{\text{tree}}(12\ldots n)$ is the full amplitude, depending on momenta k_i, helicities λ_i, and gluon gauge indices a_i, that is

$$\mathcal{A}_n^{\text{tree}}(12\ldots n) = \mathcal{A}_n^{\text{tree}}(\{k_i, \lambda_i, a_i\}), \tag{29.6}$$

g is the coupling, and σ is a permutation in the group of permutations S_n, modulo cyclic permutations Z_n (i.e. we sum over cyclically inequivalent permutations). The color-ordered amplitude A_n depends only on a single color ordering, which means that it depends only on Mandelstam variables s_{ij} for adjacent momenta s_{12}, s_{23}, s_{45}, and so on, and not on s_{ij} for non-adjacent ones like s_{13} for instance (t in the four-amplitude case). This also means that we only have one Feynman diagram, the color-ordered one, among possible diagrams for the full amplitude.

Note that the color decomposition can be performed at every loop, and also for the case of external quarks, but in each case the formula is different, there is no simple general form, hence we will not reproduce them here.

The simplest amplitude we can write, and which is in fact *useful* to write, is the three-point color-ordered amplitude for gluons.

First off, the three-point amplitudes are somewhat singular, since energy and momentum conservation means that there is a single possible nonzero case: k_1, k_2, k_3 are all parallel. Indeed, $p_1 + p_2 = -p_3$ (for all momenta in convention) implies, by squaring and using the fact that $p_1^2 = p_2^2 = p_3^2 = 0$, that $2p_1 \cdot p_2 = 0$, hence $p_2 \| p_1$, and similarly for the other products, thus all three momenta are parallel. Moreover, this means that the helicity spinors are proportional:

$$0 = 2p_1 \cdot p_2 = \langle 12 \rangle [21], \tag{29.7}$$

which means that either $|1\rangle \propto |2\rangle$, or $|1] \propto |2]$. Similarly for the other products, we find that either

$$|1\rangle \propto |2\rangle \propto |3\rangle \quad \text{or} \quad |1] \propto |2] \propto |3]. \tag{29.8}$$

For real momenta, the square and angle vectors are complex conjugate, so both are zero, which means that only for complex momenta can we have a difference.

For color-ordered amplitudes, one can calculate vertices for gluons. The three-gluon color-ordered vertex, in Gervais–Neveu gauge,[1] is found to be

$$V^{\mu\nu\rho}(k_1, k_2, k_3) = -\sqrt{2}(\eta^{\mu\nu}k_1^\rho + \eta^{\nu\rho}k_2^\mu + \eta^{\rho\mu}k_3^\nu). \tag{29.9}$$

This can easily be seen since the color ordering strips the f^{abc} factor (up to a numerical factor of $\sqrt{2}$) and introduces cyclicity in the vertex, which as we said contained $\eta^{\mu\nu}k_1$. Then the three-point color-ordered amplitude for general helicity is

$$A_3^{\text{tree}}(123) = -\sqrt{2}[(\epsilon_1 \cdot \epsilon_2)(\epsilon_3 \cdot k_1) + (\epsilon_2 \cdot \epsilon_3)(\epsilon_1 \cdot k_2) + (\epsilon_3 \cdot \epsilon_1)(\epsilon_2 \cdot k_3)]. \tag{29.10}$$

29.2 Amplitudes of Given Helicity and Parke–Taylor Formula

We can then express the result for helicities $(1^-2^-3^+)$ by expressing ϵ^+ and ϵ^- in terms of helicity spinors as above. We obtain three terms, one with $\tilde{\lambda}_3 \cdot \tilde{\lambda}_1 = [31]$, one with $\lambda_1 \cdot \lambda_2 = \langle 12 \rangle$, and one with $\lambda_2 \cdot \lambda_3 = \langle 23 \rangle$ in the numerator. In the real momentum case, when all the angle helicity spinors are proportional, and all the square helicity spinors are also proportional, the result is zero, since $[31] = \langle 12 \rangle = \langle 23 \rangle = 0$, so the amplitude vanishes.

But as we will see, it is useful to consider the amplitude at complex momenta, and then we can use only the special kinematics $|1] \propto |2] \propto |3]$, but the angle helicity spinors are not proportional. After some algebra and use of helicity spinor identities, we can show that we can write the (complex momentum) amplitude as

$$A_3[1^-2^-3^+] = i\frac{\langle 12 \rangle^3}{\langle 23 \rangle \langle 31 \rangle}. \tag{29.11}$$

We leave the proof as an exercise.

Similarly, in the complex momentum case, we can flip all the helicities of the above amplitude, which amounts to exchanging the angle with square spinors, and find by complex conjugation that

---

[1] The Gervais–Neveu gauge amounts to adding the gauge-fixing term $-\frac{1}{2}\operatorname{Tr}(\partial^\mu A_\mu - igA^\mu A_\mu)^2$, and gives the Lagrangian $\mathcal{L} = \operatorname{Tr}[-\frac{1}{2}\partial_\mu A_\nu \partial^\mu A^\nu - 2ig\partial^\mu A^\nu A_\nu A_\mu + \frac{g^2}{2}A^\mu A^\nu A_\mu A_\nu]$.

$$A_3[1^+2^+3^-] = -i\frac{[12]^3}{[23][31]},\tag{29.12}$$

or equivalently we can prove it in a similar manner.

In the three-point case, we can prove that there is no other nonzero helicity configuration, since $A_3[1^-2^-3^-] = A_3[1^+2^+3^+] = 0$, even for complex momentum.

We can now consider four-point and higher n amplitudes, which are not zero, nor do they require special kinematics. But one can prove that amplitudes with all helicities the same (the all momenta convention), or with only one different, which by cyclicity of the color-ordered amplitudes can be put to the first position, vanish:

$$A_n(1^\pm, 2^+, 3^+, \ldots, n^+) = 0.\tag{29.13}$$

A simple way to see this is as follows. In a tree Feynman diagram, there are at most $n - 2$ vertices (which is why we have factored out a g^{n-2} factor; we can see that a three-point amplitude has one three-point vertex, and then we can add each time a new three-point vertex with two propagator lines, adding one external line). Since each three-point vertex is linear in momentum, as we saw above, this procedure gives at most $n - 2$ contractions with momentum of the external helicities, which means that we must have at least one contraction of two helicities, $\epsilon_i \cdot \epsilon_j$. We can check that if we choose the arbitrary reference momentum q to be the same for all helicities, then if all helicities are the same (+ for our choice), the product $\epsilon_i^+ \cdot \epsilon_j^+ \propto \langle q_i q_j \rangle = 0$. In contrast, if one of the helicities is $-$, we can have the term $\epsilon_i^- \cdot \epsilon_j^+$, but then $\epsilon^-(k_i; q_i) \cdot \epsilon^+(k_j; q_j) \propto \langle iq_i \rangle [jq_j] = 0$ if $q_i = k_j$ or $q_j = k_i$. So if we choose for instance $q_1 = k_2, q_i = k_1, \forall i > 1$, we obtain zero for the amplitude with 1^- as well.

On the contrary, the amplitude with two different helicities, as well as ones with more than two, are nonzero. The amplitude with two different helicities is called the *maximally helicity violating* (MHV) amplitude. The reason is that, in the convention with all incoming momenta that we use, incoming particles not changing helicity with respect to outgoing ones amounts to incoming and outgoing helicities being different. So a "totally helicity violating amplitude" would have all helicities the same, for instance all pluses. But since these are zero, the "maximally helicity violating" is the one with two different helicities. The nomenclature follows with *next to maximally helicity violating* (NMHV) amplitudes, then NNMHV, and so on.

The MHV amplitudes are especially simple. In any gauge theory, it was proven by Parke and Taylor that the MHV amplitudes for n gluons take a remarkably simple form. The *Parke–Taylor formula* is

$$A_{ij}^{\text{MHV}} \equiv A_n^{\text{tree}}(1^+, \ldots, i^-, \ldots, j^-, \ldots, n^+) = i\frac{\langle ij \rangle^4}{\langle 12 \rangle \langle 23 \rangle \ldots \langle n1 \rangle}.\tag{29.14}$$

We first observe that this formula is consistent with the formula for three-point amplitudes above, since then one factor of $\langle ij \rangle$ (adjacent in the case of the three-point amplitude) cancels between the numerator and the denominator. We will prove the formula by induction, using the BCFW construction. The construction uses complex momenta, for which the three-point amplitude is nonzero.

29.3 Kleiss–Kluijf and BCJ Relations

Before continuing, we note that the amplitudes $A_n(12 \ldots n)$ are not all independent. There are $n!$ amplitudes, corresponding to the permutations of $(12 \ldots n)$. But the amplitudes are cyclically invariant, as we saw. Moreover, by taking the dagger inside the trace in the defining formula (29.5), we can show the property of *reflection invariance*:

$$A_n(12 \ldots n) = (-1)^n A_n(n \ldots 21). \tag{29.15}$$

Finally, there are further relations between the color-ordered tree amplitudes, the *Kleiss–Kluijf relations*, which are

$$A_n(1, \{\alpha\}, n\{\beta\}) = (-1)^{n_\beta} \sum_{\{\sigma\}_i \in \mathrm{OP}(\{\alpha\}, \{\beta^T\})} A_n(1, \{\sigma\}_i, n), \tag{29.16}$$

where $\{\alpha\}$, $\{\beta\}$ are orderings of external lines, $\{\sigma\}_i$ are ordered permutations (OP), where the order of $\{\alpha\}$ and $\{\beta^T\}$ is kept inside $\{\sigma\}_i$, and n_β is the number of labels in the set $\{\beta\}$.

Further, by taking one of the T^as in (29.5) to be proportional to the identity, we get the *U(1) decoupling relation*

$$A_n(1234 \ldots n) + A_n(2134 \ldots n) + A_n(2314 \ldots n) + \ldots + A_n(234 \ldots 1n) = 0. \tag{29.17}$$

By using the cyclicity and the Kleiss–Kluijf (KK) relations, we can reduce the number of independent amplitudes to $(n-2)!$, obtaining a basis of amplitudes which can be taken to be the KK basis, $A_n(1, \mathcal{P}(2, \ldots, n-1), n)$, where $\mathcal{P}$ is an arbitrary permutation of $(n-2)$ objects.

But there is one more set of relations between amplitudes, called the *Bern, Carrasco, and Johannson (BCJ) relations*. Since the color-ordered amplitudes are sums over Feynman diagrams, they are generically of the type of a numerator and a multiple pole (coming from the propagators), of the type $Q_l^2 \equiv s_l = (k_1 + \ldots + k_l)^2$, where Q_l is the momentum on the propagator, which by momentum conservation relations is a sum of various external momenta. s_l is a generalized Mandelstam variable, usually called $s_{1 \ldots l}$. The general form of the amplitude is thus

$$A_n = \sum_{a_k} \frac{n_k}{(\prod_l s_l)_{a_k}}. \tag{29.18}$$

Here, n_k are numerators. In that case, the full color amplitude $\mathcal{A}_n$ is obtained as

$$\mathcal{A}_n = \sum_i \frac{\mathrm{sgn}_i n_i c_i}{(\prod_j s_j)_i}, \tag{29.19}$$

where sgn_i are signs and c_i are color factors, obtained from "Feynman diagrams" for $\tilde{f}^{abc}$. They satisfy some color Jacobi-type identities

$$c_i + c_j + c_k = 0. \tag{29.20}$$

But then BCJ showed that one can manipulate the numerators such that they also satisfy Jacobi-type identities

$$n_i + n_j + n_k = 0, \tag{29.21}$$

and thus obtain a color-kinematic duality. The numerator relations can be solved in terms
of $(n - 2)!$ *independent* numerators. Putting the numerators into a vector $|N\rangle$, and the
amplitudes also into a vector $|A\rangle$, one finds a linear matrix relation between the two:

$$|A\rangle = M|N\rangle, \tag{29.22}$$

but one where M has $(n - 3)!$ independent eigenvectors with zero eigenvalue, $M|v\rangle = 0$.
This in turn means that the basis of independent amplitudes is in general made up of $(n-3)!$
amplitudes, though in particular cases it can be even less.

As an example, consider the four-point amplitude A_4, which can be expressed symmetrically in terms of the three Mandelstam invariants s, t, u. From the above counting, using
cyclicity, we are left with $3! = 6$ amplitudes. Reflection invariance leaves half of them, that
is three. Using the KK relations, we are left with $2! = 2$, and using the BCJ relations with
a single independent amplitude.

For completeness, we list all six KK relations, for the six amplitudes $A_4(1234), A_4(1324),$
$A_4(1243), A_4(1342), A_4(1423), A_4(1432)$:

$$A_4(1234) = A_4(1234),$$
$$A_4(1324) = A_4(1324),$$
$$A_4(1243) = -[A_4(1234) + A_4(1324)],$$
$$A_4(1342) = -[A_4(1234) + A_4(1324)] = A_4(1243),$$
$$A_4(1423) = A_4(1423),$$
$$A_4(1432) = A_4(1432). \tag{29.23}$$

But note that there is a single nontrivial KK relation. The other one is the same, due to the
identity $A_4(1342) = A_4(1243)$, obtained from reflection invariance and cyclicity. The other
two relations obtained from reflection invariance and cyclicity are $A_4(1234) = A_4(1423)$
and $A_4(1234) = A_4(1432)$.

The color factors for the four-point amplitudes are

$$c(ij; kl) \equiv \tilde{f}^{a_i a_j b} \tilde{f}^{b a_k a_l}, \tag{29.24}$$

or using an obvious notation, borrowed from the Mandelstam variables ($s = s_{12}$, $t = t_{23}$,
$u = u_{13}$):

$$c_u \equiv \tilde{f}^{a_4 a_2 b} \tilde{f}^{b a_3 a_1} , \quad c_s \equiv \tilde{f}^{a_1 a_2 b} \tilde{f}^{b a_3 a_4} , \quad c_t \equiv \tilde{f}^{a_2 a_3 b} \tilde{f}^{b a_4 a_1}. \tag{29.25}$$

These can easily be shown to satisfy

$$c_s = c_t + c_u, \tag{29.26}$$

with a convention differing from the one above by $c_s \to -c_s$, where $c_s + c_t + c_u = 0$. One
can find numerators satisfying the same formula

$$n_s = n_t + n_u, \tag{29.27}$$

again with a convention differing from above by $n_s \to -n_s$, where $n_s + n_t + n_u = 0$. Then,
the three independent amplitudes before using the KK and BCJ relations are

$$A_4(1234) = A_4(1424) = \frac{n_s}{s} + \frac{n_t}{t},$$
$$A_4(1324) = A_4(1423) = -\frac{n_t}{t} + \frac{n_u}{u},$$
$$A_4(1243) = A_4(1342) = -\frac{n_s}{s} - \frac{n_u}{u}. \tag{29.28}$$

We can immediately check the KK relation $A_4(1234) + A_4(1324) + A_4(1243) = 0$, as well as the BCJ relation

$$s_{14}A_4(1234) - s_{13}A_4(1243) \equiv tA_4(1234) - uA_4(1243) = 0. \tag{29.29}$$

Thus the only independent amplitude is $A_4(1234)$.

29.4 The BCFW Construction

The BCFW construction is a recursion relation for the on-shell amplitudes (of given helicity). The essential tool that allows one to write the BCFW relations is the fact that the amplitude is an analytical function of its momentum variables. This means that we should be able to reconstruct the function from its singular points (i.e. from the case of limiting kinematics). If we try to manipulate the full dependence on all complex momenta, that would be very difficult, since the complex analysis of multi-variables is notoriously tricky. The brilliant strategy of BCFW was to construct a single complex deformation z of the momenta, chosen such that we can completely determine a higher-point amplitude from lower-point amplitudes. Then at $z = 0$ we have the result we want to calculate, $A_n^{\text{tree}} = A_n(z = 0)$, but we can use analytic properties of the amplitude $A_n(z)$ to calculate it, and reduce it to lower amplitudes, which are supposed to be known.

In the BCFW construction, we deform two external momenta by the complex parameter z. The goal is to preserve the on-shell condition, so the deformation is taken to be at the level of the helicity spinors, and moreover involves only the spinors of the two momenta, chosen to be 1 and n. Thus, we shift

$$\lambda_1 \to \hat{\lambda}_1 = \lambda_1 + z\lambda_n , \ \tilde{\lambda}_1 \to \tilde{\lambda}_1$$
$$\lambda_n \to \lambda_n , \ \tilde{\lambda}_n \to \hat{\tilde{\lambda}}_n = \tilde{\lambda}_n - z\tilde{\lambda}_1. \tag{29.30}$$

Then the momenta are

$$(\hat{k}_1(z))_{a\dot{b}} = (\lambda_1)_a(\tilde{\lambda}_1)_{\dot{b}} + z(\lambda_n)_a(\tilde{\lambda}_1)_{\dot{b}},$$
$$(\hat{k}_n(z))_{a\dot{b}} = (\lambda_n)_a(\tilde{\lambda}_n)_{\dot{b}} - z(\lambda_n)_a(\tilde{\lambda}_1)_{\dot{b}}, \tag{29.31}$$

leading to the same momentum conservation

$$\hat{k}_1^\mu(z) + \hat{k}_n^\mu(z) = k_1^\mu + k_n^\mu. \tag{29.32}$$

Since we have deformed the helicity spinors, but we still have factorization in λ and $\tilde{\lambda}$, we are still on-shell, $\hat{k}_1^2(z) = \hat{k}_n^2(z) = 0$.

The amplitude A_n contains denominators which are of the type Q^2, which means that it is a rational function of the momenta, and the function of the deformed momenta, $\hat{A}_n(z)$, is also a rational function of the momenta, and thus of z. This function is then meromorphic, that is holomorphic except for the poles, which are simple (since no more than one propagator can have the same momentum, a given combination of the external momenta, that includes z). Moreover, one can usually prove that $\hat{A}_n(z) \to 0$ for $z \to \infty$.

Then there is no pole at infinity, which means the integral around infinity gives zero, and therefore we can use Cauchy's theorem to relate $\hat{A}_n(z = 0) = A_n$ to the residues of the function at its poles:

$$0 = \frac{1}{2\pi i} \oint_C \frac{\hat{A}_n(z)}{z} = \hat{A}_n(z = 0) + \sum_k \mathrm{Res}_{z=z_k} \frac{\hat{A}_n(z)}{z} \Rightarrow$$

$$A_n = \hat{A}_n(z = 0) = -\sum_k \mathrm{Res}_{z=z_k} \frac{\hat{A}_n(z)}{z}. \tag{29.33}$$

The pole in the denominator for each residue is for $z\hat{P}_I(z_k)^2$, where $\hat{P}_I(z)$ is the momentum on the propagator (from a Feynman diagram for A_n) that goes on-shell (has vanishing square), thus giving the pole in the amplitude. But then it is a certain sum of the external momenta, with one of them being $k_1(z)$ or $k_n(z)$. For concreteness, we choose $k_1(z)$, so $\hat{P}_I(z) = \hat{k}_1(z) + k_2 + \ldots + k_r$ is the momentum that goes on-shell at $z = z_k$, where therefore

$$0 = \hat{P}_I(z_k)^2 = (\hat{k}_1(z_k) + k_2 + \ldots + k_r)^2 = (P_I + z_k \lambda_n \tilde{\lambda}_1)^2 = P_{I,k}^2 + 2z_k P_I \cdot (\lambda_n \tilde{\lambda}_1). \tag{29.34}$$

Here, $P_{I,k}$ is the value of P_I at the pole of $\hat{P}_I(z)$, and we find the pole position

$$z_k = -\frac{P_I^2}{2P_I \cdot (\lambda_n \tilde{\lambda}_1)}. \tag{29.35}$$

But then, just near the pole, the denominator for the residue is

$$z\hat{P}_I^2(z) \simeq z_k 2(z - z_k) P_I \cdot (\lambda_n \tilde{\lambda}_1) = -(z - z_k) P_I^2, \tag{29.36}$$

where in the second equality we have used the above value for z_k. We now substitute in the Cauchy formula (29.33) and note that the tree amplitude, if the propagator goes on-shell, splits into two amplitudes, A_L and A_R (left and right), containing momenta $k_1, \ldots, k_r$ and $k_{r+1}, \ldots, k_n$, respectively, and connected by the propagator. We then obtain the *BCFW recursion relation*

$$A_n(1, 2, \ldots, n) = \sum_{h=\pm} \sum_{r=2}^{n-2} A_{r+1}(\hat{1}, 2, \ldots, r, -\hat{P}_I^{-h}) \frac{i}{P_{I,k}^2} A_{n-r+1}(\hat{P}_I^{+h}, r+1, \ldots, n-1, \hat{n}),$$

$$\tag{29.37}$$

where $\hat{P}_I^{\pm h}$ is complex momentum evaluated at $z = z_k$, and with helicity h propagating through it. The amplitudes have r external momenta plus P_I, and $n - r$ external momenta plus P_I, respectively. The sum over r is of $n - 3$ *ordered* partitions of the external momenta (since we are working with the color-ordered amplitude), but where both left and right amplitudes contain at least three external momenta ($r \geq 2, r \leq n - 2$).

29.5 Application of BCFW: Proof of the Parke–Taylor Formula

The proof is inductive. We have seen that it holds for the three-point amplitude, so we need to show that the n-point formula (together with the three-point one) gives the $(n + 1)$-point one. We consider the MHV amplitude

$$A_n = A_n^{\text{tree,MHV}}(1^+, 2^+, \ldots, i^-, \ldots, n - 1^+, n^-), \tag{29.38}$$

where by cyclicity we have put (for convenience) one of the minus helicities on the nth position, and then the other one is on the ith position.

In the recursion relation (29.37), the sum over terms is zero for all $3 \le r \le n - 3$, since otherwise we would have at least a four-point function on each side, but (since $h = +$ on one side and $h = -$ on the other), at least one of the amplitudes would have at most one $-$ helicity, thus will vanish.

So we are left with $r = 2$ and $r = n - 2$. But for $r = n - 2$, the term vanishes as well. If $i = n - 1$, then $A_{r+1} = A_{n-1}$ has at most one $-$ helicity, so vanishes. If $i < n - 1$, then for $A_{r+1} = A_{n-1}$ to be nonzero, we must have $h = +$, but then $A_{n-r+1} = A_3$ is $A_3(+ + -)$, which can be nonzero if the three λs are proportional ($A_3(- - +)$ can be nonzero for the three $\tilde{\lambda}$ proportional). But, with the shift of $\tilde{\lambda}_n$, the $\tilde{\lambda}$s are proportional instead, and $A_3(+ + -) = 0$.

Thus the only nonzero term is $r = 2$, since then $A_3(+ + -)$ can be nonzero, as the λs are proportional (we have shifted λ_1) and the BCFW formula becomes simply

$$A_n^{\text{tree,MHV}}(1^+, 2^+, \ldots, i^-, \ldots, n - 1^+, n^-)$$
$$= A_3(\hat{1}^+, 2^+, -\hat{P}^-) A_{n-1} \frac{i}{P^2} A_{n-1}(\hat{P}^+, 3^+, \ldots, i^-, \ldots, n - 1^+, n^-), \tag{29.39}$$

where $P = k_1 + k_2$, which means that $P^2 = s_{12}$. The three-point amplitude is

$$A_3(\hat{1}^+, 2^+, -\hat{P}^-) = -i \frac{[\hat{1}2]^3}{[2(-\hat{P})][(-\hat{P})\hat{1}]} = +i \frac{[12]^3}{[2\hat{P}][\hat{P}1]}. \tag{29.40}$$

Note that $[\hat{1}2] = [12]$, since $\tilde{\lambda}_1$ is not shifted.

In contrast, by the induction step, we have (note that $|\hat{n}\rangle = |n\rangle$, as λ_n is not shifted)

$$A_{n-1}(\hat{P}^+, 3^+, \ldots, i^-, \ldots, n - 1^+, n^-) = i \frac{\langle in \rangle^4}{\langle \hat{P}3 \rangle \langle 34 \rangle \ldots \langle n - 1 n \rangle \langle n\hat{P} \rangle}. \tag{29.41}$$

Then the n-point MHV amplitude is

$$A_n(1^+, 2^+, \ldots, i^-, \ldots, n - 1^+, n^-) = -i \frac{[12]^3}{[2\hat{P}][\hat{P}1]} \frac{1}{s_{12}} \frac{\langle in \rangle^4}{\langle \hat{P}3 \rangle \langle 34 \rangle \ldots \langle n - 1 n \rangle \langle n\hat{P} \rangle}. \tag{29.42}$$

We now substitute in

$$\hat{P}_{a\dot{b}} = (k_1 + k_2)_{a\dot{b}} + z_2(\lambda_n)_a (\tilde{\lambda}_1)_{\dot{b}} \tag{29.43}$$

the value of z_2:

$$z_2 = -\frac{s_{12}}{2\hat{P} \cdot (\lambda_n \tilde{\lambda}_1)}. \tag{29.44}$$

After a bit of algebra, left as an exercise, we obtain the induction step n, thus completing the proof.

Important Concepts to Remember

- We can use color decomposition, both at tree and loop level, to write the amplitude in terms of a color-ordered amplitude, which has a single ordering of the external lines, and is cyclically invariant.
- The tree color-ordered A_n amplitude in the MHV case is given by the Parke–Taylor amplitude.
- The color-ordered tree amplitude satisfies the Kleiss–Kluijf and the $U(1)$ decoupling relations.
- The numerators of the color-ordered amplitudes can be defined such that they obey $n_i + n_j + n_k = 0$, color-kinematic dual to $c_i + c_j + c_k = 0$ for the color factors, leading to BCJ relations.
- The BCFW construction is based on deforming two external momenta, more precisely two helicity spinors of opposite type, λ_1 and $\tilde{\lambda}_n$, by a complex parameter z, but respecting momentum conservation. From complex analysis, we can then derive the BCFW recursion relations.
- We can prove the Parke–Taylor formula by induction, using the BCFW recursion relation for the induction step.

Further Reading

See the review [8] and the book [9].

Exercises

1. Prove the formula for the vertex in Gervais–Neveu gauge, (29.9).
2. Show that the three-point gluon amplitudes for complex momenta are given by (29.11).
3. Check that the four-point Parke–Taylor amplitude, decomposed in color, satisfies the Kleiss–Kluijf relations.
4. Finish the algebra leading to the proof of the induction step for the Parke–Taylor formula, using the BCFW recursion relation.

Review of Path Integral and Operator Formalism and the Feynman Diagram Expansion

In this chapter, I will recapitulate the material from Part I, before moving on to Part II.

30.1 Path Integrals, Partition Functions, and Green's Functions

30.1.1 Path Integrals

In quantum field theory, the classical field $\phi(x)$ is replaced by the VEV of a quantum operator $\hat{\phi}$:

$$\langle 0|\hat{\phi}(x)|0\rangle = \int \mathcal{D}\phi e^{iS[\phi]}\phi(x). \tag{30.1}$$

Here, the path integral measure is defined as integration over the points of a discretized path, in the limit of infinite number of points on the path:

$$\mathcal{D}\phi \equiv \lim_{N\to\infty} \prod_{i=1}^{N} \int d\phi(x_i). \tag{30.2}$$

30.1.2 Scalar Field

For a scalar field, the action is typically of the type

$$\begin{aligned} S = \int d^4x \mathcal{L} &= \int d^4x \left[-\frac{1}{2}\partial_\mu\phi\partial^\mu\phi - \frac{1}{2}m^2\phi^2 - V(\phi) \right] \\ &= \int d^4x \left[\frac{1}{2}\dot{\phi}^2 - \frac{1}{2}|\vec{\nabla}\phi|^2 - \frac{1}{2}m^2\phi^2 - V(\phi) \right], \end{aligned} \tag{30.3}$$

though the kinetic term can be more complicated also. Note that the mass term was written separately, since it is quadratic, though technically it is part of the potential $V(\phi)$.

One can define objects more general than the field VEV (of which the field VEV is a particular case), which are the objects to be studied in quantum field theory, the *correlation functions*, or *Green's functions*, or *n*-point functions:

$$G_n(x_1,\dots,x_n) = \langle 0|T\{\hat{\phi}(x_1)\dots\hat{\phi}(x_n)\}|0\rangle = \int \mathcal{D}\phi e^{iS[\phi]}\phi(x_1)\dots\phi(x_n). \tag{30.4}$$

Here T denotes time ordering. Note that e^{iS} is a highly oscillatory phase, so this Green's function is hard to define rigorously, since it is not well behaved at infinity.

It is much easier to go to Euclidean space, by doing a Wick rotation, $t = -it_E$.

In quantum mechanics, one obtains the *Feynman–Kac formula* relating the transition amplitude in Euclidean space, expressed as a path integral, with the usual statistical mechanics partition function

$$Z(\beta) = \text{Tr}\{e^{-\beta\hat{H}}\} = \int dq \sum_n |\phi_n(q)|^2 e^{-\beta E_n} = \int dq \langle q, \beta | q, 0 \rangle$$

$$= \int \mathcal{D}q e^{-S_E[q]} \bigg|_{q(t_E+\beta)=q(t_E)} . \tag{30.5}$$

Here, as usual, $\beta = 1/k_B T$ and the Euclidean action S_E is defined by $iS_M = -S_E$.

In quantum field theory, where we have a space ($\vec{x}$) dependence, and one usually is interested in the vacuum functional (i.e. the transition between asymptotic vacuum states), we consider the limit of infinite periodicity, $\phi(\vec{x}, t_E + \beta) = \phi(\vec{x}, t_E)$ (or zero temperature $T = 1/\beta$), where $\beta \to \infty$. The Euclidean action for a scalar field is

$$S_E[\phi] = \int d^4x \left[\frac{1}{2} \partial_\mu \phi \partial_\mu \phi + \frac{1}{2} m^2 \phi^2 + V(\phi) \right]. \tag{30.6}$$

The Euclidean-space correlation functions are defined in a similar manner:

$$G_n^{(E)}(x_1, \ldots, x_n) = \int \mathcal{D}\phi e^{-S_E[\phi]} \phi(x_1) \ldots \phi(x_n), \tag{30.7}$$

the advantage being that now instead of a highly oscillatory phase, we have a highly decaying weight e^{-S_E}, sharply peaked on the classical action S_{cl}. The generating functional of the correlation functions is called the partition function and is given by

$$Z^{(E)}[J] = \int \mathcal{D}\phi e^{-S_E[\phi]+J\cdot\phi} \equiv {}_J\langle 0|0\rangle_J, \tag{30.8}$$

where we have defined

$$J \cdot \phi \equiv \int d^d x J(x)\phi(x). \tag{30.9}$$

The correlation functions are obtained from their generating functional as usual:

$$G_n^{(E)}(x_1, \ldots, x_n) = \frac{\delta}{\delta J(x_1)} \cdots \frac{\delta}{\delta J(x_n)} \int \mathcal{D}\phi e^{-S_E[\phi]+J\cdot\phi} \bigg|_{J=0}$$

$$= \frac{\delta}{\delta J(x_1)} \cdots \frac{\delta}{\delta J(x_n)} Z[J] \bigg|_{J=0} . \tag{30.10}$$

Note that we can define the partition function at finite temperature (finite periodicity β):

$$Z^E[\beta, J] = \text{Tr}[e^{-\beta\hat{H}_J}] = \int \mathcal{D}\phi e^{-S_E[\phi]+J\cdot\phi} \bigg|_{\phi(\vec{x},t_E+\beta)=\phi(\vec{x},t_E)} . \tag{30.11}$$

From this we can start to define quantum field theory at finite temperature, but we will not do it here.

30.2 Canonical Quantization, Operator Formalism, and Propagators

To canonically quantize a real scalar field, one expands it and its canonical conjugate momentum π in Fourier modes:

$$\phi(\vec{x}, t) = \int \frac{d^3 p}{(2\pi)^3} \frac{1}{\sqrt{2\omega_p}} (a(\vec{p}, t) e^{i\vec{p}\cdot\vec{x}} + a^\dagger(\vec{p}, t) e^{-i\vec{p}\cdot\vec{x}}),$$

$$\pi(\vec{x}, t) = \int \frac{d^3 p}{(2\pi)^3} \left(-i\sqrt{\frac{\omega_p}{2}}\right) (a(\vec{p}, t) e^{i\vec{p}\cdot\vec{x}} - a^\dagger(\vec{p}, t) e^{-i\vec{p}\cdot\vec{x}}), \tag{30.12}$$

where $\omega_p = \sqrt{\vec{p}^2 + m^2}$. From the Klein–Gordon equation of motion for the scalar field, one finds that $a(\vec{p}, t) = a_{\vec{p}} e^{-i\omega_p t}$.

Canonical quantization is achieved through the equal-time commutation relations between ϕ and its canonical conjugate momentum:

$$[\phi(\vec{x}, t), \pi(\vec{x}', t)] = i\hbar\delta^{(3)}(\vec{x} - \vec{x}'); \quad [\phi(\vec{x}, t), \phi(\vec{x}', t)] = [\pi(\vec{x}, t), \pi(\vec{x}', t)] = 0. \tag{30.13}$$

From them, we obtain the usual algebra for the creation and annihilation operator coefficients:

$$[a(\vec{p}, t), a^\dagger(\vec{p}', t)] = (2\pi)^3 \delta^{(3)}(\vec{p} - \vec{p}'). \tag{30.14}$$

For free scalars, one uses Heisenberg picture operators:

$$\phi_H(\vec{x}, t) = e^{iHt} \phi(\vec{x}) e^{-iHt}, \tag{30.15}$$

where $\phi(\vec{x})$ is a Schrödinger picture operator. As a reminder, in the Schrödinger picture, operators are time-independent and the states evolve in time with the Hamiltonian, whereas in the Heisenberg picture it is the opposite: operators evolve in time with the Hamiltonian, and the states are time-independent.

For interacting scalars, however, the useful representation is the *interaction (Dirac) picture*. One splits the Hamiltonian into a free (quadratic) part and an interaction part:

$$\hat{H} = \hat{H}_0 + \hat{H}_1. \tag{30.16}$$

Then the interaction picture operators are obtained by a canonical transformation with the free part $\hat{H}_0$:

$$\phi_I(\vec{x}, t) = e^{i\hat{H}_0(t-t_0)} \phi(\vec{x}, t_0) e^{-i\hat{H}_0(t-t_0)}. \tag{30.17}$$

Now states evolve with the interaction Hamiltonian, and operators with the free Hamiltonian:

$$i\hbar\frac{\partial}{\partial t}|\psi_I(t)\rangle = \hat{H}_{1,I}|\psi_I(t)\rangle,$$

$$i\hbar\frac{\partial}{\partial t}\hat{A}_I(t) = [\hat{A}_I(t), \hat{H}_0], \tag{30.18}$$

where $H_{1,I}$ is the interacting Hamiltonian operator in the interaction picture.

In the interacting quantum field theory, $\phi(\vec{x}, t)$ denotes the Heisenberg operator $\phi_H(\vec{x}, t)$, and is the object we are interested in, together with the true vacuum of the full theory $|\Omega\rangle$.

In contrast, for calculational purposes, we use the vacuum $|0\rangle$ of the free theory, and the interaction picture fields, since we find that

$$\phi_I(\vec{x}, t) = e^{i\hat{H}_0(t-t_0)}\phi(\vec{x}, t_0)e^{-i\hat{H}_0(t-t_0)}$$

$$= \int \frac{d^3p}{(2\pi)^3} \frac{1}{\sqrt{2E_p}}(a_{\vec{p}}e^{ipx} + a_{\vec{p}}^\dagger e^{-ipx})\bigg|_{x^0=t-t_0; p^0=E_p} . \quad (30.19)$$

The vacuum of the free theory satisfies $a_{\vec{p}}|0\rangle = 0$ for all $\vec{p}$. On the contrary, the objects we want to calculate are Green's functions of the full theory, like the two-point function (full propagator), $\langle\Omega|T\{\phi(x)\phi(y)\}|\Omega\rangle$.

We find them using perturbation theory, in terms of interaction picture objects, like the propagator. The propagator depends on a contour of complex integration, but the most commonly used one in quantum field theory is the *Feynman propagator*:

$$D_F(x-y) = \langle 0|T\{\phi_I(x)\phi_I(y)\}|0\rangle = \int \frac{d^4p}{(2\pi)^4} \frac{-i}{p^2+m^2-i\epsilon}e^{ip(x-y)}. \quad (30.20)$$

Other examples of propagators are the retarded and advanced propagators. But the Feynman propagator arises as the natural analytical continuation from Euclidean space. In Euclidean space, the propagator is uniquely defined, since there are no poles to be avoided using complex integration and contours, and it is given by

$$\Delta(x, y) = \int \frac{d^dp}{(2\pi)^d} \frac{e^{ip(x-y)}}{p^2+m^2}. \quad (30.21)$$

The relation between the full correlation functions and the interaction picture objects is given by the *Feynman theorem*, written for generality in terms of operators $\mathcal{O}$ that can specialize to the usual case of $\phi(x)$, as

$$\langle\Omega|T\{\mathcal{O}_H(x_1)\dots\mathcal{O}_H(x_n)\}|\Omega\rangle$$

$$= \lim_{T\to\infty(1-i\epsilon)} \frac{\langle 0|T\{\mathcal{O}_I(x_1)\dots\mathcal{O}_I(x_n)\exp\left[-i\int_{-T}^{T} dt H_{1,I}(t)\right]\}|0\rangle}{\langle 0|T\{\exp\left[-i\int_{-T}^{T} dt H_{1,I}(t)\right]\}|0\rangle}. \quad (30.22)$$

By expanding the right-hand side, we can calculate perturbatively in terms of the free vacuum and operators with time evolution given by $\hat{H}_0$ (i.e. interaction picture operators).

For calculations, we use the Wick theorem, which relates the time ordering with the normal ordering:

$$T\{\phi_I(x_1)\dots\phi_I(x_n)\} = N\{\phi_I(x_1)\dots\phi_I(x_n) + \text{all possible contractions}\}. \quad (30.23)$$

Since, between $\langle 0|$ and $|0\rangle$, the normal ordering of a nontrivial interaction picture operator (written in terms of as and $a^\dagger$s) gives zero, the only nonzero result is given by the full contraction, which is just a c-number.

30.3 Wick Theorem, Dyson Formula, and Free Energy in Path-Integral Formalism

In the path-integral formalism, there is an equivalent of the Wick theorem. One first writes the Dyson formula, which reads

$$
\begin{aligned}
Z[J] &= \langle 0| e^{-S_I[\hat{\phi}]} e^{\int d^d x J(x) \hat{\phi}(x)} |0\rangle \\
&= \int \mathcal{D}\phi\, e^{-S_0[\phi]+J\cdot\phi} e^{-S_I[\phi]}.
\end{aligned}
\tag{30.24}
$$

Note that the vacuum of the free theory is used, but inside the VEV we have the missing terms that allow us to make up the usual $e^{-S+J\cdot\phi}$ in the path integral, using the weight e^{-S_0}.

Then we can easily calculate the partition function of the free theory:

$$
Z_0[J] = e^{\frac{1}{2}J\cdot\Delta\cdot J} \langle 0|0\rangle_0,
\tag{30.25}
$$

where $\langle 0|0\rangle_0$ is an irrelevant normalization constant. Then the Wick theorem for path integrals is written in two forms. The first is

$$
Z[J] = e^{-\int d^d x V\left(\frac{\delta}{\delta J(x)}\right)} Z_0[J] = e^{-\int d^d x V\left(\frac{\delta}{\delta J(x)}\right)} e^{\frac{1}{2}J\cdot\Delta\cdot J},
\tag{30.26}
$$

and the second is

$$
\begin{aligned}
Z[J] &= e^{\frac{1}{2}\frac{\delta}{\delta\phi}\cdot\Delta\cdot\frac{\delta}{\delta\phi}} \left\{ e^{-\int d^d x V(\phi)+J\cdot\phi} \right\}\bigg|_{\phi=0} \\
&= \exp\left[\frac{1}{2}\int d^d x\, d^d y \Delta(x-y) \frac{\delta}{\delta\phi(x)}\frac{\delta}{\delta\phi(y)}\right] \left\{ e^{-\int d^d x V(\phi)+J\cdot\phi} \right\}\bigg|_{\phi=0}.
\end{aligned}
\tag{30.27}
$$

While it would seem that by these formulas we have solved quantum field theory, since we have a closed-form expression for the partition function, it is not so, since the expression is formal: it is understood only as a formal perturbative expansion for the exponentials. There are divergences in the terms of the expansion that need to be regulated, and so on.

To find the correlation functions of the *full* (interacting) theory, we need to get rid of the vacuum bubbles. As seen from the denominator of the Feynman theorem, this is done by dividing by the partition function (zero-point function). Diagrammatically (to be defined soon), one obtains only the connected diagrams, giving

$$
\frac{1}{Z[J]}\frac{\delta Z[J]}{\delta J(x)} = \frac{\delta(-W[J])}{\delta J(x)},
\tag{30.28}
$$

solved by

$$
Z[J] = \mathcal{N} e^{-W[J]}.
\tag{30.29}
$$

Here, $W[J]$ is called the free energy, from analogy with condensed matter, and as we see is the generating functional of the connected diagrams. $\mathcal{N}$ is an irrelevant normalization constant (it cancels out).

30.4 Feynman Rules, Quantum Effective Action, and S-Matrix

30.4.1 Feynman Rules in x Space (Euclidean)

The Feynman rules in Euclidean coordinate space are as follows:

0. Draw all Feynman diagrams.
1. A line between x and y corresponds to the Euclidean Feynman propagator $\Delta(x, y)$.
2. For an interaction potential $V = \lambda \phi^p$, we have p-legged vertices. For each vertex, we have a factor of $(-\lambda)$ and an integration over the position of the vertex, $\int d^d x$.
3. Then we obtain the value for the Feynman diagram D with external points $x_1, \ldots, x_n$ and internal points $y_1, \ldots, y_N$, integrated over:

$$F_D^{(N)} = I_D(x_1, \ldots, x_n; y_1, \ldots, y_N), \qquad (30.30)$$

and the correlation function is given as a sum over diagrams and number of vertices:

$$G_n(x_1, \ldots, x_n) = \sum_{N \geq 0} \frac{1}{N!} \sum_{\text{diags.}D} F_D^{(N)}(x_1, \ldots, x_n). \qquad (30.31)$$

30.4.2 Simplified Rules

There is a simplified version of the Feynman rules that is usually used. We rewrite the potential as $V = \lambda_p \phi^p / p!$ (i.e. $\lambda = \lambda_p / p!$). Then the vertex is $\int d^d x (-\lambda_p)$, but we write only the *topologically inequivalent diagrams* and divide by the *statistical weight factor* or *symmetry factor* S:

$$S = \frac{N! \, (p!)^N}{\# \text{ equivalent diagrams}} = \# \text{ symmetries of diagram}. \qquad (30.32)$$

30.4.3 Feynman Rules in p Space

Because of the translational invariance of the theory, which implies momentum conservation, the Fourier transform $\tilde{G}$ of the correlation function can be redefined by an overall momentum conservation delta function:

$$\tilde{G}_n(p_1, \ldots, p_n) = \int \prod_i d^d x_i e^{i \sum_j p_j x_j} G(x_1, \ldots, x_n)$$

$$= (2\pi)^d \delta^{(d)}(p_1 + \ldots + p_n) G_n(p_1, \ldots, p_n). \qquad (30.33)$$

The simplest case is the free two-point function (i.e. the propagator). Corresponding to the Feynman propagator $\Delta(x - y)$, we have the Euclidean-space propagator

$$\Delta(p) = \frac{1}{p^2 + m^2}, \qquad (30.34)$$

corresponding to G_2 above.

The rules in momentum space have:

- propagator $\Delta(p)$
- vertex $(-\lambda)$
- external line e^{-ipx}.
- Then one needs to impose momentum conservation, that is multiply by $(2\pi)^d \delta^{(d)}(\sum_j p_j)$ at each vertex and integrate $\int d^d p/(2\pi)^d$ over all *internal* momenta.
- Divide by the symmetry factor.

30.4.4 Simplified Momentum-Space Rules

We can write simplified momentum-space rules to get rid of the momentum conservation delta functions. We only introduce *independent loop momenta* $l_1, \ldots, l_L$ (integration variables). If it is not entirely obvious what the number of loops is in the diagrams, one can use the formula $L = V(p/2 - 1) - E/2 + 1$ to calculate it. Here, V is the number of vertices, p the order of vertices, and E the number of external lines. We write the momenta on each internal line in terms of the external momenta and the loop momenta l_i using momentum conservation at the vertices.

- We must integrate over these loop momenta, $\int d^d l_1/(2\pi)^d \ldots \int d^d l_L/(2\pi)^d$.
- External lines have momenta p_i, and internal lines dependent momenta q_j, depending on p_i and l_i.
- An external line then gives $1/(p_i^2 + m^2)$, and an internal line $1/(q_j^2 + m^2)$.
- As before, the vertex is $(-\lambda)$, we divide by the symmetry factor S and sum over diagrams.

30.4.5 Classical Field

As we mentioned at the beginning of the chapter, the classical field is replaced by the VEV of the quantum operator. More precisely, considering the field in the presence of a source J, one defines the *classical field* ϕ_{cl} by the VEV in the $|0\rangle_J$ vacuum, and divides by the normalization:

$$\phi_{cl} \equiv \phi(x; J) = \frac{{}_J\langle 0|\hat{\phi}|0\rangle_J}{{}_J\langle 0|0\rangle_J}. \tag{30.35}$$

Therefore, one obtains

$$\phi_{cl} = \frac{1}{Z[J]} \int \mathcal{D}\phi \, e^{-S[\phi]+J\cdot\phi} \phi(x) = \frac{\delta}{\delta J(x)} \ln Z[J] = -\frac{\delta}{\delta J(x)} W[J]. \tag{30.36}$$

30.4.6 Quantum Effective Action

Whereas classically, we have the classical action $S[\phi]$ depending on the classical field ϕ, quantum mechanically we have the *quantum effective action* that includes quantum corrections, and depends on the classical field ϕ_{cl}. It is the Legendre transform of the free energy $W[J]$:

$$\Gamma[\phi_{cl}] = W[J] + \int d^d x J(x) \phi_{cl}(x). \qquad (30.37)$$

In the same way as classically, $\phi(x)$ satisfies the classical equation of motion with a source:

$$\frac{\delta S[\phi]}{\delta \phi(x)} = J(x), \qquad (30.38)$$

quantum mechanically we have a precise analogue of it, namely

$$\frac{\delta \Gamma[\phi_{cl}]}{\delta \phi_{cl}(x)} = J(x). \qquad (30.39)$$

The quantum effective action Γ is the generator of the 1PI (one-particle irreducible) diagrams, except for the two-point function, where there is an extra term.

Therefore the partition function $Z[J]$ generates all diagrams, the free energy $W[J]$ generates the connected diagrams, the effective action $\Gamma[\phi_{cl}]$ generates the 1PI diagrams except at two-points, and we also have that the classical action S generates the tree diagrams.

30.4.7 S-Matrix

To get contact with experiments, we need to define objects that can be related to measurable quantities. Such an object is the S-matrix.

We define Heisenberg picture states (time-independent) that have well-isolated wavepackets at $t = -\infty$ (but are interacting, or mixed, at $t = +\infty$), $|\{\vec{k}_j\}\rangle_{\text{in}}$, and states that have well-isolated wavepackets at $t = +\infty$ (but are interacting, or mixed, at $t = -\infty$), $|\{\vec{p}_i\}\rangle_{\text{out}}$. We also define Schrödinger picture states (time-dependent) with the same momenta, $\langle\{\vec{p}_i\}|$ and $|\{\vec{k}_j\}\rangle$. Then the S-matrix between the Schrödinger picture states is defined as

$$\langle\{\vec{p}_i\}|S|\{\vec{k}_j\}\rangle = {}_{\text{out}}\langle\{\vec{p}_i\}|\{\vec{k}_j\}\rangle_{\text{in}}. \qquad (30.40)$$

30.4.8 Reduction Formula (LSZ)

Until now we have worked with correlation functions. But these objects are rather abstract, and we saw above that the physical objects we want to calculate are S-matrices. Fortunately, there is a relation between them, given by the *LSZ formula*. One first defines the Fourier transform of the correlation functions of the full (interacting) theory:

$$\tilde{G}_{n+m}(p_i^\mu, k_j^\mu) = \prod_{i=1}^{n} \int d^4 x_i e^{-ip_i x_i} \prod_{j=1}^{m} \int d^4 y_j e^{ik_j y_j} \langle \Omega | T\{\phi(x_1) \dots \phi(x_n)\phi(y_1) \dots \phi(y_m)\} | \Omega \rangle.$$

$$(30.41)$$

Then the S-matrix is obtained from the correlation function $\tilde{G}_{n+m}$ by going near on-shell for the external lines and dividing by the full inverse propagators for the external lines, except for the factor of Z being replaced by $\sqrt{Z}$:

$$_{\text{out}}\langle\{\vec{p}_i\}|\{\vec{k}_j\}\rangle_{\text{out}} = \lim_{p_i^2 \to -m^2; k_j^2 \to -m^2} \prod_{i=1}^{n} \frac{p_j^2 + m^2 - i\epsilon}{-i\sqrt{Z_i}} \prod_{j=1}^{m} \frac{k_j^2 + m^2 - i\epsilon}{-i\sqrt{Z_j}} \tilde{G}_{n+m}(p_i^\mu, k_j^\mu).$$

(30.42)

Diagrammatically, the S-matrix minus the trivial one (identity) is given by the sum of the connected, amputated Feynman diagrams, times a $\sqrt{Z}$ factor for each external leg, that is

$$\langle\{\vec{p}_i\}|S - 1|\{\vec{k}_j\}\rangle = \left(\sum \text{connected, amputated Feynman diagrams}\right) \times (\sqrt{Z})^{n+m}. \quad (30.43)$$

30.5 Fermions

Dirac fermions ψ_α are understood as representations of the Clifford algebra

$$\{\gamma^\mu, \gamma^\nu\} = 2g^{\mu\nu}\,\mathbb{1}, \quad (30.44)$$

that is column vector objects on which the gamma matrices act. We can also understand them as spinorial representations of the Lorentz (or Poincaré) algebra. But the Dirac fermion representation of the Lorentz algebra is not irreducible. In 3+1 dimensions, the irreducible representations (irreps) are either of the Weyl or Majorana type.

One first defines the γ_5 object as the product of the gamma matrices up to a fixed phase, in my conventions

$$\gamma_5 = -i\gamma^0\gamma^1\gamma^2\gamma^3. \quad (30.45)$$

Note that there are various conventions for the phase, but generally one chooses to have $\gamma_5^2 = 1$. Indeed, we can then define projectors $P_{L/R} = (1 \pm \gamma_5)/2$ onto the irreducible Weyl representations, defined as

$$\psi_{L/R} = \frac{1 \pm \gamma_5}{2}\psi_D. \quad (30.46)$$

Note that in the Weyl representation for the gamma matrices, $\gamma_5 = \begin{pmatrix} \mathbb{1} & 0 \\ 0 & -\mathbb{1} \end{pmatrix}$, so the irreps are the two upper components of ψ_D and the two lower components of ψ_D.

One then defines the conjugate representation $\bar{\psi}$ as $\bar{\psi} = \psi^\dagger i\gamma^0$. Note again that various conventions in the literature for $\bar{\psi}$ differ by a phase. The action in Minkowski space is

$$S_\psi = -\int d^4x\,\bar{\psi}(\gamma^\mu\partial_\mu + m)\psi, \quad (30.47)$$

and in Euclidean space without the overall minus.

The other type of irreducible representation is a real-type representation, the Majorana representation, obtained by imposing a reality constraint. Defining a C-matrix with certain properties, satisfied in 3+1 dimensions by the *choice* $C = -i\gamma^0\gamma^2$, the Majorana condition relates the Dirac conjugate $\bar{\psi}$ to the Majorana conjugate $\psi^C = \psi^T C$, that is

$$\bar{\psi} = \psi^T C. \tag{30.48}$$

Fermionic path integrals are written in terms of *Grassmann variables* (i.e. anticommuting objects ψ with $\{\psi, \psi\} = 0$). The Gaussian integral on Grassmann space gives

$$\int d^n x e^{x^T A x} = 2^{n/2}\sqrt{\det A}, \tag{30.49}$$

where for commuting objects this is $\propto 1/\sqrt{\det A}$.

The addition of fermions means new propagators and vertices in the Feynman rules.

The fermionic Euclidean space propagator is

$$\frac{1}{i\slashed{p} + m} = \frac{i}{-\slashed{p} + im} = \frac{-i\slashed{p} + m}{p^2 + m^2}, \tag{30.50}$$

whereas in Minkowski space the Feynman propagator is

$$-i\frac{-i\slashed{p} + m}{p^2 + m^2 - i\epsilon}. \tag{30.51}$$

For a Yukawa interaction

$$\int d^d x g \bar{\psi}\psi\phi, \tag{30.52}$$

the vertex is $(-g)$ in the Euclidean case and $(-ig)$ in the Minkowski case.

A fermion loop adds a minus sign to the Feynman rules.

30.6 Gauge Fields

In covariant quantization (Gupta–Bleuler), we write the expansion of the gauge field in a similar manner to the case of the scalar field:

$$A_\mu(x) = \int \frac{d^3k}{(2\pi)^3\sqrt{2E_k}} \sum_{\lambda=0}^{3} \epsilon_\mu^{(\lambda)}(k)[e^{ikx}a^{(\lambda)}(k) + a^{\dagger(\lambda)}(k)e^{-ikx}], \tag{30.53}$$

where $\epsilon_\mu^{(\lambda)}(k)$ are four polarizations. The 0 and 3 polarizations are unphysical, and the 1 and 2 are physical, transverse polarizations, satisfying $k^\mu \epsilon_\mu^{(1,2)} = 0$.

The gauge condition is imposed on the positive frequency part of the operator, that is the annihilation part of A_μ, acting on physical states:

$$\partial^\mu A_\mu^{(+)}(x)|\psi\rangle = 0, \tag{30.54}$$

which defines the physical states $|\psi\rangle$. Since $k^\mu \epsilon_\mu^{(1,2)} = 0$, on the momentum modes we obtain the condition

$$[a^{(0)}(k) + a^{(3)}(k)]|\psi\rangle = 0. \tag{30.55}$$

To define a photon propagator, we must add a gauge-fixing term to the action. If not, the action has zero modes (due to the gauge invariance), which makes it impossible to invert the kinetic operator. The gauge-fixed Lagrangian in Minkowski space is

$$\mathcal{L} = -\frac{1}{4}F_{\mu\nu}^2 - \frac{1}{2\alpha}(\partial^\mu A_\mu)^2. \tag{30.56}$$

From this, we obtain the propagator

$$G_{\mu\nu}^{(0)}(k) = \frac{1}{k^2}\left(\delta_{\mu\nu} - (1-\alpha)\frac{k_\mu k_\nu}{k^2}\right). \tag{30.57}$$

In the so-called "Feynman gauge" (it is not really a gauge, just a choice) $\alpha = 1$, the propagator becomes just the scalar propagator $G_{\mu\nu}^{(0)}(k) = \delta_{\mu\nu}/k^2$.

30.7 Quantum Electrodynamics

30.7.1 QED S-Matrix Feynman Rules

The Euclidean Lagrangian for QED is

$$\mathcal{L}^{(E)} = \frac{1}{4}F_{\mu\nu}^2 + \bar\psi(\slashed{D} + m)\psi, \tag{30.58}$$

where $D_\mu = \partial_\mu - ieA_\mu$. The Minkowski Lagrangian is

$$\mathcal{L} = -\frac{1}{4}F_{\mu\nu}^2 - \bar\psi(\slashed{D} + m)\psi. \tag{30.59}$$

- The $e^- e^+ \gamma$ vertex, with respective indices α, β, μ, is $+ie(\gamma_\mu)_{\alpha\beta}$ in Euclidean space, and $+e(\gamma_\mu)_{\alpha\beta}$ in Minkowski space.
- The incoming photon external line is $A_\mu|\vec{p}\rangle = \epsilon_\mu(p)$.
- The outgoing photon external line is $\langle\vec{p}|A_\mu = \epsilon_\mu^*(p)$.
- The incoming electron external line is $\psi|\vec{p}, s\rangle = u^s(p)$.
- The outgoing electron external line is $\langle\vec{p}, s|\bar\psi = \bar{u}^s(p)$.
- The incoming positron external line is $\bar\psi|\vec{k}, s\rangle = \bar{v}^s(k)$.
- The outgoing positron external line is $\langle\vec{k}, s|\psi = v^s(k)$.
- The fermion propagator is

$$-\frac{(\slashed{p} + im)}{p^2 + m^2 - i\epsilon}. \tag{30.60}$$

- The photon propagator is

$$-\frac{i}{k^2 - i\epsilon}\left(g_{\mu\nu} - (1-\alpha)\frac{k_\mu k_\nu}{k^2 - i\epsilon}\right). \tag{30.61}$$

- There is a minus sign for a fermion loop.

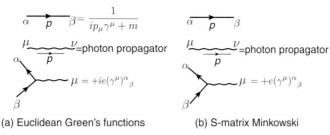

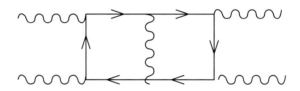

(a) Euclidean Green's functions   (b) S-matrix Minkowski

(c) S-matrix external lines

Figure 30.1   Relevant Feynman rules for Green's functions and S-matrices.

Figure 30.2   Two-loop QED Feynman diagram.

The rules are summarized in Figure 30.1.

Important Concepts to Remember

- All of them, since it is a review…

Further Reading

See the previous chapters.

Exercises

1. Use the Feynman rules to write down the integral expression for the diagram in Figure 30.2, for scattering of photons in QED (the diagram is a fermion loop, cut into two loops by a photon line, and with four external photon lines).

2. Write down the x-space and the p-space Feynman rules coming from the four-dimensional Lagrangian in Euclidean space:

$$\mathcal{L} = +\frac{1}{2}\sum_{I=1}^{N}[(\partial_\mu\phi^I)^2 + m^2(\phi^I)^2] + \left(1 + \sum_{I=1}^{N}\left(\frac{\phi^I}{M}\right)^2\right)\frac{F_{\mu\nu}F^{\mu\nu}}{4}$$

$$+ \frac{\lambda}{M^2}\sum_{I,J=1}^{N}\phi^I\phi^I\partial_\mu\phi^J\partial^\mu\phi^J. \tag{30.62}$$

3. Consider the 4-fermi interaction Lagrangian

$$\mathcal{L} = -\bar{\psi}(\not{D} + m)\psi - g_F(\bar{\psi}\gamma_\mu\psi)(\bar{\psi}\gamma^\mu\psi). \tag{30.63}$$

Write the four-point Green's function for two fermions and two antifermions up to (including) 2-loops, by writing the Feynman diagrams and the corresponding expression with the Feynman rules (without calculating the integrals explicitly).

4. Write the Dyson–Schwinger equation for the four-point Green's function in Exercise 2. Is there a Ward identity for this four-point function?

II

LOOPS, RENORMALIZATION, QUANTUM CHROMODYNAMICS, AND SPECIAL TOPICS

31

One-Loop Determinants, Vacuum Energy, and Zeta Function Regularization

In this part of the book, we will consider the effect of loops, which amounts to dealing with infinities through renormalization, after which we will consider quantum chromodynamics and other special topics.

In this chapter, we start by discussing the infinities that appear in quantum field theory, with the simplest example that we used before, namely the infinite zero-point energy. We will see the physical effect of it, in the form of the Casimir force, and in more general theories of the quantum one-loop determinants, and ways to regularize this infinity, in particular the zeta function regularization.

31.1 Vacuum Energy and Casimir Force

For a harmonic oscillator, the vacuum has a "zero-point energy" (vacuum energy) $E_0 = \hbar\omega/2$, and we have seen that analogously, for a free bosonic field, for instance for a real scalar field, we have a sum over all the harmonic oscillator modes of the field, that is (in $D-1$ space dimensions)

$$E_0 = \int \frac{d^{D-1}k}{(2\pi)^{D-1}} \frac{\hbar\omega_k}{2} (2\pi)^{D-1}\delta(0) \rightarrow E_0 = \sum_{\vec{k}} \frac{\hbar\omega_{\vec{k}}}{2} = \infty, \qquad (31.1)$$

where $\int d^{D-1}k/(2\pi)^{D-1} \rightarrow 1/V \sum_{\vec{k}}$ and $\delta(\vec{p} - \vec{k}) \rightarrow V\delta_{\vec{p}\vec{k}}$.

In the free theory, we have removed this infinity using *normal ordering*. But the issue remains, is this something physical, that has to do with quantum field theory, or just a pure artifact of the formalism? The answer is that it does in fact have a physical significance. Not the infinite result of course, but we can measure *changes* in it due to changes in geometry. This is known as the *Casimir effect*, and has already been observed experimentally, meaning that the existence of the zero-point energy has been confirmed experimentally.

Since the result is not only formally infinite (an infinite sum), but also continuous for a system of infinite size, we will consider a space that has a finite size (length) L, see Figure 31.1, and we will take $L \rightarrow \infty$ at the end of the calculation. Also, in order to consider the geometry dependence, consider two parallel perfectly conducting plates situated at a distance d. Strictly speaking, this would be for the physical case of fluctuation of the electromagnetic field, the only long-range field other than gravity. The conducting plates would

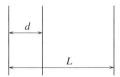

Figure 31.1 Parallel plates at a distance d, inside a box of length L.

impose constant potential on them (i.e. $A_0 = V = $ constant), thus more generally, Dirichlet boundary conditions for the fluctuating field A_μ. In our simple case then, consider a scalar field with Dirichlet boundary conditions at the plates, such that the eigenmodes between the plates have

$$k_n = \frac{\pi n}{d}. \tag{31.2}$$

One of the plates can be considered to be a boundary of the system without loss of generality, so the system is separated into d and $L - d$. (We can consider the more general case where there is space on both sides, but it will not change the final result.)

If the y and z directions are infinite, then the energy per unit area (area of x, y directions) of fluctuations *between the plates* is given by

$$\frac{E}{A} = \frac{\hbar}{2} \sum_n \int \frac{dk_y dk_z}{(2\pi)^2} \sqrt{\left(\frac{\pi n}{d}\right)^2 + k_y^2 + k_z^2}, \tag{31.3}$$

and we can do this calculation, related to the real case (experimentally measured), just that it will be more involved. It is simpler to work directly with a system in 1+1 dimensions, meaning there is no k_y, k_z.

The total vacuum energy of the system is the energy of fluctuations between the plates (boundary conditions) at distance d and the energy outside the plates. If the space is considered to be periodic, then the boundary at $x = L$ is the same (after the equivalence) as the first plate at $x = 0$, and the other boundary of the system is at $x = d$ (the second plate), just that now the space in between the plates is the "outside" one, of length $L - d$. Therefore, the energy is

$$E = f(d) + f(L - d), \tag{31.4}$$

where

$$f(d) = \frac{\hbar \pi}{2d} \sum_{n=1}^{\infty} n. \tag{31.5}$$

As we already noted, $f(d)$ is infinite, but let us consider a *regularization* that turns it into a finite quantity. This is so, since at $n \to \infty$, modes have infinite energy, so they clearly need to be suppressed. If for nothing else, when the energy reaches the Planck energy, we expect that the spacetime will start getting curved, so this naive notion of perturbation modes in flat space cannot be a good approximation anymore. In any case, it is physically clear that the contribution of modes of infinite energy should be suppressed. We consider the regulator $e^{-a\omega_n}$, so the "regulated" function is

$$\frac{\tilde{f}(d)}{\hbar} = \frac{\pi}{2d} \sum_{n \geq 1} n e^{-a\omega_n} = \frac{\pi}{2d} \sum_{n \geq 1} n e^{-a\frac{n\pi}{d}} = -\frac{1}{2}\frac{\partial}{\partial a} \sum_{n \geq 1} \left(e^{-a\frac{\pi}{d}} \right)^n$$

$$= -\frac{1}{2}\frac{\partial}{\partial a}\frac{1}{1 - e^{-a\frac{\pi}{d}}} = \frac{\pi}{2d}\frac{e^{-a\frac{\pi}{d}}}{(1 - e^{-a\frac{\pi}{d}})^2}. \tag{31.6}$$

Now, taking $a \to 0$ we obtain

$$\frac{\tilde{f}(d)}{\hbar} \simeq \frac{d}{2\pi a^2} - \frac{\pi}{24d} + \cdots \tag{31.7}$$

Adding the contributions of the two pieces (inside and outside the plates) to the Casimir effect, we obtain

$$E_0 = f(d) + f(L - d) = \frac{L}{2\pi a^2} - \frac{\pi}{24d} + \mathcal{O}\left(\frac{1}{L^2}; a^2\right), \tag{31.8}$$

and now the infinite term is *constant* (d-independent), but there is a finite d-dependent term.

In other words, by varying the geometry (i.e. by varying d), we obtain a force on the plates of

$$F = \frac{\partial E}{\partial d} = +\frac{\pi}{24d^2} + \cdots \tag{31.9}$$

31.2 General Vacuum Energy and Regularization with Riemann Zeta $\zeta(-1)$

We now consider a generalization of this case to an interacting quantum field theory for a scalar ϕ with potential $U(\phi)$, with minimum at $\phi = \phi_0$, and expand around it. We keep in the expansion of the action and the associated potential energy only the constant and quadratic terms (since the linear term is zero by the equations of motion). The potential energy of the system is then

$$V[\phi] \simeq V[\phi_0] + \int d^3x \frac{1}{2} \left\{ \delta\phi(\vec{x}, t) \left[-\vec{\nabla}^2 + \left(\frac{d^2U}{d\phi^2} \right)\bigg|_{\phi=\phi_0} \right] \delta\phi(\vec{x}, t) + \cdots \right\}. \tag{31.10}$$

We then need to find the eigenvalues (and eigenfunctions) of the operator in square brackets, namely

$$\left(-\vec{\nabla}^2 + \frac{d^2U}{d\phi^2}\bigg|_{\phi=\phi_0} \right) \eta_i(\vec{x}) = \omega_i^2 \eta_i(\vec{x}). \tag{31.11}$$

The eigenfunctions are orthonormal:

$$\int d^3x \eta_i^*(\vec{x})\eta_j(\vec{x}) = \delta_{ij}. \tag{31.12}$$

We can then expand the fluctuation in the eigenfunctions

$$\delta\phi(\vec{x}, t) = \sum_i c_i(t)\eta_i(\vec{x}).$$ (31.13)

Thus we obtain the total energy

$$E = \int d^4x \frac{1}{2}(\dot{\phi}(\vec{x}, t))^2 + \int dt V[\phi]$$
$$\simeq \int dt \sum_i \left[\frac{1}{2}\dot{c}_i(t)^2 + \frac{1}{2}\omega_i^2 c_i(t)^2\right],$$ (31.14)

which is a sum of harmonic oscillators, thus the energy of the vacuum of the system is

$$E_0 = V[\phi_0] + \frac{\hbar}{2}\sum_i \omega_i + \dots,$$ (31.15)

where we ignored corrections, which can be identified as higher-loop corrections (whereas the calculated term is only one loop, as we will see shortly). Note that this formula has many subtleties. In particular, for nonzero interactions, there will be renormalization of the parameters of the interaction (as we will show later on). But at a formal level, the above formula is correct.

As an example, consider what happens for the free theory, with $U = 0$. Then $\eta_i = e^{i\vec{k}_i \cdot \vec{x}}$, so

$$(-\vec{\nabla}^2)\eta_i = \vec{k}_i^2 \eta_i \equiv \omega_i^2 \eta_i,$$ (31.16)

and therefore

$$E_0 = \frac{\hbar}{2}\sum_i \sqrt{\vec{k}_i^2}.$$ (31.17)

Moreover, for $U = m^2\phi^2/2$ (just a mass term):

$$(-\vec{\nabla}^2 + m^2)\eta_i = (\vec{k}_i^2 + m^2)\eta_i \equiv \omega_i^2 \eta_i,$$ (31.18)

therefore

$$E_0 = \frac{\hbar}{2}\sum_i \sqrt{\vec{k}_i^2 + m^2}.$$ (31.19)

We try now to understand better the regularization that we used and its generality (or uniqueness of the nontrivial piece). The quantity we want is $\sum_{n\geq 1} n$. It is of course infinite, but in mathematics there is a well-defined finite value associated with it by analytical continuation.

Indeed, one can define the *Riemann zeta function*

$$\zeta(s) = \sum_{n\geq 1} \frac{1}{n^s}.$$ (31.20)

For real s, it is convergent for $s > 1$. However, one can define it in the complex plane $\mathbb{C}$ (as Riemann did). In this case, away from the real axis, the function $\zeta(s)$ is well defined close to -1. One can then define $\zeta(-1)$ uniquely by analytical continuation as $\zeta(s \to -1)$.[1] We have

$$\zeta(-1) = -\frac{1}{12}. \tag{31.21}$$

Then we can write directly for the Casimir effect with two plates:

$$E_0 \simeq \left(\frac{\hbar}{2d} + \frac{\hbar}{2(L-d)} \right) \sum_{n \geq 1} n = -\frac{\hbar\pi}{24d}, \tag{31.22}$$

the same as we obtained before, but with the infinite constant already removed. Therefore, the zeta function procedure is a good regularization.

31.3 Zeta Function and Heat Kernel Regularization

We now generalize the above regularization procedure using the zeta function to a general operator.

Consider an operator with positive, real, and discrete eigenvalues $a_1, \ldots, a_n$ and corresponding eigenfunctions f_n:

$$A f_n(x) = a_n f_n(x). \tag{31.23}$$

Then we define the function, associated with the operator A:

$$\zeta_A(s) = \sum_{n \geq 1} \frac{1}{a_n^s}. \tag{31.24}$$

Its derivative is

$$\frac{d}{ds} \zeta_A(s)|_{s=0} = -\sum_n \ln a_n e^{-s \ln a_n}|_{s \to 0} = -\ln \prod_n a_n. \tag{31.25}$$

So finally:

$$e^{\text{Tr} \ln A} = \det A = \prod_n a_n = e^{-\zeta_A'(0)}. \tag{31.26}$$

As we will see shortly, the object that we want to calculate is $\det A = e^{\text{Tr} \ln A}$, so we can calculate it if we know $\zeta_A(s)$.

---

[1] We can write an integral representation for the zeta function, by writing an integral representation for each term in the sum, as $\zeta(s) = \frac{1}{\Gamma(s)} \int_0^\infty t^{s-1} \sum_{n \geq 1} e^{-nt} = \frac{1}{\Gamma(s)} \int_0^\infty dt \frac{t^{s-1} e^{-t}}{1-e^{-t}}$, which form is shown to be well defined (analytic) over the whole complex plane, with the only pole at $s = +1$. For $s = -2m$, with m integer, it has in fact (trivial) poles. The Riemann hypothesis, one of the most important unsolved (unproven) problems in mathematics, is that all the nontrivial zeroes of the Riemann zeta function (other than the trivial ones above) lie on the line $Re(s) = +1/2$.

31.3.1 Heat Kernel Regularization

Sometimes we can't calculate directly $\det A$ or $\zeta_A(s)$, but it may be easier to calculate another object, called the associated "heat kernel," defined as

$$G(x, y, \tau) \equiv \sum_n e^{-a_n \tau} f_n(x) f_n^*(y). \tag{31.27}$$

This satisfies the generalized heat equation

$$A_x G(x, y, \tau) = -\frac{\partial}{\partial \tau} G(x, y, \tau), \tag{31.28}$$

as we can easily check. We can also easily check that, since the $f_n(x)$ are orthonormal, the heat kernel satisfies the boundary condition

$$G(x, y; \tau = 0) = \delta(x - y). \tag{31.29}$$

Sometimes it is easier to solve the generalized heat equation with the above boundary condition, and then we can find $\zeta_A(s)$ from it as

$$\zeta_A(s) = \frac{1}{\Gamma(s)} \int_0^\infty d\tau \, \tau^{s-1} \int d^4 x \, G(x, x, \tau), \tag{31.30}$$

since $\int d^4 x |f_n(x)|^2 = 1$.

If we find $\zeta_A(s)$, then we can calculate $\det A$, as we saw.

31.4 Saddle Point Evaluation and One-Loop Determinants

We now see how to calculate the sum over frequencies from the path integral formulation. For that, we need to make a quadratic approximation around a minimum (sometimes it is only a local minimum, not a global minimum) of the action.

This is based on the idea of saddle point evaluation of an integral. If we write the integrand as an exponential of minus something that has a minimum:

$$I \equiv \int dx \, e^{-a(x)}, \tag{31.31}$$

with

$$a(x) \simeq a(x_0) + \frac{1}{2}(x - x_0)^2 a''(x_0), \tag{31.32}$$

where $a''(x_0) > 0$, then we can approximate the integral as a Gaussian around the minimum, that is as

$$I \simeq e^{-a(x_0)} \int dx \, e^{-\frac{1}{2}(x-x_0)^2 a''(x_0)} = e^{-a(x_0)} \sqrt{\frac{2\pi}{a''(x_0)}}. \tag{31.33}$$

31.4.1 Path Integral Formulation

Consider the case of field theory in Euclidean space, with partition function

$$Z[J] = N \int \mathcal{D}\phi\, e^{-S[\phi]+J\cdot\phi} \equiv N \int \mathcal{D}\phi\, e^{-S_E[\phi,J]}. \tag{31.34}$$

We can again make a quadratic approximation around a minimum:

$$S_E[\phi,J] = S_E[\phi_0,J] + \frac{1}{2}\left\langle \frac{\delta^2 S_E}{\delta\phi_1\delta\phi_2}(\phi-\phi_0)_1(\phi-\phi_0)_2 \right\rangle_{1,2} + \ldots, \tag{31.35}$$

where $\phi_1 = (\phi-\phi_0)_1$ and $\phi_2 = (\phi-\phi_0)_2$, with 1, 2 corresponding to different variables of integration (x, y), and $<>_{1,2}$ to integrating over them.

Then the saddle point evaluation of the partition function gives

$$Z[J] \simeq N' e^{-S_E[\phi_0]+\phi_0\cdot J} \int \mathcal{D}\phi\, \exp\left\{ -\frac{1}{2}\left\langle \phi_1 \frac{\delta^2 S_E}{\delta\phi_1\delta\phi_2}\phi_2 \right\rangle_{1,2} \right\}$$

$$= N'' e^{-S_E[\phi_0]+\phi_0\cdot J} \det\left[(-\partial_\mu\partial^\mu + U''(\phi_0))\delta_{1,2}\right]^{-1/2}. \tag{31.36}$$

Since $\det A = e^{\operatorname{Tr}\ln A}$, we have

$$-W[J] = \ln Z[J] = -S[\phi_0] + \phi_0\cdot J - \frac{1}{2}\operatorname{Tr}\ln\left[(-\partial_\mu\partial^\mu + U''(\phi_0))\delta_{1,2}\right], \tag{31.37}$$

and as we saw, we can calculate

$$-\frac{1}{2}\operatorname{Tr}\ln[A] = \frac{1}{2}\left.\frac{d\zeta_A}{ds}\right|_{s=0}. \tag{31.38}$$

To make contact with what we had before, we continue to Minkowski space, where the eigenvalue–eigenfunction problem is

$$\left(-\frac{\partial^2}{\partial t^2} + \vec{\nabla}^2 - U''(\phi_0)\right)\phi_k(\vec{x},t) = a_k\phi_k(\vec{x},t). \tag{31.39}$$

We separate variables by writing

$$\phi_k(\vec{x},t) = \eta_r(\vec{x})f_n(t) \tag{31.40}$$

(i.e. $k = (rn)$), with

$$(-\vec{\nabla}^2 + U''(\phi_0))\eta_r(\vec{x}) = \omega_r^2\eta_r. \tag{31.41}$$

Then we have

$$\left(-\frac{\partial^2}{\partial t^2} - \omega_r^2\right)\sin\frac{n\pi t}{T} = \left(\frac{n^2\pi^2}{T^2} - \omega_r^2\right)\sin\frac{n\pi t}{T}, \tag{31.42}$$

so that finally the *one-loop determinant* is

$$\det\left[\partial_\mu\partial^\mu - U''(\phi_0)\right]^{-1/2} = \prod_{n\geq 1}\left(\frac{n^2\pi^2}{T^2} - \omega_r^2\right)^{-1/2}. \tag{31.43}$$

One can calculate the partition function then, and obtain (the details are left as an exercise)

$$Z = \sum_n e^{-\frac{iE_n T}{\hbar}}, \tag{31.44}$$

and as before

$$E_0 = \frac{\hbar}{2} \sum_r \omega_r. \tag{31.45}$$

The first observation is that $E_0 \propto \hbar$, and it is obtained by quantum fluctuations around the classical minimum of the action, so it is the first quantum correction (i.e. one loop). It is one loop since in the quadratic (free) approximation we used, calculating $Z_0[J]$, the only Feynman diagrams there are are vacuum bubbles (i.e. circles), which form one loop. Moreover, we have seen that the result is given by determinants of the operator acting on fluctuations. Hence one refers to these quantum corrections as *one-loop determinants*.

31.4.2 Fermions

In the case of fermions, as we know, Gaussian integration gives the determinant in the numerator instead of the denominator, that is

$$\int d^n x e^{x^T A x} = 2^{n/2} \sqrt{\det A} = N e^{+\frac{1}{2} \operatorname{Tr} \ln A}. \tag{31.46}$$

Moreover, in the operator formalism we saw that the energy is

$$H = \sum_i \sum_{n_i} \frac{\hbar \omega_i}{2} (b_i^\dagger b_i - b_i b_i^\dagger) = \sum_i \sum_{n_i} \hbar \omega \left(N_i - \frac{1}{2} \right), \tag{31.47}$$

so in both ways we obtain

$$E_0 = -\frac{\hbar}{2} \sum_i \omega_i, \tag{31.48}$$

which is opposite in sign to the bosonic corrections. In supersymmetric theories, the bosonic and fermionic zero-point energies cancel each other.

Important Concepts to Remember

- We can measure changes in the infinite zero-point energy of a field due to geometry, but not the infinite constant piece. This is known as the *Casimir effect*.
- We obtain the relevant finite piece by using a regularization, either by introducing a $e^{-a\pi n/d}$ factor with $a \to 0$, or better, by using the finite number associated with $\sum_n n$ by the zeta function, $\zeta(-1) = -1/12$.
- We generalize to an arbitrary field theory by calculating the eigenvalue–eigenfunction problem for the potential energy operator, and summing over $\hbar\omega_i/2$.

- In the path integral formulation, the saddle point evaluation gives a determinant of fluctuations around the minimum of the action. It leads to the same form of the zero-point energy. This corresponds to one-loop corrections, namely vacuum bubbles. Higher-loop corrections appear from higher-order terms in the action (neglected here).
- One can define zeta function regularization by generalizing the harmonic oscillator case via $\zeta_A(s) = \sum_{n\geq1} 1/(a_n)^s$.
- The determinant of an operator is then $\det A = e^{-\zeta_A'(0)}$.
- We can define the heat kernel associated with an operator, that satisfies the generalized heat equation, with a boundary condition $G(x,y;\tau=0) = \delta(x-y)$, which sometimes is easier to solve. Then we can find $\zeta_A(s)$ from the heat kernel.
- Fermions give $(\det A)^{+1/2}$ instead of $(\det A)^{-1/2}$, and a negative sign in front of the zero-point energy, so they can in principle cancel against the bosonic corrections.

Further Reading

See sections 3.4 and 3.5 in [4].

Exercises

1. Consider two scalar fields with potential $(\lambda_1, \lambda_2 > 0)$:
$$U(\phi, \chi) = \lambda_1(m_1\phi - \chi^2)^2 + \lambda_2(m_2\chi - \phi^2)^2. \qquad (31.49)$$
Write down the infinite sum for the zero-point energy in 1+1 dimensions *for this model* (no need to calculate the sum) around the vacuum at $\phi = \chi = 0$.
2. Write down an expression for the zeta function regularization for the above sum, and for the heat kernel.
3. Show that if the one-loop determinant is given by (31.43), then the partition function becomes (31.44).
4. Consider an operator defined by its eigenvalues $a_n = \sqrt{n}$ and eigenfunctions e^{-nx}. Calculate the zeta function associated with it, and its heat kernel.

In this chapter we will analyze possible divergences in loop integrals, in particular we will look at one loop. We will see that there are UV (high-energy) and IR (low-energy) divergences. Next, we will see how to Wick rotate a diagram whose integrands include poles, and what is the correct way to define the integral. Before calculating an integral, we will show how to determine if a theory contains divergences, by using power counting. Finally, we will define power counting renormalizability and distinguish between theories based on it.

32.1 One-Loop UV and IR Divergences

Up to now, we have only calculated explicitly tree processes, which are finite, and we have ignored the fact that loop integrals can be divergent. We have seen hints of the fact that loop integrals can be divergent, giving infinite values for Feynman diagrams.

For example, in $\lambda\phi^4$ theory in Euclidean space, consider the unique one-loop $\mathcal{O}(\lambda)$ diagram, a loop connected to the free line (propagator) at a point, see Figure 32.1. It is given by

$$-\lambda \int \frac{d^D q}{(2\pi)^D} \frac{1}{q^2 + m^2}. \tag{32.1}$$

Since the integral is

$$= -\lambda \frac{\Omega_{D-1}}{(2\pi)^D} \int q^{D-1} dq \frac{1}{q^2 + m^2} \sim \int dq \, q^{D-3}, \tag{32.2}$$

it is divergent in $D \geq 2$ and convergent only for $D < 2$. In particular, in $D = 4$ it is quadratically divergent:

$$\sim \int^\Lambda q \, dq \sim \Lambda^2. \tag{32.3}$$

We call this kind of divergence "ultraviolet," or UV divergence, from the fact that it is at large energies (4-momenta), or large frequencies.

Note also that we had one more type of divergence for loop integrals that was easily dealt with, the fact that when integrating over loop momenta in Minkowski space, the propagators can go on-shell, leading to a pole, which needed to be regulated. But the $i\epsilon$ prescription

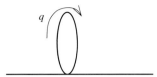

Figure 32.1 One-loop divergence in ϕ^4 theory for the two-point function.

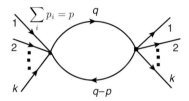

Figure 32.2 One-loop diagram in ϕ^{k+2} theory, for the $2k$-point function.

dealt with that. Otherwise, we can work in Euclidean space and then analytically continue to Minkowski space at the end of the calculation. This issue did not appear at tree level, when the propagators have fixed momenta, and are not on-shell, but it appears in loop integrals.

Let us now consider an example of a diagram that has all types of divergence, a one-loop diagram for the $2k$-point function in $\lambda \phi^{k+2}$ theory with k momenta in, then two propagators forming a loop, and then k lines out, as in Figure 32.2. The incoming momenta are called $p_1, \ldots, p_k$ and sum to $p = \sum_{i=1}^{k} p_i$. Then the two propagators have momenta q (loop variable) and $q - p$, giving for the diagram

$$\frac{\lambda^2}{2} \int \frac{d^D q}{(2\pi)^D} \frac{1}{(q^2 + m^2)((q - p)^2 + m^2)}. \tag{32.4}$$

Again, we can see that at large q, it behaves as

$$\sim \int \frac{q^{D-1} dq}{q^4}, \tag{32.5}$$

so is convergent only for $D < 4$. In particular, in $D = 4$ it is (log) divergent, and this again is UV divergence. From this example we can see that various diagrams are divergent in various dimensions.

As mentioned, the poles in the propagators when we go to Minkowski space mean that *a priori* there are these divergences as well, but they are regulated by the Feynman $i\epsilon$ prescription.

But this diagram also has another type of divergence, namely at low q ($q \to 0$). This divergence appears only *if we have $m^2 = 0$ AND $p^2 = 0$*. Thus, only if we have massless particles, and all the particles that are incoming on the same vertex sum up to something on-shell (in general, the sum of on-shell momenta is not on-shell). Then the integral is

$$\sim \int d\Omega \int \frac{q^3 dq}{q^2(q^2 - 2q \cdot p)}, \tag{32.6}$$

and in the integral over angles, there will be a point where the unit vector on q, $\hat{q}$, satisfies $\hat{q} \cdot \hat{p} = 0$ with respect to the (constant) unit vector on p, $\hat{p}$. Then we obtain

$$\int \frac{dq}{q} \tag{32.7}$$

(i.e. log divergent). We call this kind of divergence "infrared," or IR divergence, since it occurs at low energies (low 4-momenta), that is at low frequencies.

Thus we have two kinds of potential divergence, UV and IR divergence. UV divergence is an artifact of perturbation theory (i.e. of the fact that we were forced to introduce asymptotic states as states of the free theory, and calculate using Feynman diagrams). As such, it can be removed by redefining the parameters of the theory (masses, couplings, etc.), a process known as *renormalization*, which will be studied next in this book. But such UV divergences are a characteristic of the theory, hence their presence tells us that we need to do something to define the theory better.

A nonperturbative definition is not in general available, in particular for scattering processes. But for things like masses and couplings of bound states (like the proton mass in quantum chromodynamics, for instance), one can define the theory nonperturbatively, for instance on the lattice, and then we always obtain finite results. The infinities of perturbation theory manifest themselves only in something called the renormalization group, which will also be studied later in this book.

By contrast, IR divergences are genuine divergences from the point of view of the Feynman diagram (they cannot be reabsorbed by redefining the parameters). But they arise because the Feynman diagrams we are interested in, in the case of a theory with massless external states, and with external states that are on-shell at the vertex, are not quantities that can be measured experimentally. Indeed, for a massless external state ($m = 0$), of energy E, experimentally we cannot distinguish between the process with this external state and that with it and another emitted "soft and/or collinear particle," namely one of $m = 0$ and $E \simeq 0$ and/or parallel to the first. If we include the tree-level process for that second process at the same order in the coupling constant (order λ^2 for the diagram under study), and sum it together with the first (loop level), we obtain a finite differential cross-section (which can be measured experimentally), for a given cut-off in energy and/or angle of resolution between two particles. Therefore, this divergence arises because the quantity we study is not a physical one; a physical measurement always has a minimal resolution for the energy (minimal detectable energy) and the angle between emitted particles. Only by summing over processes that cannot be distinguished by the physical detector do we get a finite quantity.

Thus the physical processes are always finite, in spite of the infinities in the Feynman diagram.

32.2 Analytical Continuation of Integrals with Poles

A question which one could already have asked is the following: is Wick rotation of the final result the same as Wick rotation of the integral to Minkowski space, followed by evaluation?

Let us look at the simplest one-loop diagram in Euclidean space (in $\lambda\phi^4$, as already discussed above):

$$\int \frac{d^D q}{(2\pi)^D} \frac{1}{q^2 + m^2}. \tag{32.8}$$

In Minkowski space it becomes

$$-i \int \frac{d^D q}{(2\pi)^D} \frac{1}{q^2 + m^2 - i\epsilon} = +i \int \frac{d^{D-1}q}{(2\pi)^{D-1}} \int \frac{dq_0}{2\pi} \frac{1}{q_0^2 - \vec{q}^2 - m^2 + i\epsilon}, \tag{32.9}$$

where now $q^2 = -q_0^2 + \vec{q}^2$.

Then the poles are at $\tilde{q}_0 - i\epsilon$ and $-\tilde{q}_0 + i\epsilon$, where $\tilde{q}_0 = \sqrt{\vec{q}^2 + m^2}$. The Minkowski space integration contour is along the real axis in the q_0 plane, in the increasing direction, called C_R. On the contrary, the Euclidean-space integration contour C_I is along the imaginary axis, in the increasing direction, see Figure 32.3. As there are no poles between C_R and C_I (in the quadrants I and III of the complex plane, the poles are in the quadrants II and IV), the integral along C_I is equal to the integral along C_R (since we can close the contour at infinity, with no poles inside). Therefore, along C_I, $q_0 = iq_D$, with q_D real and increasing, and therefore $dq_0 = idq_D$, so

$$\int_{C_R} dq_0(\ldots) = \int_{C_I} dq_0(\ldots) = \int \frac{d^{D-1}q}{(2\pi)^D}(-i)i \int \frac{dq_D}{\vec{q}^2 + (q_D)^2 + m^2 - i\epsilon}, \tag{32.10}$$

which gives the same result as the Euclidean-space integral, after we drop the (now unnecessary) $-i\epsilon$.

However, in general it is not true that we can easily continue analytically. Instead, we must *define the Euclidean-space integral and Wick rotate the final result*, since in general this will seemingly be different from the continuation of the Minkowski space integral (rather, it means that the Wick rotation of the integrals is subtle). But the quantum field

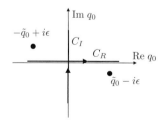

Figure 32.3   Wick rotation of the integration contour.

theory perturbation in Euclidean space is well defined, unlike the Minkowski space one, as we already saw, so it is a good starting point.

We want to see an example of this situation. Consider the second integral we analyzed, now in Minkowski space:

$$-\frac{\lambda^2}{2} \int \frac{d^D q}{(2\pi)^D} \frac{1}{q^2 + m^2 - i\epsilon} \frac{1}{(q-p)^2 + m^2 - i\epsilon}. \tag{32.11}$$

We deal with the two different propagators in the loop integral using the Feynman trick. We will study it in more detail in Chapter 33, but this time we will just use the result for two propagators. The Feynman trick for this case is the observation that

$$\frac{1}{AB} = \int_0^1 dx[xA + (1-x)B]^{-2} \tag{32.12}$$

(which can easily be checked), which allows one to turn the two propagators with different momenta into a single propagator squared. Indeed, we can now write

$$-\frac{\lambda^2}{2} \int \frac{d^D q}{(2\pi)^D} \int_0^1 dx[x(q-p)^2 + (1-x)q^2 + (x+1-x)(m^2 - i\epsilon)]^{-2}. \tag{32.13}$$

The square bracket equals $q^2 + xp^2 - 2xq\cdot p + m^2 - i\epsilon$. Changing variables to $q'^\mu = q^\mu - xp^\mu$ allows us to get rid of the term linear in q. We can change the integration variable to q', since the Jacobian for the transformation is 1, and then the square bracket becomes $q'^2 + x(1-x)p^2 + m^2 - i\epsilon$. Finally, the integral is

$$\frac{(-i\lambda)^2}{2} \int \frac{d^D q'}{(2\pi)^D} \int_0^1 dx[q'^2 + x(1-x)p^2 + m^2 - i\epsilon]^{-2}, \tag{32.14}$$

which has poles at

$$\tilde{q}_0^2 = \vec{q}^2 + m^2 - i\epsilon + x(1-x)p^2. \tag{32.15}$$

If $p^2 > 0$, this is the same as in the previous example, we just redefine m^2: the poles are outside quadrants I and III, so we can make the Wick rotation of the integral without problem. However, if $p^2 < 0$ and sufficiently large in absolute value, we can have $q_0^2 < 0$, so the poles are now in quadrants I and III, and we cannot simply rotate the contour C_R to the contour C_I, since we encounter poles along the way. So in this case, the Wick rotation is more subtle: apparently, the Minkowski-space integral gives a different result from the Euclidean-space result, Wick rotated. However, the latter is better defined, so we can use it.

32.3 Power Counting and UV Divergences

We now want to understand how we can figure out if a diagram, and more generally a theory, contains UV divergences. We do this by *power counting*. We consider here scalar $\lambda_n \phi^n$ theories.

Consider first just a (Euclidean-space) diagram, with L loops and E external lines, and I internal lines and V vertices. The loop integral will be

$$I_D(p_1, \ldots, p_E; m) = \int \prod_{\alpha=1}^{L} \frac{d^d q_\alpha}{(2\pi)^d} \prod_{j=1}^{I} \frac{1}{q_j^2 + m^2}, \tag{32.16}$$

where $q_j = q_j(p_i, q_\alpha)$ are the momenta of the internal lines (which have propagators $1/(q_j^2 + m^2)$). More precisely, they are linear combinations of the loop momenta and external momenta:

$$q_j = \sum_{\alpha=1}^{L} c_{j\alpha} q_\alpha + \sum_{i=1}^{E} c_{ji} p_i. \tag{32.17}$$

As already mentioned in Chapter 11, $L = I - V + 1$, since there are I momentum variables, constrained by V delta functions (one at each vertex), but one of the delta functions is the overall (external) momentum conservation.

If we scale the momenta and masses by the same multiplicative factor t, we can also change the integration variables (loop momenta q_α) by the same factor t, getting $\prod_{\alpha=1}^{L} d^d q \rightarrow t^{LD} \prod_{\alpha=1}^{L} d^d q$, as well as $q_j \rightarrow t q_j$, and $q^2 + m^2 \rightarrow t^2(q^2 + m^2)$, giving finally

$$I_D(t p_i; t m) = t^{\omega(D)} I_D(p_i; m), \tag{32.18}$$

where

$$\omega(D) = dL - 2I \tag{32.19}$$

is called the *superficial degree of divergence* of the diagram D, since it is the overall dimension for the scaling above. This gives rise to the following theorem:

Theorem $\omega(D) < 0$ is *necessary* for the convergence of I_D. (*Note:* but is not sufficient!)

Proof   We have

$$\prod_{i=1}^{I}(q_i^2 + m^2) \leq \left(\sum_{i=1}^{I} q_i^2 + m^2 \right)^I. \tag{32.20}$$

Then, for large enough q_α, there is a constant C such that

$$\sum_{i=1}^{I}(q_i^2 + m^2) = \sum_{i=1}^{I} \left[\left(\sum_{\alpha=1}^{L} c_{i\alpha} q_\alpha + \sum_{j=1}^{E} c_{ij} p_j \right)^2 + m^2 \right] \leq C \sum_{\alpha=1}^{L} q_\alpha^2, \tag{32.21}$$

as we can easily see. So we have

$$I_D > \frac{1}{C^I} \int_{\sum q_\alpha^2 > \Lambda^2} \prod_{\alpha=1}^{L} \frac{d^d q}{(2\pi)^d} \frac{1}{(\sum_{\alpha=1}^{L} q_\alpha^2)^I} > \int_{r > \Lambda} \frac{r^{DL-1} dr}{r^{2I}}, \tag{32.22}$$

where we used the fact that $\sum_{\alpha=1}^{L} q_\alpha^2 \equiv \sum_{M=1}^{dL} q_M^2$ is a sum of dL terms, which we can consider as a dL-dimensional space, and the condition $\sum_\alpha q_\alpha^2 > \Lambda^2$, stated before as q_α

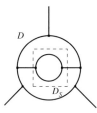

Figure 32.4 Power counting example: diagram is power-counting convergent, but subdiagram is actually divergent.

being large enough, now becomes the fact that the modulus of the dL-dimensional q_M is bounded from below. We finally see that if $\omega(D) = dL - 2I > 0$, I_D is divergent. The opposite statement is that if I_D is convergent, then $\omega(D) < 0$, that is $\omega(D) < 0$ is a necessary condition for convergence.   *q.e.d.*

As we said, the condition is necessary, but not sufficient. Indeed, we can have subdiagrams that are superficially divergent ($\omega(D_s) \geq 0$), therefore divergent, in which case the full diagram is also divergent, in spite of having $\omega(D) < 0$.

We can take as example in $\lambda\phi^3$ theory in $D = 4$ the diagram in Figure 32.4: a circle with three external lines connected to it, with a subdiagram D_s connected to the inside of the circle; a propagator line that has a loop made out of two propagators in the middle. The diagram has $I_D = 9, V_D = 7$, therefore $L_D = I_D - V_D + 1 = 9 - 7 + 1 = 3$, and then $\omega(D) = dL_D - 2I_D = 4 \times 3 - 2 \times 9 = -6 < 0$. However, the subdiagram has $I_{D_s} = 2, V_{D_s} = 2$, therefore $L_{D_s} = 2 - 2 + 1 = 1$, and then $\omega(D_s) = 4 \times 1 - 2 \times 2 = 0$, meaning that we have a logarithmically divergent subdiagram, and thus the full diagram is also logarithmically divergent.

32.4 Power-Counting Renormalizable Theories

We can then guess that we have the following theorem (which we will not prove here):

Theorem $\omega(D_s) < 0,\ \forall D_s$ 1PI subdiagrams of $D \Leftrightarrow I_D(p_1, \ldots, p_E)$ is an absolutely convergent integral.

We note that the $\Rightarrow$ implication should obviously be true, and moreover is valid for any field theory. But the $\Leftarrow$ implication is true only for scalar theories. If there are spin-1/2 and spin-1 fields, then $\omega(D) < 0$ is not even necessary, since there can be cancellations between different spins, giving a zero result for a (sum of) superficially divergent diagram(s) (hence the name *superficial* degree of divergence, it is not necessarily the *actual* degree of divergence).

We can now write a formula from which we derive a condition on the type of divergences we can have.

We note that each internal line connects to two vertices, and each external line connects to only one vertex. In a theory with $\sum_n \lambda_n \phi^n / n!$ interactions, we can have a number $n = n_v$ of legs at each vertex v, meaning that we have

$$2I + E = \sum_{v=1}^{V} n_v. \tag{32.23}$$

Then the superficial degree of divergence is

$$\omega(D) \equiv dL - 2I = (d-2)I - dV + d = d - \frac{d-2}{2}E + \sum_{v=1}^{V}\left(\frac{d-2}{2}n_v - d\right), \tag{32.24}$$

where in the second equality we used $L = I - V + 1$ and in the last equality we used $2I + E = \sum_v n_v$.

Since the kinetic term for a scalar is $-\int d^d x (\partial_\mu \phi)^2 / 2$, and it has to be dimensionless, we need the dimension of ϕ to be $[\phi] = (d-2)/2$. Then, since the interaction term is $-\int d^d x \lambda_n \phi^n / n!$, we have

$$[\lambda_{n_v}] = d - n_v[\phi] = d - \frac{d-2}{2}n_v, \tag{32.25}$$

meaning that finally

$$\omega(D) = d - \frac{d-2}{2}E - \sum_{v=1}^{V}[\lambda_v]. \tag{32.26}$$

Thus we find that if $[\lambda_v] \geq 0$, there are only a finite number (and very small) of divergent n-point functions (where $n = E$ is the number of external lines). If $[\lambda_v] > 0$, then there are also a finite number of diagrams. Indeed, first we note that by increasing the number of external lines, we get to $\omega(D) < 0$. Then, if $[\lambda_v] > 0$, we get to $\omega(D) < 0$ also by increasing V, the number of vertices.

As an example, consider the limiting case of $[\lambda_v] = 0$, and $d = 4$. Then $\omega(D) = 4 - E$, and only $E = 0, 1, 2, 3, 4$ give divergent results, irrespective of V. Since $E = 0, 1$ are not physical ($E = 0$ is a vacuum bubble, and $E = 1$ should be zero in a good theory), we have only $E = 2, 3, 4$, corresponding to three physical parameters (physical parameters are defined by the number of external lines, which define physical objects like n-point functions). For $[\lambda_v] > 0$, any vertices lower $\omega(D)$, so we could have an even smaller number of Es for divergent n-point functions, since we need at least a vertex for a loop diagram. In higher dimensions, we will have a slightly higher number of divergent n-point functions, but otherwise the same idea applies.

Such theories, where there are a finite number of n-point functions that have divergent diagrams, are called *renormalizable*, since we can absorb the infinities in the redefinition of the (finite number of) parameters of the theory, the parameters being in one-to-one correspondence with the *1PI* n-point functions.

Therefore we have three possibilities:

(A) If there is some $[\lambda_v] < 0$, we can make divergent diagrams for any n-point function (any E) with λ_v vertices just by increasing V. Therefore, there are an infinite number of divergent 1PI n-point functions that would need redefinition, so we can't make this by redefining the parameters of the theory. Such a theory is called *nonrenormalizable*. Note that a nonrenormalizable theory can only be so in perturbation theory; there exist examples of theories that are perturbatively nonrenormalizable, but the nonperturbative theory is well defined. Also note that we can work with nonrenormalizable theories in perturbation theory, just by introducing new parameters at each loop order. Therefore we can compute quantum corrections, though the degree of complexity of the theory quickly increases with the loop order.

(B) If there is some $[\lambda_v] = 0$, and all other $[\lambda_v] > 0$, this means that there are a finite number of 1PI-divergent n-point functions, which means that the theory is *renormalizable*.

(C) If all $[\lambda_v] > 0$, there are only a finite number of *diagrams* that are divergent, which means that we can actually fully renormalize the theory in practice, not just in principle. The theory is then called *super-renormalizable*.

32.4.1 Examples

Consider scalar field theories with all power laws, $\sum_{n \geq 1} \lambda_n \phi^n / n!$. Then the condition for the theory to be renormalizable is $[\lambda_n] \geq 0$, which gives

$$n \leq \frac{2d}{d-2}. \tag{32.27}$$

d=2 In two dimensions all n are renormalizable, in fact super-renormalizable, since then $[\lambda_n] = 2$, and $\omega(D) = 2 - 2V$ becomes < 0 as V is increased.

d=3 The above condition gives $n \leq 6$, with equality for $n = 6$, which means that ϕ^3, ϕ^4, ϕ^5 are super-renormalizable and ϕ^6 is just renormalizable.

d=4 The condition above gives $n \leq 4$, with equality for $n = 4$, which means that ϕ^3 is super-renormalizable and ϕ^4 is just renormalizable.

d=5 The condition above gives $n \leq 10/3$, which means that only ϕ^3 is renormalizable, and actually is super-renormalizable.

d=6 The condition gives $n \leq 3$, which means that only ϕ^3 is renormalizable.

d>6 There are no renormalizable interactions.

32.4.2 Divergent ϕ^4 1PI Diagrams in Various Dimensions

- In $d = 2$, $[\lambda_v] = 2$ and $\omega(D) = 2 - 2V$, which means that only the $V = 1$ diagram is divergent (the two-point, one-loop diagram).
- In $d = 3$, $[\lambda_v] = 1$ and $\omega(D) = 3 - E/2 - V$, which means that the only 1PI diagrams are the $V = 1, E = 2$ diagram (one loop) and the $V = 2, E = 2$ diagram (two loops).
- In $d = 4$, $[\lambda_v] = 0$, so $\omega(D) = 4 - E$, so all the diagrams of the 2, 3 and 1PI 4-point functions are divergent.

Important Concepts to Remember

- Loop diagrams can contain UV divergences (at high momenta), divergent in diagram-dependent dimensions, and IR divergences, which appear only for massless theories and for on-shell total external momenta at vertices.
- UV divergences can be absorbed in a redefinition of the parameters of the theory (renormalization), and IR divergences can be canceled by adding the tree diagrams for emission of low-momentum ($E \simeq 0$) particles, perhaps parallel to the original external particle.
- Wick rotation of the result of the Euclidean integrals can in general not be the same as Wick rotation of the Euclidean integral, since there can be poles between the Minkowskian and Euclidean contours for the loop energy integration. We can work in Euclidean space and continue the final result, since the Euclidean theory is better defined.
- Power counting gives the superficial degree of divergence of a diagram as $\omega(D) = dL - 2I$.
- In a scalar theory, $\omega(D) < 0$ is necessary for convergence of the integral I_D, but in general is not sufficient.
- In a scalar theory, $\omega(D_s) < 0$ for any 1PI subdiagram D_s of a diagram $D \Leftrightarrow I_D$ is absolutely convergent.
- Theories with couplings satisfying $[\lambda_v] \geq 0$ are renormalizable, that is one can absorb the infinities in redefinitions of the parameters of the theory, while theories with $[\lambda_v] < 0$ are nonrenormalizable, since we can't (there are an infinite number of different infinities to be absorbed).
- If all $[\lambda_v] > 0$, the theory is super-renormalizable: it has only a finite number of divergent 1PI diagrams.

Further Reading

See sections 5.1 and 5.2 in [3], section 9.1 in [2], and section 4.2 in [4].

Exercises

1. Consider the one-loop diagram in Figure 32.5, for arbitrary masses of the various lines. Check whether there are any divergences.
2. Check whether there are any UV divergences in the $D = 3$ diagram in $\lambda\phi^4$ theory in Figure 32.6.

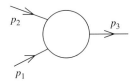

Figure 32.5   One-loop Feynman diagram.

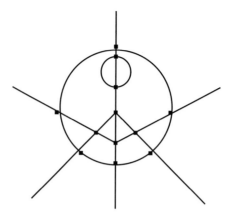

Figure 32.6   Check for UV divergences in this Feynman diagram.

3. Consider the Lagrangians

$$\mathcal{L}_1 = +\frac{1}{2}(\partial_\mu\phi)^2 + g\phi^2(\partial_\mu\phi)^2 \tag{32.28}$$

in four Euclidean dimensions, and

$$\mathcal{L}_2 = \frac{1}{2\kappa_N^2}\sqrt{g}R + \frac{1}{2}\sqrt{g}\partial_\mu\phi\partial_\nu\phi g^{\mu\nu} + \sqrt{g}\frac{m^2}{2}e^{\alpha^2\phi^2} \tag{32.29}$$

in two Euclidean dimensions, where R is the Ricci scalar for gravity.

Are these Lagrangians renormalizable, super-renormalizable, or nonrenormalizable? If they are renormalizable or super-renormalizable, write down the superficially divergent diagrams.

4. Write down the divergent diagrams for ϕ^3 theory in five dimensions.

Regularization, Definitions: Cut-off, Pauli–Villars, Dimensional Regularization, and General Feynman Parametrization

33

In this chapter we will consider methods of regularization, that will be used in the following for renormalization. We already saw a few methods of *regularization* (i.e. making finite the integrals). Here we will consider first various ways to regularize infinite sums (already started in Chapter 31), then cut-off, Pauli Villars, and dimensional regularization. The last is the most widely used, so we will show how to do general loop integrals in it, with the so-called Feynman parametrization, and we will show the algorithm for it in the general case.

33.1 Cut-off Regularization and Regularizations of Infinite Sums

The simplest is *cut-off regularization*, which means just putting upper and lower bounds on the integral over the modulus of the momenta, that is $|p|_{max} = \Lambda$ for the UV divergence, and $|p|_{min} = \epsilon$ for the IR divergence. It has to be over the modulus only (the integral over angles is not divergent), and then the procedure works best in Euclidean space (since then we don't need to worry about the fact that $-(p_0)^2 + \vec{p}^2 = \Lambda^2$ has a continuum of solutions for Minkowski space). Note that having $|p|_{max} = \Lambda$ is more or less the same as considering a lattice of size Λ^{-1} in Euclidean space, which breaks Euclidean ("Lorentz") invariance (since translational and rotational invariance are thus broken). For this reason, we very seldom consider the cut-off regularization.

There are many regularizations possible, and in general we want to consider a regularization that preserves all symmetries which play an important role at the quantum level. If there are several that preserve the symmetries we want, then all of them can in principle be used (we could even consider cut-off regularization, but then would have a hard time showing that our results are consistent with the symmetries we want to preserve).

Let us see the effect of cut-off regularization on the simplest diagram we can draw, a loop for a massless field, with no external momentum, but two external points, that is two equal loop propagators inside (see Figure 33.1(a)):

$$\int \frac{d^4p}{(2\pi)^4} \frac{1}{(p^2)^2} = \frac{\Omega_3}{(2\pi)^4} \int_\epsilon^\Lambda \frac{p^3 dp}{p^4} = \frac{1}{32\pi^2} \ln \frac{\Lambda}{\epsilon}. \tag{33.1}$$

As we said, we see that this has both UV (for $\Lambda \to \infty$) and IR (for $\epsilon \to 0$) divergences.

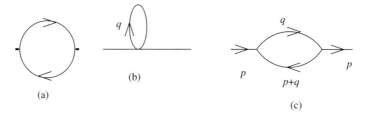

(a)

(b)

(c)

Figure 33.1   Examples of one-loop divergent diagrams.

33.1.1 Infinite Sums

As we saw in Chapter 32, sometimes we can turn integrals into infinite sums, like in the case of the zero-point energy, where a first regularization, putting the system in a box of size L, allows us to get a sum instead of an integral: $\sum_n \hbar\omega_n/2$. The sum, however, is still divergent. When one calculates the one-loop fluctuations around some classical background, the action is approximated by the classical part plus a quadratic fluctuation:

$$S \simeq S_{cl} + \frac{1}{2}\int \delta\phi \left[-\partial^\mu\partial_\mu + U''(\phi_0)\right]\delta\phi, \qquad (33.2)$$

where $\delta\phi = \phi - \phi_0$, ϕ_0 is the classical value for the field, giving the on-shell action S_{cl}, and $U(\phi)$ is the potential. Then the partition function is found to be

$$Z[J] = \mathcal{N} \det{}^{-1/2}[(\partial_\mu\partial^\mu + U''(\phi_0))\delta_{12}] = \ldots = \sum_n e^{-\frac{iE_n T}{\hbar}}, \qquad (33.3)$$

where the constant $\mathcal{N}$ includes $e^{-S_{cl}}$, and the $\ldots$ involves some calculations that will not be reproduced here. The zero-point energy E_0 is given by

$$E_0 = \frac{\hbar}{2}\sum_n \omega_n, \qquad (33.4)$$

where the ω_n are given by the eigenvalues of the spatial part of the quadratic operator:

$$(-\vec{\nabla}^2 + U''(\phi_0))\eta_r(\vec{x}) = \omega_r^2\eta_r(\vec{x}). \qquad (33.5)$$

Then, there are various ways of regularizing the sum. We can deduce that if the result depends on the method of regularization, there are two possibilities:

1. Perhaps the result is unphysical (cannot be measured). This is what happens to the *full* zero-point energy. Certainly the infinite constant (which for instance in an $e^{-a\omega_n}$ regularization is something depending only on L) is not measurable.
2. Perhaps some of the methods of regularization are unphysical, so we have to choose the physical one.

An example of a physical calculation is the *difference* of two infinite sums. This is what usually happens. In the example of the Casimir effect (the attractive force between two infinite conducting plates due to the vacuum energy), the difference between two geometries (two values for the distance d between two parallel plates) gives a physical, measurable quantity. Another example is the difference between two vacua, say a vacuum with a soliton and a trivial vacuum with no soliton. In general then, we will have

$$\sum_n \frac{\hbar \omega_n^{(1)}}{2} - \sum_n \frac{\hbar \omega_n^{(2)}}{2}. \tag{33.6}$$

For the case of the quantum corrections to masses of solitons, one considers the difference between quantum fluctuations ($\sum_n \hbar \omega_n / 2$) in the presence of the soliton, and in the vacuum (without the soliton), and this gives the physical calculation of the quantum correction to the mass of the soliton.

Note that it would seem that we could write

$$\sum_n \left(\frac{\hbar \omega_n^{(1)} - \hbar \omega_n^{(2)}}{2} \right) \tag{33.7}$$

and calculate this, but this amounts to a choice of regularization, called mode number (n) regularization. Indeed, we now have $\infty - \infty$, so unlike the case of finite integrals, now giving the $\sum_n$ operator as a common factor is only possible if we choose the *same number N as the upper bound* in both sums (if one is N, and the other $N + a$ say, with $a \sim \mathcal{O}(1)$, then obviously we obtain a different result for the difference of two sums).

This may seem natural, but there is more than one other way to calculate it: for instance, we can turn the sums into integrals in the usual way, and then take the *same upper limit in the integral (i.e. in energy)*, obtaining *energy/momentum cut-off regularization*. The difference of the two integrals gives a result differing from the mode number cut-off by a finite piece.

For $\sum_n \hbar \omega_n / 2$, there are other regularizations as well: for instance, zeta function regularization, heat kernel regularization, and $\sum_n \omega_n \to \sum_n \omega_n e^{-a\omega_n}$.

33.2 Pauli–Villars Regularization

Returning to the loop integrals appearing in Feynman diagrams, we have other ways to regulate them. One of the oldest ways used is *Pauli–Villars regularization*, and its generalizations. These fall under the category of modifications to the propagator that cut off the high-momentum modes in a smoother way than the hard cut-off $|p|_{max} = \Lambda$. The generalized case corresponds to making the substitution

$$\frac{1}{q^2 + m^2} \to \frac{1}{q^2 + m^2} - \sum_{i=1}^{N} \frac{c_i(\Lambda; m^2)}{q^2 + \Lambda_i^2}, \tag{33.8}$$

and we can adjust the c_i such that at large momentum q, the redefined propagator behaves as

$$\sim \frac{\Lambda^{2N}}{q^{2N+2}}. \tag{33.9}$$

In other words, if in a loop integral we need to have the propagator at large momentum behave as $1/q^{2N+2}$ in order for the integral to become finite, then we choose a redefinition with the desired N, and with c_is chosen for the required high-momentum behavior.

In particular, the original Pauli–Villars regularization is

$$\frac{1}{q^2+m^2} \to \frac{1}{q^2+m^2} - \frac{1}{q^2+\Lambda^2} = \frac{\Lambda^2-m^2}{(q^2+m^2)(q^2+\Lambda^2)} \sim \frac{\Lambda^2}{q^4}. \qquad (33.10)$$

We can easily see that this cannot be obtained from a normal modification of the action, because of the minus sign, however it corresponds to subtracting the contribution of a very heavy particle. Indeed, physically it is clear that a heavy particle cannot modify anything physical (e.g. a Planck mass particle cannot influence Standard Model physics). But it is equally obvious that subtracting its contribution will cancel heavy-momentum modes in the loop integral, canceling the unphysical infinities of the loop.

However, there is a simple modification that has the same result as the above Pauli–Villars subtraction at high momentum, and has a simple physical interpretation as the effect of a higher derivative term in the action. Specifically, consider the replacement of the propagator

$$\frac{1}{q^2+m^2} \to \frac{1}{q^2+m^2+q^4/\Lambda^2}. \qquad (33.11)$$

The usual propagator comes from

$$\int d^4x \frac{1}{2}(\partial_\mu\phi)^2 = \int \frac{d^4p}{(2\pi)^4} \frac{1}{2}\phi(p)p^2\phi(-p) = \int \frac{d^4p}{(2\pi)^4} \frac{1}{2}\phi(p)\Delta^{-1}(p)\phi(-p), \quad (33.12)$$

so the above replacement is obtained by adding to the action a higher derivative term

$$\int d^4x \frac{1}{2}\left[(\partial_\mu\phi)^2 + \frac{(\partial^2\phi)^2}{\Lambda^2}\right] = \frac{1}{2}\int \frac{d^4p}{(2\pi)^4}\phi(p)\left[p^2 + \frac{(p^2)^2}{\Lambda^2}\right]\phi(-p). \qquad (33.13)$$

Now consider a non-Pauli–Villars but similar modification of the loop integral, which is strictly speaking not a modification of the propagator, but of its square. Consider the same simplest loop integral, with two equal propagators, and its regularization:

$$I = \int \frac{d^4p}{(2\pi)^4} \frac{1}{(p^2+m^2)^2} \to \int \frac{d^4p}{(2\pi)^4}\left[\left(\frac{1}{p^2+m^2}\right)^2 - \left(\frac{1}{p^2+\Lambda^2}\right)^2\right] = I(m^2) - I(\Lambda^2).$$
$$(33.14)$$

The new object in the square brackets is

$$\frac{2p^2(\Lambda^2-m^2) + \Lambda^4 - m^4}{(p^2+m^2)^2(p^2+\Lambda^2)^2} \sim \frac{2\Lambda^2}{p^6}, \qquad (33.15)$$

so is now UV convergent. Since the object is UV convergent, we can use any method to calculate it. In particular, we can take a derivative $\partial/\partial m^2$ of it, and since $I(\Lambda^2)$ doesn't contribute, we get for the integral

$$\frac{\partial}{\partial m^2}[I(m^2) - I(\Lambda^2)] = \int \frac{d^4p}{(2\pi)^4} \frac{-2}{(p^2+m^2)^3} = -2\frac{\Omega_3}{(2\pi)^4}\int_0^\infty \frac{p^3 dp}{(p^2+m^2)^3}, \qquad (33.16)$$

where $\Omega_3 = 2\pi^2$ is the volume of the three-dimensional unit sphere. Considering $p^2 + m^2 = x$, so $p^3 dp = (x - m^2)dx/2$, we get

$$\frac{\partial}{\partial m^2} I(m^2, \Lambda^2) = -\frac{1}{8\pi^2} \int_{m^2}^{\infty} \frac{(x - m^2)dx}{x^3} = -\frac{1}{8\pi^2} \frac{1}{2m^2}. \tag{33.17}$$

Integrating this, we obtain $I(m^2)$, then

$$I(m^2, \Lambda^2) = I(m^2) - I(\Lambda^2) = \frac{1}{16\pi^2} \ln \frac{\Lambda^2}{m^2}. \tag{33.18}$$

This object is UV divergent as $\Lambda \to \infty$, and also divergent as $m \to 0$ (IR divergent).

33.3 Derivative Regularization

However, note that as calculated, we really introduced another type of regularization. It was implicit, since we first found a finite result by subtracting the contribution with $m \to \Lambda$, and then calculated this finite result using what was a simple trick.

However, if we keep the original integral and do the same derivative on it, after the derivative we obtain a finite result

$$\frac{\partial}{\partial m^2} \int \frac{d^4 p}{(2\pi)^4} \frac{1}{(p^2 + m^2)^2} = -2 \int \frac{d^4 p}{(2\pi)^4} \frac{1}{(p^2 + m^2)^3} = -\frac{1}{16\pi^2 m^2} = \text{finite}, \tag{33.19}$$

and now the integral (and its result) is UV convergent, despite the integral before the derivative being UV divergent. Hence the derivative with respect to the parameter m^2 is indeed a regularization. Both the initial and final results are, however, still IR divergent as $m^2 \to 0$. Integrating, we obtain

$$\int \frac{d^4 p}{(2\pi)^4} \frac{1}{(p^2 + m^2)^2} = -\frac{1}{16\pi^2} \ln \frac{m^2}{\epsilon} + \text{constant}, \tag{33.20}$$

which is still IR divergent as $\epsilon \to 0$. The UV divergence is now hidden as a possible infinite integration constant.

However, none of the regularizations analyzed so far respect a very important invariance, namely gauge invariance. Therefore, 't Hooft and Veltman introduced a new regularization to deal with the spontaneously broken gauge theories, namely *dimensional regularization* (rather late, in the early 1970s, since it is a rather strange concept).

33.4 Dimensional Regularization

Dimensional regularization means that we analytically continue in D, the dimension of spacetime, results calculated for arbitrary D. This seems like a strange thing to do, given that the dimension of spacetime is an integer, so it is not clear what a real dimension can physically mean, but we nevertheless choose $D = 4 \pm \epsilon$. The sign has some significance as well, but at this time we will just consider $D = 4 + \epsilon$.

A relevant example for us, that will not only encode the features of dimensional regularization, but will actually be the only way to obtain infinities in dimensional regularization, is the case of the Euler gamma function, which is an extension of the factorial $n!$, defined for integers, to the complex plane. Again this is done by writing an integral formula:

$$\Gamma(z) = \int_0^\infty d\alpha\, \alpha^{z-1} e^{-\alpha}. \tag{33.21}$$

Indeed, one easily shows that $\Gamma(n) = (n-1)!$, for $n \in \mathbb{N}_*$, but the integral formula can be extended to the complex plane, defining the Euler gamma function. The gamma function satisfies

$$z\Gamma(z) = \Gamma(z+1), \tag{33.22}$$

an extension of the factorial property. But this means that we can find the behavior at $z = \epsilon \to 0$, which is a simple pole, since

$$\epsilon\Gamma(\epsilon) = \Gamma(1+\epsilon) \simeq \Gamma(1) = 1 \Rightarrow \Gamma(\epsilon) \simeq \frac{1}{\epsilon}. \tag{33.23}$$

We can repeat this process:

$$(-1+\epsilon)\Gamma(-1+\epsilon) = \Gamma(\epsilon) \Rightarrow \Gamma(-1+\epsilon) \simeq -\frac{1}{\epsilon},$$

$$(-2+\epsilon)\Gamma(-2+\epsilon) = \Gamma(-1+\epsilon) \Rightarrow \Gamma(-2+\epsilon) \simeq \frac{1}{2\epsilon}, \tag{33.24}$$

and so on. We see then that the gamma function has simple poles at all $\Gamma(-n)$, with $n \in \mathbb{N}$. In fact, these poles are exactly the one we obtain in dimensional regularization, as we now show.

Consider first the simplest case, the tadpole diagram, with a single loop, of momentum q, connected at a point to a propagator line, as in Figure 33.1(b):

$$I = \int \frac{d^D q}{(2\pi)^D} \frac{1}{q^2 + m^2}. \tag{33.25}$$

We now write

$$\frac{1}{q^2 + m^2} = \int_0^\infty d\alpha\, e^{-\alpha(q^2 + m^2)}, \tag{33.26}$$

and then

$$I = \int_0^\infty d\alpha\, e^{-\alpha m^2} \int \frac{d^D q}{(2\pi)^D} e^{-\alpha q^2} = \int_0^\infty d\alpha\, e^{-\alpha m^2} \frac{\Omega_{D-1}}{(2\pi)^D} \int_0^\infty dq\, q^{D-1} e^{-\alpha q^2}$$

$$= \int_0^\infty d\alpha\, e^{-\alpha m^2} \frac{\Omega_{D-1}}{(2\pi)^D} \frac{1}{2\alpha^{D/2}} \int_0^\infty dx\, x^{\frac{D}{2}-1} e^{-x}, \tag{33.27}$$

and use the fact that $\int_0^\infty dx\, x^{D/2-1} e^{-x} = \Gamma(D/2)$, and that the volume of the D-dimensional sphere is

$$\Omega_D = \frac{2\pi^{\frac{D+1}{2}}}{\Gamma\left(\frac{D+1}{2}\right)}, \tag{33.28}$$

which we can easily test on a few examples, $\Omega_1 = 2\pi/\Gamma(1) = 2\pi$, $\Omega_2 = 2\pi^{3/2}/\Gamma(3/2) = 2\pi^{3/2}/(\sqrt{\pi}/2) = 4\pi$, $\Omega_3 = 2\pi^2/\Gamma(2) = 2\pi^2$. Then we have $\Omega_{D-1}\Gamma(D/2) = 2\pi^{D/2}$, so

$$I = \int_0^\infty d\alpha\, e^{-\alpha m^2}(4\pi\alpha)^{-\frac{D}{2}} = \frac{(m^2)^{\frac{D}{2}-1}}{(4\pi)^{\frac{D}{2}}}\Gamma\left(1-\frac{D}{2}\right). \qquad (33.29)$$

Taking derivatives $(\partial/\partial m^2)^{n-1}$ on both sides (both the definition of I and the result), we obtain in general

$$\int \frac{d^D q}{(2\pi)^D}\frac{1}{(q^2+m^2)^n} = \frac{\Gamma(n-D/2)}{(4\pi)^n\Gamma(n)}\left(\frac{m^2}{4\pi}\right)^{\frac{D}{2}-n}. \qquad (33.30)$$

We see that in $D=4$, this formula has a pole at $n=1,2$, as expected from the integral form. In these cases, the divergent part is contained in the gamma function, namely $\Gamma(-1)$ and $\Gamma(0)$. Note that, just as we continued the dimension from integer to real, we can also continue n from integer to real, since the same Euler gamma function allows us to do so analytically.

33.5 Feynman Parametrization

33.5.1 Feynman Parametrization with Two Propagators

We now move to a more complicated integral, which we will solve with Feynman parametrization, as cited in Chapter 32. Specifically, we consider the diagram for a one-loop correction to the propagator in ϕ^3 theory, with momentum p on the propagator and q and $p+q$ in the loop, as in Figure 33.1(c):

$$\int \frac{d^D q}{(2\pi)^D}\frac{1}{q^2+m^2}\frac{1}{(q+p)^2+m^2}. \qquad (33.31)$$

We now prove the Feynman parametrization in this case of two propagators. We do the trick used in the first integral (tadpole) twice, obtaining

$$\frac{1}{\Delta_1\Delta_2} = \int_0^\infty d\alpha_1\int_0^\infty d\alpha_2 e^{-(\alpha_1\Delta_1+\alpha_2\Delta_2)}. \qquad (33.32)$$

We then change variables in the integral as $\alpha_1 = t(1-\alpha)$, $\alpha_2 = t\alpha$, with Jacobian $\left|\begin{pmatrix} 1-\alpha & \alpha \\ -t & t \end{pmatrix}\right| = t$, so

$$\frac{1}{\Delta_1\Delta_2} = \int_0^1 d\alpha\int_0^\infty dt\, t\, e^{-t[(1-\alpha)\Delta_1+\alpha\Delta_2]} = \int_0^1 d\alpha\frac{1}{[(1-\alpha)\Delta_1+\alpha\Delta_2]^2}. \qquad (33.33)$$

We want to write the square bracket as a new propagator, so we redefine $q^\mu = \tilde{q}^\mu - \alpha p^\mu$, obtaining

$$(1-\alpha)\Delta_1 + \alpha\Delta_2 = (1-\alpha)q^2 + \alpha(q+p)^2 + m^2 = \tilde{q}^2 + m^2 + \alpha(1-\alpha)p^2. \qquad (33.34)$$

Finally, we obtain for the integral

$$I = \int_0^1 d\alpha \int \frac{d^D \tilde{q}}{(2\pi)^D} \frac{1}{[\tilde{q}^2 + (\alpha(1-\alpha)p^2 + m^2)]^2}$$

$$= \frac{\Gamma(2-D/2)}{(4\pi)^{D/2}} \int_0^1 d\alpha [\alpha(1-\alpha)p^2 + m^2]^{\frac{D}{2}-2}, \qquad (33.35)$$

and again we obtain the divergence as just an overall factor coming from the simple pole of the gamma function at $D = 4$, and the integral is now finite.

For completeness, we also generalize this formula to the case of an arbitrary power in the second propagator. We first define an integral with two different masses on the two propagators, and use Feynman parametrization on it, to obtain

$$I_{1,1}(p; m_1, m_2) \equiv \int \frac{d^d q}{(2\pi)^d} \frac{1}{(q^2 + m_1^2)((q+p)^2 + m_2^2)}$$

$$= \int_0^1 d\alpha \int \frac{d^d \tilde{q}}{(2\pi)^d} \frac{1}{[\tilde{q}^2 + \alpha(1-\alpha)p^2 + \alpha m_1^2 + (1-\alpha)m_2^2]^2}$$

$$= \frac{\Gamma(2-d/2)}{(4\pi)^{d/2}} \int_0^1 [\alpha(1-\alpha)p^2 + \alpha m_1^2 + (1-\alpha)m_2^2]^{\frac{d}{2}-2}. \quad (33.36)$$

Then, putting $m_1 = m_2 = 0$, we obtain

$$I_{1,1}(p) = p^{d-4} \frac{\Gamma(2-d/2)\beta(d/2-1, d/2-1)}{(4\pi)^{d/2}} = p^{d-4} \frac{\Gamma(2-d/2)[\Gamma(d/2-1)]^2}{(4\pi)^{d/2}\Gamma(d-2)}. \qquad (33.37)$$

But also, taking first $-\partial/\partial m_2^2$ and then putting $m_1 = m_2 = 0$, we obtain

$$I_{1,2}(p) \equiv \int \frac{d^d q}{(2\pi)^d} \frac{1}{q^2(q+p)^4} = \frac{\Gamma(2-d/2)(d/2-2)}{(4\pi)^{d/2}} p^{d-6} \int_0^1 d\alpha \alpha^{\frac{d}{2}-2}(1-\alpha)^{\frac{d}{2}-1}$$

$$= p^{d-6} \frac{\Gamma(2-d/2)(d/2-2)\Gamma(d/2-1)\Gamma(d/2)}{(4\pi)^{d/2}\Gamma(d-1)}$$

$$= p^{d-6} \frac{\Gamma(2-d/2)(d/2-2)[\Gamma(d/2)]^2}{(4\pi)^{d/2}(d/2-1)\Gamma(d-1)}. \qquad (33.38)$$

More generally, taking $(\partial^2/\partial m_2^2)^n$ and then putting $m_1 = m_2 = 0$, we obtain

$$I_{1,1+n}(p) \equiv \int \frac{d^d q}{(2\pi)^d} \frac{1}{q^2(p+q)^{2(n+1)}}$$

$$= \frac{\Gamma(n+2-d/2)p^{d-4-2n}}{(4\pi)^{d/2}\Gamma(n)} \int_0^1 d\alpha \alpha^{\frac{d}{2}-2-n}(1-\alpha)^{\frac{d}{2}-2}$$

$$= \frac{p^{d-4-2n}}{(4\pi)^{d/2}} \frac{\Gamma(n+2-d/2)\Gamma(d/2-1-n)\Gamma(d/2-1)}{\Gamma(n+1)\Gamma(d-2-n)}. \qquad (33.39)$$

But again this formula is valid at arbitrary n, since the Euler gamma function allows for an analytical continuation at arbitrary n. This formula can be used in various loop integrals for massless particles.

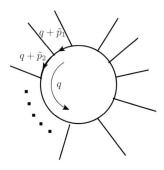

General one-loop Feynman diagram.

33.5.2 General One-Loop Integrals and Feynman Parametrization

The Feynman parametrization can in fact be generalized, to write instead of a product of propagators, a single "propagator" raised at the corresponding power, with integrals over parameters α left to do. Then we can use the formula (33.30) to calculate the momentum integral, and we are left with the integration over parameters.

Consider the general one-loop integral from Figure 33.2, obtained from a loop with n external lines coming out of it, with momenta $p_1, \ldots, p_n$. The momenta on the internal lines can be chosen to be $q + \tilde{p}_i$, with $p_i = \sum_{j=1}^{i} p_i$, for instance, but in any case we can always write them as $q + \tilde{p}_i$, with $\tilde{p}_i$ depending on the external momenta p_i. The general one-loop scalar integral is then

$$I(p_1, \ldots, p_n) = \int \frac{d^D q}{(2\pi)^D} \prod_{i=1}^{n} \frac{1}{(q + \tilde{p}_i)^2 + m^2}. \tag{33.40}$$

We first write the product of propagators in an exponential as before, as

$$\frac{1}{\Delta_1 \ldots \Delta_n} = \int_0^{\infty} \prod_{i=1}^{n} d\tilde{\alpha}_i e^{-(\tilde{\alpha}_1 \Delta_1 + \ldots + \tilde{\alpha}_n \Delta_n)}. \tag{33.41}$$

After a transformation $\tilde{\alpha}_i = \alpha_i t$, which needs to be supplemented with the constraint $\sum_i \alpha_i = 1$ in order to still have n independent integration variables, which also means that we must have $\alpha_i \leq 1$, and whose Jacobian we can easily check is $J = t^{n-1}$ (the determinant has $n - 1$ ts on the diagonal and the rest zeroes, except on the last line), the product of propagators becomes

$$\int_0^1 \prod_{i=1}^{n} d\alpha_i \delta \left(1 - \sum_i \alpha_i\right) \int_0^{\infty} dt\, t^{n-1} e^{-t(\alpha_1 \Delta_1 + \ldots + \alpha_n \Delta_n)}$$

$$= \int_0^1 \prod_{i=1}^{n} d\alpha_i \delta \left(1 - \sum_i \alpha_i\right) \frac{\Gamma(n)}{(\alpha_1 \Delta_1 + \ldots + \alpha_n \Delta_n)^n}. \tag{33.42}$$

Now we have replaced the product of different propagators with a single quadratic expression in momenta, raised to the nth power, so we can use the formula (33.30) to calculate the one-loop integral, which becomes

$$I(p_1,\ldots,p_n) = \int \prod_{i=1}^{n} d\alpha_i \delta\left(1 - \sum_i \alpha_i\right) \int \frac{d^D q}{(2\pi)^D} \frac{\Gamma(n)}{[\sum_{i=1}^{n} \alpha_i(q + \tilde{p}_i)^2 + m^2]^n}. \quad (33.43)$$

But there is still one step to do, namely to shift the momenta in order to get rid of the term linear in momenta. Define $\tilde{q}^\mu = q^\mu + \sum_i \alpha_i \tilde{p}^i$, after which the quadratic form in the denominator becomes

$$\sum_{i=1}^{n}[\alpha_i(q + \tilde{p}_i)^2 + m^2] = \tilde{q}^2 + m^2 + \sum_i \alpha_i \tilde{p}_i^2 - \left(\sum_i \alpha_i \tilde{p}_i\right)^2. \quad (33.44)$$

Since the Jacobian of the shift is 1, we finally obtain (using (33.30)) for the Feynman parametrization of the general one-loop integral

$$I(p_1,\ldots,p_n) = \frac{\Gamma(n - D/2)}{(4\pi)^{D/2}} \int_0^1 \prod_{i=1}^{n} d\alpha_i \delta\left(1 - \sum_i \alpha_i\right)$$
$$\left[m^2 + \sum_i \alpha_i \tilde{p}_i^2 - \left(\sum_i \alpha_i \tilde{p}_i\right)^2\right]^{D/2-n}. \quad (33.45)$$

The α_i are called Feynman parameters. We note that the dimensional regularization procedure described above is general, and we see that the divergence always appears as the simple pole of the gamma function at $D = 4$.

33.5.3 Alternative Version of the Feynman Parametrization

In the literature, an alternative parametrization is also sometimes used, which can be found directly by making a change of variables in (33.42), but we find it useful to start at the beginning. In (33.41), we substitute the change of variables $\tilde{\alpha}_1 = t(1 - \alpha_1)$, $\tilde{\alpha}_2 = t(\alpha_1 - \alpha_2)$, $\tilde{\alpha}_3 = t(\alpha_2 - \alpha_3)$, ..., $\tilde{\alpha}_n = t\alpha_n$. From these definitions, we see that we must have $\alpha_1 \in [0, 1]$, $\alpha_2 \in [0, \alpha_1]$, $\alpha_3 \in [0, \alpha_2]$, ... The Jacobian of the transformation is found again to be $J = t^{n-1}$, so that we get

$$\frac{1}{\Delta_1 \ldots \Delta_n} = \int_0^1 d\alpha_1 \int_0^{\alpha_1} d\alpha_2 \int_0^{\alpha_2} d\alpha_3 \ldots \int_0^{\alpha_{n-2}} d\alpha_{n-1} \int_0^\infty dt$$
$$t^{n-1} e^{-t((1-\alpha_1)\Delta_1 + (\alpha_2-\alpha_1)\Delta_2 + \ldots + (\alpha_{n-2}-\alpha_{n-1})\Delta_{n-1} + \alpha_{n-1}\Delta_n)}. \quad (33.46)$$

Finally, we have for the product of propagators

$$\frac{1}{\Delta_1 \ldots \Delta_n} = \Gamma(n) \int_0^1 d\alpha_1 \int_0^{\alpha_1} d\alpha_2 \int_0^{\alpha_2} d\alpha_3 \ldots \int_0^{\alpha_{n-2}} d\alpha_{n-1}$$
$$[\Delta_1(1 - \alpha_1) + \Delta_2(\alpha_1 - \alpha_2) + \ldots + \Delta_{n-1}(\alpha_{n-2} - \alpha_{n-1}) + \Delta_n \alpha_{n-1}]^{-n}. \quad (33.47)$$

33.6 Dimensionally Continuing Lagrangians

So we see that the loop integrals of the above type are fine to continue dimensionally, but is it fine to continue dimensionally the Lagrangians?

For scalars it is fine, but we have to be careful. The Lagrangian is

$$\mathcal{L} = \frac{1}{2}(\partial_\mu \phi)^2 + \frac{m^2}{2}\phi^2 + \frac{\lambda_n}{n!}\phi^n \qquad (33.48)$$

and since the action $S = \int d^D x$ must be dimensionless, the dimension of the scalar is $[\phi] = (D-2)/2$ and thus the dimension of the coupling is $[\lambda_n] = D - n(D-2)/2$. For instance, for $D = 4$ and $n = 4$, we have $[\lambda_4] = 0$, but for $D = 4 + \epsilon$, we have $[\lambda_4] = -\epsilon$. This means that outside $D = 4$, we must redefine the coupling with a factor μ^ϵ, where μ is some scale that appears dynamically. This process of spontaneous appearance of an arbitrary mass scale at the quantum level is called *dynamical (dimensional) transmutation*, and is related to the fact that we have a renormalization group, as we will see later on in the book.

For higher spins, however, we must be more careful. The number of components of a field depends on the dimension, which is a subtle issue. We must then use dimensional continuation of various gamma matrix formulas:

$$g^{\mu\nu}g_{\mu\nu} = D,$$
$$\gamma_\mu \not{p}\gamma^\mu = (2-D)\not{p}, \qquad (33.49)$$

and so on. In contrast, the gamma matrices still satisfy the Clifford algebra $\{\gamma^\mu, \gamma^\nu\} = 2g^{\mu\nu}$. But the dimension of the (spinor) representation of the Clifford algebra depends on dimension in an unusual way, $n = 2^{[D/2]}$, which means it is two-dimensional in $D = 2, 3$ and four-dimensional in $D = 4, 5$. This means that we cannot continue dimensionally n to $D = 4 + \epsilon$. Instead, we must still consider the gamma matrices as 4×4 even in $D = 4 + \epsilon$, and thus we still have

$$\text{Tr}[\gamma_\mu \gamma_\nu] = 4g_{\mu\nu}. \qquad (33.50)$$

This is not a problem, however there is another fact that is still a problem. The definition of γ_5 is

$$\gamma_5 = \frac{i}{4!}\epsilon_{\mu\nu\rho\sigma}\gamma^\mu \gamma^\nu \gamma^\rho \gamma^\sigma = -i\gamma^0 \gamma^1 \gamma^2 \gamma^3 = i\gamma_0 \gamma_1 \gamma_2 \gamma_3 \qquad (33.51)$$

and that cannot easily be continued dimensionally. Since chiral fermions (i.e. fermions that are eigenvalues of the chiral projectors $P_{L,R} = (1 \pm \gamma_5)/2$) appear in the Standard Model, we would need to be able to continue chiral fermions dimensionally. But that is very difficult to do.

Therefore we can say that there are no perfect regularization procedures. There is always something that does not work easily, but for particular cases we might prefer one or another.

In Chapter 34 we will see that the divergences we have regularized in this chapter can be absorbed in a redefinition of the parameters of the theory, leaving only a finite piece giving quantum corrections.

Important Concepts to Remember

- We must regularize the infinities appearing in loop integrals, and the infinite sums.
- Cut-off regularization, imposing upper and lower limits on $|p|$ in Euclidean space, regulates integrals, but is not used very much because it is related to breaking of Euclidean ("Lorentz") invariance, as well as breaking of gauge invariance.
- Often the difference in infinite sums is a physical observable, and then the result is regularization-dependent. In particular, we can have mode number cut-off (giving the sum operator as a common factor), or energy cut-off (giving a resulting energy integral as a common factor). We must choose one that is more physical.
- The choice of regularization scheme for integrals is dictated by what symmetries we want to preserve. If several respect the wanted symmetries, they are equally good.
- (Generalized) Pauli–Villars regularization removes the contribution of high-energy modes from the propagator, by subtracting a very massive particle from it. A related version is obtained from a term in the action which is a higher derivative.
- By taking derivatives with respect to a parameter (e.g. m^2), we obtain derivative regularization, which also reduces the degree of divergence of integrals.
- Dimensional regularization respects gauge invariance, and corresponds to analytically continuing the dimension, as $D = 4 + \epsilon$. This is based on the fact that we can continue $n!$ away from the integers to the Euler gamma function.
- In dimensional regularization, the divergences are the simple poles of the gamma function at $\Gamma(-n)$, and appear as a multiplicative $1/\epsilon$.
- With Feynman parametrization, we can reduce a general one-loop scalar integral to an integral over Feynman parameters only.
- For scalars, dimensional regularization of the action is fine, if we remember that couplings have extra mass dimensions away from $D = 4$. For higher spins, we must regulate the number of components, including things like $g^{\mu\nu} g_{\mu\nu}$ and gamma matrix identities.
- The dimension of gamma matrices away from $D = 4$ is still 4, so traces of gamma matrices still give a factor of 4, and the γ_5 cannot be continued away from $D = 4$, which means that analytical continuation in dimensions involving chiral fermions is very hard.

Further Reading

See section 5.3 in [3], section 4.3 in [4], and section 9.3 in [2].

Exercises

1. Calculate

$$\int \frac{d^4p}{(2\pi)^4} \frac{1}{p^2 + m_1^2} \frac{1}{p^2 + m_2^2} \tag{33.52}$$

   in Pauli–Villars regularization.

2. Calculate

$$\int \frac{d^D q}{(2\pi)^D} \frac{1}{(q^2 + m_1^2)^2} \frac{1}{(q+p)^2 + m_2^2} \tag{33.53}$$

   in dimensional regularization (it is divergent in $D = 6$).

3. Using the methods in the text, calculate

$$\int \frac{d^D q}{(2\pi)^D} \frac{1}{(q^2)^n} \frac{1}{[(q+p)^2]^m} \tag{33.54}$$

   in dimensional regularization for general real n, m.

4. Calculate

$$\int \frac{d^4 q}{(2\pi)^4} \frac{1}{q^2 + m_1^2} \frac{1}{(q+p)^2 + m_2^2} \tag{33.55}$$

   in derivative regularization.

34

One-Loop Renormalization for Scalars and Counterterms in Dimensional Regularization

In this chapter we will learn how to get rid of the infinities appearing in quantum field theory loops, by a process called renormalization. Basically, we will absorb the infinities in a redefinition of the parameters in the theory. We will do this in dimensional regularization, the most commonly used method, though of course it can be done in any regularization we want.

We start with the case of one-loop corrections in scalar theories, and specifically we consider ϕ^4 theory in $d = 4$ and ϕ^3 theory in $d = 6$. Of course, ϕ^3 theory is not well defined, since the potential is unbounded from below (it can have arbitrarily high negative value). But we will take it as a simple example, since the standard theory, ϕ^4 in $d = 4$, has no "wavefunction renormalization" at one loop (we will see what that is later), but ϕ^3 in $d = 6$ is the simplest example of a theory that does have it.

Note that both theories are in the critical dimension (meaning that for a dimension higher than the one considered, the theory would be nonrenormalizable). We can check this, since we saw that $\omega(D) = d - (d - 2)E/2 - \sum_v [\lambda_v]$ and $[\lambda_n] = d - (d - 2)n/2$. This means that for ϕ^4 in $d = 4$ we have $[\lambda_4] = 0$ and $\omega(D) = 4 - E$, and for ϕ^3 in $d = 6$ we have $[\lambda_3] = 0$ and $\omega(D) = 6 - 2E$.

As we saw in Chapter 33, in dimensional regularization we need to introduce a scale μ and redefine the coupling $\tilde{\lambda}$ in dimension d in terms of the dimensionless coupling λ in the critical dimension by

$$\tilde{\lambda} = \mu^{d - \frac{(d-2)}{2}n}\lambda \equiv \lambda_d. \tag{34.1}$$

The parameter μ is a manifestation of dimensional transmutation, the spontaneous breaking of scale invariance at the quantum level, manifest through the appearance of the arbitrary scale μ. Note that in any regularization we will find such an arbitrary scale: in the cut-off regularization it is the cut-off Λ itself, in Pauli–Villars again it is the mass parameter Λ, in the PV-like higher derivative regularization again we have a parameter Λ, and so on.

The statement is that the physics should be independent of this scale, and this will lead to the renormalization group equations, which will be presented in Chapter 35.

We now proceed to find the divergent diagrams at one loop and calculate the divergent parts. After that, we will learn how to get rid of them by a redefinition of parameters.

34.1 Divergent Diagrams in ϕ^4 Theory in $D = 4$ and its Divergences

From the above formula $\omega(D) = 4 - E$, we see that the $E = 0, 1, 2, 3, 4$-point functions are all superficially divergent. But we consider 1PI diagrams, so the $E = 0$ case (partition function) is not physical, since it just gives the normalization which cancels out in calculations. Since we are considering 1PI diagrams, at one loop in ϕ^4 theory we can convince ourselves that $\Gamma^{(1)}_{\text{one loop}} = \Gamma^{(3)}_{\text{one loop}} = 0$. So we are left with $\Gamma^{(2)}$ and $\Gamma^{(4)}$ to calculate.

We define $\epsilon = 4 - D$ ($D = 4 - \epsilon$), such that $\tilde{\lambda} = \mu^\epsilon \lambda$.

For the 1PI two-point function, the one-loop contribution (in Figure 34.1) is given by just one vertex with a loop on it, with the result

$$\delta\Gamma^{(2)}(p) = -\frac{\tilde{\lambda}}{2} \int \frac{d^D p}{(2\pi)^D} \frac{1}{p^2 + m^2}, \tag{34.2}$$

where we have used the fact that we have a symmetry factor $S = 2$. Using the formula (33.30) from Chapter 33, with $n = 1$, we obtain

$$\delta\Gamma^{(2)}(p) = -\frac{\tilde{\lambda}}{2} \frac{(m^2)^{\frac{D}{2}-1}}{(4\pi)^{\frac{D}{2}}} \Gamma\left(1 - \frac{D}{2}\right) = -\frac{\lambda m^2}{2(4\pi)^2} \Gamma\left(\frac{\epsilon}{2} - 1\right) \left(\frac{4\pi\mu^2}{m^2}\right)^{\frac{\epsilon}{2}}. \tag{34.3}$$

For the one-loop contribution to $\Gamma^{(4)}$ in Figure 34.2, we define the four momenta p_1, p_2, p_3, p_4 going in, and the Mandelstam variables $s = (p_1 + p_2)^2$, $t = (p_1 + p_4)^2 = (p_2 + p_3)^2$, $u = (p_1 + p_3)^2$, and we find three diagrams with two vertices and two external lines at each vertex. One diagram has $p_1 + p_4 \equiv \tilde{p}_1$ at one vertex (t diagram), the second has $p_1 + p_2$ at one vertex (s diagram), and the third has $p_1 + p_3$ at one vertex (u diagram), so that we have

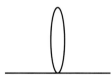

Figure 34.1   One-loop Feynman diagram for the 1PI two-point function in ϕ^4 theory.

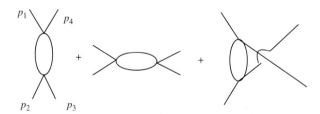

Figure 34.2   One-Loop Feynman diagram for the 1PI 4-point function in ϕ^4 theory.

$$\delta\Gamma^{(4)}(s,t,u) = \frac{\tilde{\lambda}^2}{2}[I(t) + I(s) + I(u)]. \tag{34.4}$$

Then

$$I(t) = \int \frac{d^D q}{(2\pi)^D} \frac{1}{(q^2 + m^2)((q + \tilde{p}_1)^2 + m^2)}. \tag{34.5}$$

Using the formula (33.45) from Chapter 33, for $n = 2$ and $\tilde{p}_2 = 0$, we obtain

$$\frac{\tilde{\lambda}^2}{2} I(t) = \frac{\tilde{\lambda}^2}{2} \frac{\Gamma\left(2 - \frac{D}{2}\right)}{(4\pi)^{\frac{D}{2}}} \int_0^1 d\alpha_1 d\alpha_2 \delta(1 - \alpha_1 - \alpha_2)[m^2 + \alpha_1 \tilde{p}_1^2 - \alpha_1^2 \tilde{p}_1^2]^{\frac{D}{2}-2}$$

$$= \frac{\lambda^2 \mu^{2\epsilon}}{2} \frac{\Gamma\left(\frac{\epsilon}{2}\right)}{(4\pi)^2 (4\pi)^{\frac{\epsilon}{2}}} \int_0^1 d\alpha [m^2 + \alpha(1 - \alpha)t]^{-\frac{\epsilon}{2}}. \tag{34.6}$$

Summing over s, t, u, we obtain

$$\delta\Gamma^{(4)}(s,t,u) = \frac{\lambda^2 \mu^\epsilon}{2(4\pi)^2} \Gamma\left(\frac{\epsilon}{2}\right) \left(\frac{4\pi\mu^2}{m^2}\right)^{\frac{\epsilon}{2}} \int_0^1 d\alpha \sum_{z=s,t,u} \left[1 + \alpha(1 - \alpha)\frac{z}{m^2}\right]^{-\frac{\epsilon}{2}}. \tag{34.7}$$

34.1.1 Divergent Parts

Since we are interested in getting rid of the divergent parts, we now isolate the divergent parts of $\delta\Gamma^{(2)}(p)$ and $\delta\Gamma^{(4)}(s,t,u)$.

For the two-point function, we first use the definition of the ψ function, $\psi(z) = d\Gamma/dz$, to write $\Gamma(1 + \epsilon) \simeq 1 + \epsilon\psi(1)$. Then we also have

$$\Gamma(-1 + \epsilon) = \frac{\Gamma(\epsilon)}{-1 + \epsilon} = \frac{\Gamma(1 + \epsilon)}{-\epsilon(1 - \epsilon)} \simeq -\frac{1 + \epsilon\psi(1)}{\epsilon(1 - \epsilon)} \simeq -\frac{1}{\epsilon}[1 + \epsilon(1 + \psi(1))]$$

$$= -\frac{1}{\epsilon}[1 + \epsilon\psi(2)], \tag{34.8}$$

where we have used $\psi(n + 1) = 1/n + \psi(n)$ to write $\psi(2) = 1 + \psi(1)$.

Then we have

$$\delta\Gamma^{(2)}(p) \simeq -\frac{\lambda m^2}{2(4\pi)^2} \left(-\frac{2}{\epsilon} - \psi(2)\right) \left(1 + \frac{\epsilon}{2} \ln\frac{4\pi\mu^2}{m^2}\right) + \mathcal{O}(\epsilon)$$

$$\simeq \frac{\lambda m^2}{(4\pi)^2} \left[\frac{1}{\epsilon} + \frac{\psi(2)}{2} - \frac{1}{2} \ln\frac{m^2}{4\pi\mu^2} + \mathcal{O}(\epsilon)\right]. \tag{34.9}$$

For the four-point function, we use

$$\Gamma(\epsilon) = \frac{\Gamma(1 + \epsilon)}{\epsilon} \simeq \frac{1}{\epsilon} + \psi(1) \tag{34.10}$$

to write the expansion

$$\delta\Gamma^{(4)}(s,t,u) \simeq \frac{\lambda^2 \mu^\epsilon}{2(4\pi)^2}\left(\frac{2}{\epsilon}+\psi(1)\right)\left(1+\frac{\epsilon}{2}\ln\frac{4\pi\mu^2}{m^2}\right)$$

$$\times\left(3-\frac{\epsilon}{2}\int_0^1 d\alpha\sum_{z=s,t,u}\ln\left(1+\alpha(1-\alpha)\frac{z}{m^2}\right)\right)$$

$$\simeq \frac{\lambda^2 \mu^\epsilon}{(4\pi)^2}\left(\frac{3}{\epsilon}+\frac{3\psi(1)}{2}-\frac{3}{2}\ln\frac{m^2}{4\pi\mu^2}-\frac{1}{2}\int_0^1 d\alpha\right.$$

$$\left.\sum_{z=s,t,u}\ln\left[1+\alpha(1-\alpha)\frac{z}{m^2}\right]+\mathcal{O}(\epsilon)\right).$$

(34.11)

Note that we can do the integral over α (even though here we are not interested in the finite parts):

$$\int_0^1 d\alpha\ln\left[1+\alpha(1-\alpha)\frac{z}{m^2}\right]=-2+\sqrt{1+\frac{4m^2}{z}}\ln\left(\frac{\sqrt{1+\frac{4m^2}{z}}+1}{\sqrt{1+\frac{4m^2}{z}}-1}\right).$$
(34.12)

34.2 Divergent Diagrams in ϕ^3 Theory in $D=6$ and its Divergences

We now repeat the same procedure for ϕ^3 in $d=6$. Again we want to write $\tilde{\lambda}=\mu^\epsilon\lambda$, which means that now we must take $d=6-2\epsilon$. From the formula $\omega(D)=6-2E$, we see that the $E=0,1,2,3$-point functions are superficially divergent, but $\Gamma^{(0)}$ is a normalization, and actually vanishes at one loop, and $\Gamma^{(1)}$ is trivial (it gives no term in the effective action, just a constant shift of the scalar). Therefore we have only $\Gamma^{(2)}$ and $\Gamma^{(3)}$.

For the two-point function, there is only one one-loop diagram (Figure 34.3) with two vertices connected by two propagators, each having an external line. The momenta on the two internal propagators are named q and $q+p$ (p being the external momentum), and the symmetry factor is $S=2$, giving

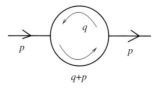

Figure 34.3   One-loop Feynman diagram for the 1PI two-point function in ϕ^3 theory.

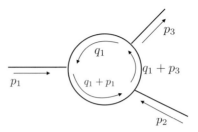

One-loop Feynman diagram for the 1PI three-point function in ϕ^3 theory.

$$\delta\Gamma^{(2)}(p) = \frac{\tilde{\lambda}^2}{2} \int \frac{d^D q}{(2\pi)^D} \frac{1}{(q^2 + m^2)((q+p)^2 + m^2)}. \tag{34.13}$$

We see that it is formally the same as the $I(t)$ integral from the ϕ^4 case, with $\tilde{p}_1$ replaced by p, so we can use that result to write

$$\delta\Gamma^{(2)}(p) = \frac{\tilde{\lambda}^2}{2} \frac{\Gamma\left(2 - \frac{D}{2}\right)}{(4\pi)^{\frac{D}{2}}} \int_0^1 d\alpha [m^2 + \alpha(1-\alpha)p^2]^{\frac{D}{2}-2}. \tag{34.14}$$

Substituting $D = 6 - 2\epsilon$, we now obtain

$$\delta\Gamma^{(2)}(p) = \frac{\lambda^2 m^2 \mu^\epsilon}{2(\pi)^3} \frac{\Gamma(-1+\epsilon)}{(4\pi)^{-\epsilon}} m^{-2\epsilon} \int_0^1 d\alpha \left[1 + \alpha(1-\alpha)\frac{p^2}{m^2}\right]^{1-\epsilon}$$

$$= \frac{\lambda^2 m^2}{2(4\pi)^3} \Gamma(-1+\epsilon) \left(\frac{4\pi\mu^2}{m^2}\right)^\epsilon \int_0^1 \left[1 + \alpha(1-\alpha)\frac{p^2}{m^2}\right]^{1-\epsilon}. \tag{34.15}$$

For the three-point function, there is also a single one-loop diagram (Figure 34.4) with three vertices connected pairwise by internal propagators. We denote the external lines by p_1, p_2 going in and p_3 coming out, and the internal lines by $q, q+p_1$, and $q+p_3$. Then we have

$$\delta\Gamma^{(3)}(p_i) = -\tilde{\lambda}^3 \int \frac{d^D q}{(2\pi)^D} \frac{1}{(q^2 + m^2)((q+p_1)^2 + m^2)((q+p_3)^2 + m^2)}. \tag{34.16}$$

We can use the result (33.45) from Chapter 33, for $n = 3$ and $\tilde{p}_1 = p_1, \tilde{p}_3 = p_3, \tilde{p}_2 = 0$ to write

$$\delta\Gamma^{(3)}(p_i) = -\tilde{\lambda}^3 \frac{\Gamma\left(3 - \frac{D}{2}\right)}{(4\pi)^{\frac{D}{2}}} \int_0^1 d\alpha_1 d\alpha_2 d\alpha_3 \delta(1 - \alpha_1 - \alpha_2 - \alpha_3)$$

$$\times [m^2 + \alpha_1 p_1^2 + \alpha_3 p_3^2 - (\alpha_1 p_1 + \alpha_3 p_3)^2]^{\frac{D}{2}-3}$$

$$= -\frac{\lambda^3 \mu^\epsilon}{(4\pi)^3} \Gamma(\epsilon) \left(\frac{4\pi\mu^2}{m^2}\right)^\epsilon \int_0^1 d\alpha_1 d\alpha_3$$

$$\left[1 + \alpha_1 \frac{p_1^2}{m^2} + \alpha_3 \frac{p_3^2}{m^2} - \left(\frac{\alpha_1 p_1 + \alpha_3 p_3}{m}\right)^2\right]^{-\epsilon}. \tag{34.17}$$

34.2.1 Divergent Parts

We now isolate the divergent parts of $\delta\Gamma^{(2)}(p)$ and $\delta\Gamma^{(3)}(p_i)$.

For the two-point function we obtain

$$
\begin{aligned}
\delta\Gamma^{(2)}(p) \simeq & \frac{\lambda^2 m^2}{2(4\pi)^3}\left(-\frac{1}{\epsilon}\right)(1+\epsilon\psi(2))\left(1-\epsilon\ln\frac{m^2}{4\pi\mu^2}\right)\left[1+\frac{p^2}{6m^2}\right. \\
& \left.-\epsilon\int_0^1 d\alpha\left[1+\alpha(1-\alpha)\frac{p^2}{m^2}\right]\ln\left[1+\alpha(1-\alpha)\frac{p^2}{m^2}\right]\right]+\mathcal{O}(\epsilon) \\
= & -\frac{\lambda^2 m^2}{2(4\pi)^3}\left(\frac{1}{\epsilon}\left(1+\frac{p^2}{6m^2}\right)+\left(1+\frac{p^2}{6m^2}\right)\left(\psi(2)-\ln\frac{m^2}{4\pi\mu^2}\right)\right. \\
& \left.-\int_0^1 d\alpha\left[1+\alpha(1-\alpha)\frac{p^2}{m^2}\right]\ln\left[1+\alpha(1-\alpha)\frac{p^2}{m^2}\right]\right)+\mathcal{O}(\epsilon),\quad(34.18)
\end{aligned}
$$

where we have used

$$
\int_0^1 d\alpha\left[1+\alpha(1-\alpha)\frac{p^2}{m^2}\right]=1+\frac{p^2}{6m^2}.\qquad(34.19)
$$

For the three-point function we obtain

$$
\begin{aligned}
\delta\Gamma^{(3)}(p_i) \simeq & -\frac{\lambda^3\mu^\epsilon}{(4\pi)^3}\frac{1}{\epsilon}(1+\epsilon\psi(1))\left(1-\epsilon\ln\frac{m^2}{4\pi\mu^2}\right)\left[1-\epsilon\int_0^1 d\alpha_1 d\alpha_3\right. \\
& \left.\ln\left[1+\alpha_1\frac{p_1^2}{m^2}+\alpha_3\frac{p_3^2}{m^2}-\left(\frac{\alpha_1 p_1+\alpha_3 p_3}{m}\right)^2\right]\right]+\mathcal{O}(\epsilon) \\
\simeq & -\frac{\lambda^3\mu^\epsilon}{(4\pi)^3}\left(\frac{1}{\epsilon}+\psi(1)-\ln\frac{m^2}{4\pi\mu^2}-\int_0^1 d\alpha_1 d\alpha_3\right. \\
& \left.\ln\left[1+\alpha_1\frac{p_1^2}{m^2}+\alpha_3\frac{p_3^2}{m^2}-\left(\frac{\alpha_1 p_1+\alpha_3 p_3}{m}\right)^2\right]\right)+\mathcal{O}(\epsilon).\quad(34.20)
\end{aligned}
$$

34.3 Counterterms in ϕ^4 and ϕ^3 Theories

We are now finally in a position to show how to get rid of the infinities calculated above. As we can check in the two examples above, in the case of renormalizable field theories, we can hide the infinities in the redefinition of the parameters of the theory. Indeed, we see that the number and type of divergences match the type of terms in the Lagrangian. In the ϕ^4 theory in $d=4$ we have infinities in $\Gamma^{(2)}$ and $\Gamma^{(4)}$, which are of the type of the propagators (quadratic in fields) and the interaction (ϕ^4 in fields), and in the ϕ^3 theory in $d=6$ we have infinities in $\Gamma^{(2)}$ and $\Gamma^{(3)}$, of the type of the propagators (quadratic in fields) and the interaction (ϕ^3 in fields).

Therefore we want to consider redefining the parameters by infinite factors, considering that the original parameters were also infinite, such that the physical, redefined parameters are finite quantities.

In a nonrenormalizable theory, it would happen that we have an infinite number of divergent terms, with new one appearing at each loop level. These can be canceled only by adding new terms to the original Lagrangian. In some sense, this means that for nonrenormalizable theories we would need to start with a Lagrangian with an infinite number of terms, most of them (except for a finite number) having zero coefficients, and then hide the infinities in the redefinition of all the coefficients (even the ones that are zero).

The procedure to get rid of infinities is then to add to the original Lagrangian new infinite terms called *counterterms*, giving *new infinite vertices*, such that the resulting amplitudes calculated with the total Lagrangian are finite.

One redefines the masses m_i, couplings λ_i, and wavefunctions for the fields ϕ_i.

ϕ^4 **in** $d = 4$. As we argued at the beginning of the chapter, for ϕ^4 theory in $d = 4$ we have no wavefunction renormalization, so only m and λ will be redefined. The counterterm Lagrangian to be added to the original one is such that the vertices coming from it cancel the divergences of $\Gamma^{(2)}$ and $\Gamma^{(4)}$ we found. This means that we must take the counterterm Lagrangian

$$\mathcal{L}_{\text{c.t.}} = \left[\frac{\lambda}{16\pi^2}\frac{1}{\epsilon}\right]\frac{1}{2}m^2\phi^2 + \left[\frac{\lambda^2}{16\pi^2}\frac{3}{\epsilon}\right]\mu^\epsilon\frac{\phi^4}{4!}. \tag{34.21}$$

Indeed, then we get two new vertices, one a two-point vertex, denoted by a line with a cross on it, as in Figure 34.5(a), with value (the vertex is minus the coupling in Euclidean space)

$$-\frac{\lambda}{16\pi^2}\frac{1}{\epsilon}m^2 \tag{34.22}$$

and one a four-point vertex, as in Figure 34.5(b), denoted by a four-point vertex with a circle on it (to distinguish it from the classical vertex), with value

$$-\mu^\epsilon\frac{\lambda^2}{16\pi^2}\frac{3}{\epsilon}. \tag{34.23}$$

Note that to identify $\mathcal{L}_{\text{c.t.}}$ above, we have considered the fact that a term $1/2[(\partial_\mu\phi)^2 + m^2\phi^2]$ in x space leads to a term $1/2\phi(p)[p^2 + m^2]\phi(-p)$ in p space (i.e. a "two-point vertex" $-(p^2 + m^2)$). The divergent 1PI two-point function was p-independent, depending only on m, hence there is no redefinition of the kinetic piece $(\partial_\mu\phi)^2/2$, which means no wavefunction renormalization (redefinition).

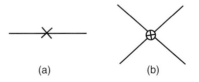

(a)   (b)

Figure 34.5   One-loop counterterm vertices in ϕ^4 theory: (a) two-point vertex; (b) four-point vertex.

Also note that the 1PI n-point functions include the classical vertices, so in the redefined theory, $\mathcal{L} + \mathcal{L}_{\text{c.t.}}$, we take both the loops of $\mathcal{L}$ and the classical vertices of $\mathcal{L}_{\text{c.t.}}$.

Then we check that we cancel the divergences coming from $\Gamma^{(2)}_{\text{one loop}}$ and $\Gamma^{(4)}_{\text{one loop}}$ with these new vertices (the "classical" tree diagram from $\mathcal{L}_{\text{c.t.}}$ and the one-loop diagram from the original Lagrangian are of the same order in λ, and the infinite parts cancel). Note that we have chosen $\mathcal{L}_{\text{c.t.}}$ to cancel just the infinite part of the one-loop diagrams. This is called minimal subtraction, and will be discussed in Chapter 35 in more detail, but note that it is not necessary to use this; we can also consider more general subtractions, where the counterterms also contain some finite parts.

ϕ^3 in $d = 6$. In the ϕ^3 theory in $d = 6$, there is a term with p^2 in $\Gamma^{(2)}(p)$, as well as a term with m^2, so now we have the counterterm Lagrangian

$$\mathcal{L}_{\text{c.t.}} = -\left[\frac{\lambda^2}{12(4\pi)^3}\frac{1}{\epsilon}\right]\frac{1}{2}(\partial_\mu\phi)^2 - \left[\frac{\lambda^2}{2(4\pi)^3}\frac{1}{\epsilon}\right]\frac{1}{2}m^2\phi^2 - \left[\frac{\lambda^3}{(4\pi)^3}\frac{1}{\epsilon}\right]\mu^\epsilon\frac{\phi^3}{3!}. \quad (34.24)$$

Note the sign difference with respect to ϕ^4 in $d = 4$. Now the new vertices coming from this Lagrangian are:

- A two-point vertex denoted by a line with a cross on it, with value

$$\frac{\lambda^3}{2(4\pi)^3}\frac{1}{\epsilon}\left(\frac{p^2}{6} + m^2\right). \quad (34.25)$$

- A three-point vertex denoted by a three-point vertex with a circle on it, with value

$$+ \mu^\epsilon\frac{\lambda^3}{(4\pi)^3}\frac{1}{\epsilon}. \quad (34.26)$$

Therefore the divergences also now cancel between the tree diagrams coming from $\mathcal{L}_{\text{c.t.}}$ and the divergent parts of the loop diagrams.

34.4 Renormalization

The general structure of $\mathcal{L}$ and $\mathcal{L}_{\text{c.t.}}$ is as follows:

$$\mathcal{L} = \frac{1}{2}(\partial_\mu\phi)^2 + \frac{1}{2}m^2\phi^2 + \frac{\mu^\epsilon\lambda}{3!}\phi^3,$$
$$\mathcal{L}_{\text{c.t.}} = C\frac{1}{2}(\partial_\mu\phi)^2 + B\frac{1}{2}m^2\phi^2 + A\frac{\mu^\epsilon\lambda}{3!}\phi^3. \quad (34.27)$$

We now define the *renormalized Lagrangian*

$$\mathcal{L}_{\text{ren}} = \mathcal{L} + \mathcal{L}_{\text{c.t.}} = \mathcal{L}_{\text{ren}}(\phi, m, \lambda\mu^\epsilon). \quad (34.28)$$

Here, ϕ, m, λ are finite quantities as $\epsilon \to 0$.

But now we can numerically identify this with a *bare Lagrangian* written in terms of *bare quantities* ϕ_0, m_0, λ_0, which are infinite:

$$\mathcal{L}_{\text{ren}}(\phi, m, \lambda\mu^\epsilon) = \mathcal{L}_{\text{bare}}(\phi_0, m_0, \lambda_0) = \frac{1}{2}(\partial_\mu \phi_0)^2 + \frac{1}{2}m_0^2\phi_0^2 + \frac{\lambda_0}{3!}\phi_0^3, \qquad (34.29)$$

which therefore looks like the classical Lagrangian as a function of ϕ_0, m_0, λ_0.

Sometimes one says that the renormalized Lagrangian is written in terms of bare (infinite) quantities, but that is not very satisfactory. A better interpretation is that the classical, bare Lagrangian, written in terms of infinite quantities, is reinterpreted as a renormalized Lagrangian, written as a sum of a finite, physical Lagrangian, plus a counterterm part that contains the infinities.

By comparison, we find that the relation between the bare and the renormalized quantities is given by

$$\phi_0 = \sqrt{1 + C}\phi \equiv \sqrt{Z_\phi}\phi,$$

$$m_0^2 = \frac{m^2(1 + B)}{Z_\phi},$$

$$\lambda_0 = \mu^\epsilon \lambda (1 + A) Z_\phi^{n/2}, \qquad (34.30)$$

where in the last line we wrote the coupling for a general $\mu^\epsilon \lambda \phi^n / n!$ case.

We see that the results we have found at one loop can be written as

$$\lambda_0 = \mu^\epsilon \left(\lambda + \frac{a_1(\lambda)}{\epsilon} \right),$$

$$m_0^2 = m^2 \left(1 + \frac{b_1(\lambda)}{\epsilon} \right),$$

$$Z_\phi = 1 + \frac{c_1(\lambda)}{\epsilon} = 1 + C. \qquad (34.31)$$

Of course, in the examples of ϕ^3 in $d = 6$ and ϕ^4 in $d = 4$, the $a_1(\lambda), b_1(\lambda), c_1(\lambda)$ are some specific powers of λ, which we will determine shortly, but in the general case of some $\lambda_n \phi^n / n!$ interactions, we will obtain nontrivial functions of the couplings.

In general, at general loops, we will find all higher-order poles in ϵ as well (increasing order in ϵ for increasing loop number), so that the general expansion takes the form

$$\lambda_0 = \mu^\epsilon \left(\lambda + \sum_{k=1}^\infty \frac{a_k(\lambda)}{\epsilon^k} \right),$$

$$m_0^2 = m^2 \left(1 + \sum_{k=1}^\infty \frac{b_k(\lambda)}{\epsilon^k} \right),$$

$$Z_\phi = 1 + \sum_{k=1}^\infty \frac{c_k(\lambda)}{\epsilon^k} = 1 + C. \qquad (34.32)$$

34.4.1 Examples

We now compute the a_1, b_1, c_1 coefficients in the two cases considered here.

(1) ϕ^4 in $d = 4$. Since $C = 0$, it follows that $c_1(\lambda) = 0$ (i.e. no wavefunction renormalization). Then $\lambda_0 = \mu^\epsilon \lambda(1 + A)$ gives

$$\lambda_0 = \lambda + \frac{\lambda^2}{16\pi^2} \frac{3}{\epsilon} \Rightarrow a_1(\lambda) = \frac{3\lambda^2}{(4\pi)^2}. \tag{34.33}$$

Then also $m_0^2 = m^2(1 + B)$ gives

$$m^2 = m_0^2 \left(1 + \frac{\lambda}{16\pi^2} \frac{1}{\epsilon} \right) \Rightarrow b_1(\lambda) = \frac{\lambda}{(4\pi)^2}. \tag{34.34}$$

(2) ϕ^3 in $d = 6$. We now have a wavefunction renormalization

$$Z_\phi = 1 + C = 1 - \frac{\lambda^2}{12(4\pi)^3} \frac{1}{\epsilon} \Rightarrow c_1(\lambda) = -\frac{\lambda^2}{12(4\pi)^3}. \tag{34.35}$$

Then from

$$Z_\phi^{3/2} \lambda_0 = \mu^\epsilon \lambda(1 + A) = \mu^\epsilon \left(\lambda - \frac{\lambda^3}{(4\pi)^3} \frac{1}{\epsilon} \right), \tag{34.36}$$

we obtain by expanding $Z_\phi^{-3/2}$ in λ

$$a_1(\lambda) = -\frac{7}{8} \frac{\lambda^3}{(4\pi)^3}. \tag{34.37}$$

Similarly, from

$$m_0^2 Z_\phi = m^2 \left(1 - \frac{\lambda^2}{2(4\pi)^3} \frac{1}{\epsilon} \right) \tag{34.38}$$

we obtain by expanding Z_ϕ^{-1} in λ

$$b_1(\lambda) = -\frac{5\lambda^2}{12(4\pi)^3}. \tag{34.39}$$

Note that here we have the first example of a calculation that is common in quantum field theory, though it is not very well defined mathematically. Even though the result is divergent as $\epsilon \to 0$, since the divergent term is multiplied by λ, which is considered small (perturbation theory), we can expand in λ as if the term is small. This treatment of divergences is bizarre, and it is not clear why it works from a mathematical point of view (there is probably a better underlying theory that we don't know yet), but it always does, so we will continue to use it. Moreover, note that at higher loops the pole order $1/\epsilon^k$ increases, so it is even more divergent, but since it is multiplied by a higher power of λ, we consider it as an even smaller term. In fact, we treat quantum calculations order by order in λ, and at each order we make the theory finite by renormalization, and calculate finite quantities without worrying that there are worse infinities at higher orders, since we know that those can be fixed as well.

Important Concepts to Remember

- The appearance of the arbitrary scale μ signals the quantum breaking of scale invariance, and independence on it will lead to renormalization group equations. This is a characteristic of all regularizations.
- ϕ^4 theory in $d = 4$ has one-loop infinities in $\Gamma^{(2)}(p)$ and $\Gamma^{(4)}(s, t, u)$, and ϕ^3 theory in $d = 6$ has one-loop infinities in $\Gamma^{(2)}(p)$ and $\Gamma^{(3)}(p_i)$. Both theories are in the critical dimension.
- These divergences are of the same type as the terms in the Lagrangian, so can be absorbed in a redefinition of the parameters.
- To the original Lagrangian, we must add a counterterm Lagrangian, made up of the divergent pieces. These give new vertices, of higher order in λ.
- When adding to the one-loop diagrams the tree-level vertices of the counterterm Lagrangian of the same order in λ, the infinities cancel.
- $\mathcal{L} + \mathcal{L}_{\text{c.t.}}$ defines the renormalized Lagrangian, written in terms of finite parameters ϕ, m, λ.
- When reinterpreted as the classical bare Lagrangian, it is written in terms of infinite bare quantities.
- The relation between bare and renormalized quantities is given by an infinite series of poles in ϵ, multiplied by functions of λ. At one loop, we have only a single pole.
- Quantities divergent in ϵ, if multiplied by λ, are nevertheless considered small, and one can expand in them.

Further Reading

See sections 5.4 and 5.5 in [3] and section 10.2 in [1].

Exercises

1. Consider ϕ^3 theory in $D = 4$. Write down the divergent diagrams and calculate the divergences.
2. Write down the counterterms and renormalized Lagrangian for the case in Exercise 1.
3. Calculate the finite parts of $\delta\Gamma^{(2)}(p)$ and $\delta\Gamma^{(3)}(p_i)$ for ϕ^3 in $D = 6$ by expanding the integral formula obtained in the text in ϵ, and then doing the integral.
4. Repeat Exercise 3 for $\delta\Gamma^{(4)}(s, t, u)$ for ϕ^4 in $D = 4$.

35

Renormalization Conditions and the Renormalization Group

In Chapter 34 we saw that we need to remove infinities by renormalization, which means adding (infinite) counterterms to the Lagrangian written in terms of ϕ, m, λ. The total Lagrangian, the sum of the classical one and the counterterm one, when written in terms of ϕ, m, λ, called the renormalized quantities (finite), is the renormalized Lagrangian. Numerically it equals the bare Lagrangian, which formally looks the same as the classical one, but is written in terms of the bare quantities ϕ_0, m_0, λ_0, which are infinite.

The relations between the bare and renormalized quantities are written as

$$\lambda_0 = \lambda_0(\lambda, \mu, \epsilon); \quad m_0 = m_0(m, \lambda, \epsilon); \quad Z = Z(\lambda, \epsilon). \tag{35.1}$$

These can be understood as follows. We can consider λ_0, m_0, ϵ as fixed quantities, defining the theory. Then we obtain $\lambda = \lambda(\mu)$, and through it, $m = m(\lambda(\mu)) = m(\mu)$ and $Z = Z(\lambda(\mu)) = Z(\mu)$. Thus masses and couplings *run with the scale* μ. An alternative viewpoint is obtained if we fix λ and μ, obtaining $\lambda_0 = \lambda_0(\epsilon)$, $m_0 = m_0(\epsilon)$, and $Z = Z(\epsilon)$.

The first viewpoint, where physical quantities depend on scale, will lead to the renormalization group equations, which are equations for the scaling behavior of the n-point functions, and will be studied next in this chapter.

35.1 Renormalization of n-Point Functions

From the fact that renormalization is a redefinition of parameters, we obtain the following scaling for the connected n-point functions:

$$G^{(n)}(p_1, \ldots, p_n; m\, \lambda, \mu, \epsilon) = Z_\phi^{-n/2} G_0^{(n)}(p_1, \ldots, p_n; m_0, \lambda_0, \epsilon). \tag{35.2}$$

Indeed, the connected n-point functions $G^{(n)}$ are obtained from the renormalized Lagrangian with sources $\mathcal{L}^{\text{ren}} - J \cdot \phi$, obtaining the partition function $Z[J]$. From this, the free energy $W[J]$ and from it, the $G^{(n)}$s are obtained from the derivation

$$G^{(n)} = -\frac{\delta^n W[J]}{\delta J_1 \ldots \delta J_n}. \tag{35.3}$$

But in order to have a well-defined action and partition function, we must have $J \cdot \phi = J_0 \cdot \phi_0$, which means that we need to define the bare sources as (since $\phi_0 = \sqrt{Z_\phi}\phi$)

$$J_0 = Z_\phi^{-1/2} J. \tag{35.4}$$

This, together with the fact explained above that $\mathcal{L}^{\text{ren}}(\phi) = \mathcal{L}_0(\phi_0)$, means that the total action is invariant under the rescaling. The path integral measure, however, transforms by

$$C_Z = \prod_{x \in \mathbb{R}^d} Z_\phi^{1/2}. \tag{35.5}$$

Then $C_Z Z[J] = Z_0[J_0]$, and since $Z = e^{-W}$, we get

$$W[J] = W_0[J_0] + \ln C_Z, \tag{35.6}$$

and so

$$G^{(n)} = -\frac{\delta^n W[J]}{\delta J_1 \ldots \delta J_n} = -Z_\phi^{-n/2} \frac{\delta^n (W_0[J_0] + \ln C_Z)}{\delta J_{01} \ldots \delta J_{0n}} = Z_\phi^{-n/2} G_0^{(n)}. \tag{35.7}$$

But we are actually more interested in 1PI n-point functions, relevant for the S-matrix. These are generated by the effective action, which is the Legendre transform of the free energy:

$$\Gamma[\phi^{cl}] = W[J] + J \cdot \phi^{cl},$$
$$\Gamma_0[\phi_0^{cl}] = W_0[J_0] + J_0 \cdot \phi_0^{cl}. \tag{35.8}$$

Here the classical field is the connected one-point function:

$$\phi^{cl} = -\frac{\delta W[J]}{\delta J},$$
$$\phi_0^{cl} = -\frac{\delta W_0[J_0]}{\delta J_0}, \tag{35.9}$$

hence we have

$$\phi_0^{cl} = Z_\phi^{1/2} \phi^{cl}, \tag{35.10}$$

which leads to a relation for Γ similar to that for W:

$$\Gamma_0[\phi_0^{cl}] = \Gamma[\phi^{cl}] - \ln C_Z. \tag{35.11}$$

In turn, for the 1PI n-point functions

$$\Gamma^{(n)} = \frac{\delta^n \Gamma}{\delta \phi_1^{cl} \ldots \delta \phi_n^{cl}}, \tag{35.12}$$

we also get a relation between the bare and renormalized quantities:

$$\Gamma^{(n)}(p_1, \ldots, p_n; m, \lambda, \mu, \epsilon) = Z_\phi^{n/2} \Gamma_0^{(n)}(p_1, \ldots, p_n; m_0, \lambda_0, \epsilon). \tag{35.13}$$

We note, therefore, both for the case of $G^{(n)}$ and of $\Gamma^{(n)}$, the multiplicative nature of renormalization, namely that all the physical quantities are renormalized by multiplying them by the infinite factors.

Coming back to the issue we started this chapter with, the renormalized quantities λ, m, ϕ depend on the scale μ, which is understood really as being related to making measurements at a chosen scale. We will obtain that masses and couplings "run" with the scale (i.e. we have scale dependence).

In order to do this, we need to fix λ_0, m_0, ϕ_0 when $\epsilon \to 0$. This is usually done in the UV (i.e. the infinite bare quantities are related to the UV of the theory). There is an issue of whether a theory can be defined in the IR, like we would need to do in the case of quantum electrodynamics or $\lambda\phi^4$ theory; it is believed that this cannot be done.

35.2 Subtraction Schemes and Normalization Conditions

35.2.1 Subtraction Schemes

In the process of renormalizing, we have subtracted only the divergences, that is the coefficient of the $1/\epsilon$ terms (the pole). This is a choice, and it is called the *minimal subtraction scheme*, or *MS* scheme, but it is only a choice. We can also choose to subtract some finite parts also. For instance, the most popular scheme is a variation of *MS* called $\overline{MS}$, which corresponds to subtracting the infinity, but also factors of $-\gamma + \ln(4\pi)$, where γ is the Euler constant. Indeed, we have seen that in all of our examples we always obtained $\psi(n) + \ln(4\pi\mu^2/m^2)$, and in the expression for $\psi(n)$ it is useful to isolate $-\gamma$ (i.e. $\psi(n) = -\gamma + \ldots$), so $-\gamma + \ln(4\pi)$ appears naturally.

35.2.2 Normalization Conditions

But we can also fix the parameters in a different way, that is however more physical. The tree values for the effective action are given by the 1PI n-point functions

$$\Gamma^{(2)}_{\text{tree}}(p) = p^2 + m^2,$$
$$\Gamma^{(4)}_{\text{tree}}(p_i) = \lambda. \tag{35.14}$$

(1) But we can imagine *defining* that at $p = 0$, the 1PI functions take their tree values, so

$$\Gamma^{(2)}(p=0) = m^2; \quad \frac{d}{dp^2}\Gamma^{(2)}(p^2)\Big|_{p^2=0} = 1,$$
$$\Gamma^{(4)}(p_i=0) = \lambda. \tag{35.15}$$

(2) The choice of $p=0$ is nothing special, so we can in fact define that more generically, at any scale $\bar\mu$ the 1PI n-point functions take their tree values, so

$$\Gamma^{(2)}(\bar\mu^2) = \bar\mu^2 + m^2; \quad \frac{d}{dp^2}\Gamma^{(2)}\Big|_{p^2=\bar\mu^2} = 1,$$
$$\Gamma^{(4)}(p_i)\Big|_{p_ip_j=\bar\mu^2(\delta_{ij}-1/4)} = \lambda. \tag{35.16}$$

The choice of $p_ip_j = \bar\mu^2(\delta_{ij} - 1/4)$ was made so as to have the Mandelstam variables $s = t = u = \bar\mu^2$, but that is not even necessary, we can generalize further and require the four-point function to be equal to its tree value at some different s, t, u related to $\bar\mu$ in another way.

Here, (1) and (2) are different *normalization conditions*, and they are a different way to fix what we renormalize, but they should in principle be related (equivalent) to the renormalization (subtraction) schemes above. However, when we try to translate, we will see that, for example, the MS scheme corresponds to some nontrivial normalization conditions.

Also, by the equivalence of the subtraction schemes (where different schemes would differ by the subtraction of different finite parts) to the normalization conditions, it follows that two renormalized theories differing just by different normalization conditions will differ by finite counterterms. Hence, the difference in counterterms (subtracted terms) must be finite.

A final observation at this point is that it is only in the MS scheme that the coefficient functions $a_k(\lambda), b_k(\lambda)$, and $c_k(\lambda)$ depend only on λ, in general they will also depend on m/μ.

35.3 Renormalization Group Equations and Anomalous Dimensions

35.3.1 Renormalization Group in MS Scheme

We note that we can trade the arbitrary parameter μ we have introduced when renormalizing (in dimensional regularization, it came from the fact that the coupling outside the critical dimension has a mass dimension) for the arbitrary subtraction point $\bar{\mu}$. That is, the freedom in μ corresponds to the freedom in defining masses and couplings at some arbitrary scale.

Then the equation for the dependence on the physical scale (like $\bar{\mu}$) at which we define parameters, of observables like the 1PI Green's functions, will be the renormalization group equation.

35.3.2 ϕ^4 in Four Dimensions

We start with the observation that in ϕ^4 theory in $d = 4$, $[\lambda_4] = 0$ and so $\omega(D) = d - (d - 2)E/2 - \sum_v [\lambda_v]$ becomes $\omega(D) = 4 - E$. Thus, for a 1PI function with n external lines, a *finite* diagram will give $\omega(D) = 4 - n$, with no necessity for renormalization, leading to the scaling law

$$\Gamma_D^{(n)}(tp_i; tm) = t^{4-n} \Gamma_D^{(n)}(p_i; m), \tag{35.17}$$

where we have scaled both momenta and masses by the same factor t, to obtain the classical scaling dimension.

But since there are divergent diagrams, we need to regularize, introducing the arbitrary scale μ (or the subtraction point $\bar{\mu}$). Then, for the fully *renormalized* 1PI Green's functions, which include the counterterms, to maintain the relation as it is, we need to scale μ also, that is

$$\Gamma^{(n)}(tp_i; tm, t\mu) = t^{4-n} \Gamma^{(n)}(p_i; m, \mu). \tag{35.18}$$

By taking td/dt on this equation, we obtain that $\Gamma^{(n)}$ satisfies

$$t\frac{d}{dt}\Gamma^{(n)}(tp_i; tm, t\mu) = (4-n)\Gamma^{(n)}(tp_i; tm, t\mu). \tag{35.19}$$

We can rewrite this by absorbing the t multiplying m and μ and trading it for derivatives with respect to m and μ:

$$\left[t\frac{\partial}{\partial t} + m\frac{\partial}{\partial m} + \mu\frac{\partial}{\partial \mu} + n - 4\right]\Gamma^{(n)}(tp_i; m, \mu) = 0. \tag{35.20}$$

We can now find another equation for $\Gamma^{(n)}$ as follows. Consider the fact that Γ_0 is independent of μ:

$$\mu\frac{d}{d\mu}\Gamma_0^{(n)}(p_i; m_0, \lambda_0, \epsilon) = 0, \tag{35.21}$$

and express it in terms of $\Gamma^{(n)}$ as

$$\mu\frac{d}{d\mu}\left[Z_\phi^{-n/2}(\lambda, \epsilon)\Gamma^{(n)}(p_i; \lambda, m, \mu, \epsilon)\right] = 0. \tag{35.22}$$

Since, as discussed, we have the dependences

$$\lambda = \lambda(\mu, \epsilon; \lambda_0, m_0); \quad m = m(\lambda(\mu), \epsilon); \quad Z_\phi = Z_\phi(\lambda(\mu), \epsilon), \tag{35.23}$$

when acting on $\Gamma^{(n)}(p_i, \lambda(\mu), m(\mu), \mu, \epsilon)$ with the μ derivative we have explicit and implicit dependence, leading to

$$\mu\frac{d}{d\mu} = \mu\frac{\partial}{\partial\mu} + \mu\frac{d\lambda}{d\mu}\frac{\partial}{\partial\lambda} + \mu\frac{dm}{d\mu}\frac{\partial}{\partial m}. \tag{35.24}$$

We now define the *beta function*

$$\beta(\lambda, \epsilon) \equiv \mu\frac{d\lambda}{d\mu}\bigg|_{m_0, \lambda_0, \epsilon}, \tag{35.25}$$

the *anomalous field dimension*

$$\gamma_d(\lambda, \epsilon) \equiv \frac{\mu}{2Z_\phi}\frac{dZ_\phi}{d\mu}\bigg|_{m_0, \lambda_0, \epsilon}, \tag{35.26}$$

and the *anomalous mass dimension*

$$\gamma_m(\lambda, \epsilon) \equiv \frac{\mu}{m}\frac{dm}{d\mu}\bigg|_{m_0, \lambda_0, \epsilon}. \tag{35.27}$$

To understand the names, note that if we have $\beta, \gamma_d, \gamma_m$ constants, then the above definitions mean that

$$\lambda \sim \lambda_0 + \beta\ln\mu; \quad Z_\phi \sim C\mu^{2\gamma_d} \Rightarrow \phi_0 \sim \sqrt{C}\mu^{\gamma_d}\phi; \quad m \sim \tilde{C}\mu^{\gamma_m}. \tag{35.28}$$

The first means that β gives the slope of λ with $\ln\mu$, γ_d the power of μ in ϕ_0, and γ_m the power of μ in m, justifying the names.

Substituting this in (35.22), we obtain

$$\Gamma^{(n)}\mu\frac{d}{d\mu}Z_\phi^{-n/2} + Z_\phi\left(\mu\frac{\partial}{\partial\mu} + \beta\frac{\partial}{\partial\lambda} + \gamma_m m\frac{\partial}{\partial m}\right)\Gamma^{(n)} = 0 \Rightarrow$$

$$\left[\mu\frac{\partial}{\partial\mu} + \beta(\lambda)\frac{\partial}{\partial\lambda} + \gamma_m(\lambda)m\frac{\partial}{\partial m} - n\gamma_d(\lambda)\right]\Gamma^{(n)}(p_i,\lambda,m,\mu) = 0. \quad (35.29)$$

Eliminating $\mu\partial/\partial\mu$ between (35.20) and (35.29), we obtain

$$\left[-t\frac{\partial}{\partial t} + \beta(\lambda)\frac{\partial}{\partial\lambda} + (\gamma_m(\lambda) - 1)m\frac{\partial}{\partial m} - n\gamma_d(\lambda) + 4 - n\right]\Gamma^{(n)}(tp_i,\lambda,m,\mu) = 0, \quad (35.30)$$

which is called the *renormalization group equation* (RGE), for the case of the ϕ^4 theory in four dimensions (for other theories, we have only the $4 - n$ term changing to the classical scaling dimension of the 1PI n-point function).

Note that this equation gives the behavior under scaling of the 1PI n-point functions, and is written only in terms of *physical* quantities, since we eliminated the μ dependence.

Solution   The formal solution of the RGE can be written as follows. Define $\bar{\lambda}(t)$ and $\bar{m}(t)$ such that they satisfy the equations

$$t\frac{\partial}{\partial t}\bar{\lambda}(t) = \beta(\bar{\lambda}(t)),$$

$$t\frac{\partial}{\partial t}\bar{m}(t) = \bar{m}(t)(\gamma_m(\bar{\lambda}(t)) - 1), \quad (35.31)$$

with the boundary conditions

$$\bar{\lambda}(t = 1) = \lambda; \quad \bar{m}(t = 1) = m. \quad (35.32)$$

Then the solution of the above will be $\bar{\lambda} = \bar{\lambda}(t,\lambda)$ and $\bar{m} = \bar{m}(t,m,\lambda)$. In turn, the formal solution to the RGE is

$$\Gamma^{(n)}(tp_i,\lambda,m;\mu) = t^{4-n}\exp\left[-n\int_1^t\frac{ds}{s}\gamma_d(\bar{\lambda}(s))\right]\Gamma^{(n)}(p_i,\bar{\lambda}(t),\bar{m}(t);\mu). \quad (35.33)$$

We will not give the proof, but one can check that this solution satisfies the equation.

It would seem that we have solved the RGE, but of course that is only in principle, not in practice. Note that in order to be able to write an explicit solution, we would need the exact formulas for $\beta(\lambda), \gamma_m(\lambda), \gamma_d(\lambda)$. We can of course plug in their values to some order in perturbation theory, but we will not obtain exact solutions.

We can compare this to the naive scaling

$$\Gamma^{(n)}(tp_i,\lambda,m;\mu) = t^{4-n}\Gamma^{(n)}(p_i,\lambda,m/t). \quad (35.34)$$

In contrast, if $\gamma_m(\bar{\lambda}(t))$ and $\gamma_d(\bar{\lambda}(t))$ are approximately constant as a function of t, which can happen only if $\bar{\lambda}(t)$ is approximately constant, which in turn requires the beta function to be zero, then we can easily integrate $\bar{\lambda}(t), \bar{m}(t)$ to obtain $\bar{\lambda}(t) \simeq \lambda$ and $\bar{m}(t) = m/t^{1-\gamma_m}$, as well as to integrate the exponent in the solution to the RGE as

$$\exp\left[-n\int_1^t\frac{ds}{s}\gamma_d(\bar{\lambda}(s))\right] \simeq t^{-n\gamma_d}, \quad (35.35)$$

such that finally the solution to the RGE is

$$\Gamma^{(n)}(tp_i, \lambda, m, \mu) = t^{4-n(1+\gamma_d)}\Gamma^{(n)}\left(p_i, \lambda, \frac{m}{t^{1-\gamma_m}}, \mu\right). \qquad (35.36)$$

We then see (by comparing with the naive scaling above) that indeed γ_m acts as an anomalous mass dimension (quantum correction for the scaling dimension of m with t) and γ_d as an anomalous field dimension (quantum correction for the scaling dimension of the field, external to the 1PI n-point function, with t).

35.4 Beta Function and Running Coupling Constant

In general, $\bar{\lambda}(t)$ is not constant (so $\lambda(\mu)$ is not constant) and we say that we have a *running coupling constant* (a coupling constant that "runs" with the energy scale), whose running is defined by the beta function $\beta(\lambda) = \mu d\lambda/d\mu|_{m_0,\lambda_0,\epsilon}$. $\bar{\lambda}(t)$ is only approximately constant near a so-called *fixed point* of the renormalization group (RG) λ_F, where $\beta(\lambda_F) = 0$.

The running of the coupling with the energy scale means that the validity of perturbation theory can depend on scale. For instance, for QCD we know that the theory is weakly coupled in the UV, a phenomenon called asymptotic freedom, whereas it becomes strongly coupled in the IR, a phenomenon called IR slavery. Therefore, QCD is perturbative only in the UV. In contrast, for QED the situation is reversed: we have a weak coupling in the IR, and the theory becomes strongly coupled in the UV. In fact, there is a *Landau pole*: namely, at some high but finite energy scale, the coupling becomes infinite in perturbation theory. Therefore, QED is perturbative only in the IR. For such theories (like QED, and ϕ^4 theory in $d = 4$ which behaves in a similar manner) it is not clear that we can consistently define the theory.

If there is a well-defined perturbation theory somewhere, we have a $\lambda = 0$ point, and $\beta(\lambda = 0) = 0$ (the validity of perturbation theory means that there is a Taylor expansion for β, given by Feynman diagrams, starting at one loop). Then $\lambda = 0$ is a universal fixed point, called the *Gaussian fixed point*, since the action is free and the path integral is Gaussian, at that point.

35.4.1 Possible Behaviors for $\beta(\lambda)$

We now analyze the most common possibilities for the behavior of $\beta(\lambda)$. We saw that $\lambda = 0$ is the Gaussian fixed point. Note that the solution to the equation $t\partial/\partial t\bar{\lambda}(t) = \beta(\bar{\lambda}(t))$ with boundary condition $\bar{\lambda}(t = 1) = \lambda$ is

$$t = \exp\int_\lambda^{\bar{\lambda}} \frac{d\lambda'}{\beta(\lambda')}. \qquad (35.37)$$

(I) Consider first the case where $\beta(\lambda)$ starts out positive at $\lambda = 0$ and keeps increasing, as in Figure 35.1(I). If, moreover, it increases faster than λ, in (35.37), the integral $\int^{\bar{\lambda}}$ is

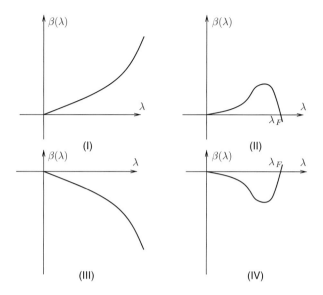

Figure 35.1 Possible behaviors of $\beta(\lambda)$.

convergent as $\bar{\lambda} \to \infty$, and we have a Landau pole, $\bar{\lambda} \to \infty$, at some large but finite t_0:

$$t_0 = \exp \int_1^\infty \frac{d\lambda'}{\beta(\lambda')}. \tag{35.38}$$

That is at some large but finite momentum. As we mentioned, QED is of this type.

The Gaussian fixed point $\lambda = 0$ is an IR-stable fixed point in this case, since for $\lambda \to 0$ and $\bar{\lambda} \to \lambda \to 0$:

$$\exp \int_0^{\bar{\lambda} \to 0} \frac{d\lambda'}{\beta(\lambda')} = \exp \int_0^{\bar{\lambda} \to 0} \frac{d\lambda'}{|\beta'(0)|\lambda'} \sim \exp[+|a| \ln(0)] \to 0. \tag{35.39}$$

We can also verify that if we start with a perturbation $\lambda > 0$, then $\mu d\lambda/d\mu = \beta(\lambda) > 0$, so decreasing μ implies that λ also decreases, towards $\lambda = 0$ (so when going to low momenta we are driven towards $\lambda = 0$).

(II) Then we can consider the case where $\beta(\lambda)$ starts out positive at $\lambda = 0$, but turns around and goes through zero, $\beta = 0$, at some $\lambda = \lambda_F$, as in Figure 35.1(II). This means that near it we have

$$\beta(\lambda) \simeq \beta'(\lambda_F)(\lambda - \lambda_F), \tag{35.40}$$

and $\beta'(\lambda_F) < 0$. In turn, it means that near λ_F, the exponent in (35.37) gives

$$\exp \int_\lambda^{\bar{\lambda} \to \lambda_F} \frac{d(\lambda' - \lambda_F)}{\beta'(\lambda_F)(\lambda' - \lambda_F)} \to +\infty. \tag{35.41}$$

Hence $t \to \infty$ as $\lambda \to \lambda_F$, so for infinite momenta (infinite t) we have $\lambda \to \lambda_F$. In other words, λ_F is a UV-stable fixed point. It is UV stable since, if we perturb $\lambda > \lambda_F$, $\mu d\lambda/d\mu = \beta(\lambda) < 0$, so as one increases μ, λ decreases towards λ_F, while if we perturb

$\lambda < \lambda_F$, $\mu d\lambda/d\mu = \beta(\lambda) > 0$, so as one increases μ, λ increases towards λ_F. Therefore either way, as we increase μ we are driven towards λ_F.

In this case the Gaussian fixed point $\lambda = 0$ is an IR-stable fixed point, exactly as in case I, for the same reason.

(III) We can consider now the mirror image of case I, that is $\beta(\lambda)$ starts out negative and keeps decreasing, and moreover that it decreases faster than $-\lambda$, as in Figure 35.1(III). Then in (35.37), the exponent behaves as

$$\exp \int_{\lambda}^{\bar{\lambda} \to 0} \frac{d\lambda'}{\beta(\lambda')} \simeq \exp \int_{\lambda}^{\bar{\lambda} \to 0} \frac{d\lambda'}{-|\beta(0)|\lambda'} = \exp[-|a| \ln(0)] \to +\infty. \qquad (35.42)$$

Therefore, as $t \to \infty$, $\bar{\lambda}(t) \to 0$, that is the Gaussian fixed point is a UV-stable fixed point, and we are driven towards it at high momenta. We can also verify this fact since $\beta'(0) < 0$, and then as in case II for λ_F, when going at high momenta we are driven towards it, even when we initially perturb away from it. This phenomenon, of $\lambda = 0$ being UV stable is called, as we said, asymptotic freedom, and it is the behavior QCD experiences.

In contrast, since $\beta(\lambda)$ decreases faster than λ, for $\bar{\lambda} \to \infty$ we obtain a Landau pole (i.e. a finite value of t), since the integral is convergent at infinity:

$$t_0 = \exp \int_{\lambda}^{\bar{\lambda} \to \infty} \frac{d\lambda'}{\beta(\lambda')}. \qquad (35.43)$$

Therefore we obtain a Landau pole (breakdown of perturbation theory at a finite energy scale) in the IR. That is fine, since for QCD we know that perturbation theory breaks down in the IR.

(IV) Finally, we can consider the mirror image case to case II, namely when the beta function starts out negative at $\lambda = 0$, but then turns around and becomes positive again at some finite $\lambda = \lambda_F$, as in Figure 35.1(IV). Then, near λ_F we can write

$$\beta(\lambda) \simeq \beta'(\lambda_F)(\lambda - \lambda_F), \qquad (35.44)$$

where $\beta'(\lambda_F) > 0$.

By the same argument as in case III, the Gaussian fixed point $\lambda = 0$ is UV stable. On the contrary, the fixed point λ_F is IR stable, since for $\bar{\lambda} \to \lambda_F$:

$$t = \exp \int^{\bar{\lambda} \to \lambda_F} \frac{d(\lambda' - \lambda_F)}{+|\beta'(\lambda_F)|(\lambda - \lambda_F)} \sim \exp[+|a| \ln(0)] \to 0. \qquad (35.45)$$

We can also verify that, since $\beta'(\lambda_F) > 0$, the same argument as for the Gaussian fixed point in case I says that if we perturb away from λ_F, when going at small μ we are driven back towards λ_F.

35.5 Perturbative Beta Function in Dimensional Regularization in MS Scheme

We now learn how to calculate the beta function in perturbation theory. We start with the formula (valid for dimensional regularization in the MS scheme) relating λ_0 and λ:

$$\lambda_0 = \mu^\epsilon \left(\lambda + \sum_{k \geq 1} \frac{a_k(\lambda)}{\epsilon^k} \right). \tag{35.46}$$

Since $\beta(\lambda, \epsilon) = \mu d/\partial \mu \lambda|_{m_0, \lambda_0, \epsilon}$, we take $d/d\mu|_{\lambda_0}$ on both sides of the equation above, and obtain (after dividing by μ^ϵ)

$$0 = \epsilon \left(\lambda + \sum_{k \geq 1} \frac{a_k(\lambda)}{\epsilon^k} \right) + \beta(\lambda, \epsilon) \left(1 + \sum_{k \geq 1} \frac{a_k'(\lambda)}{\epsilon^k} \right). \tag{35.47}$$

Consider now the ansatz $\beta(\lambda, \epsilon) = -\epsilon \lambda + \tilde{\beta}$, and plug it in the above equation. We can verify that the $\mathcal{O}(\epsilon)$ terms then cancel in the equation, and the $\mathcal{O}(\epsilon^0)$ terms give

$$a_1(\lambda) + \tilde{\beta} - \lambda a_1'(\lambda) = 0, \tag{35.48}$$

so

$$\beta(\lambda, \epsilon) = -\epsilon \lambda - a_1(\lambda) + \lambda a_1'(\lambda). \tag{35.49}$$

Taking the $\epsilon \to 0$ limit, we obtain the beta function as

$$\beta(\lambda) \equiv \lim_{\epsilon \to 0} \mu \frac{\partial \lambda}{\partial \mu} = \left(\lambda \frac{d}{d\lambda} - 1 \right) a_1(\lambda). \tag{35.50}$$

This formula is exact to all orders in perturbation theory, and note that we only need to know the coefficient of the single pole, not of any of the higher-order poles. Of course, $a_1(\lambda)$ receives corrections at each order in perturbation theory, so we can only calculate it to the order we know ($a_1(\lambda)$).

Now we can substitute $\beta(\lambda, \epsilon) = -\epsilon \lambda + \beta(\lambda)$ in the remaining equations, and at order $\mathcal{O}(\epsilon^{-k})$ we obtain the equation

$$a_{k+1}(\lambda) + \beta(\lambda) a_k'(\lambda) - \lambda a_{k+1}'(\lambda) = 0, \tag{35.51}$$

which leads to the recursion relation between the coefficients of poles

$$\left(\lambda \frac{d}{d\lambda} - 1 \right) a_{k+1}(\lambda) = \beta(\lambda) \frac{d}{d\lambda} a_k(\lambda). \tag{35.52}$$

This means that we can determine (in principle) all the $a_k(\lambda)$ in terms of $a_1(\lambda)$ only.

35.5.1 Examples

(1) ϕ^4 theory in $d = 4$. We saw that for one loop, we obtained

$$a_1(\lambda) = \frac{3\lambda^2}{(4\pi)^2} + \dots \tag{35.53}$$

Then from (35.50), we get

$$\beta(\lambda) = \frac{3\lambda^2}{(4\pi)^2} + \mathcal{O}(\lambda^4). \tag{35.54}$$

In turn, solving for $\bar{\lambda}(t)$, we obtain

$$\bar{\lambda}(t) = \frac{\lambda}{1 - \frac{3\lambda}{(4\pi)^2} \ln(t)}. \tag{35.55}$$

Thus it is IR free and there is a Landau pole, just like in QED.

(2) ϕ^3 in $d = 6$. For one loop, we obtained

$$a_1(\lambda) = -\frac{7}{8} \frac{\lambda^3}{(4\pi)^3} + \dots, \tag{35.56}$$

leading to the one-loop beta function

$$\beta(\lambda) = -\frac{7}{4} \frac{\lambda^3}{(4\pi)^3} + \mathcal{O}(\lambda^4). \tag{35.57}$$

This can then be solved for

$$\bar{\lambda}^2(t) = \frac{\lambda^2}{1 + \frac{7}{4} \frac{\lambda^2}{(4\pi)^3} \ln(t^2)}, \tag{35.58}$$

which is asymptotically free, just like in QCD.

35.6 Perturbative Calculation of γ_m and γ_d in Dimensional Regularization in the MS Scheme

We can now repeat the same kind of calculation to find γ_m. Start with

$$m_0^2 = m^2 \left(1 + \sum_{k \geq 1} \frac{b_k(\lambda)}{\epsilon^k} \right). \tag{35.59}$$

Taking $\mu d/d\mu|_{m_0, \lambda_0, \epsilon}$ on both sides and dividing by m^2, we obtain

$$0 = 2\gamma_m \left(1 + \sum_{k \geq 1} \frac{b_k(\lambda)}{\epsilon^k} \right) + \beta(\lambda, \epsilon) \sum_{k \geq 1} \frac{b_k'(\lambda)}{\epsilon^k}. \tag{35.60}$$

We then substitute $\beta(\lambda, \epsilon) = -\epsilon\lambda + \beta(\lambda)$. We can check that we cannot have a nontrivial $\gamma_m(\lambda, \epsilon)$ (a Taylor expansion in ϵ), so $\gamma_m(\lambda, \epsilon) = \gamma_m(\lambda)$. Substituting this as well, we obtain from the equation at order $\mathcal{O}(\epsilon^0)$:

$$\gamma_m(\lambda, \epsilon) = \gamma_m(\lambda) = \frac{\lambda}{2} b_1'(\lambda), \tag{35.61}$$

and substituting it back in the equation, from the order $\mathcal{O}(1/\epsilon^k)$, the recursion relation

$$\lambda b_{k+1}'(\lambda) = \beta(\lambda) b_k'(\lambda) + \lambda b_1'(\lambda) b_k(\lambda). \tag{35.62}$$

Therefore, from $b_1(\lambda)$ we can get both $\gamma_m(\lambda)$ and all higher-order poles $b_k(\lambda)$.

For γ_d, we do a similar calculation. Starting with

$$Z_\phi = 1 + \sum_{k \geq 1} \frac{c_k(\lambda)}{\epsilon^k}, \tag{35.63}$$

and taking $\mu d/d\mu|_{m_0, \lambda_0, \epsilon}$ on both sides, we obtain

$$2\gamma_d(\lambda, \epsilon) \left(1 + \sum_{k \geq 1} \frac{c_k(\lambda)}{\epsilon^k}\right) = (-\epsilon\lambda + \beta(\lambda)) \sum_{k \geq 1} \frac{c_k'(\lambda)}{\epsilon^k}. \tag{35.64}$$

Again we check that $\gamma_d(\lambda, \epsilon)$ cannot have a nontrivial Taylor expansion in ϵ (i.e. $\gamma_d(\lambda, \epsilon) = \gamma_d(\lambda)$). Then from the $\mathcal{O}(\epsilon^0)$ order in the equation, we get

$$\gamma_d(\lambda) = -\frac{\lambda}{2} c_1'(\lambda), \tag{35.65}$$

and substituting it back in the equation, from the $\mathcal{O}(\epsilon^{-k})$ order we get the recursion relations

$$\lambda c_{k+1}'(\lambda) = \beta(\lambda) c_k'(\lambda) + \lambda c_1'(\lambda) c_k(\lambda). \tag{35.66}$$

Important Concepts to Remember

- Because of the multiplicative nature of renormalization, the relation between the bare and renormalized Green's functions (connected and 1PI) is given by multiplication with Z_ϕ factors, specifically $G^{(n)} = Z_\phi^{-n/2} G_0^{(n)}$ and $\Gamma^{(n)} = Z_\phi^{n/2} \Gamma_0^{(n)}$.
- We can think of either fixing the bare quantities λ_0, m_0, ϵ, by defining the theory usually in the UV, and then we have $\lambda(\mu)$ and implicitly $m(\mu)$ and $Z(\mu)$, meaning that these physical quantities "run" with the scale, or fixing λ, m, ϕ and then λ_0, m, ϕ are functions of ϵ.
- Renormalization is defined by a subtraction scheme, defined by removing the divergent parts by counterterms (minimal subtraction), perhaps together with some finite parts. Or equivalently by a normalization condition, whereby we fix the 1PI n-point functions to take their tree-level values at some energy scale $\bar{\mu}$ (for $n \geq 2$, we need to specify also the relation of the various momenta with the scale $\bar{\mu}$).
- Two renormalized theories which differ just by normalization conditions differ by finite counterterms.

- One can trade the arbitrary parameter μ with the arbitrary subtraction point $\bar{\mu}$.
- The renormalization group equation is the equation for the variation of the renormalized 1PI n-point function under scaling, written only in terms of physical quantities (with the explicit dependence on μ solved for).
- The beta function gives the slope of scaling of λ with $\ln \mu$, the anomalous mass dimension γ_m the power of μ in m, and the anomalous field dimension the power of μ in ϕ_0. The anomalous dimensions give corresponding quantum corrections to the scaling of fields and masses in the full 1PI n-point functions.
- γ_m and γ_d are only approximately constant near a fixed point λ_F of the beta function, $\beta(\lambda_F) = 0$. If the perturbation theory is well defined, $\lambda = 0$ is a universal fixed point, the Gaussian fixed point.
- QED and ϕ^4 in $d = 4$ are perturbative in the IR and have a Landau pole in the UV; QCD and ϕ^3 in $d = 6$ are asymptotically free (perturbative in the UV) and have a pole in the IR.
- $\beta(\lambda)$, as well as $a_k(\lambda)$, are given completely in terms of $a_1(\lambda)$, the coefficient of the single pole in the coupling divergences. $\gamma_m(\lambda)$, as well as $b_k(\lambda)$, are given completely in terms of $b_1(\lambda)$. $\gamma_d(\lambda)$, as well as $c_k(\lambda)$, are given completely in terms of $c_1(\lambda)$.

Further Reading

See sections 5.6 and 5.7 in [3].

Exercises

1. Calculate the beta function for ϕ^3 theory in $d = 4$ (use the results from the exercises in Chapter 34).
2. Use $a_1(\lambda)$, $b_1(\lambda)$, and $c_1(\lambda)$ for ϕ^4 in $d = 4$ and ϕ^3 in $d = 6$ from the text to calculate explicitly $\gamma_m(\lambda), \gamma_d(\lambda), a_2, b_2, c_2$ at one loop. Then substitute in the RG equation for the divergent n-point functions in these respective theories. Write the explicit RG equation and its explicit solution at one loop.
3. Check that the formal solution (35.33) satisfies the renormalization group equations.
4. Consider a theory with an exact beta function $\beta(\lambda) = a\lambda + b\lambda^2$, with $a, b \geq 0$, and with $\gamma_m = c\lambda, c > 0$ and $\gamma_d = d\lambda, d > 0$. Solve the renormalization group equations for the n-point functions in both the $a \neq 0$ case and the $a = 0$ case.

36 One-Loop Renormalizability in QED

In this chapter we will show that quantum electrodynamics is renormalizable at one loop by doing the renormalization explicitly.

36.1 QED Feynman Rules and Power-Counting Renormalizability

We start by remembering the QED Feynman rules.

In Euclidean space, we had the following.

- Photon propagator between μ and ν:

$$G_{\mu\nu}^{(0)}(k) = \frac{1}{k^2}\left(\delta_{\mu\nu} - (1-\alpha)\frac{k_\mu k_\nu}{k^2}\right). \tag{36.1}$$

- Fermion (electron) propagator between α and β:

$$S_F^{(0)}(p) = \frac{1}{i\not{p} + m} = \frac{(-i\not{p} + m)_{\alpha\beta}}{p^2 + m^2}. \tag{36.2}$$

- Electron positron photon vertex between α, β, and μ:

$$+ ie(\gamma^\mu)_{\alpha\beta}. \tag{36.3}$$

In Minkowski space, we had the following.

- Photon propagator between μ and ν:

$$G_{\mu\nu}^{(0)}(k) = \frac{-i}{k^2 - i\epsilon}\left(g_{\mu\nu} - (1-\alpha)\frac{k_\mu k_\nu}{k^2}\right). \tag{36.4}$$

- Fermion (electron) propagator between α and β:

$$S_F^{(0)}(p) = \frac{-i(-i\not{p} + m)_{\alpha\beta}}{p^2 + m^2 - i\epsilon} = \frac{-(\not{p} + im)_{\alpha\beta}}{p^2 + m^2 + i\epsilon}. \tag{36.5}$$

- Electron positron photon vertex between α, β, and μ:

$$+ e(\gamma^\mu)_{\alpha\beta}. \tag{36.6}$$

We also remember that we have the Ward–Takahashi identity for the vertex:

$$p^\mu \Gamma_{\mu\alpha\beta}(p; q_2, q_1) = e(S_F^{-1}(q_2)_{\alpha\beta} - S_F^{-1}(q_1)_{\alpha\beta}), \tag{36.7}$$

and similarly for the n-photon case:

$$k_{(1)}^{\mu_1} \Gamma_{\mu_1 \dots \mu_n}^{(n)}(k^{(1)}, \dots, k^{(n)}) = 0. \tag{36.8}$$

Here, of course, we can contract with any of the n momenta of the vertex. These identities are a consequence of gauge invariance.

In particular, for $n = 2$ we obtain the transversality condition for the polarization, $k^\mu \Pi_{\mu\nu}(k) = 0$, which leads to the fact that

$$\Pi_{\mu\nu}(k) = (k^2 \delta_{\mu\nu} - k_\mu k_\nu) \Pi(k^2). \tag{36.9}$$

But we now note that this means we reduce the degree of the divergence by 2, since the superficial degree of divergence $\omega(D)$ was given by the scaling dimension of the Feynman integral, but the result of the integral is constrained to have the above form, which reduces its actual (effective) degree of divergence by 2 units of mass dimension, $\omega_{\text{eff}}(D) = \omega(D) - 2$.

In general, for the n-photon vertex $\Gamma^{(n)}$, we see that, because we have n conditions on it, given by contraction with each external momentum, the superficial degree of divergence will be reduced by n powers of the momenta (i.e. $\omega_{\text{eff}}(D) = \omega(D) - n$), and since for $\Gamma^{(n)}$ we have $\omega(D) = 4 - n$ in four dimensions, we have $\omega_{\text{eff}}(D) = 4 - 2n$.

We now consider a more general analysis of the superficial degree of divergence. For the case of scalar theories, we had $\omega(D) = dL - 2I$, since there were L loop integrals in d dimensions, and I propagators of $1/(p^2 + m^2)$ for the scalars. We see that the same formula is still valid for a purely bosonic theory, with all bosons having actions with two derivatives (i.e. propagators with dimension -2), $\sim 1/p^2$. But when we have fermions, the fermion propagators have dimension -1, here being $1/(i\not{p} + m)$, and therefore their contribution to $\omega(D)$ is $-I_f$, for a total of

$$\omega(D) = dL - 2I_{ph} - I_f, \tag{36.10}$$

where I_{ph} is the number of photon internal lines and I_f is the number of fermion internal lines.

For the scalar case, we also had the formula $\sum_n n V_n = 2I + E$, since at an n-vertex we have the endpoint of n lines, but each internal line has two endpoints, whereas each external line has only one. In the case of QED, we have only one vertex, on which end two fermions and a photon, therefore $n = 2$ from the point of view of the fermions, and $n = 1$ from the point of view of the photons, giving

$$2I_{ph} + E_{ph} = V; \quad 2I_f + E_f = 2V. \tag{36.11}$$

We also had the relation $L = I - V + 1$, since the number of loops was given by the number of integration variables, one for each internal line, minus the number of delta function constraints, one for each vertex, except for the overall momentum-conservation delta function. Now this still applies, with $I = I_{ph} + I_f$, so $L = I_{ph} + I_f - V + 1$. Finally, eliminating L, I_{ph}, and I_f between the four relations we wrote, we obtain for the superficial degree of divergence of QED in four dimensions

$$\omega(D) = 4 - E_{ph} - \frac{3}{2}E_f. \qquad (36.12)$$

For $\Gamma^{(n)}$, we had called E_{ph} by n before.

Then we see that *a priori* we have the following divergent Green's functions. We can have only photons, in which case $E_{ph} = 1, 2, 3, 4$ are superficially divergent, or only fermions, in which case $E_f = 1, 2$ are superficially divergent, or both fermions and photons, with $E_f = 1, 2$ and $E_{ph} = 1$, or $E_f = 1$ and $E_{ph} = 2$. But first, we note that we cannot have an odd number of fermions, since each fermion line must be uninterrupted (cannot just end in a vertex), so the $E_f = 1$ cases are ruled out. Then, for the pure photon case, at one loop we have no nonzero $E_{ph} = 1$ or $E_{ph} = 3$ diagrams, whereas the $E_{ph} = 4$ case has an *effective* degree of divergence of $\omega_{\text{eff}}(D) = -4$, so is actually convergent.

We are therefore left with only three divergences, in $E_{ph} = 2$ (photon polarization), in $E_f = 2$ (fermion self-energy), and in $E_f = 2, E_{ph} = 1$ (vertex function). We will treat these cases separately.

36.2 Dimensional Regularization of Gamma Matrices

Before that, we remember some gamma matrix identities which were derived in dimension D, and are still valid in $D = 4 - \epsilon$ dimensions, since the index μ on γ_μ now has D components (but the γ_μ matrix is still 4×4):

$$\gamma_\mu \gamma^\mu = D \, \mathbb{1},$$
$$\gamma_\nu \gamma_\mu \gamma^\nu = (2 - D)\gamma_\mu,$$
$$\gamma_\nu \gamma_{\mu_1} \gamma_{\mu_2} \gamma^\nu = 4\delta_{\mu_1\mu_2} \, \mathbb{1} + (D - 4)\gamma_{\mu_1}\gamma_{\mu_2},$$
$$\gamma_\nu \gamma_{\mu_1} \gamma_{\mu_2} \gamma_{\mu_3} \gamma^\nu = -2\gamma_{\mu_3}\gamma_{\mu_2}\gamma_{\mu_3} + (2 - D)(\delta_{\mu_1\mu_2}\gamma_{\mu_3} + \delta_{\mu_2\mu_3}\gamma_{\mu_1} - \delta_{\mu_1\mu_3}\gamma_{\mu_2}). \quad (36.13)$$

Also, we still have

$$\text{Tr}[\gamma_\mu \gamma_\nu] = 4\delta_{\mu\nu},$$
$$\text{Tr}[\gamma_\mu \gamma_\nu \gamma_\rho \gamma_\sigma] = 4(\delta_{\mu\nu}\delta_{\rho\sigma} - \delta_{\mu\rho}\delta_{\nu\sigma} + \delta_{\mu\sigma}\delta_{\nu\rho}). \qquad (36.14)$$

36.3 Case 1: Photon Polarization $\Pi_{\mu\nu}(p)$

We consider the case of an off-shell momentum p (i.e. $p^2 \neq 0$) for the photon. The diagram is a fermion loop with two external photon lines coming from it, as in Figure 36.1. The fermion momenta are k and $k + p$, the photon momenta are p, and the external photon indices are μ and ν. Then the value for the Feynman diagram is

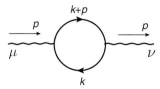

Figure 36.1   One-loop photon polarization diagram.

$$\Pi_{\mu\nu}(p) = (-1) \int \frac{d^D k}{(2\pi)^D} \, \mathrm{Tr} \left[(i\tilde{e}\gamma_\mu) \frac{(-i\slashed{k} + m)}{k^2 + m^2} (i\tilde{e}\gamma_\nu) \frac{-i(\slashed{k} + \slashed{p}) + m}{(k+p)^2 + m^2} \right]$$

$$= \tilde{e}^2 \int \frac{d^D k}{(2\pi)^D} \frac{\mathrm{Tr}[\gamma_\mu(-i\slashed{k} + m)\gamma_\nu(-i(\slashed{k} + \slashed{p}) + m)]}{(k^2 + m^2)((k+p)^2 + m^2)}. \tag{36.15}$$

Here the (-1) is because of the fermion loop, and $\tilde{e} = e\mu^{\epsilon/2}$ is the coupling in $D = 4 - \epsilon$ dimensions.

Using the Feynman trick for the two propagators $\Delta_1 = k^2 + m^2$ and $\Delta_2 = (k+p)^2 + m^2$ and shifting the momenta as $k^\mu = q^\mu - \alpha p^\mu$:

$$\frac{1}{\Delta_1 \Delta_2} = \int_0^1 \frac{1}{[q^2 + m^2 + \alpha(1 - \alpha)p^2]^2}, \tag{36.16}$$

we get

$$\Pi_{\mu\nu}(p) = e^2 \mu^\epsilon \int_0^1 d\alpha \int \frac{d^D q}{(2\pi)^D} \frac{\mathrm{Tr}[\gamma_\mu(-i(\slashed{q} - \alpha\slashed{p}) + m)\gamma_\nu(-i(\slashed{q} + (1 - \alpha)\slashed{p}) + m)]}{[q^2 + m^2 + \alpha(1 - \alpha)p^2]^2}. \tag{36.17}$$

Remembering that the trace of an odd number of gamma matrices is zero, the trace in the numerator becomes

$$\mathrm{Tr}[] = -\mathrm{Tr}[\gamma_\mu(\slashed{q} - \alpha\slashed{p})\gamma_\nu(\slashed{q} + (1 - \alpha)\slashed{p})] + m^2 \, \mathrm{Tr}[\gamma_\mu \gamma_\nu]. \tag{36.18}$$

By Lorentz invariance, we know that

$$\int d^D q \, q^\mu f(q^2) = 0, \tag{36.19}$$

since there is no Lorentz-covariant constant object with only one index. This means that the integrals of the terms in the trace with a single $\slashed{q}$ are zero, so

$$\mathrm{Tr}[] \rightarrow -\mathrm{Tr}[\gamma_\mu \slashed{q} \gamma_\nu \slashed{q}] + \alpha(1 - \alpha) \, \mathrm{Tr}[\gamma_\mu \slashed{p} \gamma_\nu \slashed{p}] + 4m^2 \delta_{\mu\nu}. \tag{36.20}$$

Using (36.14), we see that

$$\mathrm{Tr}[\gamma_\mu \slashed{q} \gamma_\nu \slashed{q}] = 4(2q_\mu q_\nu - \delta_{\mu\nu} q^2), \tag{36.21}$$

and a similar relation with p.

The integrals appearing are

$$\int \frac{d^D q}{(2\pi)^D} \frac{1}{[q^2 + \alpha(1 - \alpha)p^2 + m^2]^2} = \frac{\Gamma\left(2 - \frac{D}{2}\right)}{(4\pi)^{\frac{D}{2}} [\alpha(1 - \alpha)p^2 + m^2]^{2 - \frac{D}{2}}}, \tag{36.22}$$

where we have used the general result (33.30) for $m^2 \to m^2 + \alpha(1-\alpha)p^2$, and the integral with $q_\alpha q_\beta$, which by Lorentz invariance should be rewritten as an integral of $\delta_{\alpha\beta}q^2/D$, giving

$$
\begin{aligned}
\int \frac{d^D q}{(2\pi)^D} \frac{q_\alpha q_\beta}{[q^2 + \alpha(1-\alpha)p^2 + m^2]^2} &= \frac{\delta_{\alpha\beta}}{D} \int \frac{d^D q}{(2\pi)^D} \frac{q^2}{[q^2 + \alpha(1-\alpha)p^2 + m^2]^2} \\
&= \frac{\delta_{\alpha\beta}}{D} \int \frac{d^D q}{(2\pi)^D} \frac{1}{[q^2 + \alpha(1-\alpha)p^2 + m^2]} \\
&\quad - \frac{\delta_{\alpha\beta}}{D}(\alpha(1-\alpha)p^2 + m^2) \int \frac{d^D q}{(2\pi)^D} \frac{1}{[q^2 + \alpha(1-\alpha)p^2 + m^2]^2} \\
&= \frac{\delta_{\alpha\beta}}{D} \frac{\Gamma\left(1 - \frac{D}{2}\right)}{(4\pi)^{\frac{D}{2}}} \frac{1}{[\alpha(1-\alpha)p^2 + m^2]^{1-\frac{D}{2}}} - \frac{\delta_{\alpha\beta}}{D} \frac{\Gamma\left(2 - \frac{D}{2}\right)}{(4\pi)^{\frac{D}{2}}} \frac{1}{[\alpha(1-\alpha)p^2 + m^2]^{1-\frac{D}{2}}} \\
&= -\frac{\delta_{\alpha\beta}}{(D-2)(4\pi)^{\frac{D}{2}}} \frac{\Gamma\left(2 - \frac{D}{2}\right)}{[\alpha(1-\alpha)p^2 + m^2]^{1-\frac{D}{2}}},
\end{aligned} \tag{36.23}
$$

where we have used $\Gamma(1-D/2) = \Gamma(2-D/2)/(1-D/2)$. Also appearing is the integral of q^2, for which we just remove the factor of $\delta_{\alpha\beta}/D$ from the above integral.

Putting all the factors together, we obtain

$$
\begin{aligned}
\Pi_{\mu\nu}(p) = 4e^2 \mu^\epsilon \int_0^1 d\alpha \Bigg[&-\frac{2\delta_{\mu\nu}}{(D-2)(4\pi)^{\frac{D}{2}}} \frac{\Gamma\left(2 - \frac{D}{2}\right)}{[\alpha(1-\alpha)p^2 + m^2]^{1-\frac{D}{2}}} \\
&+ 2\alpha(1-\alpha)p_\mu p_\nu \frac{\Gamma\left(2 - \frac{D}{2}\right)}{(4\pi)^{\frac{D}{2}}[\alpha(1-\alpha)p^2 + m^2]^{2-\frac{D}{2}}} \\
&+ \frac{D\delta_{\mu\nu}}{(D-2)(4\pi)^{\frac{D}{2}}} \frac{\Gamma\left(2 - \frac{D}{2}\right)}{[\alpha(1-\alpha)p^2 + m^2]^{1-\frac{D}{2}}} - \alpha(1-\alpha)p^2 \delta_{\mu\nu} \\
&\times \frac{\Gamma\left(2 - \frac{D}{2}\right)}{(4\pi)^{\frac{D}{2}}[\alpha(1-\alpha)p^2 + m^2]^{2-\frac{D}{2}}} + m^2 \delta_{\mu\nu} \frac{\Gamma\left(2 - \frac{D}{2}\right)}{(4\pi)^{\frac{D}{2}}[\alpha(1-\alpha)p^2 + m^2]^{2-\frac{D}{2}}} \Bigg].
\end{aligned} \tag{36.24}
$$

We can simplify this by taking a common denominator $[\alpha(1-\alpha)p^2 + m^2]^{2-\frac{D}{2}}$, and then canceling terms that appear to finally obtain, replacing $D = 4 - \epsilon$:

$$
\Pi_{\mu\nu}(p) = \frac{4e^2}{(4\pi)^2}(4\pi\mu^2)^{\frac{\epsilon}{2}} \Gamma\left(\frac{\epsilon}{2}\right) 2(p_\mu p_\nu - p^2 \delta_{\mu\nu}) \int_0^1 d\alpha \frac{\alpha(1-\alpha)}{[\alpha(1-\alpha)p^2 + m^2]^{\frac{\epsilon}{2}}}. \tag{36.25}
$$

Expanding in epsilon (using $\Gamma(\epsilon/2) = 2/\epsilon - \gamma$), we find

$$
\begin{aligned}
\Pi_{\mu\nu}(p) = &-(\delta_{\mu\nu}p^2 - p_\mu p_\nu) \frac{8e^2}{(4\pi)^2}\left(\frac{2}{\epsilon} - \gamma\right) \\
&\times \left[\frac{1}{6} - \frac{\epsilon}{2} \int_0^1 d\alpha\, \alpha(1-\alpha) \ln\left(\frac{\alpha(1-\alpha)p^2 + m^2}{4\pi\mu^2}\right)\right] + \mathcal{O}(\epsilon),
\end{aligned} \tag{36.26}
$$

where we have used $\int_0^1 d\alpha\, \alpha(1-\alpha) = 1/6$ and we have put the expansion of the factor $(4\pi\mu^2)^{\epsilon/2}$ together with the expansion of the power law in $\int d\alpha$. Note that the result is indeed written as $(\delta_{\mu\nu}p^2 - p_\mu p_\nu)\Pi(p^2)$, as it should be.

The counterterm corresponds to the divergent part of $\Pi_{\mu\nu}(p)$, and since $\delta_{\mu\nu}p^2 - p_\mu p_\nu$ corresponds to $-(\delta_{\mu\nu}\partial^2 - \partial_\mu\partial_\nu)$, the kinetic operator coming from $1/4(\partial_\mu A_\nu - \partial_\nu A_\mu)^2$, the counterterm is

$$\delta\mathcal{L}_A^{(1)} = -\frac{e^2}{12\pi^2}\frac{2}{\epsilon}\frac{1}{4}(\partial_\mu A_\nu - \partial_\nu A_\mu)^2 \equiv \frac{1}{4}(Z_3 - 1)(\partial_\mu A_\nu - \partial_\nu A_\mu)^2, \qquad (36.27)$$

leading to the wavefunction renormalization factor

$$Z_3 = 1 - \frac{e^2}{6\pi^2\epsilon}. \qquad (36.28)$$

36.4 Case 2: Fermion Self-energy $\Sigma(p)$

This is the term appearing in the inverse propagator:

$$S_F^{-1} = i\not p + m + \Sigma(p). \qquad (36.29)$$

The one-loop diagram for it is given by a fermion line between indices β and α, interrupted by a photon propagator starting and ending on it, starting at index μ and ending at index ν, with momentum k, as in Figure 36.2. The fermion has external momentum p, and inside the loop has momentum $p - k$. The result of the Feynman diagram is

$$\Sigma(p) = \int \frac{d^Dk}{(2\pi)^D}\left[i\tilde{e}\gamma_\mu \frac{1}{i(\not p - \not k) + m}\frac{\delta_{\mu\nu}}{k^2}i\tilde{e}\gamma_\nu \right]_{\alpha\beta}$$

$$= -e^2\mu^\epsilon \int \frac{d^Dk}{(2\pi)^D}\frac{[\gamma_\mu(-i(\not p - \not k) + im)\gamma^\mu]_{\alpha\beta}}{k^2((p-k)^2 + m^2)}. \qquad (36.30)$$

Note that the fermion propagator is gauge-dependent, which means that it is not observable (it is not physical). Here that is obscured, since we have used the Feynman gauge $\alpha = 1$, but the result differs in other gauges.

Again we use the Feynman trick for two propagators:

$$\frac{1}{k^2((p-k)^2 + m^2)} = \int_0^1 d\alpha \frac{1}{[(1-\alpha)\Delta_1 + \alpha\Delta_2]^2} = \frac{1}{[q^2 + \alpha m^2 + \alpha(1-\alpha)p^2]^2}, \qquad (36.31)$$

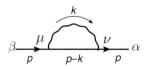

Figure 36.2   One-loop fermion self-energy diagram.

where we have shifted $k^\mu = q^\mu + \alpha p^\mu$.

Using also $\gamma_\mu \gamma^\mu = D$ and $\gamma_\mu \slashed{p} \gamma^\mu = (2 - D)\slashed{p}$ and the integral (36.22) (and again the fact that $\int d^D q\, q^\mu f(q^2) = 0$), we obtain

$$
\begin{aligned}
\Sigma(p)_{\alpha\beta} &= -e^2 \mu^\epsilon \int_0^1 \int \frac{d^D q}{(2\pi)^D} \frac{[-i\gamma_\mu((1-\alpha)\slashed{p} - \slashed{q})\gamma^\mu + m\gamma_\mu \gamma^\mu]_{\alpha\beta}}{[q^2 + \alpha(1-\alpha)p^2 + \alpha m^2]^2} \\
&= \frac{e^2 \mu^\epsilon \Gamma\left(\frac{\epsilon}{2}\right)}{(4\pi)^{2-\frac{\epsilon}{2}}} \int_0^1 d\alpha \frac{[(2-\epsilon)(1-\alpha)(-i\slashed{p}) - (4-\epsilon)m]_{\alpha\beta}}{[\alpha(1-\alpha)p^2 + \alpha m^2]^{\epsilon/2}}.
\end{aligned}
\tag{36.32}
$$

Expanding in ϵ, we obtain

$$
\begin{aligned}
\Sigma(p)_{\alpha\beta} = \frac{e^2}{(4\pi)^2} \left(\frac{2}{\epsilon} - \gamma\right) &\left[(-i\slashed{p})\left(1 - \frac{\epsilon}{2}\right) - (4-\epsilon)m \right. \\
&\left. - \frac{\epsilon}{2} \int_0^1 d\alpha [2(1-\alpha)(-i\slashed{p}) - 4m] \ln \frac{\alpha(1-\alpha)p^2 + \alpha m^2}{4\pi \mu^2} \right] + \mathcal{O}(\epsilon). \quad (36.33)
\end{aligned}
$$

Here again we have put the expansion of $(4\pi\mu^2)^{\epsilon/2}$ in the ln, so that we form the ratio $m^2/(4\pi\mu^2)$, as before.

The divergent part of the fermion self-energy is finally

$$
\Sigma(p)_{\alpha\beta} = \frac{e^2}{(4\pi)^2} \frac{2}{\epsilon}(-i\slashed{p} - 4m).
\tag{36.34}
$$

There is a term with $i\slashed{p} = \slashed{\partial}$ and a mass term, which means that we need to add a counterterm to $\bar{\psi}\slashed{\partial}\psi$ and one to $m\bar{\psi}\psi$:

$$
\delta\mathcal{L}_\psi^{(1)} = -\frac{e^2}{(4\pi)^2} \frac{2}{\epsilon} \bar{\psi}\slashed{\partial}\psi - \frac{e^2}{(4\pi)^2} \frac{8}{\epsilon} m\bar{\psi}\psi.
\tag{36.35}
$$

Defining the wavefunction renormalization of the fermions

$$
\psi_0 = \sqrt{Z_2}\psi, \quad \bar{\psi}_0 = \sqrt{Z_2}\bar{\psi}
\tag{36.36}
$$

from the identification of the first counterterm as $(Z_2 - 1)\bar{\psi}\slashed{\partial}\psi$, we obtain

$$
Z_2 = 1 - \frac{e^2}{(4\pi)^2} \frac{2}{\epsilon}.
\tag{36.37}
$$

Then, defining the renormalization of the fermion mass as

$$
m_0 = \frac{Z_m}{Z_2} m,
\tag{36.38}
$$

we obtain

$$
Z_m = 1 - \frac{e^2}{(4\pi)^2} \frac{8}{\epsilon}.
\tag{36.39}
$$

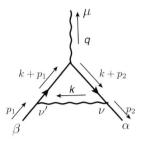

Figure 36.3 One-loop fermions–photon vertex diagram.

36.5 Case 3: Fermions–Photon Vertex $\Gamma_{\mu\alpha\beta}$

We finally consider the one-loop correction to the fermion–fermion–photon vertex $\Gamma^{(1)}_{\mu\alpha\beta}(q;p_1,p_2)$. We consider the correction to the vertex given by a photon line connecting the two fermion lines, as in Figure 36.3. The external photon has index μ and (outgoing) momentum q, the external fermion lines have index β and (incoming) momentum p_1 and index α and (outgoing) momentum p_2, the internal photon line runs between ν' on p_1 and ν on p_2, modifying the fermion momenta to $p_1 + k$ and $p_2 + k$ on the internal lines.

The result of the Feynman diagram is

$$
\begin{aligned}
\Gamma^{(1)}_{\mu\alpha\beta}(q;p_1,p_2) &= \int \frac{d^D k}{(2\pi)^D} \frac{\delta_{\nu\nu'}}{k^2} \left[i\tilde{e}\gamma_\nu \frac{1}{+i(\not{p}_2 + \not{k}) + m} i\tilde{e}\gamma_\mu \frac{1}{i(\not{p}_1 + \not{k}) + m} i\tilde{e}\gamma_{\nu'} \right]_{\alpha\beta} \\
&= -ie^3 \mu^{3\epsilon/2} \int \frac{d^D k}{(2\pi)^D} \frac{\left(\gamma_\nu[-i(\not{p}_2 + \not{k}) + m]\gamma_\mu[-i(\not{p}_1 + \not{k} + m)]\gamma_\nu\right)_{\alpha\beta}}{k^2[(p_2 + k)^2 + m^2][(p_1 + k)^2 + m^2]}.
\end{aligned}
$$

$$(36.40)$$

We use the Feynman parametrization for the three propagators in the denominator, giving

$$
\begin{aligned}
&\frac{1}{k^2[(p_2 + k)^2 + m^2][(p_1 + k)^2 + m^2]} \\
&= \int_0^1 d\alpha_1 \int_0^{1-\alpha_1} d\alpha_2 \frac{1}{\left([(p_1 + k)^2 + m^2]\alpha_1 + [(p_2 + k)^2 + m^2]\alpha_2 + k^2(1 - \alpha_1 - \alpha_2)\right)^3} \\
&= \int_0^1 d\alpha_1 \int_0^{1-\alpha_1} d\alpha_2 \frac{1}{\left(q^2 + m^2(\alpha_1 + \alpha_2) + \alpha_1 p_1^2 + \alpha_2 p_2^2 - (\alpha_1 p_1 + \alpha_2 p_2)^2\right)^3} \\
&\equiv \int_0^1 d\alpha_1 \int_0^{1-\alpha_1} d\alpha_2 \frac{1}{\left(q^2 + F(\alpha_1, \alpha_2, p_1, p_2, m)\right)^3},
\end{aligned}
$$

$$(36.41)$$

where we have eliminated $\int d\alpha_3 \delta(1 - \alpha_1 - \alpha_2 - \alpha_3)$ and also used it to set the maximum value for α_2 at $1 - \alpha_1$, and we have defined the function F to simplify notation.

Finally, the one-loop correction to the vertex is

$$\Gamma^{(1)}_{\mu\alpha\beta}(q;p_1,p_2) = -ie^3\epsilon^{3\epsilon/2} \int_0^1 d\alpha_1 \int_0^{1-\alpha_1} d\alpha_2 \int \frac{d^D q}{(2\pi)^D}$$

$$\times \frac{\{\gamma_\nu[-i\slashed{q} - i(1-\alpha_2)\slashed{p}_2 + i\alpha_1\slashed{p}_1 + m]\gamma_\mu[-i\slashed{q} - i(1-\alpha_1)\slashed{p}_1 + i\alpha_2\slashed{p}_2 + m]\gamma^\nu\}_{\alpha\beta}}{(q^2+F)^3}.$$

(36.42)

We note that, because of Lorentz invariance, the terms linear in $\slashed{q}$ give zero by integration, so we are left with terms quadratic in q in the numerator, generating a term that we will call $\Gamma^{(1,a)}_{\mu\alpha\beta}$, and a term with no q in the numerator, generating a term that we will call $\Gamma^{(1,b)}_{\mu\alpha\beta}$. Obviously $\Gamma^{(1b)}$ will be convergent (finite), since it is $\int d^4q/(q^2+F)^3$. In contrast, $\Gamma^{(1a)}_{\mu\alpha\beta}$ is UV divergent, since it is $d^4q\,q^2/(q^2+F)^3$. But it is also IR divergent, as we will see later in the book.

We concentrate on $\Gamma^{(1a)}$, since we are interested only in the divergence. It is given by

$$\Gamma^{(1a)}_{\mu\alpha\beta} = +ie^3\mu^{3\epsilon/2} \int_0^1 d\alpha_1 \int_0^{1-\alpha_1} d\alpha_2 \int \frac{d^D q}{(2\pi)^D} \frac{\{\gamma_\nu\slashed{q}\gamma_\mu\slashed{q}\gamma^\nu\}_{\alpha\beta}}{(q^2+F)^3}.$$

(36.43)

But from the relations (36.13), we find that

$$\gamma_\nu\slashed{q}\gamma_\mu\slashed{q}\gamma^\nu = -2\slashed{q}\gamma_\mu\slashed{q} - (D-4)\slashed{q}\gamma_\mu\slashed{q} = (2-D)\slashed{q}\gamma_\mu\slashed{q}.$$

(36.44)

But because $\int d^D q\, q^\alpha q^\beta f(q^2) = \int d^D q\, q^2 \delta_{\alpha\beta}/Df(q^2)$, we can replace the above with

$$\frac{(2-D)}{D}q^2\gamma_\alpha\gamma_\mu\gamma^\alpha = \frac{(2-D)^2}{D}q^2\gamma_\mu.$$

(36.45)

Then we obtain

$$\Gamma^{(1a)}_{\mu\alpha\beta} = +ie^3\mu^{3\epsilon/2}\frac{(2-D)^2}{D}(\gamma_\mu)_{\alpha\beta} \int_0^1 d\alpha_1 \int_0^{1-\alpha_1} d\alpha_2 \int \frac{d^D q}{(2\pi)^D}\frac{q^2}{(q^2+F)^3}$$

$$= +ie^3\mu^{3\epsilon/2}\frac{(2-D)^2}{D}(\gamma_\mu)_{\alpha\beta} \int_0^1 d\alpha_1 \int_0^{1-\alpha_1} d\alpha_2$$

$$\times \left[\frac{\Gamma\left(2-\frac{D}{2}\right)}{(4\pi)^{\frac{D}{2}}}(F^2)^{\frac{D}{2}-2} - F^2\frac{\Gamma\left(3-\frac{D}{2}\right)}{(4\pi)^{\frac{D}{2}}}(F^2)^{\frac{D}{2}-3}\right].$$

(36.46)

We see that the divergence comes from the first term in the square brackets, replacing $\Gamma\left(2-\frac{D}{2}\right) = 2/\epsilon + \ldots$, and so we can put $D=4$ in the rest of the integral, obtaining

$$\Gamma^{(1a)}_{\mu\alpha\beta,\text{div.}} = +ie^3(\gamma_\mu)_{\alpha\beta}\int_0^1 d\alpha_1 \int_0^{1-\alpha_1} d\alpha_2 \frac{1}{(4\pi)^2}\frac{2}{\epsilon} = +ie(\gamma_\mu)_{\alpha\beta}\frac{e^2}{(4\pi)^2\epsilon}.$$

(36.47)

This means that the counterterm is

$$\delta\mathcal{L}^{(1)} = \left[-\frac{e^2}{(4\pi)^2}\frac{1}{\epsilon}\right](-i\bar\psi\slashed{A}\psi),$$

(36.48)

and we can identify the coefficient in the square brackets as $Z_1 - 1$, the wavefunction renormalization of A, defining $A_0 = Z_1/Z_2 A$. Then we get

$$Z_1 = 1 - \frac{e^2}{(4\pi)^2}\frac{1}{\epsilon}. \tag{36.49}$$

We finally note that all the one-loop divergences were of the form of terms in the Lagrangians, so could be removed by renormalization and counterterms. Hence QED is one-loop renormalizable.

Important Concepts to Remember

- In QED, due to the Ward–Takahashi (generalized) identities, the *effective* degree of divergence of the n-photon vertex, $\Gamma^{(n)}$, is reduced by n to $\omega_{\text{eff}}(D) = 4 - 2n$.
- The superficial degree of divergence in the presence of fermions is $\omega(D) = dL - 2I_{bos} - I_f$, and for QED we obtain $\omega(D) = 4 - E_{ph} - 3E_f/2$.
- The divergent one-loop graphs in QED are in $\Pi_{\mu\nu}(p)$, $\Sigma(p)_{\alpha\beta}$, and $\Gamma_{\mu\alpha\beta}$.
- The one-loop divergences in QED can be removed by adding counterterms and renormalizing, since they have the same form as the terms in the Lagrangian.

Further Reading

See section 6.4 in [3] and section 10.3 in [1].

Exercises

1. Consider $\Gamma^{(1)}_{\alpha\beta}$ from the text. Calculate the general $\{\}_{\alpha\beta}$ matrix element. Calculate also the finite part of $\Gamma^{(1a)}_{\alpha\beta}$ *in the absence of IR divergences*.
2. Calculate the one-loop anomalous dimensions γ_m, γ_d for ψ.
3. Write explicitly the renormalization group equation and its explicit solution for the one-loop $\Sigma^{(1)}(p)_{\alpha\beta}$.
4. Calculate what are, and draw, the divergent Feynman diagrams in QED at two loops.

37 Physical Applications of One-Loop Results I: Vacuum Polarization

In this chapter, we will consider finally the first physical application of the quantum corrections arising from one-loop results and renormalization. Specifically, we will consider the vacuum polarization, and its applications for the screening of electric charge, and the pair creation rate.

But before doing that, we understand in a more systematic way the renormalization of quantum electrodynamics.

37.1 Systematics of QED Renormalization

We have seen three renormalization factors:

$$A_0 = \sqrt{Z_3}A, \quad \psi_0 = \sqrt{Z_2}\psi, \quad m_0 = \frac{Z_m}{Z_2}m, \tag{37.1}$$

and a vertex renormalization with Z_1, of the $\bar{\psi}\slashed{A}\psi$ term.

The renormalized Lagrangian is then written as

$$\mathcal{L}_{\text{ren}} = \frac{Z_3}{4}F_{\mu\nu}^2 + Z_2\bar{\psi}\slashed{\partial}\psi + Z_m\bar{\psi}\psi \\ + \frac{1}{2\alpha}(\partial^\mu A_\mu)^2 + Z_1(-ie\bar{\psi}\slashed{A}\psi), \tag{37.2}$$

where on the first line we have the renormalizations following from the three factors computed in Chapter 36. On the second line, we have a renormalization of the gauge-fixing term, considering it has no independent renormalization factor, but which, through the wavefunction renormalization of A_μ, implies still a renormalization

$$\alpha_0 = Z_3\alpha, \tag{37.3}$$

and a renormalization of the vertex function

$$e_0 = \frac{Z_1}{Z_2\sqrt{Z_3}}e, \tag{37.4}$$

defined such that the overall factor is simply Z_1.

But we first note that, at least at one loop, we have

$$Z_1^{\text{one-loop}} = Z_2^{\text{one-loop}}. \tag{37.5}$$

In fact, this is the result of the Ward–Takahashi identity:

$$p^\mu \Gamma_{\mu\alpha\beta}(p; q_1, q_2) = e(S_F^{-1}{}_{\alpha\beta}(q_2) - S_F^{-1}{}_{\alpha\beta}(q_1)), \tag{37.6}$$

since the left-hand side is related to the $\bar\psi \slashed{A} \psi$ vertex factor Z_1, and the right-hand side to the propagator term $\bar\psi \slashed{\partial} \psi$ with factor Z_2 (in the propagator there could be a momentum-independent contribution, related to Z_m, the coefficient of the mass term, but it cancels between the two S_F^{-1}'s of different momentum). Now the above relation actually implies only the result for the divergent parts, $Z_{1,div.} = Z_{2,div.}$, but since we are in the MS scheme, this is the whole factor. Then the result is actually exact to all loops:

$$Z_1 = Z_2. \tag{37.7}$$

This then implies for the vertex function

$$e_0 = \frac{e}{\sqrt{Z_3}}, \tag{37.8}$$

which to one loop gives

$$e_0 = e\mu^{\epsilon/2}\left(1 + \frac{e^2}{24\pi^2}\frac{2}{\epsilon} + \mathcal{O}(e^3)\right). \tag{37.9}$$

In turn, this gives for the beta function

$$\mu\frac{\partial}{\partial\mu}e \equiv \beta(e) = \frac{e^3}{12\pi^2} + \mathcal{O}(e^5). \tag{37.10}$$

This is solved by

$$\bar{e}^2(t) = \frac{e_0^2}{1 - \frac{e_0^2}{12\pi^2}\ln t^2}, \tag{37.11}$$

where $e_0 = e(\mu_0)$. This means that there is a Landau pole. Since $e_0^2/4\pi \simeq 1/137$ at about the eV scale (the scale of the H atom energy levels), we obtain a Landau pole at about $e^{12\pi^2/e_0^2} \sim e^{1370}$ eV.

37.2 Vacuum Polarization

Inside a nontrivial medium, the effective action is written in terms of the electric and magnetic fields as

$$S_{\text{eff}}(\vec{E}, \vec{B}) = \frac{1}{2}\int dt \int d^3x \left[\epsilon\vec{E}^2 - \frac{\vec{B}^2}{\mu}\right], \tag{37.12}$$

and the speed of light inside the medium is

$$c_{\text{medium}} = \frac{1}{\sqrt{\epsilon\mu}}. \tag{37.13}$$

The Coulomb potential associated with a static pointlike charge e is

$$eA_\mu^{\text{Coulomb}}(\vec{x}) = \frac{e^2}{4\pi\epsilon|\vec{x}|}\delta_{\mu 0}. \tag{37.14}$$

The dielectric function ϵ is in general a function, and not a constant, and is usually defined in momentum space, as the ratio of the electric induction $\vec{D}$ and the electric field $\vec{E}$:

$$\epsilon(\omega, \vec{k}) \equiv \frac{\vec{D}(\omega, \vec{k})}{\vec{E}(\omega, \vec{k})}. \tag{37.15}$$

Then really, the Coulomb potential in momentum space for a static source ($\omega = 0$) is

$$eA_\mu^{\text{Coulomb}}(\vec{k}, t) = \frac{e^2}{\vec{k}^2\epsilon(0, k)}\delta_{\mu 0}. \tag{37.16}$$

Now consider the same situation in vacuum, but with nontrivial quantum corrections (i.e. with nontrivial vacuum polarization). We can again formally consider it as a nontrivial "medium" with $\epsilon, \mu \neq 1$, but unlike a real medium, now the velocity of light must be exactly equal to 1, which means

$$\epsilon(\omega, \vec{k})\mu(\omega, \vec{k}) = 1. \tag{37.17}$$

The quantum effective action starts at quadratic order:

$$\Gamma(A_\mu^{cl}) = \frac{1}{2}\int \Gamma_{\mu\nu}^{(2)}A_\mu^{cl}A_\nu^{cl} + \mathcal{O}((A_\rho^{cl})^3), \tag{37.18}$$

where the quadratic part is written as the inverse propagator, which equals the free inverse propagator plus the vacuum polarization:

$$\Gamma_{\mu\nu}^{(2)} = G_{\mu\nu}^{-1} = G_{\mu\nu}^{(0)-1} + \Pi_{\mu\nu}, \tag{37.19}$$

so that more precisely (writing the vacuum polarization in terms of $\Pi(k^2)$):

$$\Gamma^{(2)}[A_\mu^{cl}] = \frac{1}{2}\int \frac{d^4k}{(2\pi)^4} A_\mu^{cl}(-k)G_{\mu\nu}^{-1}(k)A_\nu^{cl}(k),$$

$$G_{\mu\nu}^{-1} = (k^2\delta_{\mu\nu} - k_\mu k_\nu)(1 + \Pi(k^2)) + \frac{k_\mu k_\nu}{\alpha}. \tag{37.20}$$

It would seem like there is some gauge dependence in this effective action, due to the term with the gauge parameter α, but really there isn't, since if the classical current source is conserved (i.e. $\partial^\mu J_\mu = 0$) in momentum space $p^\mu J_\mu(p) = 0$, then

$$k^\mu J_\mu(k) = k^\mu \frac{\delta\Gamma}{\delta A_\mu^{cl}} = \frac{1}{\alpha}k^2 k_\mu A_\mu^{cl}, \tag{37.21}$$

which in turn means that the extra term in $\Gamma^{(2)}$ vanishes, so that

$$\Gamma^{(2)}[A_\mu^{cl}] = \frac{1}{2}\int \frac{d^4k}{(2\pi)^4}(1 + \Pi(k^2))A_\mu(-k)(k^2\delta_{\mu\nu} - k_\mu k_\nu)A_\nu^{cl}(k). \tag{37.22}$$

We must now Wick rotate it to Minkowski space in order to be able to extract physical information, by

$$\delta_{\mu\nu} \to g_{\mu\nu}$$
$$A^{\mu}(-k)(k^2 g_{\mu\nu} - k_{\mu}k_{\nu})A^{\nu}(k) = E_i(-k)E_i(k) - B_i(-k)B_i(k). \qquad (37.23)$$

Finally, the quantum-corrected effective action in Minkowski space is

$$\Gamma^{(2)}[A_{\mu}^{cl}] = \frac{1}{2} \int \frac{dk_0}{2\pi} \int \frac{d^3k}{(2\pi)^3} (1 + \Pi(k^2))(|\vec{E}(k_0, \vec{k})|^2 - |\vec{B}(k_0, \vec{k})|^2). \qquad (37.24)$$

From this, we can extract ϵ and μ for the vacuum:

$$\epsilon(k^2) = \frac{1}{\mu(k^2)} = 1 + \Pi(k^2). \qquad (37.25)$$

Keeping only the quadratic part of Γ, and remembering that $J_{\mu} = \delta\Gamma/\delta A_{\mu}^{cl}$, we obtain

$$J_{\mu} \simeq G_{\mu\nu}^{-1} A_{\nu}^{cl}, \qquad (37.26)$$

so that

$$A_{\mu}^{cl} \simeq G_{\mu\nu} J_{\nu}(k) = \frac{J_{\mu}}{k^2(1 + \Pi(k^2))}, \qquad (37.27)$$

where we have used the explicit form of $G_{\mu\nu}^{-1}$, and in inverting it, we have considered the fact that $J_{\nu}k^{\mu}k^{\nu} = 0$.

The Coulomb potential of a static pointlike source of

$$J_{\mu}(\vec{x}, t) = e\delta^{(3)}(\vec{x})\delta_{\mu 0} \Rightarrow J_{\mu}(k) = 2\pi e\delta(k_0)\delta_{\mu 0} \qquad (37.28)$$

for Π that depends on $k^2 = \vec{k}^2$ only is

$$eA_0^{cl}(\vec{k}, k_0) \simeq \frac{e^2}{\vec{k}^2(1 + \Pi(\vec{k}^2))} 2\pi\delta(k_0), \qquad (37.29)$$

leading to the effective coupling

$$e_{\text{eff}}^2 = \frac{e^2}{1 + \Pi(\vec{k}^2)}. \qquad (37.30)$$

Note that we have naturally $e_{\text{eff}}^2(\vec{k}^2)$, but since $|\vec{k}| \sim 1/r$, we can think of this as $e_{\text{eff}}^2(r)$. Then in the extreme IR, we have

$$e_{\text{eff}}^2(r \to \infty) = \frac{e^2}{1 + \Pi(0)}. \qquad (37.31)$$

This relation is interpreted physically as *screening of the electric charge of the pointlike source*. Indeed, we see that the effective charge is smaller than the free one. Therefore, we can interpret this in the same way as interpreting the screening of charge in a polarizable medium. The charge is effectively screened, since dipoles (charge pairs) orient themselves so as to screen the outside charge (opposite charge closer to the source). The difference is of course that in a material, it is only a local effect; globally, because of charge conservation, we have the same charge. But here, the charge is interpreted as a continuous coupling, and it can be made smaller (screened) by the interaction with the (polarizable) vacuum.

We also note that, since we are dealing with QED, which, as we argued, is defined only in the IR, it makes sense to define the physical (observed) value of the coupling in the extreme IR, at $r \to \infty$. Defining thus $e^e_{\text{eff}}(r \to \infty) = e^2$, we have the (natural) normalization condition for renormalization

$$\Pi(0) = 0. \tag{37.32}$$

Remembering that the finite part of the vacuum polarization at one loop was given by

$$\Pi(k^2) = -\frac{e^2}{2\pi^2} \left[\frac{\psi(1)}{6} + \int_0^1 d\alpha\, \alpha(1-\alpha) \ln \left[\frac{k^2\alpha(1-\alpha) + m^2 - i\epsilon}{4\pi\mu^2} \right] \right], \tag{37.33}$$

but switching now from the minimal subtraction scheme to the $\Pi(0) = 0$ normalization condition (i.e. dropping the nonzero terms at $k = 0$ in the above), we obtain

$$\Pi(k^2) = -\frac{e^2}{2\pi^2} \int_0^1 d\alpha\, \alpha(1-\alpha) \ln \frac{k^2\alpha(1-\alpha) + m^2 - i\epsilon}{m^2}. \tag{37.34}$$

It is left as an exercise to prove that, if $-k^2 \leq 4m^2$, the above integral for $k^2 = -\vec{k}^2$ reduces to

$$\Pi(\vec{k}^2) = -\frac{e^2}{12\pi^2} \vec{k}^2 \int_{4m^2}^\infty \frac{dq^2}{q^2} \frac{1}{q^2 + \vec{k}^2} \left(1 + \frac{2m^2}{q^2} \right) \sqrt{1 - \frac{4m^2}{q^2}}. \tag{37.35}$$

Since we are at small $\Pi(\vec{k}^2)$ (it is of $\mathcal{O}(e^2)$), $1/(1 + \Pi) \simeq 1 - \Pi$, so

$$A_0^{cl}(k_0, \vec{k}) \simeq \frac{e}{\vec{k}^2} (1 - \Pi(\vec{k}^2)) 2\pi\, \delta(k_0) \Rightarrow$$

$$A_0(t, \vec{x}) = e \int \frac{d^3k}{(2\pi)^3} \frac{e^{-i\vec{k}\cdot\vec{x}}}{\vec{k}^2} \left(1 + \frac{e^2}{12\pi^2} \vec{k}^2 \int_{4m^2}^\infty \frac{dq^2}{q^2} \frac{1}{q^2 + \vec{k}^2} \left(1 + \frac{2m^2}{q^2} \right) \sqrt{1 - \frac{4m^2}{q^2}} \right). \tag{37.36}$$

Using

$$\int \frac{d^3k}{(2\pi)^3} \frac{e^{-i\vec{k}\cdot\vec{x}}}{\vec{k}^2} = \frac{1}{4\pi r}; \quad \int \frac{d^3k}{(2\pi)^3} \frac{e^{-i\vec{k}\cdot\vec{x}}}{q^2 + \vec{k}^2} = \frac{e^{-qr}}{4\pi r} \tag{37.37}$$

in the two terms above, and defining $u^2 = q^2/4m^2$, we get

$$A_0^{cl}(\vec{x}, t) = \frac{e}{4\pi r} \left(1 + \frac{e^2}{6\pi^2} \int_1^\infty \frac{du}{u^2} e^{-2mru} \left(1 + \frac{1}{2u^2} \right) \sqrt{u^2 - 1} \right). \tag{37.38}$$

We consider the extreme limits of this formula. At large distances, $mr \gg 1$, we get

$$A_0^{cl}(t, \vec{x}) \simeq \frac{e}{4\pi r} \left(1 + \frac{e^2}{16} \frac{e^{-2mr}}{(\pi mr)^{3/2}} + \cdots \right), \tag{37.39}$$

whereas at small distances, $mr \ll 1$, we get

$$A_0^{cl}(t, \vec{x}) \simeq \frac{e}{4\pi r} \left(1 + \frac{e^2}{12\pi^2} \ln \frac{1}{(mr)^2} + \text{constant} + \ldots \right). \qquad (37.40)$$

In the large-distance formula we just notice again the fact mentioned previously, that this is consistent with screening, since we obtain $e_0 > e$. In the small-distance formula, however, we also notice another thing, namely that the effective charge *diverges* in the extreme UV, at $mr \to 0$. Therefore, the screening with respect to the UV is infinite: we have an infinite effective charge at $r = 0$, but it is screened down to a finite value in the IR. This is consistent with the picture of renormalization we have advocated: infinite quantities in the UV, which can be renormalized down to finite ones in the IR.

37.3 Pair Creation Rate

Previously, we considered the case of $-k^2 \leq 4m^2$, but now we can also consider the opposite case, of $-k^2 \geq 4m^2$. In this case, $k^0 \geq 2m$, so we have sufficient energy to create an electron–positron pair from the vacuum. The vacuum-to-vacuum transition amplitude in the presence of an external source J is the partition function Z ($= e^{-W}$ in Euclidean space), written in terms of the effective action ($W = \Gamma + J \cdot A^{cl}$) in Minkowski space as

$$_J\langle 0|0\rangle_J \equiv Z[J] = e^{i\Gamma[A_\mu^{cl}] + iJ \cdot A^{cl}}. \qquad (37.41)$$

If Γ has an imaginary part, we can have an absolute value different from 1, $|Z|^2 = e^{-2\mathrm{Im}[\Gamma]}$, which is interpreted as *vacuum decay*, that is the probability of the vacuum going to itself is not 1 anymore, and the difference is due, as we explained, to pair creation:

$$R = 1 - |_J\langle 0|0\rangle_J|^2 = 1 - e^{-2\mathrm{Im}[\Gamma]} \simeq 2\mathrm{Im}[\Gamma[A_\mu^{cl}] + \mathcal{O}(e^4). \qquad (37.42)$$

It remains of course to show that exactly when $-k^2 \geq 4m^2$ (i.e. when we can create pairs), we do create them (i.e. $\Gamma^{(2)}$ has an imaginary part). We will see this through explicit calculation. In the formula (37.34) the log can give us an imaginary part. Indeed, if the argument of the log is negative, we will have a term $\log(-1) = -i\pi$ added. Since $k^2 \leq -4m^2 \leq 0$, this will happen if $\alpha(1 - \alpha) + m^2/k^2$ is positive, therefore we obtain

$$\mathrm{Im}\Pi(k^2) = \frac{e^2}{2\pi} \int_0^1 d\alpha \, \alpha(1 - \alpha)\theta\left(\alpha(1 - \alpha) + \frac{m^2}{k^2}\right), \qquad (37.43)$$

where θ is the Heaviside function. Its roots are at

$$\alpha(1 - \alpha) = -\frac{m^2}{k^2} \Rightarrow \alpha_{1,2} = \frac{1 \pm \sqrt{1 + \frac{4m^2}{k^2}}}{2}, \qquad (37.44)$$

and the positivity condition of the Heaviside function for the inverted parabola is at $\alpha_1 \leq \alpha \leq \alpha_2$. We see then that the condition for this to be nonzero is indeed the condition we advocated, $1 + 4m^2/k^2 \geq 0$ (i.e. $-k^2 \geq 4m^2$). Then the integral is

$$\int_{\alpha_1}^{\alpha_2} d\alpha \, \alpha (1 - \alpha) = \left(\frac{\alpha^2}{2} - \frac{\alpha^3}{3} \right) \Big|_{\alpha_1}^{\alpha_2} = \frac{1}{4} \sqrt{1 + \frac{4m^2}{k^2}} \left(\frac{2}{3} - \frac{m^2}{3k^2} \right). \tag{37.45}$$

Finally, we obtain for the imaginary part of the vacuum polarization

$$\mathrm{Im}\,\Pi(k^2) = \frac{e^2}{12\pi} \sqrt{1 + \frac{4m^2}{k^2}} \left(1 - \frac{m^2}{2k^2} \right) \theta \left(1 + \frac{4m^2}{k^2} \right). \tag{37.46}$$

Here the condition of reality of $\mathrm{Im}\,\Pi$ would have been enough to show $-k^2 \geq 4m^2$, but we have put it explicitly with the Heaviside function for completeness.

The imaginary part of the effective action is

$$\mathrm{Im}\,\Gamma^{(2)}[A_\mu^{cl}] = \frac{1}{2} \int \frac{d^4k}{(2\pi)^4} \mathrm{Im}\,\Pi(k^2)(|\vec{E}^2(k_0, \vec{k})|^2 - |\vec{B}^2(k_0, \vec{k})|^2), \tag{37.47}$$

and the pair creation rate is $R = 2\mathrm{Im}\,\Gamma^{(2)}$.

We observe that pair creation from the vacuum is a purely electric effect, since if $k^2 \leq -4m^2 < 0$, this means we can choose a center of mass reference frame where $\vec{k} = 0$, and then the magnetic field $\vec{B} = -i\vec{k} \times \vec{A}(k) = 0$. Another way of seeing this is that the pair creation rate is positive, so only the electric field contributes, with $+|\vec{E}|^2$, whereas the magnetic field, with $-|\vec{B}|^2$, doesn't.

Important Concepts to Remember

- The renormalization of the α parameter matches that of the wavefunction of A, $\alpha_0 = Z_3 \alpha$, whereas from the Ward–Takahashi identity, $Z_1 = Z_2$, exact to all loop orders, which means that $e_0 = e/\sqrt{Z_3}$.
- The vacuum is considered like a medium with nontrivial $\epsilon(k_0, \vec{k})$ and $\mu(k_0, \vec{k})$, just that, because of relativistic invariance, signals propagate at $c = 1 = 1/\sqrt{\epsilon\mu}$, meaning that $\epsilon\mu = 1$.
- We obtain $\epsilon = 1/\mu = 1 + \Pi(k^2)$ and a screening of electric charge, $e_{\mathrm{eff}}^2 = e^2/(1 + \Pi(k^2))$.
- We can choose a normalization condition such that $\Pi(0) = 0$, identifying $e_{\mathrm{eff}}^2(r \to \infty)$ with e^2. Then, at $r \to 0$, we obtain a divergence in e_{eff}, meaning we have an infinite screening from the UV to the IR.
- An imaginary part of $\Pi(k^2)$ leads to vacuum decay through pair creation, happening when $-k^2 \geq 4m^2$, which is a purely electric effect (pair creation in an electric field).

Further Reading

See section 6.5.1 in [3].

Exercises

1. Prove the result assumed in the text, that if $-k^2 \leq 4m^2$:

$$\Pi(k^2) = -\frac{e^2}{2\pi^2} \int_0^1 d\alpha\, \alpha(1-\alpha) \ln \frac{k^2\alpha(1-\alpha) + m^2 - i\epsilon}{m^2}$$

$$= -\frac{e^2}{12\pi^2} \vec{k}^2 \int_{4m^2}^\infty \frac{dq^2}{q^2} \frac{1}{q^2 + \vec{k}^2} \left(1 + \frac{2m^2}{q^2}\right) \sqrt{1 - \frac{4m^2}{q^2}}. \qquad (37.48)$$

2. Calculate the e^+e^- pair creation rate for $\vec{B} = 0$ and an electric field

$$|\vec{E}(k)| = E_0 = \text{constant}, \qquad (37.49)$$

as well as for an electric field

$$|\vec{E}(k)| = \delta^{(3)}(\vec{k})\theta(k_0 - 2M), \qquad (37.50)$$

where $M > m$.

3. Consider a hypothetical two-loop correction to the beta function of QED, $\beta(e)$, of the form ce^5, with c a real number of either sign. If that were the full beta function, would it be possible to avoid the Landau pole?

4. At what minimum distance $r_{\min}$ does the formula (37.40) stop being valid?

Physical Applications of One-Loop Results II: Anomalous Magnetic Moment and Lamb Shift

In this chapter, we continue with the physical applications of one-loop results, describing two classic tests of radiative corrections in quantum electrodynamics. First, the anomalous magnetic moment of the electron, for which we calculate the deviation from the zeroth-order quantum-mechanical result, $g = 2$. Second, for the Lamb shift, we show how to obtain corrections coming from the Schrödinger equation, then from the Dirac equation. Finally, we just describe the quantum field theory calculation, and the corresponding Feynman diagrams that contribute, without calculating them, just stating the result.

38.1 Anomalous Magnetic Moment

The first application is related to the anomalous magnetic moment of the electron.

Classically, a particle of electric charge q, with orbital angular momentum $\vec{L}$, has a magnetic moment of

$$\vec{\mu} = \frac{q}{2m}\vec{L}. \tag{38.1}$$

But quantum mechanically, for the electron of charge e, the spin $\vec{S}$ also has a contribution to the magnetic moment:

$$\vec{\mu}_{\text{class}} = \frac{e}{m}\vec{S} = g\left(\frac{e}{2m}\right)\vec{S}, \tag{38.2}$$

where the Landé g-factor is classically (i.e. in nonrelativistic quantum mechanics)

$$g_{\text{spin,classical}} = 2. \tag{38.3}$$

But in QED, one obtains quantum corrections:

$$g = 2\left(1 + \frac{\alpha}{2\pi} + \ldots\right) = 2(1 + 0.001159652359\ldots), \tag{38.4}$$

where we have written the numerical expression including $\mathcal{O}(\alpha^3)$ corrections and more, and the 12 digits written are all verified experimentally to be correct. This is one of the most impressive tests of QED, and we will derive here the first-order term (α/π).

Consider the relativistic Dirac equation

$$(\not{D} + m)\psi = 0 \Rightarrow (-\not{D} + m)(\not{D} + m)\psi = 0. \tag{38.5}$$

But since $[D_\mu, D_\nu] = -ieF_{\mu\nu}$, $[\gamma_\mu, \gamma_\nu] = 2\gamma_{\mu\nu} \equiv -2i\sigma_{\mu\nu}$:

$$\slashed{D}\slashed{D} = D_\mu D_\nu \gamma^\mu \gamma^\nu = D_\mu D_\nu \frac{1}{2}([\gamma_\mu, \gamma_\nu] + \{\gamma_\mu, \gamma_\nu\}) = D^2 - iD_{[\mu}D_{\nu]}\sigma^{\mu\nu} = D^2 - \frac{e}{2}\sigma^{\mu\nu}F_{\mu\nu}. \tag{38.6}$$

Then the equation for ψ is

$$\left(-D^2 + m^2 + \frac{e}{2}\sigma^{\mu\nu}F_{\mu\nu}\right)\psi = 0. \tag{38.7}$$

Thus we have an extra term $+e/2\sigma^{\mu\nu}F_{\mu\nu}$, and since $F_{ij} = \epsilon_{ijk}B_k$ and $i\sigma_{ij} = i[\sigma_i, \sigma_j]/2 = -\epsilon_{ijk}\sigma_k$:

$$\frac{e}{2}\sigma^{\mu\nu}F_{\mu\nu} = -e\vec{\sigma}\cdot\vec{B} = -2e\vec{S}\cdot\vec{B}. \tag{38.8}$$

But this is a term ΔE^2 in E^2 (since $-D^2$ contains $+\partial_t^2 = -E^2$), so the difference in energy is

$$\Delta E \simeq \frac{\Delta E^2}{2E} \simeq \frac{\Delta E^2}{2m} = -\frac{e}{m}\vec{S}\cdot\vec{B} = -\vec{\mu}\cdot\vec{B} = -g\frac{e}{2m}\vec{S}\cdot\vec{B}. \tag{38.9}$$

The way to obtain $g \to g + \Delta g$ from the Lagrangian would be to add a term to the classical Lagrangian, to obtain

$$\mathcal{L} = -\bar{\psi}(\slashed{D} + m)\psi + \frac{\Delta g}{4}\frac{e}{2m}\bar{\psi}(\sigma^{\mu\nu}F_{\mu\nu})\psi. \tag{38.10}$$

Indeed, such a term would add to the Dirac equation

$$-\frac{\Delta g}{4}\frac{e}{2m}\sigma^{\mu\nu}F_{\mu\nu}\psi = +\Delta g\frac{e}{2m}\vec{S}\cdot\vec{B}\psi = \Delta\vec{\mu}\cdot\vec{B}\psi \to \Delta E\psi. \tag{38.11}$$

We will therefore search for such a term in the Lagrangian, generated by radiative (quantum loop) corrections (i.e. as a one-loop effective action correction).

But we saw that

$$\Gamma_{\mu,\alpha\beta}(p_1, p_2; q) = \Gamma_{\mu\alpha\beta}^{(0)} + \Gamma_{\mu\alpha\beta}^{(1)}(p_1, p_2; q); \quad \Gamma_{\mu\alpha\beta}^{(1)} = \Gamma_{\mu\alpha\beta}^{(1a)} + \Gamma_{\mu\alpha\beta}^{(1b)}, \tag{38.12}$$

and we saw that $\Gamma^{(1a)}$ was proportional to γ^μ (and was UV divergent), so we calculate the term

$$\bar{\psi}(p_2)\Gamma_\mu^{(1b)}\psi(p_1) \tag{38.13}$$

on-shell, that is when

$$p_1^2 = -m^2; \quad p_2^2 = -m^2; \quad 0 = q^2 = (p_1 - p_2)^2 \Rightarrow p_1 \cdot p_2 = -m^2;$$
$$(i\slashed{p}_1 + m)\psi(p_1) = 0; \quad (i\slashed{p}_2 + m)\psi(p_2) = 0. \tag{38.14}$$

But we calculated

$$\Gamma_{\mu\alpha\beta}^{(1b)}(p_1, p_2) = -i\frac{e^3}{(4\pi)^2}\int_0^1 d\alpha_1 \int_0^{1-\alpha_1} d\alpha_2$$
$$\times \frac{(\gamma_\nu[-i((1-\alpha_2)\slashed{p}_2 - \alpha_1\slashed{p}_1) + m]\gamma_\mu[-i((1-\alpha_1)\slashed{p}_1 - \alpha_2\slashed{p}_2) + m]\gamma_\nu)_{\alpha\beta}}{F},$$
$$\tag{38.15}$$

where

$$
\begin{aligned}
F &= \alpha_1(1 - \alpha_1)p_1^2 + \alpha_2(1 - \alpha_2)p_2^2 - 2\alpha_1\alpha_2 p_1 \cdot p_2 + m^2(\alpha_1 + \alpha_2) \\
&= m^2[-\alpha_1(1 - \alpha_1) - \alpha_2(1 - \alpha_2) - 2\alpha_1\alpha_2 + \alpha_1 + \alpha_2] \\
&= -m^2(\alpha_1 + \alpha_2)^2.
\end{aligned}
\tag{38.16}
$$

Then one finds (it is left as an exercise to prove it)

$$
\begin{aligned}
\bar{\psi}(p_2)\Gamma_\mu^{(1b)}(p_1, p_2)\psi(p_1) &= -i\frac{e^3}{(4\pi)^2} \int_0^1 d\alpha_1 \int_0^{1-\alpha_1} d\alpha_2 \\
\times\ \bar{\psi}(p_2)&\frac{[m^2\gamma_\mu((\alpha_1 + \alpha_2)^2 - 2(1 - \alpha_1 - \alpha_2)) + 8imq^\nu\sigma_{\mu\nu}(\alpha_1 - \alpha_2(\alpha_1 + \alpha_2))]}{F}\psi(p_1).
\end{aligned}
\tag{38.17}
$$

But we are interested only in the $\sigma_{\mu\nu}$ term, leading to

$$
\begin{aligned}
\bar{\psi}(p_2)\Gamma^{(1)}(p_1, p_2)\psi(p_1)\Big|_{\sigma_{\mu\nu}} &= \frac{e^3}{2m\pi^2}\bar{\psi}(p_2)\sigma_{\mu\nu}q^\nu\psi(p_1) \\
&\quad \int_0^1 d\alpha_1 \int_0^{1-\alpha_1} d\alpha_2 \frac{\alpha_1 - \alpha_2(\alpha_1 + \alpha_2)}{(\alpha_1 + \alpha_2)^2} \\
&= \frac{e^3}{16m\pi^2}\bar{\psi}(p_2)\sigma_{\mu\nu}q^\nu\psi(p_1).
\end{aligned}
\tag{38.18}
$$

The term in the effective action has the above multiplied by A_μ, giving

$$
\frac{e^2}{8\pi^2}\left(\frac{e}{2m}\right)\bar{\psi}(p_2)\sigma^{\mu\nu}q_\nu A_\mu(q)\psi(p_1),
\tag{38.19}
$$

but we have $F_{\mu\nu} = 2q_{[\mu}A_{\nu]}$, so

$$
\frac{\Delta g}{4} = \frac{e^2}{16\pi^2} \Rightarrow \Delta g = \frac{e^2}{4\pi} = 2\frac{\alpha}{2\pi}.
\tag{38.20}
$$

Then finally, the first quantum field theory correction to the Landé g-factor is

$$
g = 2\left(1 + \frac{\alpha}{2\pi} + \dots\right),
\tag{38.21}
$$

as advertised.

38.2 Lamb Shift

We now move on to the Lamb shift, which was the calculation that finally convinced people of the reality of quantum field theory radiative (loop) corrections. Indeed, this was the first example of a calculation that could not be obtained in any other way, but only through QED loops.

The Lamb shift is the lifting of the degeneracy of the $2S_{1/2}$ and $2P_{1/2}$ energy levels of the H atom.

We will only show the steps leading to the calculation of the Lamb shift, since the loop corrections themselves are difficult, and require the treatment of IR divergences, which will be dealt with later in the book.

Step 1. We start with the nonrelativistic Schrödinger equation analysis of the H atom, the first success of quantum mechanics. The equation

$$\left[-\frac{\Delta}{2m} + V(r) \right] \psi = E\psi \tag{38.22}$$

becomes

$$\left[-\frac{1}{2m} \left(\frac{\partial^2}{\partial r^2} + \frac{2}{r}\frac{\partial}{\partial r} - \frac{l(l+1)}{r^2} \right) - \frac{\alpha}{r} \right] \psi_{n,l}(r) = E_{n,l}\psi_{n,l}(r). \tag{38.23}$$

Here $\alpha = e^2/4\pi$ as usual, and the mass is the reduced mass of the nucleus–electron system:

$$\frac{1}{m} = \frac{1}{m_e} + \frac{1}{m_N} \simeq \frac{1}{m_e}. \tag{38.24}$$

Then the energy levels of the H atom are

$$E_{n,l} = -\frac{m\alpha^2}{2n^2}, \tag{38.25}$$

so are independent of $l = 0, 1, \ldots, n-1$, as well as of $m_z = -l, \ldots, l$, giving a degeneracy of

$$\sum_{l=0}^{n-1}(2l+1) = n^2, \tag{38.26}$$

and the energy is given by the Rydberg constant

$$R = \frac{m\alpha^2}{2} \simeq 13.6\,\text{eV}. \tag{38.27}$$

Step 2. Next, we move on to the analysis of the Dirac equation.
As we saw, $-(\not{D} - m)(\not{D} + m)\psi = 0$ gives

$$\left(-D^2 + m^2 - \frac{e}{2}\sigma^{\mu\nu}F_{\mu\nu} \right) \psi = 0. \tag{38.28}$$

We consider the Coulomb potential of a static charge:

$$eA_0 = -\frac{\alpha}{r} \Rightarrow eE_i = -\frac{\alpha\hat{r}_i}{r^2}. \tag{38.29}$$

Then, since

$$\sigma^{i0} = \frac{[\gamma^i, \gamma^0]}{2i} = i\begin{pmatrix} \sigma_i & 0 \\ 0 & -\sigma_i \end{pmatrix}, \tag{38.30}$$

we obtain

$$\frac{e}{2}\sigma^{\mu\nu}F_{\mu\nu} = e\sigma^{i0}F_{i0} = \mp i\alpha\frac{\sigma_i\hat{r}_i}{r^2}. \tag{38.31}$$

With a stationary ansatz for the wavefunction in spherical coordinates

$$\psi(x_i, t) = e^{iEt}\psi_\pm(r, \theta, \phi) \tag{38.32}$$

and using

$$D^2 = (\partial_\mu - ieA_0\delta_\mu^0)(\partial^\mu - ieA^0\delta_0^\mu) = -\partial_t^2 + \Delta - e^2A_0^2 + 2ieA_0\partial_0, \tag{38.33}$$

and considering that $\partial_0 = +iE$ on the ansatz, giving $2E\alpha/r$ for the last term, we get the equation

$$\left[-\left(\frac{\partial^2}{\partial r^2} + \frac{2}{r}\frac{\partial}{\partial r}\right) + \frac{\vec{L}^2 - \alpha^2 \pm i\alpha\sigma_i\hat{r}_i}{r^2} - \frac{2\alpha E}{r} - (E^2 - m^2)\right]\psi_\pm = 0. \tag{38.34}$$

Here $\vec{L}^2 = l(l+1)$ on the wavefunction, and with $\vec{J} = \vec{L} + \vec{S} = \vec{L} + \vec{\sigma}/2$, $[H, \vec{J}] = 0 = [\vec{L}^2, \vec{J}]$, we have also $\vec{J}^2 = j(j+1)$, with $J_z = m$ and $l = j \pm 1/2$.

Then we can prove (left as an exercise) that $\vec{L}^2 - \alpha^2 \pm i\alpha\sigma_i\hat{r}_i$ has eigenvalues $\lambda(\lambda + 1)$, where

$$\lambda_\pm = j \pm \frac{1}{2} - \delta_j,$$

$$\delta_j = j + \frac{1}{2} - \sqrt{\left(j + \frac{1}{2}\right)^2 - \alpha^2}. \tag{38.35}$$

We note that the resulting equation is formally the same as the Schrödinger equation, just with the replacements

$$\vec{L}^2 \to \vec{L}^2 - \alpha^2 \pm i\alpha\sigma_i\hat{r}_i,$$

$$l(l+1) \to \lambda(\lambda + 1),$$

$$\alpha \to \alpha\frac{E}{m},$$

$$E \to \frac{E^2 - m^2}{2m}, \tag{38.36}$$

and we also note that, because the resulting equation is an eigenvalue problem, the condition of $n - l$ to be an integer is replaced by the condition of $n - \lambda = n - l + \delta_j$ to be an integer (where $n = \sqrt{R/E}$), effectively replacing the integer n with $n - \delta_j$ in the solution to the eigenvalue problem.

Finally, we obtain the energy quantization

$$\frac{E_{nj}^2 - m^2}{2m} = -\frac{m\alpha^2}{2}\frac{E_{nj}^2}{m^2}\frac{1}{(n - \delta_j)^2}, \tag{38.37}$$

which can be solved to give

$$E_{nj} = \frac{m}{\sqrt{1 + \frac{\alpha^2}{(n - \delta_j)^2}}} = m - \frac{m\alpha^2}{2n} - \frac{m\alpha^4}{n^3(2j+1)} + \frac{3}{8}\frac{m\alpha^4}{n^4} + \mathcal{O}(\alpha^6). \tag{38.38}$$

We see that the first term is the rest mass, the second is the quantum mechanics result, and the third and fourth are new, with the third lifting the degeneracy over j (i.e. giving a *fine structure*).

With the usual notation of energy levels nl_j, the degeneracy split between the $2P_{3/2}$ and the $2P_{1/2}$ levels is now

$$E(2P_{3/2}) - E(2P_{1/2}) \simeq \frac{m\alpha^4}{32} = 4.5 \times 10^{-5}\,\mathrm{eV} = 10.9\,\mathrm{GHz}. \tag{38.39}$$

But the degeneracy over $l = j \pm 1/2$ at fixed j is not lifted, so at this point $E(2S_{1/2}) = E(2P_{1/2})$.

We need to include other effects now:

- The nucleus has a finite size, it is not a point.
- The proton recoils, since it has a finite mass, so $m \simeq m_e$ has to be corrected.
- The proton has a magnetic moment, and so it interacts with the electron spin, giving

$$\Delta E = -\frac{e}{2m}\sigma_i^{(e)}B, \tag{38.40}$$

where B is the magnetic field of the proton. This gives the *hyperfine splitting*

$$\Delta E_{\mathrm{h.f.}}(S) = 5.9 \times 10^{-6}\,\mathrm{eV} = 1.4\,\mathrm{GHz}. \tag{38.41}$$

All these three effects can be treated semiclassically, but do not account for the observations.

When taken into account, there is still a *Lamb shift* of

$$E(2S_{1/2}) - E(2P_{1/2}) \simeq 1057\,\mathrm{MHz}. \tag{38.42}$$

This comes entirely from radiative corrections. The point is that the interaction vertex of the photon with two fermions changes:

$$e\gamma_\mu A^\mu \to (e\gamma_\mu + \Gamma_\mu^{(1)} + \Pi_{\mu\nu}G^{\nu\lambda}\gamma_\lambda)A^\mu. \tag{38.43}$$

The first term gives the classical interaction with the Coulomb potential A_μ^{Coulomb}. The relevant diagrams are in Figure 38.1.

1. The $\Pi_{\mu\nu}$ term, the loop correction to the photon propagator connecting the vertex with the Coulomb source, was partly computed before, and gives a contribution of $-27\,\mathrm{MHz}$.

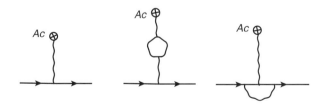

Figure 38.1   Contributions to the potential energy: Coulomb part; $\Pi_{\mu\nu}$ part; Γ_μ part.

2. The $\Gamma_\mu^{(1)}$ term contains the $\sigma_{\mu\nu}q^\nu$ piece partly computed for the anomalous magnetic moment (though one needs to take care of the IR divergences), which gives a contribution of $+68$ MHz.
3. The γ_μ piece, whose finite part we have not computed completely. This is difficult, since it contains IR divergences that need to be dealt with. It gives the largest contribution, of 1010 MHz.

In total these three contributions sum up to 1051 MHz, but by considering higher orders in α, one can go to 1057.864 ± 0.014 MHz, in perfect agreement with the experimental Lamb shift.

Important Concepts to Remember

- The corrections to the anomalous magnetic moment of the electron, specifically to $g - 2$, arise from $\bar{\psi}\sigma^{\mu\nu}F_{\mu\nu}\psi$ terms in the quantum effective action.
- The finite vertex correction $\Gamma_\mu^{(1b)}$ gives such a contribution, of $\Delta g = \alpha/\pi$.
- The Lamb shift is the lifting of the degeneracy of the energy levels $2S_{1/2}$ and $2P_{1/2}$ of the H atom.
- In nonrelativistic quantum mechanics, the energy depends only on n, but not on j or l, $E = -m\alpha^2/2n^2$.
- In relativistic quantum mechanics (i.e. the Dirac equation), the degeneracy over j is lifted at order α^4.
- The degeneracy over l is lifted only through radiative corrections in QED.
- The Lamb shift is due to a photon propagator correction, $\Pi_{\mu\nu}$, and a vertex correction, Γ_μ, splitting into a $\sigma_{\mu\nu}q^\nu$ piece and the γ_μ piece, giving the leading contribution.

Further Reading

See section 6.5.3 in [3].

Exercises

1. Fill in the omitted steps in the calculation of $\bar{\psi}(p_2)\Gamma_\mu^{(1b)}\psi(p_1)$.
2. Write down an integral expression using the Feynman rules for the three contributions to the Lamb shift, do the gamma matrix algebra, and isolate the γ_μ and $\sigma_{\mu\nu}q^\nu$ pieces of the Γ_μ diagram.
3. Show that $\vec{L}^2 - \alpha^2 \pm i\alpha\sigma_i\hat{r}_i$ has eigenvalues $\lambda(\lambda + 1)$, where λ is given by (38.35).
4. Write down the Feynman diagrams contributing to the QED anomalous magnetic moment $g - 2$, and write down the integral expression for them using the Feynman rules, without calculating it.

In this chapter, we will see how to renormalize a theory beyond one loop, by first discussing the systematics, and then giving the example of the two-loop, four-point function in ϕ^4 theory in four dimensions. We will see that we need the renormalization at one loop, which will cancel infinities in divergent subdiagrams, and then new renormalizations, that cancel infinities in the intrinsically divergent diagrams.

39.1 Types of Divergences at Two Loops and Higher

When renormalizing beyond the leading (one-loop) order, we have:

1. Divergent subdiagrams from the divergent one-loop (or, in general, $(n-1)$-loop) diagrams. They are canceled by adding diagrams with an insertion of the corresponding one-loop (or, in general, $(n-1)$-loop) counterterm vertices.
2. Intrinsically divergent two-loop (or, in general, n-loop) diagrams. They are canceled by adding *new* n-loop counterterm contributions (i.e. a two-loop (or n-loop) correction to the counterterm Lagrangian).

But the first type admits a further subdivision:

1a. *Independent subdiagrams.* For example, in quantum electrodynamics we can have a fermion loop insertion on a photon line in the diagram, canceled by the same diagram with the fermion loop substituted with the one-loop counterterm vertex (Figure 39.1). These divergences are polynomials in q^2 (though they can be logarithmic in the cutoff), which means that when Fourier transforming to x space, these divergences will be *local* (the Fourier transform of a power is a power, but the Fourier transform of a log, which is an infinite power series, is an infinite power series, i.e. nonlocal).
1b. *Nested, or overlapping divergences.* In this case, two divergent loops share a propagator, and we will see that they correspond to nonlocal divergences (non-polynomials in q^2). For instance, in ϕ^4 theory we can have the "setting sun" diagram for the propagator, with two vertices connected by three propagators, and for the same vertices coming out of the two external lines, as in Figure 39.2. Or the diagram for the four-point function, the same as for the setting sun, but with one more vertex with two external lines on one of the three propagators. In both these cases, the divergent one-loop subdiagram is one with two propagators between two vertices, and two more external legs

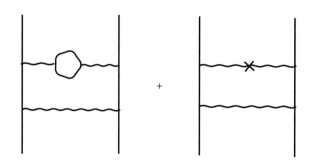

Figure 39.1   Two-loop independent subdiagram canceled by a diagram with a one-loop counterterm.

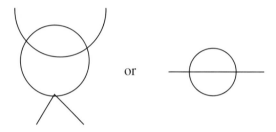

Figure 39.2   Two-loop nested divergent diagrams in ϕ^4 theory.

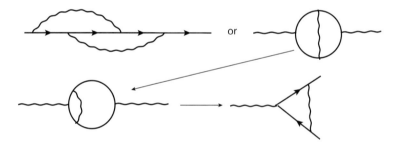

Figure 39.3   Two-loop nested divergent diagrams in QED. The photon polarization diagram can be viewed as coming from a vertex correction.

from each vertex. One of the propagators of the one-loop subdiagrams is common. In QED, we can have the fermion propagator two-loop diagram with a photon line starting off on the fermion line and returning to it, and another photon line starting between the endpoints of the first, and ending further on the fermion line, as in Figure 39.3. Or we can have a two-loop diagram for the photon propagator, with a fermion loop on the photon line, and another photon vertical line (with endpoints on both sides of the external line vertex).

Consider this last QED diagram. Its divergence includes the one-loop divergence of the photon–fermions vertex, which is of the type $(-ie\gamma_\mu)(\alpha \log \Lambda^2)$, where the first bracket isolates the classical vertex, and the second involves factors from the quantum correction. Then, when inserted inside a fermion loop correction to the photon propagator (giving in total the two-loop diagram we are describing), the classical vertex is replaced with this

Figure 39.4 Two-loop level counterterm diagrams: one-loop diagrams with the one-loop counterterm vertex in it.

quantum-corrected vertex. Since the fermion loop correction is of the type $\alpha(g^{\mu\nu}q^2 - q^\mu q^\nu)\Pi(q^2)$, and $\Pi(q^2) \sim \log \Lambda^2 + \log q^2$, we have in total

$$\sim \alpha(g^{\mu\nu}q^2 - q^\mu q^\nu)(\log \Lambda^2 + \log q^2)\alpha \log \Lambda^2. \tag{39.1}$$

This means that there is a part

$$\alpha[\log q^2 \alpha \log \Lambda^2], \tag{39.2}$$

which is divergent in Λ but nonpolynomial in q^2, since it comes from the divergent part of one divergence, times the finite part of the other divergence, which is nonpolynomial in q^2 (there is no problem in having a nonpolynomial for the finite, observable, part; the problem is for the divergent part, since all the terms in the action are local, i.e. polynomial in q^2, yet we need to remove this divergence by renormalization). This divergence is therefore nonlocal in x space, but it can be canceled by adding *loop diagrams* with the one-loop counterterm vertices, namely two one-loop fermion corrections to the photon propagator, with either one of the vertices replaced by the one-loop counterterm vertex, as in Figure 39.4.

There is of course still a part $\alpha^2(\log \Lambda^2)^2$, which however is local, so can be canceled by the addition of a new, two-loop, contribution to the local counterterm Lagrangian (local two-loop counterterm vertex).

This procedure generalizes in an obvious manner to all loops.

It is a nontrivial fact that in this way, we can cancel *all* the divergences in the theory with a finite number of local counterterms (though each term with a coefficient being an infinite series expansion in loops, i.e. in powers of the coupling), if the theory is renormalizable. There is a theorem proving this fact, that any superficially divergent theory is rendered finite by the above prescription for counterterms, called the BPHZ (Bogoliubov, Parasiuk, Hepp, and Zimmermann) theory.

39.2 Two Loops in ϕ^4 in Four Dimensions: Set-up

We consider the two-loop contributions to the four-point function of ϕ^4 theory in four dimensions, and show that we can remove all the divergence by renormalization according to the above prescription.

There are 16 diagrams corresponding to this order, 15 of which can be split into three groups, according to the channel, s, t, and u, as in Figure 39.5. In the s channel, the first diagram is a "chain" made up of two one-loop "rings," and with two external lines from a vertex at each end. The second is the diagram already described, the setting sun with an

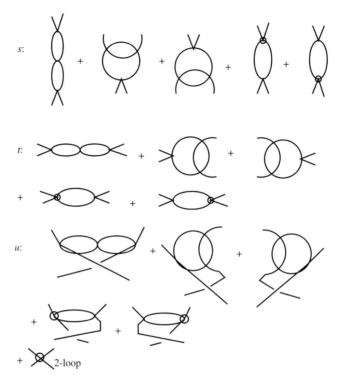

Figure 39.5 Two-loop diagrams in ϕ^4 theory, organized according to the s, t, u channel diagrams, plus the two-loop counterterm vertex diagram.

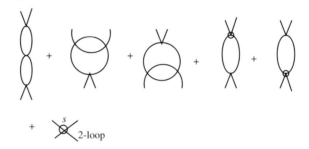

Figure 39.6 Two-loop independent diagrams in ϕ^4 theory. The other can be related by crossing. The two-loop counterterm vertex diagram contains only the s piece.

extra vertex with two external lines on one of the propagators, coming down. The third is the up–down mirror of the first. Finally, we have the one-loop diagram (with two up and two down external lines, each pair from a vertex) with one normal vertex and one vertex being the one-loop counterterm vertex. Then there are the t and u-channel versions of the same five diagrams, which can be obtained by crossing (that exchanges s, t, u). The last diagram is of course the two-loop counterterm vertex, which contains s, t, and u pieces.

Therefore, to fully renormalize the four-point vertex at two loops, we need only consider the five s-channel diagrams, and the s piece of the two-loop counterterm vertex, as in Figure 39.6, and the rest will be obtained trivially by crossing.

39.3 One-Loop Renormalization

Let us first remember the one-loop renormalization. There is a unique s-channel diagram, with momentum $p = p_1 + p_2$ coming in. The result for the diagram is

$$\frac{\tilde{\lambda}^2}{2} I_2(p^2) = \frac{\tilde{\lambda}^2}{2} \int \frac{d^D q}{(2\pi)^D} \frac{1}{q^2 + m^2} \frac{1}{(q+p)^2 + m^2}$$

$$= \frac{\tilde{\lambda}^2}{2} \frac{\Gamma\left(2 - \frac{D}{2}\right)}{(4\pi)^{\frac{D}{2}}} \int_0^1 d\alpha [\alpha(1-\alpha)p^2 + m^2]^{\frac{D}{2}-2}. \tag{39.3}$$

We consider the normalization conditions

$$\Gamma^{(2)}(p^2) = [p^2 + m^2]^{-1}; \quad \Gamma^{(4)}(s, t, u) = -\lambda \tag{39.4}$$

at $s = 4m^2$; $t = u = 0$, where $s = -(p_1 + p_2)^2$, and so on. Then we immediately obtain the one-loop counterterm vertex as

$$V^{(1)} = -\frac{\tilde{\lambda}^2}{2} [I_2(4m^2) + 2I_2(0)], \tag{39.5}$$

which splits into an s piece

$$V_s^{(1)} = -\frac{\tilde{\lambda}^2}{2} I_2(4m^2) \tag{39.6}$$

and a $t + u$ piece

$$V_{t+u}^{(1)} = -\frac{\tilde{\lambda}^2}{2} 2I_2(0). \tag{39.7}$$

Then, we can finally split the six independent s-channel diagrams into three groups, as in Figure 39.7:

I. The diagram of a chain of two one-loop rings, plus the one-loop diagrams with one of the vertices replaced by the s part of the one-loop counterterm vertex, $V_s^{(1)}$.
II. The setting sun with extra vertex down diagram, plus the one-loop diagram with normal vertex down, and the $t + u$ part of the one-loop counterterm vertex, $V_{t+u}^{(1)}$, up.
III. The same diagrams as in II, but with up and down interchanged (mirror symmetric).

Then the *momentum-dependent part of the divergence* cancels separately in I, II, and III. We also see that we only need to calculate what happens in I and II, since III is obtained from II.

(I) We start with the "chain" diagram of two one-loop rings, the first one in group I, which can be split into two independent one-loop diagrams, each with momentum p coming into it, so the result of the diagram is

$$-\frac{\tilde{\lambda}^3}{4} [I_2(p^2)]^2. \tag{39.8}$$

Note that the only nontrivial part is the vertex counting, which is $(-\tilde{\lambda})^3$ (one vertex is common to the two one-loop rings), the integral (as well as the symmetry factor of 1/2 for each ring) comes from the one-loop diagrams.

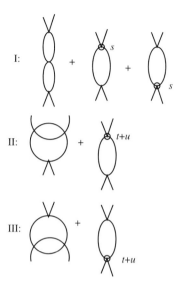

Figure 39.7   Two-loop independent diagrams in ϕ^4 theory divided into three groups. The diagrams with the one-loop counterterm vertices have been split into the s pieces and the $t + u$ pieces.

39.4 Calculation of Two-Loop Divergences in ϕ^4 in Four Dimensions and their Renormalization

Both diagrams with a $V_s^{(1)}$ insertion (up or down) (i.e. the second and third diagrams in group I) equal a $V_s^{(1)}$ factor, times a vertex $(-\tilde{\lambda})$, times the one-loop integral $I_2(p^2)$, times the 1/2 for the symmetry factor, for a total of

$$-\frac{\tilde{\lambda}}{2}I_2(p^2)\left(-\frac{\tilde{\lambda}^2}{2}\right)I_2(4m^2) = \frac{\tilde{\lambda}^3}{4}I_2(p^2)I_2(4m^2). \tag{39.9}$$

The sum of the diagrams in I is therefore

$$-\frac{\tilde{\lambda}^3}{4}[(I_2(p^2))^2 - 2I_2(p^2)I_2(4m^2)] = -\frac{\tilde{\lambda}^3}{4}\left([I_2(p^2) - I_2(4m^2)]^2 - [I_2(4m^2)]^2\right), \tag{39.10}$$

and the first term is finite, because it is the same finite term appearing in the one-loop renormalization:

$$I_2(p^2) - I_2(4m^2) = \frac{2}{\epsilon}\frac{1}{(4\pi)^2}\left(-\frac{\epsilon}{2}\right)\int_0^1 d\alpha \ln\left[\frac{m^2 + \alpha(1-\alpha)p^2}{m^2 + \alpha(1-\alpha)4m^2}\right]. \tag{39.11}$$

Note that the divergence is constant, so canceled between the two terms, and also the $\psi(1) = -\gamma$ term has canceled. In the two-loop formula, we have the square of the above, for a total of

$$\frac{\tilde{\lambda}^3}{4}\left[\frac{1}{16\pi^2}\int_0^1 d\alpha \ln[\ldots]\right]^2. \tag{39.12}$$

Therefore in I, we have canceled a product of two *independent divergences*, and we are left with only a constant divergence:

$$+\frac{\tilde{\lambda}^3}{4}[I_2(4m^2)]^2 \propto \left[\Gamma\left(2-\frac{D}{2}\right)\right]^2 \propto \left(\frac{2}{\epsilon}\right)^2, \tag{39.13}$$

which is a double pole in ϵ. This term is canceled by adding a two-loop counterterm, with vertex

$$-\frac{\tilde{\lambda}^3}{4}[I_2(4m^2)]^2. \tag{39.14}$$

Some observations are in order. The first one is that at higher loops we will see higher-order poles in ϵ, as seen here. But in all cases, at least the highest order will be a momentum-independent constant. The second observation relates to the functional form of the finite part. Note that at $p^2 \to \infty$:

$$I_2(p^2) - I_2(4m^2) \sim \log\frac{p^2}{m^2}, \tag{39.15}$$

whereas at two loops, because of the above, we can ignore the finite contribution of $I_2(4m^2)$ with respect to $I_2(p^2)$, and write that the two-loop "chain" of two rings gives

$$\sim \frac{\tilde{\lambda}^3}{4}[I_2(p^2)]^2 \propto \left(\log\frac{p^2}{m^2}\right)^2. \tag{39.16}$$

It is then easy to see that this generalizes to an arbitrary n-loop, where the "chain" diagram, with n one-loop rings, as in Figure 39.8, will give a result

$$\propto \lambda^{n+1}\left(\log\frac{p^2}{m^2}\right)^n. \tag{39.17}$$

(II) We move on to the second set of diagrams, starting with the proper two-loop diagram in Figure 39.9. Consider momenta p_1, p_2 at the classical vertex (down), connected to two internal lines of momenta k (loop momentum) and $k+p$ (where, as above, $p = p_1 + p_2$), and the other external momenta being p_3 (for the vertex connected to the k propagator) and p_4. Then we can isolate on top a one-loop diagram with total incoming momentum $k+p_3$, inserted inside an up–down one-loop diagram with total momentum $p = p_1 + p_2$ coming in. The symmetry factor of the diagram is 2, giving for the amplitude

$$\mathcal{M} = -\frac{\tilde{\lambda}^3}{2}\int\frac{d^D k}{(2\pi)^D}\frac{1}{k^2+m^2}\frac{1}{(k+p)^2+m^2}I_2((k+p_3)^2). \tag{39.18}$$

n loops

Figure 39.8 n-Loop "chain" diagram.

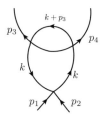

Two-loop setting sun diagram.

Replacing the two propagators with a Feynman parametrization integral over β, and writing the result of I_2 as a Feynman parametrization integral over α, we obtain

$$\mathcal{M} = -\frac{\tilde{\lambda}^3}{2} \frac{\Gamma\left(2 - \frac{D}{2}\right)}{(4\pi)^{\frac{D}{2}}} \int_0^1 d\alpha \int_0^1 d\beta \int \frac{d^D k}{(2\pi)^d} \frac{1}{[k^2 + 2\beta k \cdot p + \beta p^2 + m^2]^2}$$

$$\times \frac{1}{[\alpha(1 - \alpha)(k + p_3)^2 + m^2]^{2 - \frac{D}{2}}},$$ (39.19)

where the first denominator is $\beta \Delta_2 + (1 - \beta)\Delta_1$. But now, since the second denominator has a non-integer power, we cannot use the usual Feynman parametrization, but rather we must use the formula

$$\frac{1}{A^\alpha B^\beta} = \int_0^1 dw \frac{w^{\alpha-1}(1 - w)^{\beta-1}}{[wA + (1 - w)B]^{\alpha+\beta}} \frac{\Gamma(\alpha + \beta)}{\Gamma(\alpha)\Gamma(\beta)},$$ (39.20)

where the gamma factors together make the inverse of the beta function, $[B(\alpha, \beta)]^{-1}$.

The proof of this formula goes as follows. Consider the change of variables

$$z \equiv \frac{wA}{wA + (1 - w)B} \Rightarrow 1 - z = \frac{(1 - w)B}{wA + (1 - w)B}$$

$$\Rightarrow dz = \frac{ABdw}{[wA + (1 - w)B]^2}.$$ (39.21)

Then

$$\int_0^1 dw \frac{w^{\alpha-1}(1 - w)^{\beta-1}}{[wA + (1 - w)B]^{\alpha+\beta}} = \frac{1}{A^\alpha B^\beta} \int_0^1 dz \, z^{\alpha-1}(1 - z)^{\beta-1} = \frac{1}{A^\alpha B^\beta} B(\alpha, \beta).$$ (39.22)

$$q.e.d.$$

Then, applying it for the case of our diagram (with $\alpha = 2 - D/2$ and $\beta = 2$), we obtain

$$\mathcal{M} = -\frac{\tilde{\lambda}^3}{2} \frac{\Gamma\left(2 - \frac{D}{2}\right)}{(4\pi)^{\frac{D}{2}}} \frac{\Gamma\left(4 - \frac{D}{2}\right)}{\Gamma(2)\Gamma\left(2 - \frac{D}{2}\right)} \int_0^1 d\alpha \int_0^1 d\beta \int_0^1 dw \int \frac{d^D k}{(2\pi)^D}$$

$$\times \frac{w^{1 - \frac{D}{2}}(1 - w)}{\left(w[\alpha(1 - \alpha)(k + p_3)^2 + m^2] + (1 - w)[k^2 + 2k \cdot p + \beta p^2 + m^2]\right)^{4 - \frac{D}{2}}}.$$ (39.23)

The factor raised to $4 - D/2$ in the denominator is rewritten as

$$m^2 + k^2[1 - w + w\alpha(1 - \alpha)] + 2k \cdot [(1 - w)\beta p + w\alpha(1 - \alpha)p_3]$$

$$+ p_3^2 w\alpha(1 - \alpha) + p^2\beta(1 - w) \equiv m^2 + k'^2[1 - w + w\alpha(1 - \alpha)] + P^2, \quad (39.24)$$

where

$$k' = k + \frac{(1 - w)\beta p + w\alpha(1 - \alpha)p_3}{1 - w + w\alpha(1 - \alpha)} \quad (39.25)$$

and

$$P^2 = p_3^2 w\alpha(1 - \alpha) + p^2\beta(1 - w) - \left(\frac{(1 - w)\beta p + w\alpha(1 - \alpha)p_3}{1 - w + w\alpha(1 - \alpha)}\right)^2. \quad (39.26)$$

For future reference, we note that

$$P^2(w \to 0) = p^2\beta - \beta^2 p^2 = \beta(1 - \beta)p^2. \quad (39.27)$$

We can now do the integral over $\int d^D k = \int d^D k'$ with the formula (33.30), to obtain

$$\mathcal{M} = -\frac{\tilde{\lambda}^3}{2} \frac{\Gamma\left(4 - \frac{D}{2}\right)}{(4\pi)^{\frac{D}{2}}} \frac{\Gamma(4 - D)}{\Gamma\left(4 - \frac{D}{2}\right)(4\pi)^{\frac{D}{2}}} \int_0^1 d\alpha \int_0^1 d\beta \int_0^1 dw \frac{1}{[1 - w + w\alpha(1 - \alpha)]^{4 - \frac{D}{2}}}$$

$$\times w^{1 - \frac{D}{2}}(1 - w) \left[\frac{P^2 + m^2}{1 - w + w\alpha(1 - \alpha)}\right]^{D-4}$$

$$= -\frac{\tilde{\lambda}^3}{2} \frac{\Gamma(4 - D)}{(4\pi)^D} \int_0^1 d\alpha \int_0^1 d\beta \int_0^1 dw \frac{w^{1 - \frac{D}{2}}(1 - w)}{[1 - w + w\alpha(1 - \alpha)]^{\frac{D}{2}}} [P^2 + m^2]^{D-4}.$$

$$(39.28)$$

This diagram has of course a pole coming from $\Gamma(4 - D)$, but it also has a pole coming from the integral in w, specifically near $w = 0$. Indeed, consider the integral

$$\int_0^1 dw\, w^{1 - \frac{D}{2}} f(w) = \int_0^1 dw\, w^{1 - \frac{D}{2}} f(0) + \int_0^1 dw\, w^{1 - \frac{D}{2}}[f(w) - f(0)]. \quad (39.29)$$

Then in our case we can check that the second term gives a finite integral for $D = 4$, and the only pole comes from the overall $\Gamma(4 - D)$. This term is therefore local (there is no p-dependence at all if we set $D = 4$), and its divergence can be absorbed by an $\mathcal{O}(\lambda^3)$ counterterm (i.e. a two-loop, $1/\epsilon$ term).

The first term, with $f(0)$, gives a double pole, since

$$\int_0^1 dw\, w^{1 - \frac{D}{2}} f(0) = f(0) \left.\frac{w^{2 - \frac{D}{2}}}{2 - \frac{D}{2}}\right|_0^1 = f(0)\frac{2}{4 - D}. \quad (39.30)$$

Substituting $D = 4 - \epsilon$, we obtain

$$-\frac{\tilde{\lambda}^3 \Gamma(\epsilon)}{2(4\pi)^D} \frac{2}{\epsilon} \int_0^1 d\beta [P^2(0) + m^2]^{-\epsilon}$$

$$= -\frac{\lambda^3 \mu^\epsilon}{(4\pi)^4} \frac{1}{\epsilon} \int_0^1 d\beta \left(\frac{1}{\epsilon} - \gamma - \log \frac{\beta(1 - \beta)p^2 + m^2}{4\pi\mu^2}\right), \quad (39.31)$$

where in the last line we have grouped, as usual, the $(4\pi)^{-\epsilon}$ and $\mu^{-2\epsilon}$ terms in the expansion with the log, to make the ratio of $m^2/(4\pi\mu^2)$ manifest.

Therefore we finally have the nonlocal divergence

$$\frac{\lambda^3 \mu^\epsilon}{(4\pi)^4} \frac{1}{\epsilon} \int_0^1 d\beta \, \log \frac{\beta(1-\beta)p^2 + m^2}{4\pi\mu^2}. \tag{39.32}$$

This divergence is canceled, however, by the one-loop diagram with one one-loop $t + u$ counterterm vertex $V_{t+u}^{(1)}$, given by the product of the vertex $V_{t+u}^{(1)}$ and the $(-\tilde{\lambda}^2/2I_2(p^2))$ factor (removing one $-\tilde{\lambda}$ vertex and replacing it with $V_{t+u}^{(1)}$):

$$+ \frac{\tilde{\lambda}^3}{4} 2I_2(0)I_2(p^2)$$

$$= \frac{\tilde{\lambda}^3}{2(4\pi)^D} \int_0^1 d\alpha \frac{\left[\Gamma\left(2 - \frac{D}{2}\right)\right]^2}{[\alpha(1-\alpha)p^2 + m^2]^{2-\frac{D}{2}}(m^2)^{2-\frac{D}{2}}}$$

$$= \frac{\lambda^3 \mu^\epsilon}{2(\pi)^4} \int_0^1 d\alpha \left(\frac{2}{\epsilon} - \gamma - \log\frac{m^2}{4\pi\mu^2}\right)\left(\frac{2}{\epsilon} - \gamma - \log\frac{m^2 + \alpha(1-\alpha)p^2}{4\pi\mu^2}\right). \tag{39.33}$$

We now see that the nonlocal divergence cancels, and we are only left with a local divergence of order $1/\epsilon^2$ (constant double pole). This will be removed by adding an $\mathcal{O}(\lambda^3)$ counterterm (local two-loop counterterm).

As we mentioned, the III diagrams are obtained by symmetry, and then the t and u channels by crossing, so this means we have indeed shown the full renormalizability of the four-point function at two loops in ϕ^4 theory in four dimensions.

Important Concepts to Remember

- Intrinsically divergent two-loop diagrams are canceled by adding new two-loop (n-loop) counterterms.
- Divergences from one-loop (($n-1$)-loop) divergent subdiagrams are canceled by adding diagrams with the one-loop (($n-1$)-loop) counterterm vertices.
- Independent subdiagrams lead to local divergences (i.e. polynomial in momenta q^2), whereas nested (overlapping) divergences lead to nonlocal divergences (i.e. nonpolynomial in momenta q^2).
- The BPHZ theorem says that any superficially renormalizable theory is rendered finite by adding the diagrams with all the n-loop counterterms, which are a finite number of local counterterms (with coefficients equal to a series in loop order, or λ^n).
- The finite part of the n-loop chain diagram with n one-loop rings in ϕ^4 theory at $p^2 \to \infty$ goes like $\sim \lambda^{n+1} \log^n(p^2/m^2)$.
- At higher loops in dimensional regularization, there are higher-order poles (in ϵ), but at least the highest-order pole is a momentum-independent constant.

Further Reading

See section 10.5 in [1] and section 10.4 in [2].

Exercises

1. Write down all the two-loop divergent diagrams for $\Gamma^{(3)}$ for ϕ^3 theory in $D = 6$, paralleling ϕ^4 in $D = 4$.
2. Identify and calculate the nonlocal two-loop divergence in the above.
3. Identify the overlapping divergences among the two-loop divergent divergences in Exercise 1.
4. Draw a three-loop diagram in ϕ^4 theory that is divergent, but is renormalized by the two-loop renormalization (but not by the one-loop renormalization).

40

The LSZ Reduction Formula

In this chapter we return to the Lehmann, Symanzik, Zimmermann formula, relating correlation functions:

$$\langle \Omega | T\{\phi(x_1)\ldots\phi(x_{n+m})\} | \Omega \rangle \tag{40.1}$$

with S-matrices:

$$_{out}\langle \vec{p}_1,\ldots,\vec{p}_n | \vec{k}_1 \ldots \vec{k}_m \rangle_{in} = \langle \vec{p}_1,\ldots,\vec{p}_n | S | \vec{k}_1 \ldots \vec{k}_m \rangle. \tag{40.2}$$

We described it in Part I of the book, but now we return for a better understanding, using the knowledge of loops and renormalization that we have gained in the meantime.

40.1 The LSZ Reduction Formula and Wavefunction Renormalization

In particular, we had mentioned that there is a wavefunction renormalization factor Z that appears there. We also saw that at one loop, we defined the wavefunction renormalization factor as the factor in the renormalized Lagrangian that multiplies the p^2 part of the kinetic term, and that it comes from the calculation of the two-point function. With a bit of thought, we realize that a more formal way to define it would be as follows.

Consider the two-point function in momentum space. It will have a pole at some renormalized mass $p^2 \to -m^2$. So formally, we can say that near the pole $p^2 \to -m^2$:

$$\int d^4x e^{-ipx} \langle \Omega | T\{\phi(x)\phi(0)\} | \Omega \rangle \sim \frac{-iZ}{p^2 + m^2 - i\epsilon}. \tag{40.3}$$

To obtain the LSZ formula, we consider the Fourier transform over only one momentum of the correlation function:

$$\int d^4x e^{-ipx} \langle \Omega | T\{\phi(x)\phi(z_1)\phi(z_2)\ldots\} | \Omega \rangle, \tag{40.4}$$

and we split the integral over time as

$$\int dx^0 = \int_{T_+}^{+\infty} dx^0 + \int_{T_-}^{T_+} dx^0 + \int_{-\infty}^{T_-} dx^0, \tag{40.5}$$

and call the first term region I, the second region II, and the third region III.

Since we are interested in the behavior near a pole, we will ignore finite terms. But the integral over region II is finite, since the integrand is analytic in p^0 and the integration is over a finite interval. Therefore we can ignore this region.

Region I. Consider first the integral over region I, and let x^0 be the largest time (i.e. the time components of all the z_is are less than $T_+ < x^0$ in region I). Then we can put $\phi(x)$ to the left, and outside the time-ordering operator, and insert on its right the identity, written in terms of a complete set of states $|\lambda_{\vec{q}}\rangle$ of momentum $\vec{q}$, as

$$\mathbb{1} = \sum_\lambda \int \frac{d^3 q}{(2\pi)^3} \frac{1}{2E_{\vec{q}}(\lambda)} |\lambda_{\vec{q}}\rangle \langle \lambda_{\vec{q}}|. \tag{40.6}$$

We obtain

$$\int d^3 x \int_{T_+}^{+\infty} dx^0 e^{+ip^0 x^0 - i\vec{p}\cdot\vec{x}} \sum_\lambda \int \frac{d^3 q}{(2\pi)^3} \frac{1}{2E_{\vec{q}}(\lambda)} \langle\Omega|\phi(x)|\lambda_{\vec{q}}\rangle \langle\lambda_{\vec{q}}|T\{\phi(z_1)\ldots\phi(z_n)\}|\Omega\rangle. \tag{40.7}$$

But the first matrix element can be worked out as follows, using the fact that $\phi(x)$ is a Heisenberg operator, so

$$\langle\Omega|\phi(x)|\lambda_{\vec{q}}\rangle = \langle\Omega|e^{-i\hat{P}\cdot x}\phi(0)e^{+i\hat{P}\cdot x}|\lambda_{\vec{q}}\rangle = \langle\Omega|\phi(0)|\lambda_{\vec{q}}\rangle \, e^{+iqx}\big|_{q^0=E_{\vec{q}}}, \tag{40.8}$$

where in the last equality we have used the fact that $\hat{P}\cdot x|\lambda_{\vec{q}}\rangle = (q\cdot x)|_{q^0=E_{\vec{q}}}|\lambda_{\vec{q}}\rangle$ and $\langle\Omega|\hat{P}\cdot x = 0$.

Now, doing the integral over $d^3 x$, $\int d^3 x\, e^{i\vec{x}\cdot(\vec{q}-\vec{p})} = (2\pi)^3 \delta^3(\vec{p}-\vec{q})$, and then doing the integral over $\int d^3 q/(2\pi)^3$, we replace everywhere $\vec{q}$ with $\vec{p}$ and $E_{\vec{q}}$ with $E_{\vec{p}}$. Introducing a regularizing factor of $e^{-\epsilon x^0}$ in the usual manner, we obtain

$$\sum_\lambda \int_{T_+}^{+\infty} dx^0 \frac{1}{2E_{\vec{p}}(\lambda)} e^{+ip^0 x^0} \, e^{-iq^0 x^0}\big|_{q^0=E_{\vec{q}}=E_{\vec{p}}} e^{-\epsilon x^0} \langle\Omega|\phi(0)|\lambda_{\vec{p}}\rangle \langle\lambda_{\vec{p}}|T\{\phi(z_1)\ldots\phi(z_n)\}|\Omega\rangle. \tag{40.9}$$

The integral over t becomes

$$\int_{T_+}^{+\infty} e^{ix^0(p^0-E_p+i\epsilon)} = \frac{e^{ix^0(p^0-E_p+i\epsilon)}}{i(p^0-E_p+i\epsilon)}\bigg|_{T_+}^{+\infty} = -\frac{e^{iT_+(p^0-E_p+i\epsilon)}}{i(p^0-E_p+i\epsilon)}, \tag{40.10}$$

so our region I integral is now

$$\sum_\lambda \frac{-ie^{i(p^0-E_p+i\epsilon)T_+}}{-2E_{\vec{p}}(\lambda)(p^0-E_{\vec{p}}(\lambda)+i\epsilon)} \langle\Omega|\phi(0)|\lambda_{\vec{p}}\rangle \langle\lambda_{\vec{p}}|T\{\phi(z_1)\ldots\phi(z_n)\}|\Omega\rangle. \tag{40.11}$$

The denominator equals $p^2+m^2-i\epsilon$ and, near $p^0 \to E_{\vec{p}}$, the exponential in the numerator becomes 1. Specializing first to the case $n=1$, when the two-point function is supposed to be of the general form (40.3), we indeed find near on-shell for the unique momentum p:

$$\sim \sum_\lambda \frac{-i}{p^2+m^2-i\epsilon} \langle\Omega|\phi(0)|\lambda_{\vec{p}}\rangle \langle\lambda_{\vec{p}}|\phi(0)\Omega\rangle \sim \frac{-i}{p^2+m^2-i\epsilon} |\langle\Omega|\phi(0)|\vec{p}\rangle|^2, \tag{40.12}$$

where we have implicitly assumed that there is a single one-momentum state $|\vec{p}\rangle$, and now we can identify $\langle\Omega|\phi(0)|\vec{p}\rangle$ with the factor $\sqrt{Z}$ (whose square is the wavefunction renormalization factor Z). Substituting in the general correlation function, we find that near $p^0 \to E_{\vec{p}}$:

$$\int d^4x\, e^{-ipx} \langle\Omega|T\{\phi(x)\phi(z_1)\dots\phi(z_n)\}|\Omega\rangle \sim \frac{-i}{p^2 + m^2 - i\epsilon} \sqrt{Z}\langle\vec{p}|T\{\phi(z_1)\dots\phi(z_n)\}|\Omega\rangle.$$

(40.13)

We now repeat the procedure in region III. Assuming that x^0 is the smallest of all the times (i.e. that all the zero components of $z_1,\dots,z_n$ are greater than $T_- > x^0$), we can put x to the right, inside the time-ordering operator, and insert the identity to its left, obtaining

$$\int d^3x \int_{T_+}^{+\infty} dx^0 e^{+ip^0x^0 - i\vec{p}\cdot\vec{x}} \sum_\lambda \int \frac{d^3q}{(2\pi)^3}\frac{1}{2E_{\vec{q}}(\lambda)} \langle\Omega|T\{\phi(z_1)\dots\phi(z_n)\}|\lambda_{\vec{q}}\rangle\langle\lambda_{\vec{q}}|\phi(x)|\Omega\rangle.$$

(40.14)

The matrix element is found to be

$$\langle\lambda_{\vec{q}}|\phi(x)|\Omega\rangle = \langle\lambda_{\vec{q}}|e^{-i\hat{P}\cdot x}\phi(0)e^{+i\hat{P}\cdot x}|\Omega\rangle = \langle\lambda_{\vec{q}}|\phi(0)|\Omega\rangle \, e^{-iqx}\Big|_{q^0 = E_{\vec{q}}}.$$

(40.15)

The integral over $\vec{x}$ is therefore $\int d^3x\, e^{-i\vec{x}\cdot(\vec{p}+\vec{q})} = (2\pi)^3\delta(\vec{p}+\vec{q})$, so after doing the integral over $\vec{q}$, we substitute everywhere $\vec{q}$ with $-\vec{p}$, and E_q with E_p, and get

$$\sum_\lambda \int_{-\infty}^{T_-} dx^0 \frac{1}{2E_{\vec{p}}(\lambda)} e^{+ip^0x^0} e^{iq^0x^0}\Big|_{q^0 = E_{\vec{q}} = E_{\vec{p}}} e^{\epsilon x^0}\langle\lambda_{-\vec{p}}|\phi(0)|\Omega\rangle\langle\Omega|T\{\phi(z_1)\dots\phi(z_n)\}|\lambda_{-\vec{p}}\rangle.$$

(40.16)

The integral over x^0 gives

$$\int_{-\infty}^{T_-} e^{ix^0(p^0 + E_p + i\epsilon)} = \frac{e^{ix^0(p^0 + E_p - i\epsilon)}}{i(p^0 + E_p - i\epsilon)}\Big|_{-\infty}^{T_-} = -i\frac{e^{iT_-(p^0 + E_p - i\epsilon)}}{(p^0 + E_p - i\epsilon)},$$

(40.17)

so again we obtain the propagator, but now for $p^0 \to -E_p$, and so finally we have (near $p^0 \to -E_p$):

$$\int d^4x\, e^{-ipx} \langle\Omega|T\{\phi(x)\phi(z_1)\dots\phi(z_n)\}|\Omega\rangle \sim \frac{-i}{p^2 + m^2 - i\epsilon} \sqrt{Z}\langle\Omega|T\{\phi(z_1)\dots\phi(z_n)\}| -\vec{p}\rangle.$$

(40.18)

We can redefine $p = -k$, so that near $k^0 = E_k$ (on-shell):

$$\int d^4x\, e^{ikx} \langle\Omega|T\{\phi(x)\phi(z_1)\dots\phi(z_n)\}|\Omega\rangle \sim \frac{-i}{k^2 + m^2 - i\epsilon} \sqrt{Z}\langle\Omega|T\{\phi(z_1)\dots\phi(z_n)\}|\vec{k}\rangle.$$

(40.19)

We see that in both regions, we obtain that on-shell, we relate to a correlation function with $n \to n-1$, with a vacuum state replaced by a momentum state. If we have e^{-ipx}, we relate to an outgoing state, and if we have e^{ikx}, we relate to an incoming state. We can iteratively repeat the procedure, and obtain finally that for all momenta on-shell, we relate to a product of incoming and outgoing states. The detail is that, if we have more than one momentum, the incoming states are actually "in states," and the outgoing states are actually "out states." This is the result of the fact that we can independently consider each momentum on-shell.

40.2 Adding Wavepackets

But it remains to prove that when adding wavepackets instead of single momenta, nothing new happens. A wavepacket means that we replace

$$\int d^4x\, e^{ip^0 x^0} e^{-i\vec{p}\cdot\vec{x}} \to \int \frac{d^3k}{(2\pi)^3} \int d^4x\, e^{ip^0 x^0} e^{-i\vec{k}\cdot\vec{x}} \phi(\vec{k}). \qquad (40.20)$$

Therefore the limit $\phi(\vec{k}) \to (2\pi)^3 \delta^3(\vec{k} - \vec{p})$ takes us back to the original case. We can define $\tilde{p} = (p^0, \vec{k})$.

With this replacement, we obtain

$$\sum_\lambda \int \frac{d^3k}{(2\pi)^3} \phi(\vec{k}) \frac{-ie^{i(p^0 - E_k + i\epsilon)T_+}}{-2E_{\vec{k}}(\lambda)(p^0 - E_{\vec{k}}(\lambda) + i\epsilon)} \langle\Omega|\phi(0)|\lambda_{\vec{k}}\rangle \langle\lambda_{\vec{k}}|T\{\phi(z_1)\dots\phi(z_n)\}|\Omega\rangle, \qquad (40.21)$$

and near on-shell, $p^0 \to E_{\vec{k}}$, so $\tilde{p}^2 + m^2 \sim 0$:

$$\sim \int \frac{d^3k}{(2\pi)^3} \phi(\vec{k}) \frac{-i}{\tilde{p}^2 + m^2 - i\epsilon} \sqrt{Z} \langle \vec{k}|T\{\phi(z_1)\dots\phi(z_n)\}|\Omega\rangle. \qquad (40.22)$$

To see how this replacement works for pairs of particles, consider the scattering $n \to 2$. Then we obtain

$$\sum_\lambda \int \frac{d^3q}{(2\pi)^3} \frac{1}{2E_q} \prod_{i=1,2} \int \frac{d^3k_i}{(2\pi)^3} \int d^4x_i e^{-i\tilde{p}_i\cdot x_i} \phi_i(\vec{k}_i)$$

$$\times \langle\Omega|T\{\phi(x_1)\phi(x_2)\}|\lambda_{\vec{q}}\rangle \langle\lambda_{\vec{q}}|T\{\phi(z_1)\dots\phi(z_n)\}|\Omega\rangle. \qquad (40.23)$$

If the two outgoing particles are separated in the far future, we obtain

$$\sum_\lambda \int \frac{d^3q}{(2\pi)^3} \frac{1}{2E_{\vec{q}}} \langle\Omega|T\{\phi(x_1)\phi(x_2)\}|\lambda_{\vec{k}}\rangle \langle\lambda_{\vec{k}}|$$

$$= \sum_{\lambda_1\lambda_2} \int \frac{d^3q_1}{(2\pi)^3} \frac{1}{2E_{q_1}} \int \frac{d^3q_2}{(2\pi)^3} \frac{1}{2E_{q_2}} \langle\Omega|\phi(x_1)|\lambda_{q_1}\rangle \langle\Omega|\phi(x_2)|\lambda_{q_2}\rangle \langle\lambda_{q_1}\lambda_{q_2}|, \qquad (40.24)$$

and then we can perform the same steps independently for each particle. The same analysis can be done in the far past as well, and we can generalize to more than two particles. In the limit in which the wavepackets tend to delta functions of momenta, we get the in and out states.

All in all, we obtain the LSZ formula

$$\prod_{i=1}^n \int d^4x_i e^{-ip_i x_i} \prod_{j=1}^m d^4y_j e^{+ik_j y_j} \langle\Omega|T\{\phi(x_1)\dots\phi(x_n)\phi(y_1)\dots\phi(y_m)\}|\Omega\rangle$$

$$\underset{p_i^0 \to E_{p_i}, k_j^0 \to E_{k_j}}{\sim} \left(\prod_{i=1}^n \frac{-i\sqrt{Z}}{p_i^2 + m^2 - i\epsilon}\right) \left(\prod_{j=1}^m \frac{-i\sqrt{Z}}{k_j^2 + m^2 - i\epsilon}\right) \langle\vec{p}_1\dots\vec{p}_n|S|\vec{k}_1\dots\vec{k}_m\rangle. \qquad (40.25)$$

40.3 Diagrammatic Interpretation

The diagrammatic interpretation of the formula is as follows. In order to construct diagrams for the S-matrix from diagrams for the correlation functions (i.e. from connected diagrams), we need to perform the operation called *amputation*. We go on each external leg from the outside in until we reach the last part where it is connected with the rest of the diagram by a single leg, and cut there and excise that part, as in Figure 40.1. The reason is that we must divide out the connected correlation function by the full propagators for the external legs. Not quite the full propagators, of course, since we have a factor of $\sqrt{Z}$ instead of a factor of Z, but that is simply since a propagator has two legs instead of one, and we need to consider a factor of $\sqrt{Z}$ for each one. So, for example, for the two-point function, we would have a $\sqrt{Z}$ for each leg, but only a single free propagator factor (common to both).

This amputation procedure relates diagrams in the perturbative expansion of the correlation functions to diagrams in the perturbative expansion of the S-matrices.

Finally, we want to understand the full propagator better, and what it means to be near the on-shell pole. We saw in Part I that the connected two-point function G^c_{ij} is related to the 1PI two-point function Π_{ij} by

$$G^c = (1 + \Delta\Pi)^{-1}\Delta = \Delta - \Delta\Pi\Delta + \ldots, \tag{40.26}$$

for which we can write a diagrammatic representation, as in Figure 40.2. But denoting $\Pi(p^2) \equiv iM^2(p^2)$, and that the free propagator has bare mass m_0, we have explicitly

$$G^c = \frac{-i}{p^2 + m_0^2} - \frac{-i}{p^2 + m_0^2}iM^2\frac{-i}{p^2 + m_0^2} = \frac{-i}{p^2 + m_0^2 + M^2(p^2)}, \tag{40.27}$$

and we see that near a physical pole, we have

$$G^c \sim \frac{-iZ}{p^2 + m^2} + \text{regular}. \tag{40.28}$$

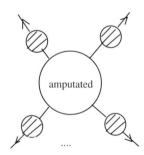

Figure 40.1   The diagrammatic amputation procedure.

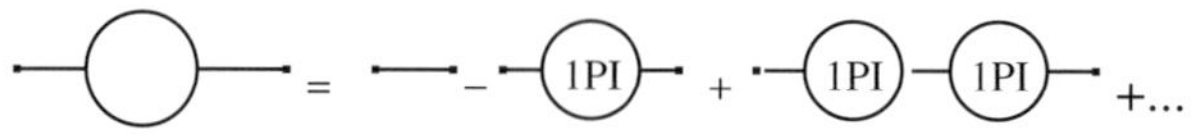

Figure 40.2   Diagrammatic expansion for the connected Green's function.

Indeed, as an example, taking $M^2(p^2) \simeq M^2 + \alpha p^2$, we would get

$$G^c = \frac{-i(1+\alpha)^{-1}}{p^2 + (m_0^2 + M^2)/(1+\alpha)}, \tag{40.29}$$

but if there are higher-order corrections in p^2 in $M^2(p^2)$, we would get other finite terms near the physical pole. We see that the addition of $M^2(p^2)$ in general both shifts the position of the physical pole, here from $-m_0^2$ to $-m^2 = -(m_0^2 + M^2)/(1+\alpha)$, and creates a Z factor, here $(1+\alpha)^{-1}$.

Important Concepts to Remember

- The LSZ formula relates correlation functions to S-matrices as follows: near the on-shell physical pole for all the external legs, removing the full propagators for the external legs (with only $\sqrt{Z}$ instead of Z), we obtain the on-shell S-matrix.
- In and out states correspond to a different sign in the Fourier transform of the correlator positions.
- The diagrammatic interpretation of going from correlators (connected diagrams) to S-matrices is of amputation, namely cut out all parts connecting external legs with the interior through a single propagator.
- For $\Pi = iM^2(p^2)$, $M^2(p^2)$ is added to the inverse propagator, shifting the physical pole and creating a Z factor.

Further Reading

See section 7.2 in [1].

Exercises

1. Write down all the three-loop divergent diagrams for the LSZ formula at four-points in ϕ^4 theory, and the associated diagrammatic amputation procedure.
2. Write down the LSZ formula for QED, and apply the diagrammatic procedure for the two-loop, six-point case.
3. Consider a hypothetical $\Pi(p^2) = i\left(M^2 e^{\alpha p^2} + p^2\right)$. Calculate the physical pole and the Z factor.
4. Consider $\Pi(p^2)$ in ϕ^3 up to three loops. To obtain the two-loop, four-point S-matrix, what order loop for the correlation function G_4 do we need?

41

The Coleman–Weinberg Mechanism for One-Loop Potential

In this chapter we will present a calculation of the one-loop effective potential in massless scalar theories. Coleman and Weinberg showed that in this case, the minimum of the one-loop corrected effective potential is shifted away from zero. However, in theories with a single coupling we need the one-loop term to be subleading to the tree term, eliminating the possibility of the minimum at one loop. In theories with more than one coupling, it is possible to maintain the validity of perturbation theory and still have the minimum away from zero.

41.1 One-Loop Effective Potential in $\lambda\phi^4$ Theory

We start with a massive theory of a real scalar field, and only later we will put $m \to 0$:

$$\mathcal{L} = -\frac{1}{2}(\partial_\mu\phi)^2 - \frac{m^2\phi^2}{2} - \lambda\frac{\phi^4}{4!}. \tag{41.1}$$

For the effective potential, we must consider (as we said in Part I of the book) a constant external field $\bar{\phi}$. The one-loop diagram for the ϕ^{2n} term is in Figure 41.1, a circle with n vertices on it, each with two external lines (and two forming propagators).

The symmetry factor for the diagram is

$$S = 2^n 2n, \tag{41.2}$$

since for each vertex we can flip the external lines, giving a factor 2 at each vertex, and cyclicity means we have n ways to define what vertex 1 is, and there is a reflection symmetry (up vs. down in the diagram).

The value of the Feynman diagram is then (keeping the factor of $\bar{\phi}^2$ of the external lines)

$$-iV^{(1)}(\bar{\phi})|_n = (-i\lambda\bar{\phi}^2)^n \frac{1}{2^n 2n} \int \frac{d^4k}{(2\pi)^4} \left(\frac{-i}{k^2 + m^2 - i\epsilon}\right)^n. \tag{41.3}$$

Summing over n, and defining the second derivative of the tree-level potential, $V''_{(0)} = \lambda\bar{\phi}^2/2!$, we find

$$-iV^{(1)}(\bar{\phi}) = -\frac{1}{2} \int \frac{d^4k}{(2\pi)^4} \ln\left[1 + \frac{V''_{(0)}}{k^2 + m^2 - i\epsilon}\right]. \tag{41.4}$$

In the sum over n, the $n = 1$ and $n = 2$ terms are UV divergent (and all terms would be IR divergent if $m \to 0$), so we need to renormalize.

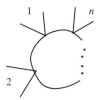

Figure 41.1   One-loop diagram for the effective potential.

41.2 Renormalization and Coleman–Weinberg Mechanism

We can try to use the renormalization conditions at zero external fields $\bar{\phi} = 0$ (which is a solution of the equation of motion for $V_{(0)}$ and, since $V^{(1)}$ has only ϕ^{2n} terms, also for the one-loop corrected V):

$$m_{\text{ren}}^2 \equiv \left. \frac{d^2 V(\bar{\phi})}{d\bar{\phi}^2} \right|_{\bar{\phi}=0} , \quad \lambda_{\text{ren}} \equiv \left. \frac{d^4 V(\bar{\phi})}{d\bar{\phi}^4} \right|_{\bar{\phi}=0} . \tag{41.5}$$

Since $V_{(0)}$ already satisfies the renormalization conditions, renormalization amounts to eliminating the $\bar{\phi}^2$ and $\bar{\phi}^4$ terms in the one-loop correction, so

$$-i V^{(1),\text{ren}}(\bar{\phi}) = -\frac{1}{2} \int \frac{d^4}{(2\pi)^4} \left\{ \ln\left[1 + \frac{V_{(0)}''}{k^2 + m^2 - i\epsilon} \right] \right.$$
$$\left. - \frac{V_{(0)}''}{k^2 + m^2 - i\epsilon} + \frac{1}{2}\left(\frac{V_{(0)}''}{k^2 + m^2 - i\epsilon} \right)^2 \right\} . \tag{41.6}$$

To evaluate the result, we can Wick rotate to Euclidean space (we could have done so from the beginning, for the calculation of the effective potential), and find

$$V^{(1),\text{ren}}(\bar{\phi}) = \frac{1}{2} \int \frac{d^4 k}{(2\pi)^4} \left\{ \ln\left[1 + \frac{V_{(0)}''}{k^2 + m^2 - i\epsilon} \right] \right.$$
$$\left. - \frac{V_{(0)}''}{k^2 + m^2} + \frac{1}{2}\left(\frac{V_{(0)}''}{k^2 + m^2} \right)^2 \right\} . \tag{41.7}$$

We write $d^4 k = \pi^2 k^2 d(k^2)$, and use $x = k^2$ as variable, obtaining

$$V^{(1),\text{ren}}(\bar{\phi}) = \frac{1}{32\pi^2} \int_0^\infty dx\, x \left[\ln\left(1 + \frac{V_{(0)}''(\bar{\phi})}{x + m_{\text{ren}}^2} \right) - \frac{V_{(0)}''(\bar{\phi})}{x + m_{\text{ren}}^2} \right.$$
$$\left. + \frac{1}{2}\left(\frac{V_{(0)}''(\bar{\phi})}{x + m_{\text{ren}}^2} \right)^2 \right] . \tag{41.8}$$

After some algebra, left as an exercise, we obtain

$$V^{(1),\text{ren}}(\bar{\phi}) = \frac{\hbar}{(8\pi)^2}\left[\left(\frac{1}{2}\lambda_{\text{ren}}\bar{\phi}^2 + m_{\text{ren}}^2\right)^2 \ln\left(1 + \frac{\lambda_{\text{ren}}\bar{\phi}^2}{2m_{\text{ren}}^2}\right)\right.$$
$$\left. - \frac{\lambda_{\text{ren}}\bar{\phi}^2}{2}\left(\frac{3}{4}\lambda_{\text{ren}}\bar{\phi}^2 + m_{\text{ren}}^2\right)\right], \tag{41.9}$$

where we have reintroduced $\hbar$ to emphasize that we have a one-loop quantum correction.

However, we want to consider massless fields, and the limit $m_{\text{ren}}^2 \to 0$ above is singular, so we need to change the renormalization conditions.

We change to defining λ_{ren} at $\bar{\phi} = M$ instead, and call the new coupling λ_M:

$$\lambda_M \equiv \left.\frac{d^4 V(\bar{\phi})}{d\bar{\phi}^4}\right|_{\bar{\phi}=M}. \tag{41.10}$$

Then we can take $m_{\text{ren}}^2 \to 0$, and it will not matter where we define it. The renormalized potential is a physical quantity, so its numerical value shouldn't depend on the renormalization conditions. We therefore take the fourth derivative of the sum of $V_{(0)}$ and $V^{(1),\text{ren}}$, evaluate it at $\bar{\phi} = M$, and equate with λ_M, to obtain

$$\lambda_M = \lambda_{\text{ren}} + \frac{\hbar\lambda_{\text{ren}}^2}{(8\pi)^2}\left[6\ln\left(1 + \frac{\lambda_{\text{ren}}\bar{\phi}^2}{2m_{\text{ren}}^2}\right) + 16 + \mathcal{O}\left(\frac{m_{\text{ren}}^2}{M^2}\right)\right]. \tag{41.11}$$

We can then substitute λ_{ren} by λ_M in the effective potential, which at tree+one loop is now

$$V^{\text{ren}}(\bar{\phi}) = \lambda_M \frac{\bar{\phi}^4}{4!} + \frac{\hbar\lambda_M^2\bar{\phi}^4}{(16\pi)^2}\left(\ln\frac{\bar{\phi}^2}{M^2} - \frac{25}{6}\right) + \mathcal{O}\left(\frac{m_{\text{ren}}^2}{M^2}\right) + \mathcal{O}(\hbar^2), \tag{41.12}$$

on which we can take the $m_{\text{ren}}^2 \to 0$ limit, obtaining the Coleman–Weinberg formula for a massless, real scalar field.

The minimum of the one-loop potential above is at $dV^{\text{ren}}/d\bar{\phi} = 0$, giving

$$\frac{32\pi^2}{3} + \hbar\lambda_M\left(-\frac{11}{3} + \ln\frac{\bar{\phi}^2}{M^2}\right) = 0. \tag{41.13}$$

It would seem that the minimum is shifted to nonzero $\bar{\phi}$, except that higher-order corrections will come with powers of $\hbar\lambda_M \ln\bar{\phi}^2/M^2$, thus must be small. Yet this contradicts the minimum condition. That was to be expected, since at the minimum we are balancing a one-loop term against a tree-level term.

41.3 Coleman–Weinberg Mechanism in Scalar-QED

We can find a consistent perturbation theory and a nontrivial minimum in the case of a theory with a few more fields, namely scalar-QED, with a complex scalar $\phi = (\phi_1 + i\phi_2)/\sqrt{2}$ minimally coupled to a gauge field, and with potential

$$V_{(0)} = \frac{\lambda}{3!}(\phi^*\phi)^2 = \frac{\lambda}{4!}(\phi_1^2 + \phi_2^2)^2. \qquad (41.14)$$

From the minimal coupling $-|D_\mu\phi|^2$ we find a vertex $-e^2\phi^*\phi A_\mu A^\mu$, which means that we must add another Feynman diagram to the one-loop potential, with the gauge field running inside the loop, instead of the scalar, but with the same external scalars. By symmetry, we can rotate such that only graphs with external ϕ_1 are considered. But then, inside the loops we have both ϕ_1, coming from the vertex $\lambda\phi_1^4/4!$, and ϕ_2, coming from the vertex $2\lambda\phi_1^2\phi_2^2/4!$.

For the loop with the photon, considering the Landau gauge propagator, we only have an extra polarization factor in front of the scalar propagator:

$$D_{\mu\nu}(k) = \left(\eta_{\mu\nu} - \frac{k_\mu k_\nu}{k^2}\right)D(k). \qquad (41.15)$$

But in the loop, this appears in products like

$$D_{\mu\nu}(k)D^\nu{}_\rho(k) = \left(\eta_{\mu\nu} - \frac{k_\mu k_\nu}{k^2}\right)\left(\delta_\rho^\nu - \frac{k^\nu k_\rho}{k^2}\right)(D(k))^2 = D_{\mu\rho}(k)D(k), \qquad (41.16)$$

until finally the polarization factor is traced, giving $\delta_\mu^\mu - k^\mu k_\mu/k^2 = 3$, times the scalar loop.

All in all, the result is just a numerical factor in front of the one-loop result for the real scalar, that is

$$V^{\mathrm{ren}}(\bar{\phi}_1) = \lambda_M\frac{\bar{\phi}_1^4}{4!} + C\frac{\hbar\bar{\phi}_1^4}{(8\pi)^2}\left(\ln\frac{\bar{\phi}_1^2}{M^2} - \frac{25}{6}\right) + \mathcal{O}\left(\frac{m_{\mathrm{ren}}^2}{M^2}\right) + \mathcal{O}(\hbar^2), \qquad (41.17)$$

where the factor C contains: the factor $(\lambda_M/2)^2$ from the ϕ_1 loop (as in the real case), plus a factor of $1/9(\lambda_M/2)^2$ from the ϕ_2 loop, since we have six ways to extract two ϕ_1s from ϕ_1^4, but only one from $2\phi_1^2\phi_2^2$, and because of the factor of 2, there is a relative $1/3$ at each λ vertex, and finally a $3e^4$, from the 3 in the Landau propagator trace, and the vertex factor of e^2. Thus

$$C = \left(\frac{\lambda_M}{2}\right)^2 + \frac{1}{9}\left(\frac{\lambda_M}{2}\right)^2 + 3e^4. \qquad (41.18)$$

Now we can assume that numerically we have λ_M of the order of $\hbar e^4$, which means that the terms with λ_M^2 can be neglected with respect to the $\hbar e^4$, and still have a consistent perturbation theory. Then the potential is approximately

$$V^{\mathrm{ren}}(\bar{\phi}_1) \simeq \lambda_M\frac{\bar{\phi}_1^4}{4!} + \frac{3\hbar e^4\bar{\phi}_1^4}{(8\pi)^2}\left(\ln\frac{\bar{\phi}_1^2}{M^2} - \frac{25}{6}\right). \qquad (41.19)$$

If we moreover choose $M = \bar{\phi}_{1,\min}$ (the value of $\bar{\phi}_1$ at the minimum), such that $\ln\bar{\phi}_{1,\min}^2/M^2 = 0$, then the minimum of the potential gives (denoting by $\tilde{\lambda}_M$)

$$\left(\tilde{\lambda}_M - \frac{11}{16\pi^2}\hbar e^4\right)\bar{\phi}_1^3 = 0, \qquad (41.20)$$

and substituting this value for λ_M in the potential, we can write

$$V^{\text{ren}} = \frac{3}{64\pi^2}\hbar e^4 \bar{\phi}_1^4 \left(\ln \frac{\bar{\phi}_1^2}{\bar{\phi}_{1,\min}^2} - \frac{1}{2} \right). \tag{41.21}$$

Thus we have obtained a minimum away from the origin, a consistent perturbation theory, and $\lambda \sim \hbar e^4$.

Important Concepts to Remember

- The one-loop effective potential has UV-divergent $n = 1$ and $n = 2$ terms, and has the formula $V = \frac{1}{2} \int \frac{d^d k}{(2\pi)^d} \ln\left[1 + \frac{V''_{(0)}}{k^2 + m^2 - i\epsilon} \right]$.
- Renormalizing, we find at one loop the Coleman–Weinberg potential, $V^{\text{ren}}(\bar{\phi}) = \lambda_M \frac{\bar{\phi}^4}{4!} + \frac{\hbar \lambda_M^2 \bar{\phi}^4}{(16\pi)^2}\left(\ln \frac{\bar{\phi}^2}{M^2} - \frac{25}{6} \right)$.
- The Coleman–Weinberg potential has a minimum balancing the one-loop term against the classical term, but that contradicts the perturbative hypothesis.
- For scalar-QED, we can choose λ_M of order $\hbar e^4$, in which case the effective potential can balance the one-loop term against the classical term, though with different couplings, $V^{\text{ren}}(\bar{\phi}_1) \simeq \lambda_M \frac{\bar{\phi}_1^4}{4!} + \frac{3\hbar e^4 \bar{\phi}_1^4}{(8\pi)^2}\left(\ln \frac{\bar{\phi}_1^2}{M^2} - \frac{25}{6} \right)$.

Further Reading

See [10].

Exercises

1. Repeat the calculation of the one-loop effective potential for $\lambda\phi^3$ theory in $d = 6$.
2. Renormalize the one-loop result for $\lambda\phi^3$ in $d = 6$ to find the equivalent of the Coleman–Weinberg potential.
3. Finish the details leading to (41.9).
4. Is it possible for the two-loop term to be balanced against the tree term in the scalar-QED effective potential?

42

Quantization of Gauge Theories I: Path Integrals and Fadeev–Popov

We now start the analysis of nonabelian gauge theories. In classical field theory we have seen how to define them, but we will review it here. After that, we will describe the Fadeev–Popov quantization procedure for the path integral of nonabelian gauge theories (we have seen in Part I how to do it for the abelian case). The result of the procedure will imply a gauge-fixing term and a ghost action in the quantum action for the gauge theory.

42.1 Review of Yang–Mills Theory and its Coupling to Matter Fields

Consider a gauge field

$$A_\mu = A_\mu^a T^a, \tag{42.1}$$

where T^a are the generators of a Lie algebra of a gauge group G, so A_μ is in the Lie algebra (i.e. in the adjoint representation). A general group element is a set of $N_R \times N_R$ matrices for the representation R:

$$U = e^{\alpha^a T^a}, \tag{42.2}$$

where $\alpha^a \in \mathbb{R}$. The generators T^a, $a = 1, \ldots, N_G$ obey the Lie algebra

$$[T^a, T^b] = f^{ab}{}_c T^c, \tag{42.3}$$

and are normalized by the relation

$$\mathrm{Tr}[T^a T^b] = T_R \delta^{ab}. \tag{42.4}$$

A note on conventions. My conventions (that I will use unless specified otherwise, or unless left free) are of anti-Hermitian generators, $(T^a)^\dagger = -T^a$, with $f^{ab}{}_c$ real and $T_R = -1/2$ in the fundamental representation of $SU(N)$. Another popular convention in the literature is with Hermitian generators, $(T^a)^\dagger = T^a$, $T_R = 1$ in the adjoint of $SU(N)$ and $[T^a, T^b] = i f^{ab}{}_c T^c$.

In a representation R, the following quadratic form is a constant (proportional to the identity over a given representation):

$$\sum_a (T_R^a)^2 = C_R \, \mathbb{1}, \tag{42.5}$$

and is called the (quadratic) Casimir of the representation R of the group G. Then

$$T_R N_G = C_R N_R. \tag{42.6}$$

In the case of the adjoint representation, defined by

$$(T^a)_{bc} = f^a{}_{bc}, \tag{42.7}$$

we have

$$T_R = C_R \equiv C_2(G) \tag{42.8}$$

as well as $N_R = N_G$.

For $G = SU(N)$, we have of course $N_G = N^2 - 1$, and in the fundamental representation (for quarks) we have $N_R = N$.

My normalization $T_R = -1/2$ in the fundamental representation also leads to $C_R = (N^2 - 1)/2N$.

The way to introduce gauge fields is by starting with some matter action invariant under the global action of the group G:

$$\psi_i(x) \rightarrow U_{ij}\psi_j(x), \tag{42.9}$$

and making the invariance local, $U_{ij} \rightarrow U_{ij}(x)$. That requires the introduction of another field, the gauge field A_μ^a, and of the minimal coupling of the matter to it, through the covariant derivative replacing the ordinary derivative:

$$\partial_\mu \psi_i(x) \rightarrow (D_\mu)_{ij}\psi_j(x); \quad (D_\mu)_{ij} \equiv \partial_\mu \delta_{ij} + g(T_R^a)_{ij}A_\mu^a(x). \tag{42.10}$$

In particular, in the adjoint representation:

$$D_\mu^{ab} = \partial_\mu \delta^{ab} + gf^{ab}{}_c A_\mu^c. \tag{42.11}$$

For an infinitesimal gauge transformation with parameter

$$U(x) = e^{g\alpha^a T^a} \simeq 1 + g\alpha^a(x)T^a + \dots, \tag{42.12}$$

the gauge field transforms as

$$A_\mu^a(x) \rightarrow A_\mu^a(x) + D_\mu^{ab}\alpha^b(x). \tag{42.13}$$

For a finite gauge transformation, $U = e^{g\alpha^a(x)T^a} = e^{g\alpha(x)}$, the gauge field transforms as

$$A_\mu(x) \rightarrow A_\mu^U(x) = U^{-1}A_\mu(x)U(x) + \frac{1}{g}\partial_\mu U(x)U^{-1}(x). \tag{42.14}$$

The field strength is defined as

$$F_{\mu\nu}^a = \partial_\mu A_\nu^a - \partial_\nu A_\mu^a + gf^a{}_{bc}A_\mu^b A_\nu^c. \tag{42.15}$$

One defines also the contraction with T^a:

$$A_\mu = A_\mu^a T^a; \quad F_{\mu\nu} = F_{\mu\nu}^a T^a, \tag{42.16}$$

such that the field strength is

$$F_{\mu\nu} = \partial_\mu A_\nu - \partial_\nu A_\mu + g[A_\mu, A_\nu], \tag{42.17}$$

as well as the form notation

$$A = A_\mu dx^\mu; \quad F = \frac{1}{2}F_{\mu\nu}dx^\mu dx^\nu \tag{42.18}$$

(note that in general $f_p = 1/p! f_{\mu_1 \ldots \mu_p} dx^{\mu_1} \wedge \ldots \wedge dx^{\mu_p}$), leading to the field strength

$$F = dA + gA \wedge A. \tag{42.19}$$

The field strength transforms *covariantly*, that is

$$F_{\mu\nu} \rightarrow F'_{\mu\nu} = U^{-1}(x) F_{\mu\nu} U(x). \tag{42.20}$$

The gauge-invariant action for the gauge field in Minkowski space is

$$S_M = \int d^4x \left[-\frac{1}{4} F^a_{\mu\nu} F^{a,\mu\nu} \right] = +\frac{1}{2} \int d^4x \, \mathrm{Tr}[F^2_{\mu\nu}]. \tag{42.21}$$

Wick rotating to Euclidean space, $x_4 = it$, so $\partial_4 = -i\partial_t$, and the same for the gauge field, which transforms as ∂_μ under Lorentz transformations, $A_4 = -iA_0$, so

$$E_i^{\text{Eucl.}} = \frac{\partial}{\partial x_4} A_i - \frac{\partial}{\partial x_i} A_4 = -iE_i^{\text{Mink.}}, \tag{42.22}$$

and so the Euclidean Lagrangian is

$$\mathcal{L}^{\text{Eucl.}} = +\frac{1}{4} F^a_{\mu\nu} F^{a\,\mu\nu} = -\frac{1}{2} \, \mathrm{Tr}[F^2_{\mu\nu}] = \frac{1}{2}((\vec{E}^a)^2 + (\vec{B}^a)^2). \tag{42.23}$$

42.2 Fadeev–Popov Procedure in Path Integrals

42.2.1 Correlation Functions

We now write the quantum correlation functions for *gauge-invariant observables* (observables in the gauge theory must be gauge invariant, by definition) in Euclidean space as path integrals:

$$\langle \mathcal{O}_1(A_\mu) \ldots \mathcal{O}_n(A_\mu) \rangle = \frac{\int \mathcal{D}A_\mu e^{-S[A_\mu]} \mathcal{O}_1(A_\mu) \ldots \mathcal{O}_n(A_\mu)}{\int \mathcal{D}A_\mu e^{-S[A_\mu]}}. \tag{42.24}$$

As in the abelian case described in Part I of the book, we "fix the gauge" by the Fadeev–Popov procedure, which amounts to dividing in the numerator and denominator by the volume of the gauge group G:

$$\prod_{x \in \mathbb{R}^d} V_x(G); \quad V(G) = \int_G dU, \tag{42.25}$$

where $\int dU$ is called the Haar measure. It is defined to be invariant under left and right multiplication by a fixed element of the gauge group:

$$U \rightarrow UU_0 \quad \text{and} \quad U \rightarrow U_0 U. \tag{42.26}$$

We are interested in covariant gauges like the Lorenz gauge, $\partial_\mu A^a_\mu = 0$, generalized to the form

$$\mathcal{F}^a(x) = c^a(x); \quad a = 1, \ldots, N. \tag{42.27}$$

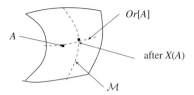

Figure 42.1   The gauge-fixed configuration is at the intersection of the orbit of A, the gauge transformations of a gauge field configuration A, and the space $\mathcal{M}$ of all possible gauge conditions.

We define the *orbit of the gauge field* $A_\mu(x)$ as the space of all possible gauge transformations of $A_\mu(x)$, that is

$$Or[A_\mu(x)] \equiv \{\tilde{A}_\mu | \exists U(x) \in G, \text{ such that } \tilde{A}_\mu(x) = {}^{U(x)}A_\mu(x)\}. \tag{42.28}$$

We also define the space of all possible gauge fields satisfying a gauge condition:

$$\mathcal{M}(A) = \{A_\mu \text{ such that } \mathcal{F}^a(A_\mu) = c^a(x)\}. \tag{42.29}$$

We assume that there is a single intersection between the two, that is $Or[A_\mu(x)] \cap \mathcal{M} = \{\text{point}\}$, or

$$\exists! \tilde{A}_\mu \text{ such that } \mathcal{F}(\tilde{A}_\mu) = c^a(x). \tag{42.30}$$

Note that this assumption is only correct for infinitesimal gauge transformations, otherwise for large gauge transformations there are *Gribov copies* which are a large distance in gauge transformation space from the identity. We will not address Gribov copies in this book. The situation above is depicted in Figure 42.1.

Then we define

$$\frac{1}{\Delta_{\mathcal{F},c}[A_\rho]} \equiv \int \prod_x dU_x \prod_{y,a} \delta(\mathcal{F}^a({}^U A_\rho) - c^a). \tag{42.31}$$

By our assumption, there is a unique $U = U^{(A)}(x)$, depending on $A_\mu(x)$, such that by transforming with it we go onto the gauge condition, that is

$$\mathcal{F}^a({}^{U^{(A)}} A_\rho) = c^a. \tag{42.32}$$

Define the matrix

$$M^{ab}(x, y; A_\rho) \equiv \frac{\partial \mathcal{F}^a}{\partial A_\mu^c(x)} D_\mu^{cb}(x; A_\rho)\delta^{(D)}(x - y), \tag{42.33}$$

where as before

$$D_\mu^{ab}(x, A_\rho) \equiv \frac{\partial}{\partial x^\mu}\delta^{ab} + gf^{ab}{}_c A_\mu^c(x). \tag{42.34}$$

The matrix M is thought of as a matrix in both the space (ab) and the space (xy). We have the following:

Lemma (Properties)

(A) $\Delta_{\mathcal{F},c}[A_\rho]$ is gauge invariant.

(B) $\Delta_{\mathcal{F},c}[A_\rho] = \det M(^{U^{(A)}}A_\rho)$.

Proof of property A   Write the definition of $\Delta[A_\rho]$ for the gauge field transformed with some U_0, $^{U_0}A_\rho$:

$$\Delta^{-1}[^{U_0}A_\rho] = \int \prod_x dU_x \prod_{x,a} \delta(\mathcal{F}^a(^U(^{U_0}A_\rho)) - c^a). \tag{42.35}$$

By invariance of the Haar measure, $dU_x = d(UU_0)$, so we write $dU_x = d(U_xU_0) \equiv d\tilde{U}_x$, and get

$$\Delta^{-1}[^{U_0}A_\rho] = \int \prod_x d\tilde{U}_x \prod_{x,a} \delta(\mathcal{F}^a(^{\tilde{U}}A_\rho) - c^a) = \Delta^{-1}[A_\rho]. \tag{42.36}$$

$$q.e.d.$$

Proof of property B   We use property A to write

$$\Delta(A_\rho) = \Delta(^{U^{(A)}}A_\rho) \equiv \Delta(\tilde{A}_\rho), \tag{42.37}$$

where by definition of $U^{(A)}$, $\mathcal{F}(\tilde{A}_\rho) = c^a$. Then

$$\Delta^{-1}[A_\rho] = \Delta^{-1}[\tilde{A}_\rho] = \int \prod_x dU_x \delta(\mathcal{F}^a(^U\tilde{A}_\rho) - c^a). \tag{42.38}$$

For infinitesimal transformations:

$$U(x) = e^{\alpha^a(x)T^a} \simeq 1 + \alpha^a(x)T^\alpha \Rightarrow dU_x = \prod_{a=1}^N d\alpha^a(x) + \mathcal{O}(\alpha^2), \tag{42.39}$$

which leads to

$$(^U\tilde{A})_\mu^a \simeq \tilde{A}_\mu^a + D_\mu^{ab}(\tilde{A}_\rho)\alpha^b + \mathcal{O}(\alpha^2), \tag{42.40}$$

and in turn to

$$\mathcal{F}^a(^U\tilde{A}_\rho)(x) \simeq \mathcal{F}(\tilde{A}_\rho)(x) + \frac{\partial \mathcal{F}(\tilde{A}_\rho(x))}{\partial \tilde{A}_\mu^c(x)} D_\mu^{cb}(x, \tilde{A}_\rho)\alpha^b(x) + \mathcal{O}(\alpha^2). \tag{42.41}$$

Substituting in $\Delta^{-1}[\tilde{A}_\rho]$, we obtain (using that $\mathcal{F}^a(\tilde{A}_\mu) = c^a$)

$$\Delta^{-1}[A_\rho] \simeq \int \prod_{x,a'} d\alpha_x^{a'} \prod_{y,a} \delta\left(\frac{\partial \mathcal{F}(\tilde{A}_\rho(y))}{\partial \tilde{A}_\mu^c} D_\mu^{cb}(y, \tilde{A}_\rho)\alpha^b(y) \right)$$

$$= \int \prod_{x,a'} d\alpha_x^{a'} \prod_{y,z} \delta\left(\int d^d z M^{ab}(y, z; \tilde{A}_\rho)\alpha^b(z) \right) \equiv \left(\det M(\tilde{A}_\rho) \right)^{-1}. \tag{42.42}$$

In the last equality, the determinant was considered both in (ab) and in (xy) space, and we have used the generalization of the relation

$$\int \prod_{i=1}^{n} d\alpha_i \prod_{j=1}^{n} \delta(M_{ij}\alpha_j) = \frac{1}{\det M}. \tag{42.43}$$

q.e.d.

Then we can take the determinant on the other side of the equation and write

$$1 = \int \prod_x dU_x \left[\det M(^{U^{(A)}}A_\rho)\right] \prod_{y,a} \delta(\mathcal{F}^a(^{U}A_\rho) - c^a(y)). \tag{42.44}$$

But the delta function enforces $U = U^{(A)}$ anyway, so we can replace it in the integral and write

$$1 = \int \prod_x dU_x \det M(^{U}A_\rho) \prod_{y,a} \delta(\mathcal{F}^a(^{U}A_\rho) - c^a(y)). \tag{42.45}$$

Now we can define

$$\text{``}1(\alpha)\text{''} \equiv \int \prod_{x,a} dc^a(x) e^{-\frac{1}{2\alpha} \int d^D x [c^a(x)]^2}, \tag{42.46}$$

which is of course a combination of πs and αs, that is irrelevant in the correlators, since it would cancel in the numerator and denominator. Now, substituting 1 in the form of (42.45) on the right-hand side of the above, and doing the integral over $dc^a(x)$, which fixes $c^a(x) = \mathcal{F}^a(x)$, we obtain

$$\text{``}1(\alpha)\text{''} = \int \prod_x dU_x \det M(^{U}A_\rho) e^{-\frac{1}{2\alpha}\mathcal{F}^a\mathcal{F}^a}. \tag{42.47}$$

Inserting this "$1(\alpha)$" in the path integral, we obtain

$$\int \mathcal{D}A_\mu e^{-S[A_\mu]} = \int \mathcal{D}[A_\mu] \int \prod_x dU_x \det M(^{U}A_\mu) e^{-S[A_\mu]-\frac{1}{2\alpha}\mathcal{F}^2[^{U}A_\mu]}$$

$$= \int \prod_x dU_x \int \mathcal{D}A_\mu \det M(^{U}A_\mu) e^{-S[A_\mu]-\frac{1}{2\alpha}\mathcal{F}^2[^{U}A_\mu]}. \tag{42.48}$$

But because of the gauge invariance of the measure $\mathcal{D}A_\mu$ and the action $S[A_\mu]$, we can replace A_μ by $^{U}A_\mu$ in the path integral, and finally rename it A_μ again everywhere. Note that it is exactly this step that can fail if there are gauge anomalies, which will be studied later, but here we will assume that there aren't.

We finally obtain that the volume of the gauge group factorizes from the integral as

$$\int \mathcal{D}A_\mu e^{-S[A_\mu]} = \left[\int \prod_x dU_x\right] \int \mathcal{D}A_\mu \det M(A_\mu) e^{-S[A_\mu]-\frac{1}{2\alpha}\mathcal{F}^2[A_\mu]}. \tag{42.49}$$

The same thing happens for the path integral with the *gauge-invariant* observables

$$\int \mathcal{D}A_\mu e^{-S[A_\mu]} \mathcal{O}_1[A_\mu] \dots \mathcal{O}_n[A_\mu]$$

$$= \left[\int \prod_x dU_x\right] \int \mathcal{D}A_\mu \det M(A_\mu) e^{-S[A_\mu]-\frac{1}{2\alpha}\mathcal{F}^2[A_\mu]} \mathcal{O}_1[A_\mu] \dots \mathcal{O}_n[A_\mu]. \tag{42.50}$$

This means that the volume of the gauge group cancels between the numerator and the denominator in the correlation functions, and we obtain

$$\langle \prod_{i=1}^{n} O_i(A_\mu) \rangle = \frac{\int \mathcal{D}A_\mu \prod_{i=1}^{n} O_i(A_\mu) e^{-S_{\text{eff}}[A_\mu]}}{\int \mathcal{D}A_\mu e^{-S_{\text{eff}}}}, \tag{42.51}$$

where S_{eff} is the sum of the classical action, a "gauge-fixing term," and an extra term:

$$S_{\text{eff}}[A_\mu] = S[A_\mu] + \frac{1}{2\alpha} \int d^D x \mathcal{F}^2[A_\mu] - \log \det M(A_\mu). \tag{42.52}$$

The last term will turn into a "ghost action" term, as we will now see.

42.3 Ghost Action

In the case of the Lorenz gauge, $\mathcal{F}^a = \partial_\mu A_\mu^a$, the matrix M becomes

$$M^{ab}(x, y; A_\rho) = \partial_\mu D_\mu^{ab} \delta^D(x - y) = (\partial_\mu^2 \delta^{ab} + g f^{ab}{}_c \partial_\mu A_\mu^c(x)) \delta^D(x - y), \tag{42.53}$$

understood as a matrix in both (ab) and (xy) space. Since we are interested in

$$\log \det M(A_\rho) = \log \frac{\det M(A_\rho)}{\det \partial^2} + \log \det \partial^2, \tag{42.54}$$

we can drop the last term, which is just a constant (albeit an infinite one), since it again cancels in correlators between the numerator and the denominator.

Note that $\partial^{-2} = \Delta(x, y)$ is the scalar (Klein–Gordon) propagator.

Therefore, we consider

$$\frac{\det M(A_\rho)}{\det \partial^2} = \det \left[\frac{M(A_\rho)}{\partial^2} \right] \equiv \det(1 + N), \tag{42.55}$$

where

$$(1 + N)^{ab}(x, y) = \delta^{ab} \delta^D(x - y) + g \int d^D z \Delta(x - z) \frac{\partial}{\partial z^\mu} f^{ab}{}_c A_\mu^c \delta^D(z - y). \tag{42.56}$$

Alternatively, this means that

$$\det(1 + N) = e^{\text{Tr} \log(1+N)} = \exp \text{Tr} \left(\sum_{n \geq 1} \frac{(-1)^{n+1}}{n} N^n \right)$$

$$= \exp \left\{ g \int d^D y_1 f^a{}_{ac} \left[\Delta(x - y_1) \frac{\partial}{\partial y_1^\mu} A_\mu^c(y_1) \delta^D(y_1 - x) \right] \right.$$

$$- \frac{g^2}{2} \int d^D y_1 d^D y_2 f^{ab}{}_c \left[\Delta(x - y_1) \frac{\partial}{\partial y_1^\mu} A_\mu^c(y_1) \delta^D(y_1 - y_2) \right] f_{bad}$$

$$\left. \times \left[\Delta(y_1 - y_2) \frac{\partial}{\partial y_2^\nu} A_\nu^d(y_2) \delta^D(y_2 - x) \right] \right\} + \dots \tag{42.57}$$

However, this gives an infinite number of vertices in the action (term $-\log\det(1+N)$), as we easily see. This is not very good.

Instead, we can use a representation in the action in terms of *fermions*, or rather anticommuting variables, that will be called ghosts.

Remember that we have the formula

$$\int \mathcal{D}\Phi \mathcal{D}\bar{\Phi} e^{-\bar{\Phi}\cdot M\cdot \Phi} \propto (\det M)^{\pm 1}, \tag{42.58}$$

where as usual the proportionality constant is not relevant, and ± 1 is for anticommuting/commuting variables, respectively.

Since we have $\det M^{+1}$, we are interested in anticommuting variables, and moreover then we have an arbitrary sign in front of the action that we can use, since

$$\int \prod_{i=1}^{N} d\eta_i d\bar{\eta}_i e^{\pm\bar{\eta}\cdot M\cdot\eta} = (\pm 1)^N \det M. \tag{42.59}$$

We choose the plus sign in the above (i.e. $e^{+\bar{\eta}\cdot M\cdot\eta}$), and as usual we choose the Lorenz gauge $\mathcal{F}^a = \partial_\mu A^a_\mu$, obtaining the gauge-fixed Yang–Mills (YM) action in Euclidean space

$$S_{\text{eff}} = \int d^D x \left[\frac{1}{4}(F^a_{\mu\nu})^2 + \frac{1}{2\alpha}(\partial_\mu A^a_\mu)^2 - \bar{\eta}^a \partial_\mu D^{ab}_\mu \eta^b \right]. \tag{42.60}$$

Note that I wrote here $\bar{\eta}^a$ and η^b, but this is misleading, since as we know from the complex integration of anticommuting objects, the integrations are really independent. That is why one usually writes b^a for $\bar{\eta}^a$ and c^a for η^a, so

$$S_{\text{eff}} = \int d^D x \left[\frac{1}{4}(F^a_{\mu\nu})^2 + \frac{1}{2\alpha}(\partial_\mu A^a_\mu)^2 - b^a \partial_\mu D^{ab}_\mu c^b \right]. \tag{42.61}$$

For a general gauge condition, we have

$$S_{\text{eff}} = \int d^D x \left[\frac{1}{4}(F^a_{\mu\nu})^2 + \frac{1}{2\alpha}(\mathcal{F}^a(A))^2 - b^a \frac{\partial \mathcal{F}^a}{\partial A^c_\mu} D^{cb}_\mu c^b \right]. \tag{42.62}$$

Other gauge conditions, generalizing the Lorenz gauge, can be written as

$$\mathcal{F}^a = \phi^{ab}_\mu A^b_\mu. \tag{42.63}$$

In particular, axial gauges (which are, however, not Lorentz-covariant) correspond to

$$\phi^{ab}_\mu = \eta_\mu \delta^{ab}. \tag{42.64}$$

Here, η_μ is a constant 4-vector.

Important Concepts to Remember

- The YM field strength transforms covariantly, $F_{\mu\nu} \rightarrow U^{-1}(x) F_{\mu\nu} U(x)$.
- The YM action in Euclidean space is $+\frac{1}{4}\int d^D x F^a_{\mu\nu} F^{a\,\mu\nu}$.

- The quantum correlators of gauge-invariant operators are written as ratios of path integrals with and without the operators, so we can use the Fadeev–Popov gauge-fixing procedure, by factorizing and canceling the volume of the gauge group from the two path integrals.
- The result of the gauge-fixing procedure for the gauge condition $\mathcal{F}^a(x) = c^a(x)$ is $S_{\text{eff}}[A] = S[A] + \int d^D x \mathcal{F}^2(A)/2\alpha - \log \det M(A)$.
- The $\log \det M(A)$ term can be written as a ghost action, $\int d^D x [-b^a \partial \mathcal{F}^a(A)/\partial A^c_\mu D^{cb}_\mu(A) c^b]$.

Further Reading

See section 7.1 in [3] and section 16.2 in [1].

Exercises

1. Consider a solution to the self-duality equation for Yang–Mills theory in Euclidean space:

$$F_{\mu\nu} = \frac{1}{2} \epsilon_{\mu\nu}{}^{\rho\sigma} F_{\rho\sigma}. \tag{42.65}$$

Show that the on-shell action is bounded by a topological term (which cannot be changed by a small transformation).

2. Consider a solution to the self-duality condition that asymptotes to flat space at $x_4 = -\infty$ and to a monopole configuration at $x_4 = +\infty$.

Wick rotate the configuration to Minkowski space, and consider the path integral centered around this configuration. What is its interpretation? Putting some reasonable numbers for the Standard Model, how relevant is this now?

3. Consider the nonabelian *Chern–Simons action* in 2+1 dimensions:

$$S = \text{Tr} \int d^3 x \epsilon^{\mu\nu\rho} \left[A_\mu \partial_\nu A_\rho + \frac{2}{3} A_\mu A_\nu A_\rho \right]. \tag{42.66}$$

Show that it is gauge invariant for gauge transformations that vanish at infinity, and write it in form language.

4. Consider a nonlinear gauge condition, $\mathcal{F}^a = (D_\mu A_\mu)^a$. Calculate the 4-vertices of the theory.

43

Quantization of Gauge Theories II: Propagators and Feynman Rules

As we have seen in Chapter 42, to calculate correlators we must fix a gauge, and then we use instead of the classical action the gauge-fixed action, with gauge-fixing term and ghost term

$$S_{\text{eff}} = \int d^D x \left[\frac{1}{4}(F_{\mu\nu}^a)^2 + \frac{1}{2\alpha}(\partial_\mu A_\mu^a)^2 - b^a \partial_\mu D_\mu^{ab} c^b \right]. \tag{43.1}$$

In order to calculate the Feynman rules, we split the action into a quadratic part, giving the propagators, and a cubic and quartic part, giving the interactions.

43.1 Propagators and Effective Action

43.1.1 Propagators

The quadratic action is

$$S^{[2]}[A, b, c] = \int d^D x \left\{ \frac{1}{2} A_\mu^a \left[\delta^{ab} \left(-\partial^2 \delta_{\mu\nu} + \partial_\mu \partial_\nu \left(1 - \frac{1}{\alpha} \right) \right) \right] A_\nu^b + b^a(-\delta^{ab}\partial^2)c^b \right\}. \tag{43.2}$$

From this, we derive the gluon propagator, which is just δ^{ab} times the abelian (photon) propagator, that is

$$\Delta_{\mu\nu}^{ab}(k) = \frac{\delta^{ab}}{k^2} \left[\delta_{\mu\nu} - (1 - \alpha)\frac{k_\mu k_\nu}{k^2} \right], \tag{43.3}$$

and the *ghost propagator*, which is (despite the anticommuting nature of the ghosts) just the scalar KG propagator

$$\Delta^{ab}(k) = \frac{\delta^{ab}}{k^2}. \tag{43.4}$$

43.1.2 Interactions

The interaction action can be rewritten as

$$S_{\text{int}}[A, b, c] = g f_{abc} \int d^D x \left[(\partial_\mu A_\nu^a) A_\mu^b A_\nu^c + (\partial_\mu b^a) A_\mu^b c^c \right]$$
$$+ \frac{g^2}{4} f^a{}_{bc} f_{ade} \int d^D x A_\mu^b A_\nu^c A_\mu^d A_\nu^e. \tag{43.5}$$

We can now define path integrals. In particular, the free energy W is defined as usual by

$$e^{-W[J,\xi_b,\xi_c]} = \int \mathcal{D}A\mathcal{D}b\mathcal{D}c \; e^{-S_{\text{eff}}[A,b,c]+\int d^D x(J\cdot A+\xi_c c+b\xi_b)}. \tag{43.6}$$

Then, as usual, the effective action, the generator of 1PI n-point functions, is the Legendre transform of the free energy:

$$\Gamma[A^{cl},b^{cl},c^{cl}] = W[J,\xi_b,\xi_c] + \int d^D x(J\cdot A^{cl} + \xi_c c^{cl} + b^{cl}\xi_b). \tag{43.7}$$

By taking derivatives of this Legendre transform relation, we obtain also

$$A_\mu^{cl}(x) = -\frac{\delta W}{\delta J(x)}; \quad c^{cl}(x) = -\frac{\delta W}{\delta \xi_c(x)}; \quad b^{cl}(x) = +\frac{\delta W}{\delta \xi_b(x)};$$

$$J_\mu(x) = +\frac{\delta \Gamma}{\delta A_\mu^{cl}(x)}; \quad \xi_c(x) = -\frac{\delta \Gamma}{\delta c^{cl}(x)}; \quad \xi_b(x) = +\frac{\delta \Gamma}{\delta b^{cl}(x)}. \tag{43.8}$$

43.2 Vertices

We can derive the vertices from the interaction action. The 3-gluon vertex comes from rewriting

$$gf_{abc} \int d^D x(\partial_\mu A_\nu^a)A_\mu^b A_\nu^c = \int \frac{d^D k d^D p d^D q}{(2\pi)^6} \frac{1}{3!} A_\mu^a(k)A_\nu^b(p)A_\rho^c(q)\tilde{\Gamma}_{\mu\nu\rho}^{abc}. \tag{43.9}$$

Then we can derive the vertex, which should have the overall momentum conservation, so

$$\tilde{\Gamma}_{\mu\nu\rho}^{abc}(k,p,q) = (2\pi)^D \delta^{(D)}(k+p+q)V_{\mu\nu\rho}^{abc}. \tag{43.10}$$

The vertex should be symmetric in the external lines. We first rewrite the interaction term as

$$(gf_{abc})\int d^D x[-A_\mu^a(\partial_\mu A_\nu^b)A_\rho^c \delta^{\nu\rho}] = (gf_{abc})\int d^D x[A_\mu^a(\partial_\rho A_\nu^b)A_\rho^c]\delta^{\mu\nu}, \tag{43.11}$$

where we have redefined the indices and used the fact that $f_{bac} = -f_{abc}$ and $f_{cab} = +f_{abc}$. Then we see that we need the terms $-(ip_\mu)\delta_{\nu\rho} + (ip_\rho)\delta_{\mu\nu}$ among the permutations (the derivative ∂ is replaced by ip), multiplied by the usual Euclidean vertex, $-gf_{abc}$. The other terms are obtained by permuting p,k,q and the external indices. In total, we have

$$V_{\mu\nu\rho}^{abc}(k,p,q) = (-igf_{abc})[(q-p)_\mu \delta_{\nu\rho} + (p-k)_\rho \delta_{\mu\nu} + (k-q)_\nu \delta_{\mu\rho}]. \tag{43.12}$$

For the 4-gluon vertex, again taking out the delta function, the vertex $V_{\mu\nu\rho\sigma}^{abcd}$, for gluons (μa), (νb), (ρc), and (σd), is momentum-independent. There are 24 terms coming from the 4! terms in the permutations of the external lines. They will give six different terms and a multiplicity of 4, canceling the 1/4 in front of the quartic interaction action. One such term is given by writing the quartic action as

$$+\frac{g^2}{4}f_{abe}f_{cd}{}^e \int d^D x A_\mu^a A_\nu^b A_\rho^c A_\sigma^d \delta^{\mu\rho}\delta^{\nu\sigma}, \tag{43.13}$$

where we have relabeled the indices and used the fact that $f^e{}_{ab}f_{ecd} = f_{abe}f^e_{cd}$. The vertex term is then $-g^2 f_{abe}f_{ad}{}^e \delta^{\mu\rho}\delta^{\nu\sigma}$, and the other five terms are found by permutations, giving in total

$$
\begin{aligned}
V^{abcd}_{\mu\nu\rho\sigma} = -\tilde{g}^2 [&f_{abe}f_{cd}{}^e (\delta_{\mu\rho}\delta_{\nu\sigma} - \delta_{\nu\rho}\delta_{\mu\sigma}) + f_{cbe}f_{ad}{}^e (\delta_{\mu\rho}\delta_{\nu\sigma} - \delta_{\mu\nu}\delta_{\rho\sigma}) \\
&+ f_{dbe}f_{ca}{}^e (\delta_{\rho\sigma}\delta_{\mu\nu} - \delta_{\nu\rho}\delta_{\mu\sigma})].
\end{aligned}
\tag{43.14}
$$

The gluon-2-ghost vertex comes from the cubic part of the interaction action. For a ghost line from a to b, where b has momentum q, and with a gluon with (μc), we have

$$
V^{abc}_\mu(q) = -\tilde{g}f_{abc}(iq_\mu).
\tag{43.15}
$$

If we also introduce fermions in a representation f with index i, ψ^i_α, so with covariant derivative

$$
D_{\mu\,ij} = \partial_\mu \delta_{ij} + g(T^a_f)_{ij}A^a_\mu,
\tag{43.16}
$$

acting on them, and $\bar{\psi}^i D^{ij}_\mu \gamma^\mu \psi^j$ kinetic term, it follows that the gluon-2-fermion vertex with the fermion going from αi to βj and the gluon with μa, is

$$
-\tilde{g}(T^a_f)_{ji}(\gamma_\mu)_{\beta\alpha}.
\tag{43.17}
$$

Note that in all the above, as usual, $\tilde{g} = g\mu^{\epsilon/2}$, where $\epsilon = 4 - D$.

43.3 Feynman Rules

All in all, we have the Feynman rules (see Figure 43.1):

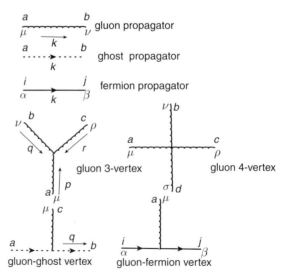

Figure 43.1   Feynman rules for nonabelian gauge theories (QCD).

- The *gluon propagator*, represented by a wiggly line from μa to νb, with momentum k, is

$$\Delta^{ab}_{\mu\nu}(k) = \frac{\delta^{ab}}{k^2}\left[\delta_{\mu\nu} - (1-\alpha)\frac{k_\mu k_\nu}{k^2}\right]. \qquad (43.18)$$

- The *ghost propagator*, represented by a dashed line with an arrow from a to b, with momentum k, is

$$\Delta^{ab}(k) = \frac{\delta^{ab}}{k^2}. \qquad (43.19)$$

- The *fermion propagator*, from $(i\alpha)$ to $(j\beta)$, with momentum k, is

$$S^{ij}_{F\alpha\beta}(k) = \delta_{ij}\left(\frac{1}{i\slashed{k}+m}\right)_{\alpha\beta}. \qquad (43.20)$$

- The *gluon 3-vertex*, represented by three wiggly lines intersecting at a point, with all momenta in (μa) with momentum p, (νb) with momentum q, and (ρc) with momentum r, is

$$V^{abc}_{\mu\nu\rho}(p,q,r) = -i\tilde{g}f_{abc}[(r-q)_\mu\delta_{\nu\rho} + (q-p)_\rho\delta_{\mu\nu} + (p-r)_\nu\delta_{\mu\rho}]. \qquad (43.21)$$

- The *gluon 4-vertex*, represented by four wiggly lines intersecting at a point, with (μa), (νb), (ρc), and (σd), is

$$\begin{aligned} V^{abcd}_{\mu\nu\rho\sigma} = &- \tilde{g}^2[f_{abe}f_{cd}{}^e(\delta_{\mu\rho}\delta_{\nu\sigma} - \delta_{\nu\rho}\delta_{\mu\sigma}) + f_{cbe}f_{ad}{}^e(\delta_{\mu\rho}\delta_{\nu\sigma} - \delta_{\mu\nu}\delta_{\rho\sigma}) \\ &+ f_{dbe}f_{ca}{}^e(\delta_{\rho\sigma}\delta_{\mu\nu} - \delta_{\nu\rho}\delta_{\mu\sigma})]. \end{aligned} \qquad (43.22)$$

- The *gluon-2-ghost vertex*, represented by a dashed line with an arrow from a to b (where b has momentum q), with a wiggly line coming out of it, ending on (μc), is

$$V^{abc}_\mu(q) = -\tilde{g}f_{abc}(iq_\mu). \qquad (43.23)$$

- The *gluon-2-fermion vertex*, represented by a continuous line with an arrow from $(i\alpha)$ to $(j\beta)$, with a wiggly line coming out of it, ending on (μc), is

$$-\tilde{g}(T^a_f)_{ji}(\gamma_\mu)_{\beta\alpha}. \qquad (43.24)$$

- The *fermion loop* has an extra (-1), but also the ghost loop, since the important thing is the anticommuting nature of the variables, not the kinetic term (which is KG for the ghost).

Observation: Note that we calculate Green's functions from derivatives of W or Γ, but while the action is gauge invariant under the nonabelian gauge transformation, the source term $\int d^D x J \cdot A$ is not, so the Green's functions are not physical observables, since they are not gauge invariant, as observables should be. But sums of Feynman diagrams could be gauge invariant. For example, in the case of quantum electrodynamics, we mentioned the fact that IR divergences will mean that we need to sum loop diagrams with tree diagrams, of the same order in the coupling, but with more external massless lines, with very small momentum. Only this combination will be related to experimentally relevant quantities like the cross-section, and will be gauge invariant (and IR safe). Now we can say a similar

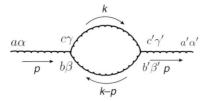

Figure 43.2 One-loop gluon diagram with 3-vertices.

thing about YM theory, where Green's functions will in general be gauge-dependent, even at a fixed order in the coupling.

43.4 Example of Feynman Diagram Calculation

As an example of a calculation, we will consider the one-loop correction to the gluon propagator with two 3-gluon vertices, as in Figure 43.2. It will also be gauge-dependent, but we will choose the Feynman gauge $\alpha = 1$. For other α, the result will change, and it will of course also change if we use instead of the Lorenz gauge some other gauge, like for instance the axial gauge.

In the diagram, a photon with momentum p and indices $(a\alpha)$ comes, and out goes a photon of momentum p and indices $(a'\alpha')$. On the internal loop, one line has momentum k and indices $(c\gamma)$ on the $(a\alpha)$ side and $(c'\gamma')$ on the $(a'\alpha')$ side, and the other line has momentum $k - p$ and indices $(b\beta)$ on the $(a\alpha)$ side and $(b'\beta')$ on the $(a'\alpha')$ side. Then the two vertices are $V_{\alpha\beta\gamma}^{abc}(p, k-p, -k)$ and $V_{\alpha'\beta'\gamma'}^{a'b'c'}(-p, p-k, k)$, and using the Feynman rules above we have the expression for the amplitude

$$\mathcal{M}_{\alpha\alpha'}^{aa'} = \int \frac{d^D k}{(2\pi)^D}(-i\tilde{g}f_{abc})[(-2k+p)_\alpha \delta_{\beta\gamma} + (p+k)_\beta \delta_{\alpha\gamma} + (k-2p)_\gamma \delta_{\alpha\beta}]\frac{\delta^{cc'}}{k^2}\delta_{\gamma\gamma'}$$

$$\times (-i\tilde{g}f_{a'b'c'})[(2k-p)_{\alpha'}\delta_{\beta'\gamma'} + (-p-k)_{\beta'}\delta_{\alpha'\gamma'} + (-k+2p)_{\gamma'}\delta_{\alpha'\beta'}]\frac{\delta^{bb'}}{(k-p)^2}\delta_{\beta\beta'}$$

$$= -\tilde{g}^2 f_{abc}f_{a'}{}^{bc}\int \frac{d^D k}{(2\pi)^D}\frac{F_{\alpha\alpha'}(k,p)}{k^2(k-p)^2}, \tag{43.25}$$

where

$$F_{\alpha\alpha'}(k,p) = -(2k-p)_\alpha(2k-p)_{\alpha'}D + (2k-p)_\alpha(p+k)_{\alpha'} + (2k-p)_\alpha(k-2p)_{\alpha'}$$

$$+ (p+k)_\alpha(2k-p)_{\alpha'} - (p+k)^2\delta_{\alpha\alpha'} - (k-2p)_\alpha(p+k)_{\alpha'}$$

$$+ (k-2p)_\alpha(2k-p)_{\alpha'} - (p+k)_{\alpha'}(k-2p)_\alpha - (k-2p)^2\delta_{\alpha\alpha'}$$

$$= (-4D+6)k_\alpha k_{\alpha'} + (-D+6)p_\alpha p_{\alpha'} + (2D-3)(k_\alpha p_{\alpha'} + p_\alpha k_{\alpha'})$$

$$- (2k^2 + 5p^2 - 2k \cdot p)\delta_{\alpha\alpha'}. \tag{43.26}$$

We note that $\tilde{g}^2 = g^2 \mu^\epsilon$ and $f_{abc} f_{a'}{}^{bc} = \delta_{aa'} C_2(G)$, where by definition $\delta_{aa'}$ is the Killing metric on the group and $C_2(G)$ is the second Casimir.

The first integral that we need is one that we already calculated:

$$\int \frac{d^D k}{(2\pi)^D} \frac{1}{k^2 (k-p)^2} = \frac{\Gamma\left(2 - \frac{D}{2}\right)}{(4\pi)^{\frac{D}{2}}} (p^2)^{\frac{D}{2}-2} \int_0^1 d\alpha [\alpha(1-\alpha)]^{\frac{D}{2}-2}$$

$$= \frac{\Gamma\left(2 - \frac{D}{2}\right)}{(4\pi)^{\frac{D}{2}}} (p^2)^{\frac{D}{2}-2} B\left(\frac{D}{2} - 1, \frac{D}{2} - 1\right)$$

$$\equiv I_{2,D}(p), \tag{43.27}$$

where we have used the Euler beta function, $B(a,b) = \int_0^1 dz \, z^{a-1}(1-z)^{b-1} = \Gamma(a + b)/\Gamma(a)\Gamma(b)$.

Then by Lorentz invariance we have

$$\int \frac{d^D k}{(2\pi)^D} \frac{k_\mu}{k^2(k-p)^2} = p_\mu \tilde{I}(p), \tag{43.28}$$

since the integral depends on p_μ (and not just on p^2). By multiplying this relation by $2p^\mu$, we obtain

$$2p^2 \tilde{I}(p) = \int \frac{d^D k}{(2\pi)^D} \frac{-(k-p)^2 + k^2 + p^2}{k^2(k-p)^2}$$

$$= -\int \frac{d^D k}{(2\pi)^D} \frac{1}{k^2} + \int \frac{d^D k}{(2\pi)^D} \frac{1}{(k-p)^2} + p^2 I_{2,D}(p), \tag{43.29}$$

where we have used the fact that the first two integrals cancel against each other, by shifting the momentum $\tilde{k} = k - p$. Actually, the integral is zero in dimensional regularization in any case, since as we saw

$$\int \frac{d^D q}{(2\pi)^D} \frac{1}{q^2 + m^2} = \frac{\Gamma\left(1 - \frac{D}{2}\right)}{(4\pi)^{\frac{D}{2}}} (m^2)^{\frac{D}{2}-1}, \tag{43.30}$$

whose $m \to 0$ limit gives zero for $D > 2$. Then we obtain

$$\tilde{I}(p) = \frac{I_{2,D}(p)}{2}. \tag{43.31}$$

Next, we need the integral

$$\int \frac{d^D k}{(2\pi)^D} \frac{k_\mu k_\nu}{k^2(k-p)^2}, \tag{43.32}$$

which by Lorentz invariance (since the integral depends on p_μ, and the only symmetric-tensor Lorentz structures available are thus $p_\mu p_\nu$ and $p^2 \delta_{\mu\nu}$) equals

$$I^a(p) p_\mu p_\nu + I^b(p) p^2 \delta_{\mu\nu}. \tag{43.33}$$

We need two relations to determine I^a and I^b. The first one is obtained by multiplying with $2p^\mu$, which gives

$$2p^2 p_\nu(I^a + I^p) = \int \frac{d^D k}{(2\pi)^D} \frac{k_\nu[-(k-p)^2 + k^2 + p^2]}{k^2(k-p)^2}$$

$$= -\int \frac{d^D k}{(2\pi)^D} \frac{k_\nu}{k^2} + \int \frac{d^D k}{(2\pi)^D} \frac{k_\nu}{(k-p)^2} + p^2 \frac{p_\nu}{2} I_{2,D}(p)$$

$$= -\int \frac{d^D k}{(2\pi)^D} \frac{k_\nu}{k^2} + \int \frac{d^D \tilde{k}}{(2\pi)^D} \frac{\tilde{k}_\nu}{\tilde{k}^2} + p_\nu \int \frac{d^D \tilde{k}}{(2\pi)^D} \frac{1}{\tilde{k}^2} + p^2 \frac{p_\nu}{2} I_{2,D}(p)$$

$$= p^2 \frac{p_\nu}{2} I_{2,D}(p), \tag{43.34}$$

where in the first equality we have used (43.31) and in the last equality we have used the fact that (43.30) vanishes. We thus obtain

$$I^a(p) + I^b(p) = \frac{I_{2,D}(p)}{4}. \tag{43.35}$$

The other relation is obtained by contracting with $\delta^{\mu\nu}$, which gives

$$p^2[I^a + DI^b] = \int \frac{d^D k}{(2\pi)^D} \frac{1}{(k-p)^2} = 0, \tag{43.36}$$

so

$$I^a = -DI^b, \tag{43.37}$$

so that finally

$$I^a(p) = \frac{D}{4(D-1)} I_{2,D}(p); \quad I^b(p) = -\frac{1}{4(D-1)} I_{2,D}(p). \tag{43.38}$$

Putting all the pieces of the amplitude together, we obtain

$$\mathcal{M}^{aa'}_{\alpha\alpha'} = -g^2 \mu^\epsilon C_2(G) \delta_{aa'} I_{2,D}(p) \left[(-4D+6) \left(\frac{D}{4(D-1)} p_\alpha p_{\alpha'} - \frac{\delta_{\alpha\alpha'}}{4(D-1)} p^2 \right) \right.$$

$$\left. + (-D+6) p_\alpha p_{\alpha'} + (2D-3) p_\alpha p_{\alpha'} - \delta_{\alpha\alpha'}(5p^2 - p^2) \right]$$

$$= -g^2 \mu^\epsilon \delta_{aa'} C_2(G) \left[p_\alpha p_{\alpha'} \frac{D + 4(D-1)(D+3)}{4(D-1)} - \delta_{\alpha\alpha'} p^2 \frac{16(D-1) - 4D + 6}{4(D-1)} \right]$$

$$\times \frac{\Gamma\left(2 - \frac{D}{2}\right)}{(4\pi)^{\frac{D}{2}}} (p^2)^{\frac{D}{2}-2} B\left(\frac{D}{2} - 1, \frac{D}{2} - 1\right). \tag{43.39}$$

From this, we can obtain the divergent part of the amplitude, using the fact that $\Gamma\left(2 - \frac{D}{2}\right) = \Gamma\left(\frac{\epsilon}{2}\right) \simeq 2/\epsilon$, and in the rest putting $D = 4$, including $B(1,1) = 1$, so that finally

$$\mathcal{M}^{aa'}_{\alpha\alpha', \text{div.}} = -\delta_{aa'} C_2(G) \frac{g^2}{(4\pi)^2} \frac{2}{\epsilon} \left(\frac{11}{3} p_\alpha p_{\alpha'} - \frac{19}{6} \delta_{\alpha\alpha'} \right). \tag{43.40}$$

Important Concepts to Remember

- The ghost propagator is the KG propagator (times $\delta_{aa'}$), and the gluon propagator is the photon propagator (times $\delta_{aa'}$).
- We have a 3-gluon vertex, a 4-gluon vertex, and a gluon-2-ghost vertex. If we add fermions, we also have a gluon-2-fermions vertex.
- A ghost loop gives a factor of (-1), the same as a fermion loop.
- Green's functions in the nonabelian gauge theory are not gauge invariant, so they cannot be directly related to observables. Sums of diagrams (for different Green's functions, even) might be gauge invariant, order by order in perturbation theory.

Further Reading

See sections 7.2 and 7.3 in [3], and section 16.1 in [1].

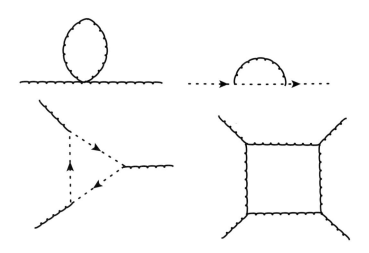

Figure 43.3 One-loop QCD diagrams.

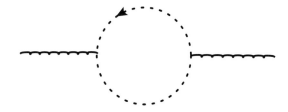

Figure 43.4 One-loop QCD diagram with ghost loop and external gluons.

Exercises

1. Write the integral expressions for the diagrams in Figure 43.3, using the Feynman rules (without calculating them).
2. Calculate the divergent part of the diagram in Figure 43.4.
3. Which of the one-loop diagrams in Figure 43.3 are *superficially* UV divergent? Are they *really* divergent? Why?
4. Calculate the UV divergences found in Exercise 43.3.

One-Loop Renormalizability of Gauge Theories

In this chapter we will study how to explicitly renormalize gauge theories (QCD) at one loop, and find that QCD is one-loop renormalizable.

44.1 Divergent Diagrams of Pure Gauge Theory

We will not derive them here, since they are too laborious (30 diagrams in total), but one can derive the result for the divergent parts of all the one-loop diagrams of pure gauge theory (Yang–Mills).

They fall into five classes, corresponding to five 1PI n-point functions which are divergent.

1. The gauge propagator self-energy $\Sigma_{\mu\nu}(p)$. There are three relevant diagrams (see Figure 44.1): the gluon loop with two 3-gluon vertices, whose divergent part we calculated in Chapter 43 (with symmetry factor 1/2); the ghost loop that was left as an exercise (with $-$ sign for the ghost loop); the gluon loop with a single 4-gluon vertex. The divergent part of the sum of these diagrams gives

$$\Sigma_{\mu\nu;\,\text{divergent}}(p) = -\frac{g^2 C_2(G)}{16\pi^2}\left(\frac{5}{3} + \frac{1}{2}(1-\alpha)\right)\frac{2}{\epsilon}[p^2\delta_{\mu\nu} - p_\mu p_\nu]\delta^{ab}. \tag{44.1}$$

2. The ghost propagator self-energy P_{ab}, with a single one-loop diagram in Figure 44.2, with a gluon line starting and ending on the ghost line. Its divergent part is

$$P_{ab;\,\text{divergent}}(p) = -\frac{g^2 C_2(G)}{16\pi^2}\left(\frac{1}{2} + \frac{1}{4}(1-\alpha)\right)\frac{2}{\epsilon}(p^2\delta^{ab}). \tag{44.2}$$

3. The gluon three-point vertex. There are now six diagrams, as in Figure 44.3: one with a gluon loop with three 3-gluon vertices attached, out of which external gluons go; another three where we contract one propagator from the first diagram to make a gluon four-point vertex (with symmetry factor 2 for each); two diagrams with a ghost loop with external gluons attached to it, and a different orientation for the ghost line differentiating between the two diagrams. The divergent part of the sum is

$$-\frac{g^2 C_2(G)}{16\pi^2}\left[\frac{2}{3} + \frac{3}{4}(1-\alpha)\right]\frac{2}{\epsilon}V^{abc}_{\mu\nu\lambda}(k,p,q), \tag{44.3}$$

where $V^{abc}_{\mu\nu\lambda}(k,p,q)$ is the classical (tree-level) vertex.

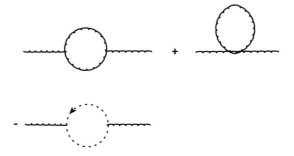

Figure 44.1   One-loop diagrams contributing to the gauge propagator self-energy $\Sigma_{\mu\nu}(p)$.

Figure 44.2   One-loop diagram contributing to the ghost propagator self-energy $P_{ab}(p)$.

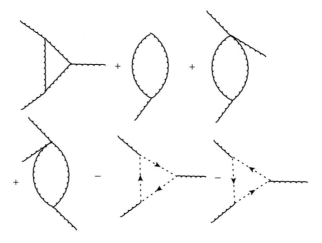

Figure 44.3   One-loop diagrams contributing to the 3-gluon vertex.

4. The gluon four-point vertex. There are 18 diagrams. The pure gluon ones are in Figure 44.4, and the ones with a ghost loop are in Figure 44.5. One is the gluon loop with four 3-gluon vertices, together with its two crossed diagrams. There are three diagrams obtained by contracting two propagators of the first diagram to obtain two 4-gluon vertices (or one diagram and two crossed ones). There are six diagrams where only one propagator has been contracted, to give one 4-gluon vertex and two 3-gluon vertices; or four diagrams obtained by contraction of the first one, and two diagrams obtained by crossing. And finally there are six diagrams with a ghost loop with four external gluon vertices: two diagrams differing by the orientation of the ghost loop, and two crossed diagrams for each of these.

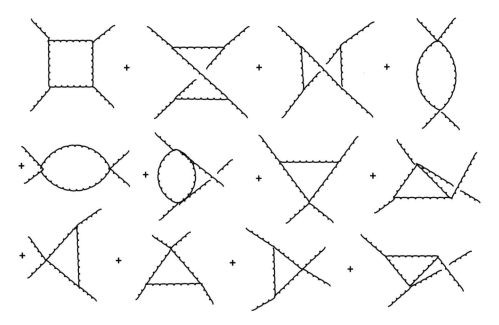

Figure 44.4   One-loop gluon diagrams contributing to the 3-gluon vertex.

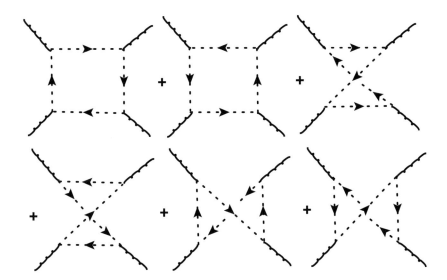

Figure 44.5   One-loop diagrams with ghost loop contributing to the 3-gluon vertex.

The divergent part of the sum of these diagrams is

$$-\frac{g^2 C_2(G)}{16\pi^2}\left[-\frac{1}{3}+(1-\alpha)\right]\frac{2}{\epsilon}V^{abcd}_{\mu\nu\rho\sigma}, \tag{44.4}$$

where $V^{abcd}_{\mu\nu\rho\sigma}$ is the classical (tree-level) vertex.

Figure 44.6   One-loop diagrams for the ghost–gluon vertex.

5. The ghost–gluon vertex. The loop is made of either two sides of ghost (with a gluon in the common vertex) and one gluon side, or two gluon sides (with the gluon coming out of the common vertex) and a single ghost side, as in Figure 44.6. The sum of these two diagrams has the divergent part

$$+\frac{g^2 C_2(G)}{16\pi^2}\frac{\alpha}{2}\frac{2}{\epsilon}V_\mu^{abc}(p), \tag{44.5}$$

where again $V_\mu^{abc}(p)$ is the classical (tree-level) vertex.

We see therefore that the five divergences all correspond to, and are proportional to, the five tree-level objects (propagators and vertices) coming from the classical Lagrangian. Therefore, renormalizability at one loop is guaranteed, since we can absorb the divergences in the redefinition of the objects in the classical Lagrangian.

But in general, to guarantee renormalizability, one would need gauge invariance, since that generates relations between the Green's functions through the Ward identities. However, the formalism we use is gauge-fixed, so it would seem we have a problem. As it turns out, we don't, because there is a *residual gauge symmetry* left after fixing the gauge (we know that fixing the gauge in general allows for the possibility of residual symmetries, meaning symmetry with a parameter that depends in a constrained way on spacetime). This symmetry is Becchi–Rouet–Stora–Tyutin (BRST) symmetry, that will be studied in Chapter 45, and we will see in Chapter 46 that it leads to relations among Green's functions similar to the Ward identities, which will allow us to prove renormalizability.

We also want to verify that the counting of the superficial degree of divergences says we have identified all the divergent 1PI n-point functions. The superficial degree of divergence in the pure gauge theory is

$$\omega(D) = 4 - E_A - \frac{3}{2}E_{b,c}, \tag{44.6}$$

where we have also introduced external ghost lines. Note the factor of 3/2 in front of the external ghost lines $E_{b,c}$, like in the case of fermions. The derivation is, however, slightly different from that for fermions. The vertex is $(\partial_\mu b^a)gf_{abc}A_\mu^b c^c$, so each b line at a vertex comes with a factor of momentum, despite the ghost propagator being Klein–Gordon, $\sim 1/p^2$. Since each propagator ends in two vertices, but in one as b and in one as c, effectively we have an extra factor of p in the propagator, giving a $\sim 1/p$ propagator, like in the fermion case.

Then we can check that indeed the five 1PI n-point functions are the divergent ones. The gluon propagator has $E_g = 2$, so $\omega(D) = 4 - 2 = 2$. The ghost propagator has $E_{bc} = 2$, so

$\omega(D) = 4 - 3 \times 2/2 = 1$. The 3-gluon vertex has $E_g = 3$, so $\omega(D) = 4 - 3 = 1$, the 4-gluon vertex has $E_g = 4$, so $\omega(D) = 4 - 4 = 0$, and the ghost–gluon vertex has $E_g = 1, E_{bc} = 2$, so $\omega(D) = 4 - 1 - 3 \times 2/2 = 0$. There are no other divergent 1PI n-point functions.

44.2 Counterterms in MS Scheme

We can now write down the counterterms in the minimal subtraction scheme, as just minus the divergent terms.

1. From the gluon propagator one-loop divergence, we obtain the counterterm

$$\delta\mathcal{L}_{A^2} = (Z_3 - 1)\frac{1}{4}(\partial_\mu A_\nu^a - \partial_\nu A_\mu^a)^2,$$

$$Z_3^{(0+1)} = 1 + \frac{g^2 C_2(G)}{16\pi^2}\left(\frac{5}{3} + \frac{1}{2}(1 - \alpha)\right)\frac{2}{\epsilon}. \tag{44.7}$$

Indeed, in p space, the gluon propagator term gives $(Z_3 - 1)\frac{1}{2}A_\mu^a(p^2\delta_{\mu\nu} - p_\mu p_\nu)A_\nu^a$, so the counterterm is indeed minus the divergence.

2. From the ghost propagator one-loop divergence, we obtain the counterterm

$$\delta\mathcal{L}_{bc} = (\tilde{Z}_3 - 1)(-b^a\partial^2 c^a),$$

$$\tilde{Z}_3^{(0+1)} = 1 + \frac{g^2 C_2(G)}{16\pi^2}\left(\frac{1}{2} + \frac{1}{4}(1 - \alpha)\right)\frac{2}{\epsilon}. \tag{44.8}$$

In p space, the gluon propagator term gives $(\tilde{Z}_3 - 1)(b^a p^2 c^a)$, so indeed the counterterm is minus the divergence.

3. From the 3-gluon vertex divergence, we obtain the counterterm

$$\delta\mathcal{L}_{A^3} = (Z_1 - 1)gf_{abc}\partial_\mu A_\nu^a A_\mu^b A_\nu^c,$$

$$Z_1^{(0+1)} = 1 + \frac{g^2 C_2(G)}{16\pi^2}\left[\frac{2}{3} + \frac{3}{4}(1 - \alpha)\right]\frac{2}{\epsilon}. \tag{44.9}$$

Since the divergence was written as a coefficient times the classical vertex, the $Z_1^{(1)}$ is just minus the coefficient, as we can check.

4. From the 4-gluon vertex divergence, we obtain the counterterm

$$\delta\mathcal{L}_{A^4} = (Z_4 - 1)\frac{g^2}{4}f_{abc}f^a{}_{de}A_\mu^b A_\nu^c A_\mu^d A_\nu^e,$$

$$Z_4^{(0+1)} = 1 + \frac{g^2 C_2(G)}{16\pi^2}\left[-\frac{1}{3} + (1 - \alpha)\right]\frac{2}{\epsilon}. \tag{44.10}$$

The same comment as above applies.

5. From the ghost–gluon vertex divergence, we obtain the counterterm

$$\delta\mathcal{L}_{bAc} = (\tilde{Z}_1 - 1)gf_{abc}(\partial_\mu b^a)A_\mu^b c^c,$$

$$\tilde{Z}_1^{(0+1)} = 1 - \frac{g^2 C_2(G)}{16\pi^2}\frac{\alpha}{2}\frac{2}{\epsilon}. \tag{44.11}$$

Again the same comment applies.

As usual, renormalization means

$$(\mathcal{L} + \delta\mathcal{L})(A, b, c, g, \alpha) = \mathcal{L}_0(A_0, b_0, c_0, g_0, \alpha_0). \tag{44.12}$$

Since we have

$$\mathcal{L} + \delta\mathcal{L} = \frac{1}{4}Z_3(\partial_\mu A_\nu^a - \partial_\nu A_\mu^a)^2 + \frac{1}{2\alpha}(\partial_\mu A_\mu^a)^2 + gf_{abc}Z_1\partial_\mu A_\nu^a A_\mu^b A_\nu^c$$
$$+ \frac{g^2}{4}f_{abc}f^a{}_{de}Z_4 A_\mu^b A_\nu^c A_\mu^d A_\nu^e + \tilde{Z}_3(\partial_\mu b^a)\partial_\mu c^a + gf_{abc}\tilde{Z}_1(\partial_\mu b^a)A_\mu^b c^c, \tag{44.13}$$

we can obtain the renormalizations of fields and couplings.

44.3 Renormalization and Consistency Conditions

But we see that we have four objects to be renormalized (since b and c must renormalize in the same way), but six coefficients (for six terms), and we will obtain also two consistency conditions.

From

$$Z_3(\partial_\mu A_\nu^a - \partial_\nu A_\mu^a)^2 = (\partial_\mu A_{0\nu}^a - \partial_\nu A_{0\mu}^a)^2, \tag{44.14}$$

we get

$$A_0 = \sqrt{Z_3}A. \tag{44.15}$$

From

$$\tilde{Z}_3(\partial_\mu b^a)\partial_\mu c^a = (\partial_\mu b_0^a)\partial_\mu c_0^a, \tag{44.16}$$

we get

$$b_0, c_0 = \sqrt{\tilde{Z}_3}b, c. \tag{44.17}$$

From

$$\frac{1}{2\alpha}(\partial_\mu A_\mu)^2 = \frac{1}{2\alpha_0}(\partial_\mu A_{0\mu})^2, \tag{44.18}$$

we get

$$\alpha_0 = Z_3\alpha. \tag{44.19}$$

From

$$gf_{abc}Z_1\partial_\mu A_\nu^a A_\mu^b A_\nu^c = g_0 f_{abc}\partial_\mu A_{0\nu}^a A_{0\mu}^b A_{0\nu}^c, \tag{44.20}$$

we get

$$g_0 = g\frac{Z_1}{Z_3^{3/2}}. \tag{44.21}$$

Now we have fixed all renormalizations, but we still have two terms to check. These will give consistency conditions. The first is the gluon 4-vertex

$$\sim g^2 Z_4 A^4 = g_0^2 A_0^4, \tag{44.22}$$

from which we obtain the consistency condition

$$\frac{Z_4}{Z_1} = \frac{Z_1}{Z_3}. \tag{44.23}$$

The second is the ghost–gluon vertex

$$\sim g\tilde{Z}_1 bAc = g_0 b_0 A_0 c_0, \tag{44.24}$$

from which we obtain the second consistency condition

$$\frac{\tilde{Z}_1}{\tilde{Z}_3} = \frac{Z_1}{Z_3}. \tag{44.25}$$

Together, these two relations form the *Slavnov–Taylor identities*.

At this point, it is not clear why they should be correct. Such relations normally appear from gauge invariance, through Ward identities that relate various 1PI n-point functions (thus their coefficients Z_i), but now we work in a gauge-fixed formalism. However, as we said, there is a residual gauge symmetry called BRST symmetry, which will allow us to write relations between the n-point functions.

We can at most check it explicitly at one loop. Remember that $Z_i = 1 + Z_i^{(1)}$ can be expanded in $Z_i^{(1)}$, so the Slavnov–Taylor identities at one loop are

$$\begin{aligned} Z_4^{(1)} &= 2Z_1^{(1)} - Z_3^{(1)}, \\ \tilde{Z}_1^{(1)} &= Z_1^{(1)} + \tilde{Z}_3^{(1)} - Z_3^{(1)}. \end{aligned} \tag{44.26}$$

The relations are indeed verified, since

$$\left[-\frac{1}{3} + (1-\alpha)\right] = +2\left[\frac{2}{3} + \frac{3}{4}(1-\alpha)\right] - \left[\frac{5}{3} + \frac{1}{2}(1-\alpha)\right],$$

$$-\frac{\alpha}{2} = \left[\frac{2}{3} + \frac{3}{4}(1-\alpha)\right] + \left[\frac{1}{2} + \frac{1}{4}(1-\alpha)\right] - \left[\frac{5}{3} + \frac{1}{2}(1-\alpha)\right]. \tag{44.27}$$

44.4 Gauge Theory with Fermions

We can couple the gauge theory to fermions with Euclidean action

$$S_{(E)} = \int d^D x \bar{\psi} (\gamma_\mu D_\mu + m)\psi, \tag{44.28}$$

where the covariant derivative is

$$D_\mu^{ij} = \partial_\mu \delta^{ij} + g(T_f^a)_{ij} A_\mu^a. \tag{44.29}$$

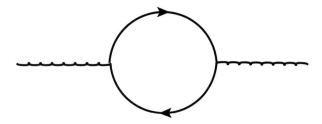

Figure 44.7   One-fermion-loop diagram for the gluon propagator.

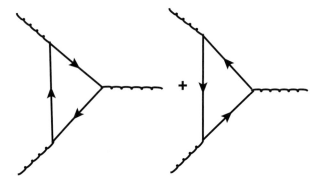

Figure 44.8   One-fermion-loop diagrams for the 3-gluon vertex.

The superficial degree of divergence in the presence of fermions is now

$$\omega(D) = 4 - E_g - \frac{3}{2}E_f - \frac{3}{2}E_{bc}. \tag{44.30}$$

Therefore, besides the previous divergent 1PI n-point functions, we also have new ones that are divergent:

- The fermion propagator $\Sigma_{\alpha\beta}(p)$, with $\omega(D) = 4 - 3 \times 2/2 = 1$.
- The fermion–gluon vertex $\Gamma_{\alpha\beta}^{a\mu}$, with $\omega(D) = 4 - 1 - 3 \times 2/2 = 0$.

But before we analyze these, we will write down the new divergent contributions to the 1PI n-point functions already considered in the pure gauge theory case, coming from a fermion loop:

- The divergent contribution to the gluon propagator (i.e. to Z_3) coming from the fermion loop with two external gluons from it in Figure 44.7, giving

$$-\frac{g^2}{16\pi^2}T_f\frac{4}{3}\frac{2}{\epsilon}. \tag{44.31}$$

- The divergent contribution to the 3-gluon vertex, (i.e. to Z_1) coming from the two diagrams in Figure 44.8, with a fermion loop (with different orientation for the arrow) and three gluons coming out of it, giving the same

$$-\frac{g^2}{16\pi^2}T_f\frac{4}{3}\frac{2}{\epsilon}. \tag{44.32}$$

Figure 44.9 One-gluon-loop fermion propagator correction.

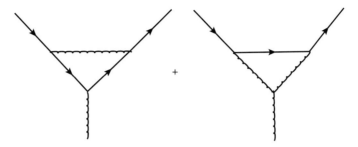

Figure 44.10 One-loop fermion–gluon vertex correction.

This is good, since from the second Slavnov–Taylor identity at one loop in (44.26), we need that (since $\tilde{Z}_1$ and $\tilde{Z}_3$ are not modified by the addition of fermions) $\delta Z_1^{(1)} = \delta Z_3^{(1)}$, which is indeed true.

- The divergent contribution to the 4-gluon vertex (i.e. to Z_4), again coming from two diagrams with a fermion loop (and different orientations for the arrow) and four gluons coming out of it, giving the same result as above for Z_1. This is again good, since now from the first Slavnov–Taylor identity at one loop in (44.26), we need that $\delta Z_4^{(1)} = \delta Z_1^{(1)}$ (since we saw already that $\delta Z_1^{(1)} = \delta Z_3^{(1)}$), which is verified.

Moving on to the new divergent 1PI n-point functions, these new divergences will be canceled by the counterterms

$$\delta \mathcal{L}(A, b, c, \psi) = (Z_{f2} - 1)\bar{\psi}\gamma_\mu \partial_\mu \psi + (Z_{f1} - 1)g\bar{\psi}\gamma_\mu A_\mu \psi + (Z_m - 1)m\bar{\psi}\psi. \quad (44.33)$$

Explicit calculations give

$$Z_{f2}^{(0+1)} = 1 - \frac{g^2 C_f}{16\pi^2}\alpha\frac{2}{\epsilon},$$

$$Z_{f1}^{(0+1)} = 1 - \frac{g^2 C_f}{16\pi^2}\left[\alpha C_f + C_2(G)\left(1 - \frac{1-\alpha}{4}\right)\right]\frac{2}{\epsilon},$$

$$Z_m^{(0+1)} = 1 - \frac{g^2 C_f}{16\pi^2}[4 - (1-\alpha)]\frac{2}{\epsilon}. \quad (44.34)$$

The correction to the fermion propagator, giving both the Z_{f2} and the Z_m terms, comes from a one-loop diagram with a fermion line out of which a gluon comes out and back, as in Figure 44.9.

The correction to the fermion–gluon vertex is given by the two diagrams equivalent to those of the ghost–gluon vertex, namely with a triangle loop two sides fermionic and one side gluonic, or two sides gluonic and one fermionic, as in Figure 44.10.

In the above, C_f and T_f are the fermion representation case of T_R and C_R defined previously, namely T_f is the normalization of the trace, and C_f the Casimir in this representation.

For the QED case (abelian group), $C_f = 1$ and $C_2(G) = 0$, and we can check that we reproduce the results we obtained for QED.

Renormalization of *the new terms* is given as usual by

$$\mathcal{L}(A, \bar{\psi}, g, m) + \delta\mathcal{L}(A, \bar{\psi}, \psi, g, m) = \mathcal{L}_0(A_0, \bar{\psi}_0, \psi_0, g_0, m_0), \tag{44.35}$$

which means that

$$Z_{f2}\bar{\psi}\gamma_\mu\partial_\mu\psi + Z_{f1}g\bar{\psi}\gamma_\mu A_\mu\psi + Z_m m\bar{\psi}\psi = \bar{\psi}_0\gamma_\mu\partial_\mu\psi_0 + g_0\bar{\psi}_0\gamma_\mu A_{0\mu}\psi_0 + m_0\bar{\psi}_0\psi_0. \tag{44.36}$$

From the first term we get the wavefunction renormalization

$$\psi_0 = \sqrt{Z_{f2}}\psi; \quad \bar{\psi}_0 = \sqrt{Z_{f2}}\bar{\psi}_0, \tag{44.37}$$

and from the last we get the mass renormalization

$$Z_{f2}m_0 = mZ_m. \tag{44.38}$$

The middle term gives a constraint, that is another Slavnov–Taylor identity

$$Z_{f1}g\bar{\psi}\gamma_\mu A_\mu\psi = g_0\bar{\psi}_0\gamma_\mu A_{0\mu}\psi_0 \Rightarrow \frac{Z_{f1}}{Z_{f2}} = \frac{Z_1}{Z_3}. \tag{44.39}$$

We can check explicitly that it is satisfied at one loop, as in the pure gauge case.

This means that, all in all, we have the Slavnov–Taylor identities

$$\frac{Z_{f2}}{Z_{f1}} = \frac{Z_1}{Z_3} = \frac{\tilde{Z}_1}{\tilde{Z}_3} = \frac{Z_4}{Z_1}. \tag{44.40}$$

Important Concepts to Remember

- Pure gauge theory is renormalizable at one loop, the divergences coming from the gauge propagator, ghost propagator, 3-gluon vertex, 4-gluon vertex, and ghost–gluon vertex, and being of the same structure as the terms in the classical Lagrangian.
- The pure gauge theory obeys the Slavnov–Taylor identities, $Z_4/Z_1 = Z_1/Z_3 = \tilde{Z}_1/\tilde{Z}_3$.
- The gauge theory with fermions has $\omega(D) = 4 - E_g - 3E_{bc}/2 - 3E_f/2$ and introduces two more divergences, in the fermion propagator and fermion–gluon vertex, being again one-loop renormalizable.
- With fermions we have one more Slavnov–Taylor identity, $Z_{f2}/Z_{f1} = Z_1/Z_3$.

Further Reading

See sections 7.3–7.5 in [3], and section 16.5 in [1].

Exercises

1. Calculate explicitly all the Z factors for a $SO(N)$ gauge theory with N_f fundamental fermions.
2. Calculate the divergent part of the one-loop graph with a fermion loop and two external gluon lines in Figure 44.7, contributing to Z_3.
3. Calculate the renormalization (divergent factor) of the fermion mass, from the fermion line with a gluon loop on it.
4. Can one have a theory in which the gluon propagator doesn't have total divergent contributions, so the fermionic loop cancels the pure gluon contributions?

In this chapter we will study two very important properties of quantum gauge theories: one is asymptotic freedom, which is the property that nonabelian gauge theories become free ($g \rightarrow 0$) in the UV. This follows from the observation that, at one loop, $\beta(g) < 0$. The second property is BRST (Becchi–Rouet–Stora–Tyutin) symmetry, which is a residual gauge symmetry, that will imply important relations for the correlators of the theory.

45.1 Asymptotic Freedom

Asymptotic freedom is a very important concept, one that gained a Nobel Prize for David Gross, Frank Wilczek, and David Politzer (in 2004). We have done most of the calculations necessary for it in Chapter 44, so now it remains to put everything together and interpret it.

We saw in Chapter 44 that renormalization of the coupling is done by $g_0 = gZ_1/Z_3^{3/2}$. However, we omitted the dimensional transmutation factor, so we actually have

$$g_0 = g\mu^{\epsilon/2}\frac{Z_1}{Z_3^{3/2}}. \tag{45.1}$$

We also saw that, in the presence of fermions, we have

$$Z_1^{(0+1)} = 1 + \frac{g^2}{16\pi^2}\left[C_2(G)\left(\frac{2}{3} + \frac{3}{4}(1-\alpha)\right) - \frac{4}{3}T_f\right]\frac{2}{\epsilon},$$

$$Z_3^{(0+1)} = 1 + \frac{g^2}{16\pi^2}\left[C_2(G)\left(\frac{5}{3} + \frac{1}{2}(1-\alpha)\right) - \frac{4}{3}T_f\right]\frac{2}{\epsilon}. \tag{45.2}$$

Then it follows that

$$g_0 = g\mu^{\frac{\epsilon}{2}}\left\{1 + \frac{g^2}{16\pi^2}\left[C_2(G)\left(\frac{2}{3} + \frac{3}{4}(1-\alpha)\right)\right.\right.$$

$$\left.\left. - \frac{3}{2}\left(\frac{5}{3} + \frac{1}{2}(1-\alpha)\right)\right) - \frac{4}{3}T_f\left(1 - \frac{3}{2}\right)\right]\frac{2}{\epsilon}\right\}$$

$$= g\mu^{\frac{\epsilon}{2}}\left\{1 - \frac{g^2}{16\pi^2}\left[\frac{11}{6}C_2(G) - \frac{2}{3}T_f\right]\frac{2}{\epsilon}\right\}. \tag{45.3}$$

Taking $\mu \partial/\partial\mu$ on both sides of this equation, since $\partial g_0/\partial\mu = 0$, we obtain

$$\mu\frac{\partial g_0}{\partial\mu} = 0 = \frac{\epsilon}{2}\mu^{\frac{\epsilon}{2}}\left(g - \frac{g^3}{16\pi^2}\left[\frac{11}{6}C_2(G) - \frac{2}{3}T_f\right]\frac{2}{\epsilon}\right)$$
$$+ \mu^{\frac{\epsilon}{2}}\left(1 - \frac{3g^2}{16\pi^2}\left[\frac{11}{6}C_2(G) - \frac{2}{3}T_f\right]\frac{2}{\epsilon}\right)\mu\frac{\partial g}{\partial\mu} + \mathcal{O}(g^5). \qquad (45.4)$$

This is solved by

$$\beta(\mu,\epsilon) \equiv \mu\frac{\partial g}{\partial\mu} = \frac{-\frac{\epsilon}{2}g + \frac{g^3}{16\pi^2}\left[\frac{11}{6}C_2(G) - \frac{2}{3}T_f\right] + \mathcal{O}(g^5)}{1 - \frac{3g^2}{16\pi^2}\left[\frac{11}{6}C_2(G) - \frac{2}{3}T_f\right]\frac{2}{\epsilon} + \mathcal{O}(g^4)}$$

$$= -\frac{\epsilon}{2}g - \frac{2g^3}{16\pi^2}\left[\frac{11}{6}C_2(G) - \frac{2}{3}T_f\right] + \mathcal{O}(g^5). \qquad (45.5)$$

Note that above we have used the usual expansion in g^2, ignoring the fact that a higher-order term might actually be divergent in ϵ, it is still considered negligible.

One also defines as usual the physical beta function as

$$\beta(g) = \beta(\mu, \epsilon \to 0) = \beta_1 g^3 + \beta_3 g^5 + \beta_5 g^7 + \dots, \qquad (45.6)$$

which leads to

$$\beta_1 = -\frac{2}{16\pi^2}\left[\frac{11}{6}C_2(G) - \frac{2}{3}T_f\right]. \qquad (45.7)$$

Again, the term in g^5 from the above would actually be divergent in ϵ, but is still considered subleading, and ignored even in the $\epsilon \to 0$ limit. Of course, the point is that the calculation is actually only to one loop. Once we do the two-loop calculation for Z_1 and Z_3 and include it in the above, one finds a finite β_2 as well.

We note here that β_1 and β_3 are actually gauge-independent (universal), nonzero, and independent of the renormalization scheme, while the other coefficients are not necessarily universal.

The most important observation is that the *nonabelian* gauge fields are actually the only ones with $\beta_1 < 0$! All other fields have $\beta_1 > 0$ contributions. In particular, we see that for fermions we have a positive β_1, proportional to T_f. For scalars, we get a similarly positive contribution.

For quantum electrodynamics, we have $C_2(G) = 0$ and $T_f = 1$, so $\beta_1 = +1/12\pi^2 > 0$.

Integrating the beta function equation at one loop:

$$\beta_1 g^3 = \mu\frac{\partial g}{\partial\mu}, \qquad (45.8)$$

by first multiplying with $2g$, we get the solution

$$g^2(\mu) = \frac{g^2(\mu_0)}{1 - \beta_1 g^2(\mu_0)\ln\frac{\mu^2}{\mu_0^2}}, \qquad (45.9)$$

as we can check explicitly. This relates the coupling constant at some fixed scale μ_0 with the coupling constant at the variable scale μ.

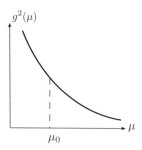

Figure 45.1 The coupling constant $g^2(\mu)$ is a decreasing function (QCD).

In particular, it means that, for $\beta_1 < 0$, the coupling constant $g^2(\mu)$ is a decreasing function of μ, as in Figure 45.1. This has two important consequences:

- IR slavery. $g^2(\mu \to 0) \to \infty$, so at large distances (small energies, i.e. infrared), we have very strong coupling, leading to confinement.
- Asymptotic freedom. $g^2(\mu \to \infty) \to 0$, so the theory is free in the UV.

For the gauge group $SU(N)$, we have $C_2(SU(N)) = N$ and $T_f = 1/2$, but we will consider N_f flavors (species) of fermions, leading to

$$\beta_{1,SU(N_c),N_f} = -\frac{1}{16\pi^2}\left(\frac{11}{3}N_c - \frac{2}{3}N_f\right). \tag{45.10}$$

In particular, for quantum chromodynamics $N_c = 3$ (three colors) and $N_f = 6$ (six flavors, u, d, c, s, t, b), so

$$\beta_{1,QCD} = -\frac{7}{16\pi^2} < 0, \tag{45.11}$$

implying that QCD is asymptotically free. It also exhibits IR slavery, leading to confinement of quarks and gluons, since the coupling becomes infinite at large distances, and we cannot separate the quarks and gluons from each other.

45.2 BRST Symmetry

To study the quantum properties of gauge theories, we will need a global symmetry called BRST symmetry, which is a remnant of gauge invariance (i.e. a residual gauge invariance), present once we fix the gauge.

It was found in a paper by Becchi, Rouet, and Stora, and independently in a paper by Tyutin, hence the name. For the ghost action, as before, we use the notation common in the BRST literature, with fields b^a and c^a. The effective gauge-fixed Lagrangian in Euclidean space is

$$\mathcal{L}_{\text{eff}}(A, b, c) = +\frac{1}{4}(F^a_{\mu\nu})^2 + \frac{1}{2\alpha}(\partial_\mu A_\mu)^2 + \partial_\mu b^a D^{ab}_\mu c^b. \tag{45.12}$$

Here, b^a, c^a are anticommuting variables, and b^a is imaginary, while c^a is real, for the reality of the action. Since the dimension of the Lagrangian must be 4, we see that we need $[b^a] + [c^a] = 2$. But since *a priori* b^a and c^a are independent fields (remember that we found them by writing the Gaussian path integral $\int \mathcal{D}b\mathcal{D}c \, e^{-bMc} = \det M$, where the two integrations are really independent), their dimensions need not be related. In fact, we can choose them of different dimension, as is done for instance in string theory. However, for simplicity, here we will choose $[b^a] = [c^a] = 1$.

Since the gauge invariance is

$$\delta_{\text{gauge},\lambda} A^a_\mu = (D_\mu \lambda)^a = \partial_\mu \lambda^a + g f^a{}_{bc} A^b_\mu \lambda^c, \tag{45.13}$$

the BRST invariance must be similar, just with the arbitrary $\lambda^a(x)$ replaced by something. That something is $c^a(x)\Lambda$, so a given x dependence (the field c^a) times an arbitrary constant, implying a global symmetry.

We can define a *ghost number* by $N_{\text{gh}}[c^a] = +1$ (since $c^a \to \eta^a$ and $b^a \to \bar{\eta}^a$, this is natural) and $N_{\text{gh}}[b^a] = -1$. This is then a symmetry of $\mathcal{L}_{\text{eff}}$, (i.e. the Lagrangian has ghost number zero). Since $N_{\text{gh}}[\lambda] = 0$, it follows that $N_{\text{gh}}[\Lambda] = -1$. Moreover, since $[\lambda] = 0$ because of $\delta A_\mu = D_\mu \lambda$, it follows that also $[\Lambda] = -1$. And since λ^a is commuting, it follows that Λ is anticommuting.

Since $c^a(x)\Lambda$ is a special case of λ^a, it trivially follows that the classical Lagrangian is BRST-invariant:

$$\delta_B \mathcal{L}_{\text{class}} = 0, \tag{45.14}$$

under the BRST transformation

$$\delta_B A^a_\mu = (D_\mu c)^a \Lambda. \tag{45.15}$$

Moreover, we can extend the BRST transformation to gauge fields coupled to matter. For instance, for scalars transforming under gauge transformations as

$$\delta_{\text{gauge}} \phi^i = -g(T_a)^i{}_j \phi^j \lambda^a(x), \tag{45.16}$$

the BRST transformation is

$$\delta_B \phi^i = -g(T_a)^i{}_j \phi^j c^a \Lambda, \tag{45.17}$$

and similarly for fermions, and so on.

Now, to find the rest of the BRST transformation laws, we require that the rest of the action be invariant. Since $\delta_B A^a_\mu$ has no terms with b^a in them, and $\mathcal{L}_{\text{eff}}$ has terms with only A_μ and a term $-b_a \partial^\mu (D_\mu c)^a$, we must have $(D_\mu c)^a$ invariant, at least when multiplied by $\partial_\mu b^a$ (if not, there would be a term in the variation $-b_a \partial^\mu \delta(D_\mu c)^a$ that could not be canceled).

We have explicitly

$$\delta_B (D_\mu c)^a = \partial_\mu (\delta_B c^a) + g f^a{}_{bc} (D_\mu c)^b \Lambda c^c + g f^a{}_{bc} A^b_\mu (\delta_B c^c). \tag{45.18}$$

We can combine the first and third terms into D_μ to find the equation

$$\delta_B (D_\mu c)^a = D_\mu (\delta_B c^a) + \frac{1}{2} D_\mu (g f^a{}_{bc} c^b \Lambda c^c) = 0, \tag{45.19}$$

but only if we have the following identity satisfied (the terms with ∂_μ are clearly the same in the two expressions, and only the terms with A_μ need to be checked):

$$\left[\frac{1}{2}f^a{}_{pq}f^q{}_{bc} - f^a{}_{qc}f^q{}_{pb}\right]A^p_\mu c^b \wedge c^c = 0, \tag{45.20}$$

where the first term comes from $1/2D_\mu(gf^a{}_{bc}c^b \wedge c^c)$ (in the final form for the variation) and the second term comes from $gf^a{}_{bc}(D_\mu c)^b \wedge c^c$ (in the initial form for the variation). The above equality follows from the Jacobi identity (coming from the double commutator of the generators of the Lie algebra):

$$f^a{}_{pq}f^q{}_{bc} + f^a{}_{bq}f^q{}_{cp} + f^a{}_{cq}f^q{}_{pb} = 0, \tag{45.21}$$

by noticing that the last two terms are equal when multiplied by an antisymmetric factor in (bc) like $c^b c^c$, and then by renaming the indices, using antisymmetry and multiplying by $1/2$ we get the needed equality. The solution to (45.19) is found by first putting $A_\mu = 0$, in which case we can peel off the ∂_μ in both terms in the equation, to find

$$\delta_B c^a = \frac{1}{2}gf^a{}_{bc}c^b c^c \Lambda, \tag{45.22}$$

and then we can check that the A_μ terms also cancel identically.

Finally, now that we fixed the variation of c^a such that $(D_\mu c)^a$ is invariant, we have that the variation of the ghost term is

$$\delta_B \mathcal{L}_{\text{ghost}} = -(\delta_B b^a)\partial^\mu (D_\mu c)^a, \tag{45.23}$$

and it needs to cancel against the variation of the gauge-fixing term (since the classical term is invariant):

$$\delta_B \mathcal{L}_{\text{gauge fix}} = \frac{1}{\alpha}(\partial^\rho A_\rho)\partial^\mu (D_\mu c)^a \Lambda. \tag{45.24}$$

This is achieved if

$$\delta_B b^a = -\frac{1}{\alpha}(\partial^\rho A_\rho)\Lambda. \tag{45.25}$$

In conclusion, the BRST transformation laws are

$$\delta_B A^a_\mu = (D_\mu c)^a \Lambda,$$
$$\delta_B c^a = \frac{1}{2}gf^a{}_{bc}c^b c^c \Lambda,$$
$$\delta_B b^a = -\frac{1}{\alpha}\partial^\mu A^a_\mu \Lambda. \tag{45.26}$$

For more general gauge-fixing terms, written as

$$\mathcal{L}_{\text{g.fix}} = +\frac{1}{2}\gamma_{ab}F^a F^b, \tag{45.27}$$

where for the Lorenz gauge we have $F^a = \partial^\mu A^a_\mu$ and $\gamma_{ab} = \delta_{ab}/\alpha$, as we saw in Chapter 11, we can write the ghost action as

$$-b_a \frac{\partial F^a}{\partial A^c_\mu}D^{cb}_\mu c^b = -b_a(\delta_B F^a)/\Lambda. \tag{45.28}$$

Here, $/\Lambda$ means we take away Λ from the right (remember that Λ is anticommuting, so taking it away from the left or the right differs by a minus sign). Then the invariance of $\delta_B F^a/\Lambda$ fixes the same $\delta_B c^a$ as before, and in turn that implies

$$\delta_B b_a = -\gamma_{ab} F^b \Lambda, \tag{45.29}$$

since

$$\delta_B\left(+\frac{1}{2}\gamma_{ab}F^b F^a\right) - (\delta_B b_a)\delta_B F^a/\Lambda = \gamma_{ab}F^b \delta_B F^a = (-\gamma_{ab}F^b\Lambda)\delta_B F^a/\Lambda = 0. \tag{45.30}$$

45.3 Nilpotency of Q_B and the Auxiliary Field Formulation

We can define a *BRST charge* Q_B that acts by $e^{Q_B\Lambda}$ (e.g. $\delta_B A_\mu^a = (Q_B A_\mu^a)\Lambda$, etc.). Then Q_B can be made nilpotent, that is

$$Q_B^2 = 0. \tag{45.31}$$

Proof We saw that $\delta_B A_\mu^a = (D_\mu c)^a \Lambda$, and $\delta_B(D_\mu c)^a = 0$, so $\delta_B^2 A_\mu^a = 0$. We also saw that $\delta c^a = \frac{1}{2}gf^a{}_{bc}c^b c^c \Lambda$, and further

$$\delta_B(f^a{}_{bc}c^b c^c) = 2f^a{}_{bc}c^b \delta c^c = gf^a{}_{bc}f^c{}_{de}c^b c^d c^e, \tag{45.32}$$

but this is zero by the Jacobi identity, since $c^b c^d c^e$ is totally antisymmetric in (bde), and

$$f^a{}_{[b|c|}f^c{}_{de]} = 0. \tag{45.33}$$

It follows that

$$\delta_B^2 c^a = 0. \tag{45.34}$$

In contrast, on b^a it is not quite nilpotent, since

$$\delta_B^2 b^a = -\frac{1}{\alpha}\partial^\mu(\delta_B A_\mu^a)\Lambda_1 = -\frac{1}{\alpha}\partial^\mu(D_\mu c)^a \Lambda_2\Lambda_1. \tag{45.35}$$

However, note that $\partial^\mu(D_\mu c)^a$ is the field equation for b^a, so $\delta_B^2 b^a = 0$ on-shell, or $Q_B^2 = 0$ on-shell.

This is a familiar situation for symmetries, in particular for supersymmetry. The algebra of charges closes only on-shell, so in order to make it close off-shell we must go to a first-order formulation, through the introduction of auxiliary fields, with field equation related to the term we want to cancel.

In our case, the problematic term arises from $\delta_B(-(\partial^\mu A_\mu)/\alpha)$, which is related to the variation of the gauge-fixing term. We will fix it by introducing an auxiliary field called the *Nakanishi–Lautrup field* d_a.

Then we write the gauge-fixing term in first-order formulation, as

$$\mathcal{L}_{\text{g.fix}} = -\frac{\alpha}{2}(d_a)^2 - d_a(\partial^\rho A_\rho^a). \tag{45.36}$$

The equation of motion for the auxiliary field d_a gives

$$d_a = -\frac{1}{\alpha}(\partial^\rho A_\rho^a), \tag{45.37}$$

which is what we have in the variation δ_B of b^a. By replacing it in the gauge-fixing term, we get back to the second-order formulation.

Therefore we write the new variations of b_a and d_a as

$$\delta_B b_a = d_a \Lambda; \quad \delta_B d_a = 0. \tag{45.38}$$

Now we see explicitly that

$$\delta_B^2 b_a = 0 = \delta_B^2 d_a \Rightarrow Q_B^2 = 0. \tag{45.39}$$

Let us verify that the above variations leave the quantum action invariant. We have $\delta_B(\alpha(d_a)^2/2) = 0$, and

$$d_a \delta_B(\partial^\rho A_\rho) + (\delta_B b_a)\partial^\mu (D_\mu c)^a = 0, \tag{45.40}$$

so indeed the action is invariant.

And now, as promised, $Q_B^2 = 0$ off-shell. Since δ_B acts by $Q_B \Lambda$, and $N_{\text{gh}}[\Lambda] = -1$, it follows that $N_{\text{gh}}[Q_B] = +1$. A more important observation is that the rules are now purely kinematical:

$$\delta_B A_\mu^a = (D_\mu c)^a \Lambda; \quad \delta_B c^a = \frac{1}{2}g f^a{}_{bc} c^b c^c;$$
$$\delta_B b_a = d_a \Lambda; \quad \delta_B d_a = 0. \tag{45.41}$$

Indeed, we see that they are completely independent of the gauge-fixing term, in particular are independent of α, so they apply to any quantum gauge theory.

The gauge-fixing term can be rewritten as

$$-\frac{1}{2}\gamma^{ab} d_a d_b - d_a F^a = -\frac{1}{2}\gamma^{ab} d_a \delta_B b_b / \Lambda - (\delta_B b_a)/\Lambda F^a, \tag{45.42}$$

and we already saw that the ghost term is

$$-b_a \delta_B F^a / \Lambda, \tag{45.43}$$

so all in all the quantum action can be written as

$$\mathcal{L}_{\text{qu}} = \mathcal{L}_{\text{class}} - \delta_B \left(b_a \left(F^a + \frac{\gamma^{ab}}{2} d_b \right) \right) / \Lambda. \tag{45.44}$$

This form is valid more generally, and it trivializes the BRST invariance of the quantum action, $\delta_B \mathcal{L}_{\text{qu}} = 0$, since it is now simply a result of the gauge invariance of the classical action, plus the nilpotency of Q_B, $Q_B^2 = 0$, which is as we saw true for the kinematical (model-independent) transformations (45.41).

The quantity

$$\psi \equiv b_a \left(F^a + \frac{\gamma^{ab}}{2} d_b \right) \tag{45.45}$$

is called the *gauge-fixing fermion*, and it can be made even more general. In terms of this, the action is

$$\mathcal{L}_{\text{qu}} = \mathcal{L}_{\text{class}} + \{Q_B, \psi\}. \tag{45.46}$$

Important Concepts to Remember

- Only nonabelian gauge fields have a negative contribution to β_1, fermions and scalars have a positive contribution, and abelian gauge fields zero.
- $SU(N_c)$ gauge theory with $N_f < 11N_c/2$ flavors is asymptotically free and IR-enslaved.
- In particular, QCD is asymptotically free and IR-enslaved, due to confinement at large distances.
- BRST symmetry is a global remnant of gauge invariance (residual gauge symmetry) in a gauge-fixed theory.
- It has an anticommuting parameter of mass dimension -1 and ghost number -1, and is defined by $\delta_B A_\mu^a = (D_\mu c)^a \Lambda$.
- The BRST charge Q_B, of ghost number $+1$, is nilpotent on-shell, $Q_B^2 = 0$, and if we introduce the Nakanishi–Lautrup auxiliary field d_a, it is nilpotent off-shell.
- With d_a, the transformation rules are purely kinematical (i.e. independent of the gauge or model), and the Lagrangian is written as the classical piece, plus the variation of a gauge-fixing fermion, $\mathcal{L}_{\text{qu}} = \mathcal{L}_{\text{class}} + \{Q_B, \psi\}$, which trivializes its invariance, since $Q_B^2 = 0$.

Further Reading

See section 7.6 in [3] and sections 16.6 and 16.7 in [1].

Exercises

1. Calculate explicitly the beta function for an $SO(N_c)$ gauge theory with N_f fundamental flavors.
2. Can a higher loop contribution to the beta function of a different sign (for instance, positive two-loop beta function term) change the property of IR slavery of gauge theories? How about UV freedom?
3. Anti-BRST invariance is obtained by interchanging b and c in the transformation rules, so

$$\delta_{\bar{B}} A_\mu^a = (D_\mu b)^a \zeta; \quad \delta_{\bar{B}} b^a = \frac{1}{2} g f^a{}_{bc} b^b b^c \zeta, \tag{45.47}$$

together with

$$\delta_{\bar{B}} c^a = -d^a \zeta + g f^a{}_{bc} b^b c^c \zeta; \quad \delta_{\bar{B}} d^a = -g f^a{}_{bc} b^b d^c \zeta. \tag{45.48}$$

Verify the nilpotency of $\delta_{\bar{B}}$ and the independence of $\delta_B, \delta_{\bar{B}}$, that is

$$\delta_B(\Lambda_1)\delta_B(\Lambda_2) = \delta_{\bar{B}}(\zeta_1)\delta_{\bar{B}}(\zeta_2) = 0 = [\delta_B(\Lambda), \delta_{\bar{B}}(\zeta)] = 0. \qquad (45.49)$$

Note: A BRST-invariant and anti-BRST-invariant model is the Curci–Ferrari model, which in Minkowski space is

$$\mathcal{L} = \mathcal{L}_{class} - \frac{1}{2\alpha}(\partial_\mu A_\mu^a)^2 + \frac{1}{2}b_a(\partial^\mu D_\mu + D_\mu \partial^\mu)c^a$$

$$+ \frac{g^2 \alpha}{8}(f^a{}_{bc}b^b c^c)^2 + \frac{\alpha}{2}\left(d^a + \frac{1}{\alpha}\partial_\mu A_\mu^a - \frac{1}{2}gf^a{}_{bc}b^b c^c\right)^2. \qquad (45.50)$$

4. Add a term $Kb_a \gamma^{ab} d_b d_c \gamma^{cd} d_d$ to the gauge-fixing fermion in the text. Write the quantum action.

46 Lee–Zinn-Justin Identities and the Structure of Divergences (Formal Renormalization of Gauge Theories)

In this chapter we will put the BRST formalism to good use, and derive from it the general structure of divergences, thus proving the renormalizability of gauge theories at one loop. The proof can be extended to all loops.

46.1 Lee–Zinn-Justin Identities

We start from the observation that the quantum action is invariant under BRST transformations. The integration measure is invariant as well. This statement needs to be tempered a bit. We will discuss anomalies later on in the book, but anomalies can arise from non-invariance of the integration measure. The potential anomaly coming from the Jacobian of gauge transformations is Q_B-exact, that is

$$\text{Anomaly} = \delta_B \Delta S. \tag{46.1}$$

This means that any actual anomaly can be removed by a local finite counterterm ΔS (the coefficient would depend on the regulator). We will not prove it here, and instead we will simply assume the invariance of the measure.

But the source terms are not invariant under BRST. The sources themselves are invariant, meaning that the source terms transform as

$$\delta_B (J \cdot A + \beta_a c^a + b_a \cdot \gamma^a) = (J \cdot Q_B A + \beta_a \cdot Q_B c^a + (Q_B b_a) \cdot \gamma^a) \Lambda. \tag{46.2}$$

Then we proceed like in the proof of Ward identities, to which these are related in spirit (they both come from invariance under a global symmetry). We make a change of variables in the path integral from the original ones to the BRST-transformed ones. The change leaves the partition function invariant. Since S_{eff} and the measure are also invariant, we obtain

$$0 = \int \mathcal{D}A \int \mathcal{D}b \int \mathcal{D}c \left\{ \int d^d x (J \cdot Q_B A + \beta_a \cdot Q_B c^a \right.$$
$$\left. + Q_B b_a \cdot \gamma^a) e^{-S_{\text{eff}}[A,b,c] + \int d^d x [J \cdot A + \beta \cdot c + b \cdot \gamma]} \right\}. \tag{46.3}$$

One could continue like this, but Lee and Zinn-Justin introduced a useful trick, to introduce new sources for $Q_B A$ and $Q_B c$, which are *nonlinear* in the fields. Note that $Q_B b$ is linear in the fields, $= -1/\alpha(\partial^\mu A_\mu)$, so it does not need an extra source. These sources are useful, since we have things like

$$\langle \delta_B A_\mu^a \rangle / \Lambda \rightarrow \langle g f^a{}_{bc} A_\mu^b c^c \rangle \neq g f^a{}_{bc} \langle A_\mu^b \rangle \langle c^c \rangle, \tag{46.4}$$

but we would like to write equations involving products of VEVs arising from single BRST transformations, as we shall see.

Therefore we add to the exponent

$$-S_{\text{extra,source}} = \int d^d x [K_\mu^a (Q_B A)_\mu^a - L^a (Q_B c)^a] = \int d^d x \left[K_\mu^a (D_\mu c)^a - L^a \frac{g}{2} f^a{}_{bc} c^b c^c \right]. \tag{46.5}$$

This term is still invariant, since the new sources are invariant, $Q_B K_\mu^a = Q_B L^a = 0$, and the rest is $Q_B(\ldots)$, and $Q_B^2 = 0$. These K_μ^a and L^a are extended to objects called "antifields" in the formalism with the same name. Here K_μ^a is anticommuting, whereas L^a is commuting (and $b_a, c^a, \beta^a, \gamma_a$ are anticommuting).

Then (46.3) can be rewritten as usual, with derivatives acting on the partition function $Z = e^{-W}$, as

$$0 = \int d^d x \left[J_\mu^a \cdot \frac{\delta}{\delta K_\mu^a} - \beta_a \frac{\delta}{\delta L^a} - \frac{1}{\alpha} \partial_\mu \left(\frac{\delta}{\delta J_\mu^a} \right) \cdot \gamma^a \right] e^{-W[J,\beta,\gamma,K,L]}, \tag{46.6}$$

that is as

$$0 = \int d^d x \left[J_\mu^a \cdot \frac{\delta}{\delta K_\mu^a} - \beta_a \frac{\delta}{\delta L^a} - \frac{1}{\alpha} \partial_\mu \left(\frac{\delta}{\delta J_\mu^a} \right) \cdot \gamma^a \right] W[J, \beta, \gamma, K, L]. \tag{46.7}$$

This is the equivalent of a Ward identity, or a Dyson–Schwinger identity, since it is an identity on n-point functions derived from the invariance of the path integral.

Like in the Ward or Dyson–Schwinger case, we can make a Legendre transform to the 1PI case:

$$\Gamma[A_\mu^{a,cl}, b_a^{cl}, c^{a,cl}, K_\mu^a, L^a] = W[J_\mu^a, \beta_a, \gamma^a, K_\mu^a, L^a] + \int d^d x (J_\mu^a A_\mu^{a,cl} + \beta_a c^{a,cl} + b_a^{cl} \gamma^a). \tag{46.8}$$

From this Legendre transform we obtain

$$\frac{\delta \Gamma}{\delta A_\mu^{a,cl}} = J_\mu^a; \quad \frac{\delta \Gamma}{\delta c^{a,cl}} = -\beta_a; \quad \frac{\delta \Gamma}{\delta b_a} = \gamma^a;$$

$$\frac{\delta \Gamma}{\delta K_\mu^a} = \frac{\delta W}{\delta K_\mu^a}; \quad \frac{\delta \Gamma}{\delta L^a} = \frac{\delta W}{\delta L^a}; \quad \frac{\delta W}{\delta J_\mu} = -A_\mu^{cl}. \tag{46.9}$$

Substituting these relations in (46.7), we obtain

$$\int d^d x \left[\frac{\delta \Gamma}{\delta A_\mu^a} \frac{\delta \Gamma}{\delta K_\mu^a} + \frac{\delta \Gamma}{\delta c^{a,cl}} \frac{\delta \Gamma}{\delta L^a} + \frac{\delta \Gamma}{\delta b_a^{cl}} \frac{1}{\alpha} (\partial_\mu A_\mu^{cl}) \right] = 0. \tag{46.10}$$

Another relation is obtained in a similar manner to the first, from the fact that the measure is invariant under a shift in the b_a field, $b_a \rightarrow b_a + \delta b_a$. Changing integration variables, using the invariance of the measure and the fact that

$$\delta(\mathcal{L}_{\text{eff}} + \mathcal{L}_{\text{source}} + \mathcal{L}_{\text{extra,source}}) = -\delta b_a [\partial^\mu (D_\mu c)^a + \gamma^a] = -\delta b_a [\partial^\mu (Q_B A_\mu)^a + \gamma^a],$$
(46.11)

we get

$$0 = \int \mathcal{D}A_\mu \mathcal{D}b_a \mathcal{D}c^a \int d^d x \, \delta b_a(x) [\partial^\mu (Q_B A_\mu)^a + \gamma^a] e^{-S_{\text{eff}}[A,b,c] + \int d^d x (J \cdot A + \beta \cdot c + b \cdot \gamma)},$$
(46.12)

which is true for any $\delta b_a(x)$, so we can write a *local* relation, and writing $Q_B A_\mu^a$ as $\delta/\delta K_\mu^a$ and extracting it outside the path integral, we obtain

$$0 = \left(\partial_\mu \frac{\delta}{\delta K_\mu^a} + \gamma^a \right) e^{-W[J,\beta,\gamma,K,L]},$$
(46.13)

giving finally

$$\partial_\mu \frac{\delta}{\delta K_\mu^a} W[J, \beta, \gamma, K, L] = \gamma^a.$$
(46.14)

Writing this relation in terms of the effective action Γ, as we did above for the first relation (46.10), we obtain

$$\partial_\mu \frac{\delta \Gamma}{\delta K_\mu^a} - \frac{\delta \Gamma}{\delta b_a^{cl}} = 0.$$
(46.15)

From the relations (46.10) and (46.15), we will then derive the Slavnov–Taylor identities.

Before that, we can simplify them a little by removing from the effective action the classical gauge-fixing term, by defining

$$\tilde{\Gamma}[A, b, c, K, L] = \Gamma[A, b, c, K, L] - \frac{1}{2\alpha} \int d^d x (\partial^\mu A_\mu^a)^2.$$
(46.16)

Indeed, note that $\int \mathcal{D}(\text{fields}) \, e^{-(S + \frac{1}{2\alpha}(\partial^\mu A_\mu^a)^2)} = Z = e^{-W}$ and $\Gamma = W + (\ldots)$, so $\Gamma^{(0)}$ contains $1/2\alpha (\partial^\mu A_\mu^a)^2$.

Then, (46.15) is rewritten in the same way with $\tilde{\Gamma}$ instead of Γ, and (46.10) is rewritten, using also (46.15) without the last term. Together, the relations are now

$$\int d^d x \left[\frac{\delta \tilde{\Gamma}}{\delta A_\mu^a} \frac{\delta \tilde{\Gamma}}{\delta K_\mu^a} + \frac{\delta \tilde{\Gamma}}{\delta c^{a,cl}} \frac{\delta \tilde{\Gamma}}{\delta L^a} \right] = 0,$$
(46.17)

$$\partial_\mu \frac{\delta \tilde{\Gamma}}{\delta K_\mu^a} - \frac{\delta \tilde{\Gamma}}{\delta b_a^{cl}} = 0.$$
(46.18)

These are called the *Lee–Zinn-Justin (LZJ) identities*.

46.2 Structure of Divergences

We now use these relations to prove the renormalizability of gauge theories at one loop (and the proof can be extended by induction to all orders). We will also derive the Slavnov–Taylor identities from the LZJ identities.

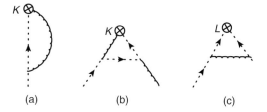

Figure 46.1 BRST divergences: (a) for $K^a_\mu \partial_\mu c^a$; (b) for $gf^a{}_{bc} K^a_\mu A^b_\mu c^c$; (c) for $gf^a{}_{bc} L^a c^b c^c$.

The new source terms

$$K(Q_B A) - L(Q_B c) = K^a_\mu(\partial_\mu c^a + gf^a{}_{bc} A^b_\mu c^c) + L^a \frac{g}{2} f^a{}_{bc} c^b c^c \qquad (46.19)$$

induce three new divergences in the quantum effective action, as seen in Figure 46.1. The sources K^a_μ and L_a are represented by crossed circles with K or L on them.

Then the one-loop divergent graph correcting the $K^a_\mu \partial_\mu c^a$ term comes in a $K^a_\mu A^b_\mu c^c$ vertex with the gluon line ending again on the gluon line. This graph is divergent, since it is the same Feynman integral (with different multiplying factors) as the ghost propagator correction at one loop (replace the K vertex with just the ghost line continuing on).

The one-loop divergent graph correcting the $gf^a{}_{bc} K^a_\mu A^b_\mu c^c$ vertex comes from the vertex itself, with a ghost line exchanged between the ghost and gluon lines, and interchanging them (turning gluon into ghost and vice versa). The graph is divergent, since it is the same Feynman integral (with different multiplying factors) as a ghost–ghost–gluon vertex correction at one loop (replace the K vertex with a ghost line continuing on).

The one-loop divergent graph correcting the $gf^a{}_{bc} L^a c^b c^c$ vertex comes from the vertex itself, with a gluon line interchanged between the two ghost lines. It is divergent, since it is the same Feynman integral as the one diagram replacing the L vertex with a gluon line continuing on, which is just the same graph as above, rotated at 90°.

Therefore, at one loop, we can isolate the divergent part that depends on K^a_μ and L^a as having these three possible divergent structures, obtaining

$$\tilde{\Gamma}_{\rm div}[A^{cl}, b^{cl}, c^{cl}, K, L] = \tilde{\Gamma}_{\rm div}[A^{cl}, b^{cl}, c^{cl}] - (\tilde{Z}_3 - 1) \int d^d x K^a_\mu \partial^\mu c^a$$
$$- (\tilde{Z}_1 - 1) \int d^d x g f_{abc} K^a_\mu A^b_\mu c^c + (X - 1) \int d^d x L_a \left(-\frac{g}{2} f^a{}_{bc} c^b c^c \right),$$
$$(46.20)$$

where we have written three renormalization factors, $\tilde{Z}_3, \tilde{Z}_1$, and X. For the first two, we have anticipated a bit that they will be the same as the previously defined $\tilde{Z}_3$ and $\tilde{Z}_1$, which we will prove shortly. Otherwise, we could have called them $\tilde{Y}_2, \tilde{Y}_1$ and show that they equal $\tilde{Z}_3, \tilde{Z}_1$.

We can now use (46.17) and (46.18) to determine $\tilde{\Gamma}_{\rm div}$, by expanding in loop order:

$$\tilde{\Gamma} = \tilde{\Gamma}^{(0)} + \hbar \tilde{\Gamma}^{(1)}_{\rm div} + \mathcal{O}(\hbar^2), \qquad (46.21)$$

and using

$$\tilde{\Gamma}^{(0)} = \int d^d x \left[\frac{1}{4}(F^a_{\mu\nu})^2 - b_a \partial^\mu D_\mu c^a - K^a_\mu Q_B A^a_\mu + L_a Q_B c^a \right]. \tag{46.22}$$

Replacing this in (46.18) and concentrating on the divergent part (to one loop), we obtain

$$\frac{\delta\tilde{\Gamma}}{\delta b_a} = \partial_\mu \frac{\delta\tilde{\Gamma}}{\delta K^a_\mu} = -(\tilde{Z}_3 - 1)\partial^\mu \partial_\mu c^a - (\tilde{Z}_1 - 1)gf^a{}_{bc}\partial^\mu(A^b_\mu c^c), \tag{46.23}$$

where in the second equality we have substituted the form of $\tilde{\Gamma}_{\mathrm{div}}$, whose K_a dependence was fixed. We can now integrate this relation with respect to b_a and obtain

$$\tilde{\Gamma}_{\mathrm{div}}[A, b, c] = \tilde{\Gamma}_{\mathrm{div}}[A] - \int d^d x \left[(\tilde{Z}_3 - 1)b_a \partial^\mu \partial_\mu c^a + (\tilde{Z}_1 - 1)b_a gf^a{}_{bc}\partial^\mu(A^b_\mu c^c) \right]. \tag{46.24}$$

Thus, we see that indeed $\tilde{Z}_3$ and $\tilde{Z}_1$ are the objects we have defined before, the renormalization of the ghost propagator and ghost–ghost–gluon vertex, respectively.

We note that we have, in two short steps, fixed the dependence of the divergent piece on all fields except A^a_μ.

To fix that as well, we look at the simple pole (single divergence) of (46.17). Indeed, there we also have a double divergence, coming from the term where both $\tilde{\Gamma}$ factors are divergent, but we want to focus instead on the terms where one $\tilde{\Gamma}$ is divergent and one is finite, so we have at one loop

$$0 = \int d^d x \left[\left(\frac{\delta\tilde{\Gamma}_{\mathrm{div}}}{\delta A^a_\mu} \frac{\delta\tilde{\Gamma}^{(0)}}{\delta K^a_\mu} + \frac{\delta\tilde{\Gamma}^{(0)}}{\delta A^a_\mu} \frac{\delta\tilde{\Gamma}_{\mathrm{div}}}{\delta K^a_\mu} \right) + \left(\frac{\delta\tilde{\Gamma}_{\mathrm{div}}}{\delta c^a} \frac{\delta\tilde{\Gamma}^{(0)}}{\delta L^a} + \frac{\delta\tilde{\Gamma}^{(0)}}{\delta c^a} \frac{\delta\tilde{\Gamma}_{\mathrm{div}}}{\delta L^a} \right) \right]. \tag{46.25}$$

46.3 Solving the LZJ and Slavov–Taylor Identities

Now we must insert in this equation the expressions for $\tilde{\Gamma}_{\mathrm{div}}$ and $\tilde{\Gamma}^{(0)}$ and identify terms. We will organize the calculation according to the types of terms in the expanded equation.

46.3.1 Terms Linear in K^a_μ

We can check that the terms coming from the second and third terms in (46.25) cancel, leaving the contributions of the first and fourth terms, giving

$$0 = \int d^d x [(\tilde{Z}_1 - 1)K^a_\mu gf^a{}_{bc}c^c]D_\mu c^b - \int d^d x (X - 1)\left[(-D_\mu K^a_\mu)\left(-\frac{g}{2}f^a{}_{bc}c^b c^c \right) \right], \tag{46.26}$$

which after a partial integration of D_μ and a rearranging of the terms gives

$$X = \tilde{Z}_1. \tag{46.27}$$

46.3.2 Terms Linear in A and Not Containing K and L, and Linear in c

We write an expansion in A of $\tilde{\Gamma}_{\text{div}}[A]$:

$$\tilde{\Gamma}_{\text{div}}[A] = \tilde{\Gamma}^{(2)}_{\text{div}}[A] + \tilde{\Gamma}^{(3)}_{\text{div}}[A] + \tilde{\Gamma}^{(4)}_{\text{div}}[A]. \tag{46.28}$$

Then, of course, the contribution to these terms from $\tilde{\Gamma}_{\text{div}}[A]$ is entirely from $\tilde{\Gamma}^{(2)}_{\text{div}}[A]$, to obtain linear terms in A after derivation. The variation of $1/4(F^a_{\mu\nu})^2$ with respect to A^a_ν gives $-D^\mu F^a_{\mu\nu}$, as we know (from the YM equation of motion). We obtain, from the first two terms in (46.25):

$$0 = \int d^d x \left\{ -\frac{\delta \tilde{\Gamma}^{(2)}_{\text{div}}[A]}{\delta A^a_\mu}(D_\mu c^a) + (D^\mu F^a_{\mu\nu})_{\text{lin.A}}[(\tilde{Z}_3 - 1)\partial_\nu c^a + (\tilde{Z}_1 - 1)gf^a_{\ bc}A^b_\nu c^c]|_{\text{no A}} \right\}.$$

$$\tag{46.29}$$

But note that by partial integration of the $(\tilde{Z}_3 - 1)$ term we obtain $\partial^\nu D^\mu F_{\mu\nu}|_{\text{lin.A}} = \partial^\mu \partial^\nu(\partial_\mu A_\nu - \partial_\nu A_\mu) = 0$, and the $(\tilde{Z}_1 - 1)$ term has too many A_μs. Therefore, we obtain

$$\int d^d x \frac{\delta \tilde{\Gamma}^{(2)}_{\text{div}}[A]}{\delta A^a_\mu}\partial_\mu c^a = 0, \tag{46.30}$$

which by partial integration gives

$$\partial_\mu \frac{\delta \Gamma^{(2)}_{\text{div}}}{\delta A^a_\mu} = 0, \tag{46.31}$$

which means that $\Gamma^{(2)}_{\text{div}}[A]$ is transverse, which uniquely selects the form of the kinetic term, that is we must have

$$\tilde{\Gamma}^{(2)}_{\text{div}}[A] = (Z_3 - 1)\int d^d x \frac{1}{2}(\partial_\mu A^a_\nu - \partial_\nu A^a_\mu)^2. \tag{46.32}$$

46.3.3 Terms Quadratic in A and Not Containing K and L

We continue with the expansion in A of the same terms as above, giving

$$0 = \int d^d x \left\{ -\frac{\delta \tilde{\Gamma}^{(3)}}{\delta A^a_\mu}\partial_\mu c^a - \frac{\delta \tilde{\Gamma}^{(2)}}{\delta A^a_\mu}gf^a_{\ bc}A^b_\mu c^c + \partial^\mu(\partial_\mu A^a_\nu - \partial_\nu A^a_\mu)(\tilde{Z}_1 - 1)gf^a_{\ bc}A^b_\nu c^c \right.$$

$$\left. + gf^a_{\ bc}A^b_\mu(\partial_\mu A^c_\nu - \partial_\nu A^c_\mu)(\tilde{Z}_3 - 1)\partial_\nu c^a \right\}, \tag{46.33}$$

which can be rewritten as

$$\int d^d x \left[\frac{\delta \tilde{\Gamma}^{(3)}}{\delta A^a_\mu} + (Z_3 - \tilde{Z}_1 - \tilde{Z}_3 - 1)gf^a_{\ bc}(\partial_\mu A^b_\nu - \partial_\nu A^b_\mu)A^c_\nu \right] \partial_\mu c^a = 0. \tag{46.34}$$

Since this relation is valid for any c^a, we can put the square bracket to zero and integrate it, to give

$$\tilde{\Gamma}^{(3)}_{\text{div}}[A] = (Z_3 + \tilde{Z}_1 - \tilde{Z}_3 - 1)\int d^d x \frac{1}{2}gf_{abc}A^a_\mu A^b_\nu(\partial_\mu A^c_\nu - \partial_\nu A^c_\mu). \tag{46.35}$$

This is again of the type of the classical 3-gluon coupling, like in the explicit one-loop renormalization. The coefficient was usually called Z_1 at one loop, so

$$Z_1^{(1)} = Z_3^{(1)} + \tilde{Z}_1^{(1)} - \tilde{Z}_3^{(1)} = Z_3 \frac{\tilde{Z}_1}{\tilde{Z}_3} + \mathcal{O}(\hbar^2). \tag{46.36}$$

46.3.4 Terms Cubic in A and Not Containing K and L

We will not do this explicitly (we will leave it as an exercise), but the analysis is similar, and the result is

$$\tilde{\Gamma}_{\text{div}}^{(4)}[A] = (Z_3 + 2\tilde{Z}_1 - 2\tilde{Z}_3 - 1)\frac{g^2}{4}f^a{}_{bc}f_{ade} \int d^d x A_\mu^b A_\nu^c A_\mu^d A_\nu^e. \tag{46.37}$$

Yet again this is of the type of the classical 4-gluon coupling, like in the explicit one-loop renormalization. The coefficient was usually called Z_4 at one loop, so

$$Z_4^{(1)} = Z_3^{(1)} + 2Z_1^{(1)} - 2\tilde{Z}_3^{(1)} = Z_3 \frac{\tilde{Z}_1^2}{\tilde{Z}_3^2} + \mathcal{O}(\hbar^2). \tag{46.38}$$

Then we see that (46.36) and (46.38) are the two Slavnov–Taylor identities at one loop, as promised.

In conclusion, from the LZJ identities (coming from BRST invariance), we have completely fixed the form of the divergent part of the quantum effective action, and derived the Slavnov–Taylor identities.

The extra source terms are

$$K(Q_B A) = K_\mu^a (\tilde{Z}_3 \partial_\mu c^a + \tilde{Z}_1 g f^a{}_{bc} A_\mu^b c^c),$$

$$L(Q_B c) = L^a X \left(\frac{g}{2} f^a{}_{bc} c^b c^c\right). \tag{46.39}$$

Using the renormalizations defined at one loop:

$$b_0, c_0 = \sqrt{\tilde{Z}_3} b, c; \quad A_0 = \sqrt{Z_3} A; \quad g_0 = g \frac{Z_1}{Z_3^{3/2}}, \tag{46.40}$$

we can write

$$K(Q_B A) = K_\mu^a \sqrt{\tilde{Z}_3}(\partial_\mu c_0^a + g_0 f^a{}_{bc} A_{0\mu}^b c_0^c) = \sqrt{\tilde{Z}_3} K(Q_B A_{0\mu}),$$

$$L(Q_B c) = \sqrt{\tilde{Z}_3} L^a \left(\frac{g_0}{2} f^a{}_{bc} c_0^b c_0^c\right) = \sqrt{\tilde{Z}_3} L(Q_B c_0), \tag{46.41}$$

therefore we deduce the renormalization of the sources

$$K_{0\mu}^a = \sqrt{\tilde{Z}_3} K_\mu^a; \quad L_0^a = \sqrt{\tilde{Z}_3} L^a. \tag{46.42}$$

In conclusion, we have the renormalization at one loop

$$\Gamma^{(1\text{-loop})}[A, b, c, K, L, g, \alpha] = \Gamma_0^{(1\text{-loop})}[A_0, b_0, c_0, K_0, L_0, g_0, \alpha_0], \tag{46.43}$$

which shows that the gauge theory is one-loop renormalizable (without calculating anything explicitly).

This proof can be extended to any n loops through induction, though we will not do it here.

One can also include fermions in the analysis. The proof of the LZJ identities in the case with fermions is left as an exercise below, as is the proof of the Slavnov–Taylor identities.

Important Concepts to Remember

- The quantum effective action is BRST-invariant, and the measure is invariant up to a possible anomaly that can be removed by a local finite counterterm.
- We derive the LZJ identities as quadratic identities on $\tilde{\Gamma}$, by adding source terms for the nonlinear terms in the BRST transformations, $K_\mu^a (Q_B A)_\mu^a$ and $-L^a (Q_B c)^a$, besides the usual source terms $\beta_a c^a + b_a \gamma^a$, and finding the equations in a similar manner to the Dyson–Schwinger identities or Ward identities.
- The introduction of K and L generates three new divergent structures, but the LZJ identities allow us to completely fix the possible structure of divergences to be exactly that in the bare Lagrangian, thus proving renormalizability at one loop.
- The proof can be generalized by induction to all loop orders, as well as to include fermions, thus a general gauge theory is renormalizable.
- The renormalization of K and L is not independent (is written in terms of the other Z factors), and the Slavnov–Taylor identities are also obtained from the LZJ identities.

Further Reading

See sections 7.7.1 and 7.7.2 in [3], and section 16.4 in [1].

Exercises

1. Introduce fermions into the gauge theory:

$$\delta_B \psi_i = -(T^a)_{ij} c^a \psi_j \Lambda, \tag{46.44}$$

and add to the action the fermionic terms

$$S_{\text{fermi}} = \int d^d x \, \bar{\psi} (\slashed{D} + m) \psi [-\bar{H}_{i\alpha} (Q_B \psi)_{i\alpha} - (Q_B \bar{\psi})_{i\alpha} H_{i\alpha}] \tag{46.45}$$

and the fermion source terms

$$\int d^d x [\bar{\zeta}_{i\alpha} \psi_{i\alpha} + \bar{\psi}_{i\alpha} \zeta_{i\alpha} + \bar{H}_{i\alpha} (Q_B \psi)_{i\alpha} + (Q_B \bar{\psi})_{i\alpha} H_{i\alpha}]. \tag{46.46}$$

Find the generalized LZJ identities, first for $W[J, \beta, \gamma, K, L, \zeta, \bar{\zeta}, H, \bar{H}]$ and then for Γ and $\tilde{\Gamma}[A, b, c, \psi, \bar{\psi}, K, L, H, \bar{H}]$.

2. Check that we get the correct Slavnov–Taylor identities at one loop.
3. Do explicitly the analysis of the terms cubic in A and not containing K and L that was omitted in the text, leading to (46.37).
4. Introduce scalars in the theory in a way similar to the fermions introduced in Exercise 1, and find the corresponding generalized LZJ identities.

47

BRST Quantization

We obtained BRST symmetry as a symmetry of gauge theories, and we saw that we could use it to get a proper understanding of the quantum theory, since it allowed a formal proof of renormalizability.

But more generally, we can turn it into a *quantization procedure* for any theory with a local invariance (like gauge invariance) that needs gauge fixing. *BRST quantization* builds upon the Dirac formalism for quantization of constrained systems, and is extended to the Batalin–Vilkovisky (BV, or field–antifield) formalism, that will not be treated here, but only mentioned at the end of the chapter.

So we have: Dirac quantization extends to BRST quantization, that extends to the BV formalism.

We will thus first review the Dirac formalism, then we will define the formalism of BRST quantization. After exemplifying it for electromagnetism in Lorenz gauge, we will write the general formalism, and then use it for pure Yang–Mills theory.

47.1 Review of the Dirac Formalism

We will quickly review the salient points of the Dirac formalism, since we will build upon it to develop BRST quantization.

One starts with a system obeying a set of *primary constraints*:

$$\phi_m(q,p) = 0; \quad m = 1, \ldots, M. \tag{47.1}$$

If a quantity is equal to zero only by using the constraints, we say we have a *weak equality*, and write ≈ 0, and we mean that we do all the derivations, and only at the end of the calculation we put $\phi_m = 0$.

Then for some function of phase space, $g(q,p)$, the time evolution will be given in weak equality by the Poisson bracket with the *total Hamiltonian* H_T:

$$H_T = H + u_m \phi_m, \tag{47.2}$$

where $H = H_0$ is the classical Hamiltonian (without the constraints), so

$$\dot{g} \approx \{g, H_T\}_{P.B.}. \tag{47.3}$$

Since the time evolution must preserve the constraints, we have $\dot{\phi}_m = 0$, which implies that we need to have the weak equality

$$\{\phi_m, H\}_{P.B.} + u_n\{\phi_m, \phi_n\}_{P.B.} \approx 0. \tag{47.4}$$

These will in general generate *secondary constraints* (we repeat the procedure of finding new constraints, then imposing that their time evolution is zero, until we don't get any more new constraints) ϕ_k, $k = M + 1, \ldots, M + K$. The set of all constraints will be denoted by

$$\phi_j \approx 0, \quad j = 1, \ldots, M + K. \tag{47.5}$$

We next define a *first-class function* of phase space $R(q, p)$ if its Poisson bracket with all the constraints is weakly zero, so

$$\{R, \phi_j\}_{P.B.} \approx 0 \quad (= r_{jj'}\phi_{j'}). \tag{47.6}$$

Of course, a particular case is of first-class constraints ϕ_a, that must therefore satisfy a kind of algebra, $\{\phi_a, \phi_j\} = f_{aj}^{j'}\phi_{j'}$.

A function of phase space is second class if there is at least one j for which $\{R, \phi_j\}$ is not weakly zero.

Since the set of ϕ_j is the full set of constraints obtained from time evolution, it follows by definition that for all of them we need to have weakly zero Poisson brackets with H_T:

$$\{\phi_j, H_T\} \approx \{\phi_j, H\} + u_m\{\phi_j, \phi_m\} \approx 0. \tag{47.7}$$

This is now thought of as a set of equations for the coefficients u_m, solved by a particular solution U_m of the inhomogeneous equation (the equation above), plus linear solutions of the homogeneous equation with arbitrary (numerical) coefficients, that is

$$u_m = U_m + v_a V_{am}, \tag{47.8}$$

where

$$V_{am}\{\phi_j, \phi_m\} \approx 0. \tag{47.9}$$

Then we can rewrite the total Hamiltonian

$$H_T = H + U_m\phi_m + v_a V_{am}\phi_m, \tag{47.10}$$

where

$$\phi_a = v_{am}\phi_m,$$
$$H' = H + U_m\phi_m. \tag{47.11}$$

Note that by definition, since $V_{am}\{\phi_m, \phi_j\}_{P.B.} \approx 0$, $V_{am}\phi_m = \phi_a$ are first-class constraints. Also, since U_m is a particular solution of the inhomogeneous equation, we have $\{\phi_j, H'\}_{P.B.} \approx 0$, so H' is also first class.

But ϕ_a are only independent first-class primary constraints, and there is nothing special about primary constraints. So we define the *extended Hamiltonian* by adding also the first-class secondary constraints, so as to have all the first-class constraints in it:

$$H_E = H_T + v_{a'}\phi_{a'}. \tag{47.12}$$

47.1.1 Dirac Brackets

To quantize, we need to define brackets that replace the Poisson brackets, to deal with the second-class constraints.

Consider χ_s, the *independent* second-class constraints, and define the matrix $c_{ss'}$ as the inverse of the bracket of constraints

$$c_{ss'}\{\chi_{s'}, \chi_{s''}\} = \delta_{ss''}. \tag{47.13}$$

Then the Dirac bracket is defined as

$$[f, g]_{D.B.} = \{f, g\}_{P.B.} - \{f, \chi_s\}_{P.B.} c_{ss'}\{\chi_{s'}, g\}_{P.B.}. \tag{47.14}$$

Then the time evolution using the Dirac brackets is the same as the one using the Poisson brackets:

$$[g, H_T]_{D.B.} \approx \{g, H_T\}_{P.B.} \approx \dot{g}, \tag{47.15}$$

but now the Dirac brackets of any function of phase space with the second-class constraints is *strongly* (identically) zero:

$$[f, \chi_{s''}]_{D.B.} = \{f, \chi_{s''}\}_{P.B.} - \{f, \chi_s\}_{P.B.} c_{ss'}\{\chi_{s'}, \chi_{s''}\}_{P.B.} = 0. \tag{47.16}$$

This means that we can impose the second-class constraints operatorially on states in order to quantize:

$$\hat{\chi}_s|\psi\rangle = 0, \tag{47.17}$$

and the Dirac brackets are quantized to the commutators, instead of the Poisson brackets.

47.2 BRST Quantization

We saw in the Lagrangian formalism that the BRST quantum action is written as

$$\mathcal{L}_{qu} = \mathcal{L}_{class} + Q_B\Psi, \tag{47.18}$$

where Ψ is the gauge-fixing fermion. More precisely, since Q_B is an operator, we really have

$$\mathcal{L}_{qu} = \mathcal{L}_{class} + \{Q_B, \Psi\}. \tag{47.19}$$

BRST quantization is a Hamiltonian formalism, since it is based on the Dirac formalism. As such, we must use Minkowski signature. In Minkowski signature:

$$\Psi = -b_a\left(F^a + \frac{\alpha}{2}d^a\right). \tag{47.20}$$

In a Hamiltonian formalism, one has states $|\psi\rangle$, for instance intial and final states $|i\rangle$ and $|f\rangle$, and observables are transition amplitudes $\langle f|i\rangle$. As such, they should be independent of the choice of gauge, thus independent of the change in gauge-fixing function F^a. Then, under a small change δF^a, such that $\delta \Psi = b_a \delta F^a$, the variation of the transition amplitude, written as a path integral

$$\langle f|i\rangle = \int \mathcal{D}(\text{fields}) \, e^{iS}, \tag{47.21}$$

is given by

$$\delta \langle f|i\rangle = \int \mathcal{D}(\text{fields}) \, e^{iS} i\delta S = i\langle f|\{Q_B, \delta\Psi\}|i\rangle, \tag{47.22}$$

and it must be zero. Assuming that Q_B is Hermitian, $Q_B^\dagger = Q_B$, putting the two terms in the anticommutator term above to zero gives $\langle f|Q_B = 0$ and $Q_B|i\rangle = 0$, so that $Q_B|f\rangle = Q_B|i\rangle = 0$. This means that *all physical states must be BRST-invariant*:

$$Q_B|\psi\rangle = 0. \tag{47.23}$$

We say that physical states are Q_B-*closed*. On the contrary, physical states that differ by Q_B-*exact* terms (i.e. terms that are Q_B of something else) are equivalent:

$$|\psi'\rangle = |\psi\rangle + Q_B|\chi\rangle \sim |\psi\rangle. \tag{47.24}$$

Note that then

$$Q_B|\psi'\rangle = Q_B|\psi\rangle + Q_B^2|\chi\rangle = 0, \tag{47.25}$$

so it is also physical. The equivalence means that matrix elements are identical. Consider the matrix element with another physical state $|\tilde\psi\rangle$ ($Q_B|\tilde\psi\rangle = 0$):

$$\langle\psi'|\tilde\psi\rangle = \langle\psi|\tilde\psi\rangle + \langle\chi|Q_B|\tilde\psi\rangle = \langle\psi|\tilde\psi\rangle. \tag{47.26}$$

These matrix elements must also be preserved under time evolution, which means that

$$\langle\psi'|H_{BRST}|\tilde\psi\rangle = \langle\psi|H_{BRST}|\tilde\psi\rangle + \langle\chi|H_{BRST}Q_B|\tilde\psi\rangle + \langle\chi|[Q_B, H_{BRST}]|\tilde\psi\rangle, \tag{47.27}$$

and this must be equal to $\langle\psi|H_{BRST}|\tilde\psi\rangle$. The middle term is zero, since $Q_B|\tilde\psi\rangle = 0$, which means that we need to have (since the states $|\chi\rangle$ and $|\tilde\psi\rangle$ are arbitrary)

$$[Q_B, H_{BRST}] = 0. \tag{47.28}$$

In order to define the BRST Hamiltonian H_{BRST}, we must consider the Hamiltonian à la Dirac, in the presence of constraints. Moreover, we need to define Q_B and H_{BRST} together, such that $Q_B^2 = 0$ and $[Q_B, H_{BRST}] = 0$.

The quantization is done by defining a Hilbert space of states. From the above considerations, physical states are states in the Q_B-*cohomology*, since we need to have Q_B-closed states, $Q_B|\psi\rangle = 0$, modulo Q_B-exact states (i.e. $|\psi\rangle \sim |\psi\rangle + Q_B|\chi\rangle$).

In general, the cohomology of an operator that squares to zero is the coset defined as closed states, modulo exact states (equivalence classes under the equivalence by exact states):

$$Q_{B\text{-cohomology}} = \frac{Q_B\text{-closed}}{Q_B\text{-exact}}, \tag{47.29}$$

or the Hilbert space is

$$\mathcal{H}_{BRST} = \frac{\mathcal{H}_{\text{closed}}}{\mathcal{H}_{\text{exact}}}. \tag{47.30}$$

The notion of cohomology should be familiar (except for the mathematical language), since the cohomology of the exterior derivative $d = dx^\mu \partial_\mu$ is nothing but the familiar one of physical states in electromagnetism. We have $d^2 = 0$, and pure gauge fields (in the absence of matter) satisfy $F = dA = 0$ (i.e. are d-closed) and equivalent gauge-field configurations differ by d-exact terms, since gauge invariance is $A \sim A + d\lambda$. Then the cohomology of d over 1-forms is the cohomology of gauge fields (i.e. the physical states for fields of vanishing field strength). The cohomology of d is naturally extended for p-forms instead of 1-forms, giving the pth cohomology of d.

47.3 Example of BRST Quantization: Electromagnetism in Lorenz Gauge

At this intermediate stage, we give a simple example for the BRST quantization formalism described so far, for abelian gauge fields (i.e. electromagnetism). The example is kind of trivial, but will show explicitly how some of the abstract ideas described above work.

The BRST transformations for the abelian gauge fields (so $f^a{}_{bc} = 0$) in Minkowski signature are

$$\delta_B A_\mu = \partial_\mu c\Lambda; \quad \delta_B b = \frac{1}{\alpha}\partial^\mu A_\mu; \quad \delta_B c = 0. \tag{47.31}$$

Consider now canonical quantization of the fields A_μ, b, c, without taking into account the gauge invariance and constraints from gauge fixing:

$$A_\mu(x) = \int \frac{d^3p}{(2\pi)^{3/2}} \frac{1}{\sqrt{2p^0}} \left[a_\mu(\vec{p})e^{ipx} + a_\mu^\dagger(\vec{p})e^{-ipx} \right],$$

$$c(x) = \int \frac{d^3p}{(2\pi)^{3/2}} \frac{1}{\sqrt{2p^0}} \left[c(\vec{p})e^{ipx} + c^\dagger(\vec{p})e^{-ipx} \right],$$

$$b(x) = \int \frac{d^3p}{(2\pi)^{3/2}} \frac{1}{\sqrt{2p^0}} \left[b(\vec{p})e^{ipx} + b^\dagger(\vec{p})e^{-ipx} \right]. \tag{47.32}$$

The BRST transformation acts by (anti)commuting with Q_B, so:

1. $\delta_B A_\mu = i[Q_B, A_\mu]\Lambda$ is compared with $\delta_B A_\mu = \partial_\mu c\Lambda$. Substituting the forms of $A_\mu(x)$ and $c(x)$, we find

$$[Q_B, a_\mu(\vec{p})] = +p_\mu c(\vec{p}); \quad [Q_B, a_\mu^\dagger(\vec{p})] = -p_\mu c^\dagger(\vec{p}). \tag{47.33}$$

2. $\delta_B b = i\{Q_B, b\}\Lambda$ is compared with $\delta_B b = 1/\alpha \partial^\mu A_\mu$. Substituting the forms of $b(x)$ and $A_\mu(x)$, we find

$$\{Q_B, b(\vec{p})\} = \frac{1}{\alpha}p^\mu a_\mu(\vec{p}); \quad \{Q_B, b^\dagger(\vec{p})\} = -\frac{1}{\alpha}p^\mu a_\mu^\dagger(\vec{p}). \tag{47.34}$$

3. $\delta c = -\{Q_B, c\}\Lambda$ is compared with $\delta_{BC} = 0$ to obtain

$$\{Q_B, c(\vec{p})\} = \{Q_B, c^\dagger(\vec{p})\} = 0. \tag{47.35}$$

We now construct the Hilbert space of physical states. They must satisfy $Q_B|\psi\rangle = 0$. Consider the state

$$|\psi\rangle = e^\mu a_\mu^\dagger(\vec{p})|\tilde{\psi}\rangle, \tag{47.36}$$

where $|\tilde{\psi}\rangle$ is a physical state (i.e. $Q_B|\tilde{\psi}\rangle = 0$). It satisfies

$$Q_B|\psi\rangle = e^\mu a_\mu^\dagger(\vec{p})Q_B|\tilde{\psi}\rangle - e^\mu p_\mu c^\dagger(\vec{p})|\tilde{\psi}\rangle = -e^\mu p_\mu c^\dagger(\vec{p})|\tilde{\psi}\rangle. \tag{47.37}$$

Therefore $Q_B|\psi\rangle = 0 \Leftrightarrow e^\mu p_\mu = 0$, so the state with one extra photon on top of $|\tilde{\psi}\rangle$ needs to be transverse.

In contrast, consider the state

$$Q_B b^\dagger(\vec{p})|\tilde{\psi}\rangle = -b^\dagger(\vec{p})Q_B|\tilde{\psi}\rangle - \frac{1}{\alpha}p^\mu a_\mu^\dagger(\vec{p})|\tilde{\psi}\rangle = -\frac{1}{\alpha}p^\mu a_\mu^\dagger(\vec{p})|\tilde{\psi}\rangle. \tag{47.38}$$

It follows that

$$|\psi(e_\mu + \beta p_\mu)\rangle = |\psi(e_\mu)\rangle + \beta p^\mu a_\mu^\dagger(\vec{p})|\tilde{\psi}\rangle = |\psi(e_\mu)\rangle - \alpha\beta Q_B b^\dagger(\vec{p})|\tilde{\psi}\rangle, \tag{47.39}$$

so we have equivalent states, $e_\mu \sim e_\mu + \beta p_\mu$, as we should, only for massless states $p^\mu p_\mu = 0$.

On the contrary, we also learn that

$$Q_B b^\dagger(\vec{p})|\tilde{\psi}\rangle = -\frac{1}{\alpha}p^\mu a_\mu^\dagger(\vec{p})|\tilde{\psi}\rangle \neq 0, \tag{47.40}$$

so there are no states with ghosts in them (i.e. with $b^\dagger(\vec{p})$) among the physical states, since any such $b^\dagger(\vec{p})$ added to a physical state turns it unphysical (Q_B on it is nonzero).

Finally, we see that

$$c^\dagger(\vec{p})|\psi\rangle = Q_B\left(\frac{-e^\mu a_\mu^\dagger(\vec{p})}{e \cdot p}\right)|\psi\rangle, \tag{47.41}$$

so it is equivalent to zero (the relation above is proven by commuting Q_B on the right-hand side with $a_\mu^\dagger(\vec{p})$ and using $Q_B|\psi\rangle = 0$).

So there are no b ghosts in physical states, and c ghosts are equivalent to zero. Therefore, the physical Hilbert space is composed only of transverse photons, as it should be.

47.4 General Formalism

We now turn to the general formalism for BRST quantization.

As mentioned, we need to define a Q_B and a H_{BRST} that satisfy $[Q_B, H_{BRST}] = 0$ and $Q_B^2 = 0$, and then we construct the Hilbert space as the states of the Q_B-cohomology. Until now $Q_B^2 = 0$ was assumed, but it is actually needed. Indeed, we want that the gauge choice does not change the commutation relation $[Q_B, H_{BRST}] = 0$, so any δH_{BRST} induced by the gauge choice should preserve it, $[Q_B, \delta H_{BRST}] = 0$. But we saw that the gauge choice in the action, and therefore in H_{BRST}, comes from the gauge-fixing fermion term, $\delta H_{BRST} = -\{Q_B, b_A \delta F^A\}$. Then we have

$$0 = [Q_B, \{Q_B, b_A \delta F^A\}] = [Q_B^2, b_A \delta F^A] \Rightarrow Q_B^2 = 0. \tag{47.42}$$

Note that there is the Fradkin–Vilkovitsky theorem, that the partition function

$$Z_\Psi = \int \mathcal{D}(\ldots) e^{iS_{qu}} \tag{47.43}$$

is Ψ-independent.

So we need to construct Q_B and H_{BRST}, which will be done by satisfying the above conditions order by order in the number of fields, but we need to start somewhere, and the starting point will be to define Q_B and H_{BRST} at the first order in the fields. We turn to that next.

47.4.1 Quantum Action

We start by writing the quantum action as

$$S_{qu} = \int dt \left[\dot{q}_i p^i + \dot{\lambda}^\mu \pi_\mu + \dot{\eta}^a p_a - H_{BRST} + \{\psi, Q_B\} \right]. \tag{47.44}$$

Here λ^μ are Lagrange multipliers, and π_μ their conjugate momenta, which are also in general first-class constraints, so the total set of first-class constraints is written as $G^a = (\phi_\alpha, \pi^\mu)$, satisfying the algebra

$$[G_a, G_b] = f_{ab}{}^c G_c. \tag{47.45}$$

We also define the ghost phase space by

$$\eta^a = (c^\alpha, -p(b)_\alpha); \quad p_a = (p(c)_\alpha, -b_\alpha). \tag{47.46}$$

We next define Poisson brackets

$$\{c^\alpha, p(c)_\alpha\} = -\delta_\beta^\alpha; \quad \{b_\alpha, p(b)^b\} = -\delta_\alpha^\beta, \tag{47.47}$$

which are written together as

$$\{\eta^a, p_b\} = -\delta_b^a. \tag{47.48}$$

Observation: in two dimensions, $p(c) = b$ and $p(b) = c$, so the story is simpler.

Then, in general, the first term in the expansion of Q_B is

$$Q_B = \eta^a G_a - \frac{1}{2} \eta^c \eta^b f_{bc}{}^a p_a(-)^c. \tag{47.49}$$

If $f_{bc}{}^a$ are constants (note that the algebra of constraints is an algebra of functions, so in general the structure "constants" can be fields, not actual constants), then the above implies $Q_B^2 = 0$ from the Jacobi identities, which is left as an exercise to prove, and the Hamiltonian in that order is

$$H_{BRST} = H_0 + \ldots \tag{47.50}$$

In general, however, we have

$$Q_B = \eta^{b_1} \ldots \eta^{b_{n+1}} U^{(n)}{}_{b_1 \ldots b_{n+1}}{}^{a_1 \ldots a_n} p_{a_1} \ldots p_{a_n} \qquad (47.51)$$

and

$$H_{BRST} = H_0 + \eta^a V_a^b p_b + \ldots \qquad (47.52)$$

Here the algebra of first-class constraints

$$\{\phi_\alpha, \phi_b\} = f_{\alpha\beta}{}^\gamma \phi_\gamma; \quad \{H_0, \phi_\alpha\} = V_\alpha^\beta \phi_\beta \qquad (47.53)$$

is generalized (extended) to

$$\{G_a, G_b\} = f_{ab}{}^c G_c; \quad \{H_0, G_a\} = V_a^b G_b. \qquad (47.54)$$

The V_a^b define the first correction to the classical Hamiltonian H_0, and the $f_{ab}{}^c$ define the second term in Q_B, since as we saw above, we have

$$U_a^{(0)} = G_a; \quad U^{(1)}{}_{b_1 b_2}{}^{a_1} = -\frac{1}{2} f_{b_1 b_2}{}^{a_1} (-)^{b_2}. \qquad (47.55)$$

Finally, note that all the general BRST formalism described until now is *classical*, quantum mechanics did not enter anywhere until now. Quantization is done by replacing the Dirac brackets with the commutator, and finding the physical Hilbert space as the space of states in the Q_B-cohomology, like we saw in the example of electromagnetism.

47.5 Example of General Formalism: Pure Yang–Mills

We now turn to a more nontrivial example to explain the general BRST formalism (though not the quantization, i.e. finding the Q_B-cohomology, which is similar to the electromagnetism case), the case of (nonabelian) Yang–Mills theory.

The action of pure Yang–Mills in Minkowski space can be written as

$$S = \int d^4x \left[-\frac{1}{4}(F_{\mu\nu}^a)^2 \right] = \int dt d^3x \left[-\frac{1}{4}(F_{ij}^a)^2 + \frac{1}{2}(\dot{A}_i^a - D_i A_0^a)^2 \right], \qquad (47.56)$$

since

$$E_i^a = F_{0i}^a = \partial_0 A_i^a - \partial_i A_0^a + g f^a{}_{bc} A_0^b A_i^c = \dot{A}_i^a - D_i A_0^a. \qquad (47.57)$$

It follows that the momentum conjugate to A_i^a is

$$(\pi_i^a \equiv) p_i^a = \dot{A}_i^a - D_i A_0^a = F_{0i}^a = E_i^a. \qquad (47.58)$$

We also define as usual

$$B_i^a = \frac{1}{2} \epsilon_{ijk} F_{jk}^a. \qquad (47.59)$$

Note that there is no momentum conjugate to A_0 (there is no $\dot{A}_0$ in the action), so the primary constraint of pure Yang–Mills is

$$p_0^a = 0. \qquad (47.60)$$

We can calculate the classical Hamiltonian by partially integrating a D_i acting on A_0^a in the action (47.56) and then writing

$$S = \int dt d^3x [\dot{A}_i^a p_i^a - H_L] = \int dt d^3x [\dot{A}_i^a E_i^a - H_0 + A_0^a (D_i E_i^a)],\qquad (47.61)$$

to obtain

$$H_0 = \int d^3x \left[\frac{1}{2}(p_i^a)^2 + \frac{1}{2}(B_i^a)^2\right] = \int d^3x \left[\frac{1}{2}(E_i^a)^2 + \frac{1}{2}(B_i^a)^2\right],\qquad (47.62)$$

where $(E_i^a)^2$ should be understood as $(p_i^a)^2$ in the Hamiltonian, and $(B_i^a)^2 = (F_{jk}^a)^2/2$. We can also define the *naive Hamiltonian* H_L as above, with

$$H_L = H_0 - \int d^3x A_0^a D_i E_i^a.\qquad (47.63)$$

We see that the time evolution of the primary constraint p_0^a with the naive Hamiltonian:

$$\dot{p}_0^a \approx \{p_0^a, H_L\} = -D_i E_i^a \approx 0,\qquad (47.64)$$

implies that the secondary constraint of the pure Yang–Mills is

$$D_i E_i^a = 0.\qquad (47.65)$$

We can check there are no other secondary constraints, since the time variation of the above vanishes:

$$\{D_i E_i^a, H_L\} = 0.\qquad (47.66)$$

From (47.61) we see that A_0^a appears as a Lagrange multiplier for the secondary constraint $D_i E_i^a$, hence the definition for H_L differing from H_0.

The algebra of constraints is found easily. We have

$$\{p_0^a(x), p_0^b(y)\}_{P.B.} = 0.\qquad (47.67)$$

In contrast,

$$\{p_0^a(x), D_i E_i^b(y)\}_{P.B.} = \{p_0^a(x), \partial_i p_i^b(x) + gf^b{}_{cd} A_i^c(y) p_i^d(y)\}_{P.B.} = 0,\qquad (47.68)$$

and

$$\{D_i E_i^a(x), D_j E_j^b(y)\}_{P.B.} = \{\partial_i p_i^a(x) + gf^a{}_{cd} A_i^c(x) p_i^d(x), \partial_j p_j^b(y) + gf^b{}_{ef} A_j^e(y) p_j^f(y)\}_{P.B.}$$
$$= gf^{ab}{}_c D_i E_i^c(x)\delta^3(x-y).\qquad (47.69)$$

Therefore, the constraints form a closed algebra (i.e. all constraints are first class) and the structure constants are indeed constant, and are $0, 0, gf^{ab}{}_c$. Since all constraints are first class, the Dirac brackets are just the Poisson brackets.

Therefore the constraints and the phase space coordinates are

$$G^a = (D_i E_i^a, p_0^a); \quad \eta_a = (c^a, -p(b)^a); \quad p_a = (p(c)_a, -b_a).\qquad (47.70)$$

Since the structure constants are actually constant, the BRST charge and Hamiltonian are given by (47.49) and (47.50), giving

$$Q_B = \int d^3x \left[c^a(D_i E_i^a) - p(b)^a p_0^a - \frac{1}{2} g c^b c^c f_{cb}{}^a p(c)_a \right],$$

$$H_{BRST} = H_0 = \int d^3x \left[\frac{1}{2}(p_i^a)^2 + \frac{1}{2}(B_i^a)^2 \right]. \tag{47.71}$$

The (first-class) constraints are

$$\phi_1 = p_0^a; \quad \phi_2 = D_i p_i^a. \tag{47.72}$$

The naive Hamiltonian is

$$H_L = H_0 - \int d^3x A_0^a \partial_i p_i^a. \tag{47.73}$$

Since all constraints are first class, $H_E = H_L$. Note that the constraint $p_0^a = 0$ does not appear with Lagrange multiplier, since it can be absorbed in the term $p_0^a \dot{A}_0^a$.

We see that in the Dirac formalism, the classical action is written as

$$S_{cl} = \int dt [\dot{q}_i p^i - H_0 + \lambda^\alpha \phi_\alpha], \tag{47.74}$$

where λ_α are Lagrange multipliers.

Now, in the BRST formalism, the quantum action (47.44) is written as

$$S_{qu} = \int d^4x \left[E_i^a \dot{A}_i^a + p_0^a \dot{A}_0^a - p(c)_a \dot{c}^a - p(b)^a \dot{b}_a - \frac{1}{2} \left\{ (E_i^a)^2 + (B_i^a)^2 \right\} + \{\psi, Q_H\} \right]. \tag{47.75}$$

Here the first two terms are $\dot{q}_i p^i$, the next two are $\dot{\eta}^a p_a$, there are no $\dot{\lambda}^\mu \pi_\mu$ terms, and we have substituted $H_{BRST} = H_0$.

We have described the classical part of the BRST formalism, but as we said, the quantization procedure replaces Dirac brackets (which now equal Poisson brackets) with commutators, and the physical Hilbert space is the space of the Q_B-cohomology. The procedure is more involved, but again we find only transverse gluons.

47.6 Batalin–Vilkovisky Formalism (Field-Antifield)

Finally, a few words about the generalization of the BRST formalism, the BV or field–antifield formalism.

When we have derived the LZJ identities and made a formal renormalization of gauge theories, we have introduced sources K_μ^a and L^a for the nonlinear terms $Q_B A$ and $Q_B c$, namely $S_{\text{extra}} = \int [K_\mu^a (Q_B A)_\mu^a - L^a (Q_B c)^a]$. These sources are called antifields for A_μ^a and c^a. In the BV formalism, one would introduce antifields also for b_a and d_a. As we argued there, for YM it is actually not needed to introduce such antifields, but in more general situations it is.

Then BV is a Lagrangian formalism (where BRST was a Hamiltonian formalism), and a Lorentz-covariant one. Moreover, there is a simplicity in it that stems from the fact that the BRST charge Q_{BRST} equals the antifield-extended quantum action S. Then we have

$$\delta_{BRST}\phi^A = (\phi^A, S\Lambda), \tag{47.76}$$

and instead of $Q_B^2 = 0$ and $[Q_B, H_{BRST}] = 0$, we have the *master equation*

$$(S, S) = 0. \tag{47.77}$$

Important Concepts to Remember

- BRST quantization is based on the Dirac formalism, and extends to the BV formalism.
- The extended Hamiltonian adds to a first-class Hamiltonian all the first-class constraints.
- Dirac brackets are introduced so as to make trivial the second-class constraints $[f, \chi_s] = 0$ strongly for any $f(q, p)$, so that we can impose χ_s on states when quantizing, $\hat{\chi}_s|\psi\rangle = 0$.
- The physical Hilbert space of BRST quantization is the Q_B-cohomology, that is Q_B-closed states, modulo Q_B-exact states.
- We need to construct the operators Q_B and H_{BRST} order by order in the fields, so as to have $Q_B^2 = 0$ and $[Q_B, H_{BRST}] = 0$.
- For electromagnetism, the BRST quantization selects a Hilbert space without b or c ghosts, and with only transverse photons, $e^\mu a_\mu^\dagger(\vec{p})$, with $e^\mu p_\mu = 0$.
- In general, $Q_B = \eta^a G_a - 1/2\eta^c\eta^b f_{bc}{}^a p_a(-)^c + \dots$ and $H_{BRST} = H_0 + \eta^a V_a^b p_b + \dots$
- In YM theory, the primary constraint is $p_0^a = 0$ and the secondary constraint is $D_i E_i^a = 0$, and both are first class. The structure constants of the algebra of constraints are constant, and are $gf_{ab}{}^c$.

Further Reading

See sections 15.7 and 15.8 Weinberg's book [11]. I have used, however, mostly various lecture notes.

Exercises

1. *The bosonic string in Dirac formalism.* Consider the action

$$S_{cl} = \int d\sigma\, dt \left[-\frac{1}{2}\tilde{g}^{\mu\nu}\partial_\mu\vec{X} \cdot \partial_\nu\vec{X} \right], \tag{47.78}$$

where $\tilde{g}^{\mu\nu} = \sqrt{-g}g^{\mu\nu}$ has $\det \tilde{g}^{\mu\nu} = -1$. Then $\tilde{g}^{00} \equiv \lambda^0$ and $\tilde{g}^{01} \equiv \lambda^1$ are independent Lagrange multipliers, as we can check, and $\tilde{g}^{11}\lambda^0 - (\lambda^1)^2 = -1$. Find the primary constraints and write H_L. From the consistency conditions $\{\phi_m, H_T\} = 0$, where ϕ_m are the primary constraints, find the secondary constraints, and the "physical constraints"= linear combinations independent of λ^i. Calculate the algebra of constraints and check that all are first class, and that the "classical Hamiltonian" H_0 vanishes, as in general relativity.

2. Check that $Q_B^2 = 0$ and $[Q_B, H_{BRST}] = 0$ to first order, for the general form of BRST quantization objects

$$Q_B = \eta^a G_a - \frac{1}{2}\eta^c\eta^b f_{bc}{}^a p_a(-)^c,$$

$$H_{BRST} = H_0 + \eta^a V_a^b p_b + \dots, \tag{47.79}$$

where the first-class algebra

$$\{\phi_\alpha, \phi_\beta\} = f_{\alpha\beta}{}^\gamma \phi_\gamma; \quad \{H_0, \phi_\alpha\} = V_\alpha^\beta \phi_\beta \tag{47.80}$$

is extended to

$$\{G_a, G_b\} = f_{ab}{}^c G_c; \quad \{H_0, G_a\} = V_a^b G_b. \tag{47.81}$$

3. Add a complex scalar to the pure Yang–Mills case, in the fundamental representation of the gauge group G. Find the new constraints, and divide them into first-class and second-class. Find Q_B and H_{BRST}.

4. Add a fermion instead of the complex scalar to the pure Yang–Mills case, and repeat Exercise 3 in this case.

48

QCD: Definition, Deep Inelastic Scattering

In this chapter we start the study of quantum chromodynamics. We have already described gauge theories, and their perturbation theory, so there is nothing new to be said there. Instead, what is of interest is the relation to experiments, which involve nonperturbative low-energy states. So, in this chapter we will learn how to combine the perturbative scattering methods studied previously with nonperturbative descriptions for the low-energy states.

After defining QCD, we will analyze the parton model and deep inelastic scattering, which is where QCD was most studied. Finally, we consider (for completeness) deep inelastic neutrino scattering, and a few ideas on hard scattering.

48.1 QCD: Definition

QCD refers to the theory of quarks and gluons. The YM gauge group is $SU(3)_c$ and there are six flavors of quarks, divided into three families (generations). The quark model was introduced by Gell-Mann and Zweig. The quarks are

$$\begin{pmatrix} u \\ d \end{pmatrix} \begin{pmatrix} c \\ s \end{pmatrix} \begin{pmatrix} t \\ b \end{pmatrix}, \tag{48.1}$$

and the elements on the top line have electric charge $Q_e = +2/3$, whereas the elements on the bottom line have $Q_e = -1/3$.

We know that there are three generations both from particle accelerator experiments (from the decay of the Z^0 for instance) and from cosmology constraints. The three families of quarks go together with three families of leptons to make the three generations of fermions:

$$\begin{pmatrix} e^- \\ \nu_e \end{pmatrix} \begin{pmatrix} \mu^- \\ \nu_\mu \end{pmatrix} \begin{pmatrix} \tau^- \\ \nu_\tau \end{pmatrix}, \tag{48.2}$$

where now the electric charge on the top line is -1, and on the bottom line is zero.

Low-energy states are gauge invariant, since the theory is confining at low energies. This means that there is a linear potential between free quarks (objects with color charge), $V \sim \sigma l$, that prevents one from breaking pairs apart. Only uncharged (gauge-invariant) objects are exempt from this potential.

The QCD states are called hadrons in general, and divide into mesons and baryons.

- Mesons are $\bar{q}q$ pairs, for instance the lightest objects of QCD, the pions, $\pi^{\pm}$ and π^{0}.
- Baryons are objects with three quarks, qqq, for instance the proton, $p = (uud)$ and the neutron, $n = (ddu)$.

In terms of color indices, the mesons are $\bar{q}^i q_i = \bar{q}^i q_j \delta^j_i$, where $i, j = 1, 2, 3$ $(1, 2, \ldots, N)$, so are constructed with the group-invariant δ^j_i that relates the fundamental $\mathbf{N}$ representation and the antifundamental $\bar{\mathbf{N}}$ representation. The baryons are $\epsilon^{ijk} q_i q_j q_k$ objects, constructed with the invariant ϵ^{ijk} of $SU(3)_c$ in the fundamental representation.

At low energies, there is also an approximate global symmetry of the Lagrangian called isospin, which acts on (ud) by

$$\begin{pmatrix} u \\ d \end{pmatrix} \to U \begin{pmatrix} u \\ d \end{pmatrix}, \tag{48.3}$$

so exchanges u and d, and correspondingly p with n.

48.2 Deep Inelastic Scattering

We begin with the most common example of scattering involving hadrons, deep inelastic scattering, see Figure 48.1. This is the scattering of an electron e^- off a proton p. We can ask, is it not simpler to start with two protons, for instance? But the proton has size and structure, which we want in fact to model, hence scattering two hadrons will make the problem harder. Instead we use the pointlike electron to probe the structure of the proton, by breaking it.

Therefore the scattering that we study is $e^- p \to e^- X$, where we have used p as a hadron, and after the collision we will have in general more than one hadron, a state that we called X.

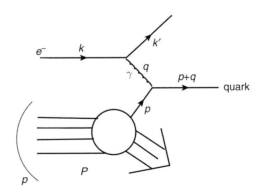

Figure 48.1   Deep inelastic scattering in QCD.

48.2.1 Parton Model

For the nonperturbative proton, Bjorken and Feynman proposed the parton model, namely that the proton is made up of constituents called *partons*. We also assume that the partons cannot exchange large transverse momentum (i.e. large q^2) through strong interactions (gluon exchange). This is valid only to leading order $\mathcal{O}(\alpha_s)$. Note that in the quark model, the proton is *classically* made up of *uud* quarks, but the parton model is more than that. It says that *quantum mechanically*, for the strong coupling state at low energy, the proton can be thought of as a combination of various quarks and gluons (real and virtual), with the total (overall) quantum numbers of the (*uud*) combination.

Then we consider an electron e^- with momentum k emitting a photon γ with momentum q and resulting in a momentum k' for the electron. Then the photon "extracts" a parton f with momentum p from the proton with overall momentum P and interacts with it, turning it into a parton with momentum $p' = p+q$. Since we work at leading order in α_s, we ignore gluon emission and exchange, so the parton will actually be a quark.

We will work in the center of mass frame and at total energies $\sqrt{s}$ much greater than the proton mass, which means that the proton is ultrarelativistic (i.e. almost lightlike). The parton constituents of the proton are assumed to have a finite fraction of the proton momentum, so they will also be almost lightlike, and almost collinear with the proton because the large p_T can come only from hard gluon exchange, which as we said is suppressed by α_s, so can be ignored at leading order.

Then the momentum of the parton constituent is written as

$$p = \xi P, \tag{48.4}$$

where ξ is called the longitudinal (momentum) fraction of the constituent.

Then the cross-section for the total process is written as

$$\sigma[e^- p \to e^- X] = \int \sum_f P_f(\xi) \sigma_{e^- q_f \text{ scatt.}}(\xi), \tag{48.5}$$

where the probability of finding a quark f at momentum fraction ξ is infinitesimal, and given by

$$P_f(\xi) = d\xi \, f_f(\xi), \tag{48.6}$$

where $f_f(\xi)$ is the *parton distribution function* for the parton f. Here the partons (constituents of the proton) are $(q, \bar{q}, g)$. The parton distribution functions cannot be computed in perturbation theory, and have to be determined from experiments. Since ξ is a momentum fraction, it takes values in $\xi \in [0, 1]$.

Then we have

$$\sigma[e^-(k)p(P) \to e^-(k')+X] = \int_0^1 d\xi \sum_f f_f(\xi)\sigma[e^-(k)q_f(\xi P) \to e^-(k')+q_f(p')]. \tag{48.7}$$

To describe the scattering, one of the parameters we use is the quantity $Q^2 \equiv -q^2$. The Mandelstam variables for the basic scattering process $e^- q \to e^- q$ are written with hats. In particular:

$$\hat{t} = -Q^2, \tag{48.8}$$

and the other independent invariant is

$$\hat{s} = (p + k)^2 = 2p \cdot k = 2\xi P \cdot k = \xi s, \tag{48.9}$$

where we have used the fact that we are in the ultrarelativistic regime for the electron, parton, and proton, so $p^2 = k^2 = P^2 \simeq 0$. Here, s is the Mandelstam invariant for the total process, $e^- p \to e^- X$. Finally, since in general $s + t + u = \sum_i m_i^2$, we have now

$$\hat{s} + \hat{t} + \hat{u} = 0. \tag{48.10}$$

Then we also have (since $(p + q)^2 \simeq 0$ for the final parton being ultrarelativistic)

$$0 \simeq (p + q)^2 = 2p \cdot q + q^2 = 2\xi P \cdot q - Q^2, \tag{48.11}$$

which means that

$$\xi = x \equiv \frac{Q^2}{2P \cdot q}. \tag{48.12}$$

For the basic scattering process, $e^- q \to e^- q$, the formula has been derived in Part I, so we will not repeat it here. The formula for the spin-averaged amplitude squared was (see (26.29) in Part I)

$$\frac{1}{4} \sum_{spins} |\mathcal{M}|^2 = \frac{8e^4 Q_f^2}{\hat{t}^2} \left(\frac{\hat{s}^2 + \hat{u}^2}{4} \right), \tag{48.13}$$

where Q_f is the electric charge of parton (quark) f.

Then, since $\alpha = e^2/4\pi$ and $\hat{u}^2 = (\hat{s} + \hat{t})^2$, we obtain for the relativistically invariant differential cross-section for the basic process

$$\frac{d\sigma[e^- q_f \to e^- q_f]}{d\hat{t}} = \frac{\frac{1}{4} \sum_{spins} |\mathcal{M}|^2}{16\pi \hat{s}^2} = \frac{2\pi \alpha^2 Q_f^2}{\hat{s}^2} \left(\frac{\hat{s}^2 + (\hat{s} + \hat{t})^2}{\hat{t}^2} \right). \tag{48.14}$$

For the total process, replacing $\hat{t}$ by $-Q^2$ and $\hat{s}$ by ξs, we have

$$\frac{d\sigma}{dQ^2} = \int_0^1 d\xi \sum_f f_f(\xi) \frac{2\pi \alpha^2 Q_f^2}{Q^4} \left[1 + \left(1 - \frac{Q^2}{\xi s} \right)^2 \right] \theta(\xi s - Q^2). \tag{48.15}$$

Note that we have introduced a Heaviside function $\theta(\xi s - Q^2)$, since we need to have $\hat{s} \geq |\hat{t}|$. Normally, that would be just a constraint on external momenta, which could be put there explicitly with a θ or not, but now there are momentum fractions integrated over, so it needs to be put in explicitly.

We can also consider the second derivative of σ, taking into account that $\xi = x$, with respect to this x, obtaining

$$\frac{d^2\sigma}{dx \, dQ^2} = \left(\sum_f f_f(x) Q_f^2 \right) \frac{2\pi \alpha^2}{Q^4} \left[1 + \left(1 - \frac{Q^2}{xs} \right)^2 \right]. \tag{48.16}$$

Note that now, being a differential formula with respect to x, there is no need to put explicitly the Heaviside function, though one could.

Finally then, we obtain *Bjorken scaling*, the scaling relation that says

$$\frac{d^2\sigma}{dx\,dQ^2}\frac{Q^4}{1+\left(1-\frac{Q^2}{xs}\right)^2} = \left(\sum_f f_f(x)Q_f^2\right)2\pi\alpha^2 \tag{48.17}$$

is independent of Q^2, and only depends on x.

Qualitatively, the scaling says that the structure of the proton looks the same to an electromagnetic probe, no matter how hard the proton is struck (how large Q^2 is). This is verified experimentally very well, but it is true *only to first order in* α_s.

Another useful variable that can be defined is

$$y \equiv \frac{2P\cdot q}{2P\cdot k} = \frac{2P\cdot q}{s}. \tag{48.18}$$

This can also be rewritten in terms of the Mandelstam variables of the basic scattering, since $(kp) \to (k'p')$ means that $\hat{s} = (p+k)^2 = 2p\cdot k$ and $\hat{u} = (p-k')^2 = -2p\cdot k'$, so

$$y = \frac{2\xi p\cdot(k-k')}{2\xi p\cdot k} = \frac{\hat{s}+\hat{u}}{\hat{s}}. \tag{48.19}$$

But since $|\hat{u}| \le \hat{s}$, we have

$$\frac{\hat{u}}{\hat{s}} = -(1-y) \Rightarrow y \le 1. \tag{48.20}$$

From $y = 2P\cdot q/s$ and $x = Q^2/2P\cdot q$, it follows that

$$xys = Q^2, \tag{48.21}$$

which also implies

$$d\xi\,dQ^2 = dx\,dQ^2 = \frac{dQ^2}{dy}dx\,dy = xs\,dx\,dy. \tag{48.22}$$

Finally then, we have for deep inelastic scattering (DIS):

$$\frac{d^2\sigma[e^-p \to e^-X]}{dx\,dy} = \left(\sum_f xf_f(x)Q_f^2\right)\frac{2\pi\alpha^2 s}{Q^4}[1+(1-y)^2]. \tag{48.23}$$

This means that

$$\frac{d^2\sigma}{dx\,dy}Q^4 \tag{48.24}$$

factorizes into two factors, one depending only on x (the Bjorken scaling factor) and one depending only on y, $[1+(1-y)^2]$, which gives the *Callan–Gross relation* for scattering of an e^- off a massless fermion (indeed, we saw that this factor originated in the calculation of the spin-averaged $|\mathcal{M}|^2$ for the e^- to scatter off a massless fermion).

One more thing to note here is that the particular form of the final hadronic states X, that come from the remnant of the original proton and the "jet" that will form out of the final quark q_f, are not part of the calculation. Of course, we observe only hadronic final states, but the effect of how these final states turn into the observed hadrons, "hadronization," does not influence too much the cross-section, and moreover, there are nonperturbative (lattice, etc.) methods to calculate these effects, and treat them as a "black box."

48.3 Deep Inelastic Neutrino Scattering

We have analyzed in DIS the effect of electromagnetic probes for the proton, but we can also consider weak interaction probes, meaning consider a W exchange instead of the γ exchange, hence the probe to be considered is a neutrino. For definiteness, we consider a ν_μ.

The weak interaction, the remnant of the broken electroweak interaction, which will not be described here, is to exchange the massive $W^\pm$ vector particles. It couples to the weak doublets (i.e. quark pairs like (ud) and lepton pairs like $(e\nu_e)$ and $(\mu\nu_\mu)$), and it turns one element of the doublet into the other.

Hence the basic interaction we consider is as follows. A ν_μ comes and emits a W^+, thus turning into a μ^-. The W^+ can now interact with a d quark parton inside the proton and turn it into a u. But remembering that the partons are not only the classical quarks, the W^+ can also interact with a $\bar{u}$ and turn it into a $\bar{d}$.

The weak interaction through W^+ exchange can be considered at low energies $E \ll m_W$ as a *4-fermi interaction*. Fermi had introduced the 4-fermi interaction as an effective model with a vertex $G_F/\sqrt{2}$ coupling four fermions (two $\bar\psi\psi$ lines), but it was later realized that this effective vertex comes from the approximation of the real quantum process, the exchange of the massive vector boson W^+. Indeed, then, in the actual diagram in Figure 48.2, we would have a massive vector propagator with momentum q coupling to two fermion lines:

$$g(\dots)^\mu \frac{\delta_{\mu\nu}}{q^2 + m_W^2} g(\dots)^\nu \simeq \frac{g^2}{m_W^2}(\dots)^\mu(\dots)_\mu, \tag{48.25}$$

where g is the coupling of the W with the fermions, and the approximation is for $q^2 \ll m_W^2$, since $q^2 = (k-k')^2$ and $|k|, |k'| \ll m_W$. This means that we can consider an effective *Fermi coupling* of

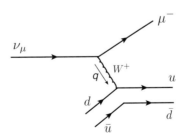

Figure 48.2   Deep inelastic ν scattering.

$$\frac{G_F}{\sqrt{2}} \equiv \frac{g^2}{8m_W^2}. \tag{48.26}$$

Then, similarly to the DIS case before, we find

$$\frac{d^2\sigma}{dx\,dy}(\nu p \to \mu^- X) = \frac{G_F^2 s}{\pi}[xf_d(x) + xf_{\bar{u}}(x)(1-y)^2]. \tag{48.27}$$

The proof is left as an exercise. We also find

$$\frac{d^2\sigma}{dx\,dy}(\bar{\nu} p \to \mu^+ X) = \frac{G_F^2 s}{\pi}[xf_u(x)(1-y)^2 + xf_{\bar{d}}(x)]. \tag{48.28}$$

48.4 Normalization of the Parton Distribution Functions

As we mentioned, classically, the proton is made up of (uud) quarks, but quantum mechanically, we expect a strongly coupled mix of states, so we will create many $q\bar{q}$ pairs, as well as gluons, inside this proton. But the overall quantum numbers of the state still have to be the numbers of the (uud) state, so this means in particular that we need to have two more u quarks than $\bar{u}$ quarks, and one more d quark than $\bar{d}$ quarks. This translates into the conditions

$$\int_0^1 dx[f_u(x) - f_{\bar{u}}(x)] = 2,$$

$$\int_0^1 dx[f_d(x) - f_{\bar{d}}(x)] = 1. \tag{48.29}$$

Of course, there can also be trace parts of other quarks, but they are negligible. The next important one is the s quark, so we need the same number of s and $\bar{s}$ quarks:

$$\int_0^1 d\xi[f_s(x) - f_{\bar{s}}(x)] = 0, \tag{48.30}$$

and so on.

A similar story holds for the other hadrons. For instance, for the neutron, we just exchange u and d from the proton, so

$$f_u^{(n)}(x) = f_d(x); \quad f_d^{(n)}(x) = f_u(x); \quad f_{\bar{u}}^{(n)}(x) = f_{\bar{d}}(x), \tag{48.31}$$

and so on.

As another example, for the antiproton, we just interchange the quarks with the antiquarks, so

$$f_u^{(\bar{p})}(x) = f_{\bar{u}}(x), \quad f_{\bar{u}}^{(\bar{p})}(x) = f_u(x), \tag{48.32}$$

and so on.

Finally, the last normalization constraint comes from the fact that the total momentum of the proton is P, so the total fraction of all the constituent partons is 1, that is

$$\int_0^1 dx \, x[f_u(x) + f_d(x) + f_{\bar{u}}(x) + f_{\bar{d}}(x) + f_g(x)] = 1. \tag{48.33}$$

Note that we could have introduced also $f_s(x), f_{\bar{s}}(x)$, and so on, but as we said, these are negligible.

The parton distribution functions are determined experimentally, and must obey the above normalization conditions. Among the many QCD experiments, one uses some to fix the distribution functions, and then uses them to predict the other cross-sections.

48.5 Hard Scattering Processes in Hadron Collisions

Finally, a few words about hard scattering, for completeness. We can now treat the next complicated case, the case of hadron–hadron scattering, as suggested at the beginning of the chapter. The basic process in leading order will be some $q_f \bar{q}_f \to Y$ process, for instance occurring electromagnetically, through an intermediate virtual photon. A parton q_f from one proton will break off and interact with the parton $\bar{q}_f$ from the second proton. All in all, we have the total process

$$\sigma(p(P_1) + p(P_2) \to Y + X) = \int_0^1 dx_1 \int_0^1 dx_2 \sum_f f_f(x_1) f_{\bar{f}}(x_2) \sigma(q_f(x_1 P_1) + \bar{q}_f(x_2 P_2) \to Y),$$

$$\tag{48.34}$$

where we have as usual that the remnants of the proton will hadronize to some state X.

Important Concepts to Remember

- QCD is YM with the color group $SU(3)_c$ and six flavors, organized in three families.
- Physical low-energy states are gauge invariant due to confinement. Mesons are $\bar{q}^i q_i$ and baryons are $\epsilon^{ijk} q_i q_j q_k$.
- In the parton model, the proton is composed at the quantum level of partons, $(q, \bar{q}, g)$.
- DIS is an electron scattering off a hadron, usually a proton, breaking out a parton from it, and interacting with it.
- The cross-sections for the total process are integrals of the cross-sections for the basic process involving the parton, with the distribution functions for momentum fraction ξ of the parton inside the proton, $p = \xi P$.
- Bjorken scaling gives the independence of some quantity on Q^2, and only dependence on $x = Q^2/2P \cdot q$, saying that the structure of the proton looks the same to an electromagnetic probe, no matter how hard the proton is struck (how high Q^2 is).

- Deep inelastic neutrino scattering occurs through W^+ exchange, which reduces to 4-fermi interaction at low energies, and turns a d into a u.
- The parton distribution functions must obey the normalization conditions, and are determined experimentally.
- Hard scattering processes (collisions of two hadrons) involve two parton distribution functions, one for each hadron.

Further Reading

See sections 17.1 and 17.3 in [1].

Exercises

1. Fill in the details of the calculation of $d^2\sigma/dx\,dy(\nu p \to \mu^- X)$.
2. Consider the $e^- p \to e^- X$ scattering, and assume that

$$f_u(x) = \frac{3}{2}f_d(x) = f_g(x) = \frac{1-x}{a(x+\epsilon)} \tag{48.35}$$

and

$$f_{\bar{u}}(x) = f_{\bar{d}}(x) = \frac{1-x}{3a(x+\epsilon)}, \tag{48.36}$$

where a and ϵ are constants and $\epsilon \ll 1$, and all possible Q_f^2 are given by their standard values ($Q_f(u) = 2/3$, etc.). From Bjorken scaling, calculate the cross-section $\sigma(e^- p \to e^- X)$.

3. Write down the explicit form of Bjorken scaling if the parton distribution functions are

$$f_u = f_g = \frac{1-x}{a(x+\epsilon)}\,, \quad f_d = \frac{2}{3}f_u\,, \quad f_{\bar{u}} = f_{\bar{d}} = \frac{1}{3}f_u, \tag{48.37}$$

and calculate a and ϵ.

4. Write down the full tree amplitude for the $\nu p \to \mu^- X$ (without the contraction to 4-fermi interaction).

49

Parton Evolution and Altarelli–Parisi Equation

In Chapter 48 we saw that we can describe the deep inelastic scattering, the collision of an electron off a hadron, via parton distribution functions $f_f(x)$, that were found to be independent of Q^2, and give Bjorken scaling. But we mentioned that this was true only to leading order in α_s. In this chapter we will consider processes subleading in α_s that will lead to violations of Bjorken scaling, through Q^2 dependence of the parton distribution functions.

The dependence on Q^2, and the subsequent *evolution* of $f_f(x)$, will be due to processes with emission of collinear quarks and gluons, also responsible for IR divergences. Therefore, the parton evolution will be linked with the regularization of IR divergences, which will be addressed in Chapter 51. Here, however, the approach will be a more practical one, so we will not deal with the formal issue of IR divergences.

We will not start with QCD, but rather with the simpler case of QED, and then see that we can import almost all the calculation to the QCD case with minimal effort.

49.1 QED Process

The process that most interests us is that of photon emission from a fermion (electron) line, as in Figure 49.1. The initial electron has momentum p, the final one momentum k, and the emitted photon momentum q. We call the momentum (energy) fraction carried by the photon z:

$$z \equiv \frac{E_\gamma}{E_{e,in}}. \tag{49.1}$$

The initial electron is ultrarelativistic (i.e. almost massless), and we choose it to be in the three direction, so

$$p = (p, 0, 0, p). \tag{49.2}$$

The kinematics of three-point scattering mean that one of the momenta of the emitted particles has to be off-shell in the presence of transverse momentum $\vec{p}_\perp$, which is small (almost collinear emission) $p_\perp \ll p$. If the photon is massless, $q^2 = 0$, to leading order in $p_\perp/p$ we have

$$q \simeq \left(zp, \vec{p}_\perp, zp - \frac{p_\perp^2}{2zp} \right), \tag{49.3}$$

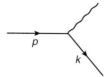

Figure 49.1   Emission of a photon off an electron line.

and then the final electron, which is also ultrarelativistic, must however be off-shell (virtual), since momentum conservation implies

$$k \simeq \left((1-z)p, -\vec{p}_\perp, (1-z)p + \frac{p_\perp^2}{2zp} \right). \tag{49.4}$$

Then we have

$$k^2 \simeq \frac{p_\perp^2}{z} \neq 0. \tag{49.5}$$

One can also consider an on-shell final electron and a virtual photon, with

$$k \simeq \left((1-z)p, -\vec{p}_\perp, (1-z)p - \frac{p_\perp^2}{2(1-z)p} \right),$$

$$q \simeq \left(zp, \vec{p}_\perp, zp + \frac{p_\perp^2}{2(1-z)p} \right), \tag{49.6}$$

which leads to $k^2 \simeq 0$ and

$$q^2 \simeq \frac{p_\perp^2}{(1-z)}. \tag{49.7}$$

We will not do the calculation here, but it is easy to calculate the amplitude, and find $|\mathcal{M}|^2$ averaged over initial polarizations and summed over final polarizations. One obtains

$$\frac{1}{2} \sum_{\text{pol.}} |\mathcal{M}|^2 = \frac{2e^2 p_\perp^2}{z(1-z)} \left[\frac{1 + (1-z)^2}{z} \right]. \tag{49.8}$$

49.2 Equivalent Photon Approximation

Now we can turn to the calculation we are interested in. We want to study QED corrections to the DIS process of an electron scattering of a hadron X to give another electron and hadron(s) Y, through interaction with an intermediate (virtual) photon γ, as in Figure 49.2. We can divide the process into the emission of a photon from the electron, $e^- \to e^- \gamma$, followed by the scattering $\gamma X \to Y$. For the total amplitude we find

$$\mathcal{M}^{\text{tot}} = \mathcal{M}^\mu \frac{\delta_{\mu\nu}}{q^2} \mathcal{M}^\nu_{\gamma X}, \tag{49.9}$$

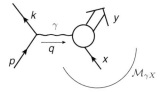

Figure 49.2 Equivalent photon approximation.

and doing the sum over the photon polarizations we find

$$\frac{1}{2}\sum_{\text{pol}}|\mathcal{M}^{\text{tot}}|^2 = \frac{1}{2}\sum_{\text{pol}}|\mathcal{M}|^2 \frac{1}{(q^2)^2}|\mathcal{M}_{\gamma X}|^2. \tag{49.10}$$

The formula for the cross-section $A + B \rightarrow \sum f$ is

$$\sigma = \frac{1}{|\vec{v}_A - \vec{v}_B|2E_A 2E_B} \int \prod_f d\Pi_f |\mathcal{M}^{\text{tot}}|^2, \tag{49.11}$$

where $d\Pi_f$ is the phase space for the final state product f. In our case, $B = X$ is the (possibly nonrelativistic) hadron with velocity v_X and energy E_X and $A = e^-$, with $E_A = p$ and $v_A = 1$ in the opposite direction to X, so

$$\sigma = \frac{1}{(1 + v_X)2p2E_X} \int \frac{d^3k}{(2\pi)^3 2k^0} \int d\Pi_Y \left[\frac{1}{2}\sum_{\text{pol}}|\mathcal{M}|^2 \right] \frac{1}{(q^2)^2}|\mathcal{M}_{\gamma X \rightarrow Y}|^2. \tag{49.12}$$

Given that the energy of the photon is zp, we can form the cross-section for $\gamma X \rightarrow Y$. Also using the fact that $k^0 = (1 - z)p$, $(q^2)^2 = p_\perp^4/(1 - z)^2$, and $d^3k = dk^0 d^2\vec{p}_\perp = pdz\pi dp_\perp^2$, we find

$$\sigma = \int \frac{pdzdp_\perp^2}{16\pi^2(1 - z)p} \left[\frac{1}{2}\sum_{\text{pol}}|\mathcal{M}|^2 \right] \frac{(1 - z)^2}{p_\perp^4} \frac{z}{(1 + v_X)2zp2E_X} \int d\Pi_y |\mathcal{M}_{\gamma X \rightarrow Y}|^2$$

$$= \int_0^1 dz \int \frac{dp_\perp^2}{p_\perp^2} \frac{\alpha}{2\pi} \left[\frac{1 + (1 - z)^2}{z} \right] \sigma(\gamma X \rightarrow Y). \tag{49.13}$$

It remains to consider the region of integration of $p_\perp^2$. $p_\perp^2$ cannot be smaller than m^2, which cuts off the potential IR divergence, and cannot be larger than the total energy squared, s, so we have $\int_{m^2}^s dp_\perp^2$. Finally, we obtain

$$\sigma(e^- X \rightarrow e^- Y) = \int_0^1 dz \frac{\alpha}{2\pi} \log \frac{s}{m^2} \left[\frac{1 + (1 - z)^2}{z} \right] \sigma(\gamma X \rightarrow Y). \tag{49.14}$$

This is the *Weizsacker–Williams equivalent photon approximation*.

We see that the formula is of the same type as in the parton model case, with the cross-section for scattering of γ integrated with a probability to find a γ inside the electron, or photon distribution function

$$f_\gamma(z) = \frac{\alpha}{2\pi} \log \frac{s}{m^2} \left[\frac{1 + (1-z)^2}{z} \right],$$ (49.15)

and the total cross-section

$$\sigma(e^- X \to Y) = \int_0^1 dz f_\gamma(z) \sigma(\gamma X \to Y).$$ (49.16)

49.3 Electron Distribution

We can now consider the case of photon emission (i.e. $e^- X \to \gamma Y$ scattering) proceeding through an intermediate e^-, so the γ is emitted from the e^-, and then we scatter $e^- X \to Y$ (Figure 49.3).

In the same way as before, we now obtain

$$\frac{1}{2} \sum_{\text{pol}} |\mathcal{M}^{\text{tot}}|^2 = \frac{1}{2} \sum_{\text{pol}} |\mathcal{M}|^2 \frac{1}{(k^2)^2} |\mathcal{M}_{e^- X \to Y}|^2,$$ (49.17)

so the total cross-section is

$$\sigma(e^- X \to \gamma Y) = \frac{1}{(1 + v_X) 2p 2E_X} \int \frac{d^3 q}{(2\pi)^3 2q^0} \int d\Pi_y \left[\frac{1}{2} \sum_{\text{pol}} |\mathcal{M}|^2 \right] \frac{1}{(k^2)^2} |\mathcal{M}_{e^- X \to Y}|^2$$

$$= \int \frac{dz dp_\perp^2}{16\pi^2 z} \left[\frac{1}{2} \sum_{\text{pol}} |\mathcal{M}|^2 \right] \frac{z^2}{p_\perp^4} (1 - z) \sigma(e^- X \to Y)$$

$$= \int_0^1 dz \int_{m^2}^s \frac{dp_\perp^2}{p_\perp^2} \frac{\alpha}{2\pi} \left[\frac{1 + (1-z)^2}{z} \right] \sigma(e^- X \to Y),$$ (49.18)

where we have used the fact that $q^0 = zp$, $d^3 q = p dz\pi dp_\perp^2$, and $(k^2)^2 = p_\perp^4/z^2$. Again we can interpret this as in the parton model, through an electron distribution function $f_e^{(1)}(x)$ at momentum fraction $x = 1 - z$, with

$$f_e^{(1)}(x) = \frac{\alpha}{2\pi} \log \frac{s}{m^2} \left[\frac{1 + x^2}{1 - x} \right],$$ (49.19)

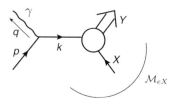

Figure 49.3   Electron distribution process.

and

$$\sigma(e^-X \to \gamma Y) = \int_0^1 dx f_e^{(1)}(x) \sigma(e^-X \to Y). \tag{49.20}$$

However, this is interpreted as a probability of finding an electron with momentum fraction x inside the electron, so we need to consider also the zeroth-order process, where we already have the original electron, that is we need to include

$$f_e^{(0)}(x) = \delta(1-x). \tag{49.21}$$

This is still not enough however, since creating an electron with momentum fraction x means we need to subtract one from momentum fraction $x = 1$ (the initial electron), so we need to subtract a term proportional to $\delta(1-x)$, giving the normalization of $f_e^{(1)}(x)$. But for that, we need to replace $1/(1-x)$, which is divergent under the integral, with a well-behaved distribution $1/(1-x)_+$, defined by

$$\int_0^1 dx \frac{f(x)}{(1-x)_+} = \int_0^1 dx \frac{f(x) - f(1)}{1-x}. \tag{49.22}$$

Then the term we have in $f_e^{(1)}(x)$ means we must consider

$$\int_0^1 dx \frac{1+x^2}{(1-x)_+} = \int_0^1 \frac{x^2-1}{1-x} = -\int_0^1 dx (1+x) = -\frac{3}{2}, \tag{49.23}$$

and subtract this normalization, multiplied by $\delta(1-x)$, to obtain finally

$$f_e(x) = \delta(1-x) + \frac{\alpha}{2\pi} \log \frac{s}{m^2} \left[\frac{1+x^2}{(1-x)_+} + \frac{3}{2}\delta(1-x) \right]. \tag{49.24}$$

49.4 Multiple Splittings

We now need to generalize the relations (49.15) and (49.24) for $f_\gamma(z)$ and $f_e(x)$ to the case of multiple splittings in Figure 49.4. Consider that we have two photons being emitted out of the electron line, with $p_{2\perp} \ll p_{1\perp}$, so the contribution is of order

$$\left(\frac{\alpha}{2\pi}\right)^2 \int_{m^2}^s \frac{dp_{1\perp}^2}{p_{1\perp}^2} \int_{m^2}^{p_{1\perp}^2} \frac{dp_{2\perp}^2}{p_{2\perp}^2} = \frac{1}{2}\left(\frac{\alpha}{2\pi}\right)^2 \log^2 \frac{s}{m^2}. \tag{49.25}$$

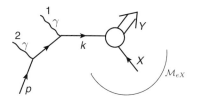

Figure 49.4 Multiple splittings from the electron line.

Note that the integral is of the type

$$\int_a^b \frac{dx}{x} \log \frac{x}{a} = \log b \log \frac{b}{a} - \int_a^b \frac{dx}{x} \log x$$

$$= \log^2 \frac{b}{a} - \int_a^b \frac{dx}{x} \log \frac{x}{a} \Rightarrow \int_a^b \frac{dx}{x} \log \frac{x}{a} = \frac{1}{2} \log^2 \frac{b}{a}. \qquad (49.26)$$

We can generalize this to the case of multiple splittings, with $p_{1\perp} \gg p_{2\perp} \gg \ldots \gg p_{n\perp}$, which gives in a similar way a contribution of

$$\frac{1}{n!} \left(\frac{\alpha}{2\pi} \right)^n \log^n \frac{s}{m^2}. \qquad (49.27)$$

This means that we can consider the splittings as independent from each other, so we can consider a continuous process of splittings. We define distribution functions $f_\gamma(x, Q)$ and $f_e(x, Q)$ for $p_\perp < Q$. Then we consider increasing Q to $Q + \Delta Q$, which means that now the electrons can emit photons γ with $Q < p_\perp < Q + \Delta Q$. The probability of a constituent e^- to emit γs with a momentum fraction z is then obtained by differentiating (49.15). We get

$$\frac{dP}{dz} = \frac{\alpha}{2\pi} \frac{dp_\perp^2}{p_\perp^2} \frac{1 + (1 - z)^2}{z}. \qquad (49.28)$$

But the constituent electron (parton) has a distribution function $f_e(x, p_\perp)$, so all in all we get

$$f_\gamma(x, Q + \Delta Q) = f_\gamma(x, Q) + \int_0^1 dx' \int_0^1 dz \left[\frac{\alpha}{2\pi} \frac{\Delta Q^2}{Q^2} \frac{1 + (1 - z)^2}{z} \right] f_e(x', p_\perp) \delta(x - zx')$$

$$= f_\gamma(x, Q) + \frac{\Delta Q}{Q} \int_x^1 \frac{dz}{z} \left[\frac{\alpha}{\pi} \frac{1 + (1 - z)^2}{z} \right] f_e\left(\frac{x}{z}, p_\perp \right), \qquad (49.29)$$

where in the first line we considered the fact that the momentum fraction x of the photon is the fraction z of the splitting times the original momentum fraction x' of the constituent electron, and in the second line we did the x' integral using $\delta(x - zx') = 1/z\delta(x' - x/z)$. We also used the fact that $1 \geq x' = x/z$, so $1 \geq z \geq x$.

We can then go to the continuum, and write the relation (for ΔQ infinitesimal)

$$\frac{d}{d \log Q} f_\gamma(x, Q) = \int_x^1 \frac{dz}{z} \left[\frac{\alpha}{\pi} \frac{1 + (1 - z)^2}{z} \right] f_e\left(\frac{x}{z}, Q \right). \qquad (49.30)$$

We can use the same logic for $f_e(x)$ and, calculating the probability of having electrons of momentum fraction x appear as a result of γ emission, it should come from (49.24) (minus the trivial delta function), times the electron distribution function itself, so

$$\frac{d}{d \log Q} f_e(x, Q) = \int_x^1 \frac{dz}{z} \left[\frac{\alpha}{\pi} \left(\frac{1 + z^2}{(1 - z)_+} + \frac{3}{2}\delta(1 - z) \right) \right] f_e\left(\frac{x}{z}, Q \right). \qquad (49.31)$$

49.4.1 Boundary Conditions

We have obtained a set of differential equations. To find a solution from them, we must give a boundary condition. The natural boundary condition is that there is only one electron, and nothing else, so $f_e = \delta(1-x)$ and $f_\gamma = 0$. Since the distribution functions f_γ and f_e in (49.15) and (49.24) have $\log s/m^2$, this means that we should define this boundary condition at $Q^2 = m^2$, so

$$f_e(x, Q)|_{Q^2=m^2} = \delta(1-x); \quad f_\gamma(x, Q)|_{Q^2=m^2} = 0. \tag{49.32}$$

With this boundary condition and the differential equations that we found, we obtain $f_e(x, Q)$ and $f_\gamma(x, Q)$, and then the cross-section for *multiple splittings* (viewed as a continuous process; note then that this does not include all possible terms, but merely resums a set of diagrams) is

$$\sigma(e^- X \to e^- + n\gamma + Y) = \int_0^1 dx f_\gamma(x, Q)\sigma(\gamma X \to Y),$$

$$\sigma(e^- X \to n\gamma + Y) = \int_0^1 dx f_e(x, Q)\sigma(e^- X \to Y). \tag{49.33}$$

49.4.2 Photon Splitting into Pairs

There is one more process that we need to consider, which is a crossed diagram for that for γ emission from an electron line, namely e^+e^- pair creation from the photon. For our DIS process, the relevant diagram is where the incoming electron emits a photon, but the photon turns into a e^+e^- pair, with e^+ emitted and e^- interacting with X as before, as in Figure 49.5.

One can easily calculate the $\gamma \to e^+e^-$ process, since it is the crossed diagram for $e^- \to e^- + \gamma$, and find (the details are left as an exercise)

$$\frac{1}{2}\sum_{\text{pol}} |\mathcal{M}|^2(\gamma \to e_L^- e_R^+) = \frac{2e^2 p_\perp^2}{z(1-z)}[z^2 + (1-z)^2]. \tag{49.34}$$

Then, in a completely similar way to the previous cases, we can consider the continuous process where the virtual photon line emits a e^+ and turns into an e^- (that emits a γ and gets out, and the γ continues on towards X and emits a pair ...), and so on. We easily see

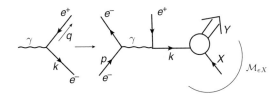

Figure 49.5   Photon splitting into pairs before interaction.

that the result will be a variation in the distribution function for e^-, and will be proportional to the new $[z^2 + (1 - z)^2]$ bracket above and the distribution function for the γ, so

$$\frac{d}{d\log Q}f_e(x, Q)\bigg|_{\text{pair cr.}} = \int_x^1 \frac{dz}{z}\left[\frac{\alpha}{\pi}(z^2 + (1-z)^2)\right]f_\gamma\left(\frac{x}{z}, Q\right). \tag{49.35}$$

But as a result, the f_γ will also be changed, since this subtracts a photon from momentum fraction $x = 1$, so as before, we must normalize, amounting to subtracting $\int_0^1 dz(z^2 + (1 - z)^2) = 2/3$ times $\delta(1 - z)$ in $df_\gamma/d\log Q$.

49.5 Evolution Equations for QED

Finally, therefore, we obtain the evolution equations for QED (i.e. the equations giving the variation of the parton distribution functions for the electron with the scale Q), found by Gribov and Lipatov. Since we have a distribution function for the electron, we must also have (due to the process of pair creation) a distribution function for the positron, which will be given by the same formula as for the electron. Indeed, the positron can turn into a positron by emitting a γ just like the electron, and a positron can be created from a γ just like an electron (the process is e^+e^- pair creation). We must also add a term to create a γ out of an e^+, equal to that creating a γ out of an e^-. We obtain

$$\frac{d}{d\log Q}f_\gamma(x, Q) = \frac{\alpha}{\pi}\int_x^1 \frac{dz}{z}\left\{P_{\gamma\leftarrow e}(z)\left[f_e\left(\frac{x}{z}, Q\right) + f_{\bar{e}}\left(\frac{x}{z}, Q\right)\right] + P_{\gamma\leftarrow\gamma}(z)f_\gamma\left(\frac{x}{z}, Q\right)\right\},$$

$$\frac{d}{d\log Q}f_e(x, Q) = \frac{\alpha}{\pi}\int_x^1 \frac{dz}{z}\left\{P_{e\leftarrow e}(z)f_e\left(\frac{x}{z}, Q\right) + P_{e\leftarrow\gamma}(z)f_\gamma\left(\frac{x}{z}, Q\right)\right\},$$

$$\frac{d}{d\log Q}f_{\bar{e}}(x, Q) = \frac{\alpha}{\pi}\int_x^1 \frac{dz}{z}\left\{P_{e\leftarrow e}(z)f_{\bar{e}}\left(\frac{x}{z}, Q\right) + P_{e\leftarrow\gamma}(z)f_\gamma\left(\frac{x}{z}, Q\right)\right\}, \tag{49.36}$$

where the *splitting functions* $P_{i\leftarrow j}(z)$ (considered as probabilities for turning parton j into parton i and calculated above) are

$$P_{e\leftarrow e}(z) = \frac{1 + z^2}{(1 - z)_+} + \frac{3}{2}\delta(1 - z),$$

$$P_{\gamma\leftarrow e}(z) = \frac{1 + (1 - z)^2}{z},$$

$$P_{e\leftarrow\gamma}(z) = z^2 + (1 - z)^2,$$

$$P_{\gamma\leftarrow\gamma}(z) = -\frac{2}{3}\delta(1 - z), \tag{49.37}$$

and the boundary conditions for integration of the differential equations are

$$f_e(x, Q)|_{Q^2=m^2} = \delta(1 - x); \quad f_{\bar{e}}(x, Q)|_{Q^2=m^2} = 0; \quad f_\gamma(x, Q)|_{Q^2=m^2} = 0. \tag{49.38}$$

The parton distribution functions also obey the same kind of normalization conditions as in the hadron (QCD) case, namely we need to have one more electron than positron, so

$$\int_0^1 dx[f_e(x,Q) - f_{\bar e}(x,Q)] = 1, \qquad (49.39)$$

and the total momentum fraction of all the partons (e^+, e^-, γ) is 1, so

$$\int_0^1 dx\, x[f_e(x,Q) + f_{\bar e}(x,Q) + f_\gamma(x,Q)] = 1. \qquad (49.40)$$

It is left as an exercise to check that these normalization conditions are respected by the evolution in the Gribov–Lipatov equations.

49.6 Altarelli–Parisi Equations and Parton Evolution

We finally return to the case of QCD. Just like in the case of QED, the evolution was due to the emission of collinear $(p_\perp/p \ll 1)$ photons and electrons, in the case of QCD, the evolution of the parton distribution functions is due to the emission of collinear gluons and quarks.

It will give a violation of Bjorken scaling, since now the DIS process will give

$$\frac{d^2\sigma}{dx\, dy}(e^- p \to e^- X) = \left(\sum_f f_f(x,Q)Q_f^2\right)\frac{2\pi\alpha_s^2}{Q^4}[1 + (1-y)^2], \qquad (49.41)$$

and the scaling is only approximate, since $f_f(x,Q)$ depends on Q as well, not just only on x.

As in the case of QED, in QCD we have two basic diagrams, for a gluon to be emitted from a quark line and the crossed diagram for a quark–antiquark pair to be emitted from a gluon (see Figure 49.6(a) and (b)). However, in the (nonabelian) QCD case, we also have a third diagram, for a gluon to split into two gluons (see Figure 49.6(c)).

The calculations resulting from diagrams (a) and (b) can be imported from QED to QCD with minimal modification. The only difference is that we sum over final polarizations and average over initial polarizations, so we must now also sum over final colors, and average over initial colors.

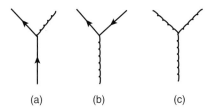

(a)   (b)   (c)

Figure 49.6   Diagrams relevant for the Altarelli–Parisi equation.

The sum over initial and final colors can easily be seen to give a factor of $\text{Tr}[t^a t^a]$. In diagram (a), the initial state is a quark, with three colors, so the averaging also gives a factor of 1/3, for a total

$$\frac{1}{3}\text{Tr}[t^a t^a] = \frac{1}{3}\frac{1}{2}\delta^a_a = \frac{4}{3}, \tag{49.42}$$

since in $SU(3)$ there are eight gluons ($N^2 - 1 = 8$ generators). In diagram (b), the initial state is a gluon, with eight states, so the averaging also gives a factor of 1/8, for a total

$$\frac{1}{8}\text{Tr}[t^a t^a] = \frac{1}{8}\frac{1}{2}\delta^a_a = \frac{1}{2}. \tag{49.43}$$

Diagram (c) needs to be done, since it doesn't appear in QED, but we will not do it here, and we will just quote the final result, for the gluon-to-gluon splitting function.

The resulting evolution equations, the *Altarelli–Parisi* evolution equations, are the same as the Gribov–Lipatov evolution equations, just replacing the electron e with quark f, $\bar{e}$ with $\bar{f}$ and γ with g, and α with $\alpha_s(Q^2)$ (which now also runs with the scale Q^2), so

$$\frac{d}{d\log Q}f_g(x, Q) = \frac{\alpha_s(Q^2)}{\pi}\int_x^1 \frac{dz}{z}\left\{ P_{g\leftarrow q}(z)\sum_f\left[f_f\left(\frac{x}{z}, Q\right) + f_{\bar{f}}\left(\frac{x}{z}, Q\right)\right] \right.$$

$$\left. + P_{g\leftarrow g}(z)f_g\left(\frac{x}{z}, Q\right) \right\},$$

$$\frac{d}{d\log Q}f_f(x, Q) = \frac{\alpha_s(Q^2)}{\pi}\int_x^1 \frac{dz}{z}\left\{ P_{q\leftarrow q}(z)f_f\left(\frac{x}{z}, Q\right) + P_{q\leftarrow g}(z)f_g\left(\frac{x}{z}, Q\right) \right\},$$

$$\frac{d}{d\log Q}f_{\bar{f}}(x, Q) = \frac{\alpha_s(Q^2)}{\pi}\int_x^1 \frac{dz}{z}\left\{ P_{q\leftarrow q}(z)f_{\bar{f}}\left(\frac{x}{z}, Q\right) + P_{q\leftarrow g}(z)f_g\left(\frac{x}{z}, Q\right) \right\}. \tag{49.44}$$

The only changes are in the splitting functions. The quark-to-quark and quark-to-gluon splitting functions come from diagram (a), so get an extra factor of 4/3, the gluon-to-quark splitting function comes from diagram (b), so gets an extra factor of 1/2, and the gluon-to-gluon splitting function is new, as it comes from diagram (c). We then obtain

$$P_{q\leftarrow q}(z) = \frac{4}{3}\left[\frac{1+z^2}{(1-z)_+} + \frac{3}{2}\delta(1-z)\right],$$

$$P_{g\leftarrow q}(z) = \frac{4}{3}\frac{1+(1-z)^2}{z},$$

$$P_{q\leftarrow g}(z) = \frac{1}{2}\left[z^2 + (1-z)^2\right],$$

$$P_{g\leftarrow g}(z) = 6\left[\frac{1-z}{z} + \frac{2}{(1-z)_+} + z(1-z) + \left(\frac{11}{2} - \frac{N_f}{18}\right)\delta(1-z)\right]. \tag{49.45}$$

- At next to leading order in α_s, Bjorken scaling is violated, by Q^2 dependence of the parton distribution functions, now $f_f(x, Q)$.
- One can describe the situation first in QED, where the DIS, with a photon emitted from an electron, interacts with X, and is understood in the Weizsacker–Williams equivalent photon approximation as being due to a photon distribution function inside the electron, so $\sigma(e^-X \to e^-Y) = \int_0^1 dz f(z) \, \sigma(\gamma X \to Y)$.
- Similarly, the interaction of an electron with X via real γ emission is described as due to an electron distribution function inside the electron.
- Then the (continuous) evolution of f_γ, through multiple emissions of γ from an e^- or e^+ line, is due to the splitting function $P_{\gamma \to e}(z)$, as well as to a $\gamma \to \gamma$ transition via $P_{\gamma \leftarrow \gamma}$, the evolution of f_e, through multiple emissions of γ from an e^- line, or pair creation e^+e^- from a γ line is due to the splitting functions $P_{e \leftarrow e}$ and $P_{e \leftarrow \gamma}$, leading to the Gribov–Lipatov equations.
- The Altarelli–Parisi equations in QCD are obtained by importing the QED calculation (Gribov–Lipatov) to QCD, changing e^- with q_f, e^+ with $\bar{q}_f$, and γ with g, with some extra color factors for the splitting functions, and with an extra diagram for a gluon to split into two gluons giving a new splitting function $P_{g \leftarrow g}(z)$.

Further Reading

See section 17.5 in [1].

Exercises

1. Prove the formula (49.8) for the amplitude of the QED process of photon emission from an electron line.

2. Verify that the Gribov–Lipatov evolution equations for QED imply that the normalization conditions for the $e^+e^-\gamma$ distribution functions

$$\int_0^1 dx[f_e(x, Q) - f_{\bar{e}}(x, Q) = 1,$$

$$\int_0^1 dx\, x[f_e(x, Q) + f_{\bar{e}}(x, Q) + f_\gamma(x, Q)] = 1 \tag{49.46}$$

are still satisfied at all Q, for the given boundary conditions at $Q = m$.

3. Consider

$$f_u(x, Q_0) = f_g(x, Q_0) = \frac{1 - x}{a(x + \epsilon)} \tag{49.47}$$

at $Q = Q_0$ and

$$\alpha_s(Q^2) = \frac{\alpha_s(Q_0^2)}{1 + \log \frac{Q^2}{Q_0^2}}. \tag{49.48}$$

Find the first correction to $f_u(x, Q)$ from the Altarelli–Parisi equations.

4. Prove the formula (49.34) for the amplitude of the pair production from a photon, $\gamma \rightarrow e^+ e^-$.

50

The Wilson Loop and the Makeenko–Migdal Loop Equation. Order Parameters; 't Hooft Loop

In Chapters 48 and 49 we have learned how to deal with the fact that QCD at low energy is strongly coupled, so nonperturbative, by introducing parton distribution functions, and using perturbation theory on top of that. In this chapter, however, we will learn how to probe truly nonperturbative physics. The most widely used tool for nonperturbative studies is the *Wilson loop*. It satisfies the Makeenko–Migdal loop equation, and defines an order parameter for a phase transition in QCD. We will also treat the 't Hooft loop, the electric–magnetic dual to the Wilson loop. These are the subjects of this chapter.

50.1 Wilson Loop

The Wilson loop is defined by external quarks, that is infinitely heavy ($m \to \infty$) sources for the gauge field, in a pure glue theory (YM). These are quarks that are not dynamical (i.e. are not in the path integral, or equivalently have decoupled because of the infinite mass).

We define the path-ordered exponential on a path $\mathcal{P}$ from x to y, namely the *Wilson line*

$$\Phi(y, x; \mathcal{P}) = P \exp\left\{ \int_x^y A_\mu(\xi) d\xi \right\} \equiv \lim_{n \to \infty} \prod_n e^{iA_\mu(\xi_n^\mu - \xi_{n+1}^\mu)}. \tag{50.1}$$

Note that $A_\mu(x) = A_\mu^a(x) T_a$ at different points do not commute in between them, so the exponential needs to be defined with an ordering along the path, that is as a limit $n \to \infty$ of a product of terms ordered along the path by an index n.

50.1.1 Abelian Case

Consider first an abelian gauge field A_μ, transforming by

$$\delta A_\mu = \partial_\mu \chi. \tag{50.2}$$

Then the objects in the definition of $\Phi(y, x; \mathcal{P})$ transform as

$$e^{iA_\mu d\xi^\mu} \to e^{iA_\mu d\xi^\mu + i\partial_\mu \chi d\xi^\mu} = e^{iA_\mu d\xi^\mu} e^{i\chi(x+dx) - i\chi(x)}. \tag{50.3}$$

It follows that the Wilson line transforms as

$$\Phi(y,x;\mathcal{P}) = \prod_x e^{iA_\mu d\xi^\mu} \to \prod_x \left(e^{iA_\mu d\xi^\mu} e^{i\chi(x+dx)-i\chi(x)} \right) = e^{i\chi(y)} \left(\prod_x e^{iA_\mu d\xi^\mu} \right) e^{-i\chi(x)}$$

$$= e^{i\chi(y)} \Phi(y,x;\mathcal{P}) e^{-i\chi(x)}. \tag{50.4}$$

Then, when acting on a charged complex scalar field $\phi(x)$, transforming under the gauge transformation as

$$\phi(x) \to e^{i\chi(x)} \phi(x), \tag{50.5}$$

the Wilson line gives

$$\Phi(y,x;\mathcal{P})\phi(x) \to e^{i\chi(y)} \Phi(y,x;\mathcal{P}) e^{-i\chi(x)} e^{i\chi(x)} \phi(x) = e^{i\chi(y)} \left(\Phi(y,x;\mathcal{P})\phi(x) \right), \tag{50.6}$$

that is it defines *parallel transport along the path $\mathcal{P}$ from x to y*. Parallel transport means that the properties of the object are preserved, just translated to a different point.

For a closed curve, $\mathcal{P} = \mathcal{C}$ for $y = x$, we have

$$\Phi(x,x;\mathcal{C}) \to e^{i\chi(x)} \Phi(x,x;\mathcal{C}) e^{-i\chi(x)} = \Phi(x,x;\mathcal{C}), \tag{50.7}$$

so the object is gauge invariant (i.e. it is a potential observable).

50.1.2 Nonabelian Case

For a nonabelian gauge field, transforming as (for coupling $g = 1$)

$$A_\mu(x) \to \Omega(x) A_\mu(x) \Omega^{-1}(x) - i(\partial_\mu \Omega)\Omega^{-1}, \tag{50.8}$$

with infinitesimal transformation for $\Omega(x) = e^{i\chi(x)}$,

$$\delta\Omega = D_\mu \chi, \tag{50.9}$$

the basic objects whose products define the Wilson line transform as

$$e^{iA_\mu d\xi^\mu} \simeq 1 + iA_\mu d\xi^\mu \to 1 + \Omega(A_\mu d\xi^\mu)\Omega^{-1} + d\xi^\mu(\partial_\mu\Omega)\Omega^{-1}$$

$$= \left[e^{i\chi(x)}(1 + iA_\mu d\xi^\mu) + d\xi^\mu \partial_\mu e^{i\chi(x)} \right] e^{-i\chi(x)}$$

$$\simeq e^{i\chi(x+dx)}(1 + iA_\mu d\xi^\mu) e^{-i\chi(x)}$$

$$\simeq e^{i\chi(x+dx)} e^{iA_\mu d\xi^\mu} e^{-i\chi(x)} + \mathcal{O}(dx^2), \tag{50.10}$$

where in the last line we have ignored quadratic terms in dx. We thus obtain that the Wilson line transforms as

$$\Phi(y,x;\mathcal{P}) \to e^{i\chi(y)} \Phi(y,x;\mathcal{P}) e^{-i\chi(x)}, \tag{50.11}$$

which is formally the same as in the abelian case, just that there the order of terms did not matter, we wrote it this way to suggest the form in the nonabelian case, but in the nonabelian case the order matters.

In particular, for closed curves $y = x$, $\mathcal{P} = \mathcal{C}$, we cannot cancel the exponentials, and the Wilson line is gaugecovariant, not invariant:

$$\Phi(x, x; \mathcal{C}) \rightarrow e^{i\chi(x)} \Phi(x, x; \mathcal{C}) e^{-i\chi(x)} \neq \Phi(x, x; \mathcal{C}). \tag{50.12}$$

But we can easily construct a gauge-invariant object, the *Wilson loop*, by taking the trace (normalized with a $1/N$, since there are N terms inside the trace for $SU(N)$):

$$W[\mathcal{C}] = \frac{1}{N} \operatorname{Tr} \Phi(x, x; \mathcal{C}). \tag{50.13}$$

Note that this object is gauge invariant, and independent of the point x (there is nothing special about x in the transformation law for W).

In the abelian case, the Wilson loop can be written in a manifestly gauge-invariant way through the use of the Stokes theorem, as

$$\Phi_{\mathcal{C}} = e^{i \int_{\mathcal{C} = \partial S} A_\mu dx^\mu} = e^{i \int_S F_{\mu\nu} d\Sigma^{\mu\nu}}. \tag{50.14}$$

In the nonabelian case, there are corrections to an explicitly invariant form. Consider a small square of sides a, in the $(\mu\nu)$ plane, so

$$\Phi_{\square\mu\nu} = e^{ia^2 F_{\mu\nu}} + \mathcal{O}(a^4). \tag{50.15}$$

Then the Wilson loop

$$W_{\square\mu\nu} = \frac{1}{N} \operatorname{Tr}[\Phi_{\square\mu\nu}] \simeq 1 - \frac{a^4}{2N} \operatorname{Tr}[F_{\mu\nu} F_{\mu\nu}] + \mathcal{O}(a^6), \tag{50.16}$$

where there is no sum over $\mu\nu$. The nontrivial object above is explicitly gauge invariant (up to a^6 terms), and moreover, by summing over $\mu\nu$, we obtain the kinetic term in the action. This is an example of why the Wilson loop contains all nonperturbative information from the gauge theory: the action, that defines the theory (in the case of the pure gauge theory, we only have the kinetic term), appears in the expansion of the Wilson loop.

50.2 Wilson Loop and the Quark–Antiquark Potential

The quantity that is the most studied from the Wilson loop is the *quark–antiquark potential*, measured for infinitely massive quarks (external quarks, fixed). Therefore we consider a contour $\mathcal{C}$ as in Figure 50.1, in the shape of a very long rectangle, made up of two long parallel lines in the time direction, of length T, one for a quark and one for an antiquark, situated at a distance R from each other, and connected by segments of length R.

Then it can be proved rigorously that the VEV of $W[\mathcal{C}]$ has the property that at large $T \rightarrow \infty$:

$$\langle W[\mathcal{C}] \rangle_0 \propto e^{-T V_{q\bar{q}}(R)}. \tag{50.17}$$

A simple (but not too rigorous) way to understand this is: add to the theory the infinitely heavy quarks, therefore appearing only as sources. The potential for the quark is $eA_0(x(q))$, and correspondingly for the antiquark $-eA_0(x(\bar{q}))$. Together, we obtain the source term

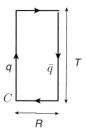

Figure 50.1 Wilson loop contour for the quark–antiquark potential.

to be added to the action, $\int d^4x j^\mu(x) A_\mu(x) = \int dt[eA_0(x(q)) - eA_0(x(\bar{q}))]$, where $j^0(x) = e\delta^{(3)}(x - x(q))$ is the quark current. This is indeed the object in the exponent of the Wilson loop, as it is $\simeq e \oint_C A_\mu d\xi^\mu$. From the interpretation as a potential, however, for a constant quark–antiquark potential, we add to e^{iS} a term $e^{iT V_{q\bar{q}}}$, so the VEV of the Wilson loop does indeed go like $e^{iT V_{q\bar{q}}(R)}$, or after a Wick rotation to Euclidean space, as $e^{-T V_{q\bar{q}}}$.

50.2.1 Area Law and Perimeter Law

For a *confining gauge theory*, the potential behaves at large distances as

$$V_{q\bar{q}}(R) \sim \sigma R, \tag{50.18}$$

where σ is called the *QCD string tension*. A linear potential means a constant force, so we cannot pull apart the quark from the antiquark, and they are confined. Confinement, however, also refers to the confinement of the electric flux lines inside a flux tube between q and $\bar{q}$ of almost constant cross-section, instead of spreading out all over space. The flux line density is proportional to the energy density, since $H \sim \frac{1}{2}[\vec{E}_a^2 + \vec{B}_a^2]$, and since the cross-section is constant, it means there is a total energy proportional to the length.

In contrast, for a *conformal gauge theory*, like QED, which doesn't have a mass scale, and is in fact conformal, the potential can only be of Coulomb type:

$$V_{q\bar{q}}(R) \sim \frac{\alpha}{R}, \tag{50.19}$$

since in that case the Wilson loop VEV scales as

$$\langle W[\mathcal{C}]\rangle_{\text{conformal}} \propto e^{-T V_{q\bar{q}}(R)} \sim e^{-\frac{\alpha T}{R}}, \tag{50.20}$$

as it should be in a conformal theory, since the only scale-invariant characterizing $\mathcal{C}$ is T/R.

In a confining theory, we obtain

$$\langle W[\mathcal{C}]\rangle_{\text{confining}} \propto e^{-\sigma TR} = e^{-\sigma \text{Area}}. \tag{50.21}$$

This is called the *area law*. But moreover, since for $C = C_1 \cup C_2$:

$$W[C = C_1 \cup C_2] = W[C_1]W[C_2], \tag{50.22}$$

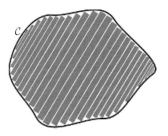

Figure 50.2 Wilson loop contour divided into infinitely thin Wilson rectangles, whose orientations cancel on the neighboring (adjacent) long lines.

as we can easily check, and in the large-N limit for an $SU(N)$ gauge group $\langle W[C_1]W[C_2]\rangle = \langle W[C_1]\rangle\langle W[C_2]\rangle$, we can extend the area law to any smooth curve C. We can approximate its area as the sum of infinitely thin rectangular contours in the T direction as in Figure 50.2, thus with $T/R \gg 1$, for each of which we have the area law, and obtain the area law for the total curve.

Therefore we have in fact that

$$\langle W[C]\rangle_0 \propto e^{-\sigma\,\mathrm{Area}[C]}, \tag{50.23}$$

for any contour C in a confining theory.

In a *Higgs phase* of a gauge theory, the quarks are screened, like in a superconductor. That is, the interaction is short-range (in a superconductor, the photon becomes effectively massive through the interaction with the medium, and the range is $a \sim 1/m$), so at large distances ($R > a$), the potential is constant:

$$V_{q\bar{q}}(R) \simeq \mathrm{const.} \equiv \mu. \tag{50.24}$$

Therefore, the Wilson loop VEV becomes

$$\langle W[C]\rangle_{\mathrm{Higgs}} \propto e^{-\mu T} \simeq e^{-\frac{\mu}{2}L[C]}, \tag{50.25}$$

where $L[C]$ is the perimeter of the contour C. We then obtain the *perimeter law*. Again, by the same argument as for the area law, we can extend the perimeter law for any smooth closed contour C.

50.3 The Makeenko–Migdal Loop Equation

For $SU(N)$ gauge theories at large N, the VEVs of gauge-invariant operators factorize (proven by Migdal) as

$$\langle \mathcal{O}_1 \ldots \mathcal{O}_n\rangle = \langle \mathcal{O}_1\rangle \ldots \langle \mathcal{O}_n\rangle + O\left(\frac{1}{N^2}\right). \tag{50.26}$$

This means that gauge-invariant operators behave like c-numbers, not as operators, so there must be a semiclassical saddle point of the path integral that allows us to write the VEV as simply the solution at the saddle point, weighted by e^{-S} at the saddle point.

We can then infer that there exists a so-called "*master field*," a colorless composite field $\Phi[A]$, with Jacobian for the gauge field

$$\left| \frac{\partial \Phi[A]}{\partial A_\mu^a} \right| \equiv e^{-N^2 J[\Phi]}. \tag{50.27}$$

In the presence of such a field, we can transform the path integral over A_μ^a into a path integral over Φ, and obtain

$$Z = \int \mathcal{D}A_\mu^a \, e^{-\frac{1}{4} \int d^4x (F_{\mu\nu}^a)^2} = \int \mathcal{D}\Phi \, \frac{1}{\left| \frac{\partial \Phi}{\partial A_\mu^a} \right|} e^{-N^2 S[\Phi]} = \int \mathcal{D}\Phi \, e^{-N^2 (S-J)}. \tag{50.28}$$

The saddle point of this is then

$$\frac{\delta S}{\delta \Phi} = \frac{\delta J}{\delta \Phi} \rightarrow \frac{\delta S}{\delta A_\mu^a} = -(\nabla_\mu F_{\mu\nu})^a = \frac{\delta J}{\delta A_\nu^a}, \tag{50.29}$$

which is called the master field equation. Note that with respect to the classical field equation, we have a nonzero term on the right-hand side, coming from the variation of the Jacobian from A_μ^a to Φ.

A natural guess for the master field (which is gauge invariant), that turns out to be correct, is the Wilson loop. But moreover, one can show that we can reformulate $SU(N)$ YM at any N (and thus QCD) in terms of $W[\mathcal{C}]$. Any observable is given by a sum over paths of Wilson loops.

Example 1　For instance, the product of two colorless vector quark currents is written as

$$\langle \bar{\psi} \gamma_\mu \psi(x_1) \bar{\psi} \gamma_\nu \psi(x_2) \rangle = \sum_{\mathcal{C} \ni x_1, x_2} J_{\mu\nu}(\mathcal{C}) \langle W[\mathcal{C}] \rangle, \tag{50.30}$$

where the sum is over paths that pass through x_1 and x_2, as in Figure 50.3.

Example 2　The connected correlator of three scalar quark currents is

$$\langle \bar{\psi} \psi(x_1) \bar{\psi} \psi(x_2) \bar{\psi} \psi(x_3) \rangle_{\text{conn.}} = \sum_{\mathcal{C} \ni x_1, x_2, x_3} J(\mathcal{C}) \langle W[\mathcal{C}] \rangle, \tag{50.31}$$

where again the sum is over paths that pass through x_1, x_2, x_3.

Figure 50.3　Paths going through x_1 and x_2 that are summed over.

If the quarks were scalars, we would have

$$J(C) = e^{-\frac{m^2}{2}\tau - \frac{m^2}{2}\int_0^\tau dt \dot{z}_\mu^2(t)} = e^{-mL[C]}, \tag{50.32}$$

where the thing in the exponent is the Lagrangian of the quark (particle), proportional to the length of the contour.

However, for the spinor quarks, things are a bit more complicated, and we get

$$J(C) = \int \mathcal{D}K_\mu(t) P \exp\left[-\int_0^\tau (iK_\mu(t)[\dot{x}^\mu(t) - \gamma^\mu(t)] + m^2)\right],$$

$$J_{\mu\nu}(C) = \int \mathcal{D}K_\mu(t) P \left\{\gamma_\mu(t_1)\gamma_\nu(t_2) \exp\left[-\int_0^\tau (iK_\mu(t)[\dot{x}^\mu(t) - \gamma^\mu(t)] + m^2)\right]\right\}, \tag{50.33}$$

where t_1 and t_2 are times of x_1, x_2.

50.3.1 Path and Area Derivatives

To write down the loop equations, we need to define some geometric objects called the path and area derivatives.

The *area derivative* of a function of a closed contour C is defined as follows. Consider the contour C with a point x singled out, C_x, as in Figure 50.4(b), and the same with an extra loop in the plane $(\mu\nu)$ and of area $\delta\sigma_{\mu\nu}$ at point x, $C_{\delta\sigma_{\mu\nu}}$, as in Figure 50.4(a). Then the area derivative is

$$\frac{\delta \mathcal{F}(C)}{\delta\sigma_{\mu\nu}(x)} \equiv \frac{1}{\delta\sigma_{\mu\nu}(x)}\left[\mathcal{F}(C_{\delta\sigma_{\mu\nu}}) - \mathcal{F}(C_x)\right]. \tag{50.34}$$

Here, $\delta\sigma_{\mu\nu} = dx_\mu \wedge dx_\nu$. For the path derivative, consider the contour $C_{\delta x_\mu}$ where at point x, the contour is shifted along δx_μ for a length δx_μ, in the μ direction, and then comes back (with zero area), as in Figure 50.5(a), so

$$\partial_\mu^x \mathcal{F}(C_x) = \frac{1}{\delta x_\mu}\left[\mathcal{F}(C_{\delta x_\mu}) - \mathcal{F}(C_x)\right]. \tag{50.35}$$

Note that the standard variational derivative can be written as a combination of the path and area derivatives as

$$\frac{\delta}{\delta x_\mu(\sigma)} = \dot{x}_\nu(\sigma)\frac{\delta}{\delta\sigma_{\mu\nu}(x(\sigma))} + \sum_{i=1}^m \partial_\mu^{x_i}\delta(\sigma - \sigma_i). \tag{50.36}$$

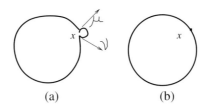

(a)   (b)

Figure 50.4 Diagrams for the definition of the area derivative: (a) $C_{\delta\sigma_{\mu\nu}}$; (b) C_x.

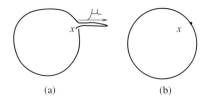

(a)   (b)

Figure 50.5   Diagrams for the definition of the path derivative: (a) $\mathcal{C}_{\delta x_\mu}$; (b) $\mathcal{C}_x$.

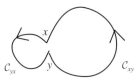

Figure 50.6   Diagram for the Makeenko–Migdal equation: $\mathcal{C}_{xy}$ and $\mathcal{C}_{yx}$.

50.3.2 Makeenko–Migdal Loop Equation

The loop equation is then written as

$$\partial_\mu^x \frac{\delta}{\delta \sigma_{\mu\nu}(x)} \langle W[\mathcal{C}] \rangle = \lambda \oint_{\mathcal{C}} dy_\nu \delta^{(D)}(x-y) \langle W[\mathcal{C}_{yx}] \rangle \langle W[\mathcal{C}_{xy}] \rangle, \qquad (50.37)$$

for $N \to \infty$, and where $\lambda = g^2 N$ is the 't Hooft coupling. Moreover, here $\mathcal{C} = \mathcal{C}_{xy} \cup \mathcal{C}_{yx}$ is closed, whereas $\mathcal{C}_{yx}$ goes from x to y (very close by points) in a long route on one side, and $\mathcal{C}_{xy}$ goes from y to x in a long route on another side, as in Figure 50.6. Note that this is a single equation for the Wilson loop VEV. This equation is the analogue of the Dyson–Schwinger equation.

At finite N, we can write a version of the above, but where now the equation does not close (i.e. it is not an equation for a single object). This is

$$\partial_\mu^x \frac{\delta}{\delta \sigma_{\mu\nu}(x)} \langle W[\mathcal{C}] \rangle = \lambda \oint_{\mathcal{C}} dy_\nu \delta^{(D)}(x-y) \left[\langle W[\mathcal{C}_{yx}] W[\mathcal{C}_{xy}] \rangle - \frac{1}{N^2} \langle W[\mathcal{C}] \rangle \right]. \qquad (50.38)$$

Therefore, we see that there is an extra term with $1/N^2$, but more importantly, the right-hand side at finite N does not factorize, and we get the VEV of a product of Wilson loops. Then we must write down an equation for the product of two Wilson loops, that will depend on the VEV of the product of three Wilson loops, and so on, obtaining an infinite chain of coupled differential equations.

50.4 Order Parameters, 't Hooft Loop, Polyakov Loop

The Wilson loop is an order parameter in the sense of Landau's theory of second-order phase transitions, namely an object that has a nonzero VEV in an ordered phase, and a zero VEV in a disordered phase, since in a confining (disordered) phase:

$$\langle W[\mathcal{C}] \rangle \sim e^{-\sigma \, \text{Area}} \to 0, \tag{50.39}$$

whereas in a Higgs (ordered) phase, we have

$$\langle W[\mathcal{C}] \rangle \sim e^{-\mu \, \text{Perimeter}} \neq 0. \tag{50.40}$$

Of course, there is a slight abuse of notation, since in the limit of large contours, really both VEVs go to zero, but the confining one goes much faster. In fact, we can of course multiply everything with $e^{+\mu \, \text{Perimeter}}$ to make it more precise, but it is still not perfectly defined.

We can define, however, another type of observable, a *disorder parameter* (i.e. one that has nonzero VEV in the disordered phase and zero in the ordered phase). This is an operator dual to the Wilson loop (in the sense that its properties are opposite to the Wilson loop), called the *'t Hooft operator* $T[\mathcal{C}]$. Unlike the Wilson loop, it can only be defined in the case where all the scalars are invariant under the center $\mathbb{Z}_N$ of $SU(N)$ (the center of a group is the subgroup that commutes with all the other elements).

50.4.1 't Hooft Loop

The 't Hooft loop is defined rather abstractly as follows. Consider a gauge transformation $\Omega^{[C]}$ that is singular along the curve C. If another curve C' winds through C with a linking number n (e.g. two links of a chain have linking number 1), and C' is parametrized by $\theta \in [0, 2\pi]$, then

$$\Omega^{[C]}(2\pi) = \Omega^{[C]}(0)e^{\frac{2\pi i n}{N}}. \tag{50.41}$$

Then the 't Hooft loop operator $T[C']$ is defined by the relation

$$W[C]T[C'] = T[C']W[C]e^{\frac{2\pi i n}{N}}. \tag{50.42}$$

As one can guess from the statement that the 't Hooft loop is dual to the Wilson loop, in a Higgs (ordered) phase we have

$$\langle T[\mathcal{C}] \rangle \sim e^{-\sigma' \, \text{Area}} \to 0, \tag{50.43}$$

that is the area law, whereas in a confining (disordered) phase we have

$$\langle T[\mathcal{C}] \rangle \sim e^{-\mu \, \text{Perimeter}} \neq 0. \tag{50.44}$$

Note, however, that 't Hooft showed there are also *mixed phases*, where

$$\langle W[\mathcal{C}] \rangle \sim e^{-\sigma \, \text{Area}}; \quad \langle T[\mathcal{C}] \rangle \sim e^{-\sigma' \, \text{Area}}. \tag{50.45}$$

50.4.2 Polyakov Loop

An important subcase of the Wilson loop, that has its own name, is the *Polyakov loop*. This is a Wilson loop in the case of a quantum field theory at finite temperature (i.e. described in Euclidean space by periodic time, with periodicity $\beta = 1/\text{temperature}$). The Polyakov loop is a loop where the rectangular contour wraps once along the periodic time direction.

In this case we obtain a better understanding of why $W[C]$ is an order parameter. Indeed, in this case, the length in time $T = \beta$ is fixed and finite. This means that now, for infinite contour C, in the Higgs (ordered) phase:

$$\langle W[C]\rangle \sim e^{-\mu T} = \text{const.} \neq 0, \tag{50.46}$$

and in the confining (disordered) phase:

$$\langle W[C]\rangle \sim e^{-(\sigma T)R} \to 0. \tag{50.47}$$

Important Concepts to Remember

- Wilson loops characterize the behavior of external quarks (infinitely massive probe quarks, not dynamical) in gauge theories.
- In an abelian theory, $\Phi = \exp[i \oint A_\mu dx^\mu]$ is gauge invariant and defines parallel transport of charged scalar fields, and in a nonabelian theory, $W[C] = 1/N \operatorname{Tr}\left[P \exp\left\{i \oint A_\mu dx^\mu\right\}\right]$ is gauge invariant and defines parallel transport of charged scalar fields.
- The Wilson loop contains all information about observables of the gauge theory. Its first nontrivial term in the expansion on a square (plaquette) is the kinetic term of the gauge action.
- The VEV of the Wilson loop on a rectangular contour infinitely long in the time direction defines the quark–antiquark potential by $\langle W[C]\rangle \propto e^{-T V_{q\bar q}(R)}$ as $T \to \infty$.
- For a confining theory (a confining phase), the potential is linear $V_{q\bar q}(R) = \sigma R$, so we obtain the area law, whereas for a Higgs phase, the potential is constant, so we obtain the perimeter law. For a conformal theory like QED, we obtain a Coulomb potential $V_{q\bar q}(R) = \alpha/R$.
- VEVs of gauge-invariant observables factorize in the large-N limit of $SU(N)$ YM, so there is a gauge invariant, composite master field Φ. In fact, the Wilson loop has its properties, and we can define (even at finite N) gauge theory observables in terms of sums over paths with the desired operator insertions.
- The Wilson loop at $N \to \infty$ satisfies the Makeenko–Migdal loop equation (the equation closes on VEVs of $W[C]$), and at finite N we obtain an infinite set of coupled equations for VEVs of products of Wilson loops.
- The Wilson loop is an order parameter, and its dual, the 't Hooft loop, is a disorder parameter.
- The Polyakov loop is a Wilson loop at finite temperature, where the infinite lines in the time directions now wrap once the periodic time.

Further Reading

See section 15.3 in [1].

Exercises

1. Consider a circular Wilson loop of radius R in Euclidean space. If the theory is confining, how will the Wilson loop VEV scale with R? How about if it is conformal (like QED)?

2. Check that the confining result for the circle satisfies the Makeenko–Migdal loop equation.

3. Calculate the $\mathcal{O}(a^6)$ term in the expansion of the Wilson loop for the plaquette (small square of size a).

4. Prove that the 't Hooft loop is a disorder operator, from the fact that the Wilson loop is an order operator, and the two are electric–magnetic dual.

51

IR Divergences in QED

In this chapter we start the study of IR divergences, focusing on the example of QED. We will see that there are two types of IR divergences, both associated with massless particles, collinear divergences and soft divergences, and in nonabelian gauge theories we have both kinds. We then move to the main example of this chapter, the QED vertex, regularizing its IR divergence by a photon mass, after which we use dimensional regularization, and show that there is a correspondence between the two calculations. We will then see that by adding another diagram, corresponding to an additional soft photon emission, which cannot be distinguished experimentally from the previous one, we eliminate the IR divergence. Finally, we will see that summing up various possible IR divergences, these exponentiate into what is known as a "Sudakov form factor."

51.1 Collinear and Soft IR Divergences

51.1.1 Collinear Divergences

We already saw in Chapter 32 that when we have massless particles in the theory, we have IR divergences in the loop diagrams. For instance, in the one-loop diagram with two $n + 2$-point vertices, two propagators, and $2n$ external lines in Figure 32.2, with Feynman diagram

$$\frac{\lambda^2}{2} \int \frac{d^D q}{(2\pi)^D} \frac{1}{q^2 + m^2} \frac{1}{(q - P)^2 + m^2}, \tag{51.1}$$

where $P = \sum_i p_i$ is the total external momentum at each of the two vertices, if $m^2 = 0$ AND $P^2 = 0$, the integral becomes

$$\frac{\lambda^2}{2} \int \frac{d^D q}{(2\pi)^D} \frac{1}{q^2} \frac{1}{q^2 - 2q \cdot P}, \tag{51.2}$$

so is divergent in $D = 4$, in the angular integration region where $\hat{q} \cdot \hat{P} = 0$ (here $\hat{q}, \hat{P}$ are the unit vectors in the directions of q, P), since then

$$\sim \frac{\lambda^2}{2} \int d\Omega \int dq \frac{q}{q^2 - 2qP\hat{q} \cdot \hat{P}} \propto \int \frac{dq}{q}. \tag{51.3}$$

Note that this divergence appears only for massless external states, since if $P^2 \neq 0$, we do not have an IR divergence. In turn, if $n > 1$, this means that the external particles must be *collinear*, $p_i \propto p_j$, so as to have $P^2 = 0$. Note also that the divergence is due to a virtual "photon" (massless particle) of momentum q being collinear with the (set of) external particle(s) $P = \sum_i p_i$, since $q \cdot P = 0 \Leftrightarrow q$ is parallel with P ($P^2 = 0$). This type of IR divergence is then called a *collinear IR divergence*. It will be canceled by an IR divergence in the amplitude to emit a (real) "photon" (massless particle) collinear with an external line from each emitted. In conclusion, a collinear IR divergence is due to a virtual or real "photon" being collinear with a massless external line. For the existence of such a divergence, we need to have a massless external state coupling to a massless internal loop (i.e. to have self-interactions of massless states), since we need $P^2 = 0$, but also $q^2 = 0$ and $(q - P)^2 = 0$. This will happen in QCD, where gluons are self-interacting, but does not happen in QED.

In dimensional regularization, using the result (33.35) at $m = 0$ (from Chapter 33), we can calculate the diagram as

$$I = \frac{\lambda^2}{2} \frac{\Gamma\left(2 - \frac{D}{2}\right)}{(4\pi)^{\frac{D}{2}}} P^{D-4} \int_0^1 d\alpha \, [\alpha(1-\alpha)]^{\frac{D}{2}-1} \propto \lambda^2 \frac{2}{\epsilon} \left(\frac{P^2}{\mu^2}\right)^{-\frac{\epsilon}{2}} \propto \lambda^2 \left[\frac{2}{\epsilon} - \log \frac{P^2}{\mu^2}\right],$$
(51.4)

where as usual $D = 4 - \epsilon$, we have introduced a dimensional transmutation parameter μ, and finally we have expanded in ϵ to obtain a term logarithmically divergent in $\mu \to 0$ (thought of as an IR cut-off).

51.1.2 Soft Divergences

Consider a part of a larger one-loop diagram with two consecutive external states, and take them to have mass m. They are taken to be continued into the loop, so two outward extending propagators, with momenta $k_1 + q$ and $k_2 - q$, will have the same mass m. In between these two lines, there is a massless propagator with momentum q. We have in mind here the application to QED, or rather the simpler version of massive scalar-QED (charged complex scalar) where the massive line would be a scalar, and the massless one a photon, but we can also consider the massless case $m = 0$, and then we have in mind a theory of massless scalars. This part of the diagram, in Figure 51.1, will give a contribution

$$\int \frac{d^D q}{(2\pi)^D} \frac{1}{(q+k_1)^2 + m^2} \frac{1}{q^2} \frac{1}{(q-k_2)^2 + m^2} = \int \frac{d^D q}{(2\pi)^D} \frac{1}{q^2 + 2q \cdot k_1} \frac{1}{q^2} \frac{1}{q^2 - 2q \cdot k_2},$$
(51.5)

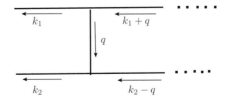

Figure 51.1 Soft divergence diagram piece.

which we see is independent of the mass m of the external states, appears at general k_1, k_2 (not necessarily collinear), and moreover the divergence is present independently of the orientation of the "photon" (i.e. of whether $q \cdot k_1 = 0$ and $q \cdot k_2 = 0$ or not, which equality is true for $m^2 = 0$, so that $k_1^2 = k_2^2 = 0$, AND $q \ll k_1$ and $q \ll k_2$). The only source of this divergence is the fact that the "photon" q is "soft" (small energy) (i.e. $q^2 \sim 0$) and moreover $|\vec{q}|^2$ small.

This divergence, for soft virtual massless particles (in a loop), and a corresponding one for soft emitted (real) massless particles, is called a *soft divergence* and is present in any theory with massless particles interacting with something, so is present both in QED and in QCD.

51.1.3 IR Divergences in Nonabelian Gauge Theories

In nonabelian gauge theories (YM), we have both soft and collinear IR divergences, and correspondingly in dimensional regularization at one loop we have a factor of $1/\epsilon$ for each: we saw that there was a $1/\epsilon$ for collinear divergences, and there is another one for soft ones. In total, in a *planar* one-loop diagram *for each pair of consecutive momenta, k_i and k_{i+1}*, with $s_{i,i+1} \equiv (k_i + k_{i+1})^2$, we have a divergent factor of

$$\sim \frac{1}{\epsilon^2} \left(\frac{-s_{i,i+1}}{\mu^2} \right)^\epsilon \simeq \frac{1}{\epsilon^2} \left[1 + \epsilon \log \frac{-s_{i,i+1}}{\mu^2} + \frac{\epsilon^2}{2} \log^2 \frac{-s_{i,i+1}}{\mu^2} + \dots \right]. \qquad (51.6)$$

We see that the term divergent as $\mu \to 0$ is $\sim \log^2(-s_{i,i+1}/\mu^2)$ in the nonabelian case. In the QED case (abelian), when we have only soft divergences, at one loop we have in dimensional regularization a term $\sim 1/\epsilon(-s_{i,i+1}/\mu^2)^\epsilon \sim 1/\epsilon + \log(s_{i,i+1}/\mu^2)$, so the term divergent as $\mu \to 0$ is $\log(-s_{i,i+1}/\mu^2)$. A useful IR regularization that will be used in the following is to introduce a photon mass μ_{ph}. When translating dimensional regularization results into photon mass regularization, we just drop the $1/\epsilon$ terms and keep only the terms divergent for $\mu \to 0$, replacing μ with μ_{ph}. Therefore, in the QED case, we expect an IR divergence of order $\log(q^2/\mu_{\text{ph}}^2)$, and we will see that this is what we obtain.

For completeness, note that at L loops, we have in nonabelian gauge theory a leading divergent term of $\mathcal{O}(1/\epsilon^{2L})$.

51.2 QED Vertex IR Divergence

Consider the full quantum-mechanical QED vertex $\Gamma^\mu_{\alpha\beta}$, with a fermion with momentum p_1 going in, a fermion with momentum p_2 coming out, and a photon with momentum $q = p_2 - p_1$ coming in, as in Figure 51.2(a). Then, in general, by Lorentz invariance we should have

$$\Gamma^\mu = A\gamma^\mu + B(p_1^\mu + p_2^\mu) + C(p_2^\mu - p_1^\mu). \qquad (51.7)$$

The vertex should satisfy the Ward–Takahashi identity:

$$q_\mu \Gamma^\mu = (p_2 - p_1)_\mu \Gamma^\mu = A(p_2 - p_1)_\mu \gamma^\mu + C(p_2 - p_1)^2 = 0, \qquad (51.8)$$

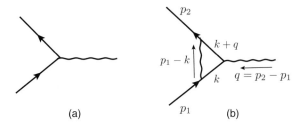

Figure 51.2 Soft divergent vertex diagram: (a) zeroth-order diagram; (b) one-loop diagram.

where we have used $p_2^2 - p_1^2 = 0$ on-shell. Since the vertex also appears in between $\bar{u}(p_2)$ and $u(p_1)$, and $(p_2 - p_1)_\mu \bar{u}(p_2)\gamma^\mu u(p_1) = 0$ on-shell (by the Dirac equation for $u(p_1)$ and $\bar{u}(p_2)$), inside physical amplitudes we can ignore A, and the Ward–Takahashi identity just says that $C = 0$.

In contrast, we have the Gordon identity, sometimes written as two equations:

$$m\bar{u}(p_2)\gamma^\mu u(p_1) = p_1^\mu \bar{u}(p_2)u(p_1) - i\bar{u}(p_2)\sigma^{\mu\nu}p_{1\nu}u(p_1),$$

$$m\bar{u}(p_2)\gamma^\mu u(p_1) = p_2^\mu \bar{u}(p_2)u(p_1) + i\bar{u}(p_2)\sigma^{\mu\nu}p_{2\nu}u(p_1), \tag{51.9}$$

and sometimes as the average of the two equations above, giving

$$\bar{u}(p_2)\gamma^\mu u(p_1) = \bar{u}(p_2)\left[\frac{p_2^\mu + p_1^\mu}{2m} + \frac{i\sigma^{\mu\nu}q_\nu}{2m}\right]u(p_1). \tag{51.10}$$

This means that we can swap the term with B (with $(p_1 + p_2)^\mu$) for a contribution to the term with A (with γ^μ) and another with $\sigma^{\mu\nu}q_\nu$, so we can always write (in between $\bar{u}(p_2)$ and $u(p_1)$)

$$\Gamma^\mu(p_2, p_1) = \gamma^\mu F_1(q^2) + \frac{i\sigma^{\mu\nu}q_\nu}{2m}F_2(q^2), \tag{51.11}$$

where we have defined the two structure functions $F_1(q^2)$ and $F_2(q^2)$.

We have treated the one-loop correction to Γ^μ, $\Gamma^{\mu(1)}$, in Chapters 36 and 38, but there we ignored the IR divergences. We will redo the calculation (some relevant steps) here, in Minkowski space, with a slightly different parametrization for the loop momentum, and considering the IR divergences.

The one-loop diagram has a photon with momentum $p_1 - k$ being emitted from the p_1 fermion line, turning it into a fermion with momentum k, and being reabsorbed into the other fermion line, initially with momentum $k + q$, to turn it into p_2, as in Figure 51.2(b). Then the diagram gives

$$\bar{u}(p_2)\Gamma^{\mu(1)}(p_1, p_2)u(p_1) = \int \frac{d^4k}{(2\pi)^4} \frac{-ig_{\nu\rho}}{(k - p_1)^2 - i\epsilon}\bar{u}(p_2)(+e\gamma^\nu)\frac{-(\slashed{k} + \slashed{q} + im)}{(k + q)^2 + m^2 - i\epsilon}$$

$$\times \gamma^\mu \frac{-(\slashed{k} + im)}{k^2 + m^2 - i\epsilon}(+e\gamma^\rho)u(p_1)$$

$$= +2ie^2 \int \frac{d^4k}{(2\pi)^4} \frac{\bar{u}(p_2)[\slashed{k}\gamma^\mu(\slashed{k} + \slashed{p}_1) - m^2\gamma^\mu + 2im(2k + q)^\mu]u(p_1)}{[(k - p_1)^2 + i\epsilon][(k + q)^2 + m^2 - i\epsilon][k^2 + m^2 - i\epsilon]}. \tag{51.12}$$

Doing the Feynman parametrization for the three propagators in the denominator, $\Delta_1, \Delta_2, \Delta_3$, with α_1 for $k^2 + m^2$, α_2 for $(k + q)^2 + m^2$, and α_3 for $(k - p_1)^2$, we get

$$\frac{1}{\Delta_1\Delta_2\Delta_3} = \int_0^1 d\alpha_1 d\alpha_2 d\alpha_3 \delta(\alpha_1 + \alpha_2 + \alpha_3 - 1)\frac{1}{(\tilde{k}^2 + F - i\epsilon)^3}, \tag{51.13}$$

where $\tilde{k} = k + \alpha_2 q - \alpha_3 p_1$. Using the formulas

$$\int \frac{d^4\tilde{k}}{(2\pi)^4}\frac{\tilde{k}^\mu}{(\tilde{k}^2 + F - i\epsilon)^3} = 0,$$

$$\int \frac{d^4\tilde{k}}{(2\pi)^4}\frac{\tilde{k}^\mu\tilde{k}^\nu}{(\tilde{k}^2 + F - i\epsilon)^3} = \int \frac{d^4\tilde{k}}{(2\pi)^4}\frac{\frac{1}{4}g^{\mu\nu}\tilde{k}^2}{(\tilde{k}^2 + F - i\epsilon)^3}, \tag{51.14}$$

after some algebra we obtain

$$\bar{u}(p_2)\Gamma^{\mu(1)}(p_1, p_2)u(p_1) = 2ie^2 \int \frac{d^4\tilde{k}}{(2\pi)^4} \int_0^1 d\alpha_1 d\alpha_2 d\alpha_3 \delta(\alpha_1 + \alpha_2 + \alpha_3 - 1)\frac{2}{(\tilde{k}^2 + F - i\epsilon)^3}$$

$$\times \bar{u}(p_2)\left[\gamma^\mu\left(-\frac{1}{2}\tilde{k}^2 - (1 - \alpha_1)(1 - \alpha_2)q^2 + (1 - 4\alpha_3 + \alpha_3^2)m^2\right)\right.$$

$$\left. + \frac{i\sigma^{\mu\nu}q_\nu}{2m}(2m^2\alpha_3(1 - \alpha_3))\right]u(p_1), \tag{51.15}$$

where

$$F = \alpha_1\alpha_2 q^2 + (1 - \alpha_3)^2 m^2. \tag{51.16}$$

One would need to UV-regulate this integral, but we will ignore it here, and instead adopt a bit later on a subtraction procedure (renormalization condition). Note that the part with $\tilde{k}^2$ in the numerator, that was called $\Gamma^{(1a)}$ in Chapter 36, is UV divergent, but is not of interest to us. The part without $\tilde{k}^2$ in the numerator, called $\Gamma^{(1b)}$ in Chapter 38, is UV finite, and will contain the relevant IR divergences, so we are interested in it.

Performing a Wick rotation on the formula (33.30) from Chapter 33, and putting $D = 4$ (for the case where the integral is UV finite) gives[1]

$$\int \frac{d^4\tilde{k}}{(2\pi)^4}\frac{1}{(\tilde{k}^2 + \Delta)^n} = \frac{i}{(4\pi)^2}\frac{1}{(n-1)(n-2)}\frac{1}{\Delta^{n-2}}, \tag{51.18}$$

and we obtain for the IR-divergent piece

$$\bar{u}(p_2)\Gamma^{\mu(1b)}u(p_1) = \frac{\alpha}{2\pi}\int_0^1 d\alpha_1 d\alpha_2 d\alpha_3 \delta(\alpha_1 + \alpha_2 + \alpha_3 - 1)$$

$$\times \bar{u}(p_2)\left[\gamma^\mu\frac{(1 - \alpha_1)(1 - \alpha_2)q^2 + (1 - 4\alpha_3 + \alpha_3^2)m^2}{F}\right.$$

$$\left. + \frac{i\sigma^{\mu\nu}q_\nu}{2m}\frac{2m^2\alpha_3(1 - \alpha_3)}{F}\right]. \tag{51.19}$$

---

[1] For use later on, note that the case relevant for us, of $n = 3$, gives, for dimensional regularization in Minkowski space:

$$\int \frac{d^D k}{(2\pi)^D}\frac{1}{(\tilde{k}^2 + \Delta)^3} = \frac{i}{(4\pi)^{D/2}}\frac{\Gamma(3 - D/2)}{\Gamma(3)}\frac{1}{\Delta^{3 - \frac{D}{2}}}. \tag{51.17}$$

The integral of the coefficient of γ^μ in the square brackets is $F_1^{(1b)}(q^2)$ and the integral of the coefficient of $\frac{i\sigma^{\mu\nu}q_\nu}{2m}$ is $F_2(q^2)$. F_2 will not contain IR divergences, but $F_1^{(1b)}$ will. To see that, we calculate $F_1^{(1b)}(q^2 = 0)$. We have

$$\int_0^1 d\alpha_1 d\alpha_2 d\alpha_3 \delta(\alpha_1 + \alpha_2 + \alpha_3 - 1)\frac{1 - 4\alpha_3 + \alpha_3^2}{F(q^2 = 0)}$$

$$= \int_0^1 d\alpha_3 \int_0^{1-\alpha_3} d\alpha_2 \frac{-2 + (1 - \alpha_3)(3 - \alpha_3)}{m^2(1 - \alpha_3)^2}$$

$$= \int^1 d\alpha_3 \frac{-2}{m^2(1 - \alpha_3)} + \text{finite}. \tag{51.20}$$

We see that we have an IR divergence, coming from the $\alpha_3 \simeq 1$ region of integration in the last Feynman parameter. We need to regulate this IR divergence. One option would be to use dimensional regularization, and we will sketch this afterwards, but here instead we introduce a small photon mass μ_{ph}.

Then the photon propagator will give $[(k - p_1)^2 + \mu_{\text{ph}}^2 - i\epsilon]^{-1}$ instead of $[(k - p_1)^2 - i\epsilon]^{-1}$, and since the inverse photon propagator Δ_3 appears multiplied by α_3 in the Feynman parametrization, the effect of the regularization is to add a term $\alpha_3\mu_{\text{ph}}^2$ to F.

We need one more ingredient. For the UV regularization of the whole diagram, a simple renormalization condition is

$$\Gamma^\mu(q^2 = 0) = \gamma^\mu. \tag{51.21}$$

This subtracts the UV divergence in $\Gamma_\mu^{(1a)}$, contributing to $F^{(1a)}(q^2)$, but it also affects the UV-finite piece $\Gamma_\mu^{(1b)}$, specifically $F_1^{(1b)}(q^2)$, by subtracting $F_1^{(1b)}(q^2 = 0)$. We finally obtain

$$F_1^{(1)}(q^2)|_{\mu_{\text{ph}}^2 \to 0} \simeq F_1^{(1b)}(q^2)|_{\mu_{\text{ph}}^2 \to 0} \simeq \frac{\alpha}{2\pi} \int_0^1 d\alpha_1 d\alpha_2 d\alpha_3 \delta(\alpha_1 + \alpha_2 + \alpha_3 - 1)$$

$$\times \left[\frac{m^2(1 - 4\alpha_3 + \alpha_3^2) - q^2(1 - \alpha_1)(1 - \alpha_2)}{q^2\alpha_1\alpha_2 + m^2(1 - \alpha_3)^2 + \mu_{\text{ph}}^2\alpha_3} - \frac{m^2(1 - 4\alpha_3 + \alpha_3^2)}{m^2(1 - \alpha_3)^2 + \mu_{\text{ph}}^2\alpha_3} \right]. \tag{51.22}$$

The IR divergence, as we saw above, comes from the $\alpha_3 \simeq 1$ ($\Rightarrow \alpha_1 \simeq \alpha_2 \simeq 0$) region of integration in the Feynman parameters, and (since the integration in the Feynman parameters is only between 0 and 1) it appears from the denominators. Therefore we can put $\alpha_3 = 1$ and $\alpha_1 = \alpha_2 = 0$ in the numerators, as well as in the regulator, so $\alpha_3\mu_{\text{ph}}^2 \to \mu_{\text{ph}}^2$.

Therefore we have

$$F_1^{(1)}(q^2)|_{\mu_{\text{ph}}^2 \to 0} \simeq \frac{\alpha}{2\pi} \int_0^1 d\alpha_3 \int_0^{1-\alpha_3} d\alpha_2 \left[\frac{-2m^2 - q^2}{m^2(1 - \alpha_3)^2 + q^2\alpha_2(1 - \alpha_3 - \alpha_2) + \mu_{\text{ph}}^2} \right.$$

$$\left. - \frac{-2m^2}{m^2(1 - \alpha_3)^2 + \mu_{\text{ph}}^2} \right]. \tag{51.23}$$

With the substitution $\alpha_2 = (1 - \alpha_3)\xi$ and $w = 1 - \alpha_3$, with Jacobian w, we obtain

$$F_1^{(1)}(q^2)|_{\mu_{\text{ph}}^2 \to 0} \simeq \frac{\alpha}{2\pi} \int_0^1 d\xi \frac{1}{2} \int_0^1 d(w^2) \left[\frac{-2m^2 - q^2}{[m^2 + q^2\xi(1-\xi)]w^2 + \mu_{\text{ph}}^2} - \frac{-2m^2}{m^2 w^2 + \mu_{\text{ph}}^2} \right]$$

$$= \frac{\alpha}{2\pi} \int_0^1 \frac{d\xi}{24} \left[\frac{-2m^2 - q^2}{m^2 + q^2\xi(1-\xi)} \log \frac{m^2 + q^2\xi(1-\xi)}{\mu_{\text{ph}}^2} + 2\log \frac{m^2}{\mu_{\text{ph}}^2} \right].$$

$$(51.24)$$

51.3 Dimensional Regularization Calculation

Before we continue, we redo this calculation in dimensional regularization. The first observation is that, with $D = 4 - 2\epsilon$, $\epsilon > 0$ gives UV regularization, but $\epsilon < 0$ gives IR regularization. A typical (limit) divergence both in the UV and in the IR is a log-divergence:

$$\int \frac{d^D k}{k^4} \propto \int_0^\infty \frac{dk}{k^{5-D}},$$

$$(51.25)$$

and we see that the UV divergence is regulated only by $4 - D = 2\epsilon > 0$:

$$\int^\infty \frac{dk}{k^{1+2\epsilon}} \sim \left. \frac{k^{-2\epsilon}}{-2\epsilon} \right|^\infty < \infty,$$

$$(51.26)$$

whereas the IR divergence is regulated only by $4 - D = 2\epsilon_{UV} = -2\epsilon_{IR} < 0$:

$$\int_0 \frac{dk}{k^{1-2\epsilon_{IR}}} \sim \left. \frac{k^{+2\epsilon_{IR}}}{\epsilon} \right|_0 < \infty.$$

$$(51.27)$$

So we consider $D = 4 + 2\epsilon$ in (51.17) and IR regularize like this instead of introducing μ_{ph}. Then the denominator in F_1 is now $F^{1-\epsilon}$ instead of F. But again the divergent integration region for the Feynman parameters is $\alpha_3 \simeq 1 \Rightarrow \alpha_1 \simeq \alpha_2 \simeq 0$, so we can repeat the same steps until we do the w^2 integration, to obtain in dimensional regularization

$$F_1^{(1)}(q^2)|_{\epsilon \to 0} \simeq \frac{\alpha}{2\pi} \int_0^1 d\xi \frac{1}{2} \int_0^1 d(w^2) \left[\frac{-2m^2 - q^2}{([m^2 + q^2\xi(1-\xi)]w^2)^{1-\epsilon}} - \frac{-2m^2}{[m^2 w^2]^{1-\epsilon}} \right]$$

$$= \frac{\alpha}{2\pi} \frac{\mu^{2\epsilon}}{\epsilon} \frac{1}{2} \int_0^1 d\xi \left[\left(\frac{m^2}{\mu^2} \right)^\epsilon \frac{-2 - q^2/m^2}{[1 + q^2\xi(1-\xi)/m^2]^{1-\epsilon}} + 2 \left(\frac{m^2}{\mu^2} \right)^\epsilon \right].$$

$$(51.28)$$

where we have introduced a dimensional transmutation scale μ, and the overall $\mu^{2\epsilon}$ should be reabsorbed, as usual, in the redefinition of the coupling. Note that we obtain the correspondence with the photon mass regularization already suggested, since

$$\frac{1}{\epsilon}\left(\frac{m^2}{\mu^2}\right)^\epsilon [1 + q^2\xi(1-\xi)/m^2]^\epsilon \simeq \frac{1}{\epsilon} + \log\frac{m^2 + q^2\xi(1-\xi)}{\mu^2},$$

$$\frac{1}{\epsilon}\left(\frac{m^2}{\mu^2}\right)^\epsilon \simeq \frac{1}{\epsilon} + \log\frac{m^2}{\mu^2}. \tag{51.29}$$

We now go back to the photon mass regularization and define

$$f_{IR}(q^2) = \int_0^1 d\xi \frac{m^2 + q^2/2}{m^2 + q^2\xi(1-\xi)}, \tag{51.30}$$

and, since generically $\log(m^2 + q^2\xi(1-\xi))/\mu_{ph}^2$ does not vary much, and in any case for $\mu_{ph} \to 0$ it doesn't matter too much what exactly we have in the log, we can write $\log(q^2 \text{ or } m^2)/\mu_{ph}^2$ and take it out of the integral in $F^{(1)}(q^2)$, and adding the tree-level part $F_1^{(0)}(q^2) = 1$, we finally have

$$F_1(q^2) = 1 - \frac{\alpha}{2\pi}f_{IR}(q^2)\log\frac{q^2 \text{ or } m^2}{\mu_{ph}^2} + \mathcal{O}(\alpha^2). \tag{51.31}$$

Note that in the $q^2 \to \infty$ limit most of the integral in f_{IR} comes from the two endpoints, $\xi \simeq 0$ and $\xi \simeq 1$, so

$$f_{IR}(q^2) \simeq \frac{1}{2}\int_0^1 d\xi \frac{q^2}{q^2\xi + m^2} + \frac{1}{2}\int^1 d\xi \frac{q^2}{q^2(1-\xi) + m^2} \simeq \log\frac{q^2}{m^2}. \tag{51.32}$$

In this limit we can write $F_1^{(1)}(q^2)$ as

$$F_1^{(1)}(q^2 \to \infty) \simeq -\frac{\alpha}{2\pi}\frac{1}{2}\int_0^1 d\xi \left[\frac{1}{\xi(1-\xi) + m^2/q^2}\log\frac{q^2}{\mu_{ph}^2}\left[\xi(1-\xi) + \frac{m^2}{q^2}\right] - \log\frac{m^2}{\mu_{ph}^2}\right], \tag{51.33}$$

and after splitting $\log[(q^2/\mu_{ph}^2)(\xi(1-\xi) + m^2/q^2)]$ into two logs, the second log term (with the overall minus removed) is small:

$$\frac{\alpha}{2\pi}\frac{1}{2}\int_0^1 d\xi \frac{1}{\xi(1-\xi) + m^2/q^2}\log\left[\xi(1-\xi) + \frac{m^2}{q^2}\right] \simeq \frac{a}{2\pi}\log^2\frac{q^2}{m^2} \ll \frac{\alpha}{2\pi}\log\frac{q^2}{\mu_{ph}^2}, \tag{51.34}$$

so can be neglected, and we obtain

$$F_1(q^2 \to \infty) \simeq 1 - \frac{\alpha}{2\pi}\log\frac{q^2}{m^2}\log\frac{q^2}{\mu_{ph}^2} + \mathcal{O}(\alpha^2) = 1 - \frac{\alpha}{2\pi}f_{IR}(q^2 \to \infty)\log\frac{q^2}{\mu_{ph}^2} + \mathcal{O}(\alpha^2). \tag{51.35}$$

The double log that we obtained is called a *Sudakov double log*. Note though that only one of the logs is IR divergent (as $\mu_{ph}^2 \to 0$), the other one is a physical log.

In dimensional regularization, we can rewrite (51.28) as

$$F_1^{(1)}(q^2)|_{\epsilon \to 0} \simeq \frac{\alpha}{2\pi}\frac{\mu^{2\epsilon}}{\epsilon}\frac{1}{2}\int_0^1 d\xi \left[\left(\frac{q^2}{\mu^2}\right)^\epsilon \frac{-1 - 2m^2/q^2}{[\xi(1-\xi) + m^2/q^2]^{1-\epsilon}} + 2\left(\frac{m^2}{\mu^2}\right)^\epsilon\right], \tag{51.36}$$

which makes it more obvious that we can always write it, after dropping the $1/\epsilon$ term, as

$$F_1^{(1)}(q^2)|_{\epsilon \to 0} \simeq -\frac{\alpha}{2\pi} f_{IR}(q^2) \log \frac{q^2}{\mu_{\rm ph}^2}, \tag{51.37}$$

where the $\log(q^2/\mu_{\rm ph}^2)$ appears from the expansion of the $(1/\epsilon)(q^2/\mu^2)^\epsilon$. Then the approximation of having $f_{IR}(q^2)$ as in (51.30) corresponds to ignoring the ϵ power of $[\xi(1 - \xi) + m^2/q^2]$, since the corresponding integral is finite, so that gives an $\mathcal{O}(\epsilon)$ correction; even though multiplication with the overall $1/\epsilon$ means we get a finite contribution also.

51.4 Cancellation of IR Divergence by Photon Emission

So we obtained an IR divergence of amplitudes that cannot be removed by renormalization. This would seem to be bad, but as we already explained, this is just a statement that the amplitude (or rather, the cross-section obtained from it) for just this process is not something physical in a theory like QED with massless particles.

That is so since we can emit the massless particles (photons) from external lines, and if the photons have sufficiently small energies, they cannot be independently detected by any physical detector, that will always have a minimal energy cut-off. Note that in QED, we only have soft divergences, so we only need to be concerned about soft emitted photons (with energies $< E_{min}$). But in a theory like QCD, which also has collinear divergences, we also need to be concerned about emitting photons that are not soft (have large energies), but instead are collinear with the particles from which they are emitted, so they cannot be distinguished from the emitted particles by a detector which has a resolution of some minimal angle θ_{min}.

Therefore we need to consider the process of emission of a soft photon off the electron lines, leading to two diagrams. Consider the tree-level diagram with amplitude $\mathcal{M}_0$, in our case the tree vertex for two fermions, one with momentum p_1 (in) and one with momentum p_2 (out) and a photon with momentum $q = p_2 - p_1$. Then the first correction diagram is for a photon with momentum k to be emitted from the (initial momentum) p_1 line, turning it into $p_1 - k$, as in Figure 51.3(a), and the second diagram is for the photon with momentum

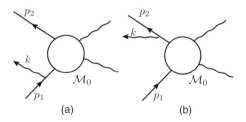

(a)   (b)

Figure 51.3   Photon emission diagrams canceling the one-loop divergences.

k to be emitted from the (final momentum) p_2 line, turning the line into $p_2 + k$ when it starts from $\mathcal{M}_0$, as in Figure 51.3(b).

We also assume that the photon is soft (i.e. $|\vec{k}| \ll |\vec{p}_2 - \vec{p}_1|$), in which case the subdiagram $\mathcal{M}_0$ for the two diagrams is the same, and the same with the tree diagram, that is

$$\mathcal{M}_0(p_2, p_1 - k) \simeq \mathcal{M}_0(p_2 + k, p_1) \simeq \mathcal{M}_0(p_2, p_1) \equiv \mathcal{M}_0. \tag{51.38}$$

Then the amplitude for emission of a photon from one of the external lines is

$$i\mathcal{M} = e\bar{u}(p_2)\left[\mathcal{M}_0 \frac{-(\not{p}_1 - \not{k} + im)}{(p_1 - k)^2 + m^2}\gamma^\mu \epsilon_\mu^*(k) + \gamma^\mu \epsilon_\mu^*(k)\frac{-(\not{p}_2 + \not{k} + im)}{(p_2 + k)^2 + m^2}\mathcal{M}_0\right]u(p_1). \tag{51.39}$$

Because p_1 and p_2 are on-shell, the two denominators are $-2p_1 \cdot k$ and $+2p_2 \cdot k$, respectively. Since the photon is soft, we can neglect $\not{k}$ in the numerator (small compared to $\not{p}_i$ and m). Also using the identities

$$(\not{p}_1 + im)\gamma^\mu \epsilon_\mu^*(k)u(p_1) = 2p_1^\mu \epsilon_\mu^*(k)u(p_1) + \gamma^\mu \epsilon_\mu^*(k)(-\not{p}_1 + im)u(p_1) = 2p_1^\mu \epsilon_\mu^*(k)u(p_1), \tag{51.40}$$

where in the first equality we have used $\{\gamma^\mu, \gamma^\nu\} = 2g^{\mu\nu}$ and in the second we have used the Dirac equation written for the spinor $u(p_1)$, and the similarly proven one

$$\bar{u}(p_2)\gamma^\mu \epsilon_\mu^*(k)(\not{p}_2 + im) = \bar{u}(p_2)2p_2^\mu \epsilon_\mu^*(k), \tag{51.41}$$

we finally write the amplitude as

$$i\mathcal{M} = -\bar{u}(p_2)\mathcal{M}_0(p_2, p_1)u(p_1)\, e\left[\frac{p_2 \cdot \epsilon^*(k)}{p_2 \cdot k} - \frac{p_1 \cdot \epsilon^*(k)}{p_1 \cdot k}\right]. \tag{51.42}$$

Integrating over the momentum of the photon and summing over its polarizations, we obtain the differential cross-section for emission of a photon as a function of the cross-section without photon emission:

$$d\sigma(p_1 \to p_2 + \gamma) = d\sigma(p_1 \to p_2)\int \frac{d^3k}{(2\pi)^3}\frac{1}{2k}\sum_{\lambda=1,2} e^2 \left|\frac{p_2 \cdot \epsilon^{(\lambda)}(k)}{p_2 \cdot k} - \frac{p_1 \cdot \epsilon^{(\lambda)}(k)}{p_1 \cdot k}\right|^2. \tag{51.43}$$

The differential probability for a photon of momentum k is then

$$d\mathcal{P}(p_1 \to p_2 + k) = \frac{d^3k}{(2\pi)^3}\sum_\lambda \frac{e^2}{2k}\left|\vec{\epsilon}_{(\lambda)} \cdot \left(\frac{\vec{p}_2}{p_2 \cdot k} - \frac{\vec{p}_1}{p_1 \cdot k}\right)\right|^2. \tag{51.44}$$

The total probability, integrating $|k|$ between the regulator μ_{ph} (photon mass) and a maximum of $|q|$ (since the condition for a soft photon was $|k| \ll |q|$, where $\vec{q} = \vec{p}_2 - \vec{p}_1$), is

$$\mathcal{P} \simeq \frac{\alpha}{\pi}\int_{\mu_{\text{ph}}}^{|q|}\frac{dk}{k}\mathcal{I} = \frac{\alpha}{2\pi}\log\frac{q^2}{\mu_{\text{ph}}^2}\mathcal{I}. \tag{51.45}$$

For the differential cross-section summed over k, we then find

$$d\sigma(p_1 \to p_2 + k) = d\sigma(p_1 \to p_2)\frac{\alpha}{2\pi}\log\frac{q^2}{\mu_{\text{ph}}^2}\mathcal{I}. \tag{51.46}$$

We will not describe here the calculation of $\mathcal{I}$ (it can be found for instance in [1]), but one finds

$$\mathcal{I}(q^2 \to \infty) \simeq 2 \log \frac{q^2}{m^2}. \qquad (51.47)$$

If follows then that the differential cross-section for $p_1 \to p_2$, taking into account the one-loop IR divergence, is written in terms of the tree-level process as (from $|F_1^{(0+1)}(q^2)|^2$)

$$\frac{d\sigma}{d\Omega}(p_1 \to p_2) = \left(\frac{d\sigma}{d\Omega}\right)_0 (p_1 \to p_2) \left[1 - \frac{\alpha}{\pi} \log \frac{q^2}{m^2} \log \frac{q^2}{\mu_{ph}^2} + \mathcal{O}(\alpha^2)\right]. \qquad (51.48)$$

In contrast, we express the differential cross-section for $p_1 \to p_1 + \gamma$ in terms of the same tree-level process (without γ, described by $\mathcal{M}_0$) as

$$\frac{d\sigma}{d\Omega}(p_1 \to p_2 + \gamma) = \left(\frac{d\sigma}{d\Omega}\right)_0 (p_1 \to p_2) \left[+\frac{\alpha}{\pi} \log \frac{q^2}{m^2} \log \frac{q^2}{\mu_{ph}^2} + \mathcal{O}(\alpha^2)\right]. \qquad (51.49)$$

This means that their sum is independent of μ_{ph}^2 (i.e. it is IR finite!). Note that this is an abstract result, because we cannot really measure the total cross-section with emission of a γ of arbitrary energy.

More physically, we can consider a detector that has an energy resolution of E_{min} (i.e. it cannot detect photons of smaller energy). Then the process with emission of a photon of smaller energy is considered as part of the process without emission. We need then to integrate $\int_{\mu_{ph}}^{E_{min}} dk/k = 1/2 \ln E_{min}^2/\mu_{ph}^2$. We also can prove (but this will not be done here) that

$$\mathcal{I}(q^2) = 2f_{IR}(q^2), \quad \forall q^2. \qquad (51.50)$$

We thus obtain

$$\frac{d\sigma}{d\Omega}(p_1 \to p_2 + \gamma(k \le E_{min})) = \left(\frac{d\sigma}{d\Omega}\right)_0 (p_1 \to p_2) \left[+\frac{\alpha}{\pi} f_{IR}(q^2) \log \frac{E_{min}^2}{\mu_{ph}^2} + \mathcal{O}(\alpha^2)\right],$$

$$\qquad (51.51)$$

and as before

$$\frac{d\sigma}{d\Omega}(p_1 \to p_2) = \left(\frac{d\sigma}{d\Omega}\right)_0 (p_1 \to p_2) \left[1 - \frac{\alpha}{\pi} f_{IR}(q^2) \log \frac{q^2 \text{ or } m^2}{\mu_{ph}^2} + \mathcal{O}(\alpha^2)\right], \qquad (51.52)$$

for a total of

$$\begin{aligned}
\left.\frac{d\sigma}{d\Omega}\right|_{\text{measured}} &= \frac{d\sigma}{d\Omega}(p_1 \to p_2) + \frac{d\sigma}{d\Omega}(p_1 \to p_2 + \gamma(\le E_{min})) \\
&= \left(\frac{d\sigma}{d\Omega}\right)_0 (p_1 \to p_2) \left[1 - \frac{\alpha}{\pi} f_{IR}(q^2) \log \frac{q^2 \text{ or } m^2}{E_{min}^2} + \mathcal{O}(\alpha^2)\right]. \qquad (51.53)
\end{aligned}$$

51.5 Summation of IR Divergences and Sudakov Factor

We now resum the one-loop diagrams on the side of virtual photons (loop corrections to the vertex), of the type of exchanging photons between the two fermion lines, but planarly ("parallel," they do not cross), and correspondingly on the photon emission side we consider the process of emission of n photons, one after another, from the external lines. In a way similar to the calculation sketched in Chapter 49 for the Alrarelli–Parisi evolution equation, we obtain that this process resums, giving a factor

$$\left[\frac{\alpha}{\pi} \log \frac{q^2}{m^2} \log \frac{q^2}{\mu_{\text{ph}}^2} \right]^n . \tag{51.54}$$

But moreover, there is a symmetry factor of $1/n!$ in front of this, because emitted photons are indistinguishable, which means that when we sum these contributions, the one-loop IR divergence exponentiates!

The exponentiation of IR divergences is rigorously proved in a theorem by Bloch and Nordsieck (1937).

Up to now, we have considered the QED vertex at one loop, but an important property of IR divergences is *factorization*, which means that in some physical process the "soft" and "hard" contributions factorize, as in Figure 51.4. We can split the process into a "hard" part (with large momenta), for scattering of two fermions with some other stuff, which we called $\mathcal{M}_0$ previously (note that we had assumed that $\mathcal{M}_0$ was the tree vertex, but we can easily check that it did not matter what $\mathcal{M}_0$ was, as long as it was a hard process) and a "soft" part (with small momenta) for scattering of two fermions to two fermions, possibly with soft photon emission.

Then the contribution to the soft part of soft *virtual* photons (i.e. of photons exchanged between the two fermion lines) gives a factor

$$\left[-\frac{\alpha}{\pi} f_{IR}(q^2) \log \frac{q^2}{\mu_{\text{ph}}^2} \right]^n \frac{1}{n!} \equiv \frac{X^n}{n!} . \tag{51.55}$$

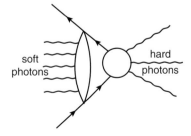

soft photons   hard photons

Figure 51.4   Factorization of the amplitude into a "hard" part and a "soft" part that governs the emission of soft photons.

In contrast, the contribution to the soft part of the soft *real* (emitted) photons (i.e. photons emitted from the two fermion lines) gives a factor of

$$\frac{1}{n!}\left[\frac{\alpha}{\pi}\mathcal{I}\log\frac{E_{min}}{\mu_{\mathrm{ph}}}\right]^n = \left[\frac{\alpha}{\pi}f_{IR}(q^2)\log\frac{E_{min}^2}{\mu_{\mathrm{ph}}^2}\right]^n\frac{1}{n!}. \tag{51.56}$$

This means that in the measured differential cross-section, we have

$$\left(\frac{d\sigma}{d\Omega}\right)_{measured} = \left(\frac{d\sigma}{d\Omega}\right)_0(p_1 \to p_2)\left(\sum_n\frac{X^n}{n!}\right)\left(\sum_m\frac{Y^m}{m!}\right)$$

$$= \left(\frac{d\sigma}{d\Omega}\right)_0(p_1 \to p_2)\exp\left[-\frac{\alpha}{\pi}f_{IR}(q^2)\log\frac{q^2}{\mu_{\mathrm{ph}}^2}\right]\exp\left[\frac{\alpha}{\pi}f_{IR}(q^2)\log\frac{E_{min}^2}{\mu_{\mathrm{ph}}^2}\right]$$

$$= \left(\frac{d\sigma}{d\Omega}\right)_0(p_1 \to p_2)\exp\left[-\frac{\alpha}{\pi}f_{IR}(q^2)\log\frac{q^2}{E_{min}^2}\right]. \tag{51.57}$$

The exponential factor is called a *Sudakov form factor*.

Once again, note that this exponential factor is only the resummation of one-loop diagrams, it is not genuinely two-loop, meaning that when expanding it in α, only the leading term is exact. In general, at each loop order we will have another power of α contribution in the exponential:

$$\exp[\alpha(\ldots) + \alpha^2(\ldots) + \alpha^3(\ldots) + \ldots]. \tag{51.58}$$

Important Concepts to Remember

- Collinear IR divergences are due to massless particles ("photons") collinear with massless external states. For virtual particles, we have IR divergences in loops, for real particles, we have IR divergences for particle emission. It appears due to the self-interaction of massless states, so is present in QCD, but not in QED.
- Soft IR divergences are due to massless particles ("photons") being soft (very small momenta). For virtual particles, we have IR divergences in loops, for real particles, we have IR divergences for particle emission. It appears independently of the external states, and is present in both QCD and QED.
- In dimensional regularization, IR divergences appear due to factors of $1/\epsilon(P^2/\mu^2)^\epsilon$ in QED (just soft divergences) or $1/\epsilon^2(P^2/\mu^2)^\epsilon$ in QCD (soft and collinear divergences), where P^2 is some relevant invariant: in the case of nonabelian YM in the planar limit, it is $-s_{i,i+1} = -(k_i+k_{i+1})^2$ for consecutive external massless lines.
- The IR divergences can be regulated by including a mass μ for the massless particles, or by dimensional regularization, with $D = 4 - 2\epsilon$, but with $\epsilon_{UV} = -\epsilon_{IR} < 0$.
- In QED, the divergence in the form factor is $F_1 = 1 - \alpha/(2\pi)f_{IR}(q^2)\log(q^2$ or $m^2)/m^2$, or at large q^2 is $F_1(q^2) = 1 - \alpha/(2\pi)\log(q^2/m^2)\log(q^2/\mu_{\mathrm{ph}}^2)$, which is the Sudakov double logarithm.

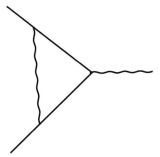

Figure 51.5   IR-divergent diagram.

- The IR divergences cancel order by order in α between processes with virtual photons (loop corrections) and processes with real photons (emission of soft or collinear photons).
- We can resum the one-loop corrections, and correspondingly the multiple emissions of photons from the external lines, and obtain that the one-loop divergences exponentiate, obtaining the Sudakov form factor.

Further Reading

See sections 12.1–12.3 in [2], and sections 6.4 and 6.5 in [1].

Exercises

1. Consider scalar-QED, that is a photon γ coupled to a massive (charged) complex scalar ϕ. Calculate $f_{IR}(q^2)$ at one loop ($\mathcal{O}(g^2)$ correction) for the vertex $V(p_2, p_1) \equiv F_1(q^2)$ in Figure 51.5.
2. Calculate (using the same formulas from the text, and ignoring the same calculations) the Sudakov form factor at one loop ($\mathcal{O}(g^2)$).
3. Isolate the diagram and region of integration that gives the double (soft collinear) IR divergence of nonabelian gauge theory at one loop.
4. Write down the QED two-loop diagrams that are IR divergent, and show which parts are the IR divergences, and how they should cancel.

52

IR Safety and Renormalization in QCD: General IR-Factorized Form of Amplitudes

In this chapter, we will analyze IR divergences in QCD. To start, however, we will give some more results about QED that easily generalize to QCD. We will consider the eikonal approximation for the QED vertex, and define better the exponentiation and factorization of IR divergences in a way that can be generalized to QCD. Then we consider the QCD process $e^+ e^- \to$ hadrons, and we will see that the cross-section is free from IR divergences (IR safe). From the renormalization group (RG) equation for this cross-section, we find a form for the running coupling of QCD, $\bar{\alpha}_s(\mu^2)$. Finally, we describe how IR factorize and exponentiate in QCD amplitudes.

52.1 QED Vertex: Eikonal Approximation, Exponentiation, and Factorization of IR Divergences

As we saw in Chapter 51, to IR-regularize we can use dimensional regularization, but in $D = 4 - 2\epsilon$, $\epsilon > 0$ for UV divergences, whereas $\epsilon < 0$ for IR divergences. We also saw how to map between photon mass regularization and dimensional regularization. We only showed how to map the divergent terms, but the exact map actually contains some finite piece as well:

$$\ln \frac{\mu_{\text{ph}}^2}{m^2} \leftrightarrow \Gamma(\epsilon) \left(\frac{4\pi \mu^2}{m^2} \right)^\epsilon = \frac{\Gamma(1+\epsilon)}{\epsilon} \left(\frac{4\pi \mu^2}{m^2} \right)^\epsilon = \frac{1}{\epsilon} - \gamma + \ln \frac{4\pi \mu^2}{m^2}. \tag{52.1}$$

This relation can be derived by computing $\Gamma_{\alpha\beta}^\mu$ in both mass and dimensional regularization, and comparing the results.

Generalizing a bit the result from Chapter 51, we saw that the same integral having both UV and IR divergences (at different endpoints) needs a different dimensional regularization for each, so we write

$$\frac{\Gamma(D/2)}{\pi^{D/2} i} \int \frac{d^D q}{(q^2)^n} = \frac{1}{D/2 - n} - \frac{1}{D'/2 - n}, \tag{52.2}$$

$D = 4 - 2\epsilon$, $D' = 4 - 2\epsilon'$ for $n = 2$, and only at the end of the calculation do we set $D = D'$.

In particular, $\Gamma_{\alpha\beta}^\mu$ still has UV divergences, that is dimensionally regularized with D, whereas the IR divergence is either regularized with mass, or dimensional regularization with D'. One finds

$$\Gamma^\mu|_{\text{mass reg.}} = \frac{\alpha}{4\pi}\left[\gamma_\mu\left(\frac{1}{\epsilon} - \gamma + \ln\frac{4\pi\mu^2}{m^2} + \frac{v^2+1}{v}\ln\frac{v+1}{v-1}\ln\frac{\mu_{\text{ph}}^2}{m^2} + F(v)\right)\right.$$
$$\left. - \frac{(p_1-p_2)_\mu}{2m}\frac{v^2-1}{v}\ln\frac{v+1}{v}\right],$$

$$\Gamma^\mu|_{\text{dim.reg.}} = \frac{\alpha}{4\pi}\left(\frac{4\pi\mu^2}{m^2}\right)^\epsilon\Gamma(1+\epsilon)\left[\gamma_\mu\left(\frac{1}{\epsilon'} + \frac{1}{\epsilon}\frac{v^2+1}{v}\ln\frac{v+1}{v-1} + F(v)\right)\right.$$
$$\left. - \frac{(p_1-p_2)_\mu}{2m}\frac{v^2-1}{v}\ln\frac{v+1}{v}\right],\tag{52.3}$$

where $v = \sqrt{1 + 4m^2/q^2}$ and $F(v)$ is a given function of v, of some complicated form, and finite.

We see that by expanding in ϵ $\left(\frac{4\pi\mu^2}{m^2}\right)^\epsilon\Gamma(1+\epsilon)$ as a coefficient of the $\ln(v+1)/(v-1)$ term, and keeping the ϵ-finite part, we have a match between the dimensional regularization and the mass regularization results only if we have (52.1).

The above was for QED, but in QCD the same calculation holds, changing external electrons for external quarks, and photons for gluons, and then we also exchange α for $\alpha_s C_F$, where $\alpha_s = g^2/(4\pi)$.

In QED, we saw that IR divergences exponentiate and factorize. We make this a bit more formal now, to compare with QCD.

At one loop, the IR-divergent part of the vertex can be expressed in the form

$$\bar{u}(p_2)\Gamma^{(IR)}_\mu u(p_1) = e\mu^\epsilon\bar{u}(p_2)\gamma_\mu u(p_1)\alpha\Gamma(\epsilon, q^2/m^2),\tag{52.4}$$

and from the calculation of Chapter 51 we, one can show (but we will not do it here, it is left as one of the exercises) that we can put the resulting Γ in the form

$$\alpha\Gamma(\epsilon, q^2/m^2) = -\frac{1}{2}\left(e\mu^\epsilon\right)^2\int\frac{d^Dk}{(2\pi)^D}\frac{-i}{k^2-i\epsilon}\left[\frac{2p_1^\alpha}{2p_1\cdot k + k^2 - i\epsilon} - \frac{2p_2^\alpha}{2p_2\cdot k + k^2 - i\epsilon}\right]^2.\tag{52.5}$$

The approximation that we made in Chapter 51, of considering for the leading IR divergence that k^2 can be neglected with respect to $2p_i\cdot k$ in the denominators, is called the *eikonal approximation*, or the *leading divergence approximation*, so

$$\alpha\Gamma^{\text{eik}}(\epsilon) = -\frac{1}{2}\left(e\mu^\epsilon\right)^2\int\frac{d^Dk}{(2\pi)^D}\frac{-i}{k^2-i\epsilon}\left[\frac{2p_1^\alpha}{2p_1\cdot k - i\epsilon} - \frac{2p_2^\alpha}{2p_2\cdot k - i\epsilon}\right]^2.\tag{52.6}$$

Note that Γ^{eik} does not depend anymore on q^2/m^2, as we can check.

We saw in Chapter 51 that IR divergences exponentiate and factorize into a hard part, containing the large momenta, and a soft part, containing the IR divergences in exponential form. Formally then

$$\Gamma_\mu(p_1, p_2) = e^{\alpha\Gamma^{\text{eik}}}\Gamma^{(H)}_\mu,\tag{52.7}$$

where $\Gamma^{(H)}_\mu$ is the hard part. Since, however, as things stand $\Gamma^{(H)}_\mu$ still has *UV divergences*, one can also write the above as

$$\Gamma_\mu(p_1, p_2) = e^{\alpha\Gamma(\epsilon, M^2/m^2)}\Gamma^{(\text{finite})}_\mu,\tag{52.8}$$

where $q^2 \equiv -M^2$ and

$$\Gamma_\mu^{\text{(finite)}} = \Gamma_\mu^{(H)} e^{\alpha \Gamma^{\text{eik}}(\epsilon) - \alpha \Gamma(\epsilon, M^2/m^2)}. \tag{52.9}$$

52.2 IR Safety in QCD for Cross-Section for $e^+ e^- \rightarrow$ Hadrons and Beta Function

In QCD, the coupling constant depends on scale and is strong at low energy, $\alpha_s \rightarrow \infty$. This means that it is very hard to define physical quantities, since besides the IR divergences we also have to deal with infinite coupling. Quantities that are free of IR divergences are called *IR safe*. As we already saw, cross-sections, that are certainly observable, are supposed to be IR safe. One way to express this is that as $m^2 \ll q^2$ (here m is the quark mass), we have

$$\sigma\left(\frac{q^2}{\mu^2}, g_s, \frac{m^2}{\mu^2}\right) = \bar{\sigma}\left(\frac{q^2}{\mu^2}, g_s\right) + \mathcal{O}\left[\left(\frac{m^2}{q^2}\right)^b\right], \tag{52.10}$$

where $b > 0$, and then $\bar{\sigma}$ is completely finite. Since in asymptotically free theories the renormalized mass satisfies $m_R(\mu \rightarrow \infty) \rightarrow 0$, the zero mass limit makes sense and moreover, the high energy ($q \rightarrow \infty$) or zero mass ($m \rightarrow 0$) limits are equivalent.

Moreover, we can use cross-sections to define the running of the coupling constant. We will explain this with the example of a simple process, very relevant experimentally.

52.2.1 Born Cross-Section for $e^+ e^- \rightarrow (q\bar{q}) \rightarrow$ Hadrons

Consider the tree diagram for $e^+ e^- \rightarrow q\bar{q}$ through an intermediate photon. The quarks have charge $Q_f e$. For the cross-section, where we have the quantity $|\mathcal{M}|^2 = \mathcal{M}\mathcal{M}^*$, we can draw diagrams with $\mathcal{M}$ and $\mathcal{M}$ flipped (time-reversed) for $\mathcal{M}^*$, forming for instance a one-loop diagram from a tree one, with the quarks in the loop being on-shell, as in Figure 52.1. It is related to the optical theorem, where cutting loop diagrams gives diagrams for the cross-section, but here we can just think of it as a useful trick.

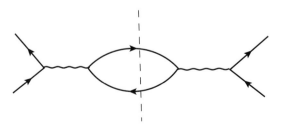

Figure 52.1 Diagram for the Born cross-section via unitarity cut.

We will not reproduce the calculation, which is left as one of the exercises, but one finds that

$$\frac{d\sigma}{d\cos\theta} = N\frac{\pi\alpha^2}{2s}\sum_f Q_f^2(1+\cos^2\theta),\tag{52.11}$$

which integrates to ($\cos\theta$ varies between -1 and $+1$)

$$\sigma_{\text{tot}} = N\frac{4\pi\alpha^2}{3s}\sum_f Q_f^2.\tag{52.12}$$

Experimentally, one defines the ratio of the process with hadrons in the final state to the one with leptons in the final state, namely $e^+e^- \to \mu^+\mu^-$:

$$R \equiv \frac{\sigma_{\text{tot}}(e^+e^- \to \text{hadrons})}{\sigma_{\text{tot}}(e^+e^- \to \mu^+\mu^-)} = N\sum_f Q_f^2,\tag{52.13}$$

where in the last equality we have considered our tree-level process. This quantity is measured experimentally very well, and is one of the most stringent tests of QCD: it depends on the fact of there being three colors and on the total number of quarks and their charges. It agrees perfectly, leaving no room for extra quarks.

We are, however, interested in corrections to this result from QCD processes (i.e. corrections of order $\alpha\alpha_s$, where $\alpha_s = g^2/(4\pi)$). These would be one-loop diagrams with a QCD loop, or two-loop cut diagrams for σ. There are eight such diagrams, and the calculation is long, so we will only parametrize the result and show the final answer. Details can be found in [2].

For the *total cross-section* for $e^+e^- \to$ hadrons, we need to consider cut diagrams with $e^+e^- \to \gamma \to$ hadrons$\to \gamma \to e^+e^-$, and the sum is over possible intermediate hadrons. This can be parametrized in the following way:

$$\sigma_{\text{tot}}(q^2) = \left[\frac{e^2\mu^{2\epsilon}}{2(q^2)^3}\right](k_1^\mu k_2^\nu + k_2^\mu k_1^\nu - k_1 \cdot k_2 g^{\mu\nu})H_{\mu\nu}(q^2),\tag{52.14}$$

where

$$H_{\mu\nu}(q^2) = e^2\mu^{2\epsilon}\sum_n \langle 0|j_\mu(0)|n\rangle\langle n|j_\nu(0)|0\rangle(2\pi)^4\delta^4(p_n - q).\tag{52.15}$$

Here the sum over n is over hadronic states, the states in the cut loop, and j_μ is the electromagnetic current. Moreover, $H_{\mu\nu}(q)$ must be transverse, which means that

$$H_{\mu\nu}(q^2) = (q_\mu q_\nu - q^2 g_{\mu\nu})H(q^2).\tag{52.16}$$

After a long calculation, where we split the contributions to $H_{\mu\nu}$, similarly to the QED case, into contributions of real and virtual gluons (i.e. real gluons are emitted gluons, thus gluons in the hadronic state $|n\rangle$, and virtual gluons are not), one finds

$$-g^{\mu\nu}H_{\mu\nu}^{\text{real}} = +2NC_2(F)Q_f^2\frac{\alpha\alpha_s}{\pi}q^2\left(\frac{4\pi\mu^2}{q^2}\right)^{2\epsilon}$$
$$\times\left[\frac{1-\epsilon}{\Gamma(2-2\epsilon)}\right]\left[\frac{1}{\epsilon^2} + \frac{3}{2\epsilon} - \frac{\pi^2}{2} + \frac{19}{4} + \mathcal{O}(\epsilon)\right],$$

$$-g^{\mu\nu}H_{\mu\nu}^{\text{virtual}} = -2NC_2(F)Q_f^2\frac{\alpha\alpha_s}{\pi}q^2\left(\frac{4\pi\mu^2}{q^2}\right)^{2\epsilon}$$

$$\times\left[\frac{1-\epsilon}{\Gamma(2-2\epsilon)}\right]\left[\frac{1}{\epsilon^2}+\frac{3}{2\epsilon}-\frac{\pi^2}{2}+4+\mathcal{O}(\epsilon)\right]. \tag{52.17}$$

We see that, while individually they are both IR divergent, in their sum the divergences cancel and, except for the overall factor, the brackets cancel except for a finite contribution $19/4 - 4 = 3/4$. Finally we obtain for the total cross-section, with the leading contribution due to $q\bar{q}$ production (and soft gluons above):

$$\sigma_{\text{tot}}(q^2) = \frac{N4\pi\alpha^2}{3q^2}\sum_f Q_f^2\left[1+\frac{\alpha_s}{\pi}\frac{3}{4}C_2(F)+\mathcal{O}(\alpha^2\alpha_s^2)\right]. \tag{52.18}$$

We can now determine α_s from σ_{tot}, which will give us the dependence $\alpha_s(\mu^2)$. At the first order treated above, however, there is no dependence on the renormalization scheme and on μ, as we can see.

At two-loop order, however, we will get inside the square bracket a contribution $(\alpha_s/\pi)^2 A_2(q^2/\mu^2)$, which is scheme-dependent and μ-dependent. Since, however, the total cross-section cannot depend on μ (which is a fictitious parameter, appearing from dimensional transmutation), it means that we can infer $\alpha_s(\mu)$ from $A_2(q^2/\mu^2)$ (the total μ dependence should cancel). A more formal way to express this is that we have a renormalization group equation (RGE) for σ_{tot} saying that the total μ dependence is zero, or

$$\left[\mu\frac{\partial}{\partial\mu}+\beta(g)\frac{\partial}{\partial g}\right]\sigma_{\text{tot}}(q^2,\mu^2,\alpha_s(\mu^2)) = 0, \tag{52.19}$$

where as usual $\beta(g) = \mu\partial g/\partial\mu$. Since μ is an arbitrary scale, we are in principle free to choose whatever we want. A useful choice is actually $\mu^2 = q^2$, which makes the running coupling depend on the physical momentum scale in the process, $\alpha_s = \alpha_s(q^2)$.

An *a priori* independent definition of α_s is from the beta function, which we will denote $\bar{\alpha}_s(\mu^2)$. From the definition of the beta function above, we have

$$\mu\frac{\partial}{\partial\mu}\frac{\bar{\alpha}_s}{\pi} = \frac{\bar{g}}{2\pi^2}\beta(\bar{g}_s) = \left(\frac{\bar{\alpha}_s}{\pi}\right)^2\frac{\beta_1}{2}+\left(\frac{\bar{\alpha}_s}{\pi}\right)^3\frac{\beta_2}{8}, \tag{52.20}$$

where one defines the numerical coefficients of the beta function as

$$\beta(g) = g\left[\frac{\alpha_s}{4\pi}\beta_1+\left(\frac{\alpha_s}{4\pi}\right)^2\beta_2+\ldots\right]. \tag{52.21}$$

We can integrate the equation from μ_0 to μ, and introducing a parameter Λ (or sometimes Λ_{QCD}), the QCD scale parameter, we find $\bar{\alpha}_s(\mu^2)$ as a function of $\ln(\mu/\Lambda)$ only.

With the nontrivial choice

$$\Lambda = \mu_0\exp\left\{\frac{2}{\beta_1}\left(\frac{\pi}{\bar{\alpha}_s(\mu_0)}-\frac{\beta_2}{4\beta_1}\ln\left[\frac{\pi}{\bar{\alpha}_s(\mu_0)}\left(1+\frac{\beta_2}{4\beta_1}\frac{\pi}{\bar{\alpha}_s(\mu_0)}\right)\right]+\frac{\beta_2}{4\beta_1}\ln\frac{-\beta_1}{4}\right)\right\}, \tag{52.22}$$

we find the somewhat simple form

$$\frac{\bar{\alpha}_s(\mu^2)}{\pi} = \frac{-2}{\beta_1 \ln(\mu/\Lambda)} - \frac{\beta_2}{\beta_1^3} \frac{\ln \ln \frac{\mu^2}{\Lambda^2}}{\ln^2 \frac{\mu}{\Lambda}} + \mathcal{O}\left(\frac{1}{\ln^3\left(\frac{\mu^2}{\Lambda^2}\right)}\right). \qquad (52.23)$$

The proof is left as one of the exercises.

52.3 Factorization and Exponentiation of IR Divergences in Gauge Theories

In QCD and nonabelian gauge theories in general we still have factorization and exponentiation of IR divergences, but the result is considerably more complicated. Here we will just give a flavor of it.

To start, we must describe the color structure of amplitudes in terms of a *color basis* $\mathcal{C}_{[i]}$. Indeed, we can convince ourselves that for all amplitudes, there are only a few color structures possible, and their coefficients are color-independent amplitudes, sometimes called color-ordered.

For instance, for amplitudes with external gluons, the external gluons are characterized only by an adjoint index a (in the case of QCD, saying which one of the eight possible gluons it belongs to), and the propagators and vertices have only delta functions for the fundamental indices $i, j, \ldots$, which means that these indices are summed over, obtaining either traces of matrices T^a with a an external gluon index, or $\delta_i^i = N$ factors. It follows that the possible color structures are all possible traces of T^a matrices.

For example, in the case of the 4-point scattering of gluons in a theory like QCD, where $\text{Tr}[T^a] = 0$, for instance for any $SU(N)$, we can have either single traces

$$\mathcal{C}_{[1]} = \text{Tr}[T^{a_1} T^{a_2} T^{a_3} T^{a_4}],$$

$$\cdots$$

$$\mathcal{C}_{[6]} = \text{Tr}[T^{a_1} T^{a_4} T^{a_3} T^{a_2}] \qquad (52.24)$$

(by cyclicity of the trace we can always put a_1 in the first position, and then there are six permutations of a_2, a_3, a_4), or double traces

$$\mathcal{C}_{[7]} = \text{Tr}[T^{a_1} T^{a_2}]\,\text{Tr}[T^{a_3} T^{a_4}],$$
$$\mathcal{C}_{[8]} = \text{Tr}[T^{a_1} T^{a_3}]\,\text{Tr}[T^{a_3} T^{a_4}],$$
$$\mathcal{C}_{[9]} = \text{Tr}[T^{a_1} T^{a_4}]\,\text{Tr}[T^{a_2} T^{a_3}]. \qquad (52.25)$$

Then the 4-point amplitude for gluons can be expanded in the above basis:

$$\mathcal{A}(1234) = \sum_{i=1}^{9} A_{[i]} \mathcal{C}_{[i]}, \qquad (52.26)$$

and this can be written as a vector with nine components:

$$|A\rangle = \begin{pmatrix} A_{[1]} \\ A_{[2]} \\ \dots \\ A_{[9]} \end{pmatrix}, \tag{52.27}$$

or rather, since we have a reflection symmetry relation:

$$A_n(12\dots n) = (-1)^n A_n(n\dots 21), \tag{52.28}$$

we can use only $A_{1234}, A_{1342}, A_{1423}$ and $A_{12;34}, A_{13;42}, A_{14;23}$.

In general, for the 4-point function for scattering of $2 \to 2$ partons (quarks and gluons), we have a similar story, with some color basis $\{\mathcal{C}_L\}$, and we expand in it as

$$\mathcal{A}^{[f]} = \sum_L A_L^{[f]}(\mathcal{C}_L) \tag{52.29}$$

and put the A_L coefficients in a vector $|A\rangle$.

In this case, we can write the *factorization* of the amplitude as

$$\left| A\left(\frac{s_{ij}}{\mu^2}, a(\mu^2), \epsilon\right) \right\rangle = J\left(\frac{Q^2}{\mu^2}, a(\mu^2), \epsilon\right) \mathbb{S}\left(\frac{s_{ij}}{Q^2}, \frac{Q^2}{\mu^2}, a(\mu^2), \epsilon\right) \left| H\left(\frac{s_{ij}}{\mu^2}, \frac{Q^2}{\mu^2}, a(\mu^2), \epsilon\right) \right\rangle, \tag{52.30}$$

where $s_{ij} = (k_i + k_j)^2$ are the possible kinematic invariants (the amplitude must depend on them) and $a(\mu^2)$ is the effective coupling:

$$a(\mu^2) = \frac{g^2 N}{8\pi^2}(4\pi e^{-\gamma})^\epsilon. \tag{52.31}$$

Here, J is called the *jet function* and is an IR-divergent scalar factor, $\mathbb{S}$ is called the *soft function* and is an IR-divergent matrix in color space, and $|H\rangle$ is called the *hard function* and contains only short-distance behavior (i.e. it is IR finite, as $\epsilon \to 0$).

The difference between the soft function and the jet function is that J contains all collinear dynamics, and as a result it contains (at one loop) all the $1/\epsilon^2$ poles (we said that one $1/\epsilon$ comes from soft divergences, and one $1/\epsilon$ from collinear divergences), whereas $\mathbb{S}$ is completely determined by the *anomalous dimension matrix* Γ, which will be defined better later on in the book, but essentially is a generalization of the anomalous dimension for scalar theories.

Note that we have introduced an arbitrary quantity Q, called the *factorization scale*. As we see, amplitudes are independent of Q, but its factorization into jet, soft, and hard functions depends on Q.

The soft function is defined from the anomalous dimension matrix Γ with components Γ_{LJ} (such that the amplitude is written as $A_L = J S_{LI} H_I$) as the renormalization group equation

$$\frac{d}{d\ln Q} S_{LI} = -\Gamma_{LJ} S_{JI}. \tag{52.32}$$

Note here that this has some similarity with the evolution equations. This is not a coincidence, and we will see in more detail in Chapter 53 that there is a connection between factorization and evolution. The solution of the above RGE (with some more input that will not be explained here) is written in the form

$$
\mathbb{S}\left(\frac{s_{ij}}{Q^2}, \frac{Q^2}{\mu^2}, a(\mu^2), \epsilon\right) = P \exp\left\{-\frac{1}{2}\int_0^{Q^2} \frac{d\tilde{\mu}^2}{\tilde{\mu}^2} \Gamma\left(\frac{s_{ij}}{Q^2}, \bar{a}\left(\frac{\mu^2}{\tilde{\mu}^2}, a(\mu^2), \epsilon\right)\right)\right\},
$$

(52.33)

where Γ is the anomalous dimension matrix, expanded in the coupling as

$$
\Gamma\left(\frac{s_{ij}}{Q^2}, a(\mu^2)\right) = \sum_{l=1}^{\infty} a(\mu^2)^l \Gamma^{(l)}\left(\frac{s_{ij}}{Q^2}\right),
$$

(52.34)

and at leading order (one loop), the coupling $\bar{a}$ is given by

$$
\bar{a}\left(\frac{\mu^2}{\tilde{\mu}^2}, a(\mu^2), \epsilon\right) = a(\mu^2)\left(\frac{\mu^2}{\tilde{\mu}^2}\right)^{\epsilon} \sum_{n=0}^{\infty}\left[\frac{\beta_1}{4\pi}\left(\left(\frac{\mu^2}{\tilde{\mu}^2}\right)^{\epsilon} - 1\right)a(\mu^2)\right]^n.
$$

(52.35)

The IR-divergent structure of gauge theories is given by:

1. $\gamma(a)$, the *cusp anomalous dimension*, or *Wilson line anomalous dimension*, or *soft anomalous dimension*. As the name suggests, this characterizes soft divergences, and also divergences arising from "cusps" (angles) in the Wilson line.
2. $\mathcal{G}_0(a)$, the *collinear anomalous dimension*, which characterizes the collinear divergences.

These functions are expanded in the coupling as

$$
\gamma(a) = \sum_{l=1}^{\infty} a^l \gamma^{(l)},
$$

$$
\mathcal{G}_0(a) = \sum_{l=1}^{\infty} a^l \mathcal{G}_0^{(l)}.
$$

(52.36)

In QCD, there is also the anomalous dimension matrix Γ, which is independent of the above two functions.

However, in a gauge theory called $\mathcal{N} = 4$ super-Yang–Mills (SYM), where many things can be computed exactly (we will not explain what $\mathcal{N} = 4$ SYM is, we use it just for purposes of illustration), there is no nontrivial Γ matrix, and in that case we can write down the jet function just in terms of $\gamma(a)$ and $\mathcal{G}_0(a)$, as

$$
J\left(\frac{Q^2}{\mu^2}, a, \epsilon\right) = \exp\left[-\frac{1}{2}\sum_{l=1}^{\infty} a^l \left(\frac{\mu^2}{Q^2}\right)^{l\epsilon}\left(\frac{\gamma^{(l)}}{(l\epsilon)^2} + \frac{2\mathcal{G}_0^{(l)}}{l\epsilon}\right)\right].
$$

(52.37)

Important Concepts to Remember

- The QED vertex is written as the exponent of the eikonal approximation vertex times a hard vertex.
- Cross-sections in QCD are IR safe.
- We can define the running $\alpha_s(\mu^2)$ from the cross-section (e.g. for e^+e^- into hadrons) by imposing that there is no dependence of μ for σ_{tot}.
- One can choose $\mu^2 = q^2$, and thus find $\alpha_s(q^2)$, but it is not necessary.
- QCD amplitudes can be decomposed into a color basis.
- QCD amplitudes factorize into a scalar jet function, a matrix soft function, and a vector hard function.
- The split (but not the amplitude) depends on an arbitrary factorization scale.
- The soft function is determined by the anomalous dimension matrix alone.
- The jet function contains all collinear dynamics, thus all double poles in ϵ, and is characterized by the cusp anomalous dimension and the collinear anomalous dimension, together with the anomalous dimension matrix.

Further Reading

See sections 12.3 and 12.4 in [2], and sections 6.1–6.3 in [12].

Exercises

1. Prove that

$$
\frac{\bar{\alpha}_s(\mu^2)}{\pi} = \frac{-2}{\beta_1 \ln(\mu/\Lambda)} - \frac{\beta_2}{\beta_1^3} \frac{\ln \ln \frac{\mu^2}{\Lambda^2}}{\ln^2 \frac{\mu}{\Lambda}} + \mathcal{O}\left(\frac{1}{\ln^3 \left(\frac{\mu^2}{\Lambda^2}\right)}\right) \tag{52.38}
$$

from

$$
\mu \frac{\partial}{\partial \mu} \frac{\bar{\alpha}_s}{\pi} = \left(\frac{\bar{\alpha}_s}{\pi}\right)^2 \frac{\beta_1}{2} + \left(\frac{\bar{\alpha}_s}{\pi}\right)^3 \frac{\beta_2}{8} \tag{52.39}
$$

for the choice of

$$
\Lambda = \mu_0 \exp\left\{\frac{2}{\beta_1}\left(\frac{\pi}{\bar{\alpha}_s(\mu_0)} - \frac{\beta_2}{4\beta_1} \ln\left[\frac{\pi}{\bar{\alpha}_s(\mu_0)}\left(1 + \frac{\beta_2}{4\beta_1}\frac{\bar{\alpha}_s(\mu_0)}{\pi}\right)\right] + \frac{\beta_2}{4\beta_1} \ln\frac{-\beta_1}{4}\right)\right\}. \tag{52.40}
$$

2. Consider the decomposition of a four-point gauge theory amplitude in coefficients c_{ijkl} related to box diagrams:

$$
\mathcal{A}(1234) = c_{1234} A_{1234} + c_{1342} A_{1342} + c_{1423} A_{1423}, \tag{52.41}
$$

Figure 52.2   Box diagram.

where

$$c_{1234} = \tilde{f}^{ea_1 b} \tilde{f}^{ba_2 c} \tilde{f}^{ca_3 d} \tilde{f}^{da_4 e}$$ (52.42)

is the color structure of the box diagram in Figure 52.2, where

$$\tilde{f}^{abc} = \mathrm{Tr}([T^a, T^b] T^c).$$ (52.43)

Express $c_{1234}, c_{1342}, c_{1423}$ in terms of $\mathcal{C}_{[i]}, i = 1, \ldots, 9$.

3. Prove that we can put the QED vertex in the form (52.5).
4. Prove that the Born differential cross-section for the $e^+ e^- \rightarrow$ hadrons is given by (52.11), using the information in the text.

Factorization and the Kinoshita–Lee–Nauenberg Theorem

In this chapter, we will first give a rather general theorem about the cancellation of IR divergences in any physical transition probability, which happens when summing over initial and final states, and then we will say some words about the relation between factorization and evolution.

53.1 The KLN Theorem

The theorem about finiteness of transition probabilities, when summing over initial and final states, is due to Kinoshita (1962) and independently Lee and Nauenberg (1964). It is somewhat more general than the specific case we are interested in, and it applies to a general quantum-mechanical system.

The formalism we will use is of quantum mechanics in the interaction picture. There, the time evolution of states is

$$i\partial_t |\psi(t)\rangle = g\hat{H}_{Ii}(t)|\psi(t)\rangle, \tag{53.1}$$

and $\hat{H}_{Ii}$ is the interaction part of the Hamiltonian, in the interaction picture, that is $H = H_0 + H_I$, and

$$\hat{H}_{Ii} = e^{iH_0 t}\hat{H}_{I,s}e^{-iH_0 t}. \tag{53.2}$$

The time evolution of states is with the evolution operator U:

$$|\psi(t)\rangle = U(t,t')|\psi(t')\rangle, \tag{53.3}$$

where

$$U(t,t') = T\exp\left[-i\int_t^{t'} dt'' g\hat{H}_{Ii}(t'')\right]. \tag{53.4}$$

The S-matrix is given by the matrix elements of

$$S = U(+\infty, -\infty) = U(0, +\infty)^\dagger U(0, -\infty). \tag{53.5}$$

Then the transition probability between state $|a\rangle$ and state $|b\rangle$ is written as

$$|\langle b|S|a\rangle|^2 = \sum_{ij}(R_{bij}^+)^* R_{aij}^-, \tag{53.6}$$

where

$$R^{\pm}_{aij} \equiv \langle i|U(0,\pm\infty)|a\rangle^* \langle j|U(0,\pm\infty)|a\rangle$$
$$= \langle j|U(0,\pm\infty)|a\rangle \langle a|U^\dagger(0,\pm\infty)|i\rangle, \tag{53.7}$$

so that

$$|\langle b|S|a\rangle|^2 = \langle j|U(0,-\infty)|a\rangle \langle a|U^\dagger(0,-\infty)|i\rangle$$
$$\langle i|U(0,+\infty)|b\rangle \langle b|U^\dagger(0,+\infty)|j\rangle. \tag{53.8}$$

We now calculate in perturbation theory (keeping only the first order in g)

$$\langle i|U(0,\pm\infty)|j\rangle = \langle i|j\rangle - ig \int_0^{\pm\infty} dt'' \langle i|e^{iH_0 t''} \hat{H}_{I,S} e^{-iH_0 t''}|j\rangle$$
$$= \delta_{ij} - ig \int_0^{\pm\infty} dt'' e^{i(E_i - E_j)t''} H_{I,ij}$$
$$= \delta_{ij} - \frac{gH_{I,ij}}{E_i - E_j \pm i\epsilon}. \tag{53.9}$$

Here we have used the fact that $|i\rangle$ are eigenstates of H_0 with energy E_i at leading order (i.e. $H_0|i\rangle = E_i|i\rangle$), we have defined

$$H_{I,ij} = \langle i|H_{I,S}|j\rangle, \tag{53.10}$$

and we have introduced a regulator in the exponential to make the term at $\pm\infty$ decay to zero, $e^{i(E_i - E_j \pm i\epsilon)(\pm\infty)} = 0$.

Therefore now to order g we obtain also

$$R^{\pm}_{aij} = \delta_{ia}\delta_{ja} - \frac{gH^*_{Iia}}{E_i - E_a \mp i\epsilon}\delta_{ja} - \frac{gH_{ja}}{E_j - E_a \pm i\epsilon}\delta_{ia} + \mathcal{O}(g^2). \tag{53.11}$$

We see the origin of the divergences we want to get rid of, corresponding to IR divergences. If there is a degeneracy between states $|a\rangle$ (external) and $|i\rangle$ or $|j\rangle$ (internal), that is $E_a = E_i$, corresponding to a degeneracy between real or virtual states, like e^- and $e^- + \gamma$ with γ having close to zero energy, then we obtain a divergence in $R^{\pm}_{aij}$. The divergence will be eliminated by summing over initial and final states.

Let $D(E)$ be the set of all states with energy E. Then we want to show that

$$\sum_{a\in D(E)} \sum_{b\in D(E)} |\langle b|S|a\rangle|^2 \tag{53.12}$$

is free of (IR) divergences. From the decomposition above in terms of $R^{\pm}$, the statement is completely equivalent to the statement that

$$R^{\pm}_{ij}(E) \equiv \sum_{a\in D(E)} R^{\pm}_{aij} \tag{53.13}$$

has no (IR) divergences.

Proof   To show this, we explicitly calculate $R_{ij}^{\pm}(E)$ case by case. We obtain

$$R_{ij}^{\pm}(E) = 0, \quad \text{for} \quad i,j \notin D(E)$$

$$= -\frac{gH_{I,ij}^*}{E_i - E \mp i\epsilon}, \quad \text{for} \quad j \in D(E), i \notin D(E) \Rightarrow E_i - E \neq 0$$

$$= -\frac{gH_{I,ji}}{E_j - E \pm i\epsilon}, \quad \text{for} \quad i \in D(E), j \notin D(E) \Rightarrow E_j - E \neq 0$$

$$= \delta_{ij} \quad \text{for} \quad i,j \in D(E). \tag{53.14}$$

As we see from the above, it is finite in all these cases.   *q.e.d.*

Next we want to generalize the proof to all orders. We will proceed by induction, since we already proved it at first order.

We first diagonalize the total Hamiltonian H by using $U(0, \pm\infty)$:

$$U^{\dagger}HU = \hat{H}_0 \tag{53.15}$$

is diagonal (but is different from H_0, the free Hamiltonian). Then

$$[U, \hat{H}_0] = U\hat{H}_0 - \hat{H}_0 U = (H - \hat{H}_0)U = (gH_I + \Delta)U, \tag{53.16}$$

where we have used the fact that $H = H_0 + gH_I$ and defined

$$\Delta \equiv H_0 - \hat{H}_0, \tag{53.17}$$

which is a diagonal operator.

We expand in perturbation theory

$$\Delta = \sum_n g^n \Delta_n; \quad U = \sum_n g^n U_n; \quad R = \sum_n g^n R_n. \tag{53.18}$$

Then we calculate

$$R_{ij}^{\pm}(E) = \sum_{a \in D(E)} R_{aij}^{\pm}(E) = \sum_{r,s} g^{r+s} \sum_{a \in D(E)} \langle i|U_r(0, \pm\infty)|a\rangle^* \langle j|U_s(0, \pm\infty)|a\rangle$$

$$\equiv \sum_n g^n R_{n,ij}^{\pm}(E), \tag{53.19}$$

so that we get

$$R_{n,ij}^{\pm}(E) = \sum_r \sum_{a \in D(E)} \langle i|U_r(0, \pm\infty)|a\rangle^* \langle j|U_{n-r}(0, \pm\infty)|a\rangle. \tag{53.20}$$

Then we have reduced the KLN theorem to the induction step for $R_{n,ij}^{\pm}(E)$, that is to a lemma described as follows.

53.2 Statement and Proof of Lemma

Lemma   If $R_{n,ij}^{\pm}(E)$ is free from IR divergences for $n \leq N$, then $R_{n,ij}^{\pm}$ is free from IR divergences for $n \leq N + 1$.

Proof   We prove it case by case.

(i) $i \notin D(E)$. Consider (53.16) in between $\langle i|$ and $|a\rangle$:

$$\langle i|[U, \hat{H}_0]|a\rangle = \langle i|(gH_I + \Delta)U|a\rangle, \qquad (53.21)$$

and consider now the energies of the total diagonalized Hamiltonian $\hat{H}_0|i\rangle = E_i|i\rangle$. Then, applying $\hat{H}_0$ on the left-hand side in the above on the states, and on the right-hand side introducing a complete set $\sum_k |k\rangle\langle k|$ in between U and $(gH_I + \Delta)$, and defining $\langle i|\Delta|i\rangle = \Delta_i$, we obtain

$$(E_a - E_i)U_{ia} = gH_{I,ik}U_{ka} + \Delta_i U_{ia}. \qquad (53.22)$$

Since $a \in D(E)$, but $i \notin D(E)$, $E_i - E_a \neq 0$, so we can divide by it and write, after expanding in powers of g both sides of the equation:

$$U_{r,ia} = \frac{1}{E_a - E_i}\left[\sum_k H_{I,ik}U_{r-1,ka} + \sum_s \Delta_{s,i}U_{r-s,ia}\right]. \qquad (53.23)$$

Then we obtain

$$
\begin{aligned}
R^{\pm}_{n,ij}(E) &= \sum_r \sum_{a \in D(E)} U^*_{r,ia}U_{n-r,ja} \\
&= \frac{1}{E - E_i}\left[\sum_r \sum_{a \in D(E)}\left(\sum_k H^*_{I,ik}U^*_{r-1,ka} + \sum_s \Delta^*_{si}U^*_{r-s,ia}\right)U_{n-r,ja}\right] \\
&= \frac{1}{E - E_i}\left[\sum_k H^*_{I,ik}R^{\pm}_{n-1,kj}(E) + \sum_{s=1}^{n-1}\Delta^*_{si}R^{\pm}_{n-s,ij}(E)\right]. \qquad (53.24)
\end{aligned}
$$

Here, after the definition in the first line, we have substituted the expansion for $U^*_{r,ia}$ and then reformed $R^{\pm}_{ij}(E)$ coefficients, so that in the final form we have $R^{\pm}_{n,ij}$ written in terms of $R^{\pm}_{m,ij}(E)$ with $m \leq n-1$, as well as $\delta_{s,i}$ and $H_{I,ik}$.

But we assume that $H_{I,ik}$ are finite (the matrix elements of the interaction Hamiltonian cannot be infinite because of unitarity), and that Δ_i is also finite (or at least is finite at each order in g, i.e. $\Delta_{r,i}$ is finite), since those are the differences between the free and interacting energies, and they must be finite (or rather, since we have chosen finite E_a, E_i, and $\Delta_i = E_i - E_i^{(0)}$, the corresponding free values for the energies, $E_i^{(0)}$, must be positive, so Δ_i must be finite).

Then indeed it follows that if all $R^{\pm}_{m,ij}(E)$ for $m \leq n$ are finite, so is $R^{\pm}_{n+1,ij}(E)$.

(ii) $j \notin D(E)$. This case is the same as case (i), because

$$(R^{\pm}_{n,ij}(E))^* = R^{\pm}_{n,ji}(E). \qquad (53.25)$$

(iii) $i, j \in D(E)$. In this case, we cannot use the same equations. Instead, from unitarity:

$$UU^\dagger = 1 \Rightarrow \sum_r U_{n-r}U^\dagger_r = 0, n \neq 0, \qquad (53.26)$$

and by sandwiching it between $\langle j|$ and $|i\rangle$, and inserting in the middle a complete set

$$
1 = \left(\sum_{a \in D(E)} + \sum_{a \notin D(E)} \right) |a\rangle\langle a|, \tag{53.27}
$$

we get

$$
\sum_r \sum_{a \in D(E)} U^*_{r,ia} U_{n-r,ja} + \sum_r \sum_{a \notin D(E)} U^*_{r,ia} U_{n-r,ja} = 0, \tag{53.28}
$$

and since on the left we form $R^\pm_{n,ij}(E)$, we get

$$
R^\pm_{n,ij}(E) = - \sum_{r=0}^n \sum_{a \notin D(E)} U^*_{r,ia} U_{n-r,ja}, \tag{53.29}
$$

so it is IR-finite, since all the matrix elements of a unitary operator (the evolution operator) are finite.

q.e.d. Lemma, thus *q.e.d.* KLN theorem.

Thus we have proved generally that IR divergences disappear in physical quantities, transition probabilities summed over all states of a given energy, which justifies the use of QFT despite the presence of IR divergences.

53.3 Factorization and Evolution

We now give some more general remarks about factorization and evolution.

The general statement of *factorization* is that the calculable, short-distance physics factorizes from the incalculable, long-distance one.

The general statement of *evolution* in momentum transfer Q is that physical quantities that characterize the long-distance (IR) behavior have an evolution in momentum transfer due to emission of soft gluons, thus are related to the IR divergences.

53.3.1 Factorization Theorem

Factorization takes the form of a theorem, but needs to be defined for a specific case. We will consider the case of hadronic structure functions $F_a^{(h)}(x, Q^2)$, $Q^2 = -q^2$, where q is the momentum transferred, defined as follows. The hadronic tensor

$$
W^{(i)}_{\mu\nu}(p, q) = \frac{1}{8\pi} \sum_{\sigma, n} \langle h(p, \sigma), in | J^{(i)\dagger}_\mu(0) | n, out \rangle \langle n, out | J^{(i)}_\nu(0) | h(p, \sigma), in \rangle (2\pi)^4 \delta^4(p_n - q - p)
$$

$$
= \frac{1}{8\pi} \sum_{\sigma} \int d^4x\, e^{iq \cdot x} \langle h(p, \sigma), in | J^{(i)\dagger}_\mu(x) J^{(i)}_\nu(0) | h(p, \sigma), in \rangle \tag{53.30}
$$

is written in terms of the electromagnetic currents, interacting with the hadronic states, and is relevant for example for DIS (with the leptonic part of the amplitude taken out). It is written as the given tensor structures times structure functions:

$$W_{\mu\nu}(p,q) = -\left(g_{\mu\nu} - \frac{q_\mu q_\nu}{q^2}\right) W_1(x,Q^2) + \left[p_\mu - q_\mu \frac{p\cdot q}{q^2}\right]\left[p_\nu - q_\nu \frac{p\cdot q}{q^2}\right] W_2(x,Q^2)$$

$$-i\epsilon_{\mu\nu}{}^{\rho\sigma} \frac{q_\nu p_\sigma}{2m} W_3(x,Q^2), \tag{53.31}$$

and

$$F_1(x,Q^2) = W_1(x,Q^2); \quad F_2(x,Q^2) = p\cdot q W_2(x,Q^2). \tag{53.32}$$

Then the factorization theorem for the structure functions (so really for the amplitudes) is

$$F_1^{(h)}(x,Q^2) = \sum_i \int_0^1 \frac{d\xi}{\xi} C_1\left(\frac{x}{\xi}, \frac{Q^2}{\mu^2}, \alpha_s(\mu^2)\right) \phi_{i/h}\left(\xi, \epsilon, \alpha_s(\mu^2)\right),$$

$$F_2^{(h)}(x,Q^2) = \sum_i \int_0^1 d\xi\, C_2\left(\frac{x}{\xi}, \frac{Q^2}{\mu^2}, \alpha_s(\mu^2)\right) \phi_{i/h}\left(\xi, \epsilon, \alpha_s(\mu^2)\right). \tag{53.33}$$

Here $\phi_{i/h}$ are the distribution functions of partons i in the hadron h and contain all long-distance dependence (including all ϵ IR dependence), and C_i are IR-safe functions independent of the external hadrons h called coefficient functions, and contain all the short-distance (Q^2) behavior. μ is like a renormalization scale, more precisely a factorization scale, such that the split depends on it, but not $F_i(x,Q^2)$.

We can define also *valence distributions* as the difference between the quark and antiquark distributions:

$$\phi_{f/h}^{(\mathrm{val})}(\xi, \epsilon, \alpha_s(\mu^2)) = \phi_{f/h}(\xi, \epsilon, \alpha_s(\mu^2)) - \phi_{\bar{f}/h}(\xi, \epsilon, \alpha_s(\mu^2)), \tag{53.34}$$

which obey an evolution equation (similar to the Altarelli–Parisi equation for the full quark and gluon distributions, but for evolution in μ):

$$\frac{d}{d\log\mu} \phi_{f/h}^{(\mathrm{val})}(x, \epsilon, \alpha_s(\mu^2)) = \int_x^1 \frac{d\xi}{\xi} P_f\left(\frac{x}{\xi}, \alpha_s(\mu^2)\right) \phi_{f/h}^{(\mathrm{val})}(\xi, \epsilon, \alpha_s(\mu^2)), \tag{53.35}$$

where P_f are related to the quark–quark splitting functions P_{qq}.

Of course, factorization and evolution are more general concepts than in this particular case, but it was shown in order to see the general principles involved.

Important Concepts to Remember

- The KLN theorem states that the transition probabilities summed over all the initial and final states of given energy are free of (IR) divergences.
- Factorization means in general that calculable short-distance physics factorizes from the incalculable long-distance one.

- Evolution means in general that physical quantities evolve in momentum transfer due to the emission of soft particles, related to IR divergences.
- The factorization theorem says that structure functions (scalar functions parametrizing amplitudes) factorize in distribution functions characterizing long-distance physics, and IR-safe coefficient functions that contain all short-distance (Q^2) behavior, and are independent of the hadron.
- The factorization depends on an arbitrary renormalization (or rather, factorization) scale μ.
- Valence distributions obey the evolution equation in μ.

Further Reading

See sections 14.3 and 14.1 in [2] for factorization and evolution, and section 6.33 in [12] for the KLN theorem.

Exercises

1. Consider the $1/N$ expansion of the IR-divergent amplitudes at L-loop, $|A^{(L)}(\epsilon)\rangle$:

$$|A^{(1)}(\epsilon)\rangle = \frac{1}{N} I^{(1)}(\epsilon) |A^{(0)}\rangle + |A^{(1f)}(\epsilon)\rangle,$$

$$|A^{(2)}(\epsilon)\rangle = \frac{1}{N^2} I^{(2)}(\epsilon) |A^{(0)}\rangle + \frac{1}{N} I^{(1)}(\epsilon) |A^{(1)}(\epsilon)\rangle + |A^{(2f)}(\epsilon)\rangle,$$

$$|A^{(3)}(\epsilon)\rangle = \frac{1}{N^3} I^{(3)}(\epsilon) |A^{(0)}\rangle + \frac{1}{N^2} I^{(2)}(\epsilon) |A^{(1)}(\epsilon)\rangle + \frac{1}{N} I^{(1)}(\epsilon) |A^{(2)}(\epsilon)\rangle + |A^{(3f)}(\epsilon),$$

(53.36)

where $I^{(L)}(\epsilon)$ are divergent and $|A^{(Lf)}(\epsilon)\rangle$ are finite.

For 4-points in $\mathcal{N} = 4$ SYM, we have

$$I^{(1)}(\epsilon) = \frac{1}{2\epsilon} \sum_{i=1}^{4} \sum_{j\neq i}^{a} T_i \cdot T_j \left(\frac{\mu^2}{-s_{ij}}\right)^\epsilon,$$

$$T_i \cdot T_j = T_i^a T_j^a; \quad s_{ij} = (k_i + k_j)^2; \quad T_i^a = (T^a)_{c_i b_i} = if_{c_i a b_i},$$

$$I^{(2)}(\epsilon) = -\frac{1}{2}\left[I^{(1)}(\epsilon)\right]^2 - N\zeta_2 c(\epsilon) I^{(1)}(2\epsilon)$$

$$+ \frac{c(\epsilon)}{4\epsilon}\left[-\frac{N\zeta_3}{2} \sum_{i=1}^{4} \sum_{j\neq i}^{4} T_i \cdot T_j \left(\frac{\mu^2}{-s_{ij}}\right)^{2\epsilon} + \hat{H}^{(2)}\right],$$

$$c(\epsilon) = 1 + \frac{\pi^2}{12}\epsilon^2 + \mathcal{O}(\epsilon^3),$$

$$\hat{H}^{(2)} = -4[T_1 \cdot T_2, T_2 \cdot T_3] \log\left(\frac{s}{t}\right) \log\left(\frac{t}{u}\right) \log\left(\frac{u}{s}\right).$$

(53.37)

Note that T_i^a is a matrix that acts on, for example $(b_1b_2b_3b_4)$, to change b_i to c_i. Prove

$$|A^{(2)}(\epsilon)\rangle = \frac{1}{2N}I^{(1)}(\epsilon)|A^{(1)}(\epsilon)\rangle - \frac{1}{N}(\zeta_2 + \epsilon\zeta_3)c(\epsilon)I^{(1)}(2\epsilon)|A^{(0)}\rangle$$
$$+\frac{1}{4N^2}\frac{c(\epsilon)}{\epsilon}\hat{H}^{(2)}|A^{(0)}\rangle + \frac{1}{2N}I^{(1)}(\epsilon)|A^{(1f)}(\epsilon)\rangle + |A^{(2f)}\rangle. \qquad (53.38)$$

2. In the basis $|A_{[i]}\rangle$, $i = 1, \ldots, 6$ (single trace) and $i = 7, 8, 9$ (double trace), expand

$$I^{(1)}(\epsilon) = -\frac{1}{\epsilon^2}\begin{pmatrix} N\alpha_\epsilon & \beta_\epsilon \\ \gamma_\epsilon & N\delta_\epsilon \end{pmatrix}. \qquad (53.39)$$

Find the $\mathcal{O}(1)$ matrices $\alpha_\epsilon(6 \times 6)$, $\beta_\epsilon(6 \times 3)$, $\gamma_\epsilon(3 \times 6)$, $\delta_\epsilon(3 \times 3)$ in terms of

$$S = \left(\frac{\mu^2}{-s}\right)^\epsilon; \quad T = \left(\frac{\mu^2}{-t}\right)^\epsilon; \quad U = \left(\frac{\mu^2}{-u}\right)^\epsilon. \qquad (53.40)$$

3. Invert the formula (53.31) to find the structure functions W_1, W_2, W_3 as a function of $W_{\mu\nu}$ and momenta q^μ, p^μ.
4. Write the normalization conditions for the parton distribution functions of parton i in hadron h, $\phi_{i/h}$.

54

Perturbative Anomalies: Chiral and Gauge

In this chapter we start the analysis of perturbative anomalies in global and local symmetries. We will first consider chiral invariance, and the anomaly associated with it, which, as we will see, is related to a bubble diagram in $d = 2$ and to a triangle diagram in $d = 4$, in both cases with fermions running in the loop. We will then study its properties, and its generalization to nonabelian gauge theories. Finally we will consider gauge anomalies, which ruin the consistency of the theory.

54.1 Chiral Invariance in Classical and Quantum Theory

For a linear symmetry

$$\phi^i(x) \to \phi^i(x) + \alpha^a (T^a)^i{}_j \phi^j(x), \tag{54.1}$$

such that the Lagrangian changes only by a boundary term

$$\mathcal{L} \to \mathcal{L} + \alpha^a \partial^\mu J^a_\mu, \tag{54.2}$$

the Noether current is

$$j^a_\mu(x) = \frac{\partial \mathcal{L}}{\partial(\partial^\mu \phi)} (T^a)^i{}_j \phi^j - J^a_\mu(x), \tag{54.3}$$

and is conserved, so

$$\partial^\mu j^a_\mu = 0. \tag{54.4}$$

The action for a massless fermion in Euclidean space:

$$\mathcal{L} = \bar{\psi} \gamma^\mu D_\mu \psi, \tag{54.5}$$

where $\bar{\psi} = \psi^\dagger \gamma^4 (= \psi^\dagger i \gamma^0)$, is invariant under the *chiral symmetry*

$$\begin{aligned}
\psi(x) &\to e^{i\alpha\gamma_5} \psi \simeq (1 + i\alpha\gamma_5)\psi, \\
\bar{\psi}(x) &\to \bar{\psi} e^{i\alpha\gamma_5} \simeq \bar{\psi}(1 + i\alpha\gamma_5).
\end{aligned} \tag{54.6}$$

This means that a mass term breaks the symmetry, since

$$m\bar{\psi}\psi \to m\bar{\psi} e^{2i\alpha\gamma_8} \psi \neq m\bar{\psi}\psi. \tag{54.7}$$

Then we have the conserved chiral current

$$j^5_\mu = \bar{\psi} \gamma_\mu \gamma_5 \psi; \quad \partial^\mu j^5_\mu = 0. \tag{54.8}$$

In quantum field theory, we expect the conservation to occur for VEVs, that is the Ward identity

$$\langle \partial^\mu j_\mu^5(x) \rangle = 0. \tag{54.9}$$

We review the derivation. For a general global invariance $\phi \to \phi'$, renaming the field in the partition function from ϕ to ϕ':

$$\int \mathcal{D}\phi' e^{-S[\phi']} = \int \mathcal{D}\phi e^{-S[\phi]}, \tag{54.10}$$

IF the Jacobian for the transformation is one (i.e., the measure is invariant, $\mathcal{D}\phi = \mathcal{D}\phi'$) then

$$0 = \int \mathcal{D}\phi \left[e^{-S[\phi']} - e^{-S[\phi]} \right] = -\int \mathcal{D}\phi \delta S[\phi] e^{-S[\phi]}. \tag{54.11}$$

But if under the global symmetry $\delta S = 0$, under a *local* version of the global symmetry, the variation of the action is

$$\delta S = \sum_a \int d^4x (\partial^\mu \epsilon^a(x)) j_\mu^a(x) = -\sum_a \int d^4x \epsilon^a(x)(\partial^\mu j_\mu^a(x)), \tag{54.12}$$

and substituting this in the above, we get

$$0 = \int d^4x \epsilon^a(x) \int \mathcal{D}\phi e^{-S[\phi]} \partial^\mu j_\mu^a(x). \tag{54.13}$$

Since $\epsilon^a(x)$ is arbitrary, we can take out the integral and write

$$0 = \int \mathcal{D}\phi e^{-S[\phi]} \partial^\mu j_\mu^a(x) \to \langle \partial^\mu j_\mu^a(x) \rangle = 0. \tag{54.14}$$

We can also repeat the same process for the partition function with insertions of $\phi(x)$, and obtain various Ward identities.

In conclusion, we obtained that the quantum nonconservation of the current (i.e. an anomaly, $\langle \partial^\mu j_\mu^a \rangle \neq 0$) appears when there is a nontrivial Jacobian for the measure.

54.2 Chiral Anomaly

A fact that will be explained a bit better further on is that anomalies are *one-loop exact*, that is there are no perturbative or nonperturbative corrections to it, and they arise in even dimensions $d = 2n$ from one-loop diagrams with $n + 1 = d/2 + 1$ vertices, one of them being the symmetry current, or more precisely $\partial^\mu j_\mu^a$, and the other n being *gauge* currents coupling to external gauge fields A_ν^{ext}.

This means that in $d = 2$, the anomalous diagram is a bubble with two vertices, in $d = 4$ it is a triangle, in $d = 6$ it is a box, in $d = 8$ a pentagon, and in $d = 10$ a hexagon, see Figure 54.1.

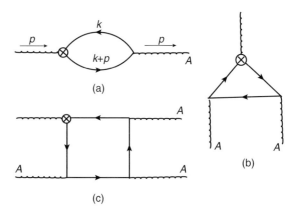

Figure 54.1 Anomalous diagrams. The crossed vertex connects to an outside current divergence, $\partial^\mu j_\mu(x)$, and the other ones to external gauge fields. Chiral fields run in the loop. (a) Anomalous bubble diagram in two dimensions. (b) Anomalous triangle diagram in four dimensions. (c) Anomalous box diagram in six dimensions.

54.2.1 Anomaly in $d = 2$ Euclidean Dimensions

We will calculate everything explicitly only in $d = 2$ Euclidean dimensions, since it is easier, and there is nothing new appearing in higher dimensions other than longer calculations.

The Lagrangian is

$$\mathcal{L} = +\bar{\psi}\gamma_\mu D_\mu \psi + \frac{1}{4}F_{\mu\nu}^2 + \text{ghosts} + \text{gauge fix}, \tag{54.15}$$

but we are interested only in the fermion part, since the gauge fields are external.

Then

$$S_0 = \int d^2 x\, \bar{\psi}\, \slashed{\partial}\psi,$$

$$S_{\text{int}} = -ie \int d^2 x\, \bar{\psi}\, \slashed{A}\psi. \tag{54.16}$$

The anomaly is

$$\langle \partial^\mu j_\mu^5 \rangle = \frac{\int \mathcal{D}\psi \mathcal{D}\bar{\psi}\, \partial^\mu j_\mu^5\, e^{+ie \int d^2 x \bar{\psi}\slashed{A}\psi}\, e^{-S_0}}{\int \mathcal{D}\psi \mathcal{D}\bar{\psi}\, e^{-S_0}}$$

$$\simeq \langle \partial^\mu j_\mu^5(x) \rangle_0 + ie \int d^2 y A_\nu^{\text{ext}}(y) \langle \bar{\psi}(y)\gamma_\nu \psi(y) \partial^\mu j_\mu^5(x) \rangle_0 + \mathcal{O}(e^2), \tag{54.17}$$

but the first term is zero, since

$$\langle \partial^\mu j_\mu^5(x) \rangle_0 = \partial_\mu \langle \bar{\psi}(x)\gamma^\mu \gamma_5 \psi(x) \rangle \propto \text{Tr}[\gamma^\mu \gamma_5] = 0, \tag{54.18}$$

which means that we have

$$\langle \partial^\mu j_\mu^5(x) \rangle = ie \partial_{x^\mu} \int d^2 y A_\nu^{\text{ext}}(y) \langle \psi_\beta(x)\bar{\psi}_{\alpha'}(y) \rangle_0 (\gamma_\nu)_{\alpha'\beta'} \langle \psi_{\beta'}(y)\bar{\psi}_\alpha(x) \rangle_0 (\gamma_\mu \gamma_5)_{\alpha\beta}$$

$$= ie \frac{\partial}{\partial x^\mu} \int d^2 y A_\nu^{\text{ext}}(y) \, \text{Tr}[S_F^0(x-y)\gamma_\nu S_F^0(y-x)\gamma_\mu \gamma_5]. \qquad (54.19)$$

The diagrammatic interpretation for this result is as a bubble diagram, with x^μ and ∂_{x^μ} at one vertex, and y^ν at the other, with the external gauge field there (i.e. exactly the diagram we said would contribute).

When going to p space, the diagram now has p_μ at both ends ($-p_\mu$ at the other, with the "all in" convention), and k and $k+p$ on the two lines of the loop, as in Figure 54.1(a), that is

$$\langle p_\mu j_\mu^5 \rangle = e p_\mu A_\nu^{\text{ext}}(-p) \int \frac{d^2 k}{(2\pi)^2} \, \text{Tr}[S_F^0(k)\gamma_\nu S_F^0(k+p)\gamma_\mu \gamma_5] = e p_\mu A_\nu^{\text{ext}}(-p) T_{\mu\nu}(p). \qquad (54.20)$$

We now regularize using dimensional regularization, introducing the parameter μ, and using that for the massless fermion $S_F^0 = -i/\slashed{p} = -i\slashed{p}/p^2$, we get

$$T_{\mu\nu}(p) = -\mu^{2-d} \int \frac{d^d k}{(2\pi)^d} \frac{\text{Tr}[(\gamma_\alpha k_\alpha)\gamma_\nu (k+p)_\beta \gamma_\beta \gamma_\mu \gamma_5]}{k^2 (k+p)^2}$$
$$= -\mu^{2-d} \, \text{Tr}[\gamma_\mu \gamma_5 \gamma_\alpha \gamma_\nu \gamma_\beta] I. \qquad (54.21)$$

Then the integral is

$$I = \int \frac{d^d k}{(2\pi)^d} \frac{k_\alpha (k+p)_\beta}{k^2 (k+p)^2} = \int_0^1 d\alpha \int \frac{d^d k}{(2\pi)^d} \frac{k_\alpha (k+p)_\beta}{(k^2 + 2\alpha k \cdot p + \alpha p^2)^2}$$
$$= \int_0^1 d\alpha \int \frac{d^d k'}{(2\pi)^d} \frac{(k' - \alpha p)_\alpha (k' + (1-\alpha)p)_\beta}{[k'^2 + p^2 \alpha(1-\alpha)]^2}$$
$$= \int_0^1 d\alpha \int \frac{d^d k'}{(2\pi)^d} \left[\frac{k'_\alpha k'_\beta}{[k'^2 + p^2 \alpha(1-\alpha)]^2} \right.$$
$$\left. - \alpha(1-\alpha) p_\alpha p_\beta \frac{1}{[k'^2 + p^2 \alpha(1-\alpha)]^2} \right], \qquad (54.22)$$

where we first wrote $k'_\mu = k_\mu + \alpha p_\mu$ and then used Lorentz invariance to put to zero the integral with a single k'_α in the numerator.

Using the results from Chapter 33 for the integrals, we get

$$I = \int_0^1 d\alpha \left[\frac{\delta_{\alpha\beta}}{(2-d)(4\pi)^{d/2}} \frac{\Gamma(2-d/2)}{[\alpha(1-\alpha)p^2]^{1-d/2}} - \frac{\alpha(1-\alpha)p_\alpha p_\beta}{(4\pi)^{d/2}} \frac{\Gamma(2-d/2)}{[\alpha(1-\alpha)p^2]^{2-d/2}} \right]$$
$$= \frac{1}{(4\pi)^{d/2}(p^2)^{1-d/2}} \left[\frac{\Gamma(2-d/2)}{2-d} \delta_{\alpha\beta} - \frac{p_\alpha p_\beta}{p^2} \Gamma(2-d/2) \right] \int_0^1 d\alpha [\alpha(1-\alpha)]^{d/2-1}. \qquad (54.23)$$

We note then that the α integral is 1 in $d=2$, and in the square bracket, the first term is divergent, and the second is finite, and proportional to $p_\alpha p_\beta$. But I is multiplied by a trace, and in $d=2$ we have

$$\text{Tr}[\gamma_\mu \gamma_5 \gamma_\alpha \gamma_\nu \gamma_\beta] p_\alpha p_\beta = 0, \qquad (54.24)$$

so we can drop the last term, and we dimensionally regularize the trace, to obtain $\propto 1/(2 - d)$ from the integral and $\propto (d - 2)$ from the trace.

Indeed, using that fact that $\gamma_\alpha \gamma_\nu \gamma_\alpha = -\gamma_\nu (\gamma_\alpha)^2 + 2\delta_{\alpha\nu}\gamma_\alpha = (2 - d)\gamma_\nu$, we have

$$\delta^{\alpha\beta} \operatorname{Tr}[\gamma_\mu \gamma_5 \gamma_\alpha \gamma_\nu \gamma_\beta] = (2 - d) \operatorname{Tr}[\gamma_\nu \gamma_\mu \gamma_5] = -2i(2 - d)\epsilon_{\mu\nu}, \tag{54.25}$$

where we have used (since "γ_5" $= \gamma_3 = -i\gamma_1\gamma_2$):

$$\gamma_\mu \gamma_\nu = i\epsilon_{\mu\nu}\gamma_5 + \delta_{\mu\nu}. \tag{54.26}$$

Then

$$T_{\mu\nu}(p) = +\mu^{2-d}(2 - d)2i\epsilon_{\mu\nu}\frac{\Gamma(2 - d/2)}{(2 - d)(4\pi)^{d/2}(p^2)^{1-d/2}} \int_0^1 d\alpha[\alpha(1 - \alpha)]^{\frac{d}{2}-1}, \tag{54.27}$$

so that when $d = 2$ we obtain

$$T_{\mu\nu}(p) \to \frac{i}{2\pi}\epsilon_{\mu\nu}. \tag{54.28}$$

Finally:

$$\langle p_\mu j_\mu^5(p) \rangle = \frac{iep_\mu}{2\pi}\epsilon_{\mu\nu}A_\nu^{\text{ext}}(-p), \tag{54.29}$$

or going back to position space:

$$\langle \partial^\mu j_\mu^5 \rangle = \frac{ie}{2\pi}\epsilon_{\mu\nu}\frac{(\partial_\mu A_\nu^{\text{ext}} - \partial_\nu A_\mu^{\text{ext}})}{2} = \frac{ie}{2\pi}\tilde{F}^{\text{ext}}, \tag{54.30}$$

where

$$\tilde{F}^{\text{ext}} \equiv \frac{1}{2}\epsilon^{\mu\nu}F_{\mu\nu}^{\text{ext}}. \tag{54.31}$$

Going back to Minkowski signature:

$$\langle \partial^\mu j_\mu^5 \rangle = \frac{e}{2\pi}\tilde{F}^{\text{ext}}. \tag{54.32}$$

54.2.2 Anomaly in $d = 4$ Dimensions

In four Euclidean dimensions, we have a triangle anomaly, and the calculation is similar, though longer. A similar expression is obtained in the end:

$$\langle \partial^\mu j_\mu^5(x) \rangle = -\frac{ie^2}{16\pi^2}\tilde{F}_{\mu\nu}^{\text{ext}}F_{\mu\nu}^{\text{ext}}, \tag{54.33}$$

where

$$\tilde{F}_{\mu\nu}^{\text{ext}} = \frac{1}{2}\epsilon_{\mu\nu\rho\sigma}F^{\rho\sigma}. \tag{54.34}$$

When going to Minkowski space:

$$\langle \partial^\mu j_\mu^5(x) \rangle = \frac{e^2}{16\pi^2}\tilde{F}_{\mu\nu}^{\text{ext}}F^{\mu\nu,\text{ext}} = \frac{e^2}{32\pi^2}\epsilon^{\mu\nu\rho\sigma}F_{\mu\nu}^{\text{ext}}F_{\rho\sigma}^{\text{ext}}. \tag{54.35}$$

54.3 Properties of the Anomaly

There is a theorem, called the Adler–Bardeen theorem, that the anomaly is one-loop only (and only comes from the given polygonal graphs). It can be proven rigorously, and in the path integral we will see it in Chapter 55, but here we give just some plausibility arguments. Note that sometimes the anomaly is called the Adler–Bell–Jackiw anomaly.

We note that in $d = 2$, the anomaly was the result of the trace of four gammas and a γ_5, being proportional to $(d - 2)\epsilon_{\mu\nu}$, times a log divergent diagram with two massless fermion propagators, $\int d^2k(1/k)(1/k) \propto 1/(d-2)$. We can see that at higher loops we will not have this near cancellation anymore. Also at higher points we will not have it anymore.

In $d = 4$, we also have a trace with six gammas and a γ_5, being proportional to $(d - 4)\epsilon_{\mu\nu\rho\sigma}$, and a log-divergent diagram with three massless fermion propagators, $\int d^4k(k_\alpha/k^2)(k'_\beta/k^2)(k''_\gamma/k^2) \propto p_\alpha \int d^4k/k^4 \propto 1/(d-4)$ (since there are no Lorentz structures with three indices, this is the only possibility). Again at higher loops or points, we don't have this cancellation anymore.

Note that we have only considered massless fermions; massive fermions, with mass term allowed by the symmetry, do not contribute to the anomaly.

The anomaly cannot be removed by a local counterterm, so it is genuine.

What we can do in four dimensions is write the anomalous contribution as a boundary term:

$$F_{\mu\nu}\tilde{F}^{\mu\nu} = 4\partial_\mu(\epsilon^{\mu\nu\rho\sigma}A_\nu\partial_\rho A_\sigma), \qquad (54.36)$$

which allows us to subtract it from the current, defining

$$\tilde{j}^5_\mu = j^5_\mu - \frac{e^2}{4\pi^2}\epsilon_{\mu\nu\rho\sigma}A^\nu\partial^\rho A^\sigma, \qquad (54.37)$$

which is conserved, and defining the conserved charge

$$\tilde{Q}_5 = \int d^3x\tilde{j}^5_0. \qquad (54.38)$$

But the new current is not gauge invariant, since under a gauge transformation $\delta A_\mu = \partial_\mu\alpha$, the current changes by

$$\delta\tilde{j}^5_\mu = -\frac{e^2}{4\pi^2}\epsilon_{\mu\nu\rho\sigma}\partial^\rho A^\sigma\partial^\nu\alpha. \qquad (54.39)$$

Yet the charge

$$\tilde{Q}_5 = Q_5 - S_{CS}[A_i], \qquad (54.40)$$

where S_{CS} is called the *Chern–Simons term* or Chern–Simons action

$$S_{CS}[A_i] = \frac{e^2}{4\pi^2}\int d^3x\epsilon^{ijk}A_i\partial_j A_k \qquad (54.41)$$

is invariant, since the variation of the current is a total derivative.

54.4 Chiral Anomaly in Nonabelian Gauge Theories

We can embed the chiral symmetry in a nonabelian theory of massless fermions. The Euclidean action

$$S = \int d^4x \, \bar{\psi} \gamma^\mu D_\mu \psi, \tag{54.42}$$

where $D_\mu = \partial_\mu + g A_\mu^a T^a$, still has an abelian chiral symmetry. Not surprisingly, it is just a trace, that is

$$\partial^\mu j_\mu^5 = \frac{g^2}{16\pi^2} \operatorname{Tr}[F^{\mu\nu} \tilde{F}_{\mu\nu}], \tag{54.43}$$

and it is usually called also chiral anomaly, or *singlet anomaly*.

Note that

$$\operatorname{Tr}[F^{\mu\nu} \tilde{F}_{\mu\nu}] = \partial_\mu \left[4\epsilon^{\mu\nu\rho\sigma} \operatorname{Tr}\left(A_\nu \partial_\rho A_\sigma + \frac{2}{3} A_\nu A_\rho A_\sigma \right) \right], \tag{54.44}$$

so that the redefined current is

$$\tilde{j}_\mu^5 = j_\mu^5 - \frac{g^2}{4\pi^2} \epsilon^{\mu\nu\rho\sigma} \operatorname{Tr}\left(A_\nu \partial_\rho A_\sigma + \frac{2}{3} A_\nu A_\rho A_\sigma \right). \tag{54.45}$$

The redefined chiral charge is

$$\tilde{Q}_5 = Q_5 - S_{CS}[A_i] = Q_5 - \frac{g^2}{4\pi^2} \int d^3x \, \epsilon^{ijk} \operatorname{Tr}\left[A_i \partial_j A_k + \frac{2}{3} A_i A_j A_k \right]. \tag{54.46}$$

But now, unlike the abelian case, under a gauge transformation

$$A_i \to U A_i U^{-1} + \frac{1}{g} \partial_i U U^{-1}, \tag{54.47}$$

the Chern–Simons term transforms by (2π times) an integer m:

$$S_{CS}[A_i] \to S_{CS}[A_i] + \frac{1}{4\pi^2} \int d^3x \, \epsilon^{ijk} \operatorname{Tr}[\partial_i U U^{-1} \partial_j U U^{-1} \partial_k U U^{-1}], \tag{54.48}$$

the extra term being a topological quantity characterizing the three-dimensional gauge transformation called a *winding number m*.

Therefore now the Chern–Simons term is not invariant anymore under "large gauge transformations," which are gauge transformations of nontrivial winding number, therefore not connected smoothly with the identity. This means that now $\tilde{Q}_5$ is not gauge invariant anymore, so the global symmetry is explicitly broken by the anomaly.

The term appearing in the anomaly gives an important topological quantity characterizing the four-dimensional gauge configuration:

$$n = \frac{g^2}{16\pi^2} \int d^4x \operatorname{Tr}[F_{\mu\nu} \tilde{F}^{\mu\nu}] \tag{54.49}$$

is called the *Pontryagin index* or *instanton number*.

54.5 Gauge Anomalies

Up to now we have discussed anomalies in global symmetries, which are good anomalies, that have physical implications, and can be measured.

But there are another type of anomalies, in gauge symmetries, that are bad anomalies, and signal the breakdown (inconsistency) of the quantum theory, so must be canceled to have a good theory.

If we have chiral fermions, $\psi_{R,L} = (1 \pm \gamma_5)/2\psi$ coupled to gauge fields, we have a potential anomaly in gauge invariance. For instance, for a ψ_L, with Euclidean action

$$S = \int d^4x \, \bar{\psi} \frac{1+\gamma_5}{2} \slashed{D} \psi, \tag{54.50}$$

we have a Noether current for the gauge symmetry

$$j_\mu^a = -\bar{\psi} \frac{1+\gamma_5}{2} \gamma_\mu T^a \psi, \tag{54.51}$$

that is covariantly conserved, $(D^\mu j_\mu)^a = 0$.

In this case, similarly to the global case, at the quantum level there is a potential anomaly that can come from a Jacobian for the transformation of the measure, or otherwise from a triangle graph, with $(D^\mu j_\mu)^a$ in one vertex, and A_ν, A_ρ in the other, similar to the case in Figure 54.1(b).

One obtains the triangle anomaly

$$\langle D^\mu j_\mu^a \rangle = \partial_\mu \left\{ \frac{1}{24\pi^2} \epsilon^{\mu\nu\rho\sigma} \, \mathrm{Tr} \left[T^a \left(A_\nu \partial_\rho A_\sigma + \frac{1}{2} A_\nu A_\rho A_\sigma \right) \right] \right\}. \tag{54.52}$$

We see that now we have a T^a inside the trace, and the anomaly is proportional to

$$d_{abc} \equiv \mathrm{Tr}[T^a(T^b T^c + T^c T^b)]. \tag{54.53}$$

However, as we said, this anomaly must cancel, so we must add up representations of fields such that the total anomaly is zero for a good theory. This is what happens for example in the Standard Model, as we will see in Chapter 56.

We also notice that the coefficients of the quadratic and cubic terms inside the trace are different from the singlet anomaly case. We will explain this in Chapter 55.

Finally, why do we need to cancel the anomaly? Since now gauge symmetry kills degrees of freedom (in four dimensions, a vector boson has three degrees of freedom, but using gauge invariance we reduce them to two), if gauge symmetry were broken at the quantum level, it would mean that there are a different number of degrees of freedom at the classical and quantum levels, which is nonsensical.

Important Concepts to Remember

- Anomalies mean nonconservation of a symmetry current at the quantum level, due to the non-invariance of the path-integral measure.
- Anomalies are one-loop exact and arise in even dimensions $d = 2n$ from polygon graphs with one $\partial^\mu j_\mu$ and n gauge currents coupled to external gauge fields.
- In two Minkowski dimensions the anomaly is $e/(4\pi)\epsilon^{\mu\nu}F_{\mu\nu}^{\text{ext}}$, and in four dimensions it is $e^2/(16\pi^2)\epsilon^{\mu\nu\rho\sigma}F_{\mu\nu}F_{\rho\sigma}$.
- The anomaly is the derivative of the Chern–Simons current.
- The Chern–Simons action changes by a winding number under large gauge transformations.
- The anomaly is given by the topological invariant Pontryagin index or instanton number.
- Gauge anomalies are bad, and must be canceled for the consistency of the theory.

Further Reading

See sections 8.1, 8.2, and 8.4 in [3], sections 19.1, 19.2, and 19.4 in [1], and sections 22.3 and 22.4 in [11].

Exercises

1. Write down (up to a coefficient, which would be calculated from the one-loop diagram) the abelian chiral anomaly $\langle\partial^\mu j_\mu^5(x)\rangle$ in $d = 6$ dimensions, and the corresponding $\tilde{j}_\mu^5$ and $\tilde{Q}_5$, justifying your work based on one-loop diagrams and symmetry arguments.
2. Consider two-dimensional conformal invariance, symmetry of flat-space field theory under the complex holomorphic transformation $z' = f(z)$, $\bar{z}' = f(\bar{z})$. Is an anomaly for it allowed, or not, and why?
3. Write down the integral expression for the chiral anomaly coming from the one-loop triangle diagram in $d = 4$, using the Feynman rules.
4. Calculate the Chern–Simons action in five dimensions, paralleling the derivation in three dimensions, from the chiral anomaly in four dimensions.

55 Anomalies in Path Integrals: The Fujikawa Method, Consistent vs. Covariant Anomalies, and Descent Equations

In this chapter we will consider anomalies from path integrals following Fujikawa, and we will explain the various nonabelian anomalies, the consistent ones (subject to the Wess–Zumino consistency conditions), and the covariant ones, and how they are related through the descent equations.

55.1 Chiral Basis vs. V–A Basis

First, some equivalent ways to express anomalies. In $d = 4$, one can consider together the anomalies by defining an axial vector current

$$j_\mu^{5,S} = \bar\psi \gamma_\mu \gamma_5 S \psi. \tag{55.1}$$

For $S = \mathbb{1}$, we have the chiral current, or singlet current, and for $S = T^a$, we have the nonabelian current.

The anomaly comes from triangle graphs with one $\mathcal{O} = D^\mu j_\mu^{5,S}$ and two As, so

$$\langle D^\mu j_\mu^{5,S} \rangle \propto \epsilon^{\mu\nu\rho\sigma} \partial_\mu \operatorname{Tr}[S(\#A_\nu \partial_\rho A_\sigma + \#A_\mu A_\rho A_\sigma)]. \tag{55.2}$$

The coefficients will be fixed in the second part of the chapter, from the descent equations.

Sometimes one considers, instead of the chiral basis $\psi_{L,R}$ for the spinors, Dirac (nonchiral) fermions, coupled to *vector* and *axial vector* gauge fields:

$$\mathcal{L} = \bar\psi_i \gamma^\mu (\partial_\mu + V_\mu^a T_a(V) + A_\mu^a T_a(A) \gamma_5) \psi^i + \frac{1}{4g_V^2}(F_{\mu\nu}^a(V))^2 + \frac{1}{4g_A^2}(F_{\mu\nu}^a(A))^2. \tag{55.3}$$

This can be rewritten in terms of the chiral basis as

$$\mathcal{L} = \bar\psi_{i,L} \gamma^\mu D_{\mu,L} \psi^i + \bar\psi_{i,R} \gamma^\mu D_{\mu,R} \psi^i + \frac{1}{4g_L^2}(F_{\mu\nu}^{(L)a})^2 + \frac{1}{4g_R^2}(F_{\mu\nu}^{(R)a})^2, \tag{55.4}$$

where

$$D_{\mu L,R} = \partial_\mu + A_{\mu L,R}^a T_a \tag{55.5}$$

and

$$V_\mu = \frac{A_\mu^L + A_\mu^R}{2}; \quad A_\mu = \frac{A_\mu^L - A_\mu^R}{2}. \tag{55.6}$$

In the V–A basis, for instance, we can write the chiral (singlet) anomaly ($S = 1$) as an AVV piece (AVV diagram) and an AAA piece (AAA diagram), with

$$\epsilon^{\mu\nu\rho\sigma} \left[F^{\text{lin}}_{\mu\nu}(V) F^{\text{lin}}_{\rho\sigma}(V) + \frac{1}{3} F^{\text{lin}}_{\mu\nu}(A) F^{\text{lin}}_{\rho\sigma}(A) \right] e^j \, \text{Tr}[T^j_b T^j_c], \tag{55.7}$$

where e^j is the abelian charge of fermion j.

55.2 Anomaly in the Path Integral: Fujikawa Method

As we already mentioned, the anomaly in the path integral appears because of the non-invariance of the path-integral measure. We want therefore to expand the ψ, $\bar{\psi}$ fields in eigenfunctions of $i\slashed{D}$ and see how this transforms under the chiral invariance.

We therefore write

$$\psi(x) = \sum_n a_n \phi_n(x),$$

$$\bar{\psi}(x) = \sum_n \phi_n^\dagger(x) \bar{b}_n, \tag{55.8}$$

where

$$i\gamma^\mu D_\mu \phi_n(x) = \lambda_n \phi_n(x), \tag{55.9}$$

and the eigenfunctions are orthonormal, that is

$$\int dx \phi_n^\dagger(x) \phi_m(x) = \delta_{nm}. \tag{55.10}$$

Then the path-integral measure is written as

$$\mathcal{D}\bar{\psi}\mathcal{D}\psi = \prod_n d\bar{b}_n \prod_m da_m. \tag{55.11}$$

Under a local chiral transformation on a_n and $\bar{b}_n$, we get the transformation

$$\psi(x) \to \sum_n a'_n \phi_n(x) = \psi'(x) = e^{i\alpha(x)\gamma_5} \psi(x) = \sum_n a_n e^{i\alpha(x)\gamma_5} \phi_n(x), \tag{55.12}$$

so, by multiplication with $\phi_n(x)$ and integration, we get

$$a'_n = \sum_m \int dx \phi_n^\dagger(x) e^{i\alpha(x)\gamma_5} \phi_m(x) a_m \equiv \sum_m C_{nm} a_m,$$

$$\bar{b}'_n = \sum_m \bar{b}_m \int dx \phi_m^\dagger(x) e^{i\alpha(x)\gamma_5} \phi_n(x) \equiv \sum_m \bar{b}_m C_{mn}. \tag{55.13}$$

Then the path-integral measure transforms as

$$\prod_n d\bar{b}'_n \prod_m da'_m = (\det C)^2 \prod_n d\bar{b}_n \prod_m da_m, \tag{55.14}$$

meaning that the Jacobian is given by (considering an infinitesimal chiral transformation)

$$C = \mathbb{1} + \hat{\alpha} + \mathcal{O}(\alpha^2),$$

$$\hat{\alpha}_{nm} \equiv \int dx \phi_n^\dagger(x) \gamma_5 \phi_m(x), \tag{55.15}$$

so that

$$(\det C)^{-1} = e^{-\operatorname{Tr}\log C} = e^{-\operatorname{Tr}\hat{\alpha} + \mathcal{O}(\alpha^2)} = 1 - \int dx \alpha \sum_n \phi_n^\dagger(x) \gamma_5 \phi_n(x) + \mathcal{O}(\alpha^2). \tag{55.16}$$

This result is formally divergent, so it needs to be regularized, with a regulator that maintains gauge invariance.

One possibility is to use *zeta function regularization* (see Chapter 31), by turning the integral into

$$\sum_n \frac{1}{\lambda_n^s} \phi_n^\dagger(x) \gamma_5 \phi_n(x), \tag{55.17}$$

and then taking $s \to 0$. This is done by analogy with Riemann's zeta function:

$$\zeta(s) = \sum_{n \geq 1} \frac{1}{n^s}. \tag{55.18}$$

But here we will instead follow the regularization used by Fujikawa, which is

$$\sum_n \phi_n^\dagger(x) \gamma_5 e^{-(\lambda_n/M)^2} \phi_n(x), \tag{55.19}$$

and then take $M \to \infty$. Note that the above sum is understood (since $i\displaystyle{\not}D \phi_n = \lambda_n \phi_n$) as

$$\sum_n \phi_n^\dagger(x) \gamma_5 e^{\frac{{\not}D^2}{M^2}} \phi_n(x) = \sum_n \langle n|x\rangle \gamma_5 e^{\frac{{\not}D^2}{M^2}} \langle x|n\rangle$$

$$= \operatorname{Tr}\left[|x\rangle \gamma_5 e^{\frac{{\not}D^2}{M^2}} \langle x| \right] = \operatorname{Tr}_\alpha \int \frac{d^d k}{(2\pi)^d} e^{-ikx} \gamma_5 e^{\frac{{\not}D^2}{M^2}} e^{ikx}, \tag{55.20}$$

where in the last form we have expressed the trace in the momentum basis, and the remaining trace is over the spinor indices.

Lemma

$$e^{-ikx} e^{\frac{{\not}D^2}{M^2}} e^{ikx} = \left(e^{-\frac{k^2}{M^2}} e^{\frac{{\not}D^2 + 2ik^\mu D_\mu}{M^2}} \right) \mathbb{1}(x), \tag{55.21}$$

where

$${\not}D^2 = D^2 - ie\sigma_{\mu\nu} F^{\mu\nu}, \tag{55.22}$$

and $\sigma_{\mu\nu} = i/4[\gamma_\mu, \gamma_\nu]$ (note that we are in Minkowski spacetime).

Proof First we note that

$${\not}D^2 = D_\mu D_\nu \gamma^\mu \gamma^\nu = D_\mu D_\nu (\delta^{\mu\nu} + \gamma^{\mu\nu}) = D^2 - i\sigma_{\mu\nu}(D_\mu D_\nu - D_\nu D_\mu) = D^2 - ie\sigma_{\mu\nu} F^{\mu\nu}. \tag{55.23}$$

Then, when acting on a function $f(x)$, we have

$$e^{-ikx} \slashed{D}^2 e^{ikx} f(x) = \slashed{D}^2 f(x) + \left(e^{-ikx} \slashed{D}^2 e^{+ikx} \right) f(x), \qquad (55.24)$$

where $\left(e^{-ikx} \slashed{D}^2 e^{+ikx} \right) = (-k^2 + 2ik^\mu D_\mu)$ (the second term is from one D_μ acting on $f(x)$ and one on e^{ikx}, and the terms with F are part of the $\slashed{D}^2 f(x)$ piece). This means that

$$e^{-ikx} e^{\frac{\slashed{D}^2}{M^2}} e^{ikx} = \left(e^{-\frac{k^2}{M^2}} e^{\frac{\slashed{D}^2 + 2ik^\mu D_\mu}{M^2}} \right) \mathbb{1}(x). \qquad (55.25)$$

q.e.d.

Then we can write our regulated sum as

$$\sum_n \phi_n^\dagger(x) \gamma_5 e^{\frac{\slashed{D}^2}{M^2}} \phi_n(x) = \int \frac{d^d k}{(2\pi)^d} e^{-\frac{k^2}{M^2}} \operatorname{Tr} \left(\gamma_5 \left[1 + \frac{\slashed{D}^2 + 2ik^\mu D_\mu}{M^2} \right. \right.$$
$$\left. \left. + \frac{(\slashed{D}^2 + eik^\mu D_\mu)^2}{2! \, M^4} + \mathcal{O}\left(\frac{1}{M^6} \right) \right] \right). \qquad (55.26)$$

But the first term, though potentially divergent, is actually zero, since $\operatorname{Tr}[\gamma_5] = 0$. The same applies to the term with $\operatorname{Tr}[\gamma_5(2ik^\mu D_\mu)]/M^2$.

d=2 We now specialize to two dimensions. Then the remaining term with $\slashed{D}^2$ is the only nonzero one, since $\int d^2k/M^2 (2\pi)^2 e^{-k^2/M^2} = \pi/(2\pi)^2 = 1/4\pi$ (double Gaussian integral) is the only finite term (the further ones are suppressed by powers of M as $M \to \infty$). But also

$$\gamma_\mu \gamma_\nu = i\gamma_5 \epsilon_{\mu\nu} + \delta_{\mu\nu} \Rightarrow \operatorname{Tr}[\gamma_5 \sigma_{\mu\nu}] = \frac{i}{4} \operatorname{Tr}[\gamma_\mu \gamma_\nu \gamma_5] - \nu \leftrightarrow \mu$$
$$= -\frac{\epsilon_{\mu\nu}}{4} \operatorname{Tr}[\mathbb{1}] - (\mu \leftrightarrow \nu) = -\epsilon_{\mu\nu}, \qquad (55.27)$$

so finally

$$\sum_n \phi_n^\dagger(x) \gamma_5 e^{\frac{\slashed{D}^2}{M^2}} \phi_n(x) = \int \frac{d^2 k/M^2}{(2\pi)^2} e^{-\frac{k^2}{M^2}} (+ie\epsilon^{\mu\nu} F_{\mu\nu}) = \frac{e}{2\pi} \left(\frac{i}{2} \epsilon^{\mu\nu} F_{\mu\nu} \right). \qquad (55.28)$$

Then the anomaly, coming from the transformation of the path-integral measure, is

$$\mathcal{D}\psi \mathcal{D}\bar{\psi} \to \mathcal{D}\psi \mathcal{D}\bar{\psi} (\det C)^{-2} \simeq \mathcal{D}\psi \mathcal{D}\bar{\psi} \left(1 - \frac{e}{\pi} \right) \int dx \alpha(x) \tilde{F}(x). \qquad (55.29)$$

d=4 In $d = 4$, because

$$\int \frac{d^4 k/M^4}{(2\pi)^4} e^{-\frac{k^2}{M^2}} = \frac{\pi^2}{(2\pi)^4}, \qquad (55.30)$$

the only nonzero term is the one with $1/M^4$. Indeed, now also $\operatorname{Tr}[\gamma_5 \sigma_{\mu\nu}] \propto \operatorname{Tr}[\epsilon_{\mu\nu\rho\sigma} \gamma^{\rho\sigma}] = 0$ as well, so the only nonzero term is from

$$\int \frac{d^4 k/M^4}{(2\pi)^4} e^{-\frac{k^2}{M^2}} \frac{1}{2} \operatorname{Tr}[\gamma_5 (\slashed{D}^2)^2] = \int \frac{d^4 k/M^4}{(2\pi)^4} e^{-\frac{k^2}{M^2}} \frac{-e^2}{2} \operatorname{Tr}[\gamma_5 \sigma_{\mu\nu} \sigma_{\rho\sigma}] F_{\mu\nu} F_{\rho\sigma}. \qquad (55.31)$$

But since

$$\mathrm{Tr}[\gamma_5 \sigma_{\mu\nu} \sigma_{\rho\sigma}] = -\frac{1}{4} \mathrm{Tr}[\gamma_5 \gamma_{\mu\nu} \gamma_{\rho\sigma}] = -\frac{i}{4}\epsilon_{\mu\nu\rho\sigma} \mathrm{Tr}[1] = -i\epsilon_{\mu\nu\rho\sigma}, \tag{55.32}$$

where we have used

$$\gamma_{\mu\nu}\gamma_{\rho\sigma} = i\epsilon_{\mu\nu\rho\sigma}\gamma_5 - (\delta_{\mu\rho}\delta_{\nu\sigma} - \delta_{\mu\sigma}\delta_{\nu\rho}) \tag{55.33}$$

(which can be proven by first putting $\mu = 1, \nu = 2, \rho = 3, \sigma = 4$ and identifying the two sides, then $\mu = \rho$, etc.), we have

$$\sum_n \phi_n^\dagger(x) \gamma_5 e^{\frac{\cancel{D}^2}{M^2}} \phi_n(x) = \frac{ie^2}{32\pi^2} \epsilon^{\mu\nu\rho\sigma} F_{\mu\nu} F_{\rho\sigma} = \frac{ie^2}{16\pi^2} F_{\mu\nu}\tilde{F}^{\mu\nu}, \tag{55.34}$$

and the anomaly is

$$\mathcal{D}\psi\mathcal{D}\bar{\psi} \rightarrow \mathcal{D}\psi\mathcal{D}\bar{\psi}\left(1 - \frac{i}{8\pi^2}\int dx\alpha(x) F_{\mu\nu}\tilde{F}^{\mu\nu}\right), \tag{55.35}$$

so

$$\langle \partial^\mu j_\mu^5 \rangle = \frac{-ie^2}{8\pi^2} F_{\mu\nu}\tilde{F}^{\mu\nu}. \tag{55.36}$$

55.3 Consistent vs. Covariant Anomaly

The nonabelian anomaly can be found up to an overall coefficient from consistency conditions found by Wess and Zumino.

The conditions come from the fact that the anomaly must be the gauge variation of the effective action $\Gamma(A)$. We define the gauge variation of the gauge field for a left-handed fermion anomaly:

$$\Delta_{\Lambda_L} A_\mu^a = (D_\mu(A)\Lambda_L)^a = \partial_\mu \Lambda_L^a + [A_\mu, \Lambda_L]^a, \tag{55.37}$$

and we introduce the operator of gauge variation which varies the action with respect to A, and then multiplies by the gauge variation of A:

$$\delta_{\Lambda_L} \equiv X_L(\Lambda_L) \equiv (\partial_\mu \Lambda_L + [A_\mu, \Lambda_L])^a \frac{\partial}{\partial A_\mu^a}. \tag{55.38}$$

If the anomaly G_a is the gauge variation of the effective action $\Gamma(A)$, it follows that

$$\delta_{\Lambda_L} \Gamma(A) = X_L(\Lambda_L)\Gamma(A) = \int d^4x \Lambda_L^a G_a. \tag{55.39}$$

But then the group algebra

$$[X_L(\Lambda_L^{(1)}), X_L(\Lambda_L^{(2)})] = X_L([\Lambda_L^{(1)}, \Lambda_L^{(2)}] \tag{55.40}$$

implies that when acting on the effective action, we get

$$\int d^4x \Lambda_L^{(2)} \delta_{\Lambda_L^{(1)a}} G_a(A) - 1 \leftrightarrow 2 = \int d^4x [\Lambda_L^{(1)}, \Lambda_L^{(2)}]^a G_a, \tag{55.41}$$

which is the *Wess–Zumino consistency condition.*

The unique solution to this equation (up to a normalization constant) is

$$G_a(A) = \text{Tr}\left[T^a d\left(A \wedge dA + \frac{1}{2}A \wedge A \wedge A\right)\right]. \tag{55.42}$$

Substituting in the consistency condition, and partially integrating d onto $\Lambda_L^{(2)}$ on the left-hand side, we obtain the condition

$$\int \text{Tr}\left[d\Lambda_L^{(2)}\delta_{\Lambda_L^{(1)}}\left(A \wedge dA + \frac{1}{2}A \wedge A \wedge A\right)\right] - (1 \leftrightarrow 2)$$

$$= \int \text{Tr}\left[\delta_{[\Lambda_L^{(1)},\Lambda_L^{(2)}]}\left(A \wedge dA + \frac{1}{2}A \wedge A \wedge A\right)\right], \tag{55.43}$$

which is left as an exercise to verify.

Then the anomaly

$$G_a = D_\mu(A)\frac{\delta\Gamma(A)}{\delta A_\mu^a} = \frac{i}{24\pi^2}\epsilon^{\mu\nu\rho\sigma}\,\text{Tr}\left[T^a\partial_\mu\left(A_\nu\partial_\rho A_\sigma + \frac{1}{2}A_\nu A_\rho A_\sigma\right)\right] \tag{55.44}$$

satisfies the consistency conditions, so is a *consistent anomaly*, but is not covariant under gauge transformations. This anomaly has physical meaning. However, theoretically, it is better to work with covariant expressions. We can add a local counterterm to the effective action or to the current J_μ^a and find a *covariant anomaly*, which is important theoretically. Then

$$\tilde{G}_a = D_\mu J_a^\mu + D_\mu X_a^\mu = G_a + D_\mu X_a^\mu = D_\mu \tilde{J}_a^\mu. \tag{55.45}$$

However, the covariant anomaly is not consistent (i.e. it does not satisfy the Wess–Zumino consistency conditions).

55.4 Descent Equations

Both the consistent and the covariant anomaly appear in the so-called (Stora–Zumino) *descent equations*, that start with the Chern form in $2n + 2$ dimensions, $F^{\wedge(n+1)}$. Here as usual $F = dA + A \wedge A = \frac{1}{2}F_{\mu\nu}dx^\mu \wedge dx^\nu$.

For instance, we start with the Chern form in six dimensions, $\omega_6 = \text{Tr}[F \wedge F \wedge F] \propto \epsilon^{\mu_1\cdots\mu_6}\,\text{Tr}[F_{\mu_1\mu_2}F_{\mu_3\mu_4}F_{\mu_5\mu_6}]$. Then $d\omega_{2n+2} = 0$, since

$$d(\text{Tr}\,F^{n+1}) = (n + 1)\,\text{Tr}[(dF)F^n] = (n + 1)\,\text{Tr}[(DF)F^n] = 0, \tag{55.46}$$

where inside the trace we can replace dF with DF, but $DF = 0$ by the Bianchi identity. This means that at least locally, though in fact globally, as we can explicitly check by exterior differentiation, $\omega_{2n+2} = d\omega_{2n+1}$, where ω_{2n+1} is called the *Chern–Simons form* in $2n + 1$ dimensions.

For instance, $\omega_6 = d\omega_5$ and we find

$$\omega_5 = \text{Tr}\left[dA \wedge dA \wedge A + \frac{3}{2}dA \wedge A \wedge A \wedge A + \frac{3}{5}A^{\wedge 5}\right]. \tag{55.47}$$

Under a general variation δA, the Chern form varies as

$$\delta\omega_{2n+2} = (n+1)\,\text{Tr}[(D\delta A)F^n] = (n+1)\,\text{Tr}[D(\delta A F^n)] = (n+1)d\,\text{Tr}[\delta A F^n], \quad (55.48)$$

where we have used the Bianchi identity $DF = 0$ to take out D, and when D is outside the trace it becomes d (since there are no more indices for it to act on). Finally we find

$$\delta d\omega_{2n+1} = d\delta\omega_{2n+1}, \quad (55.49)$$

which means that

$$\delta\omega_{2n+1} = (n+1)\,\text{Tr}[\delta A F^n] + d(\ldots). \quad (55.50)$$

As a consequence, the field equation of the Chern–Simons action $\int \omega_{2n+1}$ is $F^{\wedge n} = 0$, so is covariant.

But under a gauge variation:

$$\delta_{\text{gauge}}\int \omega_{2n+1} = (n+1)\int \text{Tr}[D\Lambda F^{\wedge n}] = (n+1)\int d(\text{Tr}[\Lambda F^{\wedge n}]) = 0, \quad (55.51)$$

where again we have taken out D using the Bianchi identity, and outside the trace, D becomes d. This means that

$$\delta_{\text{gauge}}\omega_{2n+1} = dY. \quad (55.52)$$

In fact, in our example, we find explicitly

$$\delta_{\text{gauge}}\omega_5 = \text{Tr}\left[(d\Lambda)d\left(A \wedge dA + \frac{1}{2}A \wedge A \wedge A\right)\right] = d\left[\text{Tr}\,\Lambda d\left(A \wedge dA + \frac{1}{2}A \wedge A \wedge A\right)\right]. \quad (55.53)$$

So, in general, ω_{2n+2} is the singlet anomaly in $d = 2n+2$ dimensions. $\delta_{\text{gauge}}\omega_5$ gives the *consistent nonabelian anomaly*, and the field equation of ω_5 gives the *covariant* nonabelian anomaly. In general:

$$\omega_{2n+2} = d\omega_{2n+1},$$
$$\delta_{\text{gauge}}\omega_{2n+1} = d\Lambda^a G_a(\text{cons.}),$$
$$\text{Tr}\left(T^a \frac{\delta}{\delta A}\int \omega_{2n+1}\right) = \tilde{G}_a(\text{cov.}). \quad (55.54)$$

These are the *descent equations*.

d=2　As the simplest example, we consider two dimensions. The consistent anomaly is

$$G^a = c\partial_\mu A_\nu^a \epsilon^{\mu\nu}, \quad (55.55)$$

and the extra current piece is

$$X_a^\mu = c\epsilon^{\mu\nu}A_{\nu,a}, \quad (55.56)$$

leading to

$$\tilde{G} = D_\mu J^\mu + D_\mu X^\mu = c\partial_\mu A_\nu \epsilon^{\mu\nu} + c\epsilon^{\mu\nu}D_\mu A_\nu = c(\partial_\mu A_\nu - \partial_\nu A_\mu + [A_\mu, A_\nu])\epsilon^{\mu\nu} = cF_{\mu\nu}\epsilon^{\mu\nu}. \quad (55.57)$$

d=4 The more relevant example is of four dimensions. Here the consistent anomaly is

$$G_a = c \operatorname{Tr} \left[T^a d \left(A \wedge dA + \frac{1}{2} A \wedge A \wedge A \right) \right], \tag{55.58}$$

and the extra current is

$$X_a = c \operatorname{Tr} \left[T_a \left(dA \wedge A + A \wedge dA + \frac{3}{2} A \wedge A \wedge A \right) \right], \tag{55.59}$$

leading to

$$\tilde{G}_a = G_a + D X_a = c \operatorname{Tr}[T_a 3 F \wedge F]. \tag{55.60}$$

Important Concepts to Remember

- We can write the anomaly in the chiral basis or the V–A basis.
- In the path integral, the anomaly arises because of the non-invariance of the measure. The Fujikawa method regularizes the sum over eigenfunctions of $i\slashed{D}$ with $e^{\slashed{D}^2/M^2}$.
- In the Fujikawa method, it is obvious that the anomaly is one-loop exact.
- The nonabelian anomaly must satisfy the Wess–Zumino consistency condition, leading to the consistent anomaly, but is not covariant. It has physical significance.
- By adding a local counterterm to the effective action, or to the current, we can construct a covariant anomaly that is however not consistent. It is theoretically useful.
- The various anomalies appear in the descent equation that starts from the Chern form in $2n + 2$ dimensions. The Chern form is the singlet anomaly, $\omega_{2n+2} = d\omega_{2n+1}$, and the gauge variation of the Chern–Simons form ω_{2n+1} gives the consistent anomaly. The field equation of the CS form is the covariant anomaly.

Further Reading

See section 8.3 in [3], section 19.2 in [1], and section 22.2 in [11].

Exercises

1. Regularize $\sum_n \phi_n^\dagger(x) \gamma_5 \gamma_{\mu\nu} \phi_n(x)$ with $e^{\slashed{D}^2/M^2}$ in $d = 4$ and calculate it.
2. Calculate the $d = 6$ anomaly in the Fujikawa method.
3. Prove the relation stated in the text for $d = 2$:

$$G^a = c \partial_\mu A_\nu^a \epsilon^{\mu\nu},$$

$$X_a^\mu = c\epsilon^{\mu\nu}A_{\nu,a},$$
$$\tilde{G} = c\epsilon^{\mu\nu}F_{\mu\nu} = G + D_\mu X^\mu = D_\mu J^\mu + D_\mu X^\mu. \tag{55.61}$$

4. Prove the relation stated in the text for $d = 4$:

$$G_a = c\,\mathrm{Tr}\left[T^a d\left(A \wedge dA + \frac{1}{2}A \wedge A \wedge A\right)\right],$$

$$X_a = c\,\mathrm{Tr}\left[T_a\left(dA \wedge A + A \wedge dA + \frac{3}{2}A \wedge A \wedge A\right)\right],$$

$$\tilde{G}_a = G_a + DX_a = c\,\mathrm{Tr}[T^a 3F \wedge F], \tag{55.62}$$

and find the expressions in $d = 6$.

In this chapter we present physical applications of anomalies, as well as theoretical applications, for restricting the set of consistent models. We start with physical applications of anomalies for the $\pi^0 \rightarrow \gamma\gamma$ decay, for the nonconservation of baryon number in electroweak theory, and the $U(1)$ problem. Next we treat 't Hooft's UV–IR matching conditions, which help us in finding correct effective field theories for a given theory, and finish with the (gauge) anomaly cancellation conditions, applied to the Standard Model.

56.1 $\pi^0 \rightarrow \gamma\gamma$ Decay

The most famous physical application of anomalies is to the decay of the neutral pion, into two photons.

The pions π^a, with $a = 1, 2, 3$, are related to the divergence of the axial vector current $\partial^\mu j_\mu^{5,a}$, that is

$$j_\mu^{5,a} \sim \bar{\psi}(x) \frac{\sigma^a}{2} \gamma_5 \gamma_\mu \psi(x), \tag{56.1}$$

where the σ^a are Pauli matrices for a flavor $SU(2)$ group. More precisely, in quark models, where the pions are $\bar{q}q$ objects (one quark and one antiquark), we have

$$j_\mu^{5,a}(x) = 2im_q \left(\bar{q} \gamma_5 \frac{\sigma^a}{2} \gamma_\mu q \right). \tag{56.2}$$

But, as we explained, we cannot really think of hadrons as having a fixed number of partons, but instead as having distribution functions for partons inside the handrons, due to the strong QCD interactions ("hadronization"). The claim is that under hadronization, the above relation is replaced by

$$\partial^\mu j_{\mu,\text{had.}}^{5a} = f_\pi m_\pi^2 \pi^a, \tag{56.3}$$

where f_π is the *pion decay constant*, in a sense defined by this relation, but which can be independently measured from experiment. The above relation is called the *PCAC relation*, standing for "partially conserved axial vector current."

The relation is fine for $\pi^\pm$, corresponding to the Pauli matrices $\sigma^\pm$, but is not quite correct for π^0, corresponding to the Pauli matrix σ_3, since there is a decay of the pion into two photons. More precisely, we have a decay of $\partial^\mu j_\mu^{5,3}$ into two photons through a triangle anomaly diagram, as in Figure 56.1: first the current divergence $\partial^\mu j_\mu^{5,3}$ turns (mostly) into

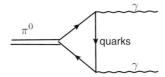

Figure 56.1 Anomaly for π^0 decaying into two photons via a quark loop.

a π^0, after which the π^0 decays through a one-loop chiral fermion triangle diagram into two γs.

Then, more precisely, we have

$$\partial^\mu j_\mu^{5(A),\pm} = f_\pi m_\pi^2 \pi^\pm(x),$$

$$\partial^\mu j_\mu^{5(A),3} = f_\pi m_\pi^2 \pi^3(x) + c\frac{e^2}{16\pi^2}\epsilon^{\mu\nu\rho\sigma}F_{\mu\nu}F_{\rho\sigma}, \tag{56.4}$$

where the extra term is the anomalous contribution. The relation is correct to all orders in α_s (it is valid for full hadronization under QCD), but is true only to first order in α (it is only one loop for QED).

The coefficient c can be calculated as follows. If it was an electron in the loop, we would have $c = 1$, since we have already calculated the $U(1)$ chiral anomaly. But for quarks in the loop, we have $c = N_c/6$, since

$$\pi^0 \sim \bar\psi\gamma_5\frac{\tau_3}{2}\psi; \quad \tau_3 = \begin{pmatrix} 1 & 0 \\ 0 & -1 \end{pmatrix}; \quad \psi = \begin{pmatrix} u \\ d \end{pmatrix}, \tag{56.5}$$

and so

$$c = \mathrm{Tr}\left[\frac{\tau_3}{2}Q^2\right] = \frac{N_c}{2}\left[\left(\frac{2}{3}\right)^2 - \left(\frac{1}{3}\right)^2\right] = \frac{N_c}{6}. \tag{56.6}$$

Note that Q^2 is the charge squared of the quarks, appearing because of the quarks coupling to each of the external photons with charge Q.

We now consider the $m_\pi \to 0$ limit, which is consistent, since the pion is the lightest state in QCD by approximately an order of magnitude.

We consider the matrix element of $j_\mu^{A,3}$ in between a photon state and the vacuum (i.e. the decay amplitude from vacuum to two photons, through the anomaly). The matrix is dominated by the intermediate state of a pion, that is

$$\langle\epsilon^1,p^1;\epsilon^2,p^2|j_\mu^{A,3}(0)|0\rangle \simeq \sum_{\vec q}\langle\epsilon^1,p^1;\epsilon^2,p^2|\pi^0,\vec q\rangle\langle\pi^0,\vec q|j_\mu^{A,3}(0)|0\rangle, \tag{56.7}$$

where ϵ^1,ϵ^2 are the polarizations of the two photons, and p^1,p^2 are their two momenta.

In contrast, the matrix element of the pion operator between the vacuum and the pion state is

$$\langle\pi^0,\vec q|\pi^0(0)|0\rangle = \frac{1}{\sqrt{2\omega_q}}, \tag{56.8}$$

which is just the relativistic wavefunction normalization. But by the PCAC relation (56.4), considering that the anomaly part has no matrix element between the pion and the vacuum, we can replace $\pi^0(0)$ with $q^\mu j_\mu^{A,3}/(f_\pi m_\pi^2)$ between the states, and given that $q_\mu^2 = m_\pi^2$, we obtain

$$\langle \pi^0, \vec{q} | j_\mu^{A,3} | 0 \rangle = \frac{q_\mu f_\pi}{\sqrt{2\omega_q}}. \tag{56.9}$$

Now we consider the PCAC relation (56.4) in between the two-photon state $\langle \epsilon^1, p^1; \epsilon^2, p^2 |$ and the vacuum $|0\rangle$.

On the right-hand side we obtain

$$\langle \epsilon^1, p^1; \epsilon^2, p^2 | \frac{N_c}{6} \frac{e^2}{16\pi^2} \left(\epsilon^{\mu\nu\rho\sigma} F_{\mu\nu} F_{\rho\sigma} \right) | 0 \rangle = \frac{N_c}{6} \frac{e^2}{16\pi^2} \left(8\epsilon^{\mu\nu\rho\sigma} \epsilon_\mu^1 \epsilon_\nu^2 p_\rho^1 p_\sigma^2 \right), \quad (56.10)$$

where $F_{\mu\nu} | \epsilon, p \rangle = 2\epsilon_{[\mu} p_{\nu]} | \epsilon, p \rangle$, and an extra factor of 2 comes about because each F can act on both (ϵp) pair. Therefore now the (56.4) relation in between the states becomes (note that $e^2/2\pi^2 = 2\alpha/\pi$)

$$\langle \epsilon^1, p^1; \epsilon^2, p^2 | \partial^\mu j_\mu^{A,3} | 0 \rangle = \left(\frac{\alpha}{\pi} \frac{N_c}{3} \right) \epsilon^{\mu\nu\rho\sigma} \epsilon_\mu^1 \epsilon_\nu^2 p_\rho^1 p_\sigma^2. \tag{56.11}$$

Finally then, using the relation (56.7) with the normalization (56.8), the above relation gives the amplitude for a pion to go to two photons:

$$\mathcal{A}(\pi^0 \to 2\gamma) \equiv \langle \epsilon^1, p^1; \epsilon^2, p^2 | \pi^0, \vec{q} \rangle = \frac{\alpha}{\pi} \frac{N_c}{3} \frac{1}{f_\pi} \epsilon^{\mu\nu\rho\sigma} \epsilon_\mu^1 \epsilon_\nu^2 p_\rho^1 p_\sigma^2, \tag{56.12}$$

since $\sum_{\vec{q}} 1/\sqrt{2\omega_q}$ gives the propagator $1/q^2$, which cancels the $q^\mu q_\mu$ coming from $\langle \pi^0, \vec{q} | \partial^\mu j_\mu^{A,3} | 0 \rangle$.

Then, using the relation between the decay probability Γ and the amplitude given in (19.49) (see Chapter 19), we have

$$\Gamma(\pi^0 \to 2\gamma) = \frac{1}{2m_\pi} \int \frac{d^3 p_1}{(2\pi)^3} \frac{1}{2\omega_1} \int \frac{d^3 p_2}{(2\pi)^3} \frac{1}{2\omega_2} \frac{1}{2}$$
$$\left(\sum_{\text{pols.}} |\mathcal{A}(\pi^0 \to 2\gamma)|^2 \right) (2\pi)^4 \delta^4(q - p^1 - p^2). \tag{56.13}$$

After some calculation that will not be reproduced here, but is left as an exercise, we find

$$\Gamma(\pi^0 \to 2\gamma) = \frac{1}{64\pi} \left(\frac{\alpha}{\pi} \frac{N_c}{3} \frac{1}{m_\pi} \right)^2 (m_\pi)^3. \tag{56.14}$$

This is tested experimentally to a high degree of accuracy, and one verifies that $N_c = 3$.

56.2 Nonconservation of Baryon Number in Electroweak Theory

The second important physical application of anomalies is to the nonconservation of baryon number. The gauge group of the Standard Model is $SU(3)_c \times SU(2) \times U(1)_Y$, where the electroweak gauge group is $SU(2) \times U(1)_Y$.

We first describe the fermion field content of the Standard Model. We have the lepton and quark left-handed $SU(2)$ doublets

$$L = \begin{pmatrix} \nu_e \\ e_L \end{pmatrix}; \quad Q_L = \begin{pmatrix} u_L \\ d_L \end{pmatrix}, \tag{56.15}$$

where the left part of the electron is

$$e_L = \frac{1 - \gamma_5}{2} e^- \tag{56.16}$$

and its right part is the full matrix

$$R = \frac{1 + \gamma_5}{2} e^-, \tag{56.17}$$

since the neutrino does not have a right-handed part in the (minimal form of the) Standard Model. A singlet right-handed neutrino is the simplest extension of the Standard Model.

We have a similar relation for the quarks, and we have three generations of fermions, which will be implicit in the notation.

Therefore, in Euclidean space, the lepton part of the action in interaction with the electroweak gauge group (ignoring the color gauge group) is

$$S_{\text{leptons}} = \int d^4x \left[\bar{R}\gamma_\mu(\partial_\mu - ig'B_\mu)R + \bar{L}\gamma_\mu \left(\partial_\mu - \frac{ig'}{2}B_\mu + \frac{ig}{2}A_\mu^a\sigma^a \right) L \right], \tag{56.18}$$

whereas the quark part of the action is

$$S_{\text{quarks}} = \int d^4x \left[\bar{Q}_L\gamma_\mu \left(\partial_\mu + \frac{ig'}{2}Y_L B_\mu + \frac{ig}{2}A_\mu^a\sigma^a \right) Q_L \right.$$

$$\left. + \sum_{i=1,2} \bar{Q}_{R(i)}\gamma_\mu \left(\partial_\mu + \frac{ig'}{2}Y_{R(i)}B_\mu \right) Q_{R(i)} \right]. \tag{56.19}$$

Note that here B_μ is the $U(1)_Y$ gauge field, so the Y_L and Y_R are hypercharges for the L and R fields, and A_μ^a are the $SU(2)$ gauge fields.

We note that the above action has as conserved quantities the baryon number B and the lepton number L (we can consider also the independent lepton numbers, for each generation, L_e, L_μ, L_τ, but these are approximate symmetries). A quark has $B = 1/3$ such that a baryon, made up of three quarks, has baryon number 1 ($B(B) = 1$). On the contrary, all leptons have lepton number $L = 1$. Then B and L are classical symmetries of the above Standard Model action.

However, we have chiral fermions, so in fact B and L are anomalous: we have an abelian (singlet) anomaly in a nonabelian theory, with

$$\partial_\mu j_\mu^B = \frac{N_{\text{gen}}g^2}{16\pi^2}\,\text{Tr}[F^{\mu\nu}\tilde{F}_{\mu\nu}]. \qquad (56.20)$$

Integrating over space and over time between t_1 and t_2, and using $\int d^4x\,\partial^\mu j_\mu = \int d^3x[j_0^B]_{t_1}^{t_2} = B(t_2) - B(t_1)$, we obtain for the difference in baryon number at times t_1 vs. t_2:

$$B(t_2) - B(t_1) = \frac{N_{\text{gen}}g^2}{16\pi^2}\int_{t_1}^{t_2} dt \int d^3x\,\text{Tr}[F^{\mu\nu}\tilde{F}_{\mu\nu}]. \qquad (56.21)$$

But there is a nontrivial field configuration called an *instanton*, obeying

$$F_{\mu\nu} = \tilde{F}_{\mu\nu} \qquad (56.22)$$

in Euclidean space, which means that for such a configuration, the topological number

$$n = \frac{g^2}{16\pi^2}\int d^4x\,\text{Tr}[F^{\mu\nu}\tilde{F}_{\mu\nu}] = \frac{g^2}{4\pi^2}\frac{1}{4}\int d^4x\,\text{Tr}[F_{\mu\nu}^2], \qquad (56.23)$$

called the *instanton number*, is proportional to the on-shell instanton action (in Euclidean space), so

$$S^{(E)} = \int d^4x\frac{1}{4}\,\text{Tr}[F_{\mu\nu}^2] = \frac{4\pi^2}{g^2}n, \qquad (56.24)$$

and the difference in baryon number is an integer:

$$B(t_2) - B(t_1) = N_{\text{gen}}n. \qquad (56.25)$$

Therefore, the difference in baryon number is defined by the instanton number, and the transition probability is given by the (classical) saddle point of the path integral with the given boundary condition (i.e. $e^{-S^{(E)}}$):

$$\langle B(t_2)|B(t_1)\rangle \simeq e^{-S^{(E)}}. \qquad (56.26)$$

However, in the vacuum corresponding to our Universe, $e^{-\frac{4\pi}{\alpha_{\text{weak}}}} \sim e^{-4\pi\cdot30}$ is negligible for the lifetime of the Universe.

On the contrary, in the high-temperature medium of the Big Bang, when the coupling is large, the probability becomes of order 1 so, by symmetry, transitions in baryon number will equalize it, resulting in $B = 0$, or an equal number of baryons and antibaryons. But we observe in our Universe a net baryon number, which means that in the initial stages of the Big Bang there was already a *baryon asymmetry*. The question is, how is this possible, given the mechanism of wiping out an initial baryon asymmetry that we just saw? This is a very important question in theoretical physics, for which there are various models, but none is perfect. Sakharov in the 1980s had already enumerated the necessary conditions to create a baryon asymmetry, but as of yet there is no perfect model.

56.3 The $U(1)$ Problem

The last of the three important physical applications of anomalies is called the $U(1)$ problem. We will not explain all the details of its resolution, among other things because it uses information that will be given later on in the book, but rather we will sketch it.

We will see in Chapter 68 that there is an effective symmetry $SU(2)_L \times SU(2)_R$, due to the near masslessness of the up and down quarks (the u and d quarks are nearly massless, for them we can use a light quark effective theory, and the c, b, and t are very heavy, for which we can use a heavy quark approximation; the s quark is intermediate). This symmetry is spontaneously broken to a diagonal $SU(2)$. We will also see later on in the book that whenever we break a symmetry spontaneously, there is a so-called Goldstone boson appearing, a massless scalar associated with the broken symmetry directions. In QCD, the $SU(2)_L \times SU(2)_R$ symmetry is approximate, so we have approximate Goldstone bosons, the three pions π^a (with masses much smaller than the masses of the other states in QCD), corresponding to the three broken generators (for a broken $SU(2)$).

But the actual symmetry of the QCD Lagrangian is $U(N_f) \times U(N_f)$ (N_f is the number of massless flavors, here two), acting on the quarks as

$$\psi \to e^{i\alpha^a T^a} \psi,$$
$$\psi \to e^{i\alpha^a T^a \gamma_5} \psi. \tag{56.27}$$

So, between the actual $U(2) \times U(2)$ and the observed $SU(2)_L \times SU(2)_R$, the difference is a $U(1) \times U(1)$. One of the $U(1)$s (the $U(1)$ is the trace of the $U(N_f)$, where we replace T^a by $\mathbb{1}$), acting as

$$\psi \to e^{i\alpha} \psi, \tag{56.28}$$

is just the hadron number, which is conserved. But then we still have the other:

$$\psi \to e^{i\alpha \gamma_5} \psi, \tag{56.29}$$

which is an abelian chiral symmetry. Before the anomaly was understood, it was thought that there could only be two possibilities: the symmetry is there, but we don't see such a symmetry in the real world; or the symmetry is spontaneously broken, but then by the Goldstone theorem we should see a Goldstone boson corresponding to this broken symmetry. However, there is no fourth pion, so that is also not true. This then was the "$U(1)$ problem," and its resolution is, of course, that the symmetry is anomalous, so is broken (though not spontaneously).

56.4 't Hooft's UV–IR Anomaly Matching Conditions

We now turn to theoretical applications of anomalies, namely applications for model building. The first such application is due to 't Hooft, and is a very useful consistency condition, which simply put states that the anomaly is independent of the energy scale, so it should give the same result, for instance in the UV and in the IR.

It is useful, since we have the effective field theory approach started by Wilson, that will be studied later on in the book, which states that for a given energy range, we can use a theory in terms of some fields, without worrying if the fields are truly fundamental, as long as we include in the Lagrangian all the possible higher-dimensional operators (even though they will in general not be renormalizable operators). This point of view allows us, say, to use the Standard Model, without worrying that there is at least a Planck scale (and maybe other scales: SUSY scale, GUT scale, etc.) at which what we think of as the fundamental degrees of freedom will change.

But in that case, the anomaly matching conditions act as an important check of the fact that we are using the right degrees of freedom at a certain scale, given knowledge about the degrees of freedom at some other scale.

As an example, consider the anomaly of a global $U(1)$ current, and in the IR consider that we can use the nearly massless fermionic degrees of freedom that may be composite (only massless chiral fermions contribute to the anomaly, as we saw). For instance, in QCD we could consider the n and the p (which are composites of three quarks) as these degrees of freedom for energies higher than the m_p, m_n, but not too high so that we need to consider perturbative QCD. On the contrary, in the UV we can use the fundamental degrees of freedom (in the case of QCD, use the quarks) to calculate the anomaly. The two calculations should match.

Proof   To prove the anomaly matching condition, couple the global $U(1)$ current to a gauge field (i.e. gauge the symmetry). Then add *free* chiral fermions that only couple to the gauge field, in such a way as to cancel the anomaly. Indeed, now that we have a local symmetry, we need to cancel this local (gauge) anomaly for consistency of the quantum theory, as we said.

But local anomaly cancellation (i.e. consistency of the quantum theory) should persist both in the UV and in the IR. Now turn off the gauge coupling, $g \to 0$, going back to the global case. Subtract the anomaly of the free chiral fermions, which is independent of the scale, since now the fermions are truly free (don't couple to anything), so they don't know what an energy scale means. The result is that the global anomaly of the original system is independent of scale.   *q.e.d.*

Note that the anomaly is purely one loop, so it can easily be calculated at an energy scale using the perturbative degrees of freedom available at that energy scale.

56.5 Anomaly Cancellation in General and in the Standard Model

The second important theoretical application is the cancellation of gauge anomalies, which is an important consistency condition for any model we might write (without it, the quantum theory is inconsistent). Here we study how it can happen that we have anomaly cancellation.

In $d = 4$, the anomaly is proportional to d^{abc}, which is proportional to $\text{Tr}[T^a\{T^b, T^c\}]$, and in turn is related to $C_3(G)$, the third Casimir of the gauge group. This means that if the Lie algebra of the group has no $C_3(G)$, there is no d^{abc}, and thus no anomaly. Such groups are called *safe groups*.

Nearly all of the classical groups are safe. In particular, the B_n series (i.e. $SO(2n+1)$), the C_n series (i.e. $Sp(2n)$), the D_n series (i.e. $SO(2n)$, except $SO(6)$), as well as the exceptional groups G_2, F_4, E_6, E_7, E_8, are all safe. The only unsafe classical groups are the A_N series (i.e. $SU(N)$, for $N \geq 3$) and $SO(6) \simeq SU(4)$. Note that $SU(2) \simeq SO(3)$ is safe, as is $SO(4) \simeq SO(3) \times SO(3)$.

Even if a group is unsafe, in some representation R we might still have $d_R^{abc} = 0$, in which case we say the representation is safe.

Otherwise, if we have an unsafe representation of an unsafe group, to cancel the anomaly we need to combine several species of fermions so as to cancel the anomaly.

56.5.1 The Standard Model

The most relevant example that we will study is the Standard Model, with gauge group $SU(3)_c \times SU(2) \times U(1)_Y$.

A potential anomaly appears for the unsafe group $SU(3)_c$, with unsafe representations. However, the $SU(3)$ anomaly in the Standard Model is canceled in a trivial way, since both left and right fermions couple in the same way with the $SU(3)_c$ gauge field, so the total anomaly vanishes.

Then, $SU(2)$ is a safe group, so there is no anomaly with an $SU(2)$ gauge field at each of the three corners of the triangle. However, the absence of d^{abc} says nothing about combining the $SU(2)$ gauge field with another in the triangle diagrams, more precisely about the $SU(2)$ contributing to the $\partial^\mu j_\mu$ anomaly of some other gauge field. But the $SU(3)_c$ couples in the same way for left and right, so that doesn't contribute to these mixed anomalies either.

Thus we need to check only the $U(1)_Y$ (hypercharge), for the pure $U(1)$ anomaly, and the mixed anomaly with $SU(2)$. The potentially anomalous diagrams are then with a $U(1)$ gauge field in $\partial^\mu j_\mu$, and two $SU(2)$ gauge fields in the others (the $SU(2)$ contribution to the $U(1)$ anomaly), and the one with three $U(1)$ gauge fields (the $U(1)$ contribution to the $U(1)$ anomaly).

Together, we can write these conditions as

$$\sum_{\text{fermion representations}} \text{Tr}\left[T^a\left(A \wedge dA + \frac{1}{2}A \wedge A \wedge A\right)\right] = 0. \tag{56.30}$$

The T^a is the generator coupling to the $\partial^\mu j_\mu^a$, so for the $U(1)$ we have $T^a = 1$.

Then the $U(1)-SU(2)-SU(2)$ anomaly has $T^a = 1$, but of course we need to multiply it by the charge Y_L, and the $A \wedge dA + 1/A \wedge A \wedge A$ is proportional to $\sigma^b \sigma^c$. Then the condition becomes

$$\sum_{\text{doublets,L}} Y_L \, \text{Tr}[\sigma^b \sigma^c] \propto \delta^{bc} \sum_{\text{doublets,L}} Y_L = 0, \qquad (56.31)$$

giving the condition

$$\sum_{\text{doublets,L}} Y_L = 0. \qquad (56.32)$$

Note that here we count doublets only (since the $SU(2)$ couples only to doublets, which are only left-handed), but we also need to count color where needed.

For the $U(1)^3$ anomaly, again we have $T^a = 1$, but this must be multiplied by Y, and $A \wedge dA$ is proportional to $\mathbb{1}$ (but again multiplied with Y_L for each gauge field), for a total condition of

$$\sum_{\text{left-handed}} (Y_L)^3 - \sum_{\text{right-handed}} (Y_R)^3 = 0. \qquad (56.33)$$

Note that here we must count each element of a doublet, and also count color.

To verify these conditions, we consider the hypercharge assignments of the Standard Model.

For *quarks*, we have

$$Y_L = 1/3; \quad Y_R(1) = 4/3; \quad Y_R(2) = -2/3, \qquad (56.34)$$

since the left quarks are doublets, but for the right quarks, we have two independent elements (not a doublet), 1 and 2.

For the *leptons*, we have

$$Y_L = -1; \quad Y_R(1) = 0; \quad Y_R(2) = -2, \qquad (56.35)$$

and the same applies, the left leptons are doublets, but the right ones are two independent elements, 1 and 2.

We now verify the two conditions. For the $U(1) - SU(2) - SU(2)$ anomaly, we get

$$N_c \times \frac{1}{3} + 1 \times (-1) = \frac{N_c}{3} - 1, \qquad (56.36)$$

and it only cancels for $N_c = 3$, as it should.

For the $U(1)^3$ anomaly, we get

$$\left[N_c \times 2 \times \left(\frac{1}{3} \right)^3 + 2 \times (-1)^3 \right] - \left[N_c \times \left(\frac{4}{3} \right)^3 + N_c \times \left(-\frac{2}{3} \right)^3 + (-2)^3 \right] = 0, \qquad (56.37)$$

giving

$$\frac{N_c}{3^3}(2 - 4^3 + 2^3) - 2 + 8 = -2(N_c - 3) = 0, \qquad (56.38)$$

which again only cancels for $N_c = 3$, as it should.

Important Concepts to Remember

- The PCAC relations relate the divergence of the axial vector $SU(2)$ current with the pions, modulo the anomaly in the σ^3 component of the current.
- The neutral pion π^0 decays into two photons because of the anomaly, via a diagram where $\partial^\mu j_\mu^{5,3}$ turns into a pion and then into two photons via a quark triangle.
- The baryon number is changed by instantons, because of the anomaly in the baryon current, and with probability $\langle B(t_2)|B(t_1)\rangle \sim e^{-S_{\text{inst}}}$.
- In the current Universe, baryon number change is irrelevant, due to the smallness of the coupling $e^{-4\pi/\alpha_{\text{weak}}}$, giving a small probability. But in the initial Big Bang it is relevant, and would wipe out any initial baryon asymmetry, hence the baryon asymmetry problem.
- The potential extra $U(1)$ does not have a Goldstone boson, since it is a chiral symmetry broken by anomalies.
- The anomaly in the UV (computed with the UV degrees of freedom) should match the anomaly in the IR (computed with the IR degrees of freedom).
- The gauge anomaly should cancel in a consistent model. There are safe groups, safe representations, and otherwise we need to combine species of fermions.
- The unsafe groups are $U(1), SU(N), N \geq 3$, and $SO(6) \simeq SU(4)$.

Further Reading

See section 8.5 in [3], section 19.3 in [1], and sections 22.1, 22.5, and 22.6 in [11].

Exercises

1. Check that the following $SU(5)$ GUT model, with fermionic field content (ignoring the Higgs sector):

   one (anti)fundamental representation $\bar{5}$ and one antisymmetric representation $5 \times 4/2 = 10$, both left-handed

   is free of gauge anomalies, given that

   $$\text{Tr}_5(T^a\{T^b, T^c\}) = d^{abc} = \text{Tr}_{10}(T^a\{T^b T^c\}). \tag{56.39}$$

2. Calculate the $U(1)$ chiral anomaly of the $SU(5)$ GUT by using the UV–IR 't Hooft matching conditions.
3. Prove that the $\pi^0 \to \gamma\gamma$ decay (56.13) gives the decay constant (56.14).
4. Given that there is an instanton solution to the Euclidean space condition $F_{\mu\nu} = \tilde{F}_{\mu\nu}$, there cannot be a solution of the same condition in Minkowski space.

The Froissart Unitarity Bound and the Heisenberg Model

In this chapter we will describe a bound, first found by Froissart in 1961, coming from unitarity, on amplitudes and the total cross-section at large center of mass energy $s \to \infty$ in a strongly coupled theory. Specifically, it will be applied to QCD and the strong interactions. We then show that a simple model by Heisenberg, proposed in 1962 (before Froissart and before QCD) is able to saturate the bound, though it is not possible in perturbative QCD.

57.1 The S-Matrix Program, Analyticity, and Partial Wave Expansions

In the late 1950s and 1960s, before the advent of QCD to describe strong interactions, it was thought that one can describe all the physics in strong interactions simply from analyticity properties of the amplitudes and S-matrices. This program didn't succeed in its original form, and it was mostly abandoned after the successes of perturbative QCD, but it led to some important results, one of them being the unitarity bound presented here. We should also mention that the program has started to reappear in recent research, as the use of Regge and multi-Regge behavior has been used to constrain amplitudes, and S-matrix constraints have been used to constrain the *a priori* list of possible quantum field theories.

The amplitudes considered are usually for two-body scattering $1 + 2 \to 3 + 4$, which we have denoted by $A(s, t)$ in Part I of the book, using Mandelstam variables. We will use the convention with all momenta incoming, so $p_1 + p_2 + p_3 + p_4 = 0$ and $s = -(p_1 + p_2)^2$, $t = -(p_2 + p_3)^2$ and $u = -(p_1 + p_3)^2$. In the center of mass system the momentum of particle 2 is $\vec{p}_{CM}$, and the momentum of particle 3 (after the collision) is $\vec{k}_{CM}$, in which case $t = -(p_{CM}^2 + k_{CM}^2 - 2p_{CM}k_{CM}\cos\theta)$, where θ is the angle made by the two momenta. If particle 2 is the same as particle 3, and all four particles have the same mass m, then $p_{CM} = k_{CM} \equiv q$, and we obtain

$$t = -2q^2(1 - \cos\theta) \Rightarrow \cos\theta = 1 + \frac{t}{2q^2}. \tag{57.1}$$

Note that then also $s = (2E)^2 = 4(p_{CM}^2 + m^2) = 4(q^2 + m^2)$, so $q^2 = (s - 4m^2)/4$. The physical region in terms of the Mandelstam variables (in the s channel) is then defined by

$$q^2 > 0 \text{ and } |\cos\theta| < 1; \text{ or } s > 4m^2, \ t \leq 0, \ u \leq 0. \tag{57.2}$$

From the analyticity properties assumed to be true for the amplitude, Mandelstam proposed a decomposition for the $A(s, \cos\theta)$ (note that we have traded t for $\cos\theta$), taking the
form of a dispersion relation at fixed s (i.e. a way of writing the amplitude in terms of an
integral over the variable of the absorbtive (imaginary) parts of the amplitude in the other
channels). Specifically, the form that Froissart starts from is

$$A(s, \cos\theta) = \frac{1}{\pi}(\cos\theta)^N \int_{x_1}^{\infty} dx \frac{\rho(s, x)}{x^N(x - \cos\theta)} + \frac{1}{\pi}(\cos\theta)^N$$

$$\int_{x_2}^{\infty} dx \frac{\rho'(s, x)}{x^N(x + \cos\theta)} + \frac{1}{\pi} \sum_{m=0}^{N-1} \rho_m \cos^m\theta. \tag{57.3}$$

Here $\rho(s, x)$ and $\rho'(s, x)$ are absorbtive parts of the amplitudes (in the t and u channels,
respectively), so that

$$x_1 = 1 + \frac{t_0}{2q^2}, \quad x_2 = 1 + \frac{u_0}{2q^2}, \tag{57.4}$$

where t_0 is the threshold (minimum value) of the absorbtive part of the amplitude in the t
channel, and u_0 of the amplitude in the u channel. Moreover, one defines

$$x_0 = \min(x_1, x_2), \quad x_0 = 1 + \frac{\kappa^2}{2q^2}, \tag{57.5}$$

so that $\kappa^2 = \min(t_0, u_0)$.

One of the uses of writing the amplitude as $A(s, \cos\theta)$ is to be able to expand it in
Legendre polynomials $P_l(\cos\theta)$, thus obtaining what are known as *partial wave amplitudes*
a_l, by

$$A(s, \cos\theta) = \frac{\sqrt{s}}{\pi q} \sum_{l=0}^{\infty} a_l(s)(2l + 1)P_l(\cos\theta). \tag{57.6}$$

The inverse relation is then

$$a_l(s) = \frac{\pi q}{2\sqrt{s}} \int_{-1}^{+1} d(\cos\theta) P_l(\cos\theta) A(s, \cos\theta). \tag{57.7}$$

57.2 The Froissart Unitarity Bound

One of the uses of the partial wave expansion, exploited by Froissart, is that the a_l form a
unitary matrix (from the statement of unitarity of the S-matrix), so in particular $|a_l| \leq 1$.
Taken together with the dispersion relation for $A(s, \cos\theta)$ (substituted in the form of $a_l(s)$
as a function of $A(s, \cos\theta)$), after some analysis, one finds that

$$A(s, \cos\theta = 1) = \frac{\sqrt{s}}{\pi q} \sum_{l=0}^{\infty} a_l(s)(2l + 1) \tag{57.8}$$

satisfies a bound:

$$|A(s, \cos\theta = 1)| < \frac{q^2}{\kappa^2} \ln^2 B(s), \qquad (57.9)$$

where $B(s)$ is a polynomial function, and as before $q^2 = (s - 4m^2)/4$. Since $\cos\theta = 1$ corresponds to forward scattering (deviation angle $\theta = 0$), and since the imaginary part of the forward amplitude is related to the total cross-section (for two particles going to anything) by the optical theorem studied in Part I:

$$\mathrm{Im}A(s, \cos\theta = 1) = E_{\mathrm{CM}}p_{\mathrm{CM}}\sigma_{\mathrm{tot}}(s) = \frac{\sqrt{s}}{2}\sqrt{\frac{s}{4} - m^2}\sigma_{\mathrm{tot}}(s), \qquad (57.10)$$

and since moreover $B(s)$ is a polynomial in s, with leading power p, at $s \to \infty$ (in particular, $s \gg m^2$), we obtain

$$\sigma_{\mathrm{tot}}(s) < \lim_{s\to\infty} \frac{4}{\sqrt{s(s - 4m^2)}} \frac{s - 4m^2}{\kappa^2} p^2 \ln^2 s = \frac{4p^2}{\kappa^2} \ln^2 s, \qquad (57.11)$$

where we remember that κ^2 is the lowest threshold (in both t and u channels), meaning effectively (possibly an integer times) the lowest available mass state in the theory.

Froissart moreover finds that the amplitude for $\cos\theta \neq \pm 1$ satisfies

$$A(s, \cos\theta) \leq s^{3/4} \ln^{3/2} s. \qquad (57.12)$$

Further, in a work in 1967 by Lukaskuk and Martin (see also previous work by Martin in 1965), it was shown that one can find an asymptotic formula for the amplitude that gives the bound

$$\sigma_{\mathrm{tot}}(s) \leq C \ln^2 s, \quad \text{where} \quad C \leq \frac{\pi}{\kappa^2}. \qquad (57.13)$$

57.2.1 Application to Strong Interactions

The relevant application is for strong interactions, described by QCD, meaning the theory of quarks and gluons. In this case, the lightest particle is the pion (that comes in three varieties, π^+, π^-, and π^0), so κ^2 will be (possibly an integer times) m_π^2. The Froissart unitarity bound in this case is therefore

$$\sigma_{\mathrm{tot}}(s) \leq C \ln^2 s, \quad \text{where} \quad C \leq \frac{\pi}{m_\pi^2}. \qquad (57.14)$$

Note that the maximum coefficient C is $\pi/m_\pi^2 \simeq 60$ milibarn.

However, recently there was some discussion that perhaps the relevant smallest mass is the mass of the glueball m_g (much larger than the pion mass), so m_π would be replaced above by m_g, though it is unclear why m_π would not be relevant, as pions can be created (and are in fact dominant) in the total cross-section.

57.3 The Heisenberg Model for Saturation of the Froissart Bound

But it is remarkable that the saturation of the Froissart unitarity bound, even though it is not yet analytically understood in QCD, was described in a simple model by Heisenberg in 1952, well before Froissart and QCD.

Heisenberg says that, as far as most of the strong interaction (QCD) total cross-section is concerned, the interaction is described by the exchange of the lowest-mass particle, the pion. Thus, one considers a pion field surrounding a nucleon, and if one considers nucleon–nucleon scattering happening at an impact parameter that is larger than the inverse QCD scale Λ_{QCD}^{-1}, we effectively have scattering of the pion fields surrounding the nucleons. Moreover, at high center of mass energy ($s \rightarrow \infty$), the nucleons become pancake-shaped due to the Lorentz contraction, and so does the pion field surrounding them. In the infinite-energy limit, the pancakes become shockwaves, and the scattering is described as a collision of two pion field shockwaves. In the collision, energy is emitted, and radiated away in the form of the lowest-mass particles (i.e. pions).

The conclusion is that we can describe the collision in a picture that combines the classical field theory involved in the collision and radiation of energy of the pion field, combined with some elements of quantum mechanics, remembering that the field is composed of quanta, the pions.

The classical pion field radiation was described in classical field theory, and will be reviewed here. One needs to start with an action for the pion field, in this nonperturbative regime (low per-pion energy, but large total energy, for a field theory limit). As we said, we aim to describe the collision of two pion field shockwaves, and nonlinearities of the pion field are expected to play a role at the shock. Therefore, Heisenberg proposes that we have an action for which the shock solution has a finite $(\partial_\mu \phi)^2$ at the shock. An alternative condition which we can impose, and which turns out to be equivalent to the previous one, is to have a saturation of the Froissart bound. We will see that the action that gives Froissart bound saturation is pretty much unique. That action is the DBI action for a (pseudo)scalar pion:

$$\mathcal{L} = l^{-4} \left[1 - \sqrt{1 + l^4 [(\partial_\mu \phi)^2 + m^2 \phi^2]} \right]. \tag{57.15}$$

Note that in reality, as mentioned, the pion is a scalar field triplet (π^+, π^-, π^0), but for simplicity we consider a single pion ϕ, with the above action.

Consider then a classical shock solution. We need a solution that is boost-invariant, which for motion in the x direction is only a function of

$$s = t^2 - x^2. \tag{57.16}$$

Now for an arbitrary classical solution in a Lorentz-invariant theory, a solution depending independently on $x^+ = t + x$ and $x^- = t - x$ would have been good, but we want moreover a solution for a shock that moves on either $x^+ = 0$ or $x^- = 0$, and then we impose $\phi(x^+ = 0) = 0$ or $\phi(x^- = 0) = 0$ in any reference frame (a boost changes

$x^+ \to e^\beta x^+$ and $x^- \to e^{-\beta} x^-$), which restricts us (from Lorentz invariance) to have $\phi = \phi(x^+ x^-) = \phi(s)$ only. Then we have

$$(\partial_\mu \phi)^2 = -(\partial_t \phi)^2 + (\partial_x \phi)^2 = -4s \left(\frac{d\phi}{ds}\right)^2, \tag{57.17}$$

and if it is constant near $s = 0$, this means that $\phi(s) \simeq A\sqrt{s}$ near $s = 0$.

If we were to choose a $\lambda \phi^4$ interaction for a massive pion with canonical kinetic term

$$\mathcal{L} = -\frac{1}{2}(\partial_\mu \phi)^2 - \frac{1}{2}m^2\phi^2 - \lambda\frac{\phi^4}{4}, \tag{57.18}$$

with the equation of motion on the ansatz $\phi = \phi(s)$ of

$$4\frac{d}{ds}\left(s\frac{d\phi}{ds}\right) + m^2\phi + \lambda\phi^3 = 0, \tag{57.19}$$

then the solution for $s \simeq 0$ would be

$$\phi(s) = a\left[1 - (m^2 + \lambda a^2)s + \frac{1}{4}(m^2 + 3\lambda a^2)s^2 + \ldots\right] \tag{57.20}$$

for $s \geq 0$, and $\phi = 0$ for $s < 0$, as we can check. Then we find that

$$(\partial_\mu \phi)^2 \simeq -4sa^2(m^2 + \lambda a^2)^2 \to 0, \tag{57.21}$$

near $s = 0$, so it doesn't satisfy our constraint. Moreover, we can easily check that any other power-law potential will not work. The modification to the action then has to be that the kinetic term becomes nonstandard, and has an infinite number of derivatives. The DBI action above is a natural candidate. Indeed, for (57.15) we find that the equation of motion for $\phi = \phi(s)$ can be manipulated to be put in the form

$$4\frac{d}{ds}\left(s\frac{d\phi}{ds}\right) + m^2\phi = 8sl^4\left(\frac{d\phi}{ds}\right)^2 \frac{\left[\frac{d\phi}{ds} + m^2\phi\right]}{1 + l^4 m^2 \phi^2}, \tag{57.22}$$

which at $m = 0$ has the exact solution

$$\phi = \frac{1}{a}\log\left(1 + \frac{a^2}{2l^4}s + \frac{a}{2l^4}\sqrt{4l^4 s + a^2 s^2}\right) \tag{57.23}$$

for $s \geq 0$ and $\phi = 0$ for $s < 0$, and at $m \neq 0$ has the perturbative solution

$$\phi = \frac{\sqrt{s}}{l^2}(1 + am^2 s + \ldots) \tag{57.24}$$

for $s \geq 0$, and $\phi = 0$ for $s < 0$. This means that now

$$(\partial_\mu \phi)^2 \simeq -l^{-4} \tag{57.25}$$

near $s = 0$, which is constant, as requested by Heisenberg.

To find the classical radiation from the collision of the two shocks, we consider the Fourier transform over x of the solution near the shock (for $a = 0$):

$$\phi(k, t) = l^{-2} \int_0^t dx \, e^{ikx} \sqrt{t^2 - x^2} \simeq l^{-2}\frac{\pi}{2}\frac{|t|}{|k|}(J_1(|k||t|) + i\mathbf{H}_1(|k||t|)), \tag{57.26}$$

where J_1 is a Bessel function and $\mathbf{H}_1$ a Struve function.

When expanding at large k, we obtain an oscillating solution, except for an additive constant:

$$\phi - l^{-2} i \frac{|t|}{|k|} \simeq \sqrt{-i} l^{-2} \sqrt{\frac{\pi}{2}} |t|^{1/2} |k|^{-3/2} e^{-i|k||t|} \left(1 + \frac{3}{8|k||t|} e^{2i|k||t|}\right). \qquad (57.27)$$

This is understood as radiation. The energy density (Hamiltonian density) coming from the Lagrangian (57.15) is

$$\mathcal{H} \equiv \frac{\partial \mathcal{L}}{\partial \dot{\phi}} \dot{\phi} - \mathcal{L} = \frac{l^{-4} + (\vec{\nabla}\phi)^2 + m^2 \phi^2}{\sqrt{1 + l^4[(\partial_\mu \phi)^2 + m^2 \phi^2]}} - l^{-4}. \qquad (57.28)$$

But then, since $(\partial_\mu \phi)^2 \to -l^{-4}$ and $\phi \to 0$ at $s \to 0$, the square root vanishes and the energy density blows up at the shock. However, Heisenberg argues that on physical grounds there must be a small perturbation that regulates the shock and keeps the square root finite (a nonvanishing constant), so that finally

$$\frac{dE}{dk} \propto k^2 \phi^2(k). \qquad (57.29)$$

From the form of the radiation piece in (57.27), $\phi(k) \propto k^{-3/2}$, so

$$\frac{dE}{dk} = \frac{\text{const.}}{k}. \qquad (57.30)$$

Note that the same calculation for a canonical kinetic term with a power-law potential would give just a constant, $dE/dk = \text{const}$.

This spectrum is not valid for any k. There is a maximum value, obtained as follows. The shockwave cannot have a width smaller than the Lorentz-contracted inverse pion mass, $(\gamma m)^{-1}$. This means that there is a minimum for $\sqrt{s}$ of $\sqrt{s_0} \equiv \sqrt{1 - v^2}/m$. That in turn translates into a maximum momentum k of γm for the radiation solution, thus for the spectrum. Moreover, there is a minimum k for the spectrum, namely the pion mass m, since quantum mechanically we cannot have an energy quantum smaller than m. Thus the spectrum is valid for $m \leq k \leq k_{0m} = \gamma m$.

Now we take the crucial step towards the quantum field theory interpretation: we consider canonical quantization of the scalar field, under which the momentum k becomes the momentum k_0 of a pion, and the energy density of the radiated field, E, becomes the energy density of the radiated pions (as the classical field is the limit of many particles of the field quanta) $\mathcal{E}$, so that finally

$$\frac{d\mathcal{E}}{dk_0} = \frac{B}{k_0}, \quad m \leq k_0 \leq k_{0m} = \gamma m, \qquad (57.31)$$

where B is a constant. This integrates to

$$E = B \log \frac{k_{0m}}{m} = B \log \gamma. \qquad (57.32)$$

Considering the number n of pions of given momentum k_0, we can write $d\mathcal{E} = k_0 dn$, so that by differentiation we have

$$\frac{dn}{dk_0} = \frac{B}{k_0^2}, \quad m \le k_0 \le k_{0m} = \gamma m, \tag{57.33}$$

integrating to

$$n = \frac{B}{m}\left(1 - \frac{m}{k_{0m}}\right). \tag{57.34}$$

Then the average, per pion, emitted energy is

$$\langle k_0 \rangle \equiv \frac{\mathcal{E}}{n} = m\frac{\log\frac{k_{0m}}{m}}{1 - \frac{m}{k_{0m}}} \simeq m\log\gamma, \tag{57.35}$$

approximately constant as a function of energy.

Finally, Heisenberg assumes that the radiated energy $\mathcal{E}$ is proportional to the center of mass energy $\sqrt{\tilde{s}}$ ($\tilde{s}$ is the Mandelstam variable, not to be confused with $s = t^2 - x^2$), with the constant of proportionality being the overlap of the pion wavefunctions. When the transverse distance $r = \sqrt{y^2 + z^2}$ (to the direction of the shock) is large, the wavefunction is small ($\phi l \ll 1$), so we can use a free massive wavefunction, $\phi(r) \sim e^{-mr}$, and the wavefunction overlap is proportional to $e^{-mr_1}e^{-mr_2} = e^{-mb}$, where b is the impact parameter. Then

$$\mathcal{E} \sim e^{-mb}\sqrt{\tilde{s}}. \tag{57.36}$$

Now in order to calculate the total cross-section, we consider the maximum impact parameter $b_{\max}$ for which we still have an interaction. That is taken to be when the radiated energy $\mathcal{E}$ equals the minimum possible value, namely the average per-pion emitted energy $\langle k_0 \rangle$:

$$e^{-mb_{\max}}\sqrt{\tilde{s}} = \langle k_0 \rangle \Rightarrow b_{\max} = \frac{1}{m}\ln\frac{\sqrt{\tilde{s}}}{\langle k_0 \rangle}. \tag{57.37}$$

Since $\langle k_0 \rangle \simeq$ constant, the total cross-section (in a classical approximation) is

$$\sigma_{\text{tot}} = \pi b_{\max}^2 = \frac{\pi}{m_\pi^2}\ln^2\frac{\sqrt{\tilde{s}}}{\langle k_0 \rangle}, \tag{57.38}$$

which saturates the Froissart bound (including for the coefficient). Note, however, that for the pion with canonical kinetic term, we get $\langle k_0 \rangle \propto \sqrt{\tilde{s}}$, so we get a constant total cross-section.

Indeed, in that case we get $\frac{d\mathcal{E}}{dk_0} = B$ (constant), from which $\frac{dn}{dk_0} = \frac{B}{k_0}$, both in the same interval $m \le k_0 \le \gamma m$. But then, integrating $\mathcal{E}$ and n, we get

$$\langle k_0 \rangle = \frac{\mathcal{E}}{n} \simeq \frac{k_{0m}}{\ln\frac{k_{0m}}{m}} = \frac{m\gamma}{\ln\gamma} \propto \frac{\sqrt{\tilde{s}}}{\ln\sqrt{\tilde{s}}}. \tag{57.39}$$

We see then that we can also reversely say that we select the DBI action *because* it is the only one that satisfies the Froissart bound (as opposed to the first argument, which restricted to DBI because of finite $(\partial_\mu\phi)^2$ at the shock, but then noticed that we obtain the Froissart bound).

Important Concepts to Remember

- Analyticity and unitarity constrain amplitudes and S-matrices.
- The partial wave expansion is an expansion of amplitudes in normalized Legendre polynomials, $(2l+1)P_l(\cos\theta)$, with coefficient a_l, the partial wave amplitude.
- The Froissart unitarity bound for the total amplitude is $\sigma_{\text{tot}} \leq C \ln^2 s$, with $C \leq \pi/\kappa^2$, and κ^2 the minimum between the t and u channel thresholds, $\kappa^2 = \min(u_0, t_0)$, so effectively the mass squared of the minimum state in the theory.
- In QCD, the natural assumption is $\kappa^2 = m_\pi^2$, which leads to $C_{\min} \simeq 60$ milibarns, though recently there were some proposals that the glueball mass $m_g \gg m_\pi$ is the relevant one.
- The Heisenberg model for saturation of the Froissart bound considers that asymptotically, both the colliding nucleons and the pion field they generate Lorentz-contract, first as pancakes, then as shockwaves, so involving collision of pion-field shockwaves sourced by nucleons.
- The correct and unique pion action that describes the classical pion shockwave field for the Froissart bound saturation in the Heisenberg model is the massive DBI action, $\mathcal{L} = l^{-4}\left[1 - \sqrt{1 + l^4[(\partial_\mu\phi)^2 + m^2\phi^2]}\right]$.
- The massive DBI action is the unique action with $\phi = \phi(s)$, $s = t^2 - x^2$, that has a finite $(\partial_\mu\phi)^2$ at the shock position $s = 0$, and gives a saturation of the Froissart bound for the classical cross-section $\sigma_{\text{tot}}(s)$.
- In the Heisenberg model $\phi(s) \simeq l^{-2}/\sqrt{s}$ near $s = 0$ (for $s > 0$) and one finds $\frac{dE}{dk} = \frac{B}{k}$, with B constant, or in the quantum theory $\frac{d\mathcal{E}}{dk_0} = \frac{B}{k_0}$ for $m_\pi \leq k_0 \leq \gamma m_\pi$, leading to average per-pion energy of $\langle k_0 \rangle m_\pi \ln \gamma$ only for the DBI action.
- In the Heisenberg model, $e^{-b_{\max}}\sqrt{s} = \langle k_0 \rangle$, with maximum classical (geometrical) cross-section of $\pi b_{\max}^2$, saturating the Froissart bound.

Further Reading

The original Froissart bound was proposed in [13]. A refinement was given in [14]. The Heisenberg model was written in [15]. A refinement and generalization of the Heisenberg model were done in [16].

Exercises

1. Expand in partial wave amplitudes the Veneziano amplitude

$$\mathcal{M}(s,t) = \frac{\Gamma(-\alpha(s))\Gamma(-\alpha(t))}{\Gamma(-\alpha(s) - \alpha(t))}. \tag{57.40}$$

2. Show that the equations of motion of the DBI action on the $\phi = \phi(s)$ ansatz, (57.22), have the solution

$$\phi = \frac{1}{a} \log \left(1 + \frac{a^2}{2l^4}s + \frac{a}{2l^4}\sqrt{4l^4s + a^2s^2} \right) \tag{57.41}$$

at $m = 0$.

3. Show (paralleling the analysis in the text for the DBI action case) that for a ϕ^4 interaction, we obtain $\frac{d\mathcal{E}}{dk_0} = $ constant, which, as we showed in the text, gives a constant cross-section, thus not saturating the Froissart bound.

4. Is the property of the DBI action that $(\partial_\mu \phi)^2$ is constant at the shock preserved by truncating the expansion of the DBI action at a given order in l^4? Why? How about the property of approximately constant $\langle k_0 \rangle$?

58 The Operator Product Expansion, Renormalization of Composite Operators, and Anomalous Dimension Matrices

In this chapter we will learn about renormalizing composite operators, a tool for describing that called the operator product expansion, and anomalous dimension matrices for composite operators. As examples, we will consider a calculation of the anomalous dimension of a composite operator, and the operator product expansion in QCD.

58.1 Renormalization of Composite Operators

Composite operators are important objects in gauge theories, since observables are gauge invariant, so need to be composite. But introducing a composite operator at a point x, $\mathcal{O}(x)$, requires additional renormalization beyond that in the Lagrangian, since putting several fields at a single point introduces new divergences.

Important examples of composite operators are the energy–momentum tensor $T_{\mu\nu}$, condensates $\bar{\psi}\psi$, and so on. We introduce the composite operators in the theory by adding a source term for them, $J_{\mathcal{O}} \cdot \mathcal{O}$ in the generating functional:

$$Z_{\mathcal{O}}[J_{\mathcal{O}}] = \int \mathcal{D}[\phi] e^{-S + \int J_{\mathcal{O}} \cdot \mathcal{O}(x)}, \tag{58.1}$$

in the same way as we did for fundamental fields in

$$Z[J] = \int \mathcal{D}\phi \, e^{-S + \int J \cdot \phi(x)}. \tag{58.2}$$

The resulting Green's functions for $\mathcal{O}$ can be thought of as Green's functions for sets of fields at $x_1, \ldots, x_n$ when we identify these points (they all converge, and are equal to x), for example for $\mathcal{O} = \phi_1 \ldots \phi_n(x)$:

$$\langle \mathcal{O}(x) \rangle \sim \langle \phi_1(x_1) \ldots \phi_n(x_n) \rangle|_{x_1 = x_2 = \ldots = x_n = x} = G^{(n)12\ldots n}(x_1, \ldots, x_n)|_{x_1 = \ldots = x_n = x}. \tag{58.3}$$

Of course, because of the divergences we mentioned, this is not so well defined. We will use the method of the operator product expansion to make better sense of this.

One can reverse the above process and consider a method of regularization for the additional divergences appearing in composite operators called "point splitting," which means pulling the constituents apart, by having each field in $\mathcal{O}$ at a different point.

For composite operators, there is also *operator mixing*, if there are Feynman diagrams that mix the operators. For this to happen, the operators must have the same charges under the symmetries respected by the Lagrangian, since then the interaction Lagrangian $\mathcal{L}_{\text{int}}$, appearing in the Feynman diagrams, also respects the symmetries.

Therefore, in general, we have the renormalization of the type

$$\mathcal{O}_n[\{\phi_j\}, g, \ldots] = \sum_m Z_n{}^m \mathcal{O}_m^{\text{ren}}[\{Z_j \phi_j^{\text{ren}}\}, Z_g g^{\text{ren}}, \ldots], \tag{58.4}$$

meaning that besides the renormalization of the fields and couplings, there is an independent renormalization of the operators, that mixes them.

In particular, in a Yang–Mills theory:

$$\mathcal{O}_j[A_\mu^a, g, \ldots] = \sum_k Z_j{}^k \mathcal{O}_k^{\text{ren}}\left[Z_3^{1/2} A_\mu^{a, \text{ren}}, \frac{Z_1}{Z_3^{3/2}} g^{\text{ren}}, \ldots \right]. \tag{58.5}$$

The matrix of renormalization factors $Z_n{}^m$ can be determined by Feynman diagrams with insertions of $\mathcal{O}_n, \mathcal{O}_m$. We will see shortly a concrete example of such a calculation.

This renormalization is also multiplicative, like the renormalization of fundamental fields. Moreover, it closes under renormalization (there are no outside operators).

In general, $Z_n{}^m$ is nontrivial, but in the particular case of conserved currents it is 1 (i.e. conserved currents do not renormalize, $Z_j = 1$).

For instance, in QED, the Ward–Takahashi identity says that

$$\partial_{z^\mu} \langle 0|T[\psi_\alpha(x)\bar\psi_\beta(y)j_\mu(z)]|0\rangle = -ie\delta(x-z)\langle 0|T[\psi_\alpha(y)\bar\psi_\beta(x)]|0\rangle$$

$$+ ie\delta(y-z)\langle 0|T[\psi_\alpha(x)\bar\psi_\beta(y)]|0\rangle. \tag{58.6}$$

The renormalization of the field is $Z_R\psi_R = \psi_0$, and of the current is $Z_j j_R = j_0$, but as we can see there is no j, so no Z_j on the right-hand side, whereas there is on the left-hand side. By matching Z factors, we see that we must have $Z_j = 1$.

58.2 Anomalous Dimension Matrix

So we see that the renormalization of operators is written as

$$\mathcal{O}_j(x) = \sum_k Z_j{}^k \mathcal{O}_k^{\text{ren}}(x). \tag{58.7}$$

We can therefore define for $\mathcal{O}$ also a notion of anomalous dimension, just that now it is an *anomalous dimension matrix*:

$$\Gamma_{ij} \equiv \left(Z^{-1} \cdot \Lambda \frac{\partial}{\partial \Lambda} Z \right)_{ij}, \tag{58.8}$$

where Z and Z^{-1} are matrices, and $\Lambda \partial/\partial \Lambda = \partial/\partial \ln \Lambda$.

Consider the *eigenvectors* $\mathcal{O}_n$ of the matrix Γ, with eigenvalue $\Delta_0 + \gamma_n$, where Δ_0 is the classical (naive) dimension, and γ_n is the anomalous dimension of the operator $\mathcal{O}_n$.

Then we must have for scalar operators (no Lorentz structure)

$$\langle \mathcal{O}_n^{\text{ren}}(x)\mathcal{O}_n^{\text{ren}}(y)\rangle = \langle (Z \cdot \mathcal{O}_n(x)(Z \cdot \mathcal{O})_n(y)\rangle = \frac{\text{const.}}{|x-y|^{2(\Delta_0+\gamma_n)}}. \tag{58.9}$$

This is so because by translational invariance the two-point function must depend on $|x - y|$ only, and then the dimension defines its power law. Strictly speaking, this result is only valid in a scale-invariant theory, without any mass scale, since otherwise we can use the mass scale to construct dimensionless quantities together with $|x - y|$.

A nontrivial example of a composite operator in QCD is the quark mass term

$$\Delta \mathcal{L}_m = m(\bar{q}q)_M, \tag{58.10}$$

with renormalization prescription at the mass scale M.

Then, for instance, the Green's functions of $\bar{q}q$ with the quarks:

$$G^{(n,k)}(x_1, \ldots, x_n; y_1, \ldots, y_n; z_1, \ldots, z_k) = \langle q(x_1) \ldots q(x_n)\bar{q}(y_1) \ldots \bar{q}(y_n)\bar{q}q(z_1) \ldots \bar{q}q(z_k) \rangle \tag{58.11}$$

obey a renormalization group equation (RGE) that is the natural extension of that for the Green's functions for fundamental fields, namely

$$\left[M\frac{\partial}{\partial M} + \beta\frac{\partial}{\partial g} + 2n\gamma + k\gamma_{\bar{q}q} \right] G^{(n,k)}(\{x_i\}, \{y_i\}, \{z_j\}, g, M) = 0. \tag{58.12}$$

We can then define a mass depending on the energy scale, $\bar{m}(Q)$, in the usual way by

$$\frac{d}{d\log(Q/M)}\bar{m} = \gamma_{\bar{q}q}(\bar{g})\bar{m}, \tag{58.13}$$

and the boundary (initial) conditions $\bar{m}(M) = m$. A perturbative calculation in QCD (that will not be reproduced here) finds

$$\bar{m}(Q) = m\left(1 + 8\frac{g^2}{(4\pi)^2}\log\frac{M^2}{Q^2} \right) + \mathcal{O}(g^4). \tag{58.14}$$

58.3 Anomalous Dimension Calculation

We can give a simple example of an anomalous dimension calculation. Consider $\lambda\phi^4/4!$ theory, and the operator $\mathcal{O}(x) = \phi^2(x)$. The two-point function $\langle\mathcal{O}(x)\mathcal{O}(0)\rangle$ is calculated as follows.

58.3.1 Tree Level: $\mathcal{O}(1)$

At the tree level, we write a Feynman diagram where we represent the operator at 0 as a line with two points at its ends (for the two ϕ fields), and the same for the operators at x, for two parallel lines, as in Figure 58.1(a). Then the free propagators connect the ϕ in the upper operator with that in the lower operator. Since in four dimensions the scalar propagator is

$$\frac{1}{4\pi^2}\frac{1}{|x|^2}, \tag{58.15}$$

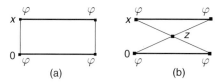

Figure 58.1 Diagrams for the anomalous dimension of an operator in ϕ^4 theory: (a) tree level; (b) one-loop level.

we obtain

$$\langle \mathcal{O}(x)\mathcal{O}(0)\rangle^{\text{tree}} = \frac{2}{(4\pi^2)^2} \frac{1}{|x|^4}, \tag{58.16}$$

for the two free propagators and two possible contractions.

58.3.2 One-Loop Level: $\mathcal{O}(\lambda)$

The Feynman diagram has a vertex at z connecting with the two fields in the upper operator and the two fields in the lower operator, as in Figure 58.1(b), for a result

$$\frac{\lambda}{(4\pi^2)^4} \int d^4z \frac{1}{|x-z|^4|z|^4} \equiv \lambda \langle \mathcal{O}(x)\mathcal{O}(0)\rangle^{\text{tree}} \mathcal{J}(x), \tag{58.17}$$

because of the two propagators from z to x and two from z to 0. From the above definition, we have

$$\mathcal{J}(x) = \frac{|x|^4}{(4\pi^2)^2} \int d^4y \frac{1}{y^4|x-y|^4}. \tag{58.18}$$

This is UV divergent, and needs to be regularized by introducing a UV cut-off Λ, after which the integral becomes

$$\mathcal{J}(x) \sim \frac{1}{4\pi^2} \log(|x|\Lambda). \tag{58.19}$$

The proof of this statement is left as an exercise.

All in all, we can write

$$\langle \mathcal{O}(x)\mathcal{O}(0)\rangle = \langle \mathcal{O}(x)\mathcal{O}(0)\rangle^{\text{tree}} \left[1 + \frac{\lambda}{4\pi^2} \log(|x|\Lambda) + \dots \right]. \tag{58.20}$$

On the contrary, in general we can expand (since $\Delta = \Delta_0 + \mathcal{O}(\lambda)$, so $\Delta - \Delta_0 = \mathcal{O}(\lambda)$)

$$\langle \mathcal{O}(x)\mathcal{O}(0)\rangle = \frac{C}{|x|^{2\Delta}} = \frac{C}{|x|^{2\Delta_0}} e^{-2(\Delta-\Delta_0)\log(|x|\Lambda)} \simeq \frac{C}{|x|^{2\Delta_0}} (1 - 2(\Delta - \Delta_0)\log(|x|\Lambda))$$
$$= \langle \mathcal{O}(x)\mathcal{O}(0)\rangle^{\text{tree}} (1 - 2(\Delta - \Delta_0)\log(|x|\Lambda) + \dots). \tag{58.21}$$

Here we have normalized the operators such that the two-point function has $C = 2/(4\pi^2)^2$ in order to reproduce the free two-point function above.

By comparing this general result with our particular Feynman diagram calculation, we obtain first $\Delta_0 = 2$, and second

$$\Delta - \Delta_0 = -\frac{\lambda}{8\pi^2} \Rightarrow \Delta = 2 - \frac{\lambda}{8\pi^2}. \tag{58.22}$$

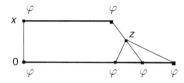

Figure 58.2 Operator mixing diagram in ϕ^4 theory.

In the same $\lambda\phi^4/4!$ theory, we can give an example of operator mixing, between the operators $\phi^2(x)$ and $\phi^4(x)$. The Feynman diagram has one free propagator coming down from $\phi^2(x)$ to $\phi^4(0)$, and one line ends in a 4-vertex, whose three other legs end up on the remaining fields in $\phi^4(0)$, as in Figure 58.2. The result for the diagram is (combinatorial factor $2 \times 4!$)

$$\langle\phi^2(x)\phi^4(0)\rangle = \frac{2 \times 4!\,\lambda}{(4\pi^2)^2|x|^4}\tilde{I}(x), \tag{58.23}$$

where

$$\tilde{I}(x) = \frac{|x|^4}{(4\pi^2)^3}\int d^4z \frac{1}{|z|^6|x-z|^2}, \tag{58.24}$$

and a calculation that will not be reproduced here gives for the integral $-2\pi^2|\Lambda|^2/|x|^2$, so we obtain

$$\tilde{I}(x) = -\frac{|x|^2\Lambda^2}{32\pi^3}. \tag{58.25}$$

58.4 The Operator Product Expansion

In general, when two composite operators $\mathcal{O}_1(x)$ and $\mathcal{O}_2(0)$ go to a point, we can create disturbances (perturbations) in the vicinity of the operators, with divergent coefficients. So the result is described in terms of local operators, times divergent coefficient functions. Under this procedure, there is a complete (closed) set of operators $\{\mathcal{O}_k\}$, consistent with the symmetries. The singularity at $x \to 0$ translates into a singularity in the coefficient functions. Therefore we have

$$\mathcal{O}_1(x)\mathcal{O}_2(0) \to \sum_n C_{12}{}^n(x)\mathcal{O}_n(0). \tag{58.26}$$

By translational invariance, the right-hand side can only depend on the difference in the positions of the two operators, so being a bit more general we can write

$$\mathcal{O}_i(x)\mathcal{O}_j(y) \to \sum_k C_{ij}{}^k(x-y)\mathcal{O}_k(y). \tag{58.27}$$

This is the *operator product expansion* (OPE). The coefficient functions $C_{ij}{}^k(x-y)$ are c-number functions that are singular in the argument going to zero.

Note that the OPE is an operator relation, which means that it must hold on any matrix element $\langle \alpha |\quad |\beta \rangle$. Dimensional analysis suggests that in a theory with no mass scales (scale-invariant):

$$C_{ij}{}^k(x-y) \to \frac{1}{|x-y|^{\Delta_i + \Delta_j - \Delta_k}}. \tag{58.28}$$

This means that, as an approximation, we can only consider the operator $\mathcal{O}_k$ of lowest dimension Δ_k, which will have the coefficient $C_{ij}{}^k$ of the highest singularity in $|x-y|$, when considering this as an expansion in $|x-y|$. That makes the OPE very useful for calculations.

The OPE is valid in all Green's functions, so for instance

$$G_{ij}(x;y;z_1,\ldots,z_m) = \langle \mathcal{O}_i(x)\mathcal{O}_j(y)\phi(z_1)\ldots\phi(z_m)\rangle = \sum_k C_{ij}{}^k(x-y)\langle \mathcal{O}_k(y)\phi(z_1)\ldots\phi(z_m)\rangle. \tag{58.29}$$

This means that we can reduce the Green's functions to lower ones, and the dependence on the absorbed point, x, is now in the coefficient functions only.

In turn, it means that knowing all the OPEs solves the theory in terms of $\mathcal{O}_j(x)$s, since we can reduce successively the number of operators in the Green's function:

$$\langle \mathcal{O}_i(x)\mathcal{O}_j(y)\mathcal{O}_l(z)\ldots\mathcal{O}_p(w)\rangle = \sum_k C_{ij}{}^k(x-y)\langle \mathcal{O}_k(y)\mathcal{O}_l(z)\ldots\mathcal{O}_p(w)\rangle$$

$$= \sum_{km} C_{ij}{}^k(x-y)C_{kl}{}^m(y-z)\langle \mathcal{O}_m(z)\ldots\mathcal{O}_p(w)\rangle = \ldots$$

$$= \sum_{k,m,\ldots} C_{ij}{}^k(x-y)C_{kl}{}^m(y-z)\ldots C_{qp}{}^r(t-w)\langle \mathcal{O}_r(w)\rangle. \tag{58.30}$$

Then, if we have the two-point function normalized as

$$\langle \mathcal{O}_q(x)\mathcal{O}_p(y)\rangle = \frac{C\delta_{qp}}{|x-y|^{2\Delta_q}}, \tag{58.31}$$

so that in particular

$$\langle \mathcal{O}_r(w)\rangle = \delta_{r1} \Rightarrow C_{qp}{}^r(t-w)\langle \mathcal{O}_r(w)\rangle = C_{qp}{}^r(t-w)\langle 1\rangle = \frac{C\delta_{qp}}{|t-w|^{2\Delta_q}}, \tag{58.32}$$

we solve completely the Green's function.

58.5 QCD Example

The example of OPE we are interested in is in QCD. In particular, the currents $J^\mu(x)J^\nu(y)$ must have an OPE that has a maximal contribution from the quark current $\bar{q}\gamma^\mu q$. In contrast, one can prove that we must expand them in the operators in irreducible representations (irreps) of the Lorentz group. That leads to the basis of gauge-invariant operators for the OPE (note that we use D^μ instead of ∂^μ for gauge invariance)

$$\mathcal{O}_f^{(n)\mu_1\ldots\mu_n} = \bar{q}_f \gamma^{\{\mu_1} D^{\mu_2} \ldots D^{\mu_n\}} q_f - \text{traces}. \tag{58.33}$$

Here f stands as usual for flavor (the type of quark). These operators start with $\bar{q}\gamma^\mu q$ for $n = 1$, have dimension $n + 2$ (since there are $n - 1$ D^μs, each with dimension 1, and two quark fields, each with dimension 3/2), and spin (i.e. Lorentz irrep) n, since there are n vector indices symmetrized and with the traces subtracted.

Then, in momentum space, the OPE of the currents is written as

$$i \int d^4 x e^{iqx} J^\mu(x) J^\nu(0) = \sum_f Q_f^2 \left[\sum_{n\geq 2} \frac{(2q^{\mu_1})\ldots(2q^{\mu_{n-2}})}{(Q^2)^{n-1}} \mathcal{O}_f^{(n)\mu\nu\mu_1\ldots\mu_{n-2}} \right.$$

$$\left. - g^{\mu\nu} \sum_{n\geq 2} \frac{(2q^{\mu_1})\ldots(2q^{\mu_n})}{(Q^2)^n} \mathcal{O}_f^{(n)\mu_1\ldots\mu_n} \right] + \ldots \tag{58.34}$$

In x space it becomes

$$T[J^\mu(x)J^\nu(0)] \sim \sum_{n\geq 2} C_n \frac{x^{\mu_1}\ldots x^{\mu_{n-2}}}{|x|^2} \mathcal{O}_f^{(n)\mu\nu\mu_1\ldots\mu_{n-2}}(0)$$

$$+ g^{\mu\nu} \sum_{n\geq 2} \tilde{C}_n \frac{x^{\mu_1}\ldots x^{\mu_n}}{(|x|^2)^2} \mathcal{O}_f^{(n)\mu_1\ldots\mu_n}(0) + \ldots, \tag{58.35}$$

where C_n and $\tilde{C}_n$ are fixed numerical coefficients.

For $\mu = \nu$ and summed over, we get a simpler expression:

$$T[J^\mu(x)J_\mu(0)] = \sum_n h_n \frac{x^{\mu_1}\ldots x^{\mu_n}}{(|x|^2)^2} \mathcal{O}_f^{(n)\mu_1\ldots\mu_n}(0), \tag{58.36}$$

where $h_n = h_n(x, \mu^2, \alpha_s(\mu^2))$ is a dimensionless function.

Note then that we have two types of OPE expansions:

- The standard one, a short distance expansion, in $|x|$, in which we can keep only the operator $\mathcal{O}^{(n)}$ of lowest dimension.
- A *lightcone expansion*, obtained by considering that the expansion is in $|x|^2$ around $|x|^2 = 0$ viewed as a lightcone (with x^μ finite), in which case the relevant dimension for the expansion is not the dimension of $\mathcal{O}_f^{(n)\mu_1\ldots\mu_n}$, but rather the dimension of the full object in the numerator (i.e. the dimension of $x^{\mu_1}\ldots x^{\mu_n}\mathcal{O}^{(n)\mu_1\ldots\mu_n}$), equal to

$$\dim[\mathcal{O}^{(n)}] - n = \dim[\mathcal{O}^{(n)}] - \text{spin}\,[\mathcal{O}] \equiv \text{twist}[\mathcal{O}^{(n)}]. \tag{58.37}$$

Here the *twist* $T = \Delta - S$. Then, in this expansion, we keep only the *leading twist operators*. But now that there is operator mixing, $\mathcal{O}_f^{(n)\mu_1\ldots\mu_n}$ mixes with

$$\mathcal{O}_g^{(n)\mu_1\ldots\mu_n} = F^{\{\mu_1\nu} D^{\mu_2} \ldots D^{\mu_{n-1}} F^{\mu_n\}}{}_\nu - \text{traces}. \tag{58.38}$$

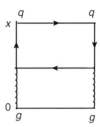

Figure 58.3   Operator mixing diagram in QCD.

Indeed, these operators also have dimension $n + 2$, since there are $n - 2$ Ds, and two Fs, each with dimension 2. They also have spin n, since there are n vector indices symmetrized and with the traces subtracted.

This means that in the OPEs, the $\mathcal{O}_g$s also appear, the same as the $\mathcal{O}_f$s. The mixing of the two operators is done through diagrams where the two quarks in the $\mathcal{O}_f$ emit quark and antiquark propagators, that join, by emitting two gluons from the quark line, that end on the two gluons in the $\mathcal{O}_g$ operators, as in Figure 58.3.

Important Concepts to Remember

- Composite operators need additional renormalization, besides that in the Lagrangian, because of extra divergences when fields come at the same point.
- Since there is also operator mixing between operators of like charges, we have in general a renormalization matrix, $\mathcal{O}_n[\{\phi_i\}, g \ldots] = \sum_m Z_n{}^m \mathcal{O}^{\text{ren}}[\{Z_j \phi_j^{\text{ren}}\}, Z_g g, \ldots]$.
- We also have an anomalous dimension matrix, $\Gamma_{ij} = (Z \cdot d/\partial \log \Lambda Z)_{ij}$.
- Conserved currents do not renormalize.
- When two composite operators go to the same point, we have an expansion in terms of other operators, with divergent c-number coefficient functions $\mathcal{O}_i(x)\mathcal{O}_j(y) \rightarrow \sum_k C_{ij}{}^k(x - y)\mathcal{O}_k(y)$, called the operator product expansion, or OPE.
- Knowing the full OPE is equivalent to solving the theory in terms of the $\mathcal{O}_i$s, since it allows us to calculate the Green's functions.
- We need to consider operators with both dimension and spin, and then we can have either a short distance expansion (usual OPE), starting with operators of minimal dimension, or a lightcone expansion, for $x^2 \rightarrow 0$, but with x^μ finite, starting with operators of leading twist $T = \Delta - S$.

Further Reading

See sections 18.1, 18.3, and 18.5 in [1], sections 11.2 and 14.5 in [1], and chapter 20 in [11].

Exercises

1. Prove that

$$\mathcal{J}(x) = \frac{|x|^4}{(4\pi^2)^2} \int \frac{d^4y}{y^4|x-y|^4} \sim \frac{1}{4\pi^2} \log(|x|\Lambda) + \text{finite}. \qquad (58.39)$$

2. Consider a set of operators $\mathcal{O}_i$, complete under the OPE, and consider their OPEs. Calculate the four-point functions of the theory.

3. Consider $\mathcal{N} = 4$ SYM, with field content: gauge fields A_μ^a, spinors $\psi_{I\alpha}^a$, scalars X_{IJ}^a, where a is an index in the adjoint of $SU(N)$, $I, J = 1, \ldots, 4$ are fundamental indices of a global $SU(4)$. What are the set of leading twist, gauge-invariant, composite ($n \geq 2$ fields) operators?

4. Show that the integral in (58.24) gives

$$\tilde{I}(x) = -\frac{|x|^2 \Lambda^2}{32\pi^3}. \qquad (58.40)$$

59

Manipulating Loop Amplitudes: Passarino–Veltman Reduction and Generalized Unitarity Cut

In this chapter we will learn to manipulate amplitudes, specifically the result of loop-order Feynman diagrams, in order to calculate them more efficiently than *ab initio*. One important tool is the use of the Passarino–Veltman reduction, which simplifies the types of integrals we have, down to a basis of bubbles, triangles, and boxes, and generalized unitarity, which is used to relate a difficult amplitude to simpler ones. Another method, that will be explored in the next chapter, is to use the properties of polylogs and what is known as the *symbol* of a result, to (1) rewrite the calculation of an integral in a simpler way and (2) calculate part of a loop amplitude "without doing the integrals."

59.1 Passarino–Veltman Reduction of One-Loop Integrals

In general, an L-loop, n-point amplitude in a quantum field theory can be written as a sum of Feynman diagrams i as

$$A_n^{L\text{-loop}} = \sum_i \int \left(\prod_{j=1}^{L} \frac{d^D l_j}{(2\pi)^D} \right) \frac{1}{S_i} \frac{n_i c_i}{\prod_k p_k^2}. \tag{59.1}$$

Here, l_j are loop momenta (integration variables), S_i are symmetry factors, n_i are numerators, which are polynomials of the Lorentz-invariant contractions of various momenta (loop l_j and external k_a) and polarization vectors ϵ_a for external states, c_i are color factors, arising from gauge theory indices (as well as coupling factors), and p_k are momenta on the propagators.

The particular case we are interested in is the one-loop, n-point case with *massless particles* (in particular, massless external lines), where there are r propagators and r vertices, forming an r-gon diagram, and $p_k = l + \sum_{i=1}^{k} K_i$, with K_i being the total external momentum at each vertex i:

$$A_n^{1\text{-loop}} = \sum_i \int \frac{d^D l}{(2\pi)^D} \frac{1}{S_i} \frac{n_i c_i}{\prod_{k=1}^{r} (l + \sum_{i=1}^{k} K_i)^2}. \tag{59.2}$$

If there is a single external line at each vertex, the momentum K_i at the vertex is "massless" (i.e. $K_i^2 = 0$), otherwise it is "massive" ($K_i^2 \neq 0$). If $r = n$, there are only massless external momenta K_i. Moreover, in a renormalizable gauge theory, the only four-point vertices are in ϕ^4 (scalars) and in the four-point gluon vertex, which has no momentum. So the presence of one of these vertices in the loop means two external lines at the vertex, but also no numerator associated with it. For a Feynman diagram then, we can at most have as many l^μs in the numerator as massless external lines (with a single external momentum).

In the above amplitude formula, the color factors c_i are l-independent, and get outside the integral, but the n_i are *a priori* polynomials in the loop momentum l^μ of some order m, $P^m(l)$. The integral to be performed is then an r-gon integral $I_r[P^m(l)]$ with a numerator formed by a polynomial $P^m(l)$. That is, a polygon, in increasing order: bubble ($m = 2$), triangle ($m = 3$), box ($m = 4$), pentagon ($m = 5$), and so on. We are mostly interested in dimensional regularization, so we use a dimension $D = 4 - 2\epsilon$.

But there is an algorithmic procedure, due to Passarino and Veltman, the so-called Passarino–Veltman reduction, that successively reduces these integrals to just scalar integrals (with no numerator) of box, triangle, and bubble type.

The first step is to reduce the rank r of the r-gon integral, together with the rank m of the polynomial $P^m(l)$. The way to do this is to contract (after we have taken out external momenta k_i^μ and polarization vectors $\epsilon_i^\mu(k)$ from the integral, and kept the integral to be calculated only with free indices on the loop momentum l^μ) the free index on l^μ with an external momentum at a vertex K_i, and write the resulting product as

$$l \cdot K_i = \frac{1}{2}\left[\left(l + \sum_{j=1}^{i} K_j\right)^2 - \left(l + \sum_{j=1}^{i-1} K_j\right)^2 - K_i^2\right],\tag{59.3}$$

and replace in the integral. In the first two terms, the corresponding propagators cancel, so effectively we "contract" a propagator line, joining together the external momenta at each of the endpoints into one. We can choose to do this for a vertex i at which K_i is massless, so $K_i^2 = 0$ (as we said before, the polynomial rank m is at most equal to the number of massless external momenta K_i). In this way, we reduce the integrals to a sum of $r - 1$-gon integrals of degree $m - 1$, as

$$I_r[P^m(l)] \to \sum_i I_{r-1}^i[P^{m-1}(l)].\tag{59.4}$$

Of course, if we did the reduction at a massive vertex, $K_i^2 \neq 0$, we would also get a $I_r[P^{m-1}(l)]$ contribution.

But we can continue this procedure repeatedly, until one of two things happen: we could get down to scalar integrals (i.e. to $I_{r'}[P^0(l)]$, or $I_{r'}[1]$ if one prefers) or we can reach box integrals (i.e. $I_4[P^{m'}(l)]$).

In the first case, we can make use of the fact that higher n-gon (n-point) scalar integrals can be reduced to lower ones, $n - 1$-gon ($n - 1$-point), as long as $n - 1 \geq 4$, a relation still valid in $D = 4 - 2\epsilon$ dimensions. Schematically, in the notation of [19]:

$$\hat{I}_n = \frac{1}{2N_n}\left[\sum_{i=1}^{n} \gamma_i \hat{I}_{n-1}^{(i)} + (n - 5 + 2\epsilon)\hat{\Delta}_i \hat{I}_n^{D=6-2\epsilon}\right],\tag{59.5}$$

where $\hat{I}_{n-1}^{(i)}$ is obtained by collapsing a single propagator in $\hat{I}_n$, N_n and $\hat{\Delta}_n$ are some specific functions of the momentum invariants $s_{ij} = -(k_i + \ldots + k_{j-1})^2/2$, and we are left with an integral in $D = 6 - 2\epsilon$ dimensions (which is UV finite as $\epsilon \to 0$).

This reduction means that we are eventually left with sums of integrals of the type $I_4^i[P^{m'}(l)]$ (if we are not using the scalar integral reduction, we have $m' = m - (r - 4)$). We can now use again the same reduction coming from (59.3), but have a massive vertex, so we obtain

$$I_4[P^m(l)] \rightarrow \sum_i I_3^i[P^{m-1}(l)] + \sum_i c_i I_4^i[1], \tag{59.6}$$

where c_i are rational functions of the Lorentz invariants. Then, moreover, the polynomial triangles reduce to polynomial bubbles plus scalar triangles:

$$I_3[P^m(l)] \rightarrow \sum_i I_2^i[P^{m-1}(l)] + \sum_i d_i I_3^i[1], \tag{59.7}$$

and finally the polynomial bubbles reduce to scalar bubbles and rational terms:

$$I_2[P^m(l)] \rightarrow \sum_i e_i I_2^i[1] + R + \mathcal{O}(\epsilon). \tag{59.8}$$

For instance, consider

$$I^{\mu\nu} = \int \frac{d^D l}{(2\pi)^D} \frac{l^\mu l^\nu}{l^2(l+p)^2} \equiv I_a p^\mu p^\nu + I_b \eta^{\mu\nu}, \tag{59.9}$$

where the two terms are the only ones allowed by Lorentz invariance and symmetry in $\mu\nu$. Then

$$(p^2 I_a + 4 I_b) \equiv I^{\mu\nu} \eta_{\mu\nu} = \int \frac{d^D l}{(2\pi)^D} \frac{1}{(l+p)^2} = \int \frac{d^D l'}{(2\pi)^D} \frac{1}{l'^2} \tag{59.10}$$

is a constant, whereas

$$(p^2 I_a + I_b)p^\nu \equiv I^{\mu\nu} p_\nu = \int \frac{d^D l}{(2\pi)^D} \frac{l^\nu}{2} \left[\frac{1}{l^2} - \frac{1}{(l+p)^2} - \frac{p^2}{l^2(l+p)^2} \right]. \tag{59.11}$$

The first term is zero (it should be a constant, but has a Lorentz index), the second and third are found again by multiplying with p^ν, and give a constant times invariants, plus a scalar bubble.

All in all, we see that the amplitude, given as a sum of Feynman diagrams, reduces to scalar boxes, triangles, and bubbles:

$$A_n^{1\text{-loop}}(1, 2, \ldots, n) = \sum_{i \in \text{F.diag}} a_i I_r^i[P^m(l; \{k_j, \epsilon_j\})]$$

$$= \sum_{i \in C} c_i I_4[1] + \sum_{j \in D} d_i I_3[1] + \sum_{k \in \mathcal{E}} e_k I_2[1] + R + \mathcal{O}(\epsilon). \tag{59.12}$$

There are particular cases, like the "maximally supersymmetric Yang–Mills" ($\mathcal{N} = 4$ SYM) and "maximally supersymmetric supergravity" ($\mathcal{N} = 8$ sugra), where in fact there are no triangles and bubbles, and the endpoint of the reduction is only written in terms of scalar box integrals. To understand this, note that in a gauge theory, the maximum degree m of the polynomial is r, since the 3-vertex is linear in the (loop) momentum, so by Passarino–Veltman reduction we arrive at $I_4[P^4(l)]$. But in a supersymmetric theory there are cancellations of loop momenta between bosons and fermions. In $\mathcal{N} = 4$ SYM, the four supersymmetries happen to cancel four-loop momenta, so we start with $I_r[P^{r-4}(l)]$, leading to $I_4[P^0(l)] = I_4[1]$ (i.e. to *scalar boxes*). More details can be found in [20].

59.2 Box Integrals

The box integrals appearing in the expansion are of five types, described in Figure 59.1, and can have one, two, three, or four "massive" external lines (with more than one massless external line at the vertex). The cases with two massive lines are split into "easy," for opposite corners, and "hard," for nearby corners.

In the case of the MHV amplitudes in $\mathcal{N} = 4$ SYM, a further simplification happens, and we have only the two-mass easy (2me) diagrams appearing.

The box integrals are defined as

$$I_4(s,t; k_1^2, k_2^2, k_3^2, k_4^2) = \mu^{4-D} \int \frac{d^D p}{(2\pi)^D} \frac{1}{p^2 (p-k_1)^2 (p-k_1-k_2)^2 (p+k_4)^2}, \quad (59.13)$$

where k_1, k_2, k_3, k_4 are the total momenta at the three corners and $s = (k_1 + k_2)^2$ and $t = (k_1 + k_4)^2$. The integral (59.13) is IR divergent if one or more of the "masses" k_i^2 vanish, but may be evaluated using dimensional regularization in $D = 4 - 2\epsilon$ dimensions.

Note that, in general, we have triangle and bubble integrals appearing as well, defined as

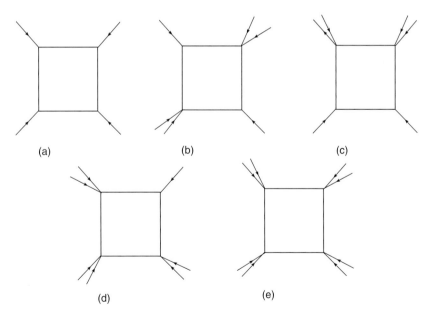

(a)   (b)   (c)

(d)   (e)

Figure 59.1 (a) One-mass (1m) scalar box; (b) two-mass "easy" (2me) scalar box; (c) two-mass "hard" (2mh) scalar box; (d) three-mass (3m) scalar box; (e) four-mass (4m) scalar box.

$$I_3(k_1, k_2, k_3) = \mu^{4-D} \int \frac{d^D p}{(2\pi)^D} \frac{1}{p^2(p-k_1)^2(p+k_3)^2},$$

$$I_2(k) = \mu^{4-D} \int \frac{d^D p}{(2\pi)^D} \frac{1}{p^2(p-k)^2}. \tag{59.14}$$

We can define the rescaled functions F by

$$F(s, t; k_1^2, k_2^2, k_3^2, k_4^2) = -\frac{2\sqrt{\det S}}{c_\gamma} I_4(s, t; k_1^2, k_2^2, k_3^2, k_4^2), \tag{59.15}$$

where the matrix S is 4×4 with elements (i, j are mod 4)

$$S_{ij} = -\frac{1}{2}(k_i + \ldots + k_{j-1})^2, \quad i \neq j, \quad S_{ii} = 0, \tag{59.16}$$

and

$$c_\Gamma = \frac{\Gamma(1+\epsilon)\Gamma^2(1-\epsilon)}{(4\pi)^{2-\epsilon}\Gamma(1-2\epsilon)}. \tag{59.17}$$

For future use, we also define, in a standard notation, the Lorentz invariants

$$t_i^{[r]} \equiv (k_i + \ldots + k_{i+r-1})^2. \tag{59.18}$$

Defining further the 2me box by

$$F^{2\text{me}}(s, t; P^2, Q^2) \equiv F(s, t; k_1^2 = 0, k_2^2 = P^2, k_3^2 = 0, k_4^2 = Q^2) \tag{59.19}$$

and the variable

$$a = \frac{P^2 + Q^2 - s - t}{P^2 Q^2 - st}, \tag{59.20}$$

the all order in epsilon 2me box can be written as [21]:

$$F^{2\text{me}}(s, t, P^2, Q^2) = -\frac{c_\Gamma}{\epsilon^2} \left[\left(\frac{\mu^2}{-s}\right)^\epsilon {}_2F_1(1, -\epsilon, 1-\epsilon, as) + \left(\frac{\mu^2}{-t}\right)^\epsilon {}_2F_1(1, -\epsilon, 1-\epsilon, at) \right.$$

$$\left. - \left(\frac{\mu^2}{-P^2}\right)^\epsilon {}_2F_1(1, -\epsilon, 1-\epsilon, aP^2) - \left(\frac{\mu^2}{-Q^2}\right)^\epsilon {}_2F_1(1, -\epsilon, 1-\epsilon, aQ^2) \right]. \tag{59.21}$$

To expand it in ϵ, we first define the standard *polylogarithms*. The dilogarithm, the one mostly appearing below, is defined by

$$\text{Li}_2(x) = -\int_0^1 \frac{dt}{t} \log(1-xt) = -\int_0^x \frac{dt}{t} \log(1-xt). \tag{59.22}$$

The polylogarithms $\text{Li}_n(z)$ can be defined inductively, by

$$\text{Li}_{n+1}(z) = \int_0^z \frac{dt}{t} \text{Li}_n(t), \tag{59.23}$$

as an integral:

$$\text{Li}_n(z) = \frac{(-1)^{n-1}}{(n-2)!} \int_0^1 \frac{dt}{t} \log^{n-2}(t) \log(1-zt), \tag{59.24}$$

or as a power series:

$$\mathrm{Li}_n(z) = \sum_{k=1}^{\infty} \frac{z^k}{k^n}. \tag{59.25}$$

Using the expansion and integral formulas

$$_2F_1(1, -\epsilon, 1 - \epsilon, z) = -\epsilon \int_0^1 \frac{t^{-1-\epsilon}}{(1 - tz)} dt = \frac{1}{1 - z} - \int_0^1 \frac{t^{-\epsilon} z}{(1 - tz)^2}$$

$$= \frac{1}{1 - z} - \sum_{k=0}^{\infty} \frac{(-\epsilon)^k}{k!} \int_0^1 \frac{z \ln^k t}{(1 - tz)^2}$$

$$= 1 - \sum_{k=1}^{\infty} \epsilon^k \mathrm{Li}_k(z), \tag{59.26}$$

we find that the expansion in ϵ of $F^{2\mathrm{me}}$ is then

$$F^{2\mathrm{me}}(s, t, P^2, Q^2) = c_\Gamma \left\{ -\frac{1}{\epsilon^2} \left[\left(\frac{-s}{\mu^2} \right)^{-\epsilon} + \left(\frac{-t}{\mu^2} \right)^{-\epsilon} - \left(\frac{-P^2}{\mu^2} \right)^{-\epsilon} - \left(\frac{-Q^2}{\mu^2} \right)^{-\epsilon} \right] \right.$$

$$\left. + \mathrm{Li}_2(1 - aP^2) + \mathrm{Li}_2(1 - aQ^2) - \mathrm{Li}_2(1 - as) - \mathrm{Li}_2(1 - at) \right\}$$

$$+ \mathcal{O}(\epsilon). \tag{59.27}$$

Note that we can expand further using the formula

$$\left(\frac{\mu^2}{-s} \right)^{\epsilon} = \sum_{k=0}^{\infty} \frac{\epsilon^k}{k!} \log^k \left(\frac{\mu^2}{-s} \right), \tag{59.28}$$

and this will in fact be useful shortly.

The other scalar box functions are given as follows.

In principle, we can derive all the box functions from the one with generic masses (i.e. the 4m function) by putting some of them to zero. However, there are subtleties, since the four-mass box is the only one that is IR finite (since IR divergences are associated with massless external states). In $D = 4$ (i.e. at $\epsilon = 0$), the four-mass box function is

$$I_4^{4m} = I_4(s, t; k_1^2, k_2^2, k_3^2, k_4^2)$$

$$= -\frac{c_\Gamma(\epsilon = 0)}{st\rho} \left[-\mathrm{Li}_2 \left(\frac{1}{2}(1 - \lambda_1 + \lambda_2 + \rho) \right) + \mathrm{Li}_2 \left(\frac{1}{2}(1 - \lambda_1 + \lambda_2 - \rho) \right) \right.$$

$$- \mathrm{Li}_2 \left(\frac{1}{2\lambda_1}(1 - \lambda_1 - \lambda_2 - \rho) \right) + \mathrm{Li}_2 \left(-\frac{1}{2\lambda_1}(1 - \lambda_1 - \lambda_2 + \rho) \right)$$

$$\left. - \frac{1}{2} \log \left(\frac{\lambda_1}{\lambda_2^2} \right) \log \left(\frac{1 + \lambda_1 - \lambda_2 + \rho}{1 + \lambda_1 - \lambda_2 - \rho} \right) \right], \tag{59.29}$$

where the quantities inside the formula are defined as

$$\rho = \sqrt{1 - 2\lambda_1 - 2\lambda_2 + \lambda_1^2 - 2\lambda_1\lambda_2 + \lambda_2^2}, \qquad \lambda_1 = \frac{k_2^2 k_4^2}{s\,t}, \qquad \lambda_2 = \frac{k_1^2 k_3^2}{s\,t}. \tag{59.30}$$

The remaining integrals are IR divergent and may be evaluated using dimensional regularization in $D = 4 - 2\epsilon$ dimensions. The three-mass box integral I_4^{3m} may be expressed for arbitrary ϵ in terms of integrals over a hypergeometric function [17]:

$$I_4^{3m} = I_4(s, t; 0, k_2^2, k_3^2, k_4^2) = \frac{\kappa}{st - k_2^2 k_4^2} (P^{3m} + Q^{3m}), \qquad \kappa = \frac{2i\Gamma(1 + \epsilon)\mu^{2\epsilon}}{(4\pi)^{2-\epsilon}}, \quad (59.31)$$

where

$$P^{3m}(s, t, k_2^2, k_3^2, k_4^2) = -\frac{1}{2\epsilon}\left[-E\left(\frac{t}{t - k_2^2}, k_2^2, k_3^2, t\right) + E\left(\frac{k_4^2}{k_4^2 - s}, s, k_3^2, k_4^2\right)\right],$$

$$Q^{3m}(s, t, k_2^2, k_3^2, k_4^2) = -\frac{1}{2\epsilon}\left[E\left(\frac{t - k_4^2}{s + t - k_2^2 - k_4^2}, k_2^2, k_3^2, t\right)\right.$$
$$\left. -E\left(\frac{t - k_4^2}{s + t - k_2^2 - k_4^2}, s, k_3^2, k_4^2\right)\right], \qquad (59.32)$$

and

$$E(z_0, k_2^2, k_3^2, t) = \int_0^1 \frac{dz}{z - z_0}\left[-zk_2^2 - (1 - z)t\right]^{-\epsilon} {}_2F_1$$
$$\times \left(1 + \epsilon, -\epsilon, 1 - \epsilon; 1 - \frac{z(1 - z)k_3^2}{zk_2^2 + (1 - z)t}\right). \qquad (59.33)$$

The remaining integrals (2mh, 2me, 1m, and 0m) may all be obtained by taking one or more of the k_i^2 in I_4^{3m} to zero. Note that this limit must be taken with ϵ arbitrary, *before* dropping any terms with positive powers of ϵ; taking k_i^2 to zero does not in general commute with taking ϵ to zero.

59.3 Generalized Unitarity Cuts

The generalized unitarity method is a method available at all loops. For a complicated amplitude, coming from complicated Feynman diagrams, we can try to utilize all possible unitarity cuts to it. But two additional factors are relevant.

The first one is that, just as in the case of one loop explained above, in general there is a basis of basic integrals, called *master integrals* $I_i^{(L)}$, into which any amplitude can be expanded with some coefficients depending on the external momenta:

$$\mathcal{A}_n^{L\text{-loop}} = \sum_i c_i I_i^{(L)}, \qquad (59.34)$$

just like in the case of the Passarino–Veltman reduction, and the basis of scalar boxes, triangles, and bubbles in (59.13) and (59.14). The specific basis depends on loop order and on the theory, and is not as easy as in the one-loop case above, so we will not give examples here. This means that we can take all the possible unitarity cuts, of any order (two-particle, three-particle, etc.) and in any possible way, and compare them with the same unitarity cuts on the basis of integrals, which can be computed. By comparing the two results, we can calculate the coefficients c_i, at the very least numerically if not analytically, and thus compute the amplitude.

The second point is that, since we are not actually interested in the unitarity relation $\text{Im}T = TT^{\dagger}$, for which the unitarity cut, as we saw in Part I, relating the imaginary part of an amplitude with a product of amplitudes would be a cut that separates the diagram into two pieces (appearing on the right-hand side of the unitarity equation), we can consider more general cuts, obtaining the *generalized unitarity method*. In particular, we don't need to restrict to a cut that separates the diagram into two, but rather to a cut that puts on-shell some propagators. We also don't need to insist on positivity of the energy on the cut lines (which was essential in the interpretation, on the right-hand side of the unitarity relation, as giving the cross-section for emission of on-shell particles), and we can even consider *complex* momenta, as considered for instance in the BCFW construction for the helicity spinor formalism (see Part I).

Since we want to put propagators on-shell, that is in the integrand

$$\frac{1}{S_i} \frac{c_i n_i}{l^2 \ldots (l + \sum_j K_j)^2 \ldots} \tag{59.35}$$

to have several of the $(l + \sum_j K_j)^2$ vanish, it follows that in D dimensions we can have a maximum of D cuts (D propagators on-shell) at one loop (a single l). Indeed, the loop momentum l^{μ} has D components, so to have a solution for generic external momenta K_i we can have at most D equations:

$$l^2 = 0 , \quad (l + K_1)^2 = 0 , \quad \ldots , \quad (l + K_1 + \ldots + K_{D-1})^2 = 0. \tag{59.36}$$

In $D = 4$, which is the case we are considering, that maximum is for a quadruple cut. We then note that the master integrals are scalar boxes, triangles, and bubbles (only boxes for $\mathcal{N} = 4$ SYM), which means that the quadruple cut will cut all the lines in the master integrals. This is not in fact a coincidence. The basis of master integrals at one loop contains up to pentagons in $D = 5$, hexagons in $D = 6$, and so on.

Thus the quadruple cut at one loop:

$$l_1^2 \equiv l^2 = 0 , \quad l_2^2 \equiv (l - K_1)^2 = 0 \Rightarrow 2l \cdot K_1 = K_1^2 ,$$
$$l_3^2 \equiv (l - K_1 - K_2)^2 = 0 \Rightarrow 2l_2 \cdot K_2 = 2(l - K_1) \cdot K_2 = K_2^2 ,$$
$$l_4^2 \equiv (l - K_1 - K_2 - K_3)^2 = 0 \Rightarrow 2l_3 \cdot K_3 = 2(l - K_1 - K_2) \cdot K_3 = K_3^2, \tag{59.37}$$

which has generically two discrete solutions for l (since we have quadratic equations), $l^{\pm}$, determines the coefficients of the scalar boxes, since the triangles and bubbles will not contribute (they don't even have four propagators to cut). Moreover, the above quadruple cut corresponds to only one possible master integral, since the cut implicitly grouped the external momenta in a specific way, into K_1, K_2, K_3, K_4, that corresponds to a single box diagram. Another quadruple cut will give the coefficient of another box (with a different grouping of external momenta into K_1, K_2, K_3, K_4), and so on.

When we are done calculating the coefficients of the boxes in this way, we move on to the triangles, by considering the triple cut (now only the bubbles will not contribute to it, not having three propagators), and finally we move to the double cut, in order to determine the coefficient of the bubbles. In these cases, however, there will be a contribution to the cut from higher point integrals, from boxes for the triple cut, and from both boxes and triangles for the double cut, which need to be isolated before we can calculate the coefficients of the triangles and bubbles, respectively. The way to deal with this involves considering complex external momenta, depending for the triple cut on a single complex parameter t, and considering the analytic structure as a function of t, but this will not be discussed further here.

Important Concepts to Remember

- One-loop massless integrals are r-gon integrals with a polynomial numerator (in the loop momentum l) $P_m(l)$.
- By Passarino–Veltman reduction, we can reduce the one-loop massless integrals to a basis of scalar integrals (no numerator), which in four dimensions is of bubbles, triangles, and boxes (and in five dimensions also has pentagons, etc.).
- The basis of boxes in four dimensions contains "masses" at corners (external lines): one, two (easy and hard, that is, opposite or nearby corners), three, or four masses.
- The basis of scalar integrals can be obtained from limits of the four massive one, but there are subtleties with the IR divergences.
- The basis of scalar integrals can be calculated in terms of polylogarithms.
- At higher loops, we can expand in a basis of master integrals, and find the coefficients by the method of generalized unitarity cuts.
- At one loop in D dimensions, we can have a maximum of D simultaneous cuts, giving the D-gon integral basis.

Further Reading

The Passarino–Veltman reduction was defined in [18]. The results of the box integrals in $D = 4-2\epsilon$ dimensions in the epsilon expansion are given in section 2 of [19] and Appendix I of [20]. The result to all orders in epsilon for the 2me box is given in Appendix A of [21]. For more details on the generalized unitarity cut method, see for instance [8].

Exercises

1. In massless scalar-QED, consider the hexagon integral with six external scalar lines with momenta $p_1, \ldots, p_6$ (all in), and alternate scalar and photon internal lines (three scalar, three photon), as in Figure 59.2. Reduce the result of the Feynman

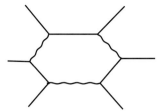

Figure 59.2   Hexagon diagram in scalar-QED.

Figure 59.3   "Theta" two-loop diagram in scalar-QED.

   integral to the scalar basis of bubbles, triangles, and boxes using Passarino–Veltman reduction.

2. Calculate the 2mh scalar box integral as a limit of the 3m scalar box integral, to all orders in ϵ, and the one-mass scalar box integral as a limit of the 2me scalar box integral.

3. Repeat Exercise 1 by using the generalized unitarity cut method to find the coefficients of the expansion in the scalar basis of integrals.

4. Write an expansion for the "theta" two-loop Feynman diagram in five dimensions with external photon lines and internal vertical photon line (and the rest scalar lines) in Figure 59.3 in terms of a basis of master integrals.

Analyzing the Result for Amplitudes: Polylogs, Transcendentality, and Symbology

In this chapter we analyze the results of the calculation for an amplitude. We will define polylogs that appear in the result, define a notion of transcendentality that organizes it, and a "symbol" to encapsulate some of its rough characteristics.

60.1 Polylogs in Amplitudes

We have already defined the standard polylogarithms $\text{Li}_n(z)$, appearing as we saw in calculations of the loop amplitudes: $\text{Li}_2(z)$ appeared at one loop, and one finds (firstly, rather "experimentally") that $\text{Li}_n(z)$ appears at $n-1$ loops in dimensional regularization.

But there exist more complicated polylogarithm functions that can also appear in loop calculations in dimensional regularization. First, there are the so-called Nielsen polylogarithms:

$$S_{n,p} = \frac{(-1)^{n+p-1}}{(n-1)!\,p!} \int_0^1 dt \frac{\log^{n-1}(t) \log^p(1-xt)}{t}, \tag{60.1}$$

which also appear in the ϵ expansion of amplitudes.

Even more generally, we have the *harmonic polylogarithms* defined in [22]. They are defined recursively by

$$H_{\vec{0}_p} = \frac{\log^p(x)}{p!}, \qquad H_{a_1\ldots a_p}(x) = \int_0^x dt f_{a_1}(t) H_{a_2\ldots a_p}(t), \tag{60.2}$$

$$f_{\pm 1}(x) = \frac{1}{1 \mp x}, \qquad f_0(x) = \frac{1}{x}, \tag{60.3}$$

$$H_{\pm 1}(x) = \mp \log(1 \mp x), \qquad H_0(x) = \log(x). \tag{60.4}$$

Here we have introduced the notation $\vec{0}_p$ for the index zero repeated p times. We are mostly interested in harmonic polylogarithms of the form $H_{a_1,a_2,\ldots,a_n}$ with $a_i = 0, 1$, however, the following relations hold true for any value of the indices.

An important property of harmonic polylogarithms is

$$H_{a_0}(x) H_{a_1\ldots a_p}(x) = H_{a_0 a_1\ldots a_p}(x) + H_{a_1 a_0 a_2\ldots a_p} + \ldots + H_{a_1\ldots a_p, a_0}(x), \tag{60.5}$$

namely the product of two harmonic polylogarithms can be expressed as a sum of single polylogarithms. In general, we can write the product of harmonic polylogarithms of degree p and q as a sum of harmonic polylogarithms of degree $p + q$, more precisely

$$H_{\vec{a}_p}(x)H_{\vec{b}_q}(x) = \sum_{\vec{c}_{p+q}\in\ \vec{a}_p\uplus\vec{b}_q} H_{\vec{c}_{p+q}}(x), \qquad (60.6)$$

where we have introduced the notation $\vec{a}_p = a_1\ldots a_p$ and $\vec{b}_q = b_1\ldots b_q$. $\vec{a}_p\uplus\vec{b}_q$ represents all mergers of $\vec{a}_p$ and $\vec{b}_q$ in which the relative orders of the elements of $\vec{a}_p$ and $\vec{b}_q$ are preserved.

There are also identities among harmonic polylogs of related arguments, in particular, identities of the type

$$H_{\vec{a}_p}(-x) = \sum_{\vec{b}_p} c_{\vec{b}_p} H_{\vec{b}_p}(x), \qquad (60.7)$$

$$H_{\vec{a}_p}(1-x) = \sum_{\vec{b}_p} c_{\vec{b}_p} H_{\vec{b}_p}(x), \qquad (60.8)$$

$$H_{\vec{a}_p}(1/x) = \sum_{\vec{b}_p} c_{\vec{b}_p} H_{\vec{b}_p}(x), \qquad (60.9)$$

$$H_{\vec{a}_p}\left(\frac{1-x}{1+x}\right) = \sum_{\vec{b}_p} c_{\vec{b}_p} H_{\vec{b}_p}(x), \qquad (60.10)$$

namely we can re-express $H_{a_1\ldots a_p}(t)$ as a homogenous sum of harmonic polylogs of degree p on x, with t and x related as above.

These harmonic polylogarithms are extensions of the normal and Nielsen ones, since we have

$$\mathrm{Li}_p(z) = H_{\vec{0}_{p-1}1}(z), \qquad S_{n,p}(z) = H_{\vec{0}_n\vec{1}_p}(z). \qquad (60.11)$$

They are relevant to the one-loop amplitudes, since for the hypergeometric function appearing in the generic IR-divergent one-loop box (the 3m box function), $_2F_1(2\epsilon,\epsilon,1+\epsilon,z)$, we have the formula

$$_2F_1(a\epsilon, b\epsilon, 1+c\epsilon, z) = \sum_{\ell=0}^{\infty}\sum_{\vec{a}} c_{\vec{a}_\ell} H_{\vec{a}_\ell}(z)\epsilon^\ell. \qquad (60.12)$$

Finally, if we define

$$G_{1,a_1\ldots a_p}(x) = -\int_0^1 \frac{dt}{t - 1/x} H_{a_1\ldots a_p}(t), \qquad (60.13)$$

then $G_{1,a_1\ldots a_p}$ is a homogeneous combination of H-functions of degree $p+1$.

60.2 Maximal and Uniform Transcendentality of Amplitudes

We can consistently define an important mathematical notion named *(degree of) transcendentality* as follows.

We assign transcendentality k to each factor of $\log^k z$, π^k, ζ_k, γ^k, where γ is Euler's constant, or any type of polylogarithm of total degree k, like the usual polylogarithm $\mathrm{Li}_k(z)$, the

Nielsen polylogarithm $S_{n,k-n}(z)$, or the harmonic polylogarithms $H_{a_1\cdots a_k}(z)$ of total degree k, with the product of two factors of transcendentality k_1 and k_2 having transcendentality $k_1 + k_2$. Then this notion of transcendentality is well defined, since the transformation properties of polylogs (all the identities between them) preserve it, as we can easily check.

As a first observation for the usefulness of this notion for amplitude calculations, note that all the terms of the IR-finite one-loop 4m box integral I_4^{4m} have a uniform (constant, in this case) transcendentality, namely 2. But this is an integral that doesn't need dimensional regularization, whereas the other boxes, appearing in one-loop amplitudes, do. We must then define a different notion, one that will be called *uniform transcendentality* of degree k, where the terms of order ϵ^n have transcendentality of degree $n + k$.

Then note that, because of (59.26) and (59.28), the ϵ^m term in the 2me one-loop box function (the only one appearing in MHV amplitudes of $\mathcal{N} = 4$ SYM) in (59.27) has transcendentality $m + 2$, which means that one-loop MHV amplitudes of $\mathcal{N} = 4$ SYM have uniform transcendentality of degree 2. These values for $\mathcal{N} = 4$ SYM of the degree of uniform transcendentality are maximal with respect to other theories, so will be called maximal transcendentality. Other theories will not in general have maximal transcendentality, or not even uniform.

For instance, in the physically relevant case of QCD, at one loop there are terms of subleading transcendentality with respect to those coming from the 2me box. In the case of the one-loop, four-point amplitude with all equal helicities, $A_4^{\text{1-loop}}(+ + ++)$ (as we said, the $\mathcal{N} = 4$ SYM counterpart vanishes, as does the pure glue amplitude), we have a uniform transcendentality, but with less than its maximal value.

More generally, an L-loop amplitude in $\mathcal{N} = 4$ SYM theory has a Laurent expansion in ϵ beginning at $1/\epsilon^{2L}$, caused by collinear and soft IR divergences. The leading term has transcendentality zero, so the maximal uniform transcendentality of the L-loop amplitude of $\mathcal{N} = 4$ SYM means that the coefficient of ϵ^m has transcendentality $m+2L$. This behavior has been observed in all $\mathcal{N} = 4$ SYM amplitudes so far calculated.

Relations between planar amplitudes of $\mathcal{N} = 4$ SYM theory at different loop orders, such as the Anastasiou–Bern–Dixon–Kosower (ABDK) relation

$$M^{(2)}(\epsilon) = \frac{1}{2}\left[M^{(1)}(\epsilon)\right]^2 - (\zeta_2 + \epsilon\zeta_3 + \epsilon^2\zeta_4)M^{(1)}(2\epsilon) - \frac{\pi^4}{72} + \mathcal{O}(\epsilon), \qquad (60.14)$$

manifestly respect the maximal uniform transcendentality of the amplitudes. The Bern–Dixon–Smirnov (BDS) conjecture for the nonperturbative, planar, MHV n-point amplitude also respects maximal uniform transcendentality for all loop orders, though since it has been explicitly shown to be incomplete for $n \geq 6$ starting at two loops, we cannot use this result to deduce the maximal transcendentality of n-point $\mathcal{N} = 4$ SYM amplitudes for all loops, and only as an indicator.

Transcendentality was first defined in [23] in a different, yet related, way, for transcendental constants (π and ζ_k). We can extend this definition to one that is completely equivalent to the above as follows.

Defining the finite harmonic sums

$$S_a(j) = \sum_{m=1}^{j} \frac{1}{m^a},$$

$$S_{a,b,c,\dots}(j) = \sum_{m=1}^{j} \frac{1}{m^a} S_{b,c,\dots}(m),$$

$$S_{-a}(j) = \sum_{m=1}^{j} \frac{(-1)^m}{m^a},$$

$$S_{-a,b,c,\dots}(j) = \sum_{m=1}^{j} \frac{(-1)^m}{m^a} S_{b,c,\dots}(m), \qquad (60.15)$$

one can analytically continue them from j to $j + \epsilon$ or a general z by writing them in terms of polygamma and related functions, for instance

$$S_i(j) = \frac{(-1)^{i-1}}{(i-1)!} [\psi^{(i-1)}(j+1) - \psi^{(i-1)}(1)], \qquad (60.16)$$

where the polygamma functions are defined by

$$\psi^{(n)}(z) = (-1)^{n+1} n! \sum_{k=0}^{\infty} \frac{1}{(k+z)^{n+1}}. \qquad (60.17)$$

This is similar to the way the digamma function at integer values

$$\psi(n+1) = -\gamma + \sum_{k=1}^{n} \frac{1}{k} \qquad (60.18)$$

is continued to general z as

$$\psi(z) = \sum_{k=1}^{\infty} \left(\frac{1}{k} - \frac{1}{k+z+1} \right) - \gamma. \qquad (60.19)$$

By expanding the harmonic sums in $j = 1+\epsilon$, we obtain series of uniform transcendentality in ϵ, as found in [23], but these are numerical series only. The degree k of the uniform transcendentality is related to the total weight $w = a + b + c + \dots$ of the harmonic sum $S_{a,b,c,\dots}(j)$.

One can find a similar definition of uniform transcendentality series that involves series of harmonic polylogs (i.e. functions) by using the "S-sums" and "Z-sums" defined below instead of harmonic sums.

The S-sums are defined by

$$S(n; m_1, \dots, m_k; x_1, \dots, x_k) = \sum_{i=1}^{n} \frac{x_1^i}{i^{m_1}} S(i; m_1, \dots, m_k; x_1, \dots, x_k) \qquad (60.20)$$

and the Z-sums are defined similarly, by restricting the sum:

$$Z(n; m_1, \ldots, m_k; x_1, \ldots, x_k) = \sum_{i=1}^{n} \frac{x_1^i}{i^{m_1}} Z(i - 1; m_1, \ldots, m_k; x_1, \ldots, x_k). \quad (60.21)$$

Then the harmonic sums in (60.15) are found by restricting to $x_i = \pm 1$. By similarly analytically continuing the sums from n to $1 + \epsilon$ and expanding in ϵ, we get series of uniform transcendentality. The degree k_0 of the series is related to the total weight $w = m_1 + \ldots + m_k$ of the sum.

60.3 Symbology

Another mathematical notion that has proven very useful for the study of amplitudes is the notion of the *symbol*. It turns out that results for the amplitudes in epsilon expansion (dimensional regularization) are often extremely complicated. A naive calculation of the integral can give many pages of special functions for a simple amplitude. But often the result can be rewritten in a simpler form, sometimes by analytical continuation, by use of complicated relations among polylogarithms. To find and/or prove those, one can first prove their validity in the much simpler symbol. Often it can be that we can reconstruct the full result, or at least parts of it, from the knowledge of its symbol.

To define the symbol, we must first define *Goncharov polynomials, or polylogarithms of one variable*, which are defined recursively by

$$G(a_1, \ldots, a_n; x) = \int_0^x \frac{dt}{t - a_1} G(a_2, \ldots, a_n; t). \quad (60.22)$$

The first one (boundary condition for the recursion relation) is

$$G(x) = G(; x) = 1; G(0) = 0. \quad (60.23)$$

These polynomials are further generalizations of the polylogarithms already defined, reducing to them for particular cases, since

$$G(\vec{0}_n; x) = \frac{1}{n!} \log^n x,$$
$$G(\vec{a}_n; x) = \frac{1}{n!} \log^n \left(1 - \frac{x}{a}\right),$$
$$G(\vec{0}_{n-1}, a; x) = -\mathrm{Li}_n \left(\frac{x}{a}\right),$$
$$G(\vec{0}_n, \vec{a}_p; x) = (-1)^p S_{n,p} \left(\frac{x}{a}\right). \quad (60.24)$$

As we can see, the Goncharov polylogarithms of one variable are defined analogously to the harmonic polylogarithms H with indices 0 and ± 1, by a similar recursion relation.

Now we can define the symbol of a Goncharov polynomial as the sum of terms of the type: tensor product of R_is, understood as $d \log R_i = dR_i/R_i$s (i.e. such that the rules the R_is satisfy follow from this $d \log$ form).

We write the tensor monomials as $R_1 \otimes \ldots \otimes R_n$, so they satisfy

$$\ldots \otimes (R_1 \cdot R_2) \otimes \ldots = \ldots \otimes R_1 \otimes \ldots + \ldots \otimes R_2 \otimes \ldots$$

$$\Rightarrow \otimes(R_1)^n\otimes = n \ldots \otimes R_1 \otimes \ldots$$

$$\ldots \otimes cR_1 \otimes \ldots = \ldots \otimes R_1 \otimes \ldots$$

$$\ldots \otimes c \otimes \ldots = 0$$

$$\Rightarrow \ldots \otimes R_1 \otimes \ldots = - \ldots \otimes 1/R_1 \otimes \ldots, \tag{60.25}$$

where $R_1, R_2, \ldots$ are variable monomials and c is a constant. The symbol of an object T_k, *a priori* a function of several variables, an extension of the simple Goncharov polynomials of one variable above, and defined recursively as

$$T_k = \int_a^b d \log R_1 \circ \ldots \circ d \log R_n = \int_a^b \left(\int_a^t d \log R_1 \circ \ldots d \log R_{n-1} \right) d \log R_n(t), \tag{60.26}$$

is

$$S[T_k] = R_1 \otimes R_2 \otimes \ldots \otimes R_n. \tag{60.27}$$

From this definition we obtain immediately the symbol of a standard Li_k polylogarithm, namely

$$S[\mathrm{Li}_k(z)] = -(1 - z) \otimes z \otimes \ldots \otimes z \tag{60.28}$$

(there are $k - 1$ factors of z), thought of as a particular case of the Goncharov polylogarithms. Note that $(1 - z)$ and $(z - 1)$ are the same, since they differ by multiplication with the constant -1. Since the overall minus sign is for the tensor monomial, it does not belong to any of the tensored factors.

We can also define the rule for multiplication of two terms with a symbol, $S[F] = \otimes_{i=1}^n R_i$ and $S[G] = \otimes_{i=n+1}^m R_i$, by

$$S[FG] = \sum_{\Pi} \otimes_{i=1}^{m+n} R_{\Pi(i)}. \tag{60.29}$$

Here the permutations Π preserve the original order of the factors in $S[F]$ and in $S[G]$ within $S[FG]$. For example, if $n = m = 2$ we get

$$S[FG] = R_1 \otimes R_2 \otimes R_3 \otimes R_4 + R_1 \otimes R_3 \otimes R_2 \otimes R_4 + R_1 \otimes R_3 \otimes R_4 \otimes R_2$$
$$+ R_3 \otimes R_1 \otimes R_2 \otimes R_4 + R_3 \otimes R_1 \otimes R_4 \otimes R_2 + R_3 \otimes R_4 \otimes R_1 \otimes R_2. \tag{60.30}$$

For logs and products of logs, we obtain

$$S[\log x] = x,$$
$$S[\log x \log y] = x \otimes y + y \otimes x. \tag{60.31}$$

For the Nielsen polylogarithms $S_{n,p}(x)$, the symbol is given by

$$S[S_{n,p}(x) = H(\vec{0}_n, \vec{1}_p; x)] = (-1)^p (1 - x) \otimes (1 - x) \otimes \ldots \otimes (1 - x) \otimes x \otimes x \ldots \otimes x, \tag{60.32}$$

where there are p $1 - x$s and n xs.

Important Concepts to Remember

- There are several notions of polylogarithms, all defined recursively: normal $\mathrm{Li}_n(z)$, Nielsen $S_{n,p}(z)$, harmonic $H_{a_1\ldots a_p}(z)$, and Goncharov $G(a_1,\ldots,a_p;z)$.
- Polylogs of total degree k appear in general in amplitudes at $k-1$ loops.
- There is a well-defined notion of transcendentality, for which $\log^k(z)$, π^k, ζ_k, γ^k, and polylogarithms of total degree k have degree of transcendentality k, respected by all functional relations.
- For uniform transcendentality of degree k in dimensional regularization of integrals, the terms of order ϵ^k have degree $n+k$ of transcendentality.
- Amplitudes in $\mathcal{N}=4$ SYM have uniform transcendentality of degree $2L$ at L loops, which is a maximal possible uniform transcendentality in loop amplitudes, respected by the ABDK relation and the BDS conjecture.
- The symbol of a Goncharov polynomial is a sum of tensor products of R_i factors defined as $d\log R_i = dR_i/R_i$, and satisfying the same rules.
- We can prove polylog relations for results of loop amplitudes by first proving the equality for the symbol, and sometimes we can deduce part of the result for the amplitude from the symbol.

Further Reading

Harmonic polylogarithms were defined in [22], and transcendentality in [23]. For more on symbols, see [24–26]. For an example of a polylog relation for the result of an amplitude that can be tested using symbology, see [26].

Exercises

1. Prove the harmonic polylogarithm composition relation (60.5) for $a_0 = 0, p = 2$ and $a_1 = 0, a_2 = 1$.
2. Do the scalar box integrals I_4^{3m}, I_4^{2mh}, and I_4^{1m} have uniform transcendentality? Why?
3. Prove that the polylogarithm relation

$$\mathrm{Li}_3(1-y) = -S_{1,2}(y) - V\mathrm{Li}_2(y) + \frac{TV^2}{2} + \frac{\pi^2}{6}V + \zeta_3, \qquad (60.33)$$

   where $V = \log(1-y)$ and $T = -\log y$, is consistent in the symbol.
4. Calculate the symbol of the finite part of the two-mass easy scalar box integral I_{2me}.

61

Representations and Symmetries for Loop Amplitudes: Amplitudes in Twistor Space, Dual Conformal Invariance, and Polytope Methods

In this chapter, I will attempt to quickly summarize an enormous amount of recent literature dealing with reformulating the loop amplitudes by using their symmetries, first in twistor and momentum twistor space, and using dual conformal invariance to write them in terms of volumes of polytopes. This program has finally led to the Grassmannian and Amplituhedron for amplitudes, which however I will not describe.

61.1 Twistor Space

Twistor space is identified with the complex projective space $\mathbb{CP}^3$, and comes from a way to satisfy a null condition. At a practical level, in the case of amplitudes of massless particles, we would like to trivialize (make explicit, so that we don't need to impose as a constraint) the null condition for external momenta, $p^2 = 0$. We have already seen in Part I how to do that: we need to introduce helicity spinors $\lambda^a, \tilde{\lambda}^{\dot{a}}$ or $|\lambda\rangle, |\tilde{\lambda}]$ such that $p_{a\dot{a}} = \lambda_a \tilde{\lambda}_{\dot{a}}$. Then automatically we have $p^2 = 0$. If, moreover, we make a Fourier transform over the angle spinor variables λ_a, thus changing

$$\langle\lambda|_a \leftrightarrow i\frac{\partial}{\partial\mu^a}, \quad \frac{\partial}{\partial\lambda^a} \leftrightarrow -i\langle\mu|_a, \tag{61.1}$$

then $(\mu_a, \tilde{\lambda}_{\dot{a}})$ forms a twistor.

To formally define twistor space, consider the conformal group in 3+1-dimensional Minkowski space, $SO(4, 2)$, and define it as the Lorentz group for a six-dimensional space with signature $(- - + + + +)$. We define the embedding of Minkowski space in six-dimensional space by the six-dimensional Lorentz-invariant constraint $X \cdot X = 0$, with X^M a six-dimensional vector. Since the constraint is invariant under rescaling by a real number λ, $X^M \to \lambda X^M$, we obtain a four-dimensional space, and by complexifying this projective constrained formalism, we see that we can identify the space with $\mathbb{CP}^3$ (three complex objects identified by multiplication with another complex object).

Since $SO(2, 4) \sim SU(2, 2)$ as we know from classical field theory, we can write X^M as a complex bispinor X^{IJ}, $I, J = 1, \ldots, 4$, in the six-dimensional antisymmetric irrep of $SU(2, 2)$, $X^{IJ} = -X^{JI}$ with a reality condition imposed. Then a null bispinor

$$X^2 = \frac{1}{2}\epsilon_{IJKL}X^{IJ}X^{KL} = 0 \tag{61.2}$$

can be written as a product of four-dimensional spinors A^I, B^I called *twistors*:

$$X^{IJ} = A^{[I}B^{J]}. \tag{61.3}$$

But if X^{IJ} is defined up to multiplication, then so are A^I, B^J, meaning that the space of A^I and of B^I is a $\mathbb{CP}^3$ space, as advertised. Then, given a point $X^M \to X^{IJ}$ in spacetime, we find a *line* in twistor space, defined by the two A^I, B^I twistors. Conversely, given a line in spacetime, it will correspond to a point in twistor space.

First, to introduce a metric in Minkowski space, we need to break conformal symmetry by introducing a reference point I^{IJ} in twistor space. Then, the separation between X_1 and X_2 is

$$d_{X_1,X_2} \equiv \frac{(X_1 \cdot X_2)}{(I \cdot X_1)(I \cdot X_2)}, \tag{61.4}$$

where $(X \cdot Y) \equiv \epsilon_{IJKL} X^{IJ} Y^{KL}$. Then, if X_1 and X_2 have a common C^I twistor (a single point in twistor space), we obtain $d_{X_1,X_2} = 0$, meaning that they are null separated (i.e. they are on a line).

Reversely, for a single point $X^M \to X^{IJ}$ in Minkowski space, the relation

$$X^{[IJ} W^{K]} = 0, \tag{61.5}$$

where W is a twistor, defines a $\mathbb{CP}^1$ line in twistor space.

We can go back to the helicity spinor formalism, and say that for $X_{a\dot{a}}$ satisfying $X^2 = 0$, we can define

$$\begin{pmatrix} \lambda \\ \mu \end{pmatrix} = Z \in \mathbb{C}^4 \tag{61.6}$$

by the condition

$$(\mu)_{\dot{a}} = (X)_{a\dot{a}}(\lambda)_a, \tag{61.7}$$

which is invariant under complex rescalings of Z (i.e. $Z \in \mathbb{CP}^3$) but more precisely, since λ_a is given by $\mu_{\dot{a}}$, it is in $\mathbb{CP}^1$, a line.

Then two Minkowski points X and Y are null separated, $d_{X,Y} = 0$, $\Leftrightarrow$ the two $\mathbb{CP}^1$s they define intersect over the same point in twistor space, defined above by W^I, standing in now for a given $Z = (\lambda, \mu)$.

61.2 Amplitudes in Twistor Space

As we said, at the practical level, we just do a half Fourier transform over the helicity spinor variables, over the angle variable, but not the bracket one, and we end up in twistor space.

As an example of the importance of the formalism, consider the Parke–Taylor formula for the *anti-MHV* amplitude (i.e. the complex conjugate of the MHV amplitude). Consider also the full amplitude, including the momentum conservation delta function, written in helicity spinor variables, so

$$\mathcal{A}_n^{\text{anti-MHV}}(1^-2^-\ldots i^+\ldots j^+\ldots n^-) = \frac{\delta^4(\sum_{i=1}^n \lambda_i^a \tilde{\lambda}_i^{\dot{a}})[ij]^4}{[12][23]\ldots[n1]}$$

$$\equiv \delta^4\left(\sum_{i=1}^n \lambda_i^a \tilde{\lambda}_i^{\dot{a}}\right) f(\tilde{\lambda}_i^{\dot{a}}). \qquad (61.8)$$

If we take the Fourier transform on λ_i, we see that it only appears in the momentum conservation delta function, which itself can be written as a Fourier transform of 1:

$$\delta^4\left(\sum_{i=1}^n \lambda_i^a \tilde{\lambda}_i^{\dot{a}}\right) = \int d^4 x e^{-i x_{a\dot{a}} \sum_{i=1}^n \lambda_i^a \tilde{\lambda}_i^{\dot{a}}}, \qquad (61.9)$$

so we can easily perform it, with the result

$$\int \left(\prod_{i=1}^n d^2\lambda_i^a e^{i\lambda_i^a \mu_a^i}\right) \mathcal{A}_n^{\text{anti-MHV}}(1^-2^-\ldots i^+\ldots j^+\ldots n^-)$$

$$= \int d^4 x \left(\prod_{i=1}^n \delta^2(\tilde{\lambda}_i^{\dot{a}} x_{a\dot{a}} + \mu_a)\right) f(\tilde{\lambda}_i^{\dot{a}}), \qquad (61.10)$$

which means we localize the amplitude on the position of the "incidence relations":

$$\tilde{\lambda}^{\dot{a}} x_{a\dot{a}} + \mu_a = 0, \qquad (61.11)$$

defining as we saw, a $\mathbb{CP}^1$ line in twistor space. The amplitude has thus been rewritten in twistor space. Note that here the Minkowski x space used to define the twistors is really configuration space, Fourier transformed from the momenta.

61.2.1 Dual Space and Momentum Twistors

If the reason for the twistor construction was to trivialize the null condition $p^2 = 0$ by the use of helicity spinors, there is one more condition that would be good to trivialize, that is make explicit in the construction, namely momentum conservation, $\sum_{i=1}^n p_i^\mu = 0$. This is the beginning of the important dual-space construction. The trivialization is obtained by introducing n points x_i instead of the n momenta p_i, such that

$$p_i = x_i - x_{i+1}, \qquad (61.12)$$

where $x_{n+1} = x_1$ (cyclic). Then we have by construction $\sum_{i=1}^n p_i = 0$. We can use these x_i^μs, converted as usual to $x_{a\dot{a}}^i = x^\mu(\sigma_\mu)_{a\dot{a}}$, as the Minkowski-space coordinates that define some twistors by the incidence relation (61.7), which are now called Hodge's *momentum twistors*.

61.3 Dual Conformal Invariance

Consider now the one-loop scalar box functions written in the dual-space x_i variables. Consider for instance the massless (zero-mass) box function

$$F_{0m}(1, 2, 3, 4) = i \int \frac{d^4 x_0}{2\pi^2} \frac{(x_1 - x_3)^2 (x_2 - x_4)^2}{(x_0 - x_1)^2 (x_0 - x_2)^2 (x_0 - x_3)^2 (x_0 - x_4)^2}, \qquad (61.13)$$

where we have written the loop momentum as $l = x_0 - x_1$, introducing a fifth coordinate x_0, and we have ignored the fact that the integral is actually IR-divergent. Observe that $s = (p_1 + p_2)^2 = (x_1 - x_2 + x_2 - x_3)^2 = (x_1 - x_3)^2$ and $t = (p_2 + p_3)^2 = (x_2 - x_3 + x_3 - x_4)^2 = (x_2 - x_4)^2$.

We note that the function now has an extra symmetry, acting on the dual space x_i^μ like a conformal invariance $SO(4, 2)$, and is hence called *dual conformal invariance*. The conformal group is generated by translations, Lorentz rotations (both of which are obvious symmetries of the integral, and in general of the dual-space representation), and the inversion. Thus to prove dual conformal invariance we only have to prove invariance of the integral under the inversion of the dual space:

$$x_i^\mu \to \frac{x_i^\mu}{x_i^2}, \quad \forall i = 0, 1, 2, 3, 4. \qquad (61.14)$$

We leave this as an exercise, though we point out the simpler scale invariance of the integral, under $x_i^\mu \to \lambda x_i^\mu$, and the fact that usually this means (together with Poincaré invariance) conformal invariance as well.

Note now that the massive boxes could not have this symmetry, since they involve scales (the external "masses"), which break scaling symmetry.

We have shown here the dual conformal invariance of a specific integral, but dual conformal invariance (and moreover dual superconformal invariance) has been proven to be a full symmetry of the perturbation theory of $\mathcal{N} = 4$ SYM. We will not insist on it here, though note that, together with the usual superconformal invariance of $\mathcal{N} = 4$ SYM, we find an infinite *Yangian* symmetry of the theory, that very drastically restricts the amplitudes of the theory.

61.4 Polytopes and Amplitudes

The construction described here is due to Mason and Skinner, though there are other constructions that lead to relevance of polytopes to amplitudes.

Consider now the $x_{a\dot{a}}^i$ dual coordinates, and construct the objects

$$X^{AB} \equiv \begin{pmatrix} -\frac{1}{2} \epsilon^{ab} x^2 & i x^a{}_{\dot{b}} \\ -i x_{\dot{a}}{}^b & \epsilon_{\dot{a}\dot{b}} \end{pmatrix}. \qquad (61.15)$$

They satisfy the conditions (we leave the proof as an exercise)

$$X^2 \equiv \frac{1}{2}\epsilon_{ABCD}X^{AB}X^{CD} = 0,$$

$$X_i \cdot X_j = -(x_i - x_j)^2. \tag{61.16}$$

Note that x^μ is in 3+1-dimensional Minkowski space with mostly plus signature. We see then that as above, the X^{AB} can be written as a bitwistor, in terms of two twistors A^A and B^B, as $X^{[AB]} = A^{[A}B^{B]}$.

Since X^{AB} are six real numbers, and the $X^2 = 0$ condition is invariant under scaling, $X^{AB} \sim \lambda X^{AB}$, these coordinates define a quadric, $X^2 = 0$, in $\mathbb{RP}^5$, the five-dimensional real projective space defined by the identification under multiplication by a real number. Moreover, we can show that, if x^μ live in Euclidean signature space, then X^{AB} live in an embedding six-dimensional space with signature $(1,5)$, that is $(-+++++)$.

Consider now a box function with vertices x_i, leading to points X_i on the quadric $X^2 = 0$ in $\mathbb{RP}^5$. Consider further $\alpha_i \in (0,1)$ Feynman parameters with $\sum_i \alpha_i = 1$, and from them construct

$$X(\alpha) = \alpha_1 X_1 + \alpha_2 X_2 + \alpha_3 X_3 + \alpha_4 X_4. \tag{61.17}$$

Except at the vertices X_i (which are excluded, since $\alpha_i \neq 0$), this will not be on the quadric $X^2 = 0$, but rather $X(\alpha) \cdot X(\alpha) \neq 0$, and we can normalize this distance to 1, by

$$Y(\alpha) \equiv \frac{X(\alpha)}{\sqrt{-X(\alpha) \cdot X(\alpha)}}, \tag{61.18}$$

such that $Y(\alpha) \cdot Y(\alpha) = -1$. The $X^{AB}(\alpha)$ vary over a tetrahedron, defined by the Feynman parameters α_i, or rather the *interior* of the tetrahedron in $\mathbb{RP}^5$, with vertices the four points X_i^{AB}, $i = 1, 2, 3, 4$. We are on the interior, since the condition $\alpha_i \neq 0, 1$ means we cannot reach either the vertices or the faces.

But $Y^{AB}(\alpha)$ are six coordinates in an embedding space with signature $(1,5)$, and the condition $Y(\alpha) \cdot Y(\alpha) = -1$ then means that the space is equivalent to the constraint

$$-Y_{-1}^2 + Y_1^2 + Y_2^2 + Y_3^2 + Y_4^2 + Y_5^2 = -1, \tag{61.19}$$

in $Mink^{(1,5)}$, which is the definition of the Euclidean AdS_5 of unit radius. Then $X(\alpha)$ on the edges and faces of the tetrahedron are straight lines in the parameters α_i, which means in X_i. That amounts to geodesics in AdS_5 (lines of minimum distance in the space, which amounts to straight lines in the embedding space of X_i). Since the faces and edges of the tetrahedron are geodesic in AdS_5, we say that the tetrahedron is *ideal*. Note that $X(\alpha)^2 \to 0$ would mean that the $Y(\alpha)$ would lie on the boundary of AdS_5 ("at infinity").

Mason and Skinner then showed that twice the volume of the ideal tetrahedron inside AdS_5 equals the fully IR-finite four-mass box.

One can also IR-regularize the other boxes, either in dimensional regularization or by a mass regularization modifying $X^2 = 0$ to $X \cdot X = 2\mu^2 X \cdot I$, with I a reference point. For the box with some massless external line, $p_i^2 = 0$, translates into $X_i \cdot X_{i+1} = 0$, which would mean that

$$(X_i + \alpha X_{i+1})^2 = 0, \quad \forall \alpha, \tag{61.20}$$

which means that an entire edge of the tetrahedron would lie on the boundary of AdS_5 ("at infinity"), thus the volume of the tetrahedron would be infinite. But the regularization above makes again the volume finite, allowing us to calculate it.

Finally, to come to the relevance of amplitudes to the story, the MHV n-point, one-loop amplitudes are given by the n-point tree amplitude, times the sum of all the one-mass and two-mass easy boxes made up of the external momenta, all with coefficient one, thus

$$R_n \equiv \frac{A_n^{\text{1-loop}}}{A_n^{\text{tree}}} = \sum_i F_i^{\text{1m}} + \sum_k F_k^{\text{2me}}. \tag{61.21}$$

The one-mass box is defined by a single index $i = 1, \ldots, n$, F_i^{1m}, such that the corners are $K_i = p_i = x_i - x_{i+1}$, $K_{i+1} = p_{i+1} = x_{i+1} - x_{i+2}$, $K_{i+2} = p_{i+2} = x_{i+2} - x_{i+3}$, and the last corner is massive, containing all the other momenta, $K_{i-1} = \sum_a p_a = x_{i+3} - x_i$. We can say that the dual-space coordinates located *between* the vertices are $x_i, x_{i+1}, x_{i+2}, x_{i+3}$. The two-mass easy box is defined by two indices, $k = (ij)$, F_{ij}^{2me}, such that there are two opposite massless momenta, $K_i = p_i = x_i - x_{i+1}$ and $K_j = p_j = x_j - x_{j+1}$, and the two in between them are massive, containing the other momenta, $K_{i+1} = x_{i+1} - x_j$ and $K_{j+1} = x_{j+1} - x_i$. Thus the dual-space coordinates located in between the vertices are then $x_i, x_{i+1}, x_j, x_{j+1}$. In a similar manner, the general four-mass box $F^{\text{4m}}(i,j,k,l)$ is defined by four random coordinates x_i, x_j, x_k, x_l located between the vertices, so $K_i = x_i - x_j$, $K_j = x_j - x_k$, $K_k = x_k - x_l$, and $K_l = x_l - x_i$.

This means that there are n tetrahedra for one-mass boxes and $n(n-5)/2$ tetrahedra for two-mass easy boxes. But the volume of each tetrahedron comes with a sign, determined by the order of the vertices x_i of the dual space in the box function $F(i,j,k,l)$. That also induces an orientation, giving a sign for the triangular faces of the tetrahedron, determined by whether the missing vertex (for the triangular face) from the set $(ijkl)$ is in an odd or even position in the ordering $(ijkl)$. This means that faces with the same vertices, but different orientation, and thus different sign, can be glued together, to give a continuous object.

Gluing together all the tetrahedra for 1m and 2me boxes in the n-point MHV amplitude, which results in *all of the faces* canceling out (being glued) in the resulting object, we thus obtain a *closed three-dimensional polytope* with n vertices, and R_n gives the volume of that polytope. Here, polytope refers to an object with flat boundaries, the generalization in higher dimensions to the polyhedron in two dimensions. Closed means that it has no boundary, due to its embedding in a higher dimension. That is a bit difficult to visualize for three-dimensional polytopes, so for ease of visualization, let us come down a dimension, and imagine instead of gluing tetrahedra (three-dimensional objects) into a closed polytope, gluing triangles into a closed polytope. The result would then be a triangulated closed surface, like for instance the surface of a tetrahedron. Similarly, in our case, by gluing tetrahedra into a closed polytope we obtain the surface of a *four-dimensional polytope*.

The construction we have considered here was for the ratio of one-loop amplitudes to tree amplitudes, but it can be generalized to the full amplitude, viewed as an expansion in the number of traces, equivalent to an expansion in the rank of the gauge group.

Before ending this topic, I will give, for completeness, the formulas for the relevant boxes in the twistor-space coordinates X^{AB} of Mason and Skinner. In the mass regularization with parameter μ, and for the choice of the reference point I as $X_i \cdot I = 1, \forall i$, we have the two-mass easy box and the one-mass box

$$
\begin{aligned}
F_{2me}(i-1,i,j-1,j) = {} & -\log\left(\frac{X_i \cdot X_j}{\mu^2}\right)\log\left(\frac{X_{i-1} \cdot X_{j-1}}{\mu^2}\right) + \frac{1}{2}\log^2\left(\frac{X_i \cdot X_{j-1}}{\mu^2}\right) \\
& + \frac{1}{2}\log^2\left(\frac{X_{i-1} \cdot X_j}{\mu^2}\right) + \mathrm{Li}_2\left(1 - \frac{X_i \cdot X_{j-1}}{X_{i-1} \cdot X_{j-1}}\right) \\
& + \mathrm{Li}_2\left(1 - \frac{X_i \cdot X_{j-1}}{X_i \cdot X_j}\right) + \mathrm{Li}_2\left(1 - \frac{X_{i-1} \cdot X_j}{X_{i-1} \cdot X_{j-1}}\right) \\
& + \mathrm{Li}_2\left(1 - \frac{X_{i-1} \cdot X_j}{X_i \cdot X_j}\right) + \mathrm{Li}_2\left(1 - \frac{X_i \cdot X_{j-1}X_{i-1} \cdot X_j}{X_i \cdot X_j X_{i-1} \cdot X_{j-1}}\right), \\
F_{1m}(i-3,i-2,i-1,i) = {} & -\log\left(\frac{X_{i-3} \cdot X_{i-1}}{\mu^2}\right)\log\left(\frac{X_{i-2} \cdot X_i}{\mu^2}\right) + \frac{1}{2}\log^2\left(\frac{X_i \cdot X_{i-3}}{\mu^2}\right) \\
& + \mathrm{Li}_2\left(1 - \frac{X_i \cdot X_{i-3}}{X_{i-3} \cdot X_{i-1}}\right) + \mathrm{Li}_2\left(1 - \frac{X_i \cdot X_{i-3}}{X_{i-2} \cdot X_i}\right) + \frac{\pi^2}{6}.
\end{aligned}
$$

$$(61.22)$$

This is finite at finite μ, but diverges as we take $\mu \to 0$.

The boxes in dimensional regularization on the dual space, but with twistor variables X_i^{AB}, are given by

$$
\begin{aligned}
F_{2me}(i-1,i,j-1,j) = {} & -\frac{1}{\epsilon^2}\left[\left(\frac{X_{i-1} \cdot X_{j-1}}{\mu^2}\right)^{-\epsilon} + \left(\frac{X_i \cdot X_j}{\mu^2}\right)^{-\epsilon} - \left(\frac{X_i \cdot X_{j-1}}{\mu^2}\right)^{-\epsilon} \right. \\
& \left. - \left(\frac{X_{i-1} \cdot X_j}{\mu^2}\right)^{-\epsilon}\right] + \frac{1}{2}\log^2\left(\frac{X_{i-1} \cdot X_{j-1}}{X_i \cdot X_j}\right) \\
& + \mathrm{Li}_2\left(1 - \frac{X_i \cdot X_{j-1}}{X_{i-1} \cdot X_{j-1}}\right) + \mathrm{Li}_2\left(1 - \frac{X_i \cdot X_{j-1}}{X_i \cdot X_j}\right) \\
& + \mathrm{Li}_2\left(1 - \frac{X_{i-1} \cdot X_j}{X_{i-1} \cdot X_{j-1}}\right) + \mathrm{Li}_2\left(1 - \frac{X_{i-1} \cdot X_j}{X_i \cdot X_j}\right) \\
& + \mathrm{Li}_2\left(1 - \frac{X_i \cdot X_{j-1}X_{i-1} \cdot X_j}{X_i \cdot X_j X_{i-1} \cdot X_{j-1}}\right) + \mathcal{O}(\epsilon), \\
F_{1m}(i-3,i-2,i-1,i) = {} & -\frac{1}{\epsilon^2}\left[\left(\frac{X_{i-1} \cdot X_{i-3}}{\mu^2}\right)^{-\epsilon} + \left(\frac{X_{i-2} \cdot X_i}{\mu^2}\right)^{-\epsilon} - \left(\frac{X_i \cdot X_{i-3}}{\mu^2}\right)^{-\epsilon}\right] \\
& + \frac{1}{2}\log^2\left(\frac{X_{i-1} \cdot X_{i-3}}{X_{i-2} \cdot X_i}\right) + \mathrm{Li}_2\left(1 - \frac{X_i \cdot X_{i-3}}{X_{i-3} \cdot X_{i-1}}\right) \\
& + \mathrm{Li}_2\left(1 - \frac{X_i \cdot X_{i-3}}{X_{i-2} \cdot X_i}\right) + \frac{\pi^2}{6} + \mathcal{O}(\epsilon),
\end{aligned}
$$

$$
\begin{aligned}
F_{2\text{mh}}(i-1,i,i+1,j) = &-\frac{1}{2\epsilon^2}\Bigg[\left(\frac{X_{i-1}\cdot X_{i+1}}{\mu^2}\right)^{-\epsilon} + 2\left(\frac{X_i\cdot X_j}{\mu^2}\right)^{-\epsilon} \\
&- \left(\frac{X_{i+1}\cdot X_j}{\mu^2}\right)^{-\epsilon} - \left(\frac{X_{i-1}\cdot X_j}{\mu^2}\right)^{-\epsilon}\Bigg] \\
&+ \text{Li}_2\left(1 - \frac{X_{i+1}\cdot X_j}{X_i\cdot X_j}\right) + \text{Li}_2\left(1 - \frac{X_{i-1}\cdot X_j}{X_i\cdot X_j}\right) \\
&- \frac{1}{2}\log\left(\frac{X_{i+1}\cdot X_j}{X_{i-1}\cdot X_{i+1}}\right)\log\left(\frac{X_{i-1}\cdot X_j}{X_{i-1}\cdot X_i + 1}\right) \\
&+ \frac{1}{2}\log^2\left(\frac{X_{i-1}\cdot X_{i+1}}{X_i\cdot X_j}\right) + \mathcal{O}(\epsilon).
\end{aligned}
\tag{61.23}
$$

Note that there is still an arbitrary mass parameter μ by dimensional transmutation, and as usual, we can go to the mass regularization by just expanding in epsilon, and keeping only the finite terms. We can indeed check that by expanding $F_{1\text{m}}$ and $F_{2\text{me}}$ in epsilon and keeping only the terms finite (in ϵ), which still diverge as $\mu \to 0$, we get the same result as in (61.22).

61.5 Leading Singularities of Amplitudes and a Conjecture for Them

We can define the notion of a *leading singularity* of an amplitude as the discontinuity of the amplitude over the singularities, where we put a maximum number of propagators on-shell. As we saw in Chapter 59*, in $D = 4$ that maximum number is four at one loop, for the four components of the loop momentum l^μ. It is then easy to generalize, and see that at L loops we have a maximum of $4L$ propagators on-shell. We can make the integral finite by regularizing the cut condition, for instance at one loop by

$$
|l^2| = \epsilon, \quad |(l-K_1)^2| = \epsilon, \quad |(l-K_1-K_2)^2| = \epsilon, \quad |(l+K_4)^2| = \epsilon,
\tag{61.24}
$$

which amounts to integrating over a contour in $\mathbb{C}^4$ that encircles each of the propagator singularities. This means that mathematically, the leading singularity is a residue. Since as we saw, there are in general two solutions to the quadruple cut condition (for $\epsilon = 0$), there are two contours of $T^4 = (S^1)^4$ type.

At L loops, we do a similar thing. The loop integral is written as

$$
\int \frac{d^4 l_1 \dots d^4 l_L}{\prod_{i=1}^{4L} P_i^2(l_1,\dots,l_{4L},p_1,..,p_n)} R(l_1,\dots,l_L),
\tag{61.25}
$$

and we take the $4L$-cut putting the corresponding propagators on-shell, or more precisely considering the contour $|P_i|^2 = \epsilon, i = 1,\dots,4L$.

In a maximally supersymmetric theory ($\mathcal{N} = 4$ SYM), the leading singularity is enough to define the amplitude itself through the use of the scalar basis of amplitudes, as mentioned in the generalized unitarity cut method.

For the leading singularity in $\mathcal{N} = 4$ SYM, there is a conjecture that has passed many tests. I will present it here, even though we have not defined supersymmetry, since it has a nice interpretation.

First, note that the Parke–Taylor formula for the MHV amplitude was generalized to an MHV *superamplitude* (a superspace version of the amplitude, encoding all of the amplitudes in the supermultiplet) by Nair, as

$$\mathcal{A}_{n,2}(12\ldots n) = \frac{\delta^4(\sum_{i=1}^n \lambda_i\tilde{\lambda}_i)\delta^8(\sum_{i=1}^n \lambda_i\tilde{\eta}^i)}{\langle 12\rangle\langle 23\rangle \ldots \langle n-1,n\rangle\langle n,1\rangle}. \tag{61.26}$$

Here, $\langle ij\rangle \equiv \epsilon^{\alpha\beta}\lambda_\alpha^{(i)}\lambda_\beta^{(j)}$, $\tilde{\eta}$ is a spinor with a suppressed index $I = 1,\ldots,4$ for supersymmetries, and the 2 in $\mathcal{A}_{n,2}$ refers to R-charge, since the N^kMHV amplitude has $m = k + 2$ R-charge, so the MHV case ($k = 0$) has R-charge 2 (R-charge is associated with an R-symmetry, a global symmetry in supersymmetric theories). The expansion in $\tilde{\eta}$ gives the various amplitudes for the supermultiplet components.

The conjecture for the leading singularity of the superamplitudes of R-charge m is

$$\mathcal{L}_{n,m} = \int \frac{d^{nm}C_{\mu i}}{\text{Vol}(Gl(m))} \prod_{\mu=1}^m \frac{\delta^2(\sum_{i=1}^n C_{\mu i}\tilde{\lambda}_i)\delta^{0|4}(\sum_{i=1}^n C_{\mu i}\tilde{\eta}_i)}{(12\ldots m)(23\ldots m+1)\ldots(n12\ldots m-1)}$$

$$\int \prod_{\mu=1}^m d^2\rho_\mu \prod_{i=1}^n \delta^2(C_{\mu i}\rho_\mu - \lambda_i)\,, \tag{61.27}$$

where $(12\ldots m)$, and so on are "Plucker coordinates on the Grassmannian $G(m,n)$," which means just the determinants of $m \times m$ minors of the matrix

$$\begin{pmatrix} C_{11} & C_{12} & \ldots & C_{1n} \\ C_{21} & C_{22} & \ldots & C_{2n} \\ .. & .. & \ldots & .. \\ C_{m1} & C_{m2} & \ldots & C_{mn} \end{pmatrix}. \tag{61.28}$$

After doing the integrals over the delta functions depending on $C_{\mu i}$, we can calculate that we are left with an $(m-2)[(n-4)-(m-2)]$-dimensional integral to be done. Then, in the MHV case ($m = 2$), there is no integral left to be done, and the formula for the leading singularity reduces to

$$\mathcal{A}_{n,2} = \int \frac{d^{2n}C_{\mu i}}{\text{Vol}(Gl(2))} \frac{\prod_{\mu=1}^2 \delta^2(\sum_{i=1}^n C_{\mu i}\tilde{\lambda}_i)\delta^{0|4}(\sum_{i=1}^n C_{\mu i}\tilde{\eta}_i)}{(12)(23)\ldots(n1)} \int \prod_{\mu=1}^2 d^2\rho_\mu \prod_{i=1}^n \delta^2(C_{\mu i}\rho_i - \lambda_i), \tag{61.29}$$

which reduces to the Nair formula, since then we can show that $(i\ i+1) = \langle i\ i+1\rangle$ after the delta functions are enforced.

A more compact form is obtained by taking a Fourier transform over λ:

$$\tilde{\mathcal{L}}_{n,m} \equiv \int \prod_{i=1}^n d^2\lambda_i e^{i<\lambda_i,\mu_i>} \mathcal{L}_{n,m} \tag{61.30}$$

and defining

$$\mathcal{W}_i = \begin{pmatrix} \tilde{\lambda}_i \\ \mu_i \\ \tilde{\eta}_i \end{pmatrix} \in C^{4|4}. \tag{61.31}$$

The physical configuration lives in $\mathbb{CP}^{3|4}$, a generalization of twistor space called supertwistor space, and the Fourier transform of the leading singularity is

$$\tilde{\mathcal{L}}_{n,m} = \int \frac{d^{nm} C_{\mu i}}{Vol(Gl(2))} \frac{\prod_{\mu=1}^{m} \delta^{4|4}(\sum_{i=1}^{n} C_{\mu i} \mathcal{W}_i)}{(12 \ldots m)(23 \ldots m+1) \ldots (n12 \ldots m)}. \tag{61.32}$$

There are other forms, including twistors and momentum twistors, but we will not give them here.

Important Concepts to Remember

- A twistor space is defined as $SO(4, 2)$ vectors X^M satisfying $X \cdot X = 0$, invariant under a real rescaling by λ, or equivalently by an $SU(2, 2)$ complex bispinor X^{IJ} with the constraint $X^2 = \frac{1}{2}\epsilon_{IJKL}X^{IJ}$ $X^{KL} = 0$.

- The twistors A^I, B^I are defined by $X^{IJ} = A^{[I}B^{J]}$, and take values in $\mathbb{CP}^3$, and trivialize the null condition $x^2 = 0$.

- The distance on twistor space is $d_{X_1, X_2} = \frac{X_1 \cdot X_2}{(I \cdot X_1)(I \cdot X_2)}$ and with it, a point (zero distance in twistor space) corresponds to a line in Minkowski space, whereas a line in twistor space corresponds to a point in Minkowski space.

- An $X_{a\dot{a}}$ satisfying $X^2 = 0$ defines a $\mathbb{CP}^1$ line in twistor space, $Z = (\lambda, \mu)$ with $\mu_{\dot{a}} = X_{a\dot{a}}\lambda_a$.

- The anti-MHV amplitude, Fourier transformed over λs, is an amplitude in twistor space, localized on the incidence relations $\tilde{\lambda}^{\dot{a}}x_{a\dot{a}} = -\mu_a$.

- The momentum twistors of Hodges, defined in the dual space $x_i^\mu \rightarrow x_i^{a\dot{a}}$, trivialize the momentum conservation condition, since $p_i = x_i - x_{i+1}$.

- Dual conformal invariance acts on the dual space x_i in massless loop integrals.

- In some theories, like $\mathcal{N} = 4$ SYM, dual conformal invariance and conformal invariance generate an infinite symmetry called the Yangian.

- From the dual coordinates x_i of the amplitudes, one can construct coordinates in Euclidean AdS_5, such that the fully IR-finite four-mass scalar box integrals correspond to twice the volume of the ideal tetrahedron with vertices corresponding to the x_i.

- The ratio of the MHV one-loop amplitude $\mathcal{A}_n$, divided by the tree amplitude, gives the volume of the closed three-dimensional polytope with n vertices, that is the surface of a four-dimensional polytope.

- The leading singularity of an amplitude is the discontinuity of the amplitude over singularities, where we put the maximum number of propagators on-shell.

- There is a conjecture for writing the leading singularity of the superamplitude of R-charge m as an amplitude on the Grassmanian, and its Fourier transform as an integral on supertwistor space.

Further Reading

The polytope picture was described in [27]. The conjecture for the leading singularities of amplitudes was defined in [28]. For more details on the twistor space construction, and on polytopes, and in particular for the scalar boxes in terms of X^{AB}, see [29].

Exercises

1. Prove that the zero-mass box function F_{0m} in (61.13) is invariant under the inversion $x^\mu \to x^\mu/x^2$, and that by composing it with scalings, translations, and Lorentz rotations, we find invariance under the conformal group $SO(4,2)$.

2. Show that the variables

$$X^{AB} \equiv \begin{pmatrix} -\frac{1}{2}\epsilon^{ab}x^2 & ix^a{}_{\dot b} \\ -ix_a{}^{\dot b} & \epsilon_{\dot a \dot b} \end{pmatrix} \tag{61.33}$$

satisfy the conditions

$$X^2 \equiv \frac{1}{2}\epsilon_{ABCD}X^{AB}X^{CD} = 0,$$
$$X_i \cdot X_j = -(x_i - x_j)^2. \tag{61.34}$$

3. Calculate $R_6 = \mathcal{A}_6/\mathcal{A}_6^{\text{tree}}$ for the six-point, one-loop MHV amplitude, and thus the volume of the corresponding closed polytope in AdS_5, as a function of the twistor-space coordinates.

4. Calculate the Fourier transform on $\tilde\lambda, \tilde\eta$ of the Nair formula for the superamplitude.

62

The Wilsonian Effective Action, Effective Field Theory, and Applications

In this chapter, we will describe a new view on the process of renormalization, one that will be continued in Chapter 63. Through the Wilsonian effective action, we will define effective field theory, and see how to apply it.

In the real world, we can always test physics only below a maximal energy scale Λ. The question is then, can we hide our ignorance about the high energy in a consistent framework? The answer turns out to be yes, by defining the Wilsonian effective action, and through it, effective field theory.

62.1 The Wilsonian Effective Action

The Wilsonian effective action is obtained in the simplest way, namely by integrating out the degrees of freedom with momenta $|k| > \Lambda$, in order to hide our ignorance about it.

62.1.1 ϕ^4 Theory in Euclidean Space

We will describe the formalism on the simplest nontrivial theory, ϕ^4 in Euclidean four dimensions. The (classical) action is

$$S_E = \int d^4x \left[\frac{1}{2}(\partial_\mu \phi)^2 + \frac{m^2}{2}\phi^2 + \frac{\lambda}{4!}\phi^4 \right], \tag{62.1}$$

and the partition function is

$$Z[J] = \int \mathcal{D}\phi\, e^{-S + \int d^4x J(x)\phi(x)}. \tag{62.2}$$

But since we want to integrate over momenta $|k| > \Lambda$, we must go to momentum space, where

$$S_E = \frac{1}{2} \int \frac{d^4k}{(2\pi)^4} \tilde\phi(-k)(k^2 + m^2)\tilde\phi(k) + \frac{\lambda}{4!}$$
$$\int \frac{d^4k_1}{(2\pi)^4} \cdots \frac{d^4k_4}{(2\pi)^4} \tilde\phi(k_1)\ldots\tilde\phi(k_4)(2\pi)^4\delta^4(k_1 + \ldots + k_4). \tag{62.3}$$

We introduce a UV cut-off Λ. Since we want to say that we cannot access energies higher than Λ by experiments, we impose that sources are zero for these momenta, $J = 0$ for $|k| > \Lambda$. Then

$$Z[J] = \int \mathcal{D}\phi_{|k|<\Lambda} e^{-S_{\text{eff}}(\phi;\Lambda)+\int J\cdot\phi}, \tag{62.4}$$

where

$$e^{-S_{\text{eff}}(\phi;\Lambda)} = \int \mathcal{D}\phi_{|k|>\Lambda} e^{-S_E[\phi]}. \tag{62.5}$$

Here, S_{eff} is called the *Wilsonian effective action*. The Wilsonian effective Lagrangian will then be (after doing the integral)

$$\mathcal{L}_{\text{eff}} = \frac{1}{2}(\partial_\mu\phi)^2 + \frac{m^2}{2}\phi^2 + \frac{\lambda}{4!}\phi^4 + \sum_{\Delta\geq 6}\sum_i c_{\Delta,i}\mathcal{O}_{\Delta,i}. \tag{62.6}$$

Here, $\mathcal{O}_{\Delta,i}$ are all higher-dimension operators (dimension higher than four, but since the classical Lagrangian has only even powers of ϕ, so should the quantum action, so the next dimension is six), organized by their dimension Δ and for given dimension by an index i.

However, in the above action, ϕ now has only momenta smaller than Λ:

$$\phi(x) = \int^\Lambda \frac{d^4k}{(2\pi)^4} e^{ikx}\tilde{\phi}(k). \tag{62.7}$$

But in reality, we have the renormalized Lagrangian

$$\mathcal{L}_{\text{ren}} = \frac{1}{2}Z_\phi(\partial_\mu\phi)^2 + \frac{Z_m m_{ph}^2}{2}\phi^2 + \frac{Z_\lambda \lambda_{ph}}{4!}\phi^4. \tag{62.8}$$

Here, λ_{ph} is defined as the 1PI vertex at $p^\mu = 0$.

Then we have, for the Wilsonian effective Lagrangian, really

$$\mathcal{L}_{\text{eff}} = \frac{1}{2}Z(\Lambda)(\partial_\mu\phi)^2 + \frac{m^2(\Lambda)}{2}\phi^2 + \frac{\lambda(\Lambda)}{4!}\phi^4 + \sum_{\Delta\geq 6}\sum_i c_{\Delta,i}\mathcal{O}_{\Delta,i}. \tag{62.9}$$

Here all coefficients are *finite* functions of Λ, since the correlation functions of renormalized fields, calculated by $\delta/\delta J(x)$ from it, are finite.

62.2 Calculation of $c_{\Delta,i}$

We now proceed to calculate explicitly $c_{\Delta,i}$ for the operators $\mathcal{O}_{2n,1} \equiv \phi^{2n}$ at one loop.

The unique one-loop diagram with $2n$ external lines is composed of a loop with n vertices on it, each having two external legs (Figure 62.1). Then, in the external lines, we have momenta $|k| < \Lambda$, but on the internal lines we integrate over $|k| > \Lambda$.

The symmetry factor for the diagram is

$$S = 2^n 2 \times n, \tag{62.10}$$

since there is a symmetry factor of 2 for the interchange of the two lines at each vertex, giving a total factor of 2^n, plus a rotation symmetry, which means that we need to define where the vertex 1 is, giving a factor of n, plus a reflection symmetry giving another factor

One-loop diagram for the effective potential in the Wilson approach.

of 2. There is also a $(2n)!$ factor for the ways to assign the momenta $p_1, \ldots, p_n$ to the external lines (permuting them).

The Feynman diagram is then identified with the effective vertex coming out as a Feynman rule out of $c_{2n,1}\mathcal{O}_{2n,1}$ in the Lagrangian (i.e. $-c_{2n,1}(2n)!$), so we have

$$- c_{2n,1}(2n)! = \frac{(2n)!}{2^n 2n}(-\lambda_{ph})^n \int_\Lambda^\infty \frac{d^4 k}{(2\pi)^4} \frac{1}{(k^2 + m_{ph}^2)^n} + \mathcal{O}(\lambda_{ph}^{n+1}). \tag{62.11}$$

Then, for $2n \geq 6$, the integral is convergent, and is

$$\int_\Lambda^\infty \frac{d^4 k}{(2\pi)^4} \frac{1}{(k^2 + m^2)^n} \simeq \frac{2\pi^2}{(2\pi)^4} \int_\Lambda^\infty \frac{k^3 dk}{k^{2n}} = \frac{1}{8\pi^2} \frac{1}{(2n-4)\Lambda^{2n-4}}, \tag{62.12}$$

where in the equality we have used the fact that $m^2 \ll \Lambda^2$, so we get for the coefficients at one loop

$$c_{2n,1}(\Lambda) = -\frac{1}{32\pi^2} \frac{(-\lambda_{ph}/2)^n}{n(n-2)\Lambda^{2n-4}} + \mathcal{O}(\lambda_{ph}^{n+1}). \tag{62.13}$$

2n=4 For $2n = 4$ (i.e. for $\lambda(\Lambda)$), we formally have the same integral, just that we need to add the tree result to the one-loop one, and we also now wrote $\lambda(\Lambda)$ instead of $c_{4,1}4!$. Since now $(2n)!/(2^n 2n) = 4!/(2^2 4) = 3/2$, we get

$$- \lambda(\Lambda) = -Z_\lambda \lambda_{ph} + \frac{3}{2}(-\lambda_{ph})^2 \int_\Lambda^\infty \frac{d^4 k}{(2\pi)^4} \frac{1}{(k^2 + m_{ph}^2)^2} + \mathcal{O}(\lambda_{ph}^3). \tag{62.14}$$

This result is UV-divergent. However, we note that in the usual renormalization, the renormalized vertex at zero momenta, which by definition is λ_{ph}, is given by

$$-\lambda_{ph} = -V_4(0,0,0,0) = -Z_\lambda \lambda_{ph} + \frac{3}{2}(-\lambda_{ph})^2 \int_0^\infty \frac{d^4 k}{(2\pi)^4} \frac{1}{(k^2 + m_{ph}^2)^2} + \mathcal{O}(\lambda_{ph}^3). \tag{62.15}$$

Note that the integration here is over *all* momenta, from 0 to ∞.

Then the difference of the two is finite:

$$- \lambda_{ph} + \lambda(\Lambda) = \frac{3}{2}(-\lambda_{ph})^2 \int_0^\Lambda \frac{d^4 k}{(2\pi)^4} \frac{1}{(k^2 + m_{ph}^2)^2} + \mathcal{O}(\lambda_{ph}^3). \tag{62.16}$$

The integral is

$$\int_0^\Lambda \frac{d^4k}{(2\pi)^4} \frac{1}{(k^2 + m_{ph}^2)^2} = \frac{1}{16\pi^2} \int_0^{\Lambda^2} \frac{k^2 dk^2}{(k^2 + m_{ph}^2)^2} = \frac{1}{16\pi^2} \int_{m_{ph}^2}^{\Lambda^2 + m_{ph}^2} \frac{(\tilde{k}^2 - m_{ph}^2)d\tilde{k}^2}{\tilde{k}^4}$$

$$= \frac{1}{16\pi^2}\left(\ln\frac{\Lambda^2 + m_{ph}^2}{m_{ph}^2} + \frac{m_{ph}^2}{\Lambda^2} - 1\right) \simeq \frac{1}{8\pi^2}\left(\ln\frac{\Lambda}{m_{ph}} - \frac{1}{2}\right),$$

$$(62.17)$$

thus it gives for the coupling

$$\lambda(\Lambda) = \lambda_{ph} + \frac{3}{16\pi^2}\lambda_{ph}^2\left[\ln\frac{\Lambda}{m_{ph}} - \frac{1}{2}\right] + \mathcal{O}(\lambda_{ph}^3).$$

$$(62.18)$$

2n=2   For $2n = 2$, the Feynman diagram is a line with a loop on it, so the integral is independent of the external momentum p. Since the p-dependent part of the Feynman diagram gives the wavefunction renormalization, in our case we have no wavefunction renormalization at one loop:

$$Z(\Lambda) = 1 + \mathcal{O}(\lambda_{ph}^2).$$

$$(62.19)$$

But there is a p-independent part, which gives the mass renormalization, so we obtain (again adding the tree contribution, and since now $(2n)!/(2^n 2n) = 2!/(2 \cdot 2) = 1/2$)

$$-m^2(\Lambda) = -Z_m m_{ph}^2 + \frac{1}{2}(-\lambda_{ph})\int_\Lambda^\infty \frac{d^4k}{(2\pi)^4}\frac{1}{k^2 + m_{ph}^2} + \mathcal{O}(\lambda_{ph}^2).$$

$$(62.20)$$

This is quadratically divergent. As before, the full renormalization gives

$$-m_{ph}^2 = -Z_m^2 m_{ph}^2 + \frac{1}{2}(-\lambda_{ph})\int_0^\infty \frac{d^4k}{(2\pi)^4}\frac{1}{k^2 + m_{ph}^2} + \mathcal{O}(\lambda_{ph}^2),$$

$$(62.21)$$

but the difference is now finite:

$$-m_{ph}^2 + m^2(\Lambda) = \frac{1}{2}(-\lambda_{ph})\int_0^\Lambda \frac{d^4k}{(2\pi)^4}\frac{1}{k^2 + m_{ph}^2} + \mathcal{O}(\lambda_{ph}^2).$$

$$(62.22)$$

The integral gives

$$\int_0^\Lambda \frac{d^4k}{(2\pi)^4}\frac{1}{k^2 + m_{ph}^2} = \frac{1}{16\pi^2}\int_{m_{ph}^2}^{\Lambda^2 + m_{ph}^2}\frac{(\tilde{k}^2 - m_{ph}^2)d\tilde{k}^2}{\tilde{k}^2}$$

$$= \frac{1}{16\pi^2}\left(\Lambda^2 - m_{ph}^2\ln\frac{\Lambda^2 + m_{ph}^2}{m^2}\right),$$

$$(62.23)$$

so the mass squared is

$$m^2(\Lambda) = m_{ph}^2 - \frac{\lambda_{ph}}{32\pi^2}\left[\Lambda^2 - m_{ph}^2\ln\frac{\Lambda^2}{m_{ph}^2}\right] + \mathcal{O}(\lambda_{ph}^2).$$

$$(62.24)$$

As a final observation, note that the nonrenormalizable operators $\mathcal{O}_{\Delta,i}$ have coefficients of order

$$c_{\Delta,i} \sim \frac{1}{\Lambda^{\Delta-4}}, \tag{62.25}$$

which is highly suppressed for large Λ (energies much smaller than it), so will not change the physics too much.

62.3 Effective Field Theory

We consider now a new set-up, closer to the physical case. In the physical case, the theory can have a true cut-off, for instance at least the Planck scalar m_{Planck} can act as such, if not the SUSY scale, GUT scale, KK scale, and so on. On top of this, we can consider also our arbitrary (variable) scale Λ below it. At this cut-off scale, the degrees of freedom of the theory change, and we have a new theory.

We also *assume* that when measured in units of the cut-off, the parameters of the theory at the cut-off scale are small:

$$\lambda(\Lambda_0) \ll 1, \quad m^2(\Lambda_0) \ll \Lambda_0^2, \quad c_{\Delta,i}(\Lambda_0) \ll \Lambda_0^{4-\Delta}. \tag{62.26}$$

So we treat the effective action as a fundamental starting point, rather than assuming a better definition of the theory in the UV.

Since the coefficients of the higher-dimension operators are small, let us assume for the moment that they actually vanish:

$$c_{\Delta,i}(\Lambda_0) = 0, \tag{62.27}$$

and see if that is consistent.

We now integrate over the region between the true cut-off and an arbitrary lower cut-off Λ (i.e. over $\Lambda < |k| < \Lambda_0$). Then we have at the lower scale

$$e^{-S_{\text{eff}}[\phi;\Lambda]} = \int \mathcal{D}\phi_{\Lambda<|k|<\Lambda_0} e^{-S_{\text{eff}}(\phi;\Lambda_0)}. \tag{62.28}$$

Using the formulas already derived for $m^2(\Lambda), \lambda(\Lambda), c_{2n,1}(\Lambda)$, we can calculate the values at Λ in terms of values at Λ_0, and obtain, for Λ not too much less than Λ_0 (i.e. for $\Lambda = b\Lambda_0, b \lesssim 1$):

$$m^2(\Lambda) = m^2(\Lambda_0) + \frac{1}{2}\lambda(\Lambda_0) \int_\Lambda^{\Lambda_0} \frac{d^4k}{(2\pi)^4} \frac{1}{k^2 + m^2(\Lambda_0)} + \cdots,$$

$$\lambda(\Lambda) = \lambda(\Lambda_0) - \frac{3}{2}\lambda^2(\Lambda_0) \int_\Lambda^{\Lambda_0} \frac{d^4k}{(2\pi)^4} \frac{1}{(k^2 + m^2(\Lambda_0))^2} + \cdots,$$

$$c_{2n,1}(\Lambda) = -\frac{(-1)^n}{2^n 2n}\lambda^n(\Lambda_0) \int_\Lambda^{\Lambda_0} \frac{d^4k}{(2\pi)^4} \frac{1}{(k^2 + m^2(\Lambda_0))^n} + \cdots \tag{62.29}$$

Then, if Λ is not too much less than Λ_0 as before, and if also $m^2(\Lambda_0) \ll \Lambda^2$, we obtain approximately (using the integrals calculated before)

$$m^2(\Lambda) = m^2(\Lambda_0) + \frac{\lambda(\Lambda_0)}{32\pi^2}(\Lambda_0^2 - \Lambda^2) + \dots,$$

$$\lambda(\Lambda) = \lambda(\Lambda_0) - \frac{3}{16\pi^2}\lambda^2(\Lambda_0)\ln\frac{\Lambda_0}{\Lambda} + \dots,$$

$$c_{2n,1}(\Lambda) = -\frac{(-\lambda(\Lambda_0)/2)^n}{32\pi^2 n(n-2)}\left(\frac{1}{\Lambda^{2n-4}} - \frac{1}{\Lambda_0^{2n-4}}\right) + \dots \qquad (62.30)$$

We note therefore that Λ_0 is not very important for $c_{2n,1}$, its effect being very small, which is why it was consistent to consider $c_{2n,1}(\Lambda_0) = 0$, whereas it is very important for m^2, so arranging for $m^2(\Lambda) \ll \Lambda^2$ is a *fine-tuning problem* that is not natural.

We also note that we can now define the beta function as usual and calculate it at one loop from the above $\lambda(\Lambda)$ as

$$\beta(\lambda) = \frac{d\lambda}{d\ln\Lambda} = \frac{3}{16\pi^2}\lambda^2(\Lambda), \qquad (62.31)$$

as it should be.

In conclusion, the picture of *effective field theory* is as follows. We define the theory with a cut-off, and correspondingly with higher-dimension operators. Then, we consider lowering the cut-off by integrating out the intermediate degrees of freedom. When we do that, the coefficients in the (Wilsonian) effective action change. This leads to a view of renormalization as a change in cut-off scale that will be developed better in Chapter 63, as Kadanoff blocking, and we will see that it leads to a connection with condensed matter theory.

62.3.1 Nonrenormalizable Theories

This picture also leads to a way to deal with nonrenormalizable theories. We can just regard them as (Wilsonian) effective actions. We impose a cut-off Λ_0, in which case the coefficients of the higher-dimension operators are

$$\frac{c_i}{\Lambda_0^{\Delta_i-4}}, \qquad (62.32)$$

with $c_i \leq 1$. We can then use this effective action below the energy scale Λ_0, as shown above, and for energies $E \ll \Lambda_0$ the coefficients of the higher-dimension operators are really small, making the theory look almost like a renormalizable one, up to powers of E/Λ_0.

62.3.2 Removing the Cut-off

An important question that arises then is, can we remove completely the cut-off Λ_0?

If we know an exact beta function, we can integrate

$$\frac{d\lambda}{d\ln\Lambda} = \beta(\lambda) \qquad (62.33)$$

to

$$\int_{\lambda(m_{ph})}^{\lambda(\Lambda_0)} \frac{d\lambda}{\beta(\lambda)} = \ln \frac{\Lambda_0}{m_{ph}}. \tag{62.34}$$

Then, as $\Lambda_0 \to \infty$, the right-hand side goes to infinity, so the left-hand side should too. In that case, it means we can remove Λ_0.

But it could happen that it does not go to infinity, but instead it goes to a constant before that. Indeed, in the case that $\beta(\lambda)$ is positive and increases at infinity faster than λ, we can see that $\int d\lambda/\beta(\lambda)$ is finite at infinity, so there must be a maximum Λ, Λ_{max}, given by

$$\ln \frac{\Lambda_{max}}{m_{ph}} = \int_{\lambda(m_{ph})}^{\infty} \frac{d\lambda}{\beta(\lambda)}. \tag{62.35}$$

For instance, using the one-loop beta function $\beta_1 = 3\lambda^2/(16\pi^2)$, which indeed goes faster than λ, we find

$$\Lambda_{max} = m_{ph} e^{\frac{16\pi^2}{3\lambda_{ph}}}. \tag{62.36}$$

This is the Landau pole that we have already explained several times.

In contrast, if $\beta(\lambda) > 0$, but it goes to infinity slower than λ, we *can* remove Λ_0, and we have a UV-fixed point λ_*.

Indeed, then for $\lambda \to \lambda_*$, $\Lambda \to \infty$ and

$$\int_{\lambda}^{\lambda_*} \frac{d\lambda}{\beta(\lambda)} \to \infty. \tag{62.37}$$

Important Concepts to Remember

- In the Wilsonian effective action approach, we hide our ignorance about high energy by integrating over momenta with $|k| > \Lambda$.
- In the Wilsonian effective action, we integrate over $|k| > \Lambda$ and obtain all the higher-dimension operators, with coefficients that go like $1/\Lambda^{\Delta-4}$, and we use it in quantum processes with energies smaller than Λ.
- The effective field theory approach is to consider the theory as fundamentally defined with a cut-off and a Wilsonian effective action, and then lower the cut-off by integrating over intermediate degrees of freedom, $\Lambda < |k| < \Lambda_0$.
- In this way, nonrenormalizable theories are thought of as effective field theories, and at energies $E \ll \Lambda_0$, the effect of the nonrenormalizable operators is very small.
- For $\beta(\lambda) > 0$, if we can remove completely the UV cut-off, we have a UV-fixed point, if not, we have a Landau pole.

Further Reading

See chapter 29 in [10] and section 12.1 in [1].

Exercises

1. Consider ϕ^3 theory in $d = 6$, with

$$\mathcal{L} = \frac{1}{2}(\partial_\mu \phi)^2 + \frac{\lambda}{3!}\phi^3 + \frac{m^2}{2}\phi^2. \qquad (62.38)$$

   Calculate the coefficients $c_{n,1}(\Lambda)$ of the $\mathcal{O}_{n,1} = \phi^n$ higher-dimension operators in Wilson's approach, at one loop.
2. In the same theory, in the effective field theory approach, calculate $\lambda(\Lambda)$ from $\lambda(\Lambda_0)$ at one loop ($\Lambda < |k| < \Lambda_0$).
3. Repeat Exercise 1 for ϕ^6 theory in $d = 3$ with a mass term.
4. Repeat Exercise 2 for ϕ^6 theory in $d = 3$ with a mass term.

Kadanoff Blocking and the Renormalization Group: Connection with Condensed Matter

In this chapter we will study a way to understand the renormalization group related to the Wilsonian method from Chapter 62, by integrating degrees of freedom, just that the way we do it is via a discretization that has a natural relation to condensed matter physics, and in particular to spin systems.

63.1 Field Theories as Classical Spin Systems

Consider a discretization of the scalar Lagrangian

$$\mathcal{L}(\phi) = \frac{1}{2}(\partial_\mu\phi)^2 + V(\phi)(+J \cdot \phi) \qquad (63.1)$$

and its path integral, on a (hyper)cubic lattice of size a, via

$$x \to x_n; \quad \phi(x) \to \phi_n = \phi(x_n); \quad \mathcal{D}\phi \to \prod_n d\phi_n;$$

$$\partial_\mu\phi \to \frac{1}{a}(\phi_{n+\mu} - \phi_n), \qquad (63.2)$$

where $n + \mu$ is the nearest neighbor on the lattice (in direction μ) to site n.

Then the discretized action is

$$S \to \sum_n a^d \left[\frac{1}{a^2} \sum_\mu \frac{1}{2}(\phi_{n+\mu} - \phi_n)^2 + V(\phi_n) + J_n\phi_n \right]. \qquad (63.3)$$

For the particular case of $V(\phi) = \lambda\phi^4/4!$, we can rescale as follows to absorb the dependence on a and put the coupling outside the action:

$$\lambda = g^{-2}a^{d-4},$$
$$\phi' = ga^{\frac{d}{2}-1}\phi,$$
$$J' = ga^{\frac{d}{2}+1}J. \qquad (63.4)$$

Then the path integral becomes

$$Z[J', g] = \mathcal{N} \int \prod_n (g^{-1}d\phi'_n)e^{-\frac{S(\phi')}{g^2}}, \qquad (63.5)$$

where the action is

$$S(\phi', J') = \sum_n \left[\frac{1}{2} \sum_\mu (\phi'_{n+\mu} - \phi_n)^2 + \frac{m^2 a^2}{2} \phi'^2 + \frac{\phi'^4}{4!} + J'_n \phi'_n \right]. \qquad (63.6)$$

This now has the form of a classical spin system. Indeed, for instance a ferromagnet has the Hamiltonian

$$H(S, h) = -\sum_{n,m} v_{n,m} S_n \cdot S_m + h \sum_m S_m, \qquad (63.7)$$

where the first term is a spin–spin interaction and the second is the interaction of the spins with an external field h. The partition function is

$$Z[h, \beta] = \int \prod_n dS_n \, \rho(S_n) e^{-\beta H(S,h)}, \qquad (63.8)$$

where $\beta = 1/(k_B T)$ and $\rho(S_n)$ is a weight describing the spin. It should depend on S_n^2 because of relativistic invariance, and it is naturally exponential, so an effective description for this measure is given phenomenologically by

$$\rho(S_n) \propto e^{-(\kappa S_n^2 + \lambda S_n^4)}. \qquad (63.9)$$

The exact form would be given by the microscopic properties of spin.

For a system with only nearest-neighbor interaction, the $\sum_{n,m}$ in the Hamiltonian turns into $\sum_n \sum_\mu$, and

$$-\sum_{n,m} v_{n,m} s_n \cdot s_m \rightarrow K \sum_n \left(\sum_\mu (s_{n+\mu} - s_n)^2 - 2ds_n^2 \right). \qquad (63.10)$$

The partition function then becomes

$$Z(K, \mu, \lambda, h) = \int \prod_n ds_n e^{-\beta H(s; K, \mu, \lambda, h)}, \qquad (63.11)$$

where the effective Hamiltonian is

$$H(s; K, \mu, \lambda, h) = \sum_n \left[K(\beta) \sum_\mu (s_{n+\mu} - s_n)^2 + \mu(\beta) s_n^2 + \lambda(\beta) s_n^4 + h s_n \right]. \qquad (63.12)$$

This is written in a more general form, but from the above considerations we would have $K(\beta) = K$, $\mu(\beta) = K/\beta - 2dK$, and $\lambda(\beta) = \lambda/\beta$.

As we see, $\mu(\beta)$ is governed by two opposing terms, so can be either positive or negative. When it is zero, $\mu = \mu(\beta_c) = 0$, we are at a phase transition.

Indeed, for $\mu(\beta) > 0$ and no external field h, we have no magnetization, since

$$\langle s \rangle = \frac{1}{V} \sum_n \langle s_n \rangle = 0, \qquad (63.13)$$

whereas at $\mu(\beta) < 0$ and no external field h, the classical minimum of the Hamiltonian is at $s_n = \sqrt{-\mu/(2\lambda)}$, so the magnetization is

$$\langle s \rangle = \frac{1}{V} \sum_n \langle s_n \rangle = \sqrt{\frac{-\mu}{2\lambda}}. \tag{63.14}$$

Moreover, near $\beta = \beta_c$, assuming smoothness of the mass term, we have

$$\mu(\beta) \simeq c_0(\beta - \beta_c), \tag{63.15}$$

which then gives

$$\langle s \rangle \sim \sqrt{\beta - \beta_c} \tag{63.16}$$

for $\beta > \beta_c$ (nonanalytic behavior).

In general, the magnetization $\langle s \rangle = \langle s \rangle (h, \beta)$, and is defined by

$$\langle s \rangle = \frac{1}{V} \sum_n \langle s_n \rangle = \frac{1}{V} \sum_n \int \prod_p ds_p e^{-\beta H} s_n. \tag{63.17}$$

Then the *magnetic susceptibility* χ is

$$\chi = \frac{\partial \langle s \rangle}{\partial h} = \frac{1}{V} \sum_n \int \prod_p ds_p e^{-\beta H} \beta s_n \sum_m s_m, \tag{63.18}$$

and can be rewritten as

$$\chi = \frac{1}{V} \sum_{n,m} \langle (s_n - \langle s \rangle)(s_m - \langle s \rangle) \rangle, \tag{63.19}$$

so it is given in terms of a two-point function. The two-point function at large distances behaves as

$$\langle (s_n - \langle s \rangle)(s_m - \langle s \rangle) \rangle \sim e^{-\frac{|x_n - x_m|}{\xi(h,\beta)}} \tag{63.20}$$

for $\xi(h, \beta) \ll |x_n - x_m|$, where ξ is called the *correlation length*, whereas at small distances (but still much larger than the lattice size) it goes as

$$\langle (s_n - \langle s \rangle)(s_m - \langle s \rangle) \rangle \sim |x_n - x_m|^{-(d-2+\eta)}, \tag{63.21}$$

for $a \ll |x_n - x_m| \ll \xi(h, \beta)$. Here η is called the anomalous dimension, for the same reason as in a general quantum field theory.

We know that the susceptibility blows up at a phase transition, and we see that this is due to the two-point function diverging.

The *scaling hypothesis* for phase transitions is that all singular behavior near the phase transition is due to the divergence of ξ. That is, $\xi \to \infty$ at the phase transition point implies all the divergences in physical (measurable) quantities. Since $\xi \to \infty$, and this is the only relevant scale for the phase transition, it means that there are no objects with dimension at the phase transition, and thus the theory is scale invariant (fixed under a scaling transformation), and thus all diverging quantities diverge as power laws, not as exponentials (which would require the existence of a scale).

In the quadratic approximation around the minimum of the spin Hamiltonian, we can calculate (though it is also clear by dimensional analysis, since μ is the only parameter with dimension, specifically dimension two) that

$$\xi(h, \beta) \sim \frac{1}{\sqrt{|\mu(h, \beta)|}}, \tag{63.22}$$

so indeed ξ diverges at the critical point.

One can define critical exponents, for instance for the susceptibility and the correlation length:

$$\chi(\beta) \sim |\beta - \beta_c|^{-\gamma},$$
$$\xi(\beta) \sim |\beta - \beta_c|^{-\nu}. \tag{63.23}$$

These critical exponents can only take a few values, that is there is a certain *universality* for them, since a large class is independent of microscopics.

63.2 Kadanoff Blocking

From now on, since we have formulated the discretized field theory like a spin system, we will talk about both on the same footing, and consider the Hamiltonian

$$H = \sum_n \left[\sum_\mu \frac{1}{2}(\phi_{n+\mu} - \phi_n)^2 + \mu\phi_n^2 + \lambda\phi_n^4 \right]. \tag{63.24}$$

We define Kadanoff blocking as follows. Divide the lattice into blocks of size s^d, where $s \in \mathbb{N}$, and average over the blocks

$$\phi'_{n'} = s^{-d} \sum_{n \in B_s(n')} \phi_n, \tag{63.25}$$

after which we rescale to the original size, by

$$x_s \equiv \frac{x}{s}. \tag{63.26}$$

If we measure ξ in lattice units, it has decreased by s:

$$\xi_s = \frac{\xi}{s}. \tag{63.27}$$

Since the two-point correlation function decays as a power law for $1 \ll n \ll \xi$:

$$\langle \phi_n \phi_0 \rangle \sim \frac{1}{n^{d-2+\eta}}, \tag{63.28}$$

this means that we need to rescale $\phi_s \equiv s^\alpha \phi$, where

$$\alpha = \frac{d - 2 + \eta}{2}. \tag{63.29}$$

This is a sort of *wavefunction renormalization* in the quantum field theory sense.

The new Hamiltonian $H'(\phi'_n)$ is found by averaging over the blocks:

$$e^{-\beta H'[\phi'_{n'}]} = \int \prod_n d\phi_n e^{-\beta H[\phi]} \prod_{n'} \delta \left(\phi'_n - s^{-d} \sum_{n \in B_s(n')} \phi_n \right). \qquad (63.30)$$

After blocking, we define the rescaled Hamiltonian by

$$H_s(\phi_s(x_s)) \equiv H'(\phi'(x')), \qquad (63.31)$$

but this procedure of blocking will generate new terms in the Hamiltonian, for instance

$$(\phi_{n+\mu} - \phi_n)^2 \phi_n^2, \quad (\phi_{n+\mu} - 2\phi_n + \phi_{n-\mu})^2, \qquad (63.32)$$

and others. This is intuitively clear from the Wilsonian picture of effective field theory from Chapter 62, which amounted also to blocking, though in momentum space (integrating from a physical momentum cut-off to a lower, variable, cut-off). Indeed, there we saw that the quantum averaging over a shell in momentum space naturally leads to all possible terms in the effective action (all terms allowed by symmetries), with coefficients given by the quantum averaging. It is left as an exercise to show this statement in a more concrete way in our case.

Then, instead of starting with a specific Hamiltonian in the UV, at the physical cut-off scale given by a, and integrating over the blocks to get new terms in the Hamiltonian, we can, like in the Wilsonian effective field theory approach from Chapter 62, start instead from an effective field theory of general type, parametrizing our ignorance about the UV.

That is, we start with the Hamiltonian

$$H(\phi) = \sum_\alpha K_\alpha S_\alpha(\phi), \qquad (63.33)$$

where $S_\alpha(\phi)$ are all possible terms in the Hamiltonian, and K_α are couplings. Then, blocking amounts to just a transformation on the space of coefficients $\{K_\alpha\}$, from the original (few) to the final (more).

This is in fact identified with the *renormalization group* (RG) transformation, since that was also a scaling transformation, just in momentum space. Formally, the transformation $T_{RG}(s)$ acts as

$$T_{RG}(s) : \{K_\alpha\} \rightarrow \{(T_{RG}(s)K)_\alpha\}. \qquad (63.34)$$

Then a fixed point of the renormalization group is one that doesn't change the couplings, so

$$(T_{RG}(s)K)_\alpha^* = K_\alpha^* \qquad (63.35)$$

for any α. We thus understand the renormalization group as the coarse graining procedure (i.e. blocking).

We define the *critical surface* as the set of all points attracted towards the fixed point by the RG transformation (i.e. the basin of attraction of the fixed point under RG):

$$(T_{RG}^n(s)K)_\alpha \rightarrow K_\alpha^* \qquad (63.36)$$

for $n \rightarrow \infty$.

A fixed point of the RG group is then identified with the critical point (i.e. the phase transition point), in critical phenomena, since as we saw that is a scale-invariant point. Then $\xi \to \infty$ there, which means that $\xi \to \infty$ on the critical surface (since as we go towards the critical point, we decrease ξ).

Then it means that the couplings of various materials $K_\alpha(\beta)$ belonging to the same critical surface ($\xi \to \infty$ at $\beta \to \beta_c$) are different, yet they lead to the same long-distance physics, defined by the critical point (since the blocking means going to larger distances, and so the fixed point is the IR behavior of the theory). This is the universality that we mentioned before.

63.3 Expansion Near a Critical Point

Expanding the couplings near the critical point K_α^*:

$$K_\alpha = K_\alpha^* + \delta K_\alpha, \tag{63.37}$$

the RG transformation is

$$(T_{RG}(s)K)_\alpha = K_\alpha^* + \sum_{\alpha'} T_{\alpha\alpha'}\delta K_{\alpha'} + \ldots \tag{63.38}$$

Then, considering a basis $v_{a\alpha}$ of eigenfunctions of $T_{\alpha\alpha'}$ with eigenvalue λ_a:

$$\sum_{\alpha'} T_{\alpha\alpha'}v_{a\alpha'} = \lambda_a v_{a\alpha}, \tag{63.39}$$

we can expand the coupling variations in it as

$$\delta K_\alpha = \sum_a h_a v_{a\alpha}. \tag{63.40}$$

The Hamiltonian is then written in this basis as (remember that $H = \sum_\alpha K_\alpha S_\alpha$)

$$H = H^* + \sum_a h_a v^a, \tag{63.41}$$

where $H^* = \sum_\alpha K_\alpha^* S_\alpha$ and

$$v^a = \sum_\alpha v_{a\alpha} S_\alpha. \tag{63.42}$$

The RG transformation then acts on the eigenfunctions as $v_{a\alpha} \to \lambda_a v_{a\alpha}$, thus $v^a \to \lambda_a v^a$, so after n steps

$$H \to H^* + \sum_a (\lambda_a)^n h_a v^a. \tag{63.43}$$

Therefore we can distinguish between:

- $\lambda_a < 1$, which multiplies an *irrelevant operator* v^a, since the term is suppressed after a few steps of the RG transformation. Note that the RG transformation along the irrelevant operator then takes us towards the fixed point ($H \to H^*$).

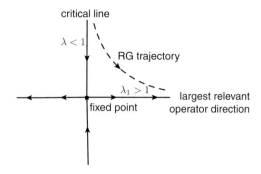

Figure 63.1 RG trajectory near fixed point.

- $\lambda_a > 1$, which multiplies a *relevant operator* v^a, as the RG transformation amplifies the effect of the operator. The RG transformation along the relevant operator takes us away from the fixed point.
- $\lambda = 1$ multiplies a *marginal operator*. This means that we need to consider higher orders in the perturbation (beyond the linear analysis) to see whether it is actually relevant or irrelevant. If it remains marginal to all orders, we say it is *exactly marginal*, and the RG transformation has no effect on it.

We then see that the critical surface is spanned by irrelevant operators (i.e. irrelevant deformations of the Hamiltonian).

In contrast, the relevant deformations of the Hamiltonian take us away from the fixed point: after a few steps, the RG trajectory is dominated by them, more precisely by the relevant deformation with the largest λ_a (the largest relevant deformation), that we will call λ_{a_1}, as we can see from Figure 63.1.

63.4 Critical Exponents (Near the Fixed Point)

We isolate this largest deformation with λ_1, writing

$$(T_{RG}(s)K)_\alpha = K_\alpha^* + \lambda_{a_1} h_{a_1}(\beta) v_{a_1\alpha} + \sum_{i\geq2} \lambda_{a_i} h_{a_i}(\beta) v_{a_i\alpha} + \mathcal{O}(h^2). \tag{63.44}$$

We also assume that $h_{a_1}(\beta)$ is smooth, and thus can be expanded near the phase transition as

$$h_{a_1}(\beta) = (\beta - \beta_c)h_{a_1}^c + \mathcal{O}((\beta - \beta_c)^2). \tag{63.45}$$

Then, after a few steps in the RG transformation, we have

$$(T_{RG}^n(s)K)_\alpha \simeq K_\alpha^* + \lambda_{a_1}^n h_{a_1}(\beta) v_{a_1\alpha}. \tag{63.46}$$

Since $\lambda_{a_1} > 1$, we can define a $\nu > 0$ by the equality

$$\lambda_{a_1} \equiv s^{1/\nu}, \tag{63.47}$$

where s is the blocking size, as before.

We can moreover define a sequence $\beta_n \to \beta_c$ of βs by the relation

$$h_{a_1}(\beta_n)\lambda_{a_1}^n = h_{a_1}(\beta_n)s^{n/\nu} = 1, \tag{63.48}$$

which implies

$$s^n \simeq \frac{1}{|(\beta_n - \beta_c)h_{a_1}^c|^\nu}. \tag{63.49}$$

It follows that

$$(T_{RG}^n(s)K)_\alpha \to K_\alpha^* + v_{a_1\alpha} \tag{63.50}$$

as $n \to \infty$, so we stay fixed away from the fixed point.

In contrast, the two-point correlation function goes like

$$\langle\phi_n\phi_0\rangle \sim \frac{1}{n^{2\alpha}}, \tag{63.51}$$

which means that under blocking it transforms as

$$G(|x|, \{K_\alpha\}) = s^{-2\alpha}G(|x|/s; \{(T_{RG}(s)K)_\alpha\}) = \ldots = s^{-2n\alpha}G(|x|/s^n; \{(T_{RG}^n(s)K)_\alpha\}), \tag{63.52}$$

but since we stay fixed away from the fixed point as $n \to \infty$, there is no nontrivial dependence on the couplings under scaling. Therefore, at $n \to \infty$ and for large x, we have

$$G = G(|x|/s^n), \tag{63.53}$$

but also in general we have

$$G \sim e^{-\frac{|x|}{\xi}}. \tag{63.54}$$

By comparing the two, we see that we need to have

$$\xi(\beta_n) \propto s^n, \tag{63.55}$$

that is

$$\xi(\beta) \propto \frac{1}{|\beta - \beta_c|^\nu}. \tag{63.56}$$

This also means that we have

$$G(|x|, \{K_\alpha(\beta)\}) = \ldots = \xi^{-2\alpha}G(|x|/\xi(\beta), \{K_\alpha^* + v_{a_1\alpha}\}), \tag{63.57}$$

where $2\alpha = d - 2 + \eta$, so by Fourier transforming to momentum space we get

$$G(p, \{K_\alpha\}) = \xi^{2-\eta}G(\xi(p)b, \{K_\alpha^* + v_{a_1\alpha}\}). \tag{63.58}$$

The momentum-space formulation is understood as the Wilsonian effective action formulation from the previous chapter.

Important Concepts to Remember

- A discretized scalar field theory can be written as a spin system.
- The two-point function at large distances decays exponentially with a correlation length and at intermediate distances it decays as a power law with an anomalous dimension.
- The scaling hypothesis states that all divergent behavior in physical quantities near a critical point (phase transition) is due to the diverging correlation length.
- The magnetic susceptibility and the correlation length have (semi-)universal critical exponents.
- Kadanoff blocking is averaging over blocks of size s in each direction. It leads to new terms in the Hamiltonian.
- Kadanoff blocking can be thought of as an RG transformation on the couplings K_α of a Hamiltonian with an infinite set of terms.
- The fixed point of the RG corresponds to the critical point of a system (at the phase transition), and the critical surface is the basin of attraction of the fixed point: various materials with various K_α all have the same long-distance physics.
- Irrelevant operators take us towards the fixed point, and they are suppressed after a few steps, while relevant operators take us away from the fixed point, and after a few steps the largest irrelevant operator (of largest eigenvalue) dominates.

Further Reading

See sections 9.2 and 9.3 in [3].

Exercises

1. Consider the discrete model with Hamiltonian

$$H = \sum_{j=1}^{J} \frac{a_j^\dagger a_j + a_j a_j^\dagger}{2} + \frac{2\lambda}{(2\pi)^2} \sum_{j=1}^{J} (\phi_j - \phi_{j+1})^2. \tag{63.59}$$

Write down the continuum version for it, find H and then the relativistic Lagrangian, and calculate the length L of the continuum system.

2. Show that by Kadanoff blocking on

$$H = \sum_n \left[\sum_\mu \frac{1}{2}(\phi_{n+\mu} - \phi_n)^2 + \mu\phi_n^2 + \lambda\phi_n^4 \right], \tag{63.60}$$

we generate terms like

$$(\phi_{n+\mu} - \phi_n)^2 \phi_n^2, \quad (\phi_{n+\mu} - 2\phi_n + \phi_{n-\mu})^2 \tag{63.61}$$

and others.

3. Consider a conserved global symmetry current $J_\mu^a(x)$ in four dimensions. Is it a relevant, irrelevant, or marginal operator?
4. If we apply the Kadanoff blocking procedure to a free theory, what happens to the Hamiltonian?

64

Lattice Field Theory

In this chapter we will see that we can define a discretization of the field theory in a consistent way, such that we recover the continuum theory in a certain limit.

It is important to put field theories on the lattice, since then we can calculate nonperturbative quantities from computer simulations (Monte Carlo calculations for the path integral).

We will first consider the continuum limit, related to the Kadanoff blocking of Chapter 63, and define the beta function in the blocking picture. Then we will move on to lattice gauge theory and its continuum limit.

64.1 Continuum Limit

An important result is that the critical point is independent of the particular RG procedure considered (the representation of the RG group; before, Kadanoff blocking).

We had described everything in terms of lattice units, but in order to define physical quantities we need to multiply by the lattice spacing a:

$$\xi_{ph} = \xi_l a; \quad x_{ph} = x_l a; \quad p_{ph} = \frac{p_l}{a}, \tag{64.1}$$

where ξ_l, x_l, p_l are in lattice units.

We need to keep ξ_{ph} fixed, but we now change the lattice spacing a by $a \to a/s$ under the RG action, thus taking $a \to 0$ in the limit instead of $\beta \to \beta_c$ $(\beta_n \to \beta_{n+1}, \ldots)$. A physical mass will be $m_{ph} = 1/\xi_{ph}$, and will be fixed as $a \to 0$.

To define continuum correlation functions, we need to make a wavefunction renormalization according to

$$G_{\mathrm{cont}}(p_{ph}, m_{ph}) = a^{-\eta} G(p_l; \{K_\alpha(a)\})|_{a \to 0}, \tag{64.2}$$

where η is the anomalous dimension, and it appears since $G \sim 1/|x|^{\Delta_{\mathrm{class}}+\eta}$ and the extra dependence must be compensated.

64.1.1 Gaussian Fixed Point

Under the RG transformation $x \to x/s$, since the engineering (classical) dimension of a scalar field is $(d-2)/2$, we have

$$\phi_s(x/s) \simeq s^{\frac{d-2}{2}} \phi(x). \tag{64.3}$$

Then, for a coupling

$$K_n \int d^d x \phi^n(x), \tag{64.4}$$

the coupling K_n transforms approximately as

$$K_n \to s^{-d_n} K_n, \tag{64.5}$$

where

$$d_n = [K_n] = -\frac{d-2}{2}n + d \tag{64.6}$$

is the engineering (classical) dimension of K_n.

That in turn means that the eigenvalue of the RG operator $T_{RG}(s)$ is

$$\lambda_n = s^{-d_n}. \tag{64.7}$$

We deduce then that near the Gaussian fixed point $\phi = 0$:

- Relevant operators, with $\lambda_n > 1$, have $d_n = [K_n] < 0$, so make the theory super-renormalizable.
- Marginal operators, with $\lambda_n = 1$, have $d_n = [K_n] = 0$, so make the theory renormalizable.
- Irrelevant operators, with $\lambda_n < 1$, have $d_n = [K_n] > 0$, so make the theory non-renormalizable.

In particular, $\lambda \int \phi^4$ has $[\lambda] = 0$, so is marginal, which means that we need to go to higher orders to see whether it is relevant, irrelevant, or exactly marginal.

Then relevant (super-renormalizable) operators are found to be UV asymptotically free, and irrelevant (nonrenormalizable) operators are IR asymptotically free. For the latter, at the Gaussian fixed point, when we take the bare coupling to zero in the IR, the renormalized coupling goes to zero even faster, so it seems impossible to define a nontrivial theory (with nonzero physical coupling on large scales). $\lambda \phi^4$ and QED are of this type, which has generated a debate on whether these theories make sense nonperturbatively (we can define the theory order by order, but what it means nonperturbatively is not clear). One possibility is that there are *other* fixed points, where one could define the theory.

For UV asymptotically free theories, however, there is no problem, since we can keep fixed the renormalized coupling at the fixed point (in the IR), and obtain a nontrivial interacting theory.

64.2 Beta Function

The general RG procedure allows us to define the beta function in an alternative way. By rescaling the correlation length by s:

$$\xi(g'^2) = s\xi(g^2), \tag{64.8}$$

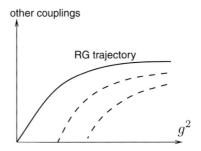

Figure 64.1 Kadanoff blocking near the RG trajectory.

the couplings are changed in general, so

$$g'^2 = g^2 - \Delta g^2. \tag{64.9}$$

But we want the long-distance physics to be unmodified by this rescaling procedure, so we need to rescale a also:

$$a(g'^2) = \frac{1}{s} a(g^2), \tag{64.10}$$

which defines $a(g^2)$ or reversely $g(a)$.

Then, after $n \to \infty$ blocking steps, we should be on the same RG trajectory, since that defines the long-distance physics, so the different blockings create various dotted lines that converge to the RG line, see Figure 64.1.

This procedure allows us to define an alternative way to define the beta function, by

$$\beta(g) = -a \frac{d}{da} g(a). \tag{64.11}$$

Note that this definition is not identical to the usual one, however one can prove that the first two coefficients in the expansion of $\beta(g)$ are the same, and of course the fixed points are the same (since the asymptotic RG trajectory is the same).

64.3 Lattice Gauge Theory

We finally come to the issue of interest, namely how to put a gauge theory on the lattice. This is of interest since QCD is a nonperturbative gauge theory in the IR, so it is hard to calculate anything at low energies, other than on the lattice.

Consider a gauge group G, so because of the local gauge invariance with G_x, we can say that we have a total group

$$G_{\text{total}} = \prod_{x \in \mathbb{R}^d} G_x. \tag{64.12}$$

We saw that we can define an observable called the Wilson loop:

$$\tilde{U}_C = \operatorname{Tr} P \exp\left[i \oint_C A_\mu dx^\mu \right], \tag{64.13}$$

where $A_\mu = A_\mu^a T_a$ and the contour C is parametrized by $t \in [0,1]$ as $C(t) : t \to x_\mu(t)$. It is the trace of a Wilson line U_C for a closed contour C, where U_C satisfies composition:

$$U_C = U_{C_n} U_{C_{n-1}} \ldots U_{C_1}, \tag{64.14}$$

and where

$$U_C(t) = \lim_{n\to\infty} \prod_{k=0}^{n} e^{idx_k^\mu A_\mu(x_k)}, \tag{64.15}$$

and $x_k \in dx_k$. We also saw that the Wilson line U_C transforms covariantly, but with different endpoints on the left and right, that is

$$U_C(t) \to V(x(t))U_C(t)V(x(0)), \tag{64.16}$$

for

$$A_\mu \to V(x)A_\mu(x)V(x)^{-1} - i\partial_\mu V(x)V^{-1}. \tag{64.17}$$

In the abelian case, we saw that we can use Stokes's theorem to write

$$e^{i \oint_C A_\mu dx^\mu} = e^{\frac{i}{2}F_{\mu\nu}da_{\mu\nu}}, \tag{64.18}$$

where $da_{\mu\nu}$ is the (infinitesimal) surface bounded by the (infinitesimal) contour C.

In the nonabelian case, we have corrections in the exponent:

$$P e^{i \oint_C A_\mu dx^\mu} = e^{\frac{i}{2}F_{\mu\nu}da_{\mu\nu}+\mathcal{O}(da^2)}. \tag{64.19}$$

By taking the trace and then the real part, and expanding the exponential, we obtain

$$\operatorname{Re} \operatorname{Tr} P e^{i \oint_C A_\mu dx^\mu} = \operatorname{Tr} I - \frac{1}{8} \operatorname{Tr}(F_{\mu\nu}da_{\mu\nu})^2 + \mathcal{O}(da_{\mu\nu}^3). \tag{64.20}$$

We are finally ready to define the gauge theory on the lattice. On a lattice, we have sites i and links (ij) connecting them, on which naturally one can define an orientation. Then it is natural to associate an element of G to each *oriented* link, $U_{ij} \in G$, which then has the properties of a Wilson line. Indeed, reversing the orientation and composing the two elements we should get back to the identity, so we should have

$$U_{ji} = U_{ij}^{-1}. \tag{64.21}$$

Moreover, the gauge transformation of U_{ij} is the same as for the Wilson line, namely

$$U_{ij} \to U_{ij}^{(V)} = V_i U_{ij} V_j^{-1}. \tag{64.22}$$

It is natural then to define along a connected path C on the links the analogue of a long Wilson line by composition:

$$U_C = U_{(i_n i_{n-1})} U_{(i_{n-1} i_{n-2})} \ldots U_{(i_3 i_2)} U_{(i_2 i_1)}, \tag{64.23}$$

and then again $\tilde{U}_C = \operatorname{Tr} U_C$ for a closed path is gauge invariant.

Note that now the total gauge group is

$$G_{\text{total}} = \prod_{i \in a\mathbf{Z}^d} G_i. \tag{64.24}$$

We associate the link variable U_{ij} with the elementary Wilson line (for the link), which can now be written as an object on the link:

$$U_{ij} \leftrightarrow Pe^{i \int_0^1 dx_\mu(t) A_\mu(x(t)) + \mathcal{O}(a^2)}, \tag{64.25}$$

where the path $x_\mu(t)$ has to be between $x_\mu(i)$ and $x_\mu(j)$, so

$$x_\mu(t) = tx_\mu(j) + (1-t)x_\mu(i), \tag{64.26}$$

and the point at which we define the object on the link is the midpoint:

$$x_\mu = \frac{x_\mu(i) + x_\mu(j)}{2}. \tag{64.27}$$

The smallest nontrivial closed loop is called a "plaquette," and is a square with nearest-neighbor vertices, $p = (ijkl)$, with area a^2. Then, from (64.20), we have for the plaquette p loop

$$\text{Tr } I - \text{Re Tr } U_p = \frac{1}{2}a^4 \text{ Tr } F^2_{\mu(p)\nu(p)} + \mathcal{O}(a^6), \tag{64.28}$$

where the plaquette square is in the directions $\mu(p)$ and $\nu(p)$.

We can define from the above the Yang–Mills action for the plaquette ($\text{Tr } F^2_{\mu\nu}/4$), and then sum over plaquettes, to obtain the *Wilson action* for the gauge theory on the lattice:

$$S_W[U] = \beta_G \sum_p \left(1 - \frac{1}{N_G} \text{Re Tr } U_p\right) \tag{64.29}$$

(note that we have absorbed a factor of $N_G = \text{Tr } I$ in the definition of the coupling β_G), which goes over in the continuum limit to

$$\frac{1}{g_0^2} \int d^d x \frac{1}{4} \text{ Tr } F^2_{\mu\nu}, \tag{64.30}$$

up to terms of order a^2 that vanish, provided we identify

$$\beta_G a^{4-d} = \frac{2N_G}{g_0^2}. \tag{64.31}$$

Here g_0 is the bare coupling constant in YM, and we have used $\int d^d x \sum_{\mu\nu} = a^d \sum_p$.

Note that the Wilson action is by no means unique. There are various actions that give rise to the same continuum limit, but the Wilson action is the simplest, so it is the one used by (almost) everyone.

Now that we have defined the lattice action, to formulate the gauge theory on the lattice we only lack a definition of the path-integral measure.

The measure for integration is the unique measure dU for integration over the group that is invariant under both left and right group multiplication (i.e. $U \to UU_0$ and $U \to U_0U$), called the *Haar measure*. It also has the property that if U is close to the identity (i.e. $U = \exp(iaA)$), then

$$dU = \prod_{i=1}^{N_G} dA_i^a (1 + \mathcal{O}(a)). \qquad (64.32)$$

Therefore the partition function on the lattice is

$$Z[\beta_G] = \int \prod_{l \in (ij) \text{ links}} dU_l e^{-S_W[U]}. \qquad (64.33)$$

64.4 Lattice Gauge Theory: Continuum Limit

The naive continuum limit of the lattice YM action would be $\beta_G \to \infty$ (or equivalently $g_0 \to 0$), since β_G is the coupling of the lattice action, and by taking it to infinity we have a small fluctuation around $\text{Tr}\, U_p = 1$.

But rather we need to do it as at the beginning of the chapter, by taking $a \to 0$ while fixing some physical length ξ. Then we have an RGE-like equation:

$$a\frac{d\xi}{da} \equiv \left(a\frac{\partial}{\partial a} - \beta(g_0)\frac{\partial}{\partial g_0} \right) \xi(a, g_0) = 0. \qquad (64.34)$$

Here, as before:

$$\beta(g_0) \equiv -a\frac{\partial}{\partial a} g_0. \qquad (64.35)$$

The solution of the RGE-like equation above is

$$\xi(a, g_0) = a \exp \int_0^{g_0} \frac{dg}{\beta(g)}, \qquad (64.36)$$

as we can easily check. This means that a divergent correlation length (and thus a fixed point) corresponds to $\beta = 0$, as it should.

In a perturbative expansion:

$$\beta(g) = -\beta_0 g^3 - \beta_1 g^5 + \mathcal{O}(g^7), \qquad (64.37)$$

and for $G = SU(N)$ we have

$$\beta_0 = \frac{11}{3}\frac{N}{16\pi^2}; \quad \beta_1 = \frac{34}{3}\frac{N^2}{16\pi^2}, \qquad (64.38)$$

leading to the solution for ξ of

$$\xi = a \cdot (\beta_0 g_0^2)^{\beta_1/2\beta_0} \exp\left(\frac{1}{2\beta_0 g_0^2} \right). \qquad (64.39)$$

Then, for fixed ξ as $a \to 0$, this defines $a(\beta_G)$, since then $g_0 \to 0 \Rightarrow \beta_G \to \infty$, which implies that we find a relation $a(g_0)$, and thus $a(\beta_G)$.

For a physical mass $m(\beta_G)$ that implies a continuum limit mass scale (for something like a glueball), $m(\beta_G)\xi(\beta_G)$ is finite as $\beta_G \to \infty$ if $a = 1$ (in lattice units) (since then the expression for $\xi(a, g_0)$ reduces to $\xi(\beta_G)$).

64.5 Adding Matter

Up to now we have seen how to describe pure YM. But for physical applications, we want to also have matter in the theory. For instance, for QCD we would need to have quarks. But unfortunately, it is difficult to put chiral fermions on the lattice while keeping the chiral symmetry. There is no perfect way to deal with chiral fermions.

So instead we will show how to put scalars on the lattice, having in mind the application to the electroweak theory, so the scalars represent the Higgs. Then the scalars $\phi(x)$ are in the fundamental representation of the gauge group.

For a scalar $\phi(x)$ defined at sites i, we now have the variables ϕ_i. The gauge invariance of the scalars is

$$\phi(x) \to V(x)\phi(x), \tag{64.40}$$

so to obtain a gauge-invariant observable we must form the composite object from two scalars and a Wilson line:

$$\phi^\dagger(y)P\, e^{i \int_{C:x \to y} A_\mu dx^\mu} \phi(x). \tag{64.41}$$

This object is easily discretized to the lattice object

$$\phi_i^\dagger U_{ij}\phi_j. \tag{64.42}$$

Next, we want to write an action, so we need to write a discrete version of the derivative, easily seen to be

$$\partial_\mu \phi(x) \to \phi_{j+\mu} - \phi_j. \tag{64.43}$$

But we actually need the covariant derivative $D_\mu \phi$, and since $U = e^{iaA} \simeq 1 + iaA$, we have

$$D_\mu \phi(x) \to \phi_{j+\mu} - U_{j+\mu,j}\phi_j. \tag{64.44}$$

Then the kinetic term of the scalar is

$$|D_\mu \phi|^2 = \sum_{j,\mu}(\phi_{j+\mu} - U_{j+\mu,j}\phi_j)^2 = 2\sum_j \phi_j^2 - 2\sum_{(ij)\text{ links}} \phi_i U_{ij}\phi_j, \tag{64.45}$$

where (ij) are links (between nearest neighbors).

In the gauge-Higgs action we need to add a mass term and a ϕ^4 term with independent coefficients besides the Wilson action for the gauge fields, for a total action

$$S[\phi, U] = -\kappa \sum_{(ij)} \mathrm{Re}\,\phi_i^\dagger U_{ij}\phi_j + \mu \sum_i \phi_i^\dagger \phi_i + \lambda \sum_i (\phi_i^\dagger \phi_i)^2 + \beta_G \sum_p \mathrm{Re}\,\mathrm{Tr}\,U_p. \tag{64.46}$$

Important Concepts to Remember

- For the continuum limit, we take $a \to 0$, while keeping physical scales like $\xi_{ph} = \xi_l a$ fixed.
- At the Gaussian fixed point, relevant interactions are super-renormalizable, marginal are renormalizable, and irrelevant are nonrenormalizable.
- Irrelevant operators at the Gaussian fixed point could lead to a trivial theory (free on physical scales): QED and ϕ^4 are in this class.
- We can vary $\xi(g)$ by $\xi(g'^2) = s\xi(g^2)$, leading to $a(g^2)$ for an invariant long-distance physics, for $a(g'^2) = a(g^2)/s$, leading to a new definition of the beta function, as $-a\, dg(a)/da$.
- In lattice gauge theory, the variables are the links $U_{ij} \in G$, acting as infinitesimal Wilson lines. The Wilson action for lattice gauge theory is written in terms of these.
- The measure on the discrete path integral is the Haar measure dU, invariant under left and right multiplications, $U \to UU_0$ and $U \to U_0 U$.
- In the continuum limit, we keep physical lengths fixed, obtaining RGE-like equations from $a\,\, d\xi/da = 0$.
- Fundamental scalars ϕ can be added, with covariant derivative $D_\mu \phi \to \phi_{j+\mu} - U_{j+\mu,j}\phi_j$.

Further Reading

See section 9.5 in [3].

Exercises

1. Calculate the number of sites, links, and plaquettes for a symmetric hypercubic lattice with periodic boundary conditions.
2. Derive the lattice version of the Yang–Mills equations (equations of motion of Yang–Mills, or nonabelian Maxwell's equations).
3. Prove that (64.39) is a solution of the RG equations for ξ, with beta function in (64.37), and extend it to find a solution for the beta function written including a g^7 term.
4. Write the lattice version of the equations of motion for the Yang–Mills–Higgs action in (64.46).

65

The Higgs mechanism

The subject of this chapter is spontaneous symmetry breaking, which is the situation when a theory is invariant under some local (gauge) symmetry, but the vacuum breaks it (i.e. the vacuum is not invariant under the symmetry). In this case, gauge fields become massive by the Higgs mechanism, and they "eat" a scalar degree of freedom along the symmetry direction in order to become massive. This works out, because a massive vector in four dimensions has three degrees of freedom, while a gauge field (massless vector) has two, the difference being supplied by the eaten scalar.

65.1 Abelian Case

We start with the simplest version of the mechanism, in the abelian case, that will be called "Abelian-Higgs," even though this case is mostly relevant for superconductivity, and not for particle physics.

The Lagrangian for a complex scalar coupled to a gauge field is

$$\mathcal{L} = -\frac{1}{4}F_{\mu\nu}^2 - |D_\mu \phi|^2 - V(|\phi|), \tag{65.1}$$

where $D_\mu = \partial_\mu - ieA_\mu$. The gauge invariance is

$$\phi(x) \to e^{i\alpha(x)}\phi(x),$$
$$A_\mu(x) \to A_\mu(x) + \frac{1}{e}\partial_\mu \alpha(x). \tag{65.2}$$

We choose the most general analytic, gauge-invariant, renormalizable potential with a symmetry-breaking term

$$V = -\mu^2 \phi^* \phi + \frac{\lambda}{2}(\phi^* \phi)^2, \tag{65.3}$$

or rather, by adding a constant (that is irrelevant as long as the model is not coupled to gravity)

$$V = \frac{\lambda}{2}\left(|\phi|^2 - \frac{\mu^2}{\lambda}\right)^2. \tag{65.4}$$

Indeed, a renormalizable potential must have powers less than or equal to 4, and a gauge-invariant and analytic potential implies only second and fourth powers. We must have

$\mu^2 = -m^2 > 0$ for symmetry breaking to occur. Then there is a VEV for ϕ and the $U(1)$ is spontaneously broken, giving a vacuum

$$\langle \phi \rangle = \phi_0 \equiv \sqrt{\frac{\mu^2}{\lambda}}. \tag{65.5}$$

The various vacua are $\phi = \phi_0 e^{i\theta_0}$, with the arbitrary phase θ_0 parametrizing the vacuum. Without loss of generality, we can set the particular vacuum we are in to be real as above, by a gauge transformation. The potential is called the "Mexican hat potential" due to its shape. Spontaneous symmetry breaking means that, starting at $\phi = 0$, a small fluctuation will lead us to a random direction, choosing some particular vacuum among the continuum of vacua, and thus breaking the invariance.

Expanding around this vacuum, we obtain

$$\phi(x) = \phi_0 + \frac{\phi_1(x) + i\phi_2(x)}{\sqrt{2}}, \tag{65.6}$$

and the potential is

$$V(\phi) = \frac{1}{2} 2\mu^2 \phi_1^2 + \mathcal{O}(\phi_1^3). \tag{65.7}$$

This means that ϕ_1 has mass $m = \sqrt{2}\mu$ and ϕ_2 is massless, and is the so-called "Goldstone boson." It is a theorem that will be proved in Chapter 66 that for every symmetry that is spontaneously broken, there is an associated massless scalar called a Goldstone boson.

Expanding the scalar kinetic term, we get

$$|D_\mu \phi|^2 = \frac{1}{2}(\partial_\mu \phi_1)^2 + \frac{1}{2}(\partial_\mu \phi_2)^2 + \frac{e^2 \mu^2}{\lambda} A_\mu A^\mu - \sqrt{2} e \phi_0 A_\mu \partial^\mu \phi_2 + \dots \tag{65.8}$$

The third term in this expansion gives a mass term for the vector, with

$$m_A^2 = 2e^2 \phi_0^2, \tag{65.9}$$

while the last term gives a mixing between the vector A_μ and the Goldstone boson ϕ_2. We can of course redefine the fields so as to get rid of this term, and in the process get rid of ϕ_2, and we will do so shortly, but before that let's keep it and see that the resulting theory after the expansion is still fine quantum mechanically.

Consider the mixing term between the scalar of momentum k and a gauge field with index μ. The Feynman rule for it is

$$i\sqrt{2} e\phi_0 (-ik^\mu) = m_A k^\mu. \tag{65.10}$$

Considering now the modification to the photon propagator, it comes from the vector mass term, and the diagram with a mixing to scalar, scalar propagation, followed by scalar mixing again. The sum of these two Feynman diagrams gives then

$$- im_A^2 g_{\mu\nu} + (m_A k^\mu) \frac{-i}{k^2} (-m_A k^\nu) = -im_A^2 \left(g_{\mu\nu} - \frac{k_\mu k_\nu}{k^2} \right), \tag{65.11}$$

which is transverse, as it should be. See Figure 65.1.

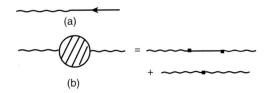

Figure 65.1 (a) Mixing term between the scalar and a gauge field. (b) The sum of the two diagrams with mixing gives the 1PI two-point function, which is correctly transverse.

65.2 Abelian Case: Unitary Gauge

The unitary, or physical gauge, is the gauge where we choose a local gauge transformation $\alpha(x)$ so as to get rid of the scalar degree of freedom in the $U(1)$ direction (i.e. we put $\phi_2 = 0$, ϕ real). Then the Lagrangian becomes

$$\mathcal{L} = -\frac{1}{4}F_{\mu\nu}^2 - (\partial_\mu\phi)^2 - e^2\phi^2 A_\mu A^\mu - V(\phi). \tag{65.12}$$

The Higgs mechanism means a mass for the gauge boson by "eating" the Goldstone boson. We can see this mechanism explicitly in the parametrization for the scalar

$$\phi = |\phi|e^{i\theta}. \tag{65.13}$$

Then the covariant derivative of the scalar is

$$D_\mu\phi = e^{i\theta}(\partial_\mu|\phi| + i\partial_\mu\theta|\phi| - ieA_\mu|\phi|). \tag{65.14}$$

By choosing the vacuum $\langle|\phi|\rangle = \phi_0 = \mu^2/\lambda$, $\theta_0 = 0$ and expanding around it as

$$|\phi| = \frac{\mu^2}{\lambda} + \frac{\delta|\phi|}{\sqrt{2}}, \tag{65.15}$$

the θ degree of freedom is removed by the redefinition

$$A_\mu \to A_\mu = A'_\mu + \frac{1}{e}\partial_\mu\theta, \tag{65.16}$$

which leaves the field strength invariant, $F'_{\mu\nu} = F_{\mu\nu}$. Then, after the redefinition, the scalar kinetic term is

$$|D_\mu\phi|^2 = (\partial_\mu|\phi|)^2 + e^2 A'_\mu A'^\mu|\phi|^2, \tag{65.17}$$

and the Lagrangian is

$$\mathcal{L} = -\frac{1}{4}F'_{\mu\nu}{}^2 - (\partial_\mu|\phi|)^2 - e^2 A'_\mu{}^2|\phi|^2 - V(|\phi|)$$
$$\simeq -\frac{1}{4}F'_{\mu\nu}{}^2 - \frac{(\partial_\mu\delta|\phi|)^2}{2} - \frac{e^2\mu^2}{\lambda}A'_\mu{}^2. \tag{65.18}$$

We see that we have removed the angle θ degree of freedom in $\phi = |\phi|e^{i\theta}$, which is the Goldstone boson (massless degree of freedom, since $V = V(|\phi|)$).

An important historical note is that the "Higgs mechanism" was discovered jointly by Higgs, Kibble, Guralnik, Hagen, Englert, and Brout. In fact, however, they explored the model, and generalized to nonabelian gauge theory, but before them the abelian model had been used in condensed matter to describe superconductivity, specifically the Meissner effect, understood as the phenomenon of the photon becoming massive due to spontaneous symmetry breaking, and thus penetrating only a distance $\mathcal{O}(1/m_{\text{photon}})$ inside the superconductor.

65.3 Abelian Case: Gauge Symmetry

What happened to the gauge symmetry that was present in the theory before the Higgsing? Before the Higgsing, the invariance was

$$\delta A_\mu = \partial_\mu \alpha,$$
$$\delta \phi = |\phi| e^{i\theta + ie\alpha} - |\phi| e^{i\theta} \Rightarrow \delta\theta = e\alpha, \tag{65.19}$$

and expanding around the vacuum does nothing to the symmetry, but the redefinition does. After the redefinition, we have

$$\delta A'_\mu = \delta \left(A_\mu - \frac{1}{e}\partial_\mu\theta \right) = 0,$$
$$\delta|\phi| = 0. \tag{65.20}$$

So the gauge invariance is simply lost (nothing happens under it), but only after the redefinition.

Note that, while it is also common to call ϕ "the Higgs," actually the real scalar $|\phi|$ (or ϕ_1) has mass $m = \sqrt{2}\mu$, and deserves the name "the Higgs," since it is the massive boson associated with the Higgs particle.

Note that in the scalar kinetic term expansion we have a term

$$e^2\sqrt{2}\phi_1\phi_0 A_\mu A^\mu = \sqrt{2}e^2\phi_0\phi_1 A'_\mu A'^\mu + \dots, \tag{65.21}$$

which leads to a vertex for 2 gluons to join and emit a Higgs, of form

$$\sqrt{2}e^2\phi_0\delta_{\mu\nu}. \tag{65.22}$$

65.4 Nonabelian Case

We consider now a nonabelian gauge group G, such that the gauge transformation is

$$\phi_i \to (e^{-i\alpha^a t_a})_{ij}\phi_j = (e^{\alpha^a T_a})_{ij}\phi_j, \tag{65.23}$$

where we have written the transformation with Hermitian and anti-Hermitian generators. The covariant derivative is

$$D_\mu \phi_i = (\partial_\mu - ig A_\mu^a t_a)_{ij} \phi_j = (\partial_\mu + g A_\mu^a T_a)_{ij} \phi_j. \tag{65.24}$$

Therefore, the kinetic term for the scalar is

$$\frac{1}{2}(D_\mu \phi_i)^2 = \frac{1}{2}(\partial_\mu \phi_i)^2 + g A_\mu^a \partial^\mu \phi_i T_{ij}^a \phi_j + \frac{g^2}{2} A_\mu^a A^{b\mu} (T^a \phi)_i (T^b \phi)_i. \tag{65.25}$$

Consider the VEV

$$\langle \phi_i \rangle = (\phi_0)_i. \tag{65.26}$$

Then from the kinetic term for the scalars we obtain the mass terms

$$\Delta \mathcal{L}_{\text{mass}} = -\frac{1}{2} m_{ab}^2 A_\mu^a A^{b\mu}, \tag{65.27}$$

where

$$m_{ab}^2 = g^2 (T^a \phi_0)_i (T^b \phi_0)_i, \tag{65.28}$$

and the diagonal elements are

$$m_{aa}^2 = g^2 (T^a \phi_0)^2 \geq 0. \tag{65.29}$$

Thus the vectors have mass squared positive or zero. Unlike the abelian case, when we had only nonzero masses, in the general nonabelian case we can have masses that are zero, corresponding to the existence of unbroken gauge fields.

Let us consider the general case of a gauge symmetry with gauge group G, broken spontaneously by the vacuum to a subgroup H that leaves the vacuum invariant. (Note that in the abelian case, the vacuum was just a point, so there was no unbroken gauge group.) Then the generators in the coset G/H take us from one vacuum to another (equivalent one). These generators give the Goldstone bosons then, which were the whole $U(1)$ in the abelian case.

The unbroken, massless generators correspond to

$$m_A^2 = 0 \Rightarrow (T^a \phi_0) = 0, \tag{65.30}$$

and these conditions define the subgroup H. We can redefine the vectors again as

$$A_\mu'^{\,a} = A_\mu^a - \partial_\mu \phi_i (T^a \phi_0)_i, \tag{65.31}$$

which means that the broken vectors (with nonzero $T^a \phi_0$) eat the Goldstone bosons and become massive.

Again the existence of the mixing vertex

$$g A_\mu^a \partial_\mu \phi_i (T^a \phi_0)_i \tag{65.32}$$

implies the transversality of the theory, so the consistency of the theory at the quantum level, since the sum of the mass term diagram and the Feynman diagram for mixing, Goldstone scalar propagation, and mixing again gives

$$-im_{ab}^2 g_{\mu\nu} + \sum_j gk^\mu (T^a \phi_0)_j \frac{-i}{k^2} (-gk^\nu (T^b \phi_0)_j) = -im_{ab}^2 \left(g_{\mu\nu} - \frac{k_\mu k_\nu}{k^2} \right), \qquad (65.33)$$

which is a transverse vacuum polarization.

65.4.1 $SU(2)$ Case

The simplest nonabelian case is the $SU(2)$ case, also relevant since it is rather close to the electroweak theory, where we have $SU(2) \times U(1)$. Consider a scalar ϕ_i that is a doublet of $SU(2)$, and

$$D_\mu \phi = (\partial_\mu - igA_\mu^a \tau^a)\phi, \qquad (65.34)$$

where $\tau_a = \sigma_a/2$ and σ_a are the Pauli matrices.

The potential is

$$V = \lambda \left(\phi^\dagger \phi - \frac{\mu^2}{\lambda} \right)^2. \qquad (65.35)$$

Choose the vacuum (as before, we can rotate it to this form by a gauge transformation)

$$\langle \phi \rangle = \frac{1}{\sqrt{2}} \begin{pmatrix} 0 \\ v \end{pmatrix}. \qquad (65.36)$$

Then the scalar kinetic term is

$$|D_\mu \phi|^2 = \frac{1}{2} g^2 \begin{pmatrix} 0 & v \end{pmatrix} \tau^a \tau^b \begin{pmatrix} 0 \\ v \end{pmatrix} A_\mu^a A^{b\mu}. \qquad (65.37)$$

Since we have symmetry in (ab), we can replace $\tau^a \tau^b$ with $\{\tau^a, \tau^b\}/2 = \delta^{ab}/4$, and obtain finally

$$\Delta \mathcal{L}_{\mathrm{mass}} = \frac{g^2 v^2}{8} A_\mu^a A^{a\mu}, \qquad (65.38)$$

that is a vector mass

$$m_A = \frac{gv}{2}. \qquad (65.39)$$

65.5 Standard Model Higgs: Electroweak $SU(2) \times U(1)$

In the case of the electroweak theory, the covariant derivative is

$$D_\mu \phi = \left(\partial_\mu - igA_\mu^a \tau^a - i\frac{g'}{2} B_\mu \right) \phi, \qquad (65.40)$$

and the last ($U(1)$) term is $-ig' Y_\phi B_\mu = -ig' B_\mu/2$.

Then we have

$$gA_\mu^a \tau^a + g' \frac{B_\mu}{2} = \frac{1}{2} \begin{pmatrix} gA_\mu^3 + g'B_\mu & g(A_\mu^1 - iA_\mu^2) \\ g(A_\mu^1 + iA_\mu^2) & -gA_\mu^3 + g'B_\mu \end{pmatrix}. \qquad (65.41)$$

The potential

$$V(\phi) = \frac{\lambda}{4}\left(\phi^\dagger\phi - \frac{v^2}{2}\right)^2 \tag{65.42}$$

has a real vacuum

$$\langle\phi\rangle = \frac{1}{\sqrt{2}}\begin{pmatrix} 0 \\ v \end{pmatrix} \tag{65.43}$$

that can be set by a gauge transformation from any other vacuum.

Then the mass term is

$$\mathcal{L}_{\text{mass}} = -\frac{v^2}{8}\begin{pmatrix} 0 & 1 \end{pmatrix}\left|\begin{pmatrix} gA_\mu^3 + g'B_\mu & g(A_\mu^1 - iA_\mu^2) \\ g(A_\mu^1 + iA_\mu^2) & -gA_\mu^3 + g'B_\mu \end{pmatrix}\right|^2\begin{pmatrix} 0 \\ 1 \end{pmatrix}$$

$$= -\frac{v^2}{8}\left[g^2(A_\mu^1)^2 + g^2(A_\mu^2)^2 + (-gA_\mu^3 + g'B_\mu)^2\right]. \tag{65.44}$$

Defining the fields

$$W_\mu^\pm = \frac{A_\mu^1 \pm iA_\mu^2}{\sqrt{2}},$$

$$Z_\mu^0 = \frac{1}{\sqrt{g^2 + g'^2}}(gA_\mu^3 - g'B_\mu),$$

$$A_\mu = \frac{1}{\sqrt{g^2 + g'^2}}(g'A_\mu^3 + gB_\mu), \tag{65.45}$$

the mass terms become

$$\Delta\mathcal{L}_{\text{mass}} = -\frac{v^2}{4}g^2 W_\mu^+ W^{-\mu} - (g^2 + g'^2)\frac{v^2}{8}Z_\mu^0 Z^{0\mu}. \tag{65.46}$$

Therefore we see that $W_\mu^\pm$ have mass

$$m_W = \frac{gv}{2}, \tag{65.47}$$

Z_μ^0 has mass

$$m_Z = \sqrt{g^2 + g'^2}\frac{v}{2}, \tag{65.48}$$

and the photon A_μ is massless, and corresponds to the unbroken electromagnetism.

Note that the photon is unbroken, since the corresponding field has no mass term in the Lagrangian, and $T^a\phi_0 = 0$ for it.

We now introduce the *weak mixing angle* or *Weinberg angle* θ_W, defined by

$$\cos\theta_W = \frac{g}{\sqrt{g^2 + g'^2}}, \quad \sin\theta_W = \frac{g'}{\sqrt{g^2 + g'^2}}. \tag{65.49}$$

We see that in terms of this, the Z and A are just a rotation of A^3 and B, that is

$$\begin{pmatrix} Z_\mu^0 \\ A_\mu \end{pmatrix} = \begin{pmatrix} \cos\theta_W & -\sin\theta_W \\ \sin\theta_W & \cos\theta_W \end{pmatrix}\begin{pmatrix} A_\mu^3 \\ B_\mu \end{pmatrix}, \tag{65.50}$$

which is why we defined Z^0_μ and A_μ in this way, in order to have an orthogonal rotation.

Using the further definitions

$$T^\pm = T^1 \pm iT^2 = \sigma^\pm,$$

$$e = \frac{gg'}{\sqrt{g^2 + g'^2}},$$

$$Q = T^3 + Y, \tag{65.51}$$

the covariant derivative is

$$D_\mu = \partial_\mu - igA^a_\mu T_a - ig'YB_\mu$$

$$= \partial_\mu - i\frac{g}{\sqrt{2}}(W^+_\mu T^+ + W^-_\mu T^-) - \frac{i}{\sqrt{g^2 + g'^2}}Z_\mu(g^2 T^3 - g'^2 Y)$$

$$- \frac{igg'}{\sqrt{g^2 + g'^2}}A_\mu(T^3 + Y), \tag{65.52}$$

and since

$$g^2 T^3 - g'^2 Y = (g^2 + g'^2)T^3 - g'^2 Q, \tag{65.53}$$

and $e = g \sin \theta_W$, we get

$$D_\mu = \partial_\mu - i\frac{g}{\sqrt{2}}(W^+_\mu T^+ + W^-_\mu T^-) - i\frac{g}{\cos\theta_W}Z_\mu(T^3 - \sin^2\theta_W Q) - ieA_\mu Q. \tag{65.54}$$

We also obtain the relation between parameters

$$m_W = m_Z \cos\theta_W. \tag{65.55}$$

In unitary gauge, we write

$$\phi(x) = \frac{1}{\sqrt{2}}\begin{pmatrix} 0 \\ v + H(x) \end{pmatrix} \tag{65.56}$$

and then $H(x)$ is the *Higgs boson* (i.e. the field associated with the Higgs particle).

Expanding the potential in terms of this, we obtain

$$V(\phi) = \frac{\lambda v^2}{4}H^2 + \frac{\lambda v}{4}H^3 + \frac{\lambda}{16}H^4, \tag{65.57}$$

so in particular

$$m_H^2 = \frac{\lambda v^2}{2}. \tag{65.58}$$

Then the electroweak bosonic Lagrangian

$$\mathcal{L} = -\frac{1}{4}(F^a_{\mu\nu})^2 - \frac{1}{4}B^2_{\mu\nu} - \frac{1}{2}|D_\mu\phi|^2 - V(\phi), \tag{65.59}$$

where $B_{\mu\nu} = \partial_\mu B_\nu - \partial_\nu B_\mu$, becomes (the proof is left as an exercise)

$$\mathcal{L} = -\frac{1}{4}F_{\mu\nu}^2 - \frac{1}{4}Z_{\mu\nu}^2 - \tilde{D}^{\dagger\mu}W^{-\nu}\tilde{D}_\mu W_\nu^+ + \tilde{D}^{\dagger\mu}W^{-\nu}\tilde{D}_\nu W_\mu^+$$

$$+ ie(F^{\mu\nu} + \cot\theta_W Z^{\mu\nu})W_\mu^+ W_\nu^- - \frac{e^2/\sin^2\theta_W}{2}(W^{+\mu}W_\mu^- W^{+\nu}W_\nu^- - W^{+\mu}W_\mu^+ W^{-\nu}W_\nu^-)$$

$$- \left(M_W^2 W^{+\mu}W_\mu^- + \frac{M_Z^2}{2}Z^\mu Z_\mu\right)\left(1 + \frac{H}{v}\right)^2$$

$$-\frac{1}{2}(\partial_\mu H)^2 - \frac{m_H^2}{2}H^2 - \frac{m_H^2}{2v}H^3 - \frac{m_H^2}{8v^2}H^4, \qquad\qquad (65.60)$$

where $Z_{\mu\nu} = \partial_\mu Z_\nu - \partial_\nu Z_\mu$ and

$$\tilde{D}_\mu = \partial_\mu - ie(A_\mu + \cot\theta_W Z_\mu). \qquad\qquad (65.61)$$

Important Concepts to Remember

- In spontaneous symmetry breaking, a local gauge group is (partially) broken ("spontaneously") by the choice of a vacuum.
- The vectors corresponding to the broken symmetry become massive by eating the massless scalars ("Goldstone bosons") corresponding to the directions of the broken symmetry in scalar-field space.
- The Goldstone bosons make the vacuum polarization of the spontaneously broken theory transverse.
- The gauge symmetry is lost after the redefinition of the vector that eats the scalar.
- In the nonabelian case, the gauge group G is broken to H, leaving G/H broken generators that give the Goldstone bosons, and H unbroken generators of zero mass, with $T^a\phi_0 = 0$.
- In the electroweak theory, the $SU(2) \times U(1)_Y$ is broken to $U(1)_{em}$, and the Z and A are an orthogonal rotation of A^3 and B.

Further Reading

See section 20.1 in [1] and chapters 85–87 in [10].

Exercises

1. Consider the abelian-Higgs Lagrangian. Expand it up to fourth order in the perturbations around the Higgs vacuum.
2. Prove that the Standard Model (electroweak) bosonic term around the Higgs vacuum takes the form

$$\mathcal{L} = -\frac{1}{4}F_{\mu\nu}^2 - \frac{1}{4}Z_{\mu\nu}^2 - \tilde{D}^{\dagger\mu}W^{-\nu}\tilde{D}_\mu W_\nu^+ + \tilde{D}^{\dagger\mu}W^{-\nu}\tilde{D}_\nu W_\mu^+$$

$$+ ie(F^{\mu\nu} + \cot\theta_W Z^{\mu\nu})W_\mu^+ W_\nu^- - \frac{e^2/\sin^2\theta_W}{2}(W^{+\mu}W_\mu^- W^{+\nu}W_\nu^- - W^{+\mu}W_\mu^+ W^{-\nu}W_\nu^-)$$

$$- \left(M_W^2 W^{+\mu}W_\mu^- + \frac{M_Z^2}{2}Z^\mu Z_\mu\right)\left(1 + \frac{H}{v}\right)^2$$

$$- \frac{1}{2}(\partial_\mu H)^2 - \frac{m_H^2}{2}H^2 - \frac{m_H^2}{2v}H^3 - \frac{m_H^2}{8v^2}H^4, \tag{65.62}$$

where $Z_{\mu\nu} = \partial_\mu Z_\nu - \partial_\nu Z_\mu$ and

$$\tilde{D}_\mu = \partial_\mu - ie(A_\mu + \cot\theta_W Z_\mu). \tag{65.63}$$

3. Calculate the vertices of the Standard Model Lagrangian in Exercise 2 that create one or more Higgs bosons.

4. Calculate the nonabelian gauge symmetry transformation after the redefinition of the gauge field, similarly to the abelian case.

In this chapter we will start to describe the quantization of spontaneously broken gauge theories, describing first the Goldstone theorem (at the quantum level), and then a gauge that is most useful for quantization of spontaneously broken gauge theories, the R_ξ gauge, a generalization of the Lorenz gauge in spontaneously broken theories in which all four unphysical states have the same mass.

66.1 The Goldstone Theorem

We already mentioned the Goldstone theorem. It states that massless states called Goldstone bosons appear whenever we break spontaneously a continuous symmetry.

Proof Consider a general Lagrangian composed of a kinetic part depending on $\partial_\mu \phi$ and a potential, that is

$$\mathcal{L} = K(\partial \phi) - V(\phi). \tag{66.1}$$

Consider the minimum of the potential, at

$$\left. \frac{\partial}{\partial \phi^a} V \right|_{\phi^a(x) = \phi_0^a} = 0, \tag{66.2}$$

and expand the potential around it:

$$V(\phi) = V(\phi_0) + \frac{1}{2}(\phi - \phi_0)^a (\phi - \phi_0)^b \left(\frac{\partial^2}{\partial \phi^a \partial \phi^b} V \right)_{\phi_0} + \ldots, \tag{66.3}$$

where

$$\left(\frac{\partial^2}{\partial \phi^a \partial \phi^b} V \right)_{\phi_0} \equiv m_{ab}^2 \tag{66.4}$$

is a mass matrix. Consider then the continuous symmetry

$$\phi^a \to \phi^a + \alpha \Delta^a(\phi). \tag{66.5}$$

In the particular case of constant ϕ^a, the invariance of the Lagrangian implies the invariance of the potential:

$$V(\phi^a) = V(\phi^a + \alpha \Delta^a(\phi)), \tag{66.6}$$

so that

$$\Delta^a(\phi) \left. \frac{\partial}{\partial \phi^a} V \right|_{\phi_0} = 0. \tag{66.7}$$

Taking a derivative $\partial/\partial \phi^b$ on the above, we get

$$0 = \left(\frac{\partial \Delta^a}{\partial \phi^b} \right)_{\phi_0} \left(\frac{\partial V}{\partial \phi^a} \right)_{\phi_0} + \Delta^a(\phi_0) \left(\frac{\partial^2}{\partial \phi^a \partial \phi^b} V \right)_{\phi_0} \Rightarrow \Delta^a(\phi_0) \left(\frac{\partial^2}{\partial \phi^a \partial \phi^b} V \right)_{\phi_0} = 0, \tag{66.8}$$

since ϕ_0 is a minimum of V, so the first term is zero.

Then there are two possibilities:

1. $\Delta^a(\phi_0) = 0$, which means that the symmetry leaves the vacuum ϕ_0^a unchanged, which is not the situation we are interested in.

2. If there is spontaneous breaking of the symmetry, then $\Delta^a(\phi_0) \neq 0$, which means that

$$(\Delta^a(\phi_0) \cdot) \left(\frac{\partial^2}{\partial \phi^a \partial \phi^b} V \right)_{\phi_0} \equiv m_{ab}^2 (\cdot \Delta^a(\phi_0)) = 0, \tag{66.9}$$

so there is a zero eigenvalue for the mass in the direction of the symmetry, which is the Goldstone boson.   *q.e.d.*

But we have actually proven the theorem only at the classical level. It is more interesting to prove that quantum corrections also don't spoil this property.

In order to consider quantum corrections, we need to consider, instead of the classical action, the quantum effective action Γ. But for constant fields ϕ^a, the effective action turns into the *effective potential*, or rather the effective potential times the volume of spacetime:

$$\Gamma[\phi_{cl}] = -(VT) V_{\text{eff}}[\phi_{cl}]. \tag{66.10}$$

But if there are no quantum anomalies, the effective potential respects the same symmetries as V, so we can repeat the same argument for the effective potential.

Then we obtain in general

$$\left. \frac{\delta^2}{\delta \phi^a \delta \phi^b} \Gamma(p) \right|_{p^2 \neq m^2} = 0. \tag{66.11}$$

If we consider the particular case of $p = 0$, corresponding to a constant classical field, Γ turns into V_{eff}, and we get

$$\frac{\partial^2}{\partial \phi^a \partial \phi^b} V_{\text{eff}} = 0, \tag{66.12}$$

as expected.

66.2 R_ξ Gauges: Abelian Case

In the analysis of the R_ξ gauges, generalizing the Lorenz gauge, we start with the abelian case.

In the case of gauge theory not spontaneously broken, the Minkowski Lagrangian is

$$\mathcal{L}_{g.f.} + \mathcal{L}_{gh} = -\frac{1}{2\xi}G^2 - b\frac{\delta G}{\delta \alpha}c, \tag{66.13}$$

where $G = 0$ is the gauge condition. The gauge transformation is

$$\delta A_\mu = -\partial_\mu \alpha; \quad \delta \phi = -ie\alpha\phi, \tag{66.14}$$

and in the case of the Lorenz (covariant) gauge $G = \partial^\mu A_\mu$, we obtain

$$\frac{\delta G}{\delta \alpha} = -\partial^2. \tag{66.15}$$

In the spontaneously broken case, with scalar field expansion, we have

$$\phi = \frac{1}{\sqrt{2}}(\phi^1 + i\phi^2) = \frac{1}{\sqrt{2}}(v + h(x) + i\varphi(x)), \tag{66.16}$$

where $h(x)$ is the Higgs boson, and $\varphi(x)$ is the Goldstone boson.

We then choose

$$G = \partial^\mu A_\mu - \xi ev\varphi, \tag{66.17}$$

which is called the R_ξ gauge, and note that for $v = 0$ it reduces to the Lorenz gauge.

We compute the kinetic term for ϕ. With $D_\mu = \partial_\mu - ieA_\mu$ as usual, we get

$$D_\mu\phi = \frac{1}{\sqrt{2}}[(\partial_\mu h + e\varphi A_\mu) + i(\partial_\mu \varphi - e(v + h)A_\mu)], \tag{66.18}$$

so that

$$
\begin{aligned}
-|D_\mu\phi|^2 &= -\frac{1}{2}(\partial_\mu h + e\varphi A_\mu)^2 - \frac{1}{2}(\partial_\mu \varphi - e(v + h)A_\mu)^2 \\
&= -\frac{1}{2}(\partial_\mu h)\partial^\mu h - \frac{1}{2}\partial_\mu \varphi \partial^\mu \varphi - \frac{e^2}{2}v^2 A_\mu A^\mu + ev A_\mu \partial^\mu \varphi \\
&\quad + eA_\mu(h\partial^\mu \varphi - \varphi\partial^\mu h) \\
&\quad - e^2 vh A_\mu A^\mu - \frac{e^2}{2}(h^2 + \varphi^2)A_\mu A^\mu.
\end{aligned} \tag{66.19}
$$

The potential expands as

$$V = \frac{\lambda}{2}\left(|\phi|^2 - \frac{v^2}{2}\right)^2 = \frac{\lambda v^2}{2}h^2 + \frac{\lambda v}{2}h(h^2 + \varphi^2) + \frac{\lambda}{8}(h^2 + \varphi^2)^2. \tag{66.20}$$

The gauge-fixing term expands as

$$\int d^dx \mathcal{L}_{g.f.} = \int d^dx \left[-\frac{1}{2\xi}(\partial^\mu A_\mu)\partial^\nu A_\nu - \frac{\xi e^2 v^2}{2}\varphi^2 + ev\varphi \partial^\mu A_\mu \right]$$

$$= \int d^dx \left[-\frac{1}{2\xi}(\partial^\mu A_\nu)\partial^\nu A_\mu - evA_\mu \partial^\mu \varphi - \frac{\xi e^2 v^2}{2}\varphi^2 \right], \quad (66.21)$$

where in the first term we have partially integrated both derivatives, and in the mixed term we have partially integrated the derivative. From the mass term for φ, we have the mass

$$m_\varphi = \sqrt{\xi}ev = \sqrt{\xi}m_A. \quad (66.22)$$

We see that the mixed term cancels between the kinetic term and the gauge-fixing term, which is one reason why the gauge-fixing term was chosen like that.

The ghost term is found as follows. The gauge transformation around the Higgs vacuum with $\phi = (v + h + \varphi)/\sqrt{2}$ is

$$\delta A_\mu = -\partial_\mu \alpha; \quad \delta h = +ea\varphi; \quad \delta\varphi = -ea(v + h). \quad (66.23)$$

Then we get

$$\frac{\delta G}{\delta\alpha} = \frac{\delta}{\delta\alpha}(\partial^\mu A_\mu - \xi ev\varphi) = -\partial^2 + \xi e^2 v(v + h), \quad (66.24)$$

leading to a ghost term

$$\int d^dx \mathcal{L}_{gh} = -\int d^dx b[-\partial^2 + \xi e^2 v(v + h)]c$$

$$= -\int d^dx [\partial^\mu b \partial_\mu c + \xi e^2 v^2 bc + \xi e^2 vhbc]. \quad (66.25)$$

From the second term we see that the ghosts b and c have masses

$$m_{b,c} = \sqrt{\xi}ev = \sqrt{\xi}m_A \quad (66.26)$$

also.

The kinetic term for A_μ contains the usual kinetic term from the $-\int F_{\mu\nu}^2/4$ action, the term from the gauge-fixing term, and the mass term from the scalar kinetic term, giving in total

$$\mathcal{L}_{A^2} = -\frac{1}{2}A_\mu \left[\partial^\mu \partial^\nu - g^{\mu\nu}\partial^2 + m_A^2 g^{\mu\nu} - \frac{\partial^\mu \partial^\nu}{\xi} \right] A_\nu, \quad (66.27)$$

or in momentum space

$$\mathcal{L}_{A^2} = -\frac{1}{2}\tilde{A}_\mu(-k)[(k^2 + m_A^2)g^{\mu\nu} - (1 - \xi^{-1})k^\mu k^\nu]\tilde{A}_\nu(k). \quad (66.28)$$

Then the kinetic matrix

$$[(k^2 + m_A^2)g^{\mu\nu} - (1 - \xi^{-1})k^\mu k^\nu] \quad (66.29)$$

can be written in terms of the projectors

$$P^{\mu\nu}(k) = g^{\mu\nu} - \frac{k^\mu k^\nu}{k^2}; \quad K^{\mu\nu}(k) = \frac{k^\mu k^\nu}{k^2}, \quad (66.30)$$

satisfying the projector, orthogonality, and completeness relations, that is schematically

$$P^2 = P; \quad K^2 = K; \quad P \cdot K = 0; \quad P + K = 1 \tag{66.31}$$

(which can easily be checked), as

$$(k^2 + m_A^2)P^{\mu\nu}(k) + \xi^{-1}(k^2 + \xi m_A^2)K^{\mu\nu}(k). \tag{66.32}$$

Thus the kinetic matrix can easily be inverted to give the photon propagator in R_ξ gauge:

$$\tilde{\Delta}_{\mu\nu}(k) = \frac{P_{\mu\nu}(k)}{k^2 + m_A^2} + \frac{\xi K_{\mu\nu}(k)}{k^2 + \xi m_A^2}. \tag{66.33}$$

We note that the transverse (physical) part of the propagator, proportional to $P_{\mu\nu}(k)$, has mass m_A, whereas the longitudinal (unphysical) part of the propagator, proportional to $K_{\mu\nu}(k)$, has mass $\sqrt{\xi}m_A$. So, in total, the quartet of unphysical states, the would-be Goldstone boson φ, the ghosts b and c, and the longitudinal photon, all have the mass $\sqrt{\xi}m_A$.

We also note that for $\xi = 1$ (the equivalent of the Feynman gauge for the unbroken theory), we have

$$\tilde{\Delta}_{\mu\nu}(k)|_{\xi=1} = \frac{g_{\mu\nu}}{k^2 + m_A^2}, \tag{66.34}$$

which is the KG propagator (for a scalar mode). We will nevertheless continue to use arbitrary ξ in order to see ξ independence for physical quantities (to test our calculation).

In conclusion, the propagators are as follows:

- For the Higgs h we have the usual scalar propagator

$$\frac{1}{k^2 + m_h^2}, \tag{66.35}$$

  where $m_h = \sqrt{\lambda}v$.
- For the would-be Goldstone boson (the unphysical scalar φ), we have the scalar propagator

$$\frac{1}{k^2 + \xi m_A^2}. \tag{66.36}$$

- The same as for the ghosts b, c.
- The vector has the propagator $\tilde{\Delta}_{\mu\nu}(k)$, with $m_A = ev$.

Putting together all the interaction terms derived above, we find the interaction Lagrangian

$$\begin{aligned}
\mathcal{L}_{\text{int}} = &-\frac{\lambda v}{2}h(h^2 + \varphi^2) - \frac{\lambda}{8}(h^2 + \varphi^2)^2 \\
&+ eA_\mu(h\partial^\mu\varphi - \varphi\partial^\mu h) \\
&- e^2 vhA_\mu A^\mu - \frac{e^2}{2}(h^2 + \varphi^2)A_\mu A^\mu \\
&- \xi e^2 vhbc.
\end{aligned} \tag{66.37}$$

66.3 R_ξ **Gauges: Nonabelian Case**

In the nonabelian case, the gauge-fixing and ghost terms in the Lagrangian are

$$\mathcal{L}_{g.f.} + \mathcal{L}_{gh} = -\frac{1}{2\xi}G_a G^a - b^a \frac{\delta G^a}{\delta \alpha^b}c^b. \tag{66.38}$$

The gauge-covariant derivative is

$$D_\mu = \partial_\mu + gA_\mu^a T_a \tag{66.39}$$

and we expand around the VEV $\langle \phi_i \rangle = v_i$ as

$$\phi_i = v_i + \chi_i. \tag{66.40}$$

Then we choose the gauge condition for the R_ξ gauge as

$$G^a = \partial^\mu A_\mu^a - \xi g(T^a)_{ij}v_j\chi_i, \tag{66.41}$$

and the gauge-fixing term becomes

$$\begin{aligned}
\int d^d x \mathcal{L}_{g.f.} &= \int d^d x \left[-\frac{1}{2\xi}(\partial^\mu A_\mu^a)\partial^\nu A_\nu^a + \xi g(T^a)_{ij}v_j\chi_i \partial^\mu A_\mu^a \right. \\
&\qquad\qquad \left. - \xi g^2((T^a)_{ik}v_k (T^a)_{jl}v_l)\chi_i\chi_j \right] \\
&= \int d^d x \left[-\frac{1}{2\xi}(\partial^\mu A_\nu^a)\partial^\nu A_\mu^a - \xi g(T^a)_{ij}v_j\partial^\mu\chi_i A_\mu^a \right. \\
&\qquad\qquad \left. - \xi g^2((T^a)_{ik}v_k (T^a)_{jl}v_l)\chi_i\chi_j \right], \tag{66.42}
\end{aligned}$$

where as before we have partially integrated the two derivatives in the first term and the derivative in the second. The last term, with $\chi_i\chi_j$, is a contribution to the mass term for scalars:

$$\xi M_{ij}^2 = \xi g^2((T^a)_{ik}v_k (T^a)_{jl}v_l). \tag{66.43}$$

The kinetic term for the scalars is $((T^a)^\dagger = -T^a)$

$$\begin{aligned}
-\frac{1}{2}(D^\mu\phi_i)^\dagger D^\mu\phi_i &= -\frac{1}{2}(\partial^\mu\chi_i)\partial_\mu\chi_i - \frac{g^2}{2}((T^a)_{ik}v_k (T^b)_{il}v_l)A_\mu^a A^{b\mu} + g(T^a)_{ik}v_k A_\mu^a \partial^\mu\chi_i \\
&\quad + gA_\mu^a \chi_i (T^a)_{ij}\partial^\mu\chi_j - g^2 A_\mu^a A^{b\mu}((T^a)_{ik}v_k)(T^b)_{ij}\chi_j \\
&\quad - \frac{g^2}{2}(T^a T^b)_{ij}\chi_i\chi_j A_\mu^a A^{b\mu}, \tag{66.44}
\end{aligned}$$

and again the mixing term between the vector and the would-be Goldstone boson cancels with the gauge-fixing term.

The ghost term is obtained by considering the gauge transformation on the fluctuation:

$$A_\mu^a \rightarrow A_\mu^a - D_\mu^{ab}\alpha^b; \quad \chi_i \rightarrow -g\alpha^a(T_a)_{ij}(v+h)_j, \tag{66.45}$$

leading to

$$\frac{\delta G^a}{\delta \alpha^b} = -\partial^\mu D^{ab}_\mu + \xi g^2 (T^a)_{ij} v_j (T^a)_{il} (v + \chi)_l$$

$$= -\partial^\mu D^{ab}_\mu + \xi g^2 (T^a)_{ij} v_j (T^b)_{il} v_l + \xi g^2 ((T^a)_{ij} v_j (T^b)_{il}) \chi_l, \tag{66.46}$$

where the middle term is written as $(M^2_{b,c})^{ab}$ and is the ghost mass term. Then the ghost term is

$$\int d^d x \mathcal{L}_{gh} = \int d^d x \left[-(\partial_\mu b^a) D^{ab}_\mu c^b - \xi (M^2_{b,c})^{ab} b^a c^b - \xi g^2 ((T^a)_{ij} v_j (T^a)_{il}) \chi_l b^a c^b \right].$$

$$\tag{66.47}$$

Important Concepts to Remember

- The Goldstone theorem says that there is a massless particle (Goldstone boson) for every spontaneously broken continuous symmetry.
- The effective potential is the effective action on constant fields, more precisely $\Gamma[\phi_{cl}] = -(VT) V_{\text{eff}}[\phi_{cl}]$.
- Quantum corrections respect the Goldstone theorem.
- The R_ξ gauge is $\partial^\mu A_\mu - \xi e v \varphi = 0$.
- In the R_ξ gauge, the quartet of unphysical states, would-be Goldstone boson φ, ghosts b and c, and longitudinal gauge boson all have the mass $\sqrt{\xi} m_A$, where $m_A = ev$; the Higgs mass is $m_h = \sqrt{\lambda} v$.
- With the choice $\xi = 1$, the photon propagator is the KG propagator.

Further Reading

See sections 11.1, 21.1, and 21.2 in [1].

Exercises

1. Consider a theory invariant under a symmetry group G, having a spontaneously breaking vacuum invariant under H sup G. How many Goldstone bosons are there? Specialize to the $SU(5) \to SU(3) \times SU(2) \times U(1)$ breaking.
2. Write down all the one-loop Feynman diagrams for the Higgs h 1PI two-point function in the spontaneously broken abelian theory in R_ξ gauge, and the integral expressions for them using the Feynman rules (without computing them).
3. Repeat Exercise 2 in the nonabelian case.
4. Argue that the loops of unphysical states cancel against each other in the case in Exercise 2.

Renormalization of Spontaneously Broken Gauge Theories II: The $SU(2)$-Higgs Model

In this chapter we will learn how to renormalize spontaneously broken gauge theories using the R_ξ gauge, for the example of the $SU(2)$-Higgs system, which is close to the electroweak theory, without being exactly that. Also, we will not do the full renormalization, but only some of the important steps.

67.1 The $SU(2)$-Higgs Model

The theory contains $SU(2)$ gauge fields, and a complex Higgs doublet. Again, the terminology is ambiguous, really we have a Higgs field h and would-be Goldstone bosons χ^a.

With respect to unbroken gauge theories, one important difference is that now there is one more Ward identity, for the fact that only the combination $v + h$ appears in the classical Lagrangian. It is of course broken by the gauge-fixing term, so we need to check explicitly that this is still satisfied at the quantum level.

We parametrize the complex field slightly differently from what we had done before, now considering the VEV in the upper component, instead of the lower component. So we start by parametrizing the complex doublet

$$\phi = \begin{pmatrix} \varphi^1 \\ \varphi^2 \end{pmatrix} \tag{67.1}$$

as

$$\phi = \frac{1}{\sqrt{2}}(\psi + i\chi^a \sigma^a)\begin{pmatrix} 1 \\ 0 \end{pmatrix} = \frac{1}{\sqrt{2}}\begin{pmatrix} \psi + i\chi^3 \\ i\chi^1 - \chi^2 \end{pmatrix}. \tag{67.2}$$

The Lagrangian is

$$\mathcal{L} = -\frac{1}{4}(F_{\mu\nu}^a)^2 - (D_\mu\phi)^\dagger D^\mu\phi - V(\phi^\dagger\phi), \tag{67.3}$$

where the potential is

$$V = -\mu^2\phi^\dagger\phi + \lambda(\phi^\dagger\phi)^2. \tag{67.4}$$

The VEV is

$$\langle \mathrm{Re}\phi^1 \rangle = \frac{v}{\sqrt{2}}, \tag{67.5}$$

so we split the scalar into VEV and fluctuations as

$$\psi = v + h. \tag{67.6}$$

The gauge-covariant derivative is

$$D_\mu \phi = \partial_\mu \phi - \frac{i}{2} g A_\mu^a \sigma^a \phi. \tag{67.7}$$

This is a particular case of the general procedure from Chapter 66, so we can write the Lagrangian that comes from the square of the covariant derivative as a sum of a kinetic piece, a piece linear in A, and a piece quadratic in A, as

$$\mathcal{L}_{(\text{kin})} = -\frac{1}{2}[(\partial^\mu h)\partial_\mu h + (\partial_\mu \chi^a)\partial^\mu \chi^a],$$

$$\mathcal{L}(A^1) = \frac{gv}{2}(A_\mu^a \partial^\mu \chi^a) + \frac{g}{2}A_\mu^a (h \overleftrightarrow{\partial}_\mu \chi^a) + \frac{1}{2}g\epsilon_{abc}(\chi^a A_\mu^b \partial^\mu \chi^c),$$

$$\mathcal{L}(A^2) = -\frac{g^2}{8}A_\mu^a A^{a\mu}[(v+h)^2 + \chi^b \chi^b]. \tag{67.8}$$

Writing the covariant derivative as

$$D_\mu \phi = \frac{1}{\sqrt{2}}(D_\mu h + i\sigma^a D_\mu \chi^a), \tag{67.9}$$

where we have defined

$$D_\mu h = \partial_\mu h + \frac{g}{2}A_\mu^a \chi_a,$$

$$D_\mu \chi^a = \partial_\mu \chi^a + \frac{g}{2}\epsilon_{abc}A_\mu^b \chi^c - \frac{g}{2}A_\mu^a(v+h), \tag{67.10}$$

we can rewrite the scalar kinetic term as

$$\mathcal{L} = -\frac{1}{2}(D_\mu h)^2 - \frac{1}{2}(D_\mu \chi^a)^2 + \frac{g}{2}v A_\mu^a \partial^\mu \chi^a. \tag{67.11}$$

To cancel the last term, we add the 't Hooft gauge-fixing term (in R_ξ gauge)

$$\mathcal{L}_{\text{g.fix}} = -\frac{1}{2\xi}\left(\partial^\mu A_\mu^a + \frac{1}{2}\xi g v \chi^a\right)^2. \tag{67.12}$$

Then the quadratic (kinetic) term for χ is

$$\mathcal{L}_{\chi^2} = -\frac{1}{2}(\partial_\mu \chi)^2 - \frac{\xi}{2}\left(\frac{gv}{2}\right)^2 (\chi^a)^2, \tag{67.13}$$

so the mass of the would-be Goldstone bosons is $m_\chi = \sqrt{\xi} m_A$.

The kinetic term for the vectors is now

$$\mathcal{L}_{YM} = -\frac{1}{4}(F_{\mu\nu}^a)^2 - \frac{1}{2}\left(\frac{gv}{2}\right)^2 (A_\mu^a)^2, \tag{67.14}$$

so $m_A = gv/2$.

The ghost Lagrangian is found, as usual, from $b_a \delta G^a / \delta \alpha^b c^b$, as

$$\mathcal{L}_{gh} = b_a \partial^\mu D_\mu c^a - \xi\left(\frac{gv}{2}\right)^2 b_a c^a - \frac{\xi}{4}g^2 v b_a (h c^a + \epsilon^a{}_{bc}\chi^b c^c). \tag{67.15}$$

As we mentioned in Chapter 66, we see that the quartet of unphysical states, the Fadeev–Popov ghosts b_a and c^a, together with the would-be Goldstone bosons χ^a and the longitudinal part of the YM field A_μ^a, all have the same mass $m = \sqrt{\xi}m_A$.

We will see why shortly, but at the quantum level we need to consider that the gauge-fixing term contains two new parameters, so we will write it as

$$\mathcal{L}_{\text{g.fix}} = -\frac{1}{2\alpha}\left(\partial^\mu A_\mu^a + \frac{1}{2}\xi gv\chi^a\right)^2, \tag{67.16}$$

where now $\alpha \neq \xi$ and equal only at the classical level. But even in this case (at the quantum level), the quartet of unphysical states will still have the same mass.

67.2 Quantum Theory and LZJ Identities

We now split the Lagrangian for the spontaneously broken theory into matter, gauge, and ghost parts:

$$\mathcal{L}_{\text{matter}} = -\frac{1}{2}(D_\mu h)^2 - \frac{1}{2}(D_\mu\chi^a)^2 + \frac{\mu^2}{2}[(v+h)^2 + (\chi^a)^2] - \frac{\lambda}{4}[(v+h)^2 + (\chi^a)^2]^2,$$

$$\mathcal{L}_{\text{gauge}} = -\frac{1}{4}\left(\partial_\mu A_\nu^a - \partial_\nu A_\mu^a + gf^a{}_{bc}A_\mu^b A_\nu^c\right)^2,$$

$$\mathcal{L}_{\text{ghost}} = b_a\left[\partial^\mu D_\mu c^a - \xi\frac{gv}{2}\left(\frac{1}{2}g(v+h)c^a + \frac{1}{2}gf^a{}_{bc}\chi^b c^c\right)\right]. \tag{67.17}$$

But we still need to add a term to the Lagrangian. We saw in the unbroken case that when we renormalize, we need to add an extra source term to the Lagrangian for the nonlinear parts of the BRST variations. We add

$$\mathcal{L}_{\text{extra}} = K_a^\mu Q_B A_\mu^a/\Lambda + KQ_B h/\Lambda + K_a Q_B \chi^a/\Lambda + L_a Q_B c^a/\Lambda$$

$$= K_a^\mu D_\mu c^a - K\left(\frac{1}{2}g\chi_a c^a\right) + K_a\left(\frac{1}{2}g(v+h)c^a + \frac{1}{2}gf^a{}_{bc}\chi^b c^c\right)$$

$$+ L_a\left(\frac{1}{2}gf^a{}_{bc}c^b c^c\right). \tag{67.18}$$

It is useful to write the linear and quadratic mass terms in the scalars as

$$-\beta vh - \frac{\beta}{2}(h^2 + (\chi^a)^2), \tag{67.19}$$

where

$$\beta = -\mu^2 + 2\lambda v^2, \tag{67.20}$$

though then β doesn't renormalize multiplicatively. Indeed, μ^2 and λv^2 both renormalize multiplicatively, however classically $\beta = 0$, so we require

$$\beta_{\text{ren}}^{(0)} = -\mu_{\text{ren}}^2 + \lambda_{\text{ren}} v_{\text{ren}}^2 = 0, \tag{67.21}$$

which means that β is renormalized additively, as

$$\beta_{\text{ren}} = 0 + \Delta\beta^{(1)}_{\text{ren}} + \dots \tag{67.22}$$

We will therefore choose to renormalize β additively as above, rather than renormalize μ^2 multiplicatively.

Another observation is that now the matter Lagrangian depends on $v + h(x)$, but the gauge-fixing and ghost terms break this, so at the quantum level we must check explicitly. (Note that we could replace v by $v + h$ in the gauge-fixing term, but then it is more complicated.)

We also consider that we will have $\alpha \neq \xi$, but we require $\alpha_{\text{ren}} = \xi_{\text{ren}} = 1$.

As in the unbroken case, we proceed to write the equivalent of the Ward identities for the effective action Γ in the BRST case, more precisely for

$$\hat{\Gamma} = \Gamma - \int \mathcal{L}_{\text{fix}} d^4 x, \tag{67.23}$$

the LZJ identities, that are now written as

$$\int d^4 x \left[\partial \hat{\Gamma} / \partial \phi^I \frac{\partial}{\partial K^I} \right] \hat{\Gamma} = 0,$$

$$\left(\partial^\mu \frac{\partial}{\partial K_a^\mu} - \xi \frac{gv}{2} \frac{\partial}{\partial K_a} - \frac{\partial}{\partial b_a} \right) \hat{\Gamma} = 0. \tag{67.24}$$

Here

$$\phi^I = \{ h, \chi^a, A_\mu^a, c^a \};$$

$$K_I = \{ K, K_a, K_a^\mu, L_a \}. \tag{67.25}$$

These can be found as in the unbroken case, and we note that restricting to A_μ^a and c^a and putting $v = 0$, we find the unbroken case.

Therefore the fields in the theory are A_μ^a, b_a, c^a, h, and χ^a, the sources are K, K^a, K_μ^a, and L_a, and the parameters are g, v, λ, α, and ξ.

67.3 Renormalization

Renormalization is then done as follows. The fields renormalize as

$$A_\mu^a = \sqrt{Z_3} A_{\mu,\text{ren}}^a,$$

$$c^a = \sqrt{Z_{gh}} c_{\text{ren}}^a,$$

$$b_a = \sqrt{Z_{gh}} b_{a,\text{ren}},$$

$$h = \sqrt{Z_h} h_{\text{ren}},$$

$$\chi^a = \sqrt{Z_\chi} \chi_{\text{ren}}^a. \tag{67.26}$$

The sources as

$$K = \sqrt{\frac{Z_3 Z_{gh}}{Z_h}} K_{\text{ren}},$$

$$K^a = \sqrt{\frac{Z_3 Z_{gh}}{Z_\chi}} K^a_{\text{ren}},$$

$$K^a_\mu = \sqrt{Z_{gh}} K^a_{\mu,\text{ren}},$$

$$L^a = \sqrt{Z_3} L_{a,\text{ren}}. \tag{67.27}$$

And the parameters as

$$g = Z_g g_{\text{ren}} \mu^{\frac{4-d}{2}},$$

$$v = \sqrt{Z_v} v_{\text{ren}},$$

$$\lambda = Z_\lambda Z_h^{-2} \lambda_{\text{ren}} \mu^{4-d},$$

$$\alpha = Z_3 \alpha_{\text{ren}},$$

$$\xi = \sqrt{\frac{Z_3}{Z_v Z_\chi}} Z_g^{-1} \xi_{\text{ren}}. \tag{67.28}$$

We must make several observations:

1. b_a, as well as K^μ_a, scale as c^a and L_a scales as A^a_μ. We have seen in the unbroken case that their renormalization is fixed by analyzing the possible divergent structures that solve the LZJ identities, and now a similar story holds. But we also see that Kh, $K_a\chi^a$, $K^\mu_a A^a_\mu$, and $L_a c^a$ all scale the same way, since the source terms are $K^a_\mu Q_B A^a_\mu$, $KQ_B h$, $K^a Q_B \chi^a$, $L_a Q_B c^a$. Therefore the renormalization of the extra source terms is completely fixed.
2. The renormalization of α and ξ is fixed by requiring that the gauge-fixing term is finite by itself. The part at $v = 0$ (from the unbroken theory) fixes in $\alpha = Z_\alpha \alpha_{\text{ren}}$ that $Z_\alpha = Z_3$, whereas the v-dependent term fixes the renormalization of ξ. The result is a different renormalization factor for α and ξ.
3. This in turn means that we need separate α and ξ as advertised, and $\alpha_{\text{ren}} = \xi_{\text{ren}} = 1$.
4. A quick one-loop calculation, left as an exercise, shows that in fact also $Z_v \neq Z_h$, even though the classical Lagrangian has them equal. One can prove renormalizability at all loops by induction. Here we will not prove the induction step, since we also didn't in the unbroken case.
5. By analyzing the solution of the LZJ identities, we find nine possible divergent structures, but we have only eight renormalization parameters: $Z_3, Z_{gh}, Z_h, Z_\chi, Z_g, Z_v, Z_\lambda$, and $\Delta\beta_{\text{ren}}$ (standing in for Z_μ). This would seem like a contradiction, but as we advertised at the beginning of the chapter, there is an extra Ward identity, coming from the fact that we only find the $v + h$ combination in the classical part (only the gauge-fixing and ghost parts break it).

The classical action is the tree-level part of the modified effective action, $\hat{\Gamma}_{\text{ren}}^{(0)} = S_{\text{ren}}$. Then the Ward identity is written as

$$\left(v\frac{\partial}{\partial h} - v\frac{\partial}{\partial v} \right) S_{\text{matter}} = 0, \tag{67.29}$$

or in terms of $\hat{\Gamma}^{(0)}$ as

$$\left(\xi_{\text{ren}}\frac{\partial}{\partial \xi_{\text{ren}}} - v_{\text{ren}}\frac{\partial}{\partial v_{\text{ren}}} + v_{\text{ren}}\frac{\partial}{\partial h_{\text{ren}}} \right) \hat{\Gamma}_{\text{ren}}^{(0)} = 0, \tag{67.30}$$

since in the ghost term and the gauge-fixing term the combination ξv appears.

The above Ward identity allows the reduction of the possible divergences to eight, equal to the number of renormalization parameters.

Another useful consistency condition is found from ghost number conservation, which implies (the coefficient of each term is their ghost number)

$$\left(b_a\frac{\partial}{\partial b^a} - c^a\frac{\partial}{\partial c^a} + K\frac{\partial}{\partial K} + K_a\frac{\partial}{\partial K_a} + K_a^\mu\frac{\partial}{\partial K_a^\mu} + 2L_a\frac{\partial}{\partial L_a} \right) \hat{\Gamma}_{\text{ren}}^{(0)} = 0. \tag{67.31}$$

Important Concepts to Remember

- The $SU(2)$-Higgs system has a gauge field and a complex scalar doublet, and the Higgs mechanism generates a quartet of unphysical states.
- To renormalize in the R_ξ ('t Hooft) gauge, we need to add sources for the nonlinear parts of the BRST transformations to the Lagrangian, K_a^μ, K, K_a, L_a for $Q_B A_\mu^a, Q_B h, Q_B \chi_a, Q_B c^a$.
- Instead of renormalizing μ^2 multiplicatively, we can renormalize $\beta = -\mu^2 + \lambda v^2$, which is 0 classically, additively.
- We must consider α and ξ renormalizations independently at the quantum level, for $G_a = \partial^\mu A_\mu^a + \xi g v/2 \chi^a$ and $(G^a)^2/2\alpha$ as gauge-fixing term.
- We must consider v and h renormalizing independently at the quantum level, but the Ward identity coming from having only $v + h$ dependence reduces the number of divergent structures to the number matching the number of parameters.

Further Reading

See section 21.3 in [1].

Exercises

1. Show that $Z_v \neq Z_h$, by a one-loop calculation.

2. Use the Ward identity (67.30) for writing the renormalized action only in terms of the eight divergences.
3. Use the LZJ identities, and show that assuming the unbroken case $v = 0$, we find the correct divergences.
4. Show that the renormalized effective action is consistent with ghost number conservation.

In this chapter we will apply the method of effective field theory from Chapter 62 to describe low-energy QCD. We will use an approximate symmetry called *chiral symmetry*, and its perturbation theory.

68.1 QCD, Chiral Symmetry Breaking, and Goldstone Theorem

We have described spontaneous symmetry breaking, and we saw that Goldstone's theorem says we obtain Goldstone bosons for the broken symmetry directions. But in reality, we never have an exact symmetry, so it is useful to know what happens when an approximate symmetry is broken. In that case, we say that we have a *pseudo-Goldstone boson*.

In QCD, we have six quarks: u, d, s, c, b, t. The last three, c, b, and t, are heavy, so a different perturbation theory is used for them. The u and d quarks are nearly massless (their masses are very small), and the s quark is intermediate in mass. Therefore, in low-energy QCD we consider the u and d quarks, and sometimes the s quark. One can consider also source terms for various currents: vector V_μ for the vector current, axial vector A_μ for the axial vector current, s for a scalar current, and p for a pseudoscalar current, for a total Lagrangian in low-energy QCD of

$$
\begin{aligned}
\mathcal{L} = &-\frac{1}{4}(F_{\mu\nu}^a)^2 - \bar{u}\gamma^\mu D_\mu u - \bar{d}\gamma^\mu D_\mu d[-\bar{s}\gamma^\mu D_\mu s] \\
&- \sum_i m_i \bar{q}_i q^i \\
&[-V_\mu \bar{q}\gamma^\mu q - A_\mu \bar{q}\gamma^\mu \gamma_5 q - s\bar{q}q - p\bar{q}\gamma_5 q],
\end{aligned}
\tag{68.1}
$$

where the first line is the massless QCD, the second line the mass terms, and the third line the source terms.

We will consider only the u and d quarks for most of the chapter, generalizing to the introduction of the s quark at the end of the chapter.

We will consider then the quark column vector $q = \begin{pmatrix} u \\ d \end{pmatrix}$, which is nearly massless ($m_u \simeq m_d \simeq 0$). Then the low-energy Lagrangian has $U(2)_L \times U(2)_R$ symmetry, as we have described in Chapter 56, composed of an $SU(2)_L \times SU(2)_R$ part and a $U(1) \times U(1)$ part.

The action of the $SU(2)_L \times SU(2)_R$ on q is

$$q = \begin{pmatrix} u \\ d \end{pmatrix} \rightarrow \exp\left[i\alpha_V^a \frac{\tau^a}{2} + i\gamma_5 \alpha_A^a \frac{\tau^a}{2}\right]\begin{pmatrix} u \\ d \end{pmatrix}, \tag{68.2}$$

where τ^a are the Pauli matrices. In terms of

$$q_{L/R} = \left(\frac{1 \pm \gamma_5}{2}\right)q,$$

$$g_{L/R} = e^{i(\alpha_V \pm \alpha_A)^a \tau_{L/R}^a},$$

$$\tau_{L/R}^a = \left(\frac{1 \pm \gamma_5}{2}\right)\tau^a, \tag{68.3}$$

the action of $SU(2)_L \times SU(2)_R$ is

$$q_{L/R} \rightarrow g_{L/R} q_{L/R},$$

$$V_\mu \pm A_\mu \rightarrow g_{L/R}(V_\mu \pm A_\mu)g_{L/R}^\dagger + i(\partial_\mu g_{L/R})g_{L/R}^\dagger,$$

$$s + ip \rightarrow g_L(s + ip)g_R. \tag{68.4}$$

Then $\tau_{L/R}^a$ generate two independent $SU(2)$ algebras (i.e. $SU(2)_L \times SU(2)_R$).

$$[\tau_L^a, \tau_L^b] = i\epsilon^{abc}\tau_L^c,$$

$$[\tau_R^a, \tau_R^b] = i\epsilon^{abc}\tau_R^c,$$

$$[\tau_L^a, \tau_R^b] = 0. \tag{68.5}$$

Besides this, we have the $U(1) \times U(1)$, acting as

$$(q) \rightarrow e^{i\alpha + i\tilde{\alpha}\gamma_5}(q). \tag{68.6}$$

The action by $e^{i\alpha}$ is (three times) the $U(1)_B$ baryon number (since the baryon number of a baryon like n and p is 1, the baryon number of the quarks u and d is 1/3).

The baryon number is conserved *in QCD!* (*in electroweak theory the baryon number is broken by anomalies through instantons*, as we said in Chapter 56, since the chiral fermions L and R couple differently to it, unlike the coupling of the fermions to $SU(3)_c$).

In contrast, $e^{i\tilde{\alpha}\gamma_5}$ is an abelian global chiral symmetry, and is broken by anomalies *in QCD*, giving the solution to the $U(1)$ problem, as we explained in Chapter 56.

The $SU(2)_L \times SU(2)_R$ is broken spontaneously to $SU(2)_V$, acting as

$$(q) \rightarrow e^{i\alpha_V^a \frac{\tau^a}{2}}(q). \tag{68.7}$$

But the other $SU(2)$ for

$$(q) \rightarrow e^{i\alpha_A^a \gamma_5 \frac{\tau^a}{2}}(q), \tag{68.8}$$

called the chiral (axial) symmetry, is spontaneously broken. That issue, of *chiral symmetry breaking*, is one of the most important of particle physics, and its exact mechanism is unknown (we don't know an exact low-energy effective action that will show the spontaneous breaking). This chapter is devoted to a phenomenological description of the phenomenon.

Since we have a spontaneous breaking of an (approximate) symmetry, by the Goldstone theorem we must have (pseudo-)Goldstone bosons for the three generators, and transforming under the unbroken $SU(2)_V$ (isospin). There is only one candidate group, the pions π^a, which are approximately massless, since $m_\pi \ll \Lambda_{QCD}$. Here Λ_{QCD} is the spontaneously generated scale for QCD, that gives the mass of the physical states. (There are various ways to define Λ_{QCD}, but we will not try to define it here.)

We note that the Wigner–Eckhart theorem, which says that states should fall under multiplets of the full symmetry group, does not apply to spontaneous symmetry breaking (SSB). Indeed, SSB can be described by the fact that the vacuum is not invariant under the symmetry (i.e. $Q|0\rangle \neq 0$), which also means that the low-energy states are not in a multiplet.

So, in our case, the pions are not in a multiplet of $SU(2)_A$, but are still in a multiplet of the unbroken group, $SU(2)_V$ or isospin. Indeed, they transform as the adjoint of this group.

We said that we have spontaneous symmetry breaking for $SU(2)_A$, but until now we have seen only SSB via a scalar field VEV, and now we have only fermions and gauge fields in the theory. But the point is that in such a theory, the fermions form *condensates* (i.e. composite scalars made up of the fermions have a VEV). Therefore, we have a *quark condensate*

$$\langle 0|\bar{q}_{Li}q_R^j|0\rangle = -v\delta_i^j. \tag{68.9}$$

Here, i, j are flavor indices. Note that there are ways to see that in a gauge theory we have a fermion condensate at low energy, and we can show this for QCD, even if we don't know the exact mechanism.

68.2 Pseudo-Goldstone Bosons, Chiral Perturbation Theory, and Nonlinear Sigma Model

Up to now we have described massless quarks leading to massless pions (Goldstone bosons). But of course, in reality, up and down quarks are not massless, and the pions also have mass, which is much smaller, however, than the mass of the hadronic states p and n, made up also of only u and d quarks. So we should make an important distinction:

- For spontaneous symmetry breaking with massless pions, we would still obtain nucleons n and p (composite fermions) with a mass. In fact, m_p and m_n are approximately independent of the quark masses m_u and m_d, as could be guessed from the fact that $m_p, m_n \gg m_\pi$.
- On the contrary, a nonzero quark mass m_q is correlated with a nonzero pion mass m_π, so in the presence of quark mass, the pions (Goldstone bosons) are not massless anymore. In fact, we will see at the end of the chapter that we have the relation

$$m_\pi^2 = \frac{2(m_u + m_d)}{f_\pi^2}v = \frac{2(m_u + m_d)}{f_\pi^2}\langle 0|\bar{q}_L q_R|0\rangle. \tag{68.10}$$

The perturbation theory that we will obtain for the pion interactions will be a perturbation theory in p/f_π, and also a perturbation theory in m_π (related to it through the above), called *chiral perturbation theory* (Ch.P.T.).

To model spontaneous symmetry breaking for $SO(4) \simeq SU(2)_L \times SU(2)_R \rightarrow SU(2)_V$, we can describe the $SO(4)$ symmetry through a bifundamental action on the set ($\mathbb{1}, \tau^a$) of generators, which is a complete set in the space of 2×2 matrices. So we define the matrix

$$\Sigma = \sigma \, \mathbb{1} + i\tau^a \pi^a, \tag{68.11}$$

which is therefore an arbitrary 2×2 matrix field (with some reality properties), and where π^a will be related to the pions, but for now is just a set of real scalars. The action of the $SO(4) \simeq SU(2)_L \times SU(2)_R$ symmetry on Σ is given by

$$\Sigma \rightarrow g_L \Sigma g_R^\dagger. \tag{68.12}$$

Then, in terms of the field Σ, we describe phenomenologically the SSB for chiral symmetry through the Lagrangian that is a simple generalization of the Higgs Lagrangian:

$$\mathcal{L}_L = -\frac{1}{4} \text{Tr}[\partial_\mu \Sigma \partial^\mu \Sigma^\dagger] + \frac{\mu^2}{4} \text{Tr}[\Sigma\Sigma^\dagger] - \frac{\lambda}{16}[\text{Tr}(\Sigma\Sigma^\dagger)]^2. \tag{68.13}$$

This is called the *linear sigma model*. The kinetic term is the standard one (note, for instance, that for σ, multiplied by the identity, the trace gives a factor of 2, so the normalization is canonical), the mass term has the spontaneous symmetry breaking sign ($m^2 = -\mu^2 < 0$), so the potential is the matrix generalization of the Higgs potential.

If we want to couple to external gauge fields A_μ and V_μ (axial vector and vector), we would do it through the covariant derivative

$$D_\mu \Sigma = \partial_\mu \Sigma - i(V_\mu + A_\mu)\Sigma + i\Sigma(V_\mu - A_\mu). \tag{68.14}$$

The theory has a spontaneously broken vacuum with VEV v, and around it the expansion of Σ is

$$\Sigma(x) = (v + s(x))U(x); \quad U(x) = e^{\frac{i\tau^a \pi'^a(x)}{v}}. \tag{68.15}$$

Here, s is a scalar with zero VEV, $\langle s \rangle = 0$, v is the VEV, and $\pi'^a(x)$ are massless, so are indentified with the pions. Indeed, there is no mass term for U, so not for π'^a, while there is one for s, which therefore is the "Higgs" (i.e. the massive mode).

In the Wilsonian effective action approach, one integrates momenta with $|k| \geq \Lambda$, which means in particular that one must integrate all the fields with masses $m \geq \Lambda$ (since they always have $|k| \geq \Lambda$). This is called *integrating out the massive modes* in the Lagrangian. When doing that, as we saw, we obtain higher-dimensional operators in the Lagrangian, but at sufficiently low energies they are negligible, since they come with inverse powers of Λ.

So now, we integrate out the massive modes, including s, and this means that at low energies, to zeroth order, we can just drop the dependence on them (on s), since the higher-dimensional operators they will generate are small.

If we do that, the Lagrangian becomes nonlinear, so we have the *nonlinear sigma model*. Indeed, the Lagrangian is now

$$\mathcal{L}_{NL} = -\frac{v^2}{4} \operatorname{Tr}[\partial_\mu U \partial^\mu U^\dagger], \tag{68.16}$$

which looks linear, however we have to remember that we now have the constraint $U^\dagger U = 1$, or $\Sigma^\dagger \Sigma = v^2$ (whereas Σ was an arbitrary matrix before the integrating out), so if we solve the constraint, we get a nonlinear action.

To relate to QCD, we must describe the matrix U. If the QCD state is the state $|U\rangle$ instead of the vacuum $|0\rangle$ in (68.9), we have a generalization of the relation:

$$\langle U | \bar{q}_{Li} q_R^j | U \rangle = -v U_i^j, \tag{68.17}$$

where U_i^j is the matrix element of the matrix U associated with the state $|U\rangle$.

The state $|U\rangle$ is a pion state so, using a new normalization that anticipates $v = f_\pi$:

$$U(x) = \exp\left[\frac{i\pi^a(x)\tau^a}{f_\pi}\right], \tag{68.18}$$

and plugging in the nonlinear sigma model action, which is now

$$\mathcal{L}_{NL} = -\frac{f_\pi^2}{4} \operatorname{Tr}[\partial_\mu U \partial^\mu U^\dagger], \tag{68.19}$$

and expanding in $1/f_\pi$, we obtain

$$\mathcal{L} = -\frac{1}{2}\partial_\mu \pi^a \partial^\mu \pi^a + \frac{f_\pi^{-2}}{6}(\pi^a \pi^a \partial^\mu \pi^b \partial_\mu \pi^b - \pi^a \pi^b \partial^\mu \pi^a \partial_\mu \pi^b) + \ldots, \tag{68.20}$$

which is indeed a perturbation in p/f_π, so is chiral perturbation theory. In fact, one can use the original normalization, and obtain a Lagrangian with v, which can be compared with predictions about the pion decay, and obtain that $v = f_\pi$. We will say more on this later in the chapter.

More generally, we call a linear sigma model a set of N scalars with a symmetry and some canonical kinetic term.

68.3 The $SO(N)$ Vector Model

The most famous example is the $SO(N)$ model, in terms of a scalar that is a vector (fundamental representation) of $SO(N)$, with spontaneous symmetry breaking. The Lagrangian is a simple generalization of the above linear sigma model Lagrangian:

$$\mathcal{L} = -\frac{1}{2}(\partial_\mu \phi^i)^2 + \frac{\mu^2}{2}(\phi^i)^2 - \frac{\lambda}{4}[(\phi^i)^2]^2, \tag{68.21}$$

and is invariant under $SO(N)$ transformations $\phi^i \rightarrow R^i{}_j \phi^j$. Note that our QCD case is $N = 4$, for $SO(4) \simeq SU(2) \times SU(2)$. The potential

$$V = -\frac{\mu^2}{2}(\phi^i)^2 + \frac{\lambda}{4}[(\phi^i)^2]^2 \tag{68.22}$$

has a spontaneously broken vacuum at

$$(\phi_0^i)^2 = \frac{\mu^2}{\lambda}, \tag{68.23}$$

and by a symmetry transformation we can orient ϕ_0 along the Nth direction:

$$\phi_0 = (0, \ldots, 0, v). \tag{68.24}$$

We expand the fields around it as

$$\phi^i(x) = (\pi^k(x), v + \sigma(x)), \tag{68.25}$$

where $k = 1, \ldots, N - 1$. Then the Lagrangian becomes

$$\mathcal{L} = -\frac{1}{2}(\partial_\mu \pi^k)^2 - \frac{1}{2}(\partial_\mu \sigma)^2 - \frac{1}{2}(2\mu^2)\sigma^2 - \sqrt{\lambda}\mu\sigma^3$$
$$-\sqrt{\lambda}\mu(\pi^k)^2\sigma - \frac{\lambda}{4}\sigma^4 - \frac{\lambda}{2}(\pi^k)^2\sigma^2 - \frac{\lambda}{4}[(\pi^k)^2]^2. \tag{68.26}$$

In general, a *nonlinear sigma model* is defined as any Lagrangian of the type

$$\mathcal{L} = f_{ij}(\{\phi^k\})\partial_\mu \phi^i \partial^\mu \phi^j. \tag{68.27}$$

That is, a model with a metric (depending on the scalars) on the space of scalars. For instance, a famous example is the two-dimensional field theory on the worldsheet of a string propagating in a general spacetime, but a more relevant example for phenomenology would be a modulus scalar field in four dimensions.

With respect to the above $O(N)$ linear sigma model, the nonlinear sigma model is

$$\mathcal{L} = -\frac{1}{2g^2}(\partial_\mu \tilde{\phi}^i)\partial^\mu \tilde{\phi}^i, \tag{68.28}$$

with the constraint

$$\sum_{i=1}^{N}(\tilde{\phi}^i(x))^2 = 1 \tag{68.29}$$

(i.e. the fields are on a unit sphere). It gives a phenomenological description of a system with $O(N)$ symmetry spontaneously broken by a VEV (e.g. by integrating out the massive (radial) mode in the above expansion around a VEV to get $\phi^i \to v\tilde{\phi}^i(x)$, so the same thing we did in the $SU(2) \times SU(2)$ case).

We can solve the constraint by $N - 1$ Goldstone bosons π^k, as

$$\tilde{\phi}^i = (\pi^1, \ldots, \pi^{N-1}, \sigma), \tag{68.30}$$

where

$$\sigma = \sqrt{1 - \vec{\pi}^2}. \tag{68.31}$$

Now the manifest $SO(N)$ symmetry turns into a manifest $SO(N - 1)$ symmetry. Then we find that the kinetic term becomes

$$(\partial_\mu \tilde{\phi}^i)^2 = (\partial_\mu \vec{\pi})^2 + \frac{(\vec{\pi} \cdot \partial_\mu \vec{\pi})^2}{1 - \vec{\pi}^2}, \tag{68.32}$$

so the nonlinear sigma model action becomes

$$\mathcal{L} = -\frac{1}{2g^2}\left[(\partial_\mu \vec{\pi})^2 + \frac{(\vec{\pi}\cdot\partial_\mu\vec{\pi})^2}{1-\vec{\pi}^2}\right]$$

$$\simeq -\frac{1}{2g^2}\left[(\partial_\mu\vec{\pi})^2 + (\vec{\pi}\cdot\partial_\mu\vec{\pi})^2 + \dots\right]. \qquad (68.33)$$

Note that here the dimension of scalars is zero, $[\phi] = 0$, where the coupling has dimension -1, $[g] = -1$, and in the second line we have expanded in the scalars π^a.

To make the connection with chiral perturbation theory, we take $N = 4$, so $SO(4) \simeq SU(2) \times SU(2)$, and is spontaneously broken to $SU(2) \simeq SO(3)$ acting on the π^ks. So initially, $SO(4)$ is manifest, but there is a constraint. When solving the constraint for $\tilde{\phi}^i$ in terms of π^k, only $SO(3)$ remains linearly realized, the other $SO(3)$ becomes nonlinearly realized.

The issue of nonlinear realizations follows the same pattern in general: what we usually call a symmetry is a linearly realized symmetry (i.e. a symmetry that acts linearly on the fields). When we have a nonlinearly realized symmetry, it is indicative of a case where there is a more fundamental representation that has the symmetry, like introducing an auxiliary field, or writing the theory with a constraint, or where we are in a spontaneously broken vacuum. But in any case, the symmetry is not *manifest* in the action. We will see that more explicitly in the last form we describe for the pions.

We can use yet another way to solve the constraint of the $SO(4)$ model. We can write the sigma model as a rotation R acting on the vector $(0,0,0,\sigma)$ parametrized only by the massive mode σ, as

$$\phi_i(x) = R_{i4}(x)\sigma(x), \qquad (68.34)$$

where the matrix R is orthogonal (in $SO(4)$), so satisfies $RR^T = 1$. Then, by squaring the above relation and using $RR^T = 1$, we get

$$\sigma(x) = \sqrt{\sum_{i=1}^{4}(\phi^i)^2}. \qquad (68.35)$$

Replacing in the linear sigma model Lagrangian, we find

$$\mathcal{L} = -\frac{1}{2}\partial_\mu\sigma\partial^\mu\sigma - \frac{\sigma^2}{2}\sum_{i=1}^{4}\partial^\mu R_{i4}\partial_\mu R_{i4} + \frac{\mu^2}{2}\sigma^2 - \frac{\lambda}{4}\sigma^4. \qquad (68.36)$$

Parametrizing the fields as

$$\zeta_a = \frac{\phi_a}{\phi_4 + \sigma},$$

$$R_{a4} = \frac{2\zeta_a}{1+\vec{\zeta}^2} = -R_{4a},$$

$$R_{44} = \frac{1-\vec{\zeta}^2}{1+\vec{\zeta}^2},$$

$$R_{ab} = \delta_{ab} - \frac{2\zeta_a\zeta_b}{1 + \vec{\zeta}^2}, \tag{68.37}$$

we obtain the Lagrangian

$$\mathcal{L} = -\frac{1}{2}\partial_\mu\sigma\,\partial^\mu\sigma - 2\sigma^2\vec{D}_\mu\vec{D}^\mu + \frac{\mu^2}{2}\sigma^2 - \frac{\lambda}{4}\sigma^4, \tag{68.38}$$

where

$$\vec{D}_\mu \equiv \frac{\partial_\mu\vec{\zeta}}{1 + \vec{\zeta}^2} \tag{68.39}$$

is often called the *covariant derivative of the pion field.*

Then the transformation rules for the scalars are:

- For isospin $SU(2)_V$, that acts as an $SO(3)$ rotation of ϕ^a, thus of ζ_a, leaving ϕ^4 and σ invariant, we have

$$\delta\vec{\zeta} = \vec{\alpha} \times \vec{\zeta}; \quad \delta\sigma = 0, \tag{68.40}$$

  which is a linear transformation, since the group is unbroken.
- For the axial vector $SU(2)_A$, which is broken, from $\delta\vec{\phi} = 2\vec{\epsilon}\phi_4$ and $\delta\phi_4 = -2\vec{\epsilon}\cdot\vec{\phi}$, we get

$$\delta\vec{D}_\mu = 2(\vec{\zeta} \times \vec{\epsilon}) \times \vec{D}_\mu, \tag{68.41}$$

  which is a nonlinear transformation (i.e. broken).

The VEV of the scalar σ is $\langle\sigma\rangle = v$, and integrating out the fluctuation in σ, we obtain the nonlinear sigma model Lagrangian

$$\mathcal{L} = -2v^2\vec{D}_\mu\vec{D}^\mu. \tag{68.42}$$

Defining the pions as

$$\vec{\pi} \equiv 2v\vec{\zeta}, \tag{68.43}$$

we get the pion Lagrangian

$$\mathcal{L} = -\frac{1}{2}\frac{\partial_\mu\vec{\pi}\cdot\partial^\mu\vec{\pi}}{\left(1 + \frac{\vec{\pi}^2}{4v^2}\right)^2}. \tag{68.44}$$

68.4 Physical Processes and Generalizations

We now observe that we have written several Lagrangians for the pions, (68.20), (68.33), and (68.44). They are all Goldstone boson Lagrangians, meaning that the interactions all involve derivatives of the pions, and the Lagrangians are an expansion in p/f_π (i.e. chiral perturbation theory). What does this mean?

Really, it means that we should use the *Wilsonian effective field theory* approach for the pions, and write down the most general Lagrangian consistent with all the symmetries (all the higher-dimension operators consistent with the symmetry), and fix the coefficients from comparison with experiments. But this will still not account for the difference between our three Lagrangians. The point is that, of course, a (possibly nonlinear) field redefinition consistent with the symmetries should not change the physics (i.e. the scattering amplitudes), for instance something like

$$\vec{\pi}' = \frac{\vec{\pi}}{1 + \vec{\pi}^2} \tag{68.45}$$

should leave the physics invariant. We can use such field redefinitions to put operators we want to zero, leading to various expressions, depending on what terms we want to keep after the redefinitions.

As an example of the comparison with experiment, expanding the Lagrangian in (68.44) in $1/v$ and comparing with experiment, we see that we need $v = f_\pi$, as already stated for (68.20).

We have not described well what f_π is until now, so we will rectify this omission. The *pion decay constant f_π* is fixed by the PCAC relation from Chapter 56, which stated that

$$\partial^\mu j_\mu^{5(A)\pm} = f_\pi m_\pi^2 \pi^\pm(x) \tag{68.46}$$

or equivalently

$$j_\mu^{5,\pm}(\text{hadronic}) = f_\pi \partial_\mu \pi^\pm(x), \tag{68.47}$$

and another relation for π^3 that contains also the anomalous part, and from which we have derived the $\pi^0 \to \gamma\gamma$ decay. But here we are interested in the decay of $\pi^\pm$, which has no anomalous component. The normalization of π^- is given by

$$\langle 0|\pi^-|\pi^-\rangle = \frac{1}{\sqrt{2m_{\pi^-}}}, \tag{68.48}$$

so we get

$$\langle 0|j_\mu^{5-}|\pi^-\rangle = q_\mu f_\pi \frac{1}{\sqrt{2m_{\pi^-}}}. \tag{68.49}$$

Then we fix f_π from the decay $\pi^- \to \mu^- + \bar{\nu}_\mu$. The relevant electroweak interaction in the 4-fermi limit is

$$\mathcal{L}_{\text{weak}} = \frac{G_F}{\sqrt{2}} \bar{\psi}_{(\mu)} \gamma^\rho (1 + \gamma_5) \psi_{(\nu_\mu)} (j_\rho^{V,-} + j_\rho^{A,-}), \tag{68.50}$$

which (after a calculation that will not be reproduced here, but is left as an exercise) gives for the decay rate

$$\Gamma = \frac{1}{8\pi} \left(\frac{m_\pi^2 - m_\mu^2}{m_\pi^2}\right)^2 (G_F m_\mu m_\pi)^2 \left(\frac{f_\pi}{m_\pi}\right)^2 \tag{68.51}$$

and then experimentally from the decay, we can fix $f_\pi \simeq 93\,\text{MeV}$.

68.4.1 Generalization

We can also generalize the mechanism of spontaneous symmetry breaking for a chiral symmetry to any $G \rightarrow H$, though we will not do any calculation here. Then the Goldstone bosons parametrize the coset G/H and transform linearly under H and nonlinearly under G/H. Again we can find the most general form allowed by symmetries, do field redefinitions, and fix coefficients from experiment.

68.4.2 Generalization to $SU(3)$

We now show how to include the s quark, which has a slightly larger mass, so its chiral perturbation theory is somewhat less useful (the corrections are large). The Goldstone boson matrix U is now written as

$$U = \begin{pmatrix} \frac{\pi^0}{\sqrt{2}} + \frac{\eta^0}{\sqrt{6}} & \pi^+ & K^+ \\ \pi^- & -\frac{\pi^0}{\sqrt{2}} + \frac{\eta^0}{\sqrt{6}} & K^0 \\ K^- & \bar{k}^0 & -\frac{2}{\sqrt{6}}\eta^0 \end{pmatrix} \qquad (68.52)$$

otherwise we have the same idea, so the construction will not be repeated here. We just note that now the adjoint of $SU(3)$ has eight components, and these are parametrized by $\pi^+, \pi^-, \pi^0, \eta^0, K^+, K^-, K^0, \bar{K}^0$.

As mentioned, the light quarks are u and d, possibly together with the s quark, whereas the c, b, t quarks are heavy.

68.5 Heavy Quark Effective Field Theory

For the c, b, t quarks we have another type of effective field theory, one that takes into account the fact that for many things the quarks can be treated as nonrelativistic. So we write something like nonrelativistic quantum mechanics (NRQM), but with Lorentz indices.

We start with the quark Lagrangian

$$\bar{Q}_i i\gamma^\mu \partial_\mu Q_i - M\bar{Q}_i Q_i - \bar{Q}_i g\slashed{A} Q_i \qquad (68.53)$$

and consider momenta

$$p_{(i)} = Mv + k_{(i)}, \qquad (68.54)$$

where $v \ll 1$ and k is small. The *4-vector* velocity v satisfies $v^2 = 1 \Rightarrow \slashed{v}^2 = 1$, and we define the projectors

$$P_\pm = \frac{1}{2}(1 \pm \slashed{v}) \qquad (68.55)$$

and split the fields according to it into positive and negative "chirality," $\slashed{v}h = h$ and $\slashed{v}\chi = -\chi$, and throw away the "negative chirality," really negative energy, states (such as going from field theory to NRQM).

Thus we define

$$Q = e^{-iMv \cdot x}(h + \chi), \tag{68.56}$$

where x^μ is a 4-vector like v^μ, and $h(x)$ now contains only the small variation k^μ, and throw away χ, arriving at the HQET Lagrangian to zeroth order

$$\mathcal{L} = \sum_j \bar{h}_j (iv \cdot D) h_j. \tag{68.57}$$

Keeping the next order also, we get from general considerations

$$\mathcal{L} = \sum_j \left[\bar{h}_j (iv \cdot D) h + \frac{1}{2M_Q} \bar{h}_j [\alpha (i\slashed{D})^2 + \beta (v \cdot D)^2] h_j \right]. \tag{68.58}$$

But by reparametrization invariance we can put $\alpha = 1$, and the second term doesn't contribute to physical processes, so can be dropped.

68.6 Coupling to Nucleons

Until now we have described only the pions, but from a phenomenological perspective it is even more interesting to describe how they couple to the nucleons that compose matter.

The free nucleon Lagrangian is written as

$$- \bar{\mathcal{N}} \slashed{\partial} \mathcal{N} - m_N \bar{\mathcal{N}} \mathcal{N}. \tag{68.59}$$

The interaction is written as follows. From Lorentz invariance it must have $\bar{\mathcal{N}}$ and $\mathcal{N}$. From the pions point of view it must be a derivative interaction, so $\partial_\mu \pi^a$, which means that we need $\gamma^\mu \tau^a$ in the action as well. The coupling is an axial coupling, so it will have a γ_5 as well, and by dimensional analysis we need a g_A/f_π, for a total interaction

$$- i \frac{g_A}{f_\pi} \partial_\mu \vec{\pi} \bar{\mathcal{N}} \gamma^\mu \gamma_5 \frac{\vec{\tau}}{2} \mathcal{N}, \tag{68.60}$$

where g_A is an axial vector coupling, found experimentally to be about 1.27. By partially integrating the derivative, using the free equation of motion for the nucleon to replace it with m_N, we get

$$g_{\pi NN} = \frac{m_N g_A}{f_\pi}. \tag{68.61}$$

Defining the fields

$$u = \exp \left[\frac{i\pi^a \tau^a}{2f_\pi} \right],$$

$$A_\mu = \frac{i}{2} (u^\dagger \partial_\mu u - u \partial_\mu u^\dagger),$$

$$V_\mu = \frac{i}{2} (u^\dagger \partial_\mu u + u \partial_\mu u^\dagger) \tag{68.62}$$

(note that $u = \sqrt{U}$), we can complete the interaction term to the full terms

$$\bar{\mathcal{N}} V_\mu \gamma^\mu \mathcal{N} - g_A \bar{\mathcal{N}} A_\mu \gamma^\mu \gamma_5 \mathcal{N}. \tag{68.63}$$

Then with the field redefinition

$$\mathcal{N} = \left(u^\dagger \frac{1 + \gamma_5}{2} + u \frac{1 - \gamma_5}{2} \right) N, \tag{68.64}$$

we can rewrite the Lagrangian as

$$\mathcal{L} = -\bar{N} \slashed{\partial} N - m_N \bar{N} \left(U^\dagger \frac{1 + \gamma_5}{2} + U \frac{1 - \gamma_5}{2} \right) N$$
$$- \frac{1}{2} (g_A - 1) \bar{N} \gamma^\mu \left(U \partial_\mu U^\dagger \frac{1 + \gamma_5}{2} + U^\dagger \partial_\mu U \frac{1 - \gamma_5}{2} \right) N. \tag{68.65}$$

Note that it is now written in terms of U, not u.

68.7 Mass Terms

By an $SU(2)_L \times SU(2)_R$ transformation, we can put the quark mass matrix in the diagonal form

$$M = \begin{pmatrix} m_u & 0 \\ 0 & m_d \end{pmatrix} e^{-i\theta/2}. \tag{68.66}$$

Then the mass term is

$$\mathcal{L}_{\text{mass}} = - \operatorname{Tr}[\bar{q}_L M q_R], \tag{68.67}$$

where the trace is over the flavor indices i, j. Since in the vacuum we have the fermion condensate (68.9), and in the physical state we have (68.17), we replace the mass term by

$$\mathcal{L}_{\text{mass}} = v \operatorname{Tr}[M U + M^\dagger U^\dagger]. \tag{68.68}$$

Considering a real mass matrix, $M = M^\dagger$, and expanding in $1/f_\pi$, we get

$$\mathcal{L}_{\text{mass}} = - \frac{v}{f_\pi^2} (\operatorname{Tr} M) \pi^a \pi^a + \dots, \tag{68.69}$$

which implies

$$m_\pi^2 = \frac{2(m_u + m_d) v}{f_\pi^2}, \tag{68.70}$$

as advertised at the beginning of the chapter. This is called the *Gell-Mann–Oakes–Renner relation*.

Note that sometimes one replaces the mass term by

$$- m_\pi^2 \operatorname{Tr}[U + U^\dagger - 2], \tag{68.71}$$

considering that $m_u \simeq m_d$ and subtracting the constant term from the Lagrangian.

Important Concepts to Remember

- When we have an approximate symmetry spontaneously broken, we get pseudo-Goldstone bosons.
- Chiral symmetry is the $U(2) \times U(2)$ approximate symmetry of low-energy QCD with just u and d quarks. The two $U(1)$s are the conserved baryon number and an abelian chiral symmetry broken by anomalies, and the $SU(2)_L \times SU(2)_R$ is spontaneously broken to $SU(2)_V$.
- The VEV that breaks the symmetry is a fermion (quark) condensate (i.e. a composite field).
- The exact mechanism of chiral symmetry breaking is unknown, but we make models for it.
- The quark masses do not affect the nucleon masses much (they are nonzero because of confinement, not because of quark masses), but the pion mass squared is proportional to the quark masses.
- Chiral perturbation theory is the effective theory for pions, which is an expansion in p/f_π and m_π.
- The linear sigma model is a phenomenological model for the SSB of chiral symmetry, in terms of the pions π^a and the massive σ field, with canonical (linear) kinetic term. There are various descriptions for it.
- The nonlinear sigma model is the nonlinear model in terms of only the pions, obtained by integrating out the massive modes. In general, it is a model with a metric on scalar field space.
- In the $SO(N)$ model, we have N scalars in a vector representation of $SO(N)$. The nonlinear sigma model corresponds to the scalars on a unit sphere.
- In general, for $G \rightarrow H$ breaking, the Goldstone bosons live in the coset G/H and transform linearly under H and nonlinearly under G/H.
- Chiral perturbation theory is understood from the Wilsonian effective field theory approach: write the most general Lagrangian consistent with the symmetries, and use field redefinitions to put various terms to zero.
- One can include the s quark in chiral perturbation theory, and the Goldstone boson matrix includes now $K^\pm, K^0, \bar{K}^0$, and η^0, but is less useful since it has larger corrections.
- For the heavy quarks c, b, t, one can use heavy quark effective field theory, which is like nonrelativistic quantum mechanics with Lorentz indices.

Further Reading

See chapter 83 in [10], sections 11.1 and and 13.3 in [1], and sections 19.4 and 19.5 in [11].

Exercises

1. Calculate the pion action in chiral perturbation theory up to order f_π^{-4}, by substituting the matrix U in (68.18) into (68.19) to obtain (68.20) plus higher-order terms.

2. Calculate the pion Lagrangian (68.33) under the redefinition (68.45). Show that the corresponding three-point scalar (tree-level, i.e. classical) amplitude is the same for the two versions, using the equations of motion of the Lagrangian.

3. Show that the pion decay $\pi^- \to \mu^- + \bar{\nu}_\mu$ with the electroweak interaction in the 4-fermi limit (68.50) has a decay rate (68.51).

4. Show that the heavy quark fermion Lagrangian, in the nonrelativistic large mass limit, gives (68.58) for $\alpha = 1$ and $\beta = 0$.

The Background Field Method

In this chapter we will consider a method to deal mostly with gauge theories, by keeping a kind of gauge invariance at the quantum level. In the usual formalism, at the quantum level gauge invariance is lost, and we are left with the global BRST invariance, its remnant, as well as with Ward identities that come from gauge invariance.

But what one can do is split the gauge field $\tilde{A}_\mu^a$ into a classical, background part A_μ^a and a quantum part Q_μ^a, and integrate only over Q_μ^a. This leaves the classical gauge invariance (coming from A_μ^a) intact, and it turns out to be useful for calculations. This formalism was introduced at one loop by Bryce DeWitt in 1964.

69.1 General Method and Quantum Partition Function

The action is just the usual action for $\tilde{A}_\mu^a$, so at the classical level

$$S_{cl} = S_{cl}, \text{YM}[A_\mu^a + Q_\mu^a] = -\int d^4x \frac{1}{4} \operatorname{Tr} F_{\mu\nu}[A_\mu^a + Q_\mu^a]^2. \tag{69.1}$$

It is invariant under

$$\delta_{\text{gauge}}(A_\mu^a + Q_\mu^a) = \partial_\mu \lambda^a + g f^a{}_{bc}(A_\mu^b + Q_\mu^b)\lambda^c \equiv D_\mu(A+Q)\lambda^a. \tag{69.2}$$

We now decompose this invariance as a transformation of A_μ and Q_μ in two different ways. One as an invariance under a *classical (background) gauge transformation*, under which A_μ^a transforms as a gauge field and the quantum field Q_μ^a is viewed as a kind of vector field, namely

$$\delta_{\text{bgr}}A_\mu^a = D_\mu(A)\lambda^a \equiv \partial_\mu \lambda^a + g f^a{}_{bc}A_\mu^b \lambda^c,$$
$$\delta_{\text{bgr}}Q_\mu^a = g f^a{}_{bc}Q_\mu^b \lambda^c, \tag{69.3}$$

and another as a *quantum gauge transformation*, under which the background A_μ^a is invariant, and the quantum field transforms as

$$\delta_{\text{qu}}Q_\mu^a = D_\mu(A+Q)\lambda^a \equiv \partial_\mu \lambda^a + g f^a{}_{bc}(A_\mu^b + Q_\mu^b)\lambda^c,$$
$$\delta_{\text{qu}}A_\mu^a = 0. \tag{69.4}$$

Under the classical (background) gauge transformation, A_μ is a connection and Q_μ is an adjoint matter field, which means that their sum, $\tilde{A}_\mu = A_\mu + Q_\mu$, is also a connection, as it should be. Note then that $S_{cl}, \text{SYM}[A_\mu]$ is also background gauge invariant.

Under the quantum gauge transformation, the classical field is invariant. We gauge-fix this symmetry in the usual way, with the standard-gauge fixing term modified by the addition of the background field A_μ^a:

$$\mathcal{L}_{\text{g.fix}} = -\frac{1}{2\alpha}(D^\mu(A)Q_\mu^a)^2. \tag{69.5}$$

After gauge fixing, the remnant of the quantum gauge symmetry is the global BRST symmetry, that changes the parameter by the local field $c^a(x)$ times a global parameter, so

$$\delta_{\text{BRST}}A_\mu = 0,$$
$$\delta_{\text{BRST}}Q_\mu = D_\mu(A+Q)c^a\Lambda. \tag{69.6}$$

But the gauge-fixing term was chosen so that it is still invariant under the *background* gauge transformation, under which Q_μ^a transforms as a gauge group vector (matter field in the adjoint), and so does $D^\mu(A)Q_\mu^a$. Also notice that

$$D^\mu(A)Q_\mu^a = D^\mu(A+Q)Q_\mu^a. \tag{69.7}$$

As usual, the ghost action is obtained from the gauge-fixing function $F^a = D^\mu(A)Q_\mu^a$ (that gives the gauge-fixing term as $-\frac{1}{2}\gamma_{ab}F^aF^b$) by turning Q_μ^a into its BRST variation and multiplying with b^a:

$$\mathcal{L}_{\text{ghost}} = b_a\delta_{\text{BRST}}F^a/\Lambda = b_aD^\mu(A)\delta_{\text{BRST}}Q_\mu^a/\Lambda = b^aD^\mu(A)D_\mu(A+Q)c^a. \tag{69.8}$$

The construction above guarantees BRST invariance, which besides the above variations of A_μ and Q_μ also has the usual ($\gamma_{ab} = \delta_{ab}/\alpha$)

$$\delta_{\text{BRST}}c^a = \frac{1}{2}gf^a{}_{bc}c^bc^c\Lambda,$$
$$\delta_{\text{BRST}}b_a = -\gamma_{ab}F^b\Lambda = -\frac{1}{\alpha}\delta_{ab}D^\mu(A)A_\mu^b\Lambda. \tag{69.9}$$

But now we also have background gauge invariance, which is an invariance (by construction) for the classical part $\mathcal{L}_{cl,\text{YM}}$ and the gauge-fixing part. In order for it to also be an invariance for the ghost part, we need to define the background gauge transformations of the ghost fields as transformations of matter (vector) fields in the adjoint representation:

$$\delta_{\text{bgr}}b^a = gf^a{}_{bc}b^b\lambda^c,$$
$$\delta_{\text{bgr}}c^a = gf^a{}_{bc}c^b\lambda^c. \tag{69.10}$$

To complete the quantum partition function, we must add external sources. As usual, we add sources β_a and γ^a for the fields c^a and b_a, respectively. For the gauge field, we have in principle two choices: to add a source for the full gauge field $\tilde{A}_\mu^a = A_\mu^a + Q_\mu^a$, or just for the quantum gauge field Q_μ^a. We will opt for the latter, since the source is supposed to be for a quantum field; a classical component has no need for a source in a quantum theory. Thus, we add the source part to the action

$$S_{\text{source}} = \int d^4x(J_a^\mu Q_\mu^a + \beta_ac^a + b_a\gamma^a). \tag{69.11}$$

However, as we have seen before, it is useful to also add a source term for the BRST variations of Q_μ^a and c^a (to derive LZJ identities, as we saw in Chapter 46):

$$S_{\text{extra,source}} = \int d^4x[K_a^\mu(Q_{\text{BRST}}Q)_\mu^a - L_a(Q_{\text{BRST}}c)^a]$$

$$= \int d^4x \left[K_a^\mu D_\mu(A+Q)c^a - L_a \frac{1}{2}gf^a{}_{bc}c^bc^c \right], \tag{69.12}$$

for a total partition function of

$$Z[J_a^\mu, \beta_a, \gamma^a, K_a^\mu, L_a, A_\mu^a] = \int \mathcal{D}Q_\mu^a \mathcal{D}b_a \mathcal{D}c^a,$$

$$\exp \left[iS_{cl,\text{YM}}(A+Q) + iS_{\text{g.fix}} + iS_{\text{ghost}} + iS_{\text{source}} + iS_{\text{extra,source}} \right]. \tag{69.13}$$

Since Q_μ^a, c^a, b_a and $Q_{\text{BRST}}c^a, Q_{\text{BRST}}Q_\mu^a$ transform as vectors (matter in the adjoint representation) under *background* gauge transformations, if we also impose that $J_a^\mu, \beta_a, \gamma^a, K_a^\mu, L_a$ transform as vectors, that is

$$\delta_{\text{bgr}}J^{\mu a} = gf^a{}_{bc}J^{\mu b}\lambda^c,$$
$$\delta_{\text{bgr}}\beta^a = gf^a{}_{bc}\beta^b\lambda^c,$$
$$\delta_{\text{bgr}}\gamma^a = gf^a{}_{bc}\gamma^b\lambda^c,$$
$$\delta_{\text{bgr}}K^{\mu a} = gf^a{}_{bc}K^{\mu b}\lambda^c,$$
$$\delta_{\text{bgr}}L^a = gf^a{}_{bc}L^b\lambda^c, \tag{69.14}$$

then the full *partition function is background gauge invariant*.

69.2 Scalar Field Analysis for Effective Action

Before we continue with the gauge field, we explore the simpler case of a scalar field, in order to understand better the logic and manipulations we will do to arrive at our goal, to obtain the effective action of usual QFT, from the background field formalism. For simplicity of later references, we denote the fields by the same names as in the gauge field case: $\tilde{A} = A + Q$ for the split of the real scalar field into a background A and a quantum piece Q.

Unlike the gauge field case, now there is no gauge fix and ghost pieces, just the classical action $S_{cl}(A+Q)$ and the source piece $\int d^4xJQ$, with partition function

$$Z[J,A] = \int \mathcal{D}Qe^{i[S_{cl}(A+Q)+\int d^4xJQ]}e^{iW[J,A]}. \tag{69.15}$$

The classical field, or VEV of the quantum field Q in the presence of a source, is

$$Q_{cl}[J,A] = \frac{\langle 0|Q|0 \rangle}{\langle 0|0 \rangle} = \frac{1}{Z[J,A]} \int \mathcal{D}Qe^{i[S_{cl}(A+Q)+\int d^4xJQ]}Q = \left. \frac{\delta}{\delta J}W[J,A] \right|_A. \tag{69.16}$$

In the usual formulation of quantum field theory, we need to impose that there are no tadpoles (i.e. that the classical field in the absence of sources is zero, $Q_{cl}[J = 0] = 0$). But in the presence of a background field A, that is not necessarily so anymore:

$$Q_{cl}[J = 0, A] \neq 0. \tag{69.17}$$

The effective action, which is the generator of the 1PI Green's functions, except for an inverse propagator in the two-point function, is the Legendre transform of the free energy, which now still depends on the background field:

$$\Gamma[Q_{cl}, A] = W[J, A] - \int d^4 x J Q_{cl}. \tag{69.18}$$

Our first goal now is to establish the relation to the usual QFT formalism, in the absence of the background field.

The first step is to shift the integration variable in $Z[J, A]$ from $Q \to \tilde{Q} = Q - A$, obtaining

$$Z[J, A] = \int \mathcal{D}Q e^{i[S_{cl}(Q) + \int d^4 x J Q]} e^{-i \int d^4 x J A} = Z_{QFT}[J] e^{-i \int d^4 x J A}, \tag{69.19}$$

where $Z_{QFT}[J]$ is the usual QFT partition function, and as usual

$$Z_{QFT}[J] = e^{i W_{QFT}[J]},$$

$$Q_{cl}[J] = \frac{\delta}{\delta J} W_{QFT}[J],$$

$$\Gamma_{QFT}[Q_{cl}] = W_{QFT}[J] - \int d^4 x J Q_{cl}. \tag{69.20}$$

Then we further obtain, first

$$W[J, A] = W_{QFT}[J] - \int d^4 x J A, \tag{69.21}$$

as well as, by differentiation:

$$\tilde{Q}_{cl} \equiv Q_{cl}[J; A] = \frac{\delta}{\delta J} W[J; A] \bigg|_A = Q_{cl}[J] - A \equiv Q_{cl} - A. \tag{69.22}$$

Second, we use these relations in the definition of the effective action, to obtain

$$\Gamma[\tilde{Q}_{cl}, A] = W[J, A] - \int d^4 x J (Q_{cl} - A) = W_{QFT}[J] - \int d^4 x J A - \int d^4 x J (Q_{cl} - A)$$

$$= W_{QFT}[J] - \int d^4 x J Q_{cl} \equiv \Gamma_{QFT}[Q_{cl}] = \Gamma_{QFT}[\tilde{Q}_{cl} + A]. \tag{69.23}$$

So the effective action in the background field formalism only depends on the combination $\tilde{Q}_{cl} + A$. In particular, we can consider putting

$$\tilde{Q}_{cl} = Q_{cl}[J, A] = 0 \Rightarrow J = J[A], \tag{69.24}$$

and then

$$\Gamma[0, A] = \Gamma_{QFT}[A], \tag{69.25}$$

which means that differentiation $\Gamma[0, A]$ with respect to A gives the 1PI Green's functions of usual QFT (whose generating functional is $\Gamma_{\text{QFT}}[A]$). We will see that this implies we can obtain S-matrices from the background field formalism by this procedure.

69.3 Gauge Theory Analysis

We now generalize the above analysis to gauge theory. The new elements are the gauge-fixing term and the ghost term, both defined by the gauge-fixing function

$$F^a(A, Q) \equiv D^\mu(A) Q_\mu^a, \tag{69.26}$$

where we have emphasized the dependence on A and Q for later use, and the source and extra source terms. Emphasizing the dependence on this gauge-fixing function $F^a(A, Q)$ in the partition function, we write

$$Z[J, A] \equiv Z[J_a^\mu, \beta_a, \gamma^a, K_a^\mu, L_a, A_\mu^a; F^a(A, Q)] = \int \mathcal{D}Q_\mu^a \mathcal{D}b_a \mathcal{D}c^a$$

$$\exp\left[i \int d^4x \left\{ \mathcal{L}_{cl, \text{YM}}(A + Q) - \frac{1}{2\alpha} (D^\mu(A) Q_\mu^a)^2 + b_a D_\mu(A) D_\mu(A + Q) c^a \right.\right.$$

$$\left.\left. + J_a^\mu Q_\mu^a + \beta_a c^a + b_a \gamma^a + K_a^\mu D_\mu(A + Q) c^a - L_a \frac{1}{2} g f^a{}_{bc} c^b c^c \right\} \right]. \tag{69.27}$$

We now shift as before the integration variable, $Q \to \tilde{Q} = Q - A$, obtaining

$$Z[J, A] \equiv Z[J_a^\mu, \beta_a, \gamma^a, K_a^\mu, L_a, A_\mu^a; F^a(A, Q - A)] = \int \mathcal{D}Q_\mu^a \mathcal{D}b_a \mathcal{D}c^a$$

$$\exp\left[i \int d^4x \left\{ \mathcal{L}_{cl, \text{YM}}(Q) - \frac{1}{2\alpha} (D^\mu(A) Q_\mu^a - \partial^\mu A_\mu^a)^2 + b_a D_\mu(A) D_\mu(Q) c^a \right.\right.$$

$$\left.\left. + J_a^\mu (Q_\mu^a - A_\mu^a) + \beta_a c^a + b_a \gamma^a + K_a^\mu D_\mu(Q) c^a - L_a \frac{1}{2} g f^a{}_{bc} c^b c^c \right\} \right]. \tag{69.28}$$

Here we have used the fact that

$$F^a(A, Q - A) = D^\mu(A)(Q_\mu^a - A_\mu^a) = D^\mu(A) Q_\mu^a - \partial^\mu A_\mu^a. \tag{69.29}$$

But then note that under the *usual* formalism of QFT:

$$\delta_{\text{BRST}} F^a(A, Q - A)/\Lambda = D^\mu(A) \delta_{\text{BRST}} Q_\mu^a / \Lambda = D^\mu(A) D_\mu(Q) c^a, \tag{69.30}$$

so the partition function in the *usual* QFT formalism, yet with the gauge-fixing function $F^a(A, Q - A)$ (that depends on a classical gauge field A_μ^a):

$$Z_{\text{QFT}}[J; A, F^a(A, Q - A)] \equiv Z_{\text{QFT}}[J_a^\mu, \beta_a, \gamma^a, K_a^\mu, L_a; F^a(A, Q - A)]$$

$$= \int \mathcal{D}Q_\mu^a \mathcal{D}b_a \mathcal{D}c^a \exp\left[i \int d^4x \left\{ \mathcal{L}_{cl,\text{YM}}(Q) - \frac{1}{2\alpha}(D^\mu(A)Q_\mu^a - \partial^\mu A_\mu^a)^2 \right.\right.$$

$$+ b_a D_\mu(A) D_\mu(Q) c^a + J_a^\mu Q_\mu^a + \beta_a c^a$$

$$\left.\left. + b_a \gamma^a + K_a^\mu D_\mu(Q) c^a - L_a \frac{1}{2} g f^a{}_{bc} c^b c^c \right\}\right]. \qquad (69.31)$$

Finally, we have a relation between the partition function of the background field formalism and the usual QFT, yet with A_μ^a-dependent gauge-fixing function (and thus depending nevertheless on A_μ^a) given by

$$Z[J, A; F^a(A, Q)] = Z_{\text{QFT}}[J; A, F^a(A, Q - A)] e^{-i \int d^4x J_a^\mu A_\mu^a}, \qquad (69.32)$$

or for the free energies

$$W[J, A; F^a(A, Q)] = W_{\text{QFT}}[J; A, F^a(A, Q - A)] - \int d^4x J_a^\mu A_\mu^a. \qquad (69.33)$$

As in the scalar case, we have

$$\tilde{Q}_{\mu,cl}^a \equiv Q_{\mu,cl}^a[J, A; F^a(A, Q)] = \left. \frac{\delta}{\delta J_a^\mu} W[J, A; F^a(A, Q)] \right|_A, \qquad (69.34)$$

and moreover now for the usual QFT with the unusual gauge-fixing function:

$$Q_{\mu,cl}^a \equiv Q_{\mu,cl}^a[J; A, F^a(A, Q - A)] = \left. \frac{\delta}{\delta J_a^\mu} W_{\text{QFT}}[J; A, F^a(A, Q - A)] \right|_A, \qquad (69.35)$$

so that as before we obtain

$$\tilde{Q}_{\mu,cl}^a = Q_{\mu,cl}^a - A_\mu^a. \qquad (69.36)$$

We now construct the effective action in the usual way, as the Legendre transform of the free energy, just that now we have more sources (for b_a and c^a as well):

$$\Gamma[\tilde{Q}_{cl}, A; F^a(A, Q)] \equiv \Gamma[\tilde{Q}_{\mu,cl}^a, c_{cl}^a, b_{a,cl}, K_a^\mu, L_a; F^a(A, Q)]$$

$$= W[J_a^\mu, \beta_a, \gamma^a, K_a^\mu, L_a, A_\mu^a; F^a(A, Q)]$$

$$- \int d^4x (J_a^\mu \tilde{Q}_{\mu,cl}^a + \beta_a c_{cl}^a + b_{a,cl} \gamma^a)$$

$$= W[J, A; F^a(A, Q)] - \int d^4x (J\tilde{Q}_{cl} + \beta c_{cl} + b_{cl}\gamma). \qquad (69.37)$$

Similarly, we construct the usual QFT effective action, but with the unusual gauge-fixing term:

$$\Gamma_{\text{QFT}}[Q_{cl}; A, F^a(A, Q - A)] \equiv \Gamma_{\text{QFT}}[Q^a_{\mu,cl}, c^a_{cl}, b_{a,cl}, K^\mu_a, L_a; F^a(A, Q - A)]$$
$$= W_{\text{QFT}}[J^\mu_a, \beta_a, \gamma^a, K^\mu_a, L_a; F^a(A, Q - A)]$$
$$- \int d^4x (J^\mu_a Q^a_{\mu,cl} + \beta_a c^a_{cl} + b_{a,cl}\gamma^a)$$
$$= W_{\text{QFT}}[J; A, F^a(A, Q - A)] - \int d^4x (JQ_{cl} + \beta c_{cl} + b_{cl}\gamma).$$
$$(69.38)$$

We finally perform the same manipulations as in the scalar field case, replacing $\tilde{Q}$ as a function of Q, and obtaining

$$\Gamma[\tilde{Q}_{cl}, A; F^a(A, Q)] = W[J, A; F^a(A, Q)] - \int d^4x (J\tilde{Q}_{cl} + \beta c_{cl} + b_{cl}\gamma)$$
$$= W_{\text{QFT}}[J; A, F^a(A, Q - A)] - \int d^4x JA$$
$$- \int d^4x [J(Q_{cl} - A) + \beta c_{cl} + b_{cl}\gamma]$$
$$= W_{\text{QFT}}[J; A, F^a(A, Q - A)] - \int d^4x [JQ_{cl} + \beta c_{cl} + b_{cl}\gamma]$$
$$= \Gamma_{\text{QFT}}[Q_{cl}; A, F^a(A, Q - A)]$$
$$= \Gamma_{\text{QFT}}[\tilde{Q}_{cl} + A; A, F^a(A, Q - A)].$$
$$(69.39)$$

We then put $\tilde{Q}^a_{\mu,cl} = 0$, which implies that $J^\mu_a = J^a_\mu[A^a_\mu, K^\mu_a, L_a, \beta_a, \gamma^a]$, and obtain

$$\Gamma[0, A; F^a(A, Q)] = \Gamma_{\text{QFT}}[A; A, F^a(A, Q - A)].$$
$$(69.40)$$

This means that by differentiating the effective field action in the background field formalism with respect to A^a_μ, we obtain something different than the usual 1PI n-point functions of the QFT with the unusual gauge-fixing term (obtained by differentiating with respect to the first A in Γ_{QFT}, or more precisely with respect to Q, and then putting $Q = A$), since now we have two dependences on A. However, since the gauge-fixing term is unphysical, and S-matrices are physical, the claim is that the S-matrices are independent of A (which can be proved more rigorously), and hence one obtains the usual QFT S-matrices in the background field formalism.

The advantage is that the background field formalism effective action $\Gamma[0, A; F^a(A, Q)]$ (from which the usual S-matrices are constructed) is actually background gauge field invariant!

Indeed, $Z[J, A; F^a(A, Q)]$ is by construction background gauge field invariant, and therefore so is $W[J, A; F^a(A, Q)]$. Then $Q^a_{\mu,cl} = \frac{\delta}{\delta J^\mu_a} W[J, A; F^a(A, Q)]$ transforms as a vector (since J^μ_a does), so the term $\int d^4x J^\mu_a Q^a_{\mu,cl}$ is invariant, and $\Gamma[\tilde{Q}_{cl}, A; F^a(A, Q)]$ is background gauge invariant. The effective action is at least quadratic in $\tilde{Q}_{cl}$, which means that $\tilde{Q}_{cl} = 0$ is a "consistent truncation" (respects the equations of motion).

As an example of the usefulness of this formalism with background gauge invariance, consider renormalization of the theory. As usual, the formalism is α-dependent (i.e. gauge choice (F^a)-dependent). But then background gauge invariance imposes constraints on the renormalization parameters.

Consider the gauge-fixing term

$$-\frac{1}{2\alpha}(D^\mu(A)Q^a_\mu)^2 = -\frac{1}{2\alpha}(\partial^\mu Q^a_\mu - gf^a{}_{bc}A^{b\mu}Q^c_\mu)^2, \qquad (69.41)$$

and impose that it does not renormalize, as in usual gauge theory (since there is no corresponding divergent term, for instance). Then, by writing the usual renormalization relations

$$A^a_\mu = \sqrt{Z_A}A^{a,\mathrm{ren}}_\mu,$$

$$Q^a_\mu = \sqrt{Z_Q}Q^{a,\mathrm{ren}}_\mu,$$

$$g = Z_g\mu^{\epsilon/2}g_{\mathrm{ren}},$$

$$\alpha = Z_\alpha\alpha_{\mathrm{ren}}, \qquad (69.42)$$

from the leading term we obtain the usual relation

$$Z_\alpha = Z_Q \qquad (69.43)$$

(both were called Z_3 in Chapter 44), but now on top of it we get from the subleading term the extra condition

$$Z_g\sqrt{Z_A} = 1, \qquad (69.44)$$

which means that the coupling constant renormalization, defining the beta function ($\beta(g) = \mu\partial g/\partial\mu$), is found just from the renormalization of the background field A^a_μ, for which we only need the two-point function (not the three-point function, or vertex, as for g), simplifying calculations.

Important Concepts to Remember

- The background field method splits the gauge field $\tilde{A}^a_\mu$ into a classical part A^a_μ and a quantum part Q^a_μ, and integrate only the latter.
- In the background field method we have classical (background) gauge invariance, under which Q^a_μ transforms as an adjoint vector, as $\delta_{\mathrm{bgr}}Q^a_\mu = gf^a{}_{bc}Q^b_\mu\lambda^c$, and a quantum gauge transformation, under which A^a_μ is invariant, but Q^a_μ transforms as $\delta_{\mathrm{qu}}Q^a_\mu = D_\mu(Q+A)\lambda^a$.
- The gauge-fixing term in the background formalism is $-(D^\mu(A)Q^a_\mu)^2/2\alpha$, so for a gauge-fixing function $F^a = D^\mu(A)Q^a_\mu$ is invariant under background gauge transformations.
- The quantum action is found by adding source terms $\int(J^\mu_a Q^a_\mu + \beta_a c^a + b_a\gamma^a)$ and extra source terms $\int[K^\mu_a(\delta_{\mathrm{BRST}}Q)^a_\mu - L_a(\delta_{\mathrm{BRST}}c)^a]$, with all the sources transforming as adjoint vectors under background gauge transformations, so that the quantum action and the partition function are all background gauge invariant.

- Defining the classical field in the presence of the classical background, in the scalar field case, $\tilde{Q}_d \equiv Q_d[J;A] = \frac{\delta}{\delta J} W[J;A]\big|_{A}$, we obtain for the effective action $\Gamma[\tilde{Q}_d, A] = \Gamma_{QFT}[\tilde{Q}_d + A]$, so that in particular $\Gamma[0, A] = \Gamma_{QFT}[A]$.
- In the gauge theory case we have $\tilde{Q}^a_{\mu,d} = Q^a_{\mu,d} - A^a_\mu$ and $\Gamma[\tilde{Q}_d, A; F^a(A,Q)] = \Gamma_{QFT}[\tilde{Q}_d + A; A, F^a(A, Q-A)]$, in particular $\Gamma[0, A; F^a(A,Q)] = \Gamma_{QFT}[A; A, F^a(A, Q-A)]$.
- In the background field formalism, the S-matrices are the usual QFT ones, but we obtain them in a background gauge field invariant way, which allows us to obtain extra relations when renormalizing.

Further Reading

See [10] and [11].

Exercises

1. Differentiate twice (69.40) with respect to A^a_μ, to obtain an expression for the propagator in the background field formalism, in terms of similar quantities in the usual QFT formalism.
2. Derive LZJ identities for the gauge theory in the background field formalism, with the partition function (69.13).
3. Write full renormalization relations (using the relations between renormalization factors obtained in the background field formalism) for the full quantum partition function with sources (69.13).
4. Add fermionic matter in the fundamental representation to the gauge theory background field formalism. Write the background and quantum gauge transformations, add sources, and write the partition function.

In this chapter, we will start to learn how to introduce a finite temperature in quantum field theory. This is necessary, since we know for instance that in Brookhaven's Relativistic Heavy Ion Collider and LHC's ALICE experiments, heavy ions are collided resulting in an almost thermal plasma, that cools off into a large number of pions and other particles, which behave thermally, while obeying quantum field theory. An astrophysical situation which has both temperature and the need for quantum field theory is the interior of a neutron star, where matter is crushed to the point of having only (highly quantum) nuclear matter, and the high pressure is also associated with a large temperature. Moreover, at the beginning of the Universe, there was a very high temperature in which quantum fields existed, so to understand the beginning phase of cosmology, we also need to introduce temperature in quantum field theory.

We have seen in Part I a basic way to deal with temperature, by going to Euclidean theory and compactifying time with period $\beta = 1/T$. Therefore in this formalism, in terms of the original Minkowskian theory, the time is *imaginary* (and periodic). This is therefore called the *imaginary-time formalism*, and was introduced by Matsubara. It also has many applications in the context of condensed matter, where we work at finite temperature by definition; in this case, we are in a nonrelativistic situation, so we need to specialize the formalism a bit (in fact, the finite-temperature formalism started as a "manybody" non-relativistic formalism, with applications in condensed matter, and was later generalized to relativistic field theories). But in this formalism, we have no (real) time dependence (time is gone, and it has been "converted" into temperature), hence we can only calculate static quantities. However, it is clear from the previous paragraph that we want to calculate dynamics also (things like scatterings), so we need a formalism with time dependence.

The solution is to consider the fact that time has become imaginary, and the periodicity is imposed as a boundary condition on the quantum fields on a line. We can then generalize the formalism to consider time (x_0) defined on the complex plane, and define formalisms by the specific contours in the complex plane we use, and the corresponding boundary conditions for the fields. In this classification, besides the imaginary-time formalism above, we also have the "real-time formalism," with its variant, the formalism of the "Schwinger–Keldysh contour" and the "thermofield dynamics" formalism. In these two latter cases, we have both an imaginary time leading to temperature, and a real time that describes real-time evolution.

Specifically, in this chapter we will treat the nonrelativistic or "manybody" case of interest to condensed matter, and in Chapter 71* we will move on to the relativistic case.

70.1 Review of Thermodynamics of Quantum Systems (Quantum Statistical Mechanics)

Before plunging into the quantum field theory formalism, it is worth making a quick review of quantum statistical mechanics, the way to deal with temperature in the quantum mechanics of large numbers of particles, with a view towards quantum field theory generalization.

The average in this case is a double one: quantum mechanical and temperature average, so we write

$$\langle\langle\hat{A}\rangle_{\text{QM}}\rangle_\beta = \text{Tr}[\hat{\rho}\hat{A}] = \frac{\text{Tr}[e^{-\beta\hat{H}}\hat{A}]}{\text{Tr}[e^{-\beta\hat{H}}]}, \tag{70.1}$$

where

$$\hat{\rho} \equiv \frac{e^{-\beta\hat{H}}}{Z}, \quad Z(\beta) \equiv \text{Tr}[e^{-\beta\hat{H}}] \tag{70.2}$$

is the quantum-mechanical density operator, with Z the partition function. More precisely, we are here in the canonical ensemble (for the partition function) or, if we replace $\hat{H} \rightarrow \hat{H} - \mu\hat{N}$, in the grand-canonical ensemble.

We can apply this idea to the Green's functions (n-point function), for an operator $\hat{A}$ that is the time-ordered (**T**) product of n fields. In the scalar field case, and dropping the hats for simplicity of notation, we can thus write the Green's function at finite temperature as

$$G_\beta(x_1, \ldots, x_n) = \frac{\text{Tr}[e^{-\beta H}\mathbf{T}(\phi(x_1)\ldots\phi(x_n))]}{\text{Tr}[e^{-\beta H}]}. \tag{70.3}$$

In this canonical formalism, we can calculate for instance the entropy

$$S = -\sum_n p_n \ln p_n = -\beta^2 \frac{\partial}{\partial\beta} \frac{1}{\beta} \ln Z(\beta) \tag{70.4}$$

and the average energy as

$$\langle E\rangle_\beta = -\frac{\partial}{\partial\beta} \ln Z(\beta). \tag{70.5}$$

We also have as usual the free energy

$$F(\beta) = -\frac{1}{\beta} \ln Z(\beta) \Rightarrow Z(\beta) = e^{-\beta F(\beta)}, \tag{70.6}$$

which easily generalizes to quantum field theory, and as we know, the free energy $F(\beta)$ is the generating functional of the connected Green's functions.

For bosons, the Hamiltonian is a sum of bosonic oscillators:

$$H = \sum_{\vec{k}} \hbar\omega_{\vec{k}} \left(a_{\vec{k}}^\dagger a_{\vec{k}} + \frac{1}{2}\right), \tag{70.7}$$

which gives (by the above) for the average energy of a single oscillator of frequency ω the Bose–Einstein distribution

$$\langle E \rangle_\beta = \frac{\hbar\omega}{2} + \frac{\hbar\omega}{e^{\beta\hbar\omega} - 1}. \tag{70.8}$$

For fermions, the Hamiltonian is a sum of fermionic oscillators:

$$H = \sum_{\vec{k},r} \hbar\omega_{\vec{k}} \left(b^\dagger_{\vec{k},r} b_{\vec{k},r} - \frac{1}{2} \right), \tag{70.9}$$

leading to the average energy of a single oscillator of frequency ω of the Fermi–Dirac distribution form:

$$\langle E \rangle_\beta = -\frac{\hbar\omega}{2} + \frac{\hbar\omega}{e^{\beta\hbar\omega} + 1}. \tag{70.10}$$

We can consider a generalized canonical formalism, where we add an interaction term to the energy in the exponent, by having a fixed external field like the electric field $\vec{E}$ (or magnetic field $\vec{H}$), and a varying response field like the polarization $\vec{\mathcal{P}}$ or the induction $\vec{D} = \epsilon_0 \vec{E} + \vec{\mathcal{P}}$ (or magnetic induction field $\vec{B} = \mu_0(\vec{H} + \vec{M})$). In this case, the work for polarization is $\vec{E} \cdot d\vec{\mathcal{P}}$ or, if including the energy of the electromagnetic field itself, $\vec{E} \cdot d(V\vec{D})$; or $\vec{H} \cdot d(V\vec{B})$ in the magnetic case, so the Hamiltonian is

$$H(\vec{E}) = H_0 - \int \vec{E} \cdot d\vec{\mathcal{P}}. \tag{70.11}$$

This leads to a partition function and thermodynamic potential

$$Z = Z(\beta, \vec{E}), \quad -\frac{1}{\beta} \ln Z = \bar{U}(\beta, \vec{E}) \tag{70.12}$$

of the *generalized Gibbs potential* type. In a more general setting, this is

$$G^*(\beta, P_1, \ldots, P_n, X_{n+1}, \ldots, X_r) = U - TS - \sum_{l=1}^{n} P_l X_l, \tag{70.13}$$

where P_l are intensive quantities, like T, P and $\vec{E}, \vec{H}$, and X_l are extensive quantities, like S, V and $\vec{\mathcal{P}}, \vec{M}$. In particular, $G = U - TS - PV$ is the proper (usual) Gibbs potential, or free enthalpy.

In the case of field theory, we usually have a partition function of the type

$$Z(\beta, \mu, J) = \mathrm{Tr}\, e^{-\beta(\hat{H} - \mu\hat{N} - J\hat{A})}, \tag{70.14}$$

where $\hat{N}$ is a variable number of particles (as always in QFT), μ is a fixed chemical potential, and J is an external source for the quantum field $\hat{A}$. It is slightly a matter of interpretation, but one considers J as the extensive quantity, since it is proportional to a volume of the system and A as the intensive one, as it characterizes the response of the system to the external stimulus. In that case, we traditionally call the thermodynamical potential

$$F(\beta, \mu, J) = -\frac{1}{\beta} \ln Z \tag{70.15}$$

the free energy, though grand-canonical potential would be more accurate, since it depends on μ. Then its Legendre transform

$$G(\beta, \mu, A) = F(\beta, \mu, J) + JA \tag{70.16}$$

is a Gibbs potential, and by its definition (A is what can be called the classical field ϕ_{cl}), it is actually the effective action (or effective potential, if we consider no spacetime dependence for the fields). This means that the Gibbs potential is the generating function of 1PI graphs.

As we saw, we can consider the (connected) Green's functions at finite temperature, and from the above, we can also consider the effective action at finite temperature, in order to characterize the system. But to define the perturbation theory, one must first define propagators, then vertices and the perturbative expansion.

70.2 Nonrelativistic QFT at Finite Temperature: "Manybody" Theory

We start with a nonrelativistic example, of use in condensed matter, the case of "manybody" theory. We start by better defining Matsubara's imaginary-time formalism, which involves analytical continuation to Euclidean time, and then periodic time with period $\beta = 1/T$.

Define the time translation operator in the grand-canonical ensemble, $\hat{K} = \hat{H} - \mu\hat{N}$.

Since we use a QFT formalism, we define the Euclidean time "Heisenberg" picture, for a Schrödinger picture operator $\hat{A} = \hat{A}(\vec{r})$, namely the time-dependent operator

$$\hat{A}_{\hat{K}}(\tau) = e^{\frac{\tau}{\hbar}\hat{K}}\hat{A}e^{-\frac{\tau}{\hbar}\hat{K}}, \tag{70.17}$$

depending on the imaginary (Euclidean) time τ.

The "Dirac" picture operator is instead

$$\hat{A}_{\hat{K}_0}(\tau) = e^{\frac{\tau}{\hbar}\hat{K}_0}\hat{A}e^{-\frac{\tau}{\hbar}\hat{K}_0}, \tag{70.18}$$

and then the evolution operator is

$$\hat{U}(\tau_1, \tau_2) = e^{\frac{\tau_1}{\hbar}\hat{K}_0}e^{-\frac{\tau_1-\tau_2}{\hbar}\hat{K}}e^{-\frac{\tau_2}{\hbar}\hat{K}_0}. \tag{70.19}$$

The Green–Matsubara (imaginary-time) two-point function for a field ψ_σ (thought of as a fermion, or electron, of spinor index σ) is defined as

$$\mathcal{G}_{\sigma\sigma'}(\vec{r}, \tau; \vec{r}', \tau') \equiv \mathrm{Tr}\left[\hat{\rho}\mathbf{T}_\tau\left(\hat{\psi}_{\sigma\hat{K}}(\vec{r}, \tau)\psi_{\sigma'}^\dagger(\vec{r}, \tau)\right)\right]. \tag{70.20}$$

We must impose (anti)periodicity in time with period $\hbar\beta$ in *both* τ and τ', that is

$$\mathcal{G}_{\sigma,\sigma'}(\vec{r}, 0; \vec{r}', \tau') = \pm\mathcal{G}_{\sigma,\sigma'}(\vec{r}, \hbar\beta; \vec{r}', \tau'),$$

$$\mathcal{G}_{\sigma,\sigma'}(\vec{r}, \tau; \vec{r}', 0) = \pm\mathcal{G}_{\sigma,\sigma'}(\vec{r}, \tau; \vec{r}', \hbar\beta), \tag{70.21}$$

where the plus sign is for bosons and the minus for fermions. Then, if we take both bosons and fermions together, the system has period $2\hbar\beta$.

Moreover, if the system is time translation invariant, that is time independent (for the imaginary-time formalism, it has to be, as we said), then

$$\mathcal{G}_{\sigma,\sigma'}(\vec{r},\tau;\vec{r}',\tau') = \mathcal{G}_{\sigma,\sigma'}(\vec{r},\tau-\tau';\vec{r}',0), \tag{70.22}$$

and we can expand the periodic functions of $\tau - \tau'$ in Fourier series, with even or odd frequencies, for bosons and fermions, respectively:

$$\mathcal{G}_{\sigma,\sigma'}(\vec{r},\tau;\vec{r}',\tau') = \frac{1}{\hbar\beta}\sum_{n\in\mathbb{Z}} e^{-i\omega_n(\tau-\tau')}\tilde{\mathcal{G}}_{\sigma,\sigma'}(\vec{r},\vec{r}';\omega_n), \tag{70.23}$$

where the *Matsubara frequencies* are

$$\omega_n = \frac{\pi}{\hbar\beta}2n \quad \text{for bosons,}$$

$$\omega_n = \frac{\pi}{\hbar\beta}(2n+1) \quad \text{for fermions.} \tag{70.24}$$

Inversely, we have

$$\tilde{G}_{\sigma,\sigma'}(\vec{r},\vec{r}';\omega_n) = \int_0^{\hbar\beta} d(\tau-\tau') e^{i\omega_n(\tau-\tau')}\mathcal{G}_{\sigma,\sigma'}(\vec{r},\tau;\vec{r}',\tau'). \tag{70.25}$$

70.3 Parenthesis: Condensed Matter Calculations

In the context of condensed matter, we usually have particles with a two-body interaction potential $V(\vec{r},\vec{r}')$. In this case, we cannot simply write an equation satisfied by the two-point Green's function $\mathcal{G}_{\sigma,\sigma'}$, but rather one that relates it to a certain four-point function. Thus the differential equation of the Green–Matsubara function in this nonrelativistic context (similar to a Schrödinger equation) is

$$\left(-\hbar\frac{\partial}{\partial\tau} + \frac{\hbar^2}{2m}\vec{\nabla}^2 + \mu\right)\mathcal{G}_{\sigma,\sigma'}(\vec{r},\tau;\vec{r}',\tau')$$

$$\pm \int d^3\vec{r}'' \sum_{\sigma''} V(\vec{r},\vec{r}'')\mathcal{G}_{\sigma\sigma'',\sigma'\sigma''}(\vec{r},\tau;\vec{r}'',\tau'|\vec{r}',\tau';\vec{r}'',\tau+\epsilon)$$

$$= \hbar\delta_{\sigma\sigma'}\delta^3(\vec{r}-\vec{r}')\delta(\tau-\tau'). \tag{70.26}$$

Note first that the form is due to the nonrelativistic approach, and the related fact that we have a *two-particle* potential, instead of the QFT notion of a *potential for the fields*. Second, this is the analytical continuation ($t = -i\tau$, $iV_{(M)} = -V_{(E)}$) of the $T = 0$ formalism, where the differential equation, also with an external field (interaction term) $u_{\sigma\sigma'}(\vec{r})$ and with a spin-dependent potential, is

$$\left(i\hbar \frac{\partial}{\partial t} + \frac{\hbar^2}{2m} \vec{\nabla}^2 + \mu \right) \mathcal{G}_{\sigma\sigma'}(\vec{r},t;\vec{r}',t') - \sum_{\sigma''} u_{\sigma\sigma''}(\vec{r}) \mathcal{G}_{\sigma''\sigma}(\vec{r},t;\vec{r}',t')$$

$$+ i \int d^3\vec{r}'' \sum_{\sigma_1',\sigma_2,\sigma_2'} V_{\sigma\sigma_1',\sigma_2,\sigma_2'}(\vec{r},\vec{r}'') \mathcal{G}_{\sigma_2\sigma_2',\sigma'\sigma_1'}(\vec{r},t;\vec{r}'',t|\vec{r}',t';\vec{r}'',t')$$

$$= -\hbar \delta_{\sigma\sigma'} \delta^3(\vec{r}-\vec{r}') \delta(t-t'). \tag{70.27}$$

Note that the causal Green's functions at $T=0$ are defined as:

- The one-particle one is ($|\Omega\rangle$ is the Heisenberg vacuum state of the interacting system and ψ_H refers to the Heisenberg picture)

$$G_{\sigma,\sigma'}(\vec{r},t;\vec{r}',t') \equiv -i \frac{\langle \Omega | \mathbf{T} \left(\hat{\psi}_{H,\sigma}(\vec{r},t) \hat{\psi}_{H,\sigma'}^\dagger(\vec{r}',t') \right) |\Omega\rangle}{\langle \Omega|\Omega\rangle}. \tag{70.28}$$

- The two-particle one is

$$G_{\sigma_1\sigma_2,\sigma_1'\sigma_2'}(\vec{r}_1,t_1;\vec{r}_2,t_2|\vec{r}_1',t_1';\vec{r}_2',t_2')$$

$$\equiv (-i)^2 \frac{\langle \Omega | \mathbf{T} \left(\hat{\psi}_{H,\sigma_1}(\vec{r}_1,t_1) \hat{\psi}_{H,\sigma_1'}(\vec{r}_1',t_1') \hat{\psi}_{H,\sigma_2'}^\dagger(\vec{r}_2',t_2') \hat{\psi}_{H,\sigma_2}^\dagger(\vec{r}_2',t_2') \right) |\Omega\rangle}{\langle \Omega|\Omega\rangle}, \tag{70.29}$$

and at $T \neq 0$ in the same way, except for the $(-i)$ factors and $G \to \mathcal{G}$, fields defined with K instead of H, and the insertion of the thermal density $\hat{\rho}$, thus for instance the two-point function is

$$\mathcal{G}_{\sigma,\sigma'}(\vec{r},\tau;\vec{r}',\tau') \equiv -\operatorname{Tr}\left[\hat{\rho}\mathbf{T}_\tau \left(\hat{\psi}_{K,\sigma}(\vec{r},\tau) \hat{\psi}_{K,\sigma'}^\dagger(\vec{r}',\tau') \right) \right]. \tag{70.30}$$

Then i times the Green's functions has the interpretation as amplitudes for propagation from

$$|\psi_I(\vec{r}',\sigma',t')\rangle = \hat{\psi}_{I,\sigma'}^\dagger(\vec{r}',t')|\psi_{I,0}(t')\rangle = \hat{U}_I(t',0)\hat{\psi}_{H,\sigma}^\dagger(\vec{r}',t')|\Omega\rangle \tag{70.31}$$

to

$$|\psi_I(\vec{r},\sigma,t)\rangle = \hat{\psi}_{I,\sigma}^\dagger(\vec{r},t)|\psi_{I,0}(t)\rangle = \hat{U}_I(t,0)\hat{\psi}_{H,\sigma}^\dagger(\vec{r},t)|\Omega\rangle. \tag{70.32}$$

From the Green's functions we can find observables, by tracing them with the quantum operator.

In the $T=0$ case, for a one-particle operator $\hat{a}(\vec{r})$, the average is (introducing an identity, expressed as $I = \sum_n |n\rangle\langle n|$, and using the fact that $a(\vec{n})$ is diagonal in the coordinate representation)

$$\langle a(\vec{r})\rangle = \sum_{\sigma,\sigma'} \lim_{\vec{r}'\to\vec{r},t\to t'+\epsilon} \langle \psi_I(\vec{r},\sigma,t)|a(\vec{r})|\psi_I(\vec{r},\sigma',t)\rangle \langle \psi_I(\vec{r},\sigma',t)|\psi_I(\vec{r}',\sigma,t)\rangle, \tag{70.33}$$

and identifying the first factor as $a(\vec{r})$ in the coordinate representation, $\tilde{a}_{\sigma,\sigma'}(\vec{r})$, we have

$$\langle a(\vec{r})\rangle = -i \sum_{\sigma,\sigma'} \lim_{\vec{r}'\to\vec{r},t'\to t+\epsilon} \tilde{a}_{\sigma,\sigma'}(\vec{r}) G_{\sigma',\sigma}(\vec{r},t;\vec{r}',t') = -i \operatorname{Tr}_\sigma[\tilde{a} \cdot G], \tag{70.34}$$

where the trace is over spinor matrix indices, and is a limit in the coordinate space $(\vec{r}, t)$. Integrating over $\vec{r}$, we find the average of $A = \int d^3 r a(\vec{r})$ as a trace over $\vec{r}$ as well.

For a two-particle operator $b(\vec{r}, \vec{r}')$, we find similarly, dividing by a factor of 2 for symmetry:

$$\langle B \rangle = \int d^3 \vec{r} \int d^3 \vec{r}' \langle b(\vec{r}, \vec{r}') \rangle$$

$$= -\frac{1}{2} \int d^3 \vec{r} d^3 \vec{r}' \sum_{\sigma_1, \sigma_1', \sigma_2, \sigma_2'} \tilde{b}_{\sigma_1 \sigma_1', \sigma_2, \sigma_2'}(\vec{r}, \vec{r}') \lim_{t' \to t + \epsilon} G_{\sigma_2 \sigma_2', \sigma_1 \sigma_1'}(\vec{r}, t; \vec{r}', t' | \vec{r}, t + \epsilon; \vec{r}', t').$$

$$(70.35)$$

Similarly, in the $T \neq 0$ case, we find

$$\langle A \rangle = \mp \int d^3 \vec{r} \sum_{\sigma, \sigma'} \lim_{\vec{r}' \to \vec{r}, \tau' \to \tau + \epsilon} \tilde{a}_{\sigma, \sigma'}(\vec{r}) G_{\sigma', \sigma}(\vec{r}, \tau; \vec{r}', \tau'),$$

$$\langle B \rangle = \frac{1}{2} \int d^3 \vec{r} d^3 \vec{r}' \sum_{\sigma_1, \sigma_1', \sigma_2, \sigma_2'} \tilde{b}_{\sigma_1 \sigma_1', \sigma_2, \sigma_2'}(\vec{r}, \vec{r}') \lim_{\tau' \to \tau + \epsilon} G_{\sigma_2 \sigma_2', \sigma_1 \sigma_1'}(\vec{r}, \tau; \vec{r}', \tau' | \vec{r}, \tau + \epsilon; \vec{r}', \tau').$$

$$(70.36)$$

Then, in particular, the kinetic energy average is

$$\langle T \rangle = \int d^3 \vec{r} \lim_{\vec{r}' \to \vec{r}, \tau' \to \tau + \epsilon} -\frac{\hbar^2}{2m} \vec{\nabla}^2 \operatorname{Tr} \tilde{G}(\vec{r}, t; \vec{r}', t'). \qquad (70.37)$$

70.4 Free Green's Function

The free vacuum for nonrelativistic fermionic systems at $T = 0$ is an occupied state until the Fermi surface:

$$|\Omega_0\rangle = \left(\prod_{\vec{k}, \sigma, k < k_F} \hat{a}^\dagger_{\vec{k}, \sigma} \right) |0\rangle. \qquad (70.38)$$

Near $T = 0$, we consider the "Bogoliubov transformation" (or change of basis for the creation and annihilation operators) from fermions to particles and holes (which is a canonical transformation), by

$$\hat{c}_{\vec{k}, \sigma} = \hat{a}_{\vec{k}, \sigma}, \quad \text{particle annihilation}, \quad k > k_F$$

$$= \hat{b}^\dagger_{-\vec{k}, -\sigma}, \quad \text{hole creation}, \quad k < k_F. \qquad (70.39)$$

Thus, a particle is an addition of a fermion outside the Fermi surface, and a hole is the absence of a fermion inside the Fermi surface (Fermi sea). The canonical transformation leads to the vacuum

$$|\Omega_0\rangle = \prod_{\vec{k}, \sigma, k < k_F} \hat{c}^\dagger_{\vec{k}, \sigma} |0\rangle, \qquad (70.40)$$

generically with a vacuum energy different from the ($T = 0$) Fermi energy:

$$\epsilon_k^0 = \epsilon_F^0 + \hbar\Omega_k. \tag{70.41}$$

At $T \neq 0$, in order to calculate the free two-point Green's function according to (70.30), we need the free vacuum $|\Omega_0\rangle$ defined above, and also the free field operator in the free Heisenberg picture (Dirac). In the particle–hole representation, the free time translation operator is

$$\hat{K}_0 = \hat{H}_0 - \mu\hat{N} = \sum_{\vec{k},\sigma}(\epsilon_k^0 - \mu)\hat{c}_{\vec{k},\sigma}^\dagger \hat{c}_{\vec{k},\sigma} \equiv \sum_{\vec{k},\sigma} \hbar\overline{\omega}_k^0 \hat{c}_{\vec{k},\sigma}^\dagger \hat{c}_{\vec{k},\sigma}, \tag{70.42}$$

where we have defined the frequency

$$\overline{\omega}_k^0 \equiv \frac{\epsilon_k^0 - \mu}{\hbar}. \tag{70.43}$$

Then the (nonrelativistic) free field operator is a plane wave with a $\hat{c}_{\vec{k},\sigma}$ annihilation operator, time-evolved with $\hat{K}_0$, of frequency $\overline{\omega}_k^0$, so

$$\hat{\psi}_{K_0,\sigma}(\vec{r},\tau) = \sum_{\vec{k}} \frac{e^{i\vec{k}\cdot\vec{r}-\overline{\omega}_k^0\tau}}{\sqrt{V}}\hat{c}_{\vec{k},\sigma}. \tag{70.44}$$

We now substitute in (70.30) and obtain first

$$\mathrm{Tr}\left[\hat{\rho}_0\hat{c}_{\vec{k},\sigma}^\dagger \hat{c}_{\vec{k},\sigma}\right] = \langle\hat{c}_{\vec{k},\sigma}^\dagger \hat{c}_{\vec{k},\sigma}\rangle_0 = \delta_{\sigma,\sigma'}\delta_{\vec{k},\vec{k}'}\langle n_{\vec{k},\sigma}\rangle_0 = \delta_{\sigma,\sigma'}\delta_{\vec{k},\vec{k}'}\frac{1}{e^{\beta(\epsilon_k^0-\mu)}\mp 1}, \tag{70.45}$$

where we have used the BE or FD distributions

$$\langle n_{\vec{k},\sigma}\rangle_0 = n_k^0 = \frac{1}{e^{\beta(\epsilon_k^0-\mu)}\mp 1}. \tag{70.46}$$

Then we find a homogeneous, time-translation-invariant Green's function:

$$\mathcal{G}_{\sigma,\sigma'}^0(\vec{r},\tau;\vec{r}',\tau') = \frac{\delta_{\sigma,\sigma'}}{V}\sum_{\vec{k}} e^{i\vec{k}\cdot(\vec{r}-\vec{r}')-\overline{\omega}_k^0(\tau-\tau')}[-(1\pm n_k^0)\theta(\tau-\tau')\mp n_k^0\theta(\tau'-\tau)], \tag{70.47}$$

for which we can define not only the Fourier transform over time in (70.23), but also one over space:

$$\mathcal{G}_{\sigma,\sigma'}(\vec{r},\tau;\vec{r}',\tau') = \frac{1}{\hbar\beta}\sum_{n\in\mathbb{Z}} e^{-i\omega_n(\tau-\tau')}\frac{1}{V}\sum_{\vec{k}} e^{i\vec{k}\cdot(\vec{r}-\vec{r}')}\tilde{\mathcal{G}}_{\sigma,\sigma'}(\vec{k},\omega_n), \tag{70.48}$$

so that finally we find (after doing the integral over $\tau-\tau'$ in (70.25))

$$\tilde{\mathcal{G}}_{\sigma,\sigma'}^0(\vec{k},\omega_n) = \delta_{\sigma,\sigma'}\frac{1}{i\omega_n - \overline{\omega}_k^0}. \tag{70.49}$$

70.5 Perturbation Theory and Dyson Equations

Having settled the free case, we now define perturbation theory. We first define, as in the $T = 0$ formalism, Wick contractions

$$\overline{A(\tau)B(\tau')} \equiv \langle \mathbf{T}_\tau \left(\hat{A}(\tau)\hat{B}(\tau') \right) \rangle_0 = \mathrm{Tr}\left[\hat{\rho}_0 \mathbf{T}_\tau \left(\hat{A}(\tau)\hat{B}(\tau') \right) \right]. \tag{70.50}$$

We then have the basic free field contractions

$$\overline{\psi_{K_0,\sigma}(\vec{r},\tau)\psi^\dagger_{K_0,\sigma'}}(\vec{r}',\tau') = -\mathcal{G}^0_{\sigma,\sigma'}(\vec{r},\tau;\vec{r}',\tau'),$$
$$\overline{\psi^\dagger_{K_0,\sigma}(\vec{r},\tau)\psi_{K_0,\sigma'}}(\vec{r}',\tau') = \mp\mathcal{G}^0_{\sigma,\sigma'}(\vec{r},\tau;\vec{r}',\tau'). \tag{70.51}$$

Note then that

$$\overline{\hat{c}_{K_0,\vec{k},\sigma}(\tau)\hat{c}^\dagger_{K_0,\vec{k}',\sigma'}}(\tau') = \frac{[\hat{c}_{K_0,\vec{k},\sigma}(\tau),\hat{c}^\dagger_{K_0,\vec{k},\sigma}(\tau')]_\mp}{1 \mp e^{-\beta(\epsilon^0_k - \mu)}} = \frac{\delta_{\sigma\sigma'}\delta_{\vec{k},\vec{k}'}e^{-\overline{\omega}^0_k(\tau-\tau')}\mathbf{1}}{1 \mp e^{-\beta(\epsilon^0_k - \mu)}}. \tag{70.52}$$

Then, just as for the usual Wick theorem, one can prove the *thermal Wick theorem, or Bloch–De Dominicis theorem*: If $A(\tau), B(\tau)$ are linear combinations of $\hat{c}_{K_0,\vec{k},\sigma}(\tau)$ and $\hat{c}^\dagger_{K_0,\vec{k},\sigma}(\tau)$, we have the equality of the VEV of time-ordered products to the sum of all total contractions:

$$\langle \mathbf{T}_\tau \left(\hat{A}(\tau_a)\hat{B}(\tau_b)\dots\hat{Z}(\tau_z) \right) \rangle_0 = \sum \text{total contractions} \left(\hat{A}(\tau_a)\dots\hat{Z}(\tau_z) \right). \tag{70.53}$$

Then we can find the thermal analogue of the Feynman theorem for perturbation theory:

$$\mathcal{G}_{\sigma,\sigma'}(\vec{r},\tau;\vec{r}',\tau') = -e^{\beta(\Omega-\Omega_0)}\sum_{n\geq 0}\frac{1}{n!}\left(-\frac{1}{\hbar}\right)^n\int_0^{\hbar\beta}d\tau_1\dots\int_0^{\hbar\beta}d\tau_n$$
$$\langle \mathbf{T}_\tau \left(\hat{K}'_{K_0}(\tau_1)\dots\hat{K}'_{K_0}(\tau_n)\hat{\psi}_{K_0,\sigma}(\vec{r},\tau)\hat{\psi}^\dagger_{K_0,\sigma'}(\vec{r}',\tau') \right) \rangle_0 e^{-\beta(\Omega-\Omega_0)}$$
$$= \sum_{n\geq 0}\frac{1}{n!}\left(-\frac{1}{\hbar}\right)^n\int_0^{\hbar\beta}d\tau_1\dots\int_0^{\hbar\beta}d\tau_n$$
$$\langle \mathbf{T}_\tau \left(\hat{K}'_{K_0}(\tau_1)\dots\hat{K}'_{K_0}(\tau_n)\hat{\psi}_{K_0,\sigma}(\vec{r},\tau)\hat{\psi}^\dagger_{K_0,\sigma'}(\vec{r}',\tau') \right) \rangle_0, \tag{70.54}$$

where $\hbar(\Omega - \Omega_0)$ is the shift in the energy of the vacuum due to the interactions and $\langle \rangle_0$ is the VEV in the *unperturbed vacuum*. Here we have defined the (condensed matter) interaction Hamiltonian (potential term) $\hat{K}'_{K_0}(\tau)$ ($\hat{V}_I(t)$ in the $T = 0$ formalism) out of the two-body interaction potential $V(\vec{r},\vec{r}')$ as (usual in condensed matter)

$$\hat{K}'_{K_0}(\tau) = \frac{1}{2}\int d^3\vec{r}\int d^3\vec{r}'\int_0^{\hbar\beta}d\tau'\sum_{\lambda\mu\lambda'\mu'}V_{\lambda\lambda',\mu\mu'}(\vec{r},\vec{r}')\delta(\tau-\tau')$$
$$\hat{\psi}^\dagger_{K_0\lambda}(\vec{r},\tau)\hat{\psi}^\dagger_{K_0,\lambda'}(\vec{r}',\tau')\hat{\psi}_{K_0,\mu}(\vec{r}',\tau')\hat{\psi}_{K_0,\mu'}(\vec{r},\tau). \tag{70.55}$$

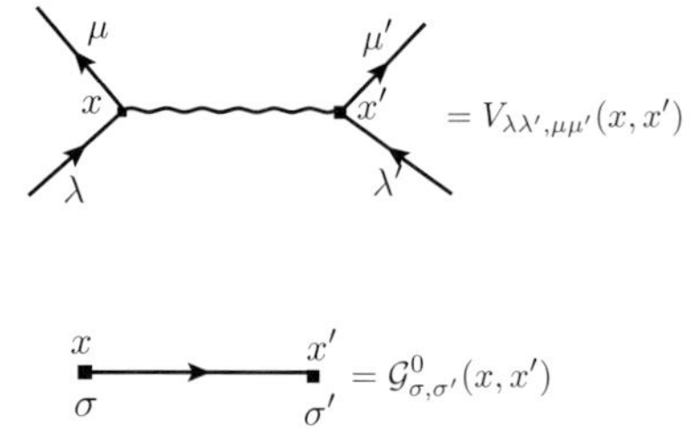

$$= V_{\lambda\lambda',\mu\mu'}(x,x')$$

$$= \mathcal{G}^0_{\sigma,\sigma'}(x,x')$$

Figure 70.1 Feynman rules for the x-space Green's functions.

Combining the Wick and Feynman formulas, we find also

$$\mathcal{G}_{\sigma,\sigma'}(\vec{r},\tau;\vec{r}',\tau') = -\sum_{n\geq 0}\frac{1}{n!}\left(-\frac{1}{\hbar}\right)^n \int_0^{\hbar\beta} d\tau_1 \ldots \int_0^{\hbar\beta} d\tau_n$$

connected total contractions $\left(\hat{K}'_{K_0}(\tau_1)\ldots\hat{K}'_{K_0}(\tau_n)\hat{\psi}_{K_0,\sigma}(\vec{r},\tau)\hat{\psi}^\dagger_{K_0,\sigma'}(\vec{r}',\tau')\right).$

$$(70.56)$$

70.5.1 Feynman Rules in x Space

For the Green's functions above, we have Feynman rules (see Figure 70.1):

1. Draw all the connected diagrams.
2. For a free propagator $\mathcal{G}^0_{\sigma,\sigma'}(x,x')$, draw an oriented line from (x',σ') to (x,σ). For a two-body interaction potential $V_{\lambda\lambda',\mu\mu'}(x,x') = \mathcal{V}_{\lambda\lambda',\mu\mu'}(\vec{r},\vec{r}')\delta(\tau-\tau')$, draw a wave line between x and x', with two lines with indices λ,μ and two with indices λ',μ' going out of the two points.
3. Introduce a factor $f_j^{(n)} = \left(-\frac{1}{\hbar}\right)^n (-1)^{L_j}$ for the diagram, where L_j is the number of loops of the diagram.
4. Finally, sum over indices and integrate over loop positions, to obtain

$$\mathcal{G}^{(n)}_{\sigma,\sigma'}(x,x') = \sum_j f_j^{(n)} \int d^{8n}\mathbf{x} \sum_{\lambda,\mu} \mathcal{D}^{(n,j)}_{\lambda,\mu}(\mathbf{x},\mathbf{x}'). \qquad (70.57)$$

70.5.2 Feynman Rules in Momentum Space

For $\tilde{\mathcal{G}}_{\sigma,\sigma'}(\vec{k},\omega_n)$, we have Feynman rules (see Figure 70.2):

1. Draw all the connected diagrams.
2. Draw all $\vec{k},\omega_n$, with conservation at each vertex.
3. For a line of momentum $\vec{k}$ and frequency ω_n, we have the propagator $\tilde{\mathcal{G}}^0(\vec{k},\omega_n)$, represented by a directed line with momentum on it. For a momentum-space potential $\tilde{\mathcal{V}}(\vec{k})$, draw a wavy line with an arrow on it.
4. Sum and integrate over loops, with $\int d^3k_j \frac{1}{\hbar\beta}\sum_n$, and add a factor $\left(-\frac{1}{\hbar}\right)^n [\pm(2s+1)]^{L_j}$.
5. Add a factor $e^{i\omega_n \epsilon}$ for a propagator, or propagator–potential, loop.

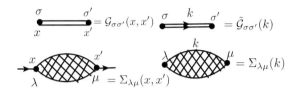

Figure 70.2   Feynman rules for the p-space Green's functions.

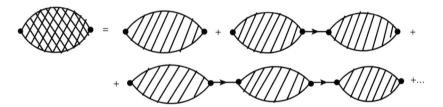

Figure 70.3   Feynman rules for the Dyson equation: full propagator $\mathcal{G}$, self-energy Σ, and 1PI self-energy Σ^*.

Figure 70.4   Diagrammatic equation for the self-energy Σ in terms of the 1PI self-energy Σ^*.

70.5.3 Dyson Equation

To complete the perturbation theory, consider the Dyson's equations.

(a) For the momentum-space propagator $\tilde{\mathcal{G}}(\vec{k}, \omega_n)$, see the double (thick) line with an arrow on it in Figure 70.3

Define also the (connected) self-energy Σ, represented as a full circle with two dots at the end, crossed by left and right diagonals, and the 1PI self-energy Σ^*, represented by the same, but with only right diagonals (Figure 70.3). Then, from diagrammatics (see Figure 70.4), we have

$$\Sigma = \Sigma^* + \Sigma^* \tilde{\mathcal{G}}^0 \Sigma^* + \Sigma^* \tilde{\mathcal{G}}^0 \Sigma^* \tilde{\mathcal{G}}^0 \Sigma^* + \dots \tag{70.58}$$

Writing the diagrammatically obvious relation

$$\tilde{\mathcal{G}} = \tilde{\mathcal{G}}^0 + \tilde{\mathcal{G}}^0 \Sigma \tilde{\mathcal{G}}^0 \tag{70.59}$$

Figure 70.5 Diagrammatically obvious equation, and Dyson equation for the propagator.

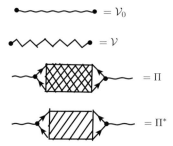

Figure 70.6 Feynman rules for the free interaction $\mathcal{V}_0$, effective interaction $\mathcal{V}$, total connected polarization Π, and 1PI polarization Π^*.

and then expanding Σ in terms of Σ^* as in the above, and finally identifying with the left-hand side ($\mathcal{G}$), we obtain the *Dyson equation for the propagator*, as in Figure 70.5:

$$\tilde{\mathcal{G}} = \tilde{\mathcal{G}}^0 + \tilde{\mathcal{G}}^0 \Sigma \tilde{\mathcal{G}}^0 = \tilde{\mathcal{G}}^0 + \tilde{\mathcal{G}}^0 \Sigma^* \tilde{\mathcal{G}}. \tag{70.60}$$

This can be solved to give

$$\tilde{\mathcal{G}} = \tilde{\mathcal{G}}^0 (1 + \Sigma^* \tilde{\mathcal{G}}^0)^{-1}, \tag{70.61}$$

or finally

$$\tilde{\mathcal{G}} = \frac{1}{\tilde{\mathcal{G}}_0^{-1} + \Sigma^*}, \tag{70.62}$$

or more precisely

$$\tilde{\mathcal{G}}(\vec{k}, \omega_n) = \frac{1}{i\omega_n - \frac{\epsilon_k^0 - \mu}{\hbar} - \tilde{\Sigma}^*(\vec{k}, \omega_n)}. \tag{70.63}$$

(b) For the momentum-space effective interaction $\mathcal{V}$, represented by a zigzagging line, define also the free interaction $\mathcal{V}_0$ as a wavy line (the one defined previously), the total (connected) polarization Π, represented by a box with crossed left and right diagonal lines and two fermionic (fermi–antifermi) lines on the left and on the right, to be attached at a point each, and the proper (1PI) polarization Π^*, the same but with only right diagonal lines, as in Figure 70.6.

Then again, we have the diagrammatically obvious relation (see Figure 70.7)

$$\Pi = \Pi^* + \Pi^* \mathcal{V}_0 \Pi^* + \Pi^* \mathcal{V}_0 \Pi^* \mathcal{V}_0 \Pi^* + \ldots = \Pi^* + \Pi^* \mathcal{V}_0 \Pi. \tag{70.64}$$

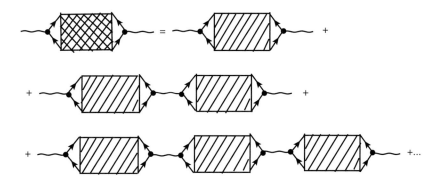

Figure 70.7　Diagrammatic relation for the polarization Π in terms of the proper (1PI) polarization Π^*.

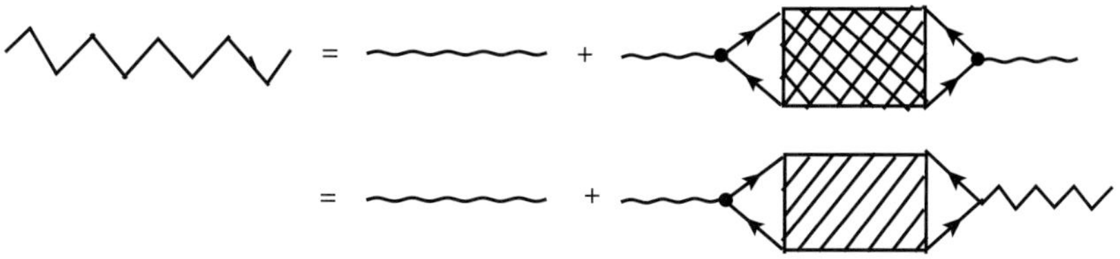

Figure 70.8　Diagrammatically obvious relation and Dyson equation for the effective interaction.

We can also write the diagrammatically obvious relation

$$\mathcal{V} = \mathcal{V}_0 + \mathcal{V}_0 \Pi \mathcal{V}_0, \tag{70.65}$$

and then expand Π as above and reidentify the left-hand side, to finally obtain the Dyson equation for the effective interaction, as in Figure 70.8

$$\mathcal{V} = \mathcal{V}_0 + \mathcal{V}_0 \Pi^* \mathcal{V}. \tag{70.66}$$

Again this can be solved to give the effective interaction in terms of the proper polarization as

$$\mathcal{V} = \mathcal{V}_0 (1 - \mathcal{V}_0 \Pi^*)^{-1}, \tag{70.67}$$

or more precisely

$$\tilde{\mathcal{V}}(\vec{k}, \omega_n) = \frac{\tilde{\mathcal{V}}_0(\vec{k})}{1 - \tilde{\mathcal{V}}_0(\vec{k}) \tilde{\Pi}^*(\vec{k}, \omega_n)}. \tag{70.68}$$

70.6 Lehmann Representation and Dispersion Relations

In the $T = 0$ case, we can use the definition (70.28) of the two-point function, where the full vacuum $|\Omega\rangle$ is a vacuum of N particles, so can be denoted by $|\Omega_{N,0}\rangle$, write $\mathbf{T}(\psi(t)\psi^\dagger(t')) = \theta(t-t')\psi(t)\psi^\dagger(t') + \theta(t'-t)\psi^\dagger(t')\psi(t)$, and then insert the identity written as $\mathbf{1} = \sum_n |n\rangle\langle n|$, but remember that ψ only contains c, which annihilates one particle, and $\psi^\dagger$ only $c^\dagger$, which creates one particle, so for each of the two terms, only a single term contributes, and then we finally write $\psi_{H,\sigma}(\vec{r},t) = e^{\frac{i}{\hbar}\hat{H}t}\psi_\sigma(\vec{r})e^{-\frac{i}{\hbar}\hat{H}t}$ and act with the Hamiltonian on the corresponding state (at left or right), to obtain its energy, and find for the causal Green's function

$$G_{\sigma,\sigma'}(\vec{r},t;\vec{r}',t') = -i \sum_{(N+1,\hbar\vec{k}_\alpha)} \theta(t-t')e^{\frac{i}{\hbar}(E_{N,0}-E_{N+1,\alpha})t}$$

$$\frac{\langle\Omega_{N,0}|\hat{\psi}_\sigma(\vec{r},0)|N+1,\hbar\vec{k}_\alpha\rangle\langle N+1,\hbar\vec{k}_\alpha|\hat{\psi}^\dagger_{\sigma'}(\vec{r}',0)|\Omega_{N,0}\rangle}{\langle\Omega_{N,0}|\Omega_{N,0}\rangle} e^{\frac{i}{\hbar}(E_{N+1,\alpha}-E_{N,0})t'}$$

$$- i \sum_{(N-1,\hbar\vec{k}_\alpha)} \theta(t'-t)e^{\frac{i}{\hbar}(E_{N,0}-E_{N-1,\alpha})t}$$

$$\frac{\langle\Omega_{N,0}|\hat{\psi}^\dagger_{\sigma'}(\vec{r}',0)|N-1,-\hbar\vec{k}_\alpha\rangle\langle N-1,-\hbar\vec{k}_\alpha|\hat{\psi}_\sigma(\vec{r},0)|\Omega_{N,0}\rangle}{\langle\Omega_{N,0}|\Omega_{N,0}\rangle} e^{\frac{i}{\hbar}(E_{N-1,\alpha}-E_{N,0})t}.$$

$$(70.69)$$

Next we take the Fourier transform over $\vec{r}-\vec{r}'$ and $t-t'$, use the relation

$$\theta(\tau) = \lim_{\epsilon\to 0} \frac{i}{2\pi} \int_{-\infty}^{+\infty} d\tilde{\omega}\frac{e^{i\tilde{\omega}\tau}}{\tilde{\omega}+i\epsilon}, \qquad (70.70)$$

which can be proven by complex integration over a contour that closes by a semicircle either in the upper half plane, or the lower half plane (as in Part I), which further means that

$$\int_{-\infty}^{+\infty} d\tau\,\theta(\pm\tau)e^{i\omega t} = \int_{-\infty}^{+\infty} \frac{d\tau}{2\pi} \int d\tilde{\omega}\frac{ie^{i(\tilde{\omega}-\omega)\tau}}{\tilde{\omega}-i\epsilon} = \lim_{\epsilon\to 0}\frac{\pm i}{\omega\pm i\epsilon}, \qquad (70.71)$$

define $\mu \equiv E_{N+1,0} - E_{N,0}$ and the excitation energy $\epsilon_\alpha^{N+1} \equiv N_{N+1,\alpha} - E_{N+1,0}$, to find

$$E_{N+1,\alpha} - E_{N,0} = \epsilon_\alpha^{(N+1)} + \mu^{(N)}, \quad E_{N-1,\alpha} - E_{N,0} = \epsilon_\alpha^{(N-1)} - \mu^{(N-1)}, \qquad (70.72)$$

and finally find the *Lehmann representation of the causal Green's function in momentum space*:

$$G_{\sigma,\sigma'}^C(\vec{k},\omega) = \lim_{\epsilon\to 0}\left[\frac{1}{V} \sum_{\alpha,(N+1,\hbar\vec{k}_\alpha)} \frac{a_{\sigma,\sigma'}(\vec{k},\alpha)}{\omega - \omega_\alpha^{(N+1)} - \frac{\mu}{\hbar} + i\epsilon} \right.$$

$$\left. + \frac{1}{V} \sum_{\alpha,(N-1,-\hbar\vec{k}_\alpha)} \frac{b_{\sigma,\sigma'}(\vec{k},\alpha)}{\omega + \omega_\alpha^{(N-1)} - \frac{\mu}{\hbar} - i\epsilon} \right]. \qquad (70.73)$$

Here we have denoted

$$a_{\sigma,\sigma'}(\vec{k},\alpha) \equiv \frac{V\langle\Omega_{N,0}|\hat{\psi}_\sigma(\vec{k},0)|N+1,\hbar\vec{k},\alpha\rangle\langle N+1,\hbar\vec{k},\alpha|\hat{\psi}^\dagger_{\sigma'}(\vec{k},0)|\Omega_{N,0}\rangle}{\langle\Omega_{N,0}|\Omega_{N,0}\rangle}$$

$$= V\frac{\left|\langle\Omega_{N,0}|\hat{\psi}_\sigma(\vec{k},0)|N+1,\hbar\vec{k},\alpha\rangle\right|^2}{\langle\Omega_{N,0}|\Omega_{N,0}\rangle}$$

$$
b_{\sigma,\sigma'}(\vec{k},\alpha) \equiv \frac{V\langle\Omega_{N,0}|\hat{\psi}_{\sigma'}^{\dagger}(\vec{k},0)|N-1,-\hbar\vec{k},\alpha\rangle\langle N-1,-\hbar\vec{k},\alpha|\hat{\psi}_{\sigma}(\vec{k},0)|\Omega_{N,0}\rangle}{\langle\Omega_{N,0}|\Omega_{N,0}\rangle}
$$

$$
= V\frac{\left|\langle\Omega_{N,0}|\hat{\psi}_{\sigma'}^{\dagger}(\vec{k},0)|N-1,-\hbar\vec{k},\alpha\rangle\right|^2}{\langle\Omega_{N,0}|\Omega_{N,0}\rangle}, \tag{70.74}
$$

and note that $\frac{1}{V}\sum$ becomes an integral in the thermodynamic limit, with a density of states $dN(\omega)/d\omega$:

$$
\frac{1}{V}\sum_{\alpha}F(\epsilon_{\alpha}) = \int_0^{\infty} d\omega' \frac{dN(\omega')}{d\omega'}F(\hbar\omega'). \tag{70.75}
$$

We have written only the causal Green's function, but for the retarded (R) or advanced (A) one, we can replace the $\mathbf{T}(A(t)B(t'))$ by $\theta(t-t')[A(t),B(t')]$ or $-\theta(t'-t)[A(t),B(t')]$, and obtain similarly

$$
G_{\sigma,\sigma'}^{R,A}(\vec{k},\omega) = \lim_{\epsilon\to0}\left[\frac{1}{V}\sum_{\alpha,(N+1,\hbar\vec{k}_{\alpha})}\frac{a_{\sigma,\sigma'}(\vec{k},\alpha)}{\omega-\omega_{\alpha}^{(N+1)}-\frac{\mu}{\hbar}\pm i\epsilon}\right.
$$

$$
\left. +\frac{1}{V}\sum_{\alpha,(N-1,-\hbar\vec{k}_{\alpha})}\frac{b_{\sigma,\sigma'}(\vec{k},\alpha)}{\omega+\omega_{\alpha}^{(N-1)}-\frac{\mu}{\hbar}\pm i\epsilon}\right]. \tag{70.76}
$$

The usefulness of the Lehmann representation is that it manifests the analytic properties of the Green's function. Further using the fact that

$$
\lim_{\epsilon\to0}\frac{1}{x\pm i\epsilon} = \mathcal{P}\left(\frac{1}{x}\right)\mp i\pi\delta(x), \tag{70.77}
$$

where $\mathcal{P}$ stands for the principal part, and translating sums into integrals as above (in the thermodynamic limit), we obtain for the retarded and advanced Green's functions

$$
G^{R,A}(\vec{k},\omega) = \mathcal{P}\int_{-\infty}^{+\infty}d\omega'\frac{\theta(\omega'-\mu/\hbar)A(\vec{k},\omega'-\mu/\hbar)+\theta(\mu/\hbar-\omega')B(\vec{k},\mu/\hbar-\omega')}{\omega-\omega'}
$$

$$
\mp i\pi\left[\theta\left(\omega-\frac{\mu}{\hbar}\right)A\left(\vec{k},\omega-\frac{\mu}{\hbar}\right)+\theta\left(\frac{\mu}{\hbar}-\omega\right)B\left(\vec{k},\frac{\mu}{\hbar}-\omega'\right)\right], \tag{70.78}
$$

from which we deduce the *dispersion relations*

$$
\mathrm{Re}\,G^{R,A}(\vec{k},\omega) = \mp\frac{1}{\pi}\mathcal{P}\int_{-\infty}^{+\infty}d\omega'\frac{\mathrm{Im}\,G^{R,A}(\vec{k},\omega')}{\omega-\omega'} \tag{70.79}
$$

that just encapsulate the analytic properties of the retarded and advanced Green's functions.

70.7 Real-Time Formalism

We are finally in a position to describe the basics of the real-time formalism in the non-relativistic case, which was defined by Zubarev. In this nonrelativistic case, it is used for scatterings in a heat bath, for instance in a condensed matter system at finite temperature.

As we said, we need to introduce a real time, that can define the evolution of the system, besides the heat bath.

Then we define Heisenberg operators depending on the real time t as

$$\hat{A}_K(t) \equiv e^{\frac{i}{\hbar}\hat{K}t}\hat{A}e^{-\frac{i}{\hbar}\hat{K}t}, \tag{70.80}$$

where as usual $\hat{A}$ is in the Schrödinger picture.

Next we define the causal Green–Zubarev functions at finite temperature (with real time) in the same way as in the imaginary-time formalism, namely with a Boltzmann weight factor, but with the real-time operators $\hat{A}_K(t)$:

$$\mathbf{G}_{\sigma,\sigma'}(\vec{r},t;\vec{r}',t') = -i\,\mathrm{Tr}\left[\hat{\rho}\mathbf{T}_\tau[\hat{\psi}_{K,\sigma}(\vec{r},t)\hat{\psi}_{K,\sigma'}^\dagger(\vec{r}',t')]\right], \tag{70.81}$$

and similarly for the retarded and advanced Green–Zubarev functions:

$$\mathbf{G}_{\sigma,\sigma'}^{R,A}(\vec{r},t;\vec{r}',t') = \mp i\theta[\pm(t-t')]\,\mathrm{Tr}\left[\hat{\rho}[\hat{\psi}_{K,\sigma}(\vec{r},t),\hat{\psi}_{K,\sigma'}^\dagger(\vec{r}',t')]\right]. \tag{70.82}$$

70.7.1 Lehmann Representation and Dispersion Relations

We can calculate the same Lehmann representation as in the $T = 0$ case, now in the real-time formalism. Note that at finite temperature, we sum (trace) over states in the Green–Zubarev function, so we replace $|\Omega\rangle_{N,0}$ with a sum over states $|\alpha\rangle$. We also don't assume anymore that only the $|N + 1, \alpha'\rangle$ and $|N - 1, \alpha'\rangle$ states contribute to the sum, as before, for generality. Then we replace also $\omega_\alpha^{(N+1)} + \mu/\hbar = (E_{N+1,\alpha} - E_{N,0})/\hbar$ with

$$\frac{K_{\alpha'} - K_\alpha}{\hbar} = \frac{E_{\alpha'} + \mu_{\alpha'} - E_\alpha - \mu_\alpha}{\hbar}, \tag{70.83}$$

and $-\omega_\alpha^{(N-1)} + \mu/\hbar = (E_{N,0} - E_{N-1})/\hbar$ with the same, and introduce the Boltzmann factors, to obtain

$$\tilde{\mathbf{G}}(\vec{k},\omega) = \frac{\mathcal{V}}{Z}\sum_{\alpha,\alpha'}\delta_{\vec{P}_{\alpha'}-\vec{P}_\alpha,\hbar\vec{k}}\frac{1}{2S+1}\sum_\sigma |\langle\alpha|\hat{\psi}_\sigma(0)|\alpha'\rangle|^2$$

$$\times \lim_{\epsilon\to 0+}\left[\frac{e^{-\beta K_\alpha}}{\omega - \frac{K_{\alpha'}-K_\alpha}{\hbar} + i\epsilon} \mp \frac{e^{-\beta K_{\alpha'}}}{\omega - \frac{K_{\alpha'}-K_\alpha}{\hbar} - i\epsilon}\right]. \tag{70.84}$$

Here $\mathcal{V}$ is the volume and Z the partition function. Similarly, we obtain in the retarded and advanced cases

$$\tilde{\mathbf{G}}^{R,A}(\vec{k},\omega) = \frac{\mathcal{V}}{Z}\sum_{\alpha,\alpha'}\delta_{\vec{P}_{\alpha'}-\vec{P}_\alpha,\hbar\vec{k}}\frac{1}{2S+1}\sum_\sigma |\langle\alpha|\hat{\psi}_\sigma(0)|\alpha'\rangle|^2 \lim_{\epsilon\to 0+}\left[\frac{e^{-\beta K_\alpha}\mp e^{-\beta K_{\alpha'}}}{\omega - \frac{K_{\alpha'}-K_\alpha}{\hbar} \pm i\epsilon}\right].$$

$$\tag{70.85}$$

To identify the analytic properties of the Green–Zubarev functions, and to make a connection with the imaginary-time (Matsubara) formalism, we define a quantity usually called the *spectral function*, $\rho(\vec{k},\omega)$, as follows:

$$\rho(\vec{k}, \omega) = \frac{\mathcal{V}}{Z} \sum_{\alpha, \alpha'} \delta_{\vec{P}_{\alpha'} - \vec{P}_\alpha, \hbar \vec{k}} \frac{1}{2S+1} \sum_\sigma |\langle \alpha | \hat{\psi}_\sigma(0) | \alpha' \rangle|^2 (e^{-\beta K_\alpha} \mp e^{-\beta K_{\alpha'}})$$

$$\times 2\pi \delta \left(\omega - \frac{K_{\alpha'} - K_\alpha}{\hbar} \right). \tag{70.86}$$

We also define the quantity

$$\Gamma(\vec{k}, \omega) \equiv \mathcal{P} \int_{-\infty}^{+\infty} \frac{d\omega'}{2\pi} \frac{\rho(\vec{k}, \omega)}{\omega - \omega'}, \tag{70.87}$$

where $\mathcal{P}$ refers to the principal part of the integral, as before.

Then the retarded and advanced Green–Zubarev functions are

$$\tilde{\mathbf{G}}^{R,A}(\vec{k}, \omega) = \lim_{\epsilon \to 0+} \int_{-\infty}^{+\infty} \frac{d\omega'}{2\pi} \frac{\rho(\vec{k}, \omega)}{\omega - \omega' \pm i\epsilon} = \lim_{\epsilon \to 0+} \Gamma(\vec{k}, \omega \pm i\epsilon)$$

$$= \lim_{\epsilon \to 0+} \mathcal{P} \int_{-\infty}^{+\infty} \frac{d\omega'}{2\pi} \frac{\rho(\vec{k}, \omega)}{\omega - \omega' \pm i\epsilon} \mp \frac{i}{2} \rho(\vec{k}, \omega), \tag{70.88}$$

and the causal Green–Zubarev function is

$$\tilde{\mathbf{G}}(\vec{k}, \omega) = \int_{-\infty}^{+\infty} \frac{d\omega'}{2\pi} \rho(\vec{k}, \omega) \left[\mathcal{P} \left(\frac{1}{\omega - \omega'} \right) - i\pi \left(\coth \left(\frac{\beta \hbar \omega}{2} \right) \right)^{\pm 1} \delta(\omega - \omega') \right]. \tag{70.89}$$

Finally, we can write dispersion relations coming from the above analytic structure. The relation for the retarded and advanced functions is the analogue of (70.79) in the $T = 0$ case, namely

$$\mathrm{Re}\tilde{\mathbf{G}}^{R,A}(\vec{k}, \omega) = \mp \frac{1}{\pi} \mathcal{P} \int_{-\infty}^{+\infty} d\omega' \frac{\mathrm{Im}\tilde{\mathbf{G}}^{R,A}(\vec{k}, \omega)}{\omega - \omega'}. \tag{70.90}$$

There is now also a simple relation for the causal Green's function:

$$\mathrm{Re}\tilde{\mathbf{G}}(\vec{k}, \omega) = \mp \frac{1}{\pi} \mathcal{P} \int_{-\infty}^{+\infty} d\omega' \frac{\mathrm{Im}\tilde{\mathbf{G}}(\vec{k}, \omega)}{\omega - \omega'} \left(\tanh \frac{\beta \hbar \omega}{2} \right)^{\pm 1}. \tag{70.91}$$

If the spectral function is normalized in the usual way:

$$\int_{-\infty}^{+\infty} \frac{d\omega}{2\pi} \rho(\vec{k}, \omega) = 1, \tag{70.92}$$

we find that at infinity the quantity Γ behaves as

$$\Gamma(\vec{k}, \omega) \sim \frac{1}{|\omega|}, \quad \text{as } |\omega| \to \infty. \tag{70.93}$$

70.7.2 Relation with Green–Matsubara Functions

We can now relate to the imaginary-time formalism, since directly from the definitions of the Green–Matsubara (imaginary-time) and Green–Zubarev (real-time) functions, or from comparing the explicit Lehmann representations, we find that at the Matsubara frequencies, the Γ function is the Wick rotation of the Green–Matsubara one:

$$\tilde{\mathcal{G}}(\vec{k}, \omega_n) = \Gamma(\vec{k}, i\omega_n), \qquad (70.94)$$

where we had $\mathbf{G}^{R,A}(\vec{k}, \omega) = \Gamma(\vec{k}, \omega \pm i\epsilon)$.

Moreover, the spectral function is the same, and is the discontinuity of the imaginary part of the Green–Matsubara one at the Matsubara frequencies, or

$$\rho(\vec{k}, \omega) = i \lim_{\epsilon \to 0+} \left[\tilde{\mathcal{G}}(\vec{k}, \omega_n) \Big|_{\omega_n = \omega + i\epsilon} - \tilde{\mathcal{G}}(\vec{k}, \omega_n) \Big|_{\omega_n = \omega - i\epsilon} \right]. \qquad (70.95)$$

We might wonder that on one side we have a function of a continuous variable, and on the right a function of a discrete variable, but the analytic continuation of the Γ function from the $z = \omega_n$ points is actually unique, provided we impose the asymptotic condition noticed before:

$$\Gamma(z) \simeq \frac{1}{|z|}, \quad \text{for } |z| \to \infty. \qquad (70.96)$$

70.7.3 Free Green–Zubarev Function

The spectral function representation is good because it allows us, among other things, to make assumptions directly about $\rho(\vec{k}, \omega)$. In particular, in the free case, we expect that all the factors in the definition of ρ become trivial except for the delta functions, so we obtain

$$\rho^0(\vec{k}, \omega) = 2\pi \delta(\omega - \bar{\omega}_k^0). \qquad (70.97)$$

Then the retarded and advanced functions are found to be

$$\tilde{\mathbf{G}}^{R,A}(\vec{k}, \omega) = \frac{1}{\omega - \bar{\omega}_k^0 \pm i\epsilon}, \qquad (70.98)$$

and the causal function is

$$\tilde{\mathbf{G}}(\vec{k}, \omega) = \frac{1}{1 \mp e^{-\beta\hbar\omega}} \frac{1}{\omega - \bar{\omega}_K^0 + i\epsilon} + \frac{1}{1 \mp e^{\beta\hbar\omega}} \frac{1}{\omega - \bar{\omega}_K^0 - i\epsilon}. \qquad (70.99)$$

70.7.4 Correlation Functions and Scattering

To consider scattering, we must consider first correlation functions. Then, through a version of the LSZ formalism, we can relate them to scattering amplitudes.

Consider two operators $\hat{A}_K(\vec{r}, \tau), \hat{B}_K(\vec{r}, \tau)$ in the Matsubara formalism and $\hat{A}_K(\vec{r}, t)$, $\hat{B}_K(\vec{r}, t)$ in the Zubarev formalism. Then we can define a causal two-point correlation function in the imaginary case, and a retarded two-point correlation function in the real case:

$$C_{AB}(\vec{r}, \tau; \vec{r}', \tau') = -\text{Tr}\left[\hat{\rho} \mathbf{T}_\tau [\hat{A}_K(\vec{r}, \tau) \hat{B}_K(\vec{r}, \tau; \vec{r}', \tau')] \right],$$

$$\mathbf{C}_{AB}^R(\vec{r}, t; \vec{r}', t') = -i\theta(t - t') \text{Tr}\left[\hat{\rho} [\hat{A}_K(\vec{r}, t), \hat{B}(\vec{r}', t')] \right]. \qquad (70.100)$$

Note that $\hat{A}_K = \hat{A}_H$ in the case that the number operator is conserved, $[\hat{N}, \hat{H}] = 0$ and $[\hat{N}, \hat{A}] = 0$.

We can also define a Lehmann representation for these correlation functions, and as before find (for the Fourier transform in time)

$$
\mathcal{C}_{AB}(\vec{r}, \vec{r}\,'; \omega_n) = \int_{-\infty}^{+\infty} \frac{d\omega'}{2\pi} \frac{\hbar \Delta_{AB}(\vec{r}, \vec{r}\,'; \omega')}{i\omega_n - \omega'},
$$

$$
\mathbf{C}_{AB}^R(\vec{r}, \vec{r}\,'; \omega) = \int_{-\infty}^{+\infty} \frac{d\omega'}{2\pi} \frac{\hbar \Delta_{AB}(\vec{r}, \vec{r}\,'; \omega')}{\omega - \omega' + i\epsilon}, \tag{70.101}
$$

where we denoted

$$
\hbar \Delta_{AB}(\vec{r}, \vec{r}\,'; \omega') = \frac{1}{Z} \sum_{\alpha, \alpha'} \langle \alpha | \hat{A}(\vec{r}) | \alpha' \rangle \langle \alpha' | \hat{B}(\vec{r}\,') | \alpha \rangle e^{-\beta K_\alpha} (1 - e^{-\beta \hbar \omega}) 2\pi \delta \left(\omega' - \frac{K_{\alpha'} - K_\alpha}{\hbar} \right).
$$

$$\tag{70.102}$$

The useful and relevant correlation function is that for the (local) number density operator itself, $\hat{A} = \hat{B} = \delta \hat{n}$, when we denote

$$
\mathcal{C}_n(\vec{r}, \vec{r}\,'; \omega_n) = \mathcal{C}_{\delta \hat{n}, \delta \hat{n}}(\vec{r}, \vec{r}\,'; \omega_n), \quad \mathbf{C}_n^R(\vec{k}, \omega) = \mathbf{C}_{\delta \hat{n}, \delta \hat{n}}^R(\vec{k}, \omega). \tag{70.103}
$$

Then we can prove a **theorem**, saying that this correlation function is given by the (connected) polarization function in both Green–Zubarev and Green–Matsubara cases:

$$
\mathcal{C}_n(\vec{r}, \vec{r}\,'; \omega_n) = \hbar \Pi(\vec{r}, \vec{r}\,'; \omega_n), \quad \tilde{\mathbf{C}}_n^R(\vec{k}, \omega) = \hbar \tilde{\mathbf{\Pi}}^R(\vec{k}, \omega). \tag{70.104}
$$

Important Concepts to Remember

- At finite temperature T, the free-energy/grand-canonical potential is the generator of the connected graphs, and the effective action/Gibbs potential is the generator of 1PI graphs.
- In Matsubara's imaginary-time formalism we have periodic Euclidean time with period $\beta = 1/T$, but otherwise we define Heisenberg picture and Dirac picture operators in the same way, with $e^{\frac{\tau}{\hbar}\hat{K}}$.
- For time-translation-invariant Green–Matsubara functions in Fourier (energy) space, we have Matsubara frequencies $\omega_n = \frac{\pi}{\hbar \beta} 2n$ for bosons and $\frac{\pi}{\hbar \beta}(2n + 1)$.
- In the nonrelativistic (manybody) context, the Green–Matsubara function satisfies a Schrödinger-like equation, with a two-particle potential.
- The VEV of a one-particle operator integrated over the volume, $\langle A \rangle$, is given by the trace, in spin and coordinate space, of the operator a with the one-particle causal Green's function, and the VEV of a two-particle operator integrated over both volumes, $\langle B \rangle$, is given by the trace, over both spin and coordinate space, of the operator b with the two-particle causal Green's function.
- The free Green–Matsubara two-point function in momentum space is $\tilde{G}_{\sigma, \sigma'}^0(\vec{k}, \omega_n) = \delta_{\sigma, \sigma'} \frac{1}{i\omega_n - \bar{\omega}_k^0}$.
- The thermal Wick theorem (Bloch–De Dominicis theorem) is valid for (time-ordered products of) $\hat{A}(\tau)$ operators that are combinations of $\hat{c}_{K_0, \vec{k}, \sigma}$ and $\hat{c}_{K_0, \vec{k}, \sigma}^\dagger$, and then we also have a thermal version of the Feynman theorem.

- We can write Feynman rules for the Green–Matsubara functions, both in coordinate and in momentum space, for $\tilde{G}_{\sigma,\sigma'}(\vec{k}, \omega_n)$.
- We can write a Dyson equation for the propagator, solved by $\tilde{\mathcal{G}}(\vec{k}, \omega_n) = \dfrac{1}{\tilde{\mathcal{G}}_0^{-1}(\vec{k},\omega_n) - \tilde{\Sigma}^*(\vec{k},\omega_n)} = \dfrac{1}{i\omega_n - \frac{\epsilon_k^0 - \mu}{\hbar} - \tilde{\Sigma}^*(\vec{k},\omega_n)}$, and a Dyson equation for the effective interaction, solved by $\tilde{\mathcal{V}}(\vec{k}, \omega_n) = \dfrac{\tilde{\mathcal{V}}_0(\vec{k})}{1 - \tilde{\mathcal{V}}_0(\vec{k})\tilde{\Pi}^*(\vec{k},\omega_n)}$.
- We can write a Lehmann representation for the causal Green–Matsubara functions in momentum space $G^C_{\sigma,\sigma'}(\vec{k}, \omega)$, and dispersion relations (coming from the analytic structure) for the retarded and advanced ones, $\text{Re } G^{R,A}(\vec{k}, \omega) = \mp\frac{1}{\pi}\mathcal{P}\int_{-\infty}^{+\infty} d\omega' \frac{\text{Im } G^{R,A}(\vec{k},\omega')}{\omega - \omega'}$.
- In the real-time formalism, we define Green–Zubarev functions with the evolution operator $e^{\frac{i}{\hbar}\hat{K}t}$.
- We can write a Lehmann representation for the Green–Zubarev functions (causal, advanced, and retarded) in momentum space, and all of them can be written in terms of the spectral function $\rho(\vec{k}, \omega)$.
- The retarded and advanced Green–Zubarev functions satisfy the usual dispersion relation (coming from the analytic properties), whereas the causal one satisfies another relation.
- Defining $\Gamma(\vec{k}, \omega) \equiv \mathcal{P}\int_{-\infty}^{+\infty}\frac{d\omega'}{2\pi}\frac{\rho(\vec{k},\omega)}{\omega - \omega'}$, satisfying $\Gamma(\vec{k}, \omega) \sim \frac{1}{|\omega|}$, as $|\omega| \to \infty$, we can relate Green–Matsubara with Green–Zubarev functions via $\tilde{\mathcal{G}}(\vec{k}, \omega_n) = \Gamma(\vec{k}, i\omega_n)$, where $\mathbf{G}^{R,A}(\vec{k}, \omega) = \Gamma(\vec{k}, \omega \pm i\epsilon)$.
- The free Green–Zubarev functions are: retarded and advanced $\tilde{\mathbf{G}}^{R,A}(\vec{k}, \omega) = \frac{1}{\omega - \tilde{\omega}_k^0 \pm i\epsilon}$ and causal $\tilde{\mathbf{G}}(\vec{k}, \omega) = \frac{1}{1 \mp e^{-\beta\hbar\omega}}\frac{1}{\omega - \tilde{\omega}_K^0 + i\epsilon} + \frac{1}{1 \mp e^{\beta\hbar\omega}}\frac{1}{\omega - \tilde{\omega}_K^0 - i\epsilon}$.
- We can define correlation functions for operators in Matsubara and Zubarev formalisms, relate them to basic quantities in the imaginary or real-time formalisms, and calculate scattering from them.

Further Reading

See any book on manybody theory.

Exercises

1. Write an expression for the total energy of electrons in a material, given the two-point interaction potential $V_{\sigma,\sigma'}(\vec{r}, \vec{r}')$ and the causal Green's functions.
2. Write down (no need to calculate) the one-loop integral for $\mathcal{G}_{\sigma,\sigma'}(\vec{r}, \tau; \vec{r}', \tau')$, given a two-body interaction potential $\mathcal{V}_{\lambda\lambda',\mu\mu'}(\vec{r}, \vec{r}')$, using the Feynman rules.
3. In the case in Exercise 2, write a formula for the one-loop propagator in momentum space $\tilde{\mathcal{G}}(\vec{k}, \omega_n)$ and for the one-loop effective interaction $\tilde{\mathcal{V}}(\vec{k}, \omega_n)$.
4. Prove the missing steps in the Lehmann representation for the causal and advanced/retarded Green–Zubarev function (70.84) and (70.85).

71

Finite-Temperature Quantum Field Theory II: Imaginary and Real-Time Formalisms

We are now in a position to describe a more general formalism that is simpler in the relativistic case, but can also be used in the nonrelativistic case from the previous chapter. As I described at the beginning of Chapter 70, one of the ways we can distinguish the various formalisms is by the use of different integration contours in the complex-time plane. The imaginary-time formalism uses a vertical contour, for the temperature, and various real-time formalisms use a combination of the vertical piece and a real piece, for real-time evolution, that can involve two parallel lines. One starts with the Euclidean formulation (imaginary-time formalism), which has a natural Minkowski counterpart (real-time formalism) by Wick rotation, but then we encounter various ways to solve for the need to have time evolution different from the temperature.

In thermofield dynamics (studied in Chapter 72), we use different fields (a "doubling" formalism) on the two lines, and in the Schwinger–Keldysh case, we go back on almost the same line (Figure 71.1).

71.1 The Imaginary-Time Formalism

The simplest case is the imaginary-time formalism, described very briefly in Chapter 8 (in Part I). The *Euclidean* field theory Heisenberg field operators are defined via a Wick rotation from Minkowski time, by $t \equiv x^0 = -i\tau$, so

$$\hat{\phi}(\vec{x}, x^0) = e^{\frac{i}{\hbar}\hat{H}x^0}\phi(\vec{x}, 0)e^{-\frac{i}{\hbar}\hat{H}x^0} = e^{\frac{1}{\hbar}\hat{H}\tau}\phi(\vec{x}, 0)e^{-\frac{1}{\hbar}\hat{H}\tau}, \qquad (71.1)$$

where now we assume that $\tau \in \mathbb{R}$.

Then, as we saw there, we can write a path integral for propagation of a state for a finite Euclidean time τ, and naturally obtain the partition function as the Euclidean path integral over periodic τ, with periodicity $\beta = 1/T$, since schematically, first we write propagation as ($i\int dt\pi \frac{d}{dt}\phi = i\int d\tau\pi \frac{d}{d\tau}$ and $i\int dt\mathcal{H} = -\int d\tau\mathcal{H}$)

$$\langle \phi_1 | e^{-\frac{1}{\hbar}\hat{H}\tau} | \phi_0 \rangle = \int \mathcal{D}\phi\mathcal{D}\pi \, e^{\frac{1}{\hbar}\int_0^\tau d\tau \int d^3\vec{x}[i\pi\dot{\phi} - \mathcal{H}(\pi)]}, \qquad (71.2)$$

and then we write the partition function as a trace, thus identifying the initial and final states, and obtaining periodic fields:

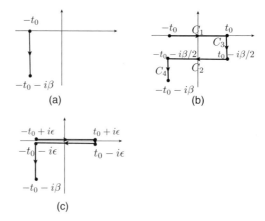

Figure 71.1 Contours for finite temperature: (a) contour for imaginary-time formalism; (b) contour for imaginary-time and thermofield dynamics (TFD) formalisms; (c) contour for Schwinger–Keldysh formalism.

$$Z_\beta = \text{Tr}[e^{-\beta\hat{H}}] = \ldots = N(\beta) \int_{\phi(0)=\phi(\hbar\beta)} \mathcal{D}\phi e^{\int_0^\beta d\tau \int d^3\vec{x}\left[-\frac{1}{2}(\partial_\tau\phi)^2 - \frac{1}{2}(\partial_i\phi)^2 - \frac{m^2}{2}\phi^2 - V_{\text{int}}(\phi)\right]}$$

$$\equiv N(\beta) \int_{\phi(0)=\phi(\hbar\beta)} \mathcal{D}\phi e^{-S}. \tag{71.3}$$

In Part I we didn't explore this too much, beyond saying that the $T \to 0$ limit gives the natural formulation of QFT.

Now we can add sources, and define the partition function with sources, first abstractly in Minkowski space:

$$Z_\beta[J] = \langle\langle\mathbf{T}e^{\frac{i}{\hbar}\int dt \int d^3\vec{x}J(x)\hat{\phi}(x)}\rangle\rangle_{\text{QM}}\rangle_\beta \equiv \frac{\text{Tr}\left[e^{-\beta\hat{H}}\mathbf{T}e^{\frac{i}{\hbar}\int dt \int d^3\vec{x}J(x)\hat{\phi}(x)}\right]}{\text{Tr}\left[e^{-\beta\hat{H}}\right]}, \tag{71.4}$$

and then by Wick rotation to Euclidean space:

$$Z_\beta^{(E)}[J] = \text{Tr}\left[e^{-\beta\hat{H}'}\right] = N(\beta) \int \mathcal{D}\phi e^{\frac{-S+\int J\hat{\phi}}{\hbar}}, \tag{71.5}$$

where we have defined the sourced Hamiltonian

$$H' \equiv H - \int d^3x J(x)\hat{\phi}(x). \tag{71.6}$$

Next, as in Chapter 70, we define the n-point functions in the same way, as

$$G_\beta(x_1,\ldots,x_n) = \langle\langle\mathbf{T}(\phi(x_1)\ldots\phi(x_n))\rangle\rangle_{\text{QM}}\rangle_\beta = \text{Tr}\left[\hat{\rho}\mathbf{T}(\phi(x_1)\ldots\phi(x_n))\right]$$

$$= \frac{\text{Tr}\left[e^{-\beta\hat{H}}\mathbf{T}(\phi(x_1)\ldots\phi(x_n))\right]}{\text{Tr}\left[e^{-\beta\hat{H}}\right]}, \tag{71.7}$$

which can be found from the partition function via differentiation:

$$G_\beta(x_1, \ldots, x_n) = \frac{\delta^n}{\delta J(x_1) \ldots \delta J(x_n)} Z_\beta^{(E)}[J]\bigg|_{J=0}. \tag{71.8}$$

Then the Feynman rules come as usual from the Dyson equation:

$$Z_\beta[J] = e^{-\int_{\text{period }\beta} V_{\text{int}}\left(\hbar\frac{\delta}{\delta J}\right)} Z_{0,\beta}[J]. \tag{71.9}$$

The only difference is now the discreteness of the integration over k_0: it is a sum over n, so

$$\int \frac{d^4\vec{k}}{(2\pi)^4} \to \int \frac{d^3\vec{k}}{(2\pi)^3} \left(\frac{1}{\beta}\sum_n\right), \quad \delta\left(\sum_i k_i\right) \to \delta\left(\sum_i \vec{k}_i\right) \beta\delta_{\sum_i \omega_i, 0}. \tag{71.10}$$

But note that the above definition (71.7) for the thermal n-point functions does not distinguish between real and imaginary time ($\phi(x)$ can be taken to mean $\phi(\vec{x}, x^0)$ or $\phi(\vec{x}, -i\tau)$), that is Minkowski or Euclidean space.

71.2 Imaginary-Time Formalism: Propagators

In particular, the free two-point function, or Feynman propagator, is

$$D_{F,\beta}(x_1, x_2) = \frac{\text{Tr}\left[e^{-\beta\hat{H}_0}\mathbf{T}(\phi(x_1)\phi(x_2))\right]}{\text{Tr}\left[e^{-\beta\hat{H}_0}\right]}. \tag{71.11}$$

Since the *Euclidean* (imaginary-time) propagator must be periodic in both τ_1 and τ_2 with period $\hbar\beta$ (as both field insertions are periodic), the propagator itself is periodic, and can be expanded in a Fourier series of *Matsubara frequencies* $\omega_n = (2\pi n)/\beta$, $n \in \mathbb{Z}$.

Note that, as usual, the Feynman propagator is a combination of Wightman functions:

$$D_{F,\beta}(x - y) = \theta(x^0 - y^0)D_\beta(x - y) + \theta(y^0 - x^0)D_\beta(x - y), \tag{71.12}$$

where the Wightman function at finite temperature is

$$D_\beta(x, y) = \langle\langle\phi(x)\phi(y)\rangle_{\text{QM}}\rangle_\beta = \frac{\text{Tr}\left[e^{-\beta\hat{H}_0}\phi(x)\phi(y)\right]}{\text{Tr}\left[e^{-\beta\hat{H}_0}\right]} \equiv D_{+,\beta}(x, y),$$

$$D_\beta(y, x) \equiv D_{-,\beta}(x, y). \tag{71.13}$$

We can proceed a long way towards calculating the propagator without explicitly stating whether we have real or imaginary time. We can use, for instance, canonical quantization to calculate the Feynman propagator, calculating first

$$\langle a(\vec{k})a(\vec{k}')^\dagger \rangle_\beta = \frac{\text{Tr}\left[e^{-\beta\hat{H}}a(\vec{k})a(\vec{k}')^\dagger\right]}{\text{Tr}\left[e^{-\beta\hat{H}}\right]} = \dots$$

$$= (1 + n_{BE}(\omega_k))\delta^3(\vec{k} - \vec{k}'),$$

$$n_{BE}(\omega) \equiv \frac{1}{e^{\beta\omega} - 1},$$

$$\langle a(\vec{k}')^\dagger a(\vec{k}) \rangle_\beta = \dots = n_{BE}(\omega_k)\delta^3(\vec{k} - \vec{k}'), \tag{71.14}$$

to finally find

$$D_{F,\beta}(x - y) = \int \frac{d^3k}{(2\pi)^3} \frac{1}{2\omega} \left[\theta(x^0 - y^0)e^{ik(x-y)} + \theta(y^0 - x^0)e^{-ik(x-y)}\right.$$

$$\left. + n_{BE}(\omega)(e^{ik(x-y)} + e^{-ik(x-y)})\right]. \tag{71.15}$$

At this point, the expressions can still contain either real or imaginary time (either Minkowski space or Euclidean space).

Then, we can calculate the imaginary-time propagator, either by rewriting the explicit integrals, or by expanding first the fields in Matsubara frequencies as

$$\phi(\vec{x}, \tau) = \sum_{n\in\mathbb{Z}} \int \frac{d^3k}{(2\pi)^{3/2}} e^{i(\vec{k}\cdot\vec{x} + \omega_n\tau)}\phi_n(\vec{k}), \tag{71.16}$$

or by directly inverting the kinetic operator (solving the KG equation), subject to the periodicity conditions for the fields. One finds the obvious formula for the propagator:

$$D_{F,\beta}(\vec{x}, -i\tau) = \int \frac{d^3k}{(2\pi)^3} \left(\frac{1}{\beta}\sum_n\right) \frac{-i\hbar e^{i(\vec{k}\cdot\vec{x} + \omega_n\tau)}}{\omega_n^2 + \vec{k}^2 + m^2}. \tag{71.17}$$

71.3 KMS (Kubo–Martin–Schwinger) Relation

The propagators obey a certain periodicity relation for the Wightman functions, the KMS relation, which we now derive. We write the Wightman function

$$D_{+,\beta}(\vec{x}, x^0; \vec{y}, y^0) = Z_\beta^{-1} \text{Tr}\left[e^{-\beta\hat{H}}e^{\frac{i}{\hbar}\hat{H}x^0}\phi(\vec{x}, 0)e^{-\frac{i}{\hbar}\hat{H}(x^0 - y^0)}\phi(\vec{y}, 0)e^{-\frac{i}{\hbar}\hat{H}y^0}\right]$$

$$= Z_\beta^{-1} \text{Tr}\left[\left(e^{\frac{i}{\hbar}H(x^0 + i\beta)}\phi(\vec{x}, 0)e^{-\frac{i}{\hbar}H(x^0 + i\beta)}\right)e^{-\beta H}\right.$$

$$\left. \times \left(e^{+\frac{i}{\hbar}Hy^0}\phi(\vec{x}, 0)e^{-\frac{i}{\hbar}Hy^0}\right)\right]$$

$$= Z_\beta^{-1} \text{Tr}\left[e^{-\beta H}\left(e^{\frac{i}{\hbar}Hy^0}\phi(\vec{y}, 0)e^{-\frac{i}{\hbar}Hy^0}\right)\left(e^{\frac{i}{\hbar}H(x^0 + i\beta)}\phi(\vec{x}, 0)e^{-\frac{i}{\hbar}H(x^0 + i\beta)}\right)\right]$$

$$= D_{+,\beta}(\vec{y}, y^0; \vec{x}, x^0 + i\beta) \equiv D_{-,\beta}(\vec{x}, x^0 + i\beta; \vec{y}, y^0), \tag{71.18}$$

where in the third line we used cyclicity of the trace, and then reformed the Heisenberg operator fields. Simplifying the notation, we thus have the KMS relation:

$$D_{+,\beta}(\vec{x} - \vec{y}; x^0 - y^0) = D_{-,\beta}(\vec{x} - \vec{y}; x^0 - y^0 + i\beta). \tag{71.19}$$

We can then explicitly check that it is satisfied by the propagator formulas both in the real/imaginary-time formulation (71.15) and in the imaginary-time one (71.17). Specifically, for imaginary time, $x^0 = -i\tau$, and inserting $1 = \theta(x^0) + \theta(-x^0)$ and grouping the factors, we find for the imaginary-time propagator coming from (71.15):

$$D_{F,\beta}(\vec{x}, -i\tau) = \int \frac{d^3k}{(2\pi)^3} \frac{1}{2\omega} \frac{1}{e^{\beta\omega} - 1} \Big[\theta(\tau)(e^{i\vec{k}\cdot\vec{x} - \omega\tau + \beta\omega} + e^{-i\vec{k}\cdot\vec{x} + \omega\tau})$$
$$+ \theta(-\tau)(e^{-i\vec{k}\cdot\vec{x} + \omega\tau + \beta\omega} + e^{i\vec{k}\cdot\vec{x} - \omega\tau}) \Big]$$
$$= \int \frac{d^3k}{(2\pi)^3} \frac{1}{2\omega} \frac{e^{i\vec{k}\cdot\vec{x}}}{e^{\beta\omega} - 1} \Big[e^{-\omega(|\tau| - \beta)} + e^{\omega|\tau|} \Big], \tag{71.20}$$

where in the second equality we have converted all $e^{-i\vec{k}\cdot\vec{x}}$ to $e^{i\vec{k}\cdot\vec{x}}$, and where $\omega = \sqrt{\vec{k}^2 + m^2}$, and this is in fact equal to (71.15), as

$$\frac{-i}{\beta} \sum_n \frac{e^{i\omega_n\tau}}{\omega^2 + \omega_n^2} = \frac{1}{2\omega} \frac{1}{e^{\beta\omega} - 1} \Big[e^{-\omega(|\tau| - \beta)} + e^{\omega|\tau|} \Big], \tag{71.21}$$

as we can check (the proof is left as an exercise), exactly as in the nonrelativistic case from Chapter 70. Note that the form of the propagator is the obvious generalization of the quantum-mechanical case in QFT (Chapter 8).

71.4 Real-Time Formalism

As we saw, in the imaginary (Euclidean)-time formalism, in the thermal Green's functions we had fields defined on the imaginary line, through $\phi(\vec{x}, t) = e^{\frac{1}{\hbar}H\tau}\phi(\vec{x}, 0)e^{-\frac{1}{\hbar}H\tau}$, with time ordering $\mathbf{T}$ along this line, and the trace, together with the Boltzmann factor of $e^{-\beta H}$, introduced a periodicity for the correlators. In particular, we showed that for the propagator (two-point function), we have the KMS relation. Considering that the Schrödinger fields are defined at a time t_0, the line goes from t_0 to $t_0 - i\beta$. At the level of the path integral, from the trace with the Boltzmann factor we find a periodicity in imaginary time:

$$\phi_{\text{in}} = \phi(t_0) = \phi(t_0 - i\beta) = \phi_{\text{fin}}. \tag{71.22}$$

Note that the Wick rotation from Minkowski spacetime implicit in the imaginary-time formalism means that the Feynman propagator contains Heaviside functions defined *on this contour* (imaginary line):

$$D_{F,\beta}(x_1 - x_2) = \theta_C(t_1 - t_2)D_{+,\beta}(x_1 - x_2) + \theta_C(t_2 - t_1)D_{-,\beta}(x_1 - x_2), \tag{71.23}$$

where θ_C is defined on the contour.

But for the real-time formalism, we need time coordinates for the field operators defined on the real line, yet we still have the same $\text{Tr}\, e^{-\beta H}$, which means again a periodicity of β in imaginary time, from some t_0 to $t_0 - i\beta$.

This means that we need to consider a complex-time contour, and define the general form (71.23) along it. But in $D_{+,\beta}$ we have factors of $e^{\frac{i}{\hbar}H(t_1 - t_2 + i\beta)}$ and $e^{-\frac{i}{\hbar}H(t_1 - t_2)}$, meaning that it is analytic in $-\beta \leq \mathrm{Im}(t_1 - t_2) \leq 0$, and in $D_{-,\beta}$ we have factors of $e^{-\frac{i}{\hbar}(t_1 - t_2 - i\beta)}$ and $e^{\frac{i}{\hbar}(t_1 - t_2)}$, so it is analytic in $0 \leq \mathrm{Im}(t_1 - t_2) \leq \beta$. This means that the contour must always go down or stay horizontal in the complex-time plane, and moreover it must be limited by

$$-\beta \leq \mathrm{Im}(t_1 - t_2) \leq \beta. \tag{71.24}$$

Moreover, we have also established that the contour must start at some t_0 and end up at $t_0 - i\beta$, for periodicity, and that it must cover most of the real line (whatever times we need in our calculation). It is conventional to replace $t_0 \to -t_0$. Then one such contour starts at $-t_0$ and goes on to $+t_0$, then down to $t_0 - i\beta/2$, then left to $-t_0 - i\beta/2$, then down to $-t_0 - i\beta$, as in Figure 71.1(b).

More generally, we can consider a contour C that goes on the real line (slightly above it) from $-t_0$ to $+t_0$ (C_1), then down to $t_0 - i\beta'$ (C_3), then left to $-t_0 - i\beta'$ (C_2), then down to $-t_0 - i\beta$ (C_4). The choice above, $\beta' = \beta/2$, is a symmetric one that is good for the thermofield dynamics formalism, and the choice $\beta' = \epsilon$ (one contour slightly above, another slightly below the real axis) is the Schwinger–Keldysh contour, used in the formalism of the same name.

Then the partition function is found in the usual way, by integrating out the momenta in the Hamiltonian formalism, and using a Dyson equation. Thus, first we write generically

$$Z_\beta[J] = \int \mathcal{D}\phi \mathcal{D}\pi \, e^{\frac{i}{\hbar} \int_C d\tau_C d^3 x (\pi \partial_C \phi - H[J])}, \tag{71.25}$$

where C is the contour, π is the canonical conjugate to ϕ, τ_C is a real coordinate along the contour C, $\phi = \phi(\tau_C, \vec{x})$, and $\partial_C \phi \equiv \partial \phi(\tau_C, \vec{x})/\partial \tau_C$. Doing the integral over the momenta to go to the Lagrangian formulation, and then isolating the quadratic part and completing squares for it, we find the Dyson equation

$$
\begin{aligned}
Z_\beta[J] &= \int \mathcal{D}\phi \exp\left[\frac{i}{\hbar} S_{\mathrm{int}}\left(\frac{\hbar}{i} \frac{\delta}{\delta J(x)}\right)\right] \\
&\quad \times \exp\left[-\frac{i}{2\hbar} \int_C d\tau_x \int_C d\tau_y \int d^3 x \int d^3 y \, J(x) \left(\frac{i}{\hbar} D_{F,\beta}(x - y)\right) J(y)\right] \\
&\equiv \int \mathcal{D}\phi \exp\left[\frac{i}{\hbar} S_{\mathrm{int}}\left(\frac{\hbar}{i} \frac{\delta}{\delta J(x)}\right)\right] Z_{0,\beta}[J].
\end{aligned}
\tag{71.26}
$$

Here $D_{F,\beta}$ is the Feynman propagator *on the contour*, satisfying the differential equation on the contour

$$(-\partial_C^2 + \partial_k^2 - m^2) D_{F,\beta}(x - y) = \delta_C(x - y) = \delta_C(\tau_x - \tau_y)\delta^3(\vec{x} - \vec{y}), \tag{71.27}$$

and satisfying the periodic boundary conditions defined by the KMS relation:

$$D_{+,\beta}(\vec{x} - \vec{y}, \tau_{C,x} - \tau_{C,y}) = D_{-,\beta}(\vec{x} - \vec{y}, \tau_{C,x} - \tau_{C,y} + i\beta). \tag{71.28}$$

71.5 Interpretation of Green's Functions

Before proceeding further, let us take stock, and understand the meaning of the various contours, for the partition function and for Green's functions.

For imaginary time, the partition function is

$$Z_\beta[J] = \sum_n \langle \phi_n | e^{-\beta(H - \int J\phi)} | \phi_n \rangle, \tag{71.29}$$

and is turned into a path integral with periodic boundary, since $\phi_{\text{in}} = \phi_n = \phi_{\text{final}}$, but $\phi_{\text{final}} = \phi(t + i\beta\hbar)$. For two-point functions:

$$\begin{aligned} G(x, y) &= \sum_n \langle \phi_n | e^{-\beta(H - \int J\phi)} \mathbf{T}(\phi(\vec{x}, \tau_x)\phi(\vec{y}, \tau_y)) | \phi_n \rangle \\ &= \frac{1}{\beta^2} \frac{\delta}{\delta J(x)} \frac{\delta}{\delta J(y)} Z_\beta[J] \bigg|_{J=0}, \end{aligned} \tag{71.30}$$

we obtain the same.

This means that for these Euclidean Green's functions, the path integral is taken for periodic fields on the complex-time contour going from t_0 to $t_0 - i\beta\hbar$, and since also $\int_C J\phi$ is taken on the same contour, τ is on the contour also for the Green's functions.

The limit $\beta \to \infty$ ($T = 0$) gives the vacuum-to-vacuum functional of the $T = 0$ QFT in Euclidean space, which is the Wick-rotated theory from Minkowski space, which however corresponds to an extreme case of the real-time formalism.

Indeed, Wick rotation in this infinite periodicity case means considering in complex time the closed contour going from $t = 0$ to $t = +\infty$ (on the real line), then a quarter circle at infinity down to $-i\infty$, then up again to $t = 0$, as in Figure 71.2. Since there are no poles in this quadrant, the total integral is zero, and since the quarter circle is at infinity in the lower half plane, it gives no contribution. That results in the Wick rotation (i.e. the integral over the real line is equal to the integral over the imaginary line).

But in the case of the finite periodicity, and therefore finite time contour, the integral on the quarter circle at infinity is not zero anymore, so we cannot do the same. Then we are forced to consider the contours discussed here. This way we have a contour that still starts at t_0 and ends at $t_0 - i\beta$, but there are no parts that need to vanish for the contour to make sense. The equality is now between the contour from t_0 to $t_0 - i\beta$ (imaginary-time formalism) and the real-time formalism contour C, at the price of having two parts that

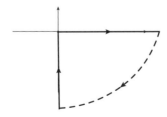

Figure 71.2   Wick rotation of the finite-temperature contour, for $\beta \to \infty$ ($T \to 0$) can be done without problem.

go in the real direction, C_1 and C_2 (and two that go in the imaginary direction, C_3 and C_4): one for physical fields (on $\mathbb{R}$) and one for the heat bath (on $\mathbb{R} - i\beta/2$, or $\mathbb{R} - i\beta'$ in general). Note that the contour needs to be monotonically decreasing in the imaginary direction, which corresponds to the Feynman $i\epsilon$ prescription (consider the limiting case of $\beta \to \infty$, where the real contour is also slightly decreasing).

Then the partition function of this real-time formalism becomes

$$Z_\beta[J] \equiv Z_C[J] = \sum_n \langle \phi_n | \mathbf{T}_C \left[e^{-i \int_C d\tau_C H[J]} \right] | \phi_n \rangle$$

$$= \mathrm{Tr} \left[\mathbf{T}_C \left[e^{-i \int_{C_1} d\tau_C H[J]} \right] \mathbf{T}_C \left[e^{-i \int_{C_3} d\tau_C H[J]} \right] \right.$$

$$\left. \mathbf{T}_C \left[e^{-i \int_{C_2} d\tau_C H[J]} \right] \mathbf{T}_C \left[e^{-i \int_{C_4} d\tau_C H[J]} \right] \right]$$

$$= \mathrm{Tr} \left[U[J; C_1] U[J; C_3] U[J; C_2] U[J; C_4] \right], \qquad (71.31)$$

where $H[J] = H - \int d^4x J(x)\phi(x)$ and

$$U[J; C] \equiv \mathbf{T}_C \left[e^{-i \int_C d\tau_C H[J]} \right] \qquad (71.32)$$

is the time-evolution operator on the contour.

Note that if $\beta' = 0$ and $J[C_4] = J[C_3] = 0$, then $U[J; C_3] = 1$, $U[J; C_4] = e^{-\beta H}$, and $U[J; C_2] = (U[J; C_1])^\dagger$, and we obtain

$$Z_\beta[J] = \mathrm{Tr} \left[U[J; C_1]^\dagger e^{-\beta \hat{H}} U[J; C_1] \right]. \qquad (71.33)$$

This will be of use in Chapter 72.

The Green's functions are again obtained simply by differentiation with respect to $J(x)$, as before.

71.6 Propagators and Field Doubling

From the form of the Feynman propagator in (71.20), we find that the Wightman functions in momentum space for the spatial directions are given by

$$D_{+,\beta}(\vec{k}, \tau_C) = f(E) \left[e^{-iE\tau_C} + e^{iE(\tau_C + i\beta)} \right] \Rightarrow$$

$$D_{-,\beta}(\vec{k}, \tau_C) = f(E) \left[e^{iE\tau_C} + e^{-iE(\tau_C - i\beta)} \right],$$

$$f(E) \equiv \frac{1}{2E} \frac{1}{1 - e^{-\beta E}},$$

$$E \equiv \sqrt{\vec{k}^2 + m^2}. \qquad (71.34)$$

Here we have written τ_C, since the same formula is obtained on an arbitrary contour, and in fact by solving the differential equation for $D_{F,\beta}$ with the KMS boundary conditions, we arrive at the above results, just with a relative factor α between the two terms, that can be fixed to 1 by hermiticity of $D_{F,\beta}$.

For the contour C considered before, if $\tau_{C,1}$ is on C_1 or C_2 (horizontal pieces) at finite positions, and $\tau_{C,2}$ on C_3 or C_4 (vertical pieces), and $t_0 \to \infty$, so $\tau_C = \tau_{C,2} - \tau_{C,1} \to \infty$ as well. Then the two factors in $D_{\pm,\beta}(\vec{k}, \tau_C)$ both oscillate very fast (have $e^{\pm iEt_0}$ factors), and by the Riemann–Lebesgue lemma their Fourier transform, and thus $D_{F,\beta}(\vec{x}, \tau_C)$, vanishes. This means that there is no interaction between the C_1, C_2 contours and the C_3, C_4 contours in the free partition function $Z_{0,\beta}$, and

$$Z_{0,\beta}[J] = Z_{0,\beta}[J; C_1, C_2] Z_{0,\beta}[J; C_3, C_4] = Z_{0,\beta}[J; C_1, C_2], \qquad (71.35)$$

where in the second equality we have assumed the vanishing of the sources at temporal infinity, $J(\vec{x}, t) \to 0$ for $t = \pm t_0 \to \pm\infty$. Then finally

$$Z_{0,\beta}[J] = Z_{0,\beta}[J; C_1, C_2] = \exp\left[-\frac{i}{2\hbar} \int_{C_1, C_2} dx\, dy\, J(x) \frac{D_{F,\beta}(x-y)}{i\hbar} J(y) \right]. \qquad (71.36)$$

But moreover now, since we have fields defined on two parallel lines ($\mathbb{R}$ and $\mathbb{R} - i\beta/2$), we can define two independent *fields* as the fields on the two contours (i.e. *field doubling*):

$$J_1(\vec{x}, t) \equiv J(\vec{x}, t), \quad J_2(\vec{x}, t) \equiv J\left(\vec{x}, t - \frac{i}{2}\beta\right), \qquad (71.37)$$

and write a matrix formula

$$Z_{0,\beta}[J_1, J_2] = \exp\left[-\frac{i}{2\hbar} \int_{C_1} dx\, dy\, J_a(x) \frac{D^{ab}_{F,\beta}(x-y)}{i\hbar} J_b(y) \right] \qquad (71.38)$$

as a function of the matrix propagator $D^{ab}_{F,\beta}(x-y)$, with $a = 1$ corresponding to the usual fields and $a = 2$ to the contour C_2, *with opposite direction of integration*, corresponding to the heat bath. Because of this, D^{11} is the normal propagator, D^{22} is the propagator with reversed endpoints and D^{12} and D^{21} have a complex time shift, specifically:

$$
\begin{aligned}
D^{11}_{F,\beta} &= D_{F,\beta}(x-y), \\
D^{22}_{F,\beta} &= D_{F,\beta}(y-x) = D^*_{F,\beta}(x-y), \\
D^{12}_{F,\beta} &= D_{-,\beta}\left(\vec{x} - \vec{y}, x^0 - y^0 + \frac{i\beta}{2}\right), \\
D^{21}_{F,\beta} &= D_{+,\beta}\left(\vec{x} - \vec{y}, x^0 - y^0 - \frac{i\beta}{2}\right).
\end{aligned}
\qquad (71.39)
$$

Then, the matrix in momentum space is

$$D_{F,\beta}(k) = \begin{pmatrix} D_{F,\beta}(k) & 0 \\ 0 & D^*_{F,\beta}(k) \end{pmatrix} + \frac{2\pi \delta(k^2 + m^2)}{e^{\beta E} - 1} \begin{pmatrix} 1 & e^{\frac{\beta E}{2}} \\ e^{\frac{\beta E}{2}} & 1 \end{pmatrix}, \qquad (71.40)$$

and we only have a symmetric matrix because we chose C_2 to be at $\beta' = \beta/2$ (otherwise we would have different $e^{\beta' E}$ and $e^{(\beta - \beta')E}$ components in the second term).

One can similarly calculate the propagators for photons and fermions. For completeness, we write here the propagator for fermions:

$$S_{F,\beta}(k) = \begin{pmatrix} S_{F,\beta}(k) & 0 \\ 0 & S_{F,\beta}^*(k) \end{pmatrix} + \frac{2\pi\,\delta(k^2 + m^2)}{e^{\beta E} - 1} \begin{pmatrix} 1 & \epsilon(k_0)e^{\frac{\beta E}{2}} \\ -\epsilon(k_0)e^{\frac{\beta E}{2}} & 1 \end{pmatrix}. \quad (71.41)$$

We can also express the partition function completely in terms of these field doubles, just that because of the different direction of integration, we get a relative minus sign between the two interaction terms:

$$Z_\beta[J] = Z_\beta[J_1, J_2] = \exp\left[\frac{i}{\hbar}\int_C d\tau_C L_{\text{int}}\left(\frac{\hbar}{i}\frac{\delta}{\delta J}\right)\right] Z_{0,\beta}[J]$$

$$= \exp\left[\frac{i}{\hbar}\int_{C_1} d\tau_C \left\{ L_{\text{int}}\left(\frac{\hbar}{i}\frac{\delta}{\delta J_1}\right) - L_{\text{int}}\left(\frac{\hbar}{i}\frac{\delta}{\delta J_2}\right)\right\}\right]$$

$$= \int \mathcal{D}\phi_1 \int \mathcal{D}\phi_2 \exp\left[\frac{i}{\hbar}\int d^4x \int d^4y\ \phi_a(x)D_{F,\beta}^{-1,ab}(x-y)\phi^b(y)\right.$$

$$\left. + \frac{i}{\hbar}\int d^4x\, [\mathcal{L}_{\text{int}}(\phi_1) - \mathcal{L}_{\text{int}}(\phi_2) + J_a\phi_a]\right], \quad (71.42)$$

where $J_a\phi_a \equiv J_1\phi_1 - J_2\phi_2$.

Important Concepts to Remember

- As in the nonrelativistic case, the various formalisms are defined by contours in complex time, on which we consider path integrals.
- The imaginary-time formalism has a contour on the imaginary axis (for the temperature), the real-time formalism has vertical pieces for the temperature and contours parallel to the real line for time evolution.
- In the imaginary-time formalism, one can expand in Matsubara frequencies, $\omega_n = 2\pi n/\beta$ for bosons.
- Wightman functions (propagators) in the imaginary-time formalism obey the KMS relation, $D_+(x^0 - y^0) = D_-(x^0 - y^0 + i\beta)$, and in a real-time formalism, a similar equation defined on the contour C, $D_{+,\beta}(\tau_{C,x} - \tau_{C,y}) = D_{-,\beta}(\tau_{C,x} - \tau_{C,y} + i\beta)$.
- To obtain Green's functions, we consider source terms on the contour C, $\int_C J\phi$, where the contour C is monotonically decreasing in the imaginary direction, corresponding to Feynman's $i\epsilon$ prescription.
- For a contour formed of two horizontal pieces C_1 and C_2 (with $i\beta/2$ difference in the symmetric case), and two vertical pieces C_3 and C_4, we can define field doubling, a field on each of C_1 and C_2, with opposite directions of integration, one for the physical fields and one for the heat bath.
- Because of field doubling, the propagator becomes a matrix in the 1,2 space.

Further Reading

See [30] for the thermofield double formalism.

Exercises

1. Prove the relation

$$\frac{-i}{\beta} \sum_n \frac{e^{i\omega_n \tau}}{\omega^2 + \omega_n^2} = \frac{1}{2\omega} \frac{1}{e^{\beta\omega} - 1} \left[e^{-\omega(|\tau|-\beta)} + e^{\omega|\tau|} \right], \qquad (71.43)$$

   where ω_n are Matsubara frequencies.

2. Show that the differential equation (71.27) for the Wightman functions, with the KMS periodicity condition, is solved by (71.34), imposing also hermiticity.

3. Calculate the free four-point function in the real-time formalism, as a matrix in C_1, C_2 space.

4. Write the (formal) Dyson equation for the partition function on the C_1, C_2, C_3, C_4 contour, in the field-doubling formalism.

72

Finite-Temperature Quantum Field Theory III: Thermofield Dynamics and Schwinger–Keldysh "In–In" Formalism for Thermal and Nonequilibrium Situations. Applications

In this chapter, we first present the thermofield dynamics, which is a rewriting in canonical formalism of the field-doubling formalism from Chapter 71, and then move on to the Schwinger–Keldysh "in–in" formalism, which can be used not only in thermal situations, but also in nonequilibrium ones, where we don't know the final point of the evolution.

72.1 Thermofield Dynamics

This formalism, defined by Umesawa, is a canonical (operatorial) formulation corresponding to the field-doubling mechanics observed at the end of Chapter 71.

What we want is to have a *thermal vacuum* $|0, \beta\rangle$ and thermal creation and annihilation operators, and do calculations in them, instead of the sums of expectation values in different states, with Boltzmann weights, as in the case of previous chapters. This will correspond to a *Bogoliubov transformation of basis*, defined by some angle, different for bosons and fermions.

We want to rewrite the thermal average as a VEV in the thermal vacuum state:

$$\langle A \rangle_\beta = \frac{\sum_n e^{-\beta E_n} \langle n|A|n\rangle}{\sum_m e^{-\beta E_m}} \equiv \langle 0, \beta|A|0, \beta\rangle, \qquad (72.1)$$

which can be achieved if we take the thermal vacuum state

$$|0, \beta\rangle \equiv \frac{\sum_{n=\tilde{n}} e^{-\frac{\beta E}{2}} |n\rangle \otimes |\tilde{n}\rangle}{\left(\sum_m e^{-\beta E_m}\right)^{1/2}}. \qquad (72.2)$$

Note that here we need the second Hilbert space, $|\tilde{n}\rangle$, since it allows us to project only onto *diagonal* matrix elements $\langle n|A|n\rangle$ (because we have orthonormality in the second Hilbert space, $\langle \tilde{n}|\tilde{m}\rangle$, otherwise we would have $\langle n|A|m\rangle$. This doubling of Hilbert spaces corresponds to the field doubling obtained in the path integral, with the $|n\rangle$ space being the original Hilbert space, on the real line, and $|\tilde{n}\rangle$ corresponding to the heat bath. Also note that we don't need to have only two Hilbert spaces, more would do as well, which in the path-integral formalism would correspond to having more lines parallel to the real line, going back and forth until reaching $-t_0 - i\beta$.

72.1.1 Thermal Fermionic Harmonic Oscillator

The simplest case is that of the fermionic harmonic oscillator, with only two states (unoccupied, $|0\rangle$ and occupied, $1\rangle$). In the field doubling, we still obtain two states, $|0,0\rangle$ and $|1,1\rangle$, so now

$$|0,\beta\rangle_F = \frac{1}{\sqrt{1+e^{-\beta\omega}}}\left(|0,0\rangle + e^{-\frac{\beta\omega}{2}}|1,1\rangle\right) \equiv \cos\theta_F(\beta)|0,0\rangle + \sin\theta_F(\beta)|1,1\rangle, \quad (72.3)$$

where we have defined

$$\cos\theta_F = \frac{1}{\sqrt{1+e^{-\beta\omega}}} \Rightarrow \tan\theta_F = e^{-\frac{\beta\omega}{2}}. \quad (72.4)$$

We can also define this transformation as a "unitary" transformation, from one particle vacuum to another, and one set of creation and annihilation states to another, or *Bogoliubov transformation*, by

$$|0,\beta\rangle_F \equiv U_F(\beta)|0,0\rangle_F \equiv e^{-iQ_F(\theta)}|0,0\rangle_F,$$
$$Q_F(\theta) \equiv -i\theta_F(\beta)(\tilde{a}_F a_F - a_F^\dagger \tilde{a}_F^\dagger). \quad (72.5)$$

The proof of this relation is left as an exercise.

We can formalize this transformation by defining first thermal creation and annihilation operators:

$$a_F(\theta) = U_F(\beta)a_F U_F^{-1}(\beta) \,, \quad \tilde{a}_F(\theta) = U_F(\beta)\tilde{a}_F U_F^{-1}(\beta), \quad (72.6)$$

and then the column matrix annihilation operator:

$$A_F \equiv \begin{pmatrix} a_F \\ \tilde{a}_F^\dagger \end{pmatrix}, \quad A_F(\beta) \equiv \begin{pmatrix} a_F(\beta) \\ \tilde{a}_F^\dagger(\beta) \end{pmatrix}, \quad A_F' \equiv \begin{pmatrix} a_F^\dagger \\ \tilde{a}_F \end{pmatrix}, \quad A_F'(\beta) \equiv \begin{pmatrix} a_F^\dagger(\beta) \\ \tilde{a}_F(\beta) \end{pmatrix}. \quad (72.7)$$

Then the Bogoliubov transformation acts by a rotation matrix on the A_F:

$$A_F(\beta) = U_F(\beta)A_F U_F^{-1}(\beta) = \begin{pmatrix} \cos\theta_F(\beta) & -\sin\theta_F(\beta) \\ \sin\theta_F(\beta) & \cos\theta_F(\beta) \end{pmatrix} A_F,$$
$$A_F'(\beta) = U_F(\beta)A_F' U_F^{-1}(\beta) = \begin{pmatrix} \cos\theta_F(\beta) & -\sin\theta_F(\beta) \\ \sin\theta_F(\beta) & \cos\theta_F(\beta) \end{pmatrix} A_F'. \quad (72.8)$$

The new thermal vacuum is a vacuum of the thermal annihilation operators:

$$a_F(\theta)|0,\beta\rangle = U_F(\beta)a_F U_F^{-1}(\beta)U_F(\beta)|0,0\rangle_F$$
$$= \cos\theta_F(\beta)\sin\theta_F(\beta)(a_F|1,1\rangle - \tilde{a}_F^\dagger|0,0\rangle) = 0,$$
$$\tilde{a}_F(\theta)|0,\beta\rangle = U_F(\beta)\tilde{a}_F U_F^{-1}(\beta)U_F(\beta)|0,0\rangle$$
$$= \cos\theta_F(\beta)\sin\theta_F(\beta)(\tilde{a}_F|1,1\rangle + a^\dagger|0,0\rangle) = 0. \quad (72.9)$$

Note that the new thermal creation and annihilation operators are then dressed ordinary particles $(a, a^\dagger)$ and dressed particles in the heat bath $(\tilde{a}, \tilde{a}^\dagger)$.

Since $A'_F \cdot A_F - 1 = a^\dagger_F a_F + \tilde{a}_F \tilde{a}^\dagger_F - 1 = a^\dagger_F a_F - \tilde{a}^\dagger_F \tilde{a}_F$ is rotationally $(U_F(\beta))$-invariant, we have

$$H - \tilde{H} = \hbar\omega(a^\dagger_F a_F - \tilde{a}^\dagger_F \tilde{a}_F) = \hbar\omega(a^\dagger_F(\beta)a_F(\beta) - \tilde{a}^\dagger_F(\beta)\tilde{a}_F(\beta)), \tag{72.10}$$

which means that (excited) thermal states

$$|n, \tilde{n}\rangle \propto (a^\dagger_F(\beta))^n (\tilde{a}^\dagger_F(\beta))^{\tilde{n}}|0, 0\rangle \tag{72.11}$$

are eigenstates of $H - \tilde{H}$, but not of $H = \omega(a^\dagger_F a_F - 1/2)$ itself.

72.1.2 Bosonic Harmonic Oscillator

Now we can move on to the bosonic case, and similarly define

$$\langle A\rangle_\beta = \frac{\sum_{n\geq 0} e^{-\beta\omega}\langle n|A|n\rangle}{\sum_n e^{-\beta\omega}} = (1 - e^{-\beta\omega})\sum_{n\geq 0} e^{-\beta\omega}\langle n|A|n\rangle \equiv {}_B\langle 0, \beta|A|0, \beta\rangle_B, \tag{72.12}$$

which is solved by

$$|0, \beta\rangle_B \equiv \sqrt{1 - e^{-\beta\omega}}\sum_{n\geq 0} e^{-\frac{\beta\omega}{2}}|n, \tilde{n}\rangle_B, \tag{72.13}$$

where $|n, \tilde{n}\rangle_B \equiv |n\rangle \otimes |\tilde{n}\rangle$. Moreover, we can also write this as a Bogoliubov transformation on the vacuum state:

$$|0, \beta\rangle_B = U_B(\beta)|0, 0\rangle = e^{-iQ_B(\beta)}|0, 0\rangle,$$
$$Q_B(\beta) = -i\theta_B(\beta)(\tilde{a}_B a_B - a^\dagger_B \tilde{a}^\dagger_B). \tag{72.14}$$

We can also define the Bogoliubov transformation as a Lorentz rotation on creation and annihilation operators:

$$A(\beta) = \begin{pmatrix} a_B(\beta) \\ \tilde{a}^\dagger_B(\beta) \end{pmatrix} = U_B(\beta)AU_B^{-1}(\beta) = \begin{pmatrix} \cosh\theta_B(\beta) & -\sinh\theta_B(\beta) \\ -\sinh\theta_B(\beta) & \cosh\theta_B(\beta) \end{pmatrix} \begin{pmatrix} a_B \\ \tilde{a}^\dagger_B \end{pmatrix},$$
$$A'(\beta) = \begin{pmatrix} a^\dagger_B(\beta) \\ \tilde{a}_B(\beta) \end{pmatrix} = U_B(\beta)A'U_B^{-1}(\beta) = \begin{pmatrix} \cosh\theta_B(\beta) & -\sinh\theta_B(\beta) \\ -\sinh\theta_B(\beta) & \cosh\theta_B(\beta) \end{pmatrix} \begin{pmatrix} a^\dagger_B \\ \tilde{a}_B \end{pmatrix}.$$
$$\tag{72.15}$$

which we leave to prove as an exercise. Then, as before, the relative Hamiltonian is left invariant by the Bogoliubov transformation:

$$H - \tilde{H} = U(\beta)(H - \tilde{H})U^{-1}(\beta) = H(\beta) - \tilde{H}(\beta), \tag{72.16}$$

so thermal states are not eigenstates of the Hamiltonian, but of $H - \tilde{H}$.

But we have not defined what $\theta_B(\beta)$ is. To do this, we first define the thermal expectation value of the number operator:

$$\langle a_B^\dagger a_B \rangle_\beta = \frac{\sum_n n e^{-\beta E_n}}{\sum_n e^{-\beta E_n}} = (1 - e^{-\beta\omega}) \left(\sum_n n e^{-n\beta\omega} \right)$$

$$= (1 - e^{-\beta\omega}) \frac{\partial}{\partial(-\beta\omega)} \frac{1}{1 - e^{-\beta\omega}} = \frac{1}{1 - e^{-\beta\omega}}, \tag{72.17}$$

then alternatively, relating a_B to $a_B(\beta)$ via the Lorentz rotation, and thus obtaining

$$\langle a_B^\dagger a_B \rangle_\beta = \cosh^2(\theta_B(\beta)), \tag{72.18}$$

so that finally

$$\cosh \theta_B(\beta) = \frac{1}{\sqrt{1 - e^{-\beta\omega}}}. \tag{72.19}$$

72.2 The Schwinger–Keldysh Formalism at $T = 0$

Until now, in the generic real-time formalism, and in the thermofield dynamics one above, we have mostly assumed that the complex-time contour is the one with two horizontal lines situated symmetrically, at $\beta' = \beta/2$. This was useful for the case of the field doubling having a symmetric propagator matrix. But there is another useful contour, the one with $\beta' = 2\epsilon$, so where we go on $\mathbb{R} + i\epsilon$, and then go back on $\mathbb{R} - i\epsilon$, before going down to $-t_0 - i\beta$, as in Figure 71.1(c). But then, the temperature is reduced to act as in the imaginary-time formalism, by the imaginary line from $-t_0$ to $-t_0 - i\beta$.

We can use the same formalism from Chapter 71, with field doubling, for this contour with $\beta' = 2\epsilon$. We will restate it here, for this particular contour, where, as we will see, there are some peculiarities, and we obtain more information. Moreover, we don't need to describe the temperature piece (imaginary line to $-t_0 - i\beta$) yet, we will introduce it later. As it is, we obtain a new formalism, that can be used at $T = 0$, called the "in–in" formalism, in that it doesn't concern itself with the future evolution of the system; everything is described from the point of view of its past (the "in" modes, at $-t_0$). This formalism can then be used also for nonequilibrium physics.

First, a two-point function on the contour, more generally for an operator $\mathcal{O}_H(x)$ in the Heisenberg picture, rather than just a field $\phi(x)$, is defined as usual, just with time ordering *on the contour*, as

$$G_C(x_1, x_2) = \langle \Omega | \mathbf{T}_C \left(\mathcal{O}_H(x_1) \mathcal{O}_H(x_2) \right) | \Omega \rangle, \tag{72.20}$$

and similarly for higher n-point functions.

We are then interested in rewriting things in terms of the unperturbed vacuum $|0\rangle$ via Feynman's theorem (see Chapter 5 in Part I), in order to set up perturbation theory. The interaction picture evolution operator *on the contour* is

$$U_{I,C}(t, t_0) = \mathbf{T}_C \exp\left[-i \int_{C, t_0}^{t} d\tau H_{\text{int},I}(\tau) \right]. \tag{72.21}$$

Normally, we would like to write a Feynman theorem for the case $t_0 \to -\infty$ and $t \to +\infty$, with a convenient $i\epsilon$ prescription (that ∞ is replaced by $\infty(1 - i\epsilon)$):

$$G(x_1, x_2) = \lim_{t_0 \to -\infty(1-i\epsilon), t \to +\infty(1-i\epsilon)} \frac{\langle 0|\mathbf{T}\left(U_I(t, t_0)\mathcal{O}_I(x_1)\mathcal{O}_I(x_2)\right)|0\rangle}{\langle 0|U_I(t, t_0)|0\rangle}. \tag{72.22}$$

But now, because of the contour, which comes back to $-t_0$, albeit at -2ϵ below the starting point, we actually have

$$U_{I,C} \equiv U_{I,C}(-t_0 + i\epsilon, -t_0 - i\epsilon) \Rightarrow U_{I,C}|0\rangle = |0\rangle, \tag{72.23}$$

since we come back to an "in" state, which is taken to be an *equilibrium* one (even if the future evolution of the system is out of equilibrium), specifically the instantaneous vacuum of the free Hamiltonian at $-t_0 \to -\infty$, $|0\rangle$. This means that we don't have a denominator anymore, and we have simply

$$G_C(x_1, x_2) = \lim_{t_0 \to \infty} \langle 0|\mathbf{T}_C\left(U_{I,C}\mathcal{O}(x_1)\mathcal{O}(x_2)\right)|0\rangle. \tag{72.24}$$

We can then consider "field doubling," as in Chapter 71, with fields on the upper branch of the contour C denoted by the subscript 1 (sometimes one uses R) and fields on the lower branch denoted by the subscript 2 (sometimes one uses L).

The Feynman propagator for the contour (free two-point function on the contour) becomes a matrix, corresponding to where (on which branch) the field insertion is:

$$D^{ab}_{F,C}(x_1, x_2) = \begin{pmatrix} D^{11}_F(x_1, x_2) & D^{12}(x_1, x_2) \\ D^{21}_F(x_1, x_2) & D^{22}_F(x_1, x_2) \end{pmatrix}, \tag{72.25}$$

and we have (note that we are in the *formal* case $\beta = 0$ compared to last time, meaning we don't have the vertical branch of the contour, and $\beta' \simeq 0$):

$$D^{11}_F(x_1, x_2) = D_F(x_1 - x_2),$$
$$D^{12}_F(x_1, x_2) = D_-(x_1 - x_2),$$
$$D^{21}_F(x_1, x_2) = D_+(x_1, x_2),$$
$$D^{22}_F(x_1, x_2) = D^*_F(x_1 - x_2). \tag{72.26}$$

As usual:

$$D_{F,C}(x - y) = \langle 0|\mathbf{T}_C\left(\mathcal{O}(x)\mathcal{O}^\dagger(y)\right)|0\rangle$$
$$= \theta_C(t_x - t_y)D_{+,C}(x - y) + \theta_C(t_y - t_x)D_{-,C}(x - y),$$
$$D_F(x - y) = \langle 0|\mathbf{T}\left(\mathcal{O}(x)\mathcal{O}^\dagger(y)\right)|0\rangle = \theta(t_x - t_y)D_+(x - y) + \theta(t_y - t_x)D_-(x - y). \tag{72.27}$$

Similar relations hold for the full Feynman two-point functions $G^{ab}_{F,C}(x, y) \equiv G^{ab}(x, y)$.

One can also define the retarded (R), advanced (A), and Keldysh (K) two-point functions:

$$D_R(x_1, x_2) = \theta(t_1 - t_2)\langle 0| \left[\mathcal{O}(x_1), \mathcal{O}^\dagger(x_2) \right] |0\rangle = D_F - D_-,$$

$$D_A(x_1, x_2) = -\theta(t_2 - t_1)\langle 0| \left[\mathcal{O}(x_1), \mathcal{O}^\dagger(x_2) \right] |0\rangle = D_F - D_+,$$

$$D_K(x_1, x_2) = \langle 0| \left\{ \mathcal{O}(x_1), \mathcal{O}^\dagger(x_2) \right\} |0\rangle = D_F + D_F^*. \tag{72.28}$$

We can also generalize to n-point functions $G^{a_1 \cdots a_n}(x_1, \ldots, x_n)$, where $a_i = 1, 2$.

Then we can write down a partition function Z_{SK}, for the Schwinger–Keldysh contour, similarly to what we did in Chapter 71. Define

$$J_1(\vec{x}, t) = J(\vec{x}, t + i\epsilon) , \quad J_2(\vec{x}, t) = J(\vec{x}, t - i\epsilon), \tag{72.29}$$

and then the free partition function is

$$Z_{0,SK}[J] = Z_0[J; C] = \exp\left[-\frac{i}{2\hbar} \int_C dx\, dy\, J(x) \frac{D_{F,C}(x - y)}{i\hbar} J(y) \right]$$

$$= \exp\left[-\frac{i}{2\hbar} \int_{C_1} dx\, dy\, J_a(x) \frac{D_F^{ab}(x - y)}{i\hbar} J_b(y) \right] \equiv Z_{0,SK}[J_1, J_2]. \tag{72.30}$$

We can then write a Dyson equation:

$$Z_{SK}[J] = \int \mathcal{D}\phi \exp\left[\frac{i}{\hbar} S_{\text{int}}\left(\frac{\hbar}{i} \frac{\delta}{\delta J(x)} \right) \right] Z_{0,SK}[J], \tag{72.31}$$

and with the field-doubling formalism rewrite it as

$$Z_{SK}[J] = Z_{SK}[J_1, J_2] = \exp\left[\frac{i}{\hbar} \int_C d\tau_C L_{\text{int}}\left(\frac{\hbar}{i} \frac{\delta}{\delta J} \right) \right] Z_{0,SK}[J]$$

$$= \exp\left[\frac{i}{\hbar} \int_{C_1} d\tau_C \left\{ L_{\text{int}}\left(\frac{\hbar}{i} \frac{\delta}{\delta J_1} \right) - L_{\text{int}}\left(\frac{\hbar}{i} \frac{\delta}{\delta J_2} \right) \right\} \right] Z_{0,SK}[J]$$

$$= \int \mathcal{D}\phi_1 \int \mathcal{D}\phi_2 \exp\left[\frac{i}{\hbar} \int d^4x \int d^4y\, \phi_a(x) D_F^{-1,ab}(x - y)\phi^b(y) \right.$$

$$\left. + \frac{i}{\hbar} \int d^4x \left[\mathcal{L}_{\text{int}}(\phi_1) - \mathcal{L}_{\text{int}}(\phi_2) + J_a\phi_a \right] \right], \tag{72.32}$$

where $J_a\phi_a = J_1\phi_1 - J_2\phi_2$.

From this, we obtain as usual the Green's functions by differentiation:

$$G^{a_1 \cdots a_n}(x_1, \ldots, x_n) = \left. \frac{\delta^n}{\delta J_{a_1}(x_1) \ldots \delta J_{a_n}(x_n)} Z_{SK}[J_1, J_2] \right|_{J_1 = J_2 = 0}. \tag{72.33}$$

An observation to be made is that, because of the time ordering on the contour, we obtain only correlators where the operator insertions on the lower contour all come before the insertions on the upper contour.

The next observation is that we can introduce a general initial density matrix for the system $\hat{\rho}_{\text{initial}}$ in the partition function and correlation functions, and define the partition function according to an obvious generalization of the thermal case in (71.33), as

$$Z_{SK}[J_1, J_2] = \text{Tr}\left[U[J_1]\hat{\rho}_{\text{initial}} U^\dagger[J_2] \right], \tag{72.34}$$

where

$$U[J; t_i, t_f] = \mathbf{T}\left[e^{-i \int_{C;t_i}^{t_f} d\tau H_I(\tau; J)}\right], \quad H[J] = H - \int d^4 x J(x)\phi(x), \tag{72.35}$$

and the $\hat{\rho}_{\text{initial}}$ is normalized to one, $\text{Tr}[\hat{\rho}_{\text{initial}}] = 1$.

Note that in the absence of the sources J_1, J_2, U and $U^\dagger$ would annihilate (given the cyclicity of the trace), and Z_{SK} would be $\text{Tr}[\hat{\rho}_{\text{initial}}] = 1$.

It is also of use to consider the change of basis from $\mathcal{O}_1, J_1$ and $\mathcal{O}_2, J_2$ to difference and average operators:

$$\mathcal{O}_{\text{diff}} = \mathcal{O}_1 - \mathcal{O}_2, \ J_{\text{diff}} = J_1 - J_2,$$
$$\mathcal{O}_{\text{av}} = \frac{\mathcal{O}_1 + \mathcal{O}_2}{2}, \ J_{\text{av}} = \frac{J_1 + J_2}{2}. \tag{72.36}$$

Then we can rewrite the object appearing in the exponent of the partition function:

$$J_1 \phi_1 - J_2 \phi_2 \rightarrow \mathcal{O}_1 J_1 - \mathcal{O}_2 J_2 = \mathcal{O}_{\text{diff}} J_{\text{av}} + \mathcal{O}_{\text{av}} J_{\text{diff}}. \tag{72.37}$$

This means in turn that we can calculate correlators of $\mathcal{O}_{\text{diff}}$ and $\mathcal{O}_{\text{av}}$, obtained from differentiation with respect to J_{av} and J_{diff}, respectively. But then we find that the correlator of only $\mathcal{O}_{\text{diff}}$s vanishes:

$$\langle \Omega | \mathbf{T}_{SK} \left(\mathcal{O}_{\text{diff}}(x_1) \dots \mathcal{O}_{\text{diff}}(x_n)\right) | \Omega \rangle = \langle \Omega | \mathbf{T}_{SK} \left((\mathcal{O}_1 - \mathcal{O}_2)(x_1) \dots (\mathcal{O}_1 - \mathcal{O}_2)(x_n)\right) | \Omega \rangle = 0. \tag{72.38}$$

Indeed, this correlator is obtained by differentiation of $Z_{SK}[J]$ with respect to J_{av}. For this calculation, we can put $J_{\text{diff}} = J_1 - J_2 = 0$, $J_1 = J_2 = J = J_{\text{av}}$. But $Z_{SK}[J, J] = \text{Tr}[\hat{\rho}_{\text{initial}}] = 1$ by cyclicity of the trace, so the result of differentiation with respect to $J = J_{\text{av}}$ vanishes, as stated. Moreover, we can also find the same result for the correlators of $\mathcal{O}_{\text{diff}}$s and $\mathcal{O}_{\text{av}}$s, as long as the operator at the latest time is an $\mathcal{O}_{\text{diff}}$:

$$\langle \Omega | \mathbf{T}_{SK} \left(\mathcal{O}_{\text{diff}}(x_p) \mathcal{O}_{I_1} \dots \mathcal{O}_{I_n}(x_n)\right) | \Omega \rangle = 0, \tag{72.39}$$

if $t_p > t_i$, $\forall i = 1, \dots, n$, and $I_1, \dots, I_n = $ diff or av. The proof follows in the same way, and is left as an exercise.

72.3 Schwinger–Keldysh Formalism at Nonzero T

We can now easily extend the formalism to a thermal state, by replacing the initial density matrix $\hat{\rho}_{\text{initial}}$ with a thermal one, $\hat{\rho}_T = e^{-\beta \hat{H}} / \text{Tr}[e^{-\beta \hat{H}}]$. As we noted before, this amounts to an extra contribution to the contour in complex time, going vertically from $-t_0$ to $-t_0 - i\beta$.

The fields in Green's functions must obey periodicity for bosons, and antiperiodicity for fermions under time translation by $i\beta$, resulting in the KMS condition.

But then, in the partition function, as seen in (71.31), we must have an extra propagation in time by the evolution operator $e^{-\beta H}$, corresponding to the imaginary-time translation $-t_0 \rightarrow -t_0 - i\beta$, acting on the fields in the second contour C_2. This means that in the

Green's functions in the SK basis $\mathcal{O}_{\text{diff}}$ and $\mathcal{O}_{\text{av}}$, obtained by differentiation with respect to J_{av} and J_{diff}, we have also this extra imaginary-time propagation. Then the same condition (72.38) is obtained, just with $\mathcal{O}_2$s propagated by $-i\beta$:

$$\langle\Omega|\mathbf{T}_{SK}\left((\mathcal{O}_1(\vec{x}_1,t_1)-\mathcal{O}_2(\vec{x}_1,t_1-i\beta))\ldots(\mathcal{O}_1(\vec{x}_n;t_n)-\mathcal{O}_n(\vec{x}_n;t_n-i\beta))\right)||\Omega\rangle=0,\tag{72.40}$$

and similarly vanishes if we have only the operator at largest time $(\mathcal{O}_1(\vec{x}_1,t_1)-\mathcal{O}_2(\vec{x}_1,t_1-i\beta))$.

72.4 Application of Thermal Field Theory: Finite-Temperature Effective Potential

The effective potential is the effective action at constant fields, or generator of zero-momentum 1PI Green's functions. In this case, we can also consider constant sources, and write the partition function at finite temperature and constant sources as

$$Z_\beta[J]=\text{Tr}[e^{-\beta H[J]}]\,,\quad H[J]=H-VJ\phi(t),\tag{72.41}$$

where V is the volume. The free energy is

$$W_\beta[J]=-\frac{1}{\beta}\ln Z_\beta[J],\tag{72.42}$$

and the effective potential is its Legendre transform:

$$F_\beta[\phi]=W_\beta[J]-VJ\phi,\tag{72.43}$$

such that

$$-VJ=\frac{\partial F_\beta[\phi]}{\partial\phi}.\tag{72.44}$$

We will use the imaginary-time formalism to calculate the one-loop potential.

Note that we can either calculate the effective potential with the Feynman rules at finite temperature, or with the background field method. Also note that the loop expansion is no longer an $\hbar$ expansion, since the periodicity depends on $\hbar$ (it is $\hbar\beta$).

However, we can still formally repeat the steps in the Coleman–Weinberg case, and write the Euclidean version (41.4), just that with a sum over frequencies, due to the finite periodicity, instead of an integral:

$$V^{(1)}=\frac{\hbar}{2}\int\frac{d^3k}{(2\pi)^3}\left(\frac{1}{\beta}\sum_n\right)\ln\left(1+\frac{V''}{\vec{k}^2+\left(\frac{2\pi n}{\beta}\right)^2+m^2}\right).\tag{72.45}$$

To calculate this, we can use the formula

$$\sum_{n\geq1}\frac{2y}{n^2+y^2}=-\frac{1}{y}+\pi\coth\pi y,\tag{72.46}$$

which can be obtained by calculating the integral in the complex plane:

$$\oint_C \frac{\coth \pi z}{z - y} dz \tag{72.47}$$

on a contour close to infinity, by the residue theorem (the proof is left as an exercise).

Then, defining the function

$$f(E) \equiv \sum_n \ln \left[E^2 + \left(\frac{2\pi n}{\beta} \right)^2 \right], \tag{72.48}$$

where $E^2 \equiv \vec{k}^2 + m^2 + V''$, we find by differentiation

$$\frac{\partial f(E)}{\partial E} = \sum_{n \in \mathbb{Z}} \frac{2E}{\frac{(2\pi n)^2}{\beta^2} + E^2} = \beta \left[1 + \frac{2}{e^{\beta E} - 1} \right]. \tag{72.49}$$

Note that in fact differentiation by E amounts to a regularization procedure, as before it, we had a divergent result. Now we obtain

$$f(E) = \beta E + 2 \ln \left(1 - e^{-\beta E} \right) + E_0, \tag{72.50}$$

where E_0 is a constant that we will drop (renormalization procedure).

Also note that now we are left with the integral over $\vec{k}$, which gives a constant piece, and a finite-temperature dependent piece:

$$V^{(1)}(\phi) = \frac{1}{2} \int \frac{d^3 k}{(2\pi)^3} E + \frac{1}{\beta} \int \frac{d^3 k}{(2\pi)^3} \ln \left(1 - 2e^{-\beta E} \right), \tag{72.51}$$

where

$$E^2(\phi) = \vec{k}^2 + m^2 + V''(\phi) = \vec{k}^2 + m^2 + \frac{\lambda \phi^2}{2}. \tag{72.52}$$

Again, the divergence is in the temperature-independent piece, and the temperature-dependent piece is (UV and IR) finite (including for $m \to 0$).

Then renormalization follows exactly as at $T = 0$, since there are no new, T-dependent divergences, and thus the renormalization of the $T = 0$ theory also renormalizes the finite-T one.

One can calculate more explicitly the large-T ($\beta \to 0$) expansion of the above effective potential, by evaluating the integral

$$I_B(y) = \int_0^\infty dr \, r^2 \ln \left(1 - e^{-\sqrt{r^2 + y^2}} \right) = I_0 + y^2 I_2 + \dots \tag{72.53}$$

as an expansion in $y^2 = \beta^2 (V'' + m^2)$ ($r = \beta k$).

After a calculation, one finds

$$V(\phi, \beta) = \frac{1}{2} m^2 \phi^2 + \frac{\lambda \phi^4}{4!} + \frac{\hbar}{\beta^4} \frac{4\pi}{(2\pi)^3} \left[-\frac{\pi^4}{45} + \frac{\pi^2 \beta^2}{12} \left(\frac{\lambda \phi^2}{2} + m^2 \right) \right] + \mathcal{O} \left(\frac{\hbar}{\beta} \right)$$
$$+ \mathcal{O}(\hbar^2). \tag{72.54}$$

We can isolate the ϕ-independent terms V_0, and in the case $m^2 < 0$ (symmetry breaking at $T = 0$), write

$$V(\phi, \beta) = V_0 + \frac{m^2 \phi^2}{2} \left(1 - \hbar \frac{T^2}{T_C^2} \right) + \frac{\lambda \phi^4}{4!} + \dots, \tag{72.55}$$

where the critical temperature is

$$\kappa T_C = \sqrt{ -\frac{24 m^2}{\lambda} }. \tag{72.56}$$

We see then that, since $m^2 < 0$, there is a symmetry breaking at $T = 0$, but for $T \geq T_C$, $m^2(T)$ changes sign, and we find no symmetry breaking, therefore a second-order phase transition occurs here.

This means that at sufficiently high temperature, symmetry is restored. Note that here $\langle \phi \rangle = -6m^2/\lambda$, so

$$kT_C = 2\langle \phi \rangle, \tag{72.57}$$

and taking the scalar VEV in the Standard Model, of about 250 GeV, one obtains a critical temperature for symmetry restoration of about 500 GeV, though this was a simplistic model. In reality, we must consider all of the fields in the model, not just the scalar.

For the fermions, we must consider *antiperiodic modes*, and the Matsubara frequencies are $\omega_n = (2n + 1)\pi/\beta$, and one finds

$$V_F^{(1)}(\phi) = V_{T=0}^{(1)}(\phi) - \frac{4\hbar}{\beta} \int \frac{d^3 \vec{k}}{(2\pi)^3} \ln \left(1 + e^{-\beta E(k)} \right). \tag{72.58}$$

For gauge fields, one finds that they act at the free level as a gas of massless bosons with two polarizations, and the unphysical degrees of freedom cancel each other in their effects.

We can then consider various theories, and find the one-loop effective potential in them. One finds for instance that supersymmetry is broken by finite-temperature effects, but grand unified theories (GUTs) have the symmetry restored at high energies.

Important Concepts to Remember

- Thermofield dynamics defines a thermal vacuum $|0, \beta\rangle$ such that thermal averages are VEVs in the thermal vacuum, $\langle A \rangle_\beta = \langle 0, \beta | A | 0, \beta \rangle$.
- The thermal vacuum is defined in terms of field-double Hilbert states, $|n\rangle \otimes |\tilde{n}\rangle$.
- The thermal vacua of the bosonic and fermionic harmonic oscillators are Bogoliubov transformations of the vacuum $|0, 0\rangle$, and the oscillators are the same transformations (rotations) of the zero-temperature oscillators.
- The Schwinger–Keldysh contour is a field-doubling contour, which has the physical states on $\mathbb{R} + i\epsilon$ (C_1) and the heat bath on $\mathbb{R} - i\epsilon$ (C_2), so going forwards and backwards near the real line, and then going down on the imaginary line.

- The field-doubling matrix propagators in the Schwinger–Keldysh case are written as $D_F^{11} = D_F$, $D_F^{22} = D_F^*, D_F^{12} = D_-, D_F^{21} = D_+$, and are related to the retarded, advanced, and Keldysh propagators by $D_R = D_F - D_-, D_A = D_F - D_+$, and $D_K = D_F + D_F^*$.
- The Schwinger–Keldysh partition function is defined in terms of the contour C integral, or equivalently in terms of matrix propagators in C_1, C_2, and the Green's function by differentiation.
- The Schwinger–Keldysh formalism, or "in–in formalism," can be used at $T = 0$ for nonequilibrium situations, where everything is described in terms of the past, at t_0, or "in" states.
- Defining $\mathcal{O}_{\text{diff}} = \mathcal{O}_1 - \mathcal{O}_2, \mathcal{O}_{\text{av}} = \mathcal{O}_1 + \mathcal{O}_2/2$, and similarly for the sources, we find $\langle \Omega | \mathbf{T}_{SK} \left(\mathcal{O}_{\text{diff}}(x_p) \mathcal{O}_{I_1} \ldots \mathcal{O}_{I_n}(x_n) \right) | \Omega \rangle = 0$, where $t_p > t_i, \forall i = 1, \ldots, n$, and $I_1, \ldots, I_n = \text{diff or av}$.
- At finite temperature, the Schwinger–Keldysh formalism only involves an extra translation in imaginary time by $-i\beta$.
- The Coleman–Weinberg effective potential at finite temperature has a correction that changes symmetry breaking, providing a phase transition at $\kappa T_C = \sqrt{-24m^2/\lambda}$.

Further Reading

See [30] for a more detailed review of the Schwinger–Keldysh formalism in (relativistic or not) quantum field theories.

Exercises

1. Prove the relation (72.5), where the thermal vacuum is defined by (72.3), for the fermionic harmonic oscillator.
2. Prove that the Bogoliubov transformation on the thermal vacuum of the bosonic harmonic oscillator can be written as in (72.15).
3. Prove that the correlators of $\mathcal{O}_{\text{diff}}$ and $\mathcal{O}_{\text{av}}$, with the latest time being $\mathcal{O}_{\text{diff}}$, vanish:

$$\langle \Omega | \mathbf{T}_{SK} \left(\mathcal{O}_{\text{diff}}(x_p) \mathcal{O}_{I_1} \ldots \mathcal{O}_{I_n}(x_n) \right) | \Omega \rangle = 0, \tag{72.59}$$

   where $t_p > t_i, \forall i = 1, \ldots, n$, and $I_1, \ldots, I_n = \text{diff or av}$.
4. Prove the relation (72.46).

References

[1] M. E. Peskin and D. V. Schroeder, *An Introduction to Quantum Field Theory*, Westview Press, Boulder, CO, 1995.

[2] G. Sterman, *An Introduction to Quantum Field Theory*, Cambridge University Press, Cambridge, 1993.

[3] J. Ambjorn and J. L. Petersen, *Quantum Field Theory*, Niels Bohr Institute Lecture Notes, 1994.

[4] P. Ramond, *Field Theory: A Modern Primer* (2nd edn), Westview Press, Boulder, CO, 2001.

[5] P. A. M. Dirac, *Lectures on Quantum Mechanics*, Dover, New York, 2001.

[6] T. Banks, *Modern Quantum Field Theory*, Cambridge University Press, Cambridge, 2008.

[7] M. J. G. Veltman, "Unitarity and causality in a renormalizable field theory with unstable particles," *Physica* **29**, 186 (1963).

[8] L. J. Dixon, "A brief introduction to modern amplitude methods," arXiv:1310.5353 [hep-ph].

[9] H. Elvang and Y. t. Huang, *Scattering Amplitudes in Gauge Theory and Gravity*, Cambridge University Press, Cambridge, 2015.

[10] M. Srednicki, *Quantum Field Theory*, Cambridge University Press, Cambridge, 2007.

[11] S. Weinberg, *The Quantum Theory of Fields*, vol. II: *Modern Applications*, Cambridge University Press, Cambridge, 1996.

[12] T. Muta, *Foundation of Quantum Chromodynamics: An Introduction to Perturbative Methods in Gauge Theories*, World Scientific, Singapore, 1998.

[13] M. Froissart, "Asymptotic behavior and subtractions in the Mandelstam representation," *Phys. Rev.* **123**, 1053 (1961).

[14] L. Lukaszuk and A. Martin, "Absolute upper bounds for pi pi scattering," *Nuovo Cim.* A **52**, 122 (1967).

[15] W. Heisenberg, "Mesonenerzeugung als Stosswellenproblem," *Z. Phys.* **133**, 65 (1952).

[16] H. Nastase and J. Sonnenschein, "More on Heisenberg's model for high energy nucleon-nucleon scattering," *Phys. Rev. D* **92**, 105028 (2015) (arXiv:1504.01328 [hep-th]).

[17] G. Duplancic and B. Nizic, "Dimensionally regulated one loop box scalar integrals with massless internal lines," *Eur. Phys. J. C* **20**, 357 (2001) [hep-ph/0006249].

[18] G. Passarino and M. J. G. Veltman, "One loop corrections for e+ e− annihilation into mu+ mu− in the Weinberg model," *Nucl. Phys. B* **160**, 151 (1979).

[19] Z. Bern, L. J. Dixon, and D. A. Kosower, "Dimensionally regulated pentagon integrals," *Nucl. Phys. B* **412**, 751 (1994) [hep-ph/9306240].

[20] Z. Bern, L. J. Dixon, D. C. Dunbar, and D. A. Kosower, "One loop *n* point gauge theory amplitudes, unitarity and collinear limits," *Nucl. Phys. B* **425**, 217 (1994) [hep-ph/9403226].

[21] A. Brandhuber, B. Spence, and G. Travaglini, "From trees to loops and back," *JHEP* **0601**, 142 (2006) [hep-th/0510253].

[22] E. Remiddi and J. A. M. Vermaseren, "Harmonic polylogarithms," *Int. J. Mod. Phys. A* **15**, 725 (2000) [hep-ph/9905237].

[23] A. V. Kotikov, L. N. Lipatov, A. I. Onishchenko, and V. N. Velizhanin, "Three loop universal anomalous dimension of the Wilson operators in $N = 4$ SUSY Yang–Mills model," *Phys. Lett. B* **595**, 521 (2004). Erratum: *Phys. Lett. B* **632**, 754 (2006) [hep-th/0404092].

[24] D. Gaiotto, J. Maldacena, A. Sever, and P. Vieira, "Pulling the straps of polygons," *JHEP* **1112**, 011 (2011) (arXiv:1102.0062 [hep-th]).

[25] C. Duhr, H. Gangl, and J. R. Rhodes, "From polygons and symbols to polylogarithmic functions," *JHEP* **1210**, 075 (2012) (arXiv:1110.0458 [math-ph]).

[26] S. G. Naculich, H. Nastase, and H. J. Schnitzer, "All-loop infrared-divergent behavior of most-subleading-color gauge-theory amplitudes," *JHEP* **1304**, 114 (2013) (arXiv:1301.2234 [hep-th]).

[27] L. Mason and D. Skinner, "Amplitudes at weak coupling as polytopes in AdS_5," *J. Phys. A* **44**, 135401 (2011) (arXiv:1004.3498 [hep-th]).

[28] N. Arkani-Hamed, F. Cachazo, C. Cheung, and J. Kaplan, "A duality for the S matrix," *JHEP* **1003**, 020 (2010) (arXiv:0907.5418 [hep-th]).

[29] H. Nastase and H. J. Schnitzer, "Twistor and polytope interpretations for subleading color one-loop amplitudes," *Nucl. Phys. B* **855**, 901 (2012) (arXiv:1104.2752 [hep-th]).

[30] F. M. Haehl, R. Loganayagam, and M. Rangamani, "Schwinger–Keldysh formalism. Part I: BRST symmetries and superspace," *JHEP* **1706**, 069 (2017) (arXiv:1610.01940 [hep-th]).

Index